Response Feature Technology for High-Frequency Electronics. Optimization, Modeling, and Design Automation

Anna Pietrenko-Dabrowska • Slawomir Koziel

Response Feature Technology for High-Frequency Electronics. Optimization, Modeling, and Design Automation

Springer

Anna Pietrenko-Dabrowska
Faculty of Electronics,
Telecommunications and Informatics
Gdansk University of Technology
Gdansk, Poland

Slawomir Koziel
Department of Engineering
Reykjavik University
Reykjavik, Iceland

ISBN 978-3-031-43844-8 ISBN 978-3-031-43845-5 (eBook)
https://doi.org/10.1007/978-3-031-43845-5

This Springer imprint is published by the registered company Springer Nature Switzerland AG
The registered company address is: Gewerbestrasse 11, 6330 Cham, Switzerland

Preface

Contemporary high-frequency electronics, including design of antennas and microwave systems, is heavily reliant on computational tools, especially full-wave electromagnetic (EM) analysis. While offering evaluation reliability and versatility, EM-driven design entails considerable computational expenses, which manifest themselves whenever repetitive EM simulations are required. Examples include parametric optimization, multi-criterial design, or uncertainty quantification. More often than not, the employment of conventional numerical modeling or optimization techniques is impractical if not prohibitive. Needless to say, the literature offers a plethora of approaches developed to alleviate these difficulties. This book offers a yet another methodological perspective on simulation-driven design of high-frequency structures, which is the response feature technology as well as its applications to modeling, optimization, and computer-aided design of antenna and microwave components. The method capitalizes on the exploration of the specific structure of the system outputs, which facilitates simulation-driven design procedures, both in terms of improving computational efficiency and reliability. These advantages can be attributed to weakly nonlinear dependence between the feature point coordinates and designable parameters of the system at hand. In the context of optimization, another benefit is inherent regularization of the objective function leading to faster convergence and enabling quasi-global search capability while using formally local search techniques. The book overviews a definition and extraction of characteristic points, as well as feature-based design problem reformulation. It also outlines algorithms developed to handle local, global, and multi-criterial design, surrogate modeling, as well as uncertainty quantification. The discussed frameworks are extensively illustrated using examples of real-world microwave and antenna structures along with numerous design cases. Furthermore, the book provides an introductory material on simulation-driven design, numerical optimization, but also behavioral and physics-based surrogate modeling. Practical aspects of feature-based surrogate modeling and optimization are discussed along with recommendations concerning particular techniques. The book will be useful for the readers working in the area of high-frequency electronics, including microwave engineering, antenna design, and microwave photonics.

Gdansk, Poland
Reykjavik, Iceland
July 2023

Anna Pietrenko-Dabrowska
Slawomir Koziel

Acknowledgments

We would like to acknowledge the efforts of all those students, researchers, and colleagues who have helped us during the research work presented in this book. We would also like to thank Dassault Systemes, France, for making CST Microwave Studio available for our research purposes, and Keysight Technologies for ADS.

Contents

1 Introduction 1
References 12

2 Simulation-Driven Design of High-Frequency Structures 25
2.1 Overview. Notation and Terminology 25
2.2 Local and Global Parameter Tuning 26
2.3 Multi-objective Optimization 28
2.4 Surrogate Modeling 34
2.5 Uncertainty Quantification 37
References 40

3 Fundamentals of Numerical Optimization 45
3.1 Optimization Problem Formulation 45
3.2 Gradient-Based Optimization 47
3.2.1 Optimization by Descent Methods 48
3.2.2 Newton and Quasi-Newton Methods 52
3.2.3 Qualitative Comparison of Descent Methods 54
3.2.4 Remarks on Constrained Optimization 55
3.3 Derivative-Free Optimization 59
3.3.1 Pattern Search Methods 60
3.3.2 Hooke-Jeeves Direct Search 61
3.3.3 Nelder-Mead Algorithm 61
3.4 Global Optimization: Nature-Inspired Population-Based Algorithms 62
3.4.1 Fundamentals of Population-Based Metaheuristics 64
3.4.2 Evolution Strategies 67
3.4.3 Genetic Algorithms 68
3.4.4 Evolutionary Algorithms 73
3.4.5 Particle Swarm Optimization 74

3.4.6 Differential Evolution 75
3.4.7 Firefly Algorithm 76
3.4.8 Other Methods 77
3.5 Summary 78
References 79

4 Fundamentals of Surrogate Modeling and Surrogate-Assisted Optimization 85
4.1 Surrogate-Assisted Optimization 85
4.2 Surrogate Modeling: Data-Driven Surrogates 88
4.2.1 Modeling Flow for Data-Driven Surrogates 89
4.2.2 Design of Experiments 90
4.2.3 Data-Driven Modeling Techniques 92
4.2.4 Model Validation 97
4.3 Surrogate Modeling: Physics-Based Surrogates 98
4.4 Optimization with Data-Driven Surrogates 103
4.4.1 Optimization by Means of Response Surfaces 103
4.4.2 Sequential Approximate Optimization 104
4.4.3 SBO with Kriging Surrogates: Exploration Versus Exploitation 106
4.4.4 Summary 108
4.5 SBO Using Physics-Based Surrogates 108
4.5.1 Space Mapping 108
4.5.2 Approximation Model Management Optimization (AMMO) 111
4.5.3 Manifold Mapping (MM) 111
4.5.4 Shape-Preserving Response Prediction (SPRP) 112
4.5.5 Adaptive Response Scaling 113
4.5.6 Summary 118
References 119

5 Basics of Response Feature Technology 127
5.1 Case Study: Antenna Parameter Tuning 128
5.1.1 Problem Statement 128
5.1.2 Response Features 129
5.1.3 Feature-Based Antenna Optimization 131
5.1.4 Illustration Example 1: Planar Inverted-F Antenna (PIFA) 133
5.1.5 Single-Brick Dielectric Resonator Antenna 136
5.1.6 Discussion 137
5.2 Case Study: Feature-Based Optimization of Dual-B and Coupler 139
5.2.1 Compact Dual-Band Microstrip Coupler: Geometry and Design Task 140
5.2.2 Response Features 141

5.2.3 Objective Function and Constraints: Optimization Algorithm 143
5.2.4 Numerical Results and Experimental Validation 144
5.3 Summary 146
References 147

6 Response Features for Local Optimization 149
6.1 Conventional Feature-Based Optimization 149
6.1.1 Matching Enhancement of Multi-band Antennas 150
6.1.2 Feature-Based Optimization of an Impedance Matching Transformer 159
6.1.3 Feature-Based Optimization of a SIW Filter 163
6.1.4 FBO for the Design of Linear Microstrip Antenna Array Apertures 167
6.1.5 Conventional Feature-Based Optimization: Summary . . 174
6.2 FBO with Variable Number of Features 174
6.2.1 Response Features for UWB Antennas 175
6.2.2 Case Studies 177
6.3 FBO with Circuit Decomposition 184
6.3.1 Conventional and Compact Circuit. Structure Decomposition 185
6.3.2 Cell Optimization Using Response Features 185
6.3.3 Simulation-Based Transformer Tuning 187
6.3.4 Optimization Algorithm 188
6.3.5 Results 188
6.4 Feature-Based Size Reduction 190
6.4.1 Problem Formulation 191
6.4.2 Controlling Reflection Response Using Response Features 191
6.4.3 Objective Function for FBO Design 193
6.4.4 Results 193
6.5 Summary 199
References 202

7 Accelerated Feature-Based Local Optimization with Variable-Fidelity EM Simulations 205
7.1 Variable-Fidelity Feature-Based Optimization of Antennas 206
7.1.1 Problem Formulation 206
7.1.2 Automated Low-Fidelity Model Selection 208
7.1.3 Bandwidth Enhancement of a Quasi-Isotropic DRA . . . 211
7.1.4 Matching Improvement of a Slot-Ring Coupled Patch Antenna 213
7.2 Variable-Fidelity Feature-Based Optimization of Microwave Structures 217
7.2.1 Response Features for Multi-Band Couplers 218

7.2.2 Design Objectives 220
7.2.3 Automated Low-Fidelity Model Selection 221
7.2.4 Results for Compact Dual-Band Coupler I 222
7.2.5 Results for Compact Dual-Band Coupler II 227
7.3 Summary 232
References 232

8 Expedited Feature-Based Local Optimization: Adjoint Sensitivities and Sparse Sensitivity Updates 233
8.1 Feature-Based Optimization with Adjoint Sensitivities 233
8.1.1 Problem Formulation 234
8.1.2 Optimization Algorithm with Adjoint Sensitivities 236
8.1.3 Case Studies 238
8.2 Feature-Based Optimization with Sparse Sensitivity Updates 241
8.2.1 Globalized Feature-Based Optimization with Sparse Sensitivity Updates 242
8.2.2 Verification Case Studies 248
8.3 Summary 254
References 255

9 Response Features for Reliability Improvement of Local Optimization Procedures 257
9.1 Design Specification Adaptation for Reliability Improvement 257
9.1.1 Design Specification Adaptation: Concept and Prerequisites 258
9.1.2 Optimization Engine: Trust-Region Algorithm 264
9.1.3 Optimization Algorithm 265
9.1.4 Results: Dual-Band Branch-Line Coupler 265
9.1.5 Results: Dual-Band Power Divider 266
9.2 Design Specification Management with Variable-Fidelity Models 272
9.2.1 Problem Formulation: Multi-Band Antennas 272
9.2.2 Improving Computational Efficiency: Variable-Resolution Models 274
9.2.3 Complete Optimization Framework 276
9.2.4 Results: Dual-Band Dipole 277
9.2.5 Results: Triple-Band Dipole 279
9.3 Feature-Based Frequency-Based Regularization 284
9.3.1 Optimization by Frequency-Based Regularization 284
9.3.2 Results: Quasi-Yagi Planar Antenna 287
9.3.3 Results: Ultra Wide-Band Antenna 288
9.3.4 Summary 292
9.4 Regularization with Multi-Fidelity Models and Sparse Sensitivity Updates 292
9.4.1 Restricted Sensitivity Updates 293

9.4.2 Optimization Framework . . . 294
9.4.3 Results . . . 294
9.5 Summary . . . 302
References . . . 302

10 Response Features for Global and Multi-objective Optimization . . . 305
10.1 Global Optimization Using Feature-Based Inverse Surrogates . . 305
10.1.1 Design Problem Formulation . . . 306
10.1.2 Response Features for Initial Design Quality Assessment . . . 307
10.1.3 Global Search Using Inverse Surrogates . . . 309
10.1.4 Local Refinement by Trust-Region Gradient Search . . . 316
10.1.5 Optimization Framework . . . 316
10.1.6 Demonstration Case Study I: Triple-Band Antenna . . . 318
10.1.7 Demonstration Case Study II: Rat-Race Coupler . . . 321
10.2 Global Optimization Using Simplex-Based Regression Surrogates . . . 324
10.2.1 Problem Formulation . . . 325
10.2.2 Simplex-Based Predictors . . . 327
10.2.3 Global Search by Means of Simplex-Based Predictors . . . 332
10.2.4 Local Tuning Using Gradient-Based Trust-Region Search . . . 337
10.2.5 Optimization Procedure . . . 337
10.2.6 Verification Case I: Quasi-Yagi Antenna . . . 338
10.2.7 Verification Case II: Power Divider . . . 343
10.3 Multi-objective Optimization with Features . . . 345
10.3.1 Multi-objective Design Problem Formulation . . . 346
10.3.2 Optimization Algorithm . . . 346
10.3.3 Pareto Front Exploration Using Response Features . . . 347
10.3.4 Verification Case: Ultra-wideband Monopole Antenna . . . 349
10.4 Summary . . . 352
References . . . 352

11 Response Features for Warm-Start Optimization and Rapid Database Acquisition . . . 355
11.1 Expedited Parameter Tuning Using Inverse/Forward Surrogates and Response Features . . . 355
11.1.1 Problem Formulation . . . 356
11.1.2 Response Features . . . 358
11.1.3 Inverse Model: Feature-Based Forward Surrogate—Initial Design . . . 359
11.1.4 Design Refinement by Fast Gradient Search . . . 362
11.1.5 Complete Procedure . . . 362

11.1.6 Demonstration Examples—Case I: Ring-Slot Antenna . . . 363
11.1.7 Demonstration Examples—Case II: Quasi-Yagi Antenna . . . 365
11.1.8 Demonstration Examples—Case III: Rat-Race Coupler . . . 370
11.2 Computationally-Efficient Reference Design Acquisition for Reduced-Cost Constrained Modeling and Redesign of Compact Microwave Passives . . . 371
11.2.1 Design Optimality: Database Design Acquisition Problem . . . 373
11.2.2 Assessing Design Quality Using Response Features . . . 374
11.2.3 Inverse Sensitivity . . . 377
11.2.4 Rapid Database Design Acquisition by Inverse Sensitivity and Response Features . . . 379
11.2.5 Demonstration Example: Miniaturized Branch-Line Coupler . . . 383
11.3 Expedited Microwave Design Closure Using Kriging Surrogates and Iterative Correction . . . 386
11.3.1 Preexisting Designs and Kriging Surrogates . . . 387
11.3.2 Initial Design . . . 388
11.3.3 Basic Design Refinement: Fast Gradient Search . . . 390
11.3.4 Iterative Design Refinement . . . 391
11.3.5 Demonstration Case Study: Three-Section Impedance-Matching Transformer . . . 392
11.4 Automated Inverse Design of Bandpass Filters Through Linear Approximation of Physical Dimensions and Response Features . . . 393
11.4.1 Filter Optimization Using Response Features . . . 394
11.4.2 Coupling Coefficients as Filter Features . . . 397
11.4.3 Filter Implementation Using Inverse Model of Design Parameters . . . 398
11.4.4 Illustration and Application Example . . . 399
References . . . 402

12 Response Features for Low-Cost Behavioral Modeling . . . 403
12.1 Feature-Based Modeling Using Characteristic Point Allocation . . . 404
12.1.1 Modeling Methodology . . . 404
12.1.2 Case Study I: Dielectric Resonator Antenna . . . 407
12.1.3 Case Study II: Patch Antenna Coupled to Rectangular Slot . . . 410
12.1.4 Case Study III: Dual-Band Uniplanar Dipole Antenna . . . 412

12.2 Feature-Based Modeling of Microwave Circuits 415
12.2.1 Modeling Methodology 416
12.2.2 Demonstration Case Studies 421
12.2.3 Application Case Studies 423
12.3 Variable-Fidelity Feature-Based Modeling 424
12.3.1 Modeling Procedure 425
12.3.2 Verification Examples 429
References 433

13 Constrained Modeling with Response Features 435
13.1 Feature-Based Nested Kriging 436
13.1.1 Nested Kriging 436
13.1.2 Nested Kriging with Response Features 439
13.1.3 Verification Case: Dual-Band Antenna 441
13.2 Feature-Based Nested Kriging with Dimensionality Reduction 445
13.2.1 Nested Kriging with Explicit Dimensionality Reduction 445
13.2.2 PCA-Based Constrained Modeling with Response Features 448
13.2.3 Verification Case: Triple-Band Antenna 454
13.3 Two-Stage Observable-Based Constrained Modeling 457
13.3.1 Inverse Regression Surrogate 458
13.3.2 Domain Definition and Final Surrogate 461
13.3.3 Complete Two-Stage Observable-Based Modeling Framework 464
13.3.4 Verification Case Study I: Quasi-Yagi Antenna 466
13.3.5 Verification Case Study II: Compact Branch-Line Coupler 467
13.4 Two-Stage Feature-Based Modeling 474
13.4.1 Verification Case Study I: Ring-Slot Antenna 476
13.4.2 Verification Case Study II: Quasi-Yagi Antenna 480
13.5 Summary 484
References 485

14 Feature-Based Uncertainty Quantification 487
14.1 Motivation 488
14.2 Fabrication Yield and Design Centering: Problem Formulation 490
14.3 Yield Optimization Using Response Features 493
14.3.1 Yield Optimization Task Using Response Features 493
14.3.2 Design Centering by Feature-Based Surrogates and Trust Regions 495
14.3.3 Demonstration Examples 496

14.4 Yield Optimization of Microwave Filters Using Feature-Based Linear Regression Models 501
14.4.1 Feature-Based Approximation Model 502
14.4.2 Yield Estimation Examples 506
14.4.3 Yield Optimization Examples 509
14.5 Expedited Feature-Based Uncertainty Quantification Using Variable-Fidelity EM Simulations 511
14.5.1 Variable-Fidelity Yield Optimization Using Second-Order Feature-Based Surrogates 511
14.5.2 Variable-Fidelity Yield Optimization Using Linear Feature-Based Surrogates and Trust Regions 516
14.6 Feature-Based Tolerance Optimization 522
14.6.1 Feature-Based Regression Model 523
14.6.2 Evaluating Objective Function: Tolerance Optimization Algorithm 525
14.6.3 Demonstration Examples 1: Antenna Design 527
14.6.4 Demonstration Examples 1: Microwave Design 529
14.7 Design Centering Using Response Features and Inverse Regression 535
14.7.1 Task Formulation and Models 535
14.7.2 Design Centering Algorithm 539
14.7.3 Demonstration Case Studies 540
14.8 Summary 545
References 545

15 Generalized Response Features 549
15.1 Motivation 550
15.2 Response Features: Recollection of the Concept and Advantages 551
15.3 Generalized Response Features 556
15.3.1 Concept and Feature Point Definition 556
15.3.2 Objective Function Definition 560
15.3.3 Optimization Algorithm Flow 565
15.4 Illustration Examples 565
15.4.1 Case I: Ring Slot Antenna 569
15.4.2 Case II: Triple-Band Dipole Antenna 572
15.4.3 Case III: Bandwidth-Enhanced Patch Antenna 574
15.4.4 Case IV: Ultra-Wideband Antenna 576
15.4.5 Discussion 578
15.5 Summary 579
References 581

16 Summary 583

Index 591

Chapter 1
Introduction

The reliance of all branches of engineering on computational (or simulation) models has never been as strong as it is today. Their fundamental advantages over traditional methods of evaluating the system performance, e.g., analytical or semiempirical, include versatility and reliability. On the one hand, simulation environments allow for representing structures of almost arbitrary complexity, embed them in larger systems, and extract information about all sorts of characteristics, e.g., scattering parameters for microwave components (Pozar, 2012), or integral quantities such as gain or efficiency for antennas (Balanis, 2005). On the other hand, computer simulation permits quantification of physical phenomena affecting system operation, even those that are difficult or impossible to be obtained otherwise, e.g., mutual coupling in antenna arrays (Koziel & Ogurtsov, 2019a), or dielectric/radiation losses (Tang & Luo, 2021). The last two decades witnessed tremendous developments in simulation methods and computing hardware. Commercial simulation packages have evolved in all engineering disciplines, including mechanical and structural engineering (Fusion 360; Autodesk, 2022), aerospace engineering (Inventor; Autodesk, 2019), electrical engineering (Cadence, 2016; Microwave Office, 2019; National Instruments, 2019), or multi-physics and multidisciplinary simulations (Multiphysics Simulation; ANSYS, 2022; COMSOL Multiphysics, COMSOL Inc., 2022). Nowadays, it is possible to accurately simulate complex and large-scale systems such as allocation of antennas on military vehicles (Byun et al., 2013), full-scale aircraft (Liersch & Hepperle, 2011), laminar and turbulent flow through wind turbines (Rosenberg et al., 2014), weather and climate phenomena (hurricanes, concentration of pollutants in the atmosphere or the ocean; Smith & Fortin, 2022; Cebe & Balas, 2018), or chemical processes (Upreti, 2017). Practical limitations of simulation-based modeling capacity are mainly related to the reliability of available input data (e.g., boundary conditions, material parameters) and computational resources. Notwithstanding, simulation models have had major effects on engineer-

A. Pietrenko-Dabrowska, S. Koziel, *Response Feature Technology for High-Frequency Electronics. Optimization, Modeling, and Design Automation*,
https://doi.org/10.1007/978-3-031-43845-5_1

ing practice. Accurate system evaluation obtained therefrom allows us to replace costly prototyping, investigate larger numbers of alternative designs, and reduce time to market as well as the overall cost of design procedures.

The focus of this book is high-frequency electronics. Therein, the two fundamental types of computational models are circuit simulators (ADS, Keysight Technologies, 2021) and full-wave electromagnetic (EM) analysis (Davidson, 2010; Sullivan, 2013; Sevgi, 2014; Swanson & Hoefer, 2003; White, 2004). The former allows for fast (circuit theory-based) evaluation of electrical characteristics; however, it becomes grossly inaccurate at the presence of strong cross-coupling effects (Kurgan & Koziel, 2019), dielectric/radiation losses (Rajab et al., 2005), or when the geometry of the structure at hand is of high level of complexity. On the other hand, EM analysis is conducted by numerical solving of Maxwell equations established over a selected computational domain. It enables identification of the distribution of the electric and magnetic fields, and, through appropriate post-processing, acquisition of the relevant system characteristics such as scattering parameters (Nikolova et al., 2006) or radiation patterns (Paulotto et al., 2008). Over the years, a large variety of general-purpose and specialized analysis techniques have been developed that are suitable for various purposes (e.g., finite-element analysis; Jin, 2002; finite-difference time domain analysis; Webb, 2004; methods of moments; Gibson, 2007, etc.). EM-driven design including design closure (in particular, adjustment of geometry and/or material parameters to fine-tune the system performance) has become an academic and industry standard. Nowadays, it is possible to conduct the entire design process within the simulation environment. Many commercial simulation software packages are available, including the general-purpose EM solvers such as HFSS (2021), Altair FEKO (2021), CST Microwave Studio (2022), XFDTD (2021), Momentum (Keysight, 2020), SEMCAD (2022), or Sonnet *em* (Sonnet, 2020), but also specific design tools, e.g., ADS (Advanced Design System, Keysight 2021) and Antenna Magus (2023), for antenna design.

The times when computational models were mainly used for design verification are long gone. Their larger-scale employment was fundamentally hindered by available computers and inefficiency of early simulation software. Rapid advancements in simulation technologies but also the developments of hardware, along with dramatically increased computational throughput (that including parallelization and hardware acceleration), have changed this picture entirely. In many cases, it is possible to carry out the entire design cycle with the aid of simulation tools. In fact, this is a practical necessity for a growing number of components and systems. In particular, traditional approaches, largely based on design-ready theoretical models are no longer adequate. One of the reasons is the increasing level of complexity of engineering systems; another is various system- and component-level interactions that have to be taken into account in the design process (Fang & Lin, 2016; Cheung & Long, 2006; Ittipiboon et al., 1998; Taflove & Hagness, 2005; Kuo et al., 1997). Simulation-driven design tasks can be formulated in various ways, depending on the type (geometry, material ones, etc.) and the number of adjustable parameters that have to be handled, the number of objectives, as well as possible constraints. Undoubtedly, the most common design task is parametric optimization, where the values of selected variables are adjusted in order to improve the system performance (Bubnicki, 2005; Zaslavski, 2010; Pistikopoulos et al., 2007; Koziel & Bandler, 2015a; Koziel et al., 2016; Cao et al., 2011; Chakravorty & Mandal, 2016; Sadrossadat et al., 2013). The latter is quantified by means of appropriately defined objective function, which is typically scalar-valued (Koziel et al., 2013; Sobester & Forrester, 2015). In some cases, simultaneous handling of multiple goals is required, leading to multi-criterial problems (Mirjalili & Dong, 2019; Mandal et al., 2018), where the objective function is vector-valued (Koziel & Pietrenko-Dabrowska, 2021a; Taran et al., 2018). Multi-criterial design typically seeks for a family of designs (a so-called Pareto set, Toktas et al., 2019), representing the best possible trade-offs between conflicting goals (Lv et al., 2019; Bramerdorfer & Zavoianu, 2017). Here, a representative example is a reduction of the physical dimensions of a microwave or antenna component, which is normally detrimental to electrical performance figures (Mahrokh & Koziel, 2021; Koziel & Pietrenko-Dabrowska, 2022). Other tasks include statistical design (e.g., Monte Carlo analysis; Styblinski & Opalski, 1986) or, in more general terms, uncertainty quantification (Hosder, 2012; Allaire & Willcox, 2014). One of the frequently considered problems in this field is maximization of the fabrication yield under the assumed manufacturing tolerances (Pietrenko-Dabrowska et al., 2020; Kouassi et al., 2016). A wide range of specialized algorithms have been developed to perform each of these tasks (Nocedal & Wright, 2000; Conn et al., 2000; Gorissen et al., 2010; Yang, 2010). Some of the methods are generic, i.e., applicable to a number of problems in different areas (Nocedal & Wright, 2000); others are problem-specific (e.g., Koziel et al., 2013; Koziel & Leifsson, 2013a).

Despite the mentioned advancements in computing hardware and software, simulation-driven design poses considerable challenges for a majority of real-world situations. The fundamental issue is the high cost of evaluating the computational models. The CPU cost of simulation highly depends on the model complexity. It may be just a few seconds per frequency for simple electromagnetic (EM) analysis of two-dimensional models (e.g., planar microwave filters; Hazdra et al., 2005), a few minutes for computational fluid dynamics (CFD) analysis of two-dimensional airfoil profiles (Siegler et al., 2016) or EM analysis of compact planar antennas (Koziel & Pietrenko-Dabrowska, 2021b), and up to several hours (e.g., CFD analysis of three-dimensional structures such as aircraft wings, or EM analysis of integrated photonic components; Krause & Jäger, 2005; Fakhfakh et al., 2015). Particularly involved structures (a full aircraft, a ship, climate models) may require many hours or even days of the analysis time (Wehner et al., 2010; Dennis et al., 2012; Yondo et al., 2018). At the same time, availability of better computational resources does not alleviate the cost-related difficulties to a large extent because it is counterbalanced by a growing complexity of the systems considered in all fields of engineering and science. Simulation-driven design procedures require repetitive and sometimes massive evaluations of the system at hand, which entail computational costs that might be impractical or even prohibitive. The problem is more pronounced for high-dimensional parameter spaces but also when global or multi-objective optimization is necessary. Nowadays, the most popular global search procedures involve nature-inspired population-based algorithms (Li et al., 2019; Jia & Lu, 2019), known for poor computational efficiency: the typical number of objective function evaluations ranges from a few thousands to many thousands per algorithm run. Multi-criterial problems are also routinely handled using population-based methods that allow generation of the entire Pareto set representation in a single run (Tang et al., 2018; Yang & Kiang, 2015). Uncertainty quantification problems are even more expensive due to the necessity of numerical integration of the underlying probability density functions describing parameter deviations (e.g., Monte Carlo analysis; Ma et al., 2018), which has to be carried out repetitively when optimizing statistical figures of merit (yield, Zhang et al., 2021; tolerance hypervolume, Wu et al., 2021). Another issue pertinent to computational models is the numerical noise, which may be a result of terminating the simulation process before full convergence, or it may be related to adaptive meshing employed by modern solvers. The latter manifests itself through noticeable changes of the simulated system responses due to even very small changes of the structure geometry (these leading to considerable changes of the mesh topology). The noise may affect the operation of gradient-based optimization routines that normally require the objective function to be smooth. Yet another problem is reliability of simulation-based design procedures. For example, the most widely used procedure is local parameter tuning, the performance of which is highly dependent on availability of a sufficiently good starting point. In many cases, this is problematic, e.g., in size reduction of compact microwave or antenna components, where miniaturization techniques employing slow-wave phenomenon (Qin & Xue, 2013) or geometrical modifications (stubs, Wei et al., 2015; slots, Zhang et al., 2019; defected ground structures, Sen & Moyra, 2019) increase the number of variables and result in nonintuitive relations between geometry parameters and the system outputs (Pietrenko-Dabrowska & Koziel, 2020a). Also, many problems are of multimodal nature (i.e., feature multiple local optima), and the employment of local algorithms typically leads to inferior results, whereas resorting to global methods generates excessive computational costs.

From the perspective of practical design applications of the simulation tools, the challenges outlined in the previous paragraphs are combined with inefficiency of conventional optimization/modeling algorithms, which are normally incorporated into commercial simulation software packages (e.g., CST, 2022; HFSS, 2021; Altair FEKO, 2021). This resulted in the development of a variety of informal and interactive ways of utilizing simulation models for design purposes, which are still ubiquitous in many engineering fields, including high-frequency design. The key factor is the engineering insight, which allows for making reasonable predictions about the promising changes of the system parameters. Practical workflows involve parameter sweeping (usually, one parameter at a time). Experienced designers are often capable of identifying satisfactory parameter setups using an acceptable number of simulations even though the interactive procedures are rather laborious and do not guarantee optimum results. However, the adequacy of parameter sweeping-based and similar methods has become questionable with the increasing complexity of the engineering systems, the need of simultaneous tuning of many variables, as well as the presence of multiple objectives and constraints.

Needless to say, the last two decades or so witnessed considerable research efforts directed toward mitigating the common issues of simulation-driven design, particularly, its high cost. In the realm of local optimization, perhaps the most serious bottleneck is evaluation of the system response gradients, which—when carried out using finite differentiation (Levy & Lessman, 1992)—generates considerable computational overhead proportional to the dimensionality of the parameter space. This can be alleviated using adjoint sensitivities (Director & Rohrer, 1969; Pironneau, 1984; Jameson, 1988; El Sabbagh et al., 2006; Papadimitriou & Giannakoglou, 2008; Toivanen et al., 2009), or automated differentiation (Griewank, 2000; Bischof et al., 2008)). These methods allow for a fast evaluation of gradients of the figures of interest at small extra computational effort (often only one additional simulation), regardless of the number of designable parameters. Consequently, the benefits of adjoints are particularly evident for higher-dimensional problems. Adjoint sensitivities are currently available in some commercial simulation packages in various areas (e.g., computational fluid dynamics solvers: ANSYS Fluent, 2015; Star-CCM+, 2015), as well as some noncommercial codes (e.g., Stanford University Unstructured; Palacios et al., 2013). In the context of high-frequency modeling, they are supported by CST Microwave Studio (CST, 2022) and ANSYS HFSS (HFSS, 2021). Other available techniques include sparse sensitivity updating schemes, where finite differentiation-based gradient estimation is limited to selected parameters based on various metrics (e.g., Koziel & Pietrenko-Dabrowska, 2019a, b; Pietrenko-Dabrowska & Koziel, 2019, 2020a). Local search procedures can also be expedited using parallelization (Zhang et al., 2020b), utilization of fast dedicated solvers (Arndt, 2012), or techniques exploiting specific shapes of the system responses (e.g., cognition-driven design, Zhang et al., 2015; feature-based optimization, Koziel, 2015).

One of the most promising approaches to alleviating the challenges of simulation-driven design is utilization of fast replacement models, referred to as surrogates (Simpson et al., 2001; Queipo et al., 2005; Bandler et al., 2008; Forrester & Keane, 2009; Koziel & Pietrenko-Dabrowska, 2020b; Yondo et al., 2019). The underlying idea is to shift most of the computational burden into a cheap surrogate while only making occasional references to the original (high-fidelity) simulation model. Although a construction of the surrogate is also associated with certain computational expenses, the benefits often outweigh those costs for well-working procedures (Koziel & Ogurtsov, 2019b; Van Nechel et al., 2018). This is particularly the case for design tasks incurring massive system evaluations (uncertainty quantification, Koziel & Ogurtsov, 2019b; global search, Easum et al., 2018) but may be advantageous also for local optimization (e.g., space mapping, Li et al., 2020). For the surrogate to be effectively used instead of the original simulation model, it needs to be fast, sufficiently accurate, and, preferably, analytically tractable (e.g., smooth) (Alexandrov & Lewis, 2001; Bandler et al., 2008; Jin, 2011; Koziel & Leifsson, 2013a). The surrogates can be constructed locally (e.g., in the vicinity of the optimization path for the purpose of local optimization (Koziel & Unnsteinsson, 2018) in the neighborhood of the nominal design for the purpose of statistical analysis (Rayas-Sanchez et al., 2021)), or over the entire parameter space pertinent to a given problem (e.g., for the purpose of global optimization (Koziel & Abdullah, 2021)). Although surrogate models have been around for several decades, their importance has been steadily growing over the last two decades or so because of the increasing role of computational models themselves. The literature on surrogate models, also referred to as replacement models or metamodels, is replete (Simpson et al., 2001; Muller & Shoemaker, 2014; Bilicz, 2016; Declercq et al., 2013; Leary et al., 2003; Mendes et al., 2013; Du & Roblin, 2018; Wang et al., 2018; Yang et al., 2019). On the other hand, the surrogates entered the field of high-frequency engineering relatively recently, i.e., in the late 1990s and early 2000s (Rayas-Sanchez, 2016; Koziel et al., 2008), whereas their widespread use originated about a decade ago (Liu et al., 2014; Zhang et al., 2021; Koziel & Ogurtsov, 2019a).

Surrogate models can be categorized into two major groups, data-driven (Simpson et al., 2001) and physics-based ones (Koziel & Ogurtsov, 2014). Data-driven or approximation models are constructed from sampled high-fidelity simulation data (Queipo et al., 2005). Some of the popular methods include polynomial regression (Jin et al., 2001), radial basis functions, RBF (Wild et al., 2008), kriging (Jones, 2001; Forrester & Keane, 2009; Kleijnen, 2009), Gaussian process regression, GPR (Angiulli et al., 2007; Jacobs, 2012), support vector regression, SVR (Smola & Schölkopf, 2004), rational approximation (Shaker et al., 2009), polynomial chaos expansion, PCE (Zhang et al., 2018a), as well as artificial neural networks, ANN (Haykin, 1998). The latter come in many variations, including convolutional neural networks (Liang et al., 2020), recurrent neural networks (Shen & Wang, 2008) and deep neural networks (DNN) (Koziel et al., 2021b), and found many applications in high-frequency electronics (Zhang et al., 2020a; Belen et al., 2021; Zhao & Wu, 2020). Approximation models are constructed exclusively using data acquired from the system of interest (no engineering insight is required), which makes them versatile and easily transferrable between the problem domains. At the same time, they are cheap to evaluate because their underlying structure is based on analytical equations (e.g., linear combinations of appropriate basis functions; Rozhenko, 2018). For the purpose of design optimization but also general-purpose modeling, data-driven surrogates are often incorporated into machine learning frameworks (Xiao et al., 2018; Wu et al., 2020), where the model is iteratively updated using sequential sampling methods (Couckuyt et al., 2012). Various ways of incorporating new training points into the model (so-called infill criteria) have been developed, including exploitative models (i.e., models oriented toward improving the design in the vicinity of the current one), explorative models (i.e., models aiming at improving global accuracy), as well as models with balanced exploration and exploitation (Forrester & Keane, 2009; Couckuyt et al., 2010). Generally, these techniques are often referred to as efficient global optimization (EGO) methods (Jones et al., 1998) or surrogate-assisted evolutionary algorithms (SAEAs) (Gorissen et al., 2009; Lim et al., 2010; Jin, 2011; Yang et al., 2019).

Physics-based models constitute the second major class of surrogates. The name refers to the fundamental structure of these surrogates which exploit, to a certain extent, the system-specific knowledge (Cervantes-González et al., 2016; Koziel &

Leifsson, 2016). This, in turn, is most often a simplified physical description of the system in the form of an underlying low-fidelity model (Robinson et al., 2008; Koziel & Leifsson, 2013b; Sarkar et al., 2019). The low-fidelity model is subsequently corrected using a limited amount of high-fidelity simulation data, typically through linear or nonlinear regression (Bandler et al., 2004). A representative example of a low-fidelity model in the area of microwave engineering is an equivalent network of the structure (e.g., a filter), with the high-fidelity model being evaluated through a full-wave electromagnetic analysis (Zhu, 2002). The fundamental benefit of this type of arrangement is that sharing the same physics by the low- and high-fidelity models normally results in a better generalization capability of the surrogate, its validity over wider ranges of parameters, and limited vulnerability to dimensionality-related issues. Consequently, the number of high-fidelity training samples required to construct physics-based models is lower than for the data-driven ones. Thus, physics-based modeling solves some of the issues of data-driven surrogates, which contribute to their growing popularity (Pantoja et al., 2007; Salleh et al., 2008; Crevecoeur et al., 2010; Koziel & Leifsson, 2013a; Cervantes-González et al., 2016; Baratta et al., 2018; Zhang et al., 2018b). Unfortunately, the same factors that contribute to the attractiveness of the physics-based surrogates also imply their limitations. The first issue is the low-fidelity model itself. It is normally problem-specific (Bandler et al., 2004; Koziel & Leifsson, 2013a); therefore, physics-based surrogates lack the versatility of data-driven models. In high-frequency engineering, low-fidelity models are typically obtained by simulating the system at a different level of physical description (e.g., equivalent network representation evaluated using circuit theory rules; Bandler et al., 2004), or lower resolution (e.g., simulation with coarser discretization of the structure and/or relaxed convergence criteria; Koziel & Ogurtsov, 2014). Because the low-fidelity models typically involve computer simulation, their evaluation time cannot be neglected. The aggregated cost of the low-fidelity model simulations may be significant in certain applications, e.g., parametric optimization (Zhou et al., 2007; Koziel & Leifsson, 2016). Another issue is a trade-off between the low-fidelity model speed and accuracy. While this can be easily adjusted in many situations, e.g., by changing the discretization density in coarse-mesh simulation models, selection of a particular model setup might not be a trivial task (Koziel & Ogurtsov, 2012).

Physics-based surrogates are typically used for local optimization. One of the most popular physics-based SBO methods in high-frequency electronics is space mapping (SM) (Bandler et al., 2004; Koziel et al., 2008; Rayas-Sanchez, 2016) which comes in many variations (aggressive SM, Bandler et al., 1995; implicit SM, Koziel et al., 2011; manifold-mapping, Echeverria et al., 2006; neural SM, Gutiérrez-Ayala & Rayas-Sánchez, 2010; output SM, Ayed et al., 2012; input SM, Khalatpour et al., 2011). Because the most straightforward way of constructing the surrogate model from the underlying low-fidelity model is correcting its response, many physics-based SBO algorithms employ such mechanisms. Some of the techniques include approximation model management optimization (AMMO; Alexandrov & Lewis, 2001), multipoint correction (Toropov, 1989), manifold mapping (Echeverria & Hemker, 2005), adaptive response correction (Koziel et al.,

2009), shape-preserving response prediction (Koziel, 2010), or adaptive response scaling (Koziel & Unnsteinsson, 2018). Apart from design optimization, surrogate models are used for many other purposes as well. An important area is uncertainty quantification (e.g., yield estimation; Bandler et al., 2002; Biernacki et al., 2012) and tolerance-aware (or robust) design (Ko et al., 2011; Koziel & Bandler, 2015a; Kouassi et al., 2016; Aubry et al., 2016). Statistical analysis is typically performed in order to find the effects of manufacturing tolerances or uncertainties concerning operating conditions on the system performance. Traditional methods such as Monte Carlo (MC) simulation (Hu et al., 2016; Liu, 2017) are computationally heavy, and fast surrogates seem to be an ideal way of accelerating the process. Certain types of models, such as polynomial chaos expansion (PCE), are particularly suitable in this context due to their capability of directly assessing the statistical moments of the output probability distributions without the necessity of running MC (Sudret, 2008; Xiu, 2009; Du & Roblin, 2017; Manfredi et al., 2017). Another important application of surrogates is inverse modeling, which allows for directly yielding the optimum values of design variables corresponding to the required values of performance figures (Akkaram et al., 2007; Kabir et al., 2008; Liu et al., 2016; Zhang et al., 2018b).

Despite their potential efficacy, surrogate-assisted methods are affected by a number of issues. The major bottleneck is the curse of dimensionality (Wu et al., 2019), which is commonly referred to as a rapid growth of the number of training data samples necessary to render a reliable model. As a matter of fact, a construction of the surrogates within highly dimensional spaces (say, 20 or more dimensions) is only possible if the system responses are weakly nonlinear. In the case of high-frequency electronics, practical data-driven modeling of components, such as multi-band antennas (characterized by sharp, resonant-like responses), filters (featuring multiple poles and transmission zeros), or compact microwave components (e.g., couplers), is limited to a few parameters. Meanwhile, in the case of the mentioned structures, modeling within wide ranges of parameters is even more important for the surrogates to be of any utility for design purposes (Feng et al., 2019; Yelten et al., 2012; Koziel et al., 2013, 2016). This poses even more challenges than the dimensionality issue because characteristic features of the responses (e.g., the resonances) relocate considerably along the frequency spectrum (Ullah & Koziel, 2018; Rossi et al., 2014). In practice, globally accurate approximation modeling is usually justified in the case of multiple-use library models of components described by a limited number of parameters. There have been many attempts and techniques developed to mitigate these problems. One is sequential sampling (Xiong et al., 2009; Mukhopadhyay, 2011; Wei et al., 2012, Mackman et al., 2013; Xu et al., 2014), nowadays commonly used as a design of experiments approach for data-driven surrogates. The idea is to replace a one-shot (typically uniform) sample allocation by an iterative process in which additional (or infill) samples are distributed based on the feedback obtained from the current sample distribution and/or the current surrogate. In the case of explorative design of experiments schemes (Mandal et al., 2012), the infill samples are added without referring to the system output but merely based on the data points allocated so far, with the purpose of improving uniformity of the set. In general, especially for the exploitative sampling schemes (Koziel & Ogurtsov, 2019a), a suitable infill criterion may be maximization of the mean square error, i.e., identifying locations where the error (as predicted by the surrogate, typically, kriging) is the highest (Jiang et al., 2018). In these cases, finding the infill points requires global optimization, typically realized using population-based metaheuristics (Beheshti & Shamsuddin, 2013; Mehmani et al., 2015; Park et al., 2018; Liu et al., 2018a), which makes the procedure slow. The advantage is that more samples may be allocated in the regions corresponding to higher nonlinearity of the system responses, which improves the model accuracy. Unfortunately, this sort of approach rarely pays off for many high-frequency structures where "nonlinear" regions are almost everywhere within the model domain (San et al., 2004; Hajjaj et al., 2017; Shitvov et al., 2014). Co-kriging (Forrester et al., 2007) and gradient kriging (Han et al., 2013) are other ways of reducing the computational cost of the training data acquisition. Co-kriging uses densely sampled low-fidelity model data blended together with a limited number of high-fidelity points which allows for constructing the surrogate of the accuracy similar to that obtained solely from (densely sampled) high-fidelity simulations. This is of course under the assumption that the low- and high-fidelity models are sufficiently well correlated. Gradient kriging, on the other hand, incorporates sensitivity data into the surrogate, which also potentially reduces the number of necessary training points; however, this approach is only practical if the gradient information can be obtained in a computationally efficient matter (e.g., through adjoints; Giles & Pierce, 2000; Allaire, 2015). A different approach to handle dimensionality issues is high-dimensional model representation (HDMR) (Foo & Karniadakis, 2010; Ma & Zabaras, 2010; Shan & Wang, 2011; Cai et al., 2017; Liu et al., 2018b; Wu et al., 2019), where the system responses are represented as a linear combination of functions that account for lower order effects, specifically, cooperative effects of individual variables, pairs of variables, etc. Model order reduction (MOR) is another class of methods that generally refers to various ways of reducing the complexity of large-scale dynamical systems, while preserving (as much as possible) their input-output behavior (Baur et al., 2014; Henneron & Clénet, 2014). The reduced models mimic (from the point of view of the input and output) the behavior of the large-scale system so that they can be efficiently used for design automation, parametric optimization, or sensitivity analysis. Particular numerical techniques utilized by MOR include, among others, Krylov subspace methods (Lin et al., 2007), proper orthogonal decomposition (POD; Willcox & Peraire, 2002), principal component analysis (PCA; Dray, 2008), solution space projection (SSP; Lee & Jin, 2007), or rational approximation (Deschrijver et al., 2007). Examples of MOR in high-frequency engineering include parameterized MOR models of complex electromagnetic systems (Burgard et al., 2013; Sato et al., 2015), reduced order macromodels of high-speed microstrip structures (Zhu & Cangellaris, 2001), or accelerated frequency sweeps in FEM analysis (de la Rubia et al., 2009). Several methods have also been developed to handle situations when the regression problem (e.g., in the construction of response surface approximation models; Khuri & Mukhopadhyay,

2010) is heavily underdetermined. This essentially means that the system at hand is characterized by a large number of internal degrees of freedom (e.g., analog or mixed-signal circuits, analog-to-digital converters, or RF front ends), whereas the regression model has to be built using a relatively small number of samples due to the computational budget issues or simply numerical problems pertinent to handling large data sets. Orthogonal matching pursuit (Tropp, 2004; Tao et al., 2016; Li, 2010; Bishop, 2006) is a technique that identifies a small set of basis functions that approximate the model (or function) of interest. In order to ensure fast convergence, a set of basis functions that are normalized and orthogonal is normally adopted (Needell & Tropp, 2009). Another method, Bayesian model fusion (Wang et al., 2013, 2016) can be classified as a physics-based surrogate technique because it uses the so-called early-stage data (e.g., schematic-level simulation data) in order to fit the late-stage model (e.g., a post-layout one with extracted parasitic components) (Li et al., 2012). By blending these two-level data, realized using Bayesian inference, the overall cost of model construction can be greatly reduced because very few late-stage samples are typically used (Tao et al., 2019).

The last few paragraphs outlined the methodologies developed to mitigate the common issues of simulation-driven design, especially those related to excessive costs of the optimization or modeling procedures. While keeping the formulation of the underlying design problem intact, they focused on reducing the number of evaluation of the high-fidelity model (e.g., by reducing the sensitivity updating schemes in the gradient-based algorithms) or incorporating cheaper representations (surrogate models), delegated to absorb most of the computational burden incurred by a given algorithm. Many of these methods exhibit excellent efficiency when implemented properly, although generality and automation are often questionable. Also, acceleration techniques do not address reliability of the design procedures. For example, the performance of local (e.g., gradient- or stencil-based algorithms; Nocedal & Wright, 2000; Kolda et al., 2003), whether conventional or expedited, is still reliant on the availability of a sufficiently good initial design. In the recent years, a conceptually distinct approach has been proposed, which is referred to as the response feature technology (Koziel, 2015). In the context of design optimization, it is often termed feature-based optimization (FBO) (Koziel & Pietrenko-Dabrowska, 2020b). Its major purpose is to improve both the computational efficiency and the reliability of the modeling and optimization procedures in high-frequency engineering. The response feature approach operates by reformulating the underlying design tasks in terms of the appropriately defined characteristic (feature) points of the system at hand. The characteristic points are problem-dependent and assigned to account for the relevant features of the system outputs, e.g., frequency and level coordinates of antenna resonances (Koziel & Pietrenko-Dabrowska, 2021b), frequencies corresponding to specific levels of reflection or transmission characteristics (Koziel and Bandler, 2015b), local maxima of the in-band return loss response of a filter (Koziel & Bandler, 2015a), etc. The type and the number of response features are decided upon so that the information carried therein is sufficient to quantify the design objectives.

The major advantage of the response feature approach is that the dependence between the characteristic point coordinates and the system parameters (both geometry and material) is normally weakly nonlinear, as opposed to a similar relationship for the complete response. Consequently, problem reformulation leads to "flattening" of the functional landscape to be handled, which has several advantages: (i) accelerating convergence of the optimization procedures (Koziel, 2015); (ii) enabling quasi-global search capabilities (Pietrenko-Dabrowska et al., 2021); (iii) reducing the number of training data sets required for surrogate model construction (Pietrenko-Dabrowska et al., 2022); (iv) improving reliability of EM-driven design processes, in particular, reducing sensitivity of local tuning algorithms to the quality of the initial design (Koziel & Pietrenko-Dabrowska, 2021b); and (v) facilitating inverse modeling procedures (Koziel & Pietrenko-Dabrowska, 2020c), which also carries over to forward modeling (Koziel et al., 2021a). Response features can be incorporated into the majority of existing optimization and modeling procedures because the employment is related to reformulating the design task rather than modifying the procedure itself. However, feature-based algorithms can also be developed from ground up, particularly in relation to modeling methods (both inverse and forward, Koziel & Pietrenko-Dabrowska, 2021c) but also optimization (Koziel & Pietrenko-Dabrowska, 2021d) and uncertainty quantification (Pietrenko-Dabrowska & Koziel, 2021).

The primary purpose of this book is an exposition of the response feature technology and its application for modeling, optimization, and design automation of high-frequency structures, including antennas and microwave devices. In order to be self-contained, in Chap. 2, we formulate common simulation-driven tasks, including local and global optimization, multi-objective design, statistical analysis, as well as tolerance-aware optimization, whereas, Chaps. 3 and 4 overview local and global optimization methods and provide the fundamentals of surrogate modeling and surrogate-assisted optimization. Chapter 5 introduces the response feature technology, discusses the definition of characteristic points and their fundamental properties, as well as illustrates their employment for optimization of antenna and microwave components. Chapter 6 provides an overview of local feature-based optimization. The main purpose is to discuss the various ways of defining the characteristic points, depending on a particular design scenario, as well as the underlying type (shape) of the system characteristics. Chapters 7 and 8 discuss accelerated feature-based optimization strategies, whereas Chap. 9 outlines feature-inspired methods for reliability improvement of local tuning algorithms. The two representative techniques discussed therein are frequency-based regularization and adaptively adjusted design specification approach. Chapter 10 overviews utilization of response features for global optimization of antenna and microwave components. Feature-based methodologies for warm-start optimization and rapid redesign of high-frequency components are discussed in Chap. 11. The next two chapters cover feature-based surrogate modeling, both in conventional (Chap. 12) and confined domains (using a so-called performance-driven paradigm; Chap. 13). Chapter 14 elaborates on utilization of response features for rapid statistical analysis and tolerance optimization of high-frequency devices. The concept of generalized features is discussed in Chap. 15, whereas Chap. 16 concludes the work. The book is illustrated with a large number of practical design cases from various areas of high-frequency electronics, especially microwave and antenna engineering. Despite this particular focus, some of the optimization and modeling methods discussed here are of a generic nature and can be adopted for solving problems in other engineering disciplines. The authors believe that the presented material may be helpful for engineers and researchers interested in methods for improving computational efficiency and reliability of simulation-driven procedures, especially in the context of design optimization, constructing fast replacement models, or other projects that require massive evaluations of expensive computer simulation models as well as handling highly nonlinear system outputs.

References

ADS (Advanced Design System). (2021). Keysight Technologies, Fountaingrove Parkway 1400, Santa Rosa, CA 95403–1799.

Akkaram, S., Beeson, D., Agarwal, H., & Wiggs, G. (2007). Inverse modeling technology for parameter estimation. *Structural and Multidisciplinary Optimization, 34*(2), 151–164.

Alexandrov, N. M., & Lewis, R. M. (2001). An overview of first-order model management for engineering optimization. *Optical Engineering, 2*(4), 413–430.

Allaire, G. (2015). A review of adjoint methods for sensitivity analysis, uncertainty quantification, and optimization in numerical codes. *Ingenieurs de l'Automobile, SIA, 836*, 33–36.

Allaire, D., & Willcox, K. (2014). A mathematical and computational framework for multifidelity design and analysis with computer models. *International Journal for Uncertainty Quantification, 4*, 1–20.

Altair FEKO. (2021). Altair HyperWorks, 1820 E Big Beaver Rd, Troy, MI 48083, USA.

Angiulli, G., Cacciola, M., & Versaci, M. (2007). Microwave devices and antennas modelling by support vector regression machines. *IEEE Transactions on Magnetics, 43*(4), 1589–1592.

Antenna Magus. (2023). Magus (Pty) Ltd, Unit 9B Octo Place, 5 Electron Rd., Technopark Stellenbosch 7600, South Africa.

Arndt, F. (2012). WASP-NET: Recent advances in fast EM CAD and optimization of waveguide components, feeds and aperture antennas. In *IEEE international symposium on antennas and propagation* (pp. 1–2), Chicago, IL, USA, 8–14 July, 2012.

Aubry, A., De Maio, A., Huang, Y., & Piezzo, M. (2016). Robust design of radar Doppler filters. *IEEE Transactions on Signal Processing, 64*(22), 5848–5860.

Ayed, R. B., Gong, J., Brisset, S., Gillon, F., & Brochet, P. (2012). Three-level output space mapping strategy for electromagnetic design optimization. *IEEE Transactions on Magnetics, 48*(2), 671–674.

Balanis, C. A. (2005). *Antenna theory analysis and design* (3rd ed.). John Wiley & Sons.

Bandler, J. W., Biernacki, R. M., Chen, S. H., Hemmers, R. H., & Madsen, K. (1995). Electromagnetic optimization exploiting aggressive space mapping. *IEEE Transactions on Microwave Theory and Techniques, 41*(12), 2874–2882.

Bandler, J. W., Rayas-Sánchez, J. E., & Zhang, Q. J. (2002). Yield-driven electromagnetic optimization via space mapping-based neuromodels. *International Journal of RF and Microwave Computer-Aided Engineering, 12*, 79–89.

Bandler, J. W., Cheng, Q. S., Dakroury, S. A., Mohamed, A. S., Bakr, M. H., Madsen, K., & Søndergaard, J. (2004). Space mapping: The state of the art. *IEEE Transactions on Microwave Theory and Techniques, 52*(1), 337–361.

Bandler, J. W., Koziel, S., & Madsen, K. (2008). Editorial—Surrogate modeling and space mapping for engineering optimization. *Optimization and Engineering, 9*(4), 307–310.

Baratta, I. A., de Andrade, C. B., de Assis, R. R., & Silva, E. J. (2018). Infinitesimal dipole model using space mapping optimization for antenna placement. *IEEE Antennas and Wireless Propagation Letters, 17*(1), 17–20.

Baur, U., Benner, P., & Feng, L. (2014). Model order reduction for linear and nonlinear systems: A system-theoretic perspective. *Archives of Computational Methods in Engineering, 21*(4), 331–358.

Beheshti, Z., & Shamsuddin, S. M. H. (2013). A review of population-based meta-heuristic algorithm. *International Journal of Advances in Soft Computing and its Applications, 5*(1), 1–35.

Belen, M. A., Mahouti, P., Koziel, S., Caliskan, A., & Szczepanski, S. (2021). On decomposition-based surrogate-assisted design of leaky wave antenna input characteristics for beam scanning applications. *IEEE Access, 9*, 161318–161325.

Biernacki, R., Chen, S., Estep, G., Rousset, J., & Sifri, J. (2012). Statistical analysis and yield optimization in practical RF and microwave systems. In *IEEE/MTT-S International Microwave Symposium Digest, Montreal, Canada* (pp. 1–3).

Bilicz, S. (2016). Sparse grid surrogate models for electromagnetic problems with many parameters. *IEEE Transactions on Magnetics, 52*(3), 1–4.

Bischof, C., Bücker, H. M., Hovland, P. D., Naumann, U., & Utke, J. (Eds.). (2008). *Advances in automatic differentiation, lecture notes in computational science and engineering*. Springer.

Bishop, C. (2006). *Pattern recognition and machine learning*. New York, NY, USA, Springer.

Bramerdorfer, G., & Zavoianu, A. C. (2017). Surrogate-based multi-objective optimization of electrical machine designs facilitating tolerance analysis. *IEEE Transactions on Magnetics, 53, no. 8*, paper 8107611.

Bubnicki, Z. (2005). Parametric optimization. In *Modern control theory*. Springer.

Burgard, S., Farle, O., & Edlinger, R. D. (2013). A novel parametric model order reduction approach with applications to geometrically parameterized microwave devices. *International Journal for Computation and Mathematics in Electrical and Electronic Engineering, 32*(5), 1525–1538.

Byun, G., Choo, H., & Ling, H. (2013). Optimum placement of DF antenna elements for accurate DOA estimation in a harsh platform environment. *IEEE Transactions on Antennas and Propagation, 61*(9), 4783–4791.

Cadence Allegro. (2016). Cadence Design Systems, 2655 Seely Ave, San Jose, CA 95134, USA.

Cai, X., Qiu, H., Gao, L., & Shao, X. (2017). Metamodeling for high dimensional design problems by multi-fidelity simulations. *Structural and Multidisciplinary Optimization, 56*(1), 151–166.

Cao, Y., Reitzinger, S., & Zhang, Q. (2011). Simple and efficient high-dimensional parametric modeling for microwave cavity filters using modular neural network. *IEEE Microwave and Wireless Components Letters, 21*(5), 258–260.

Cebe, K., & Balas, L. (2018). Monitoring and modeling land-based marine pollution. *Regional Studies in Marine Science, 24*, 23–39.

Cervantes-González, J. C., Rayas-Sánchez, J. E., López, C. A., Camacho-Pérez, J. R., Brito-Brito, Z., & Chávez-Hurtado, J. L. (2016). Space mapping optimization of handset antennas considering EM effects of mobile phone components and human body. *International Journal of RF and Microwave Computer-Aided Engineering, 26*(2), 121–128.

Chakravorty, P., & Mandal, D. (2016). Radiation pattern correction in mutually coupled antenna arrays using parametric assimilation technique. *IEEE Transactions on Antennas and Propagation, 64*(9), 4092–4095.

Cheung, T. S. D., & Long, J. R. (2006). Shielded passive devices for silicon-based monolithic microwave and millimeter-wave integrated circuits. *IEEE Journal of Solid-State Circuits, 41*(5), 1183–1200.

COMSOL Multiphysics. (2022). COMSOL Inc., 1 New England Executive Park, Burlington, MA 01803, USA.

Conn, A. R., Gould, N. I. M., & Toint, P. L. (2000). *Trust region methods* (MPS-SIAM series on optimization). MPS-SIAM.

Couckuyt, I., Declercq, F., Dhaene, T., Rogier, H., & Knockaert, L. (2010). Surrogate-based infill optimization applied to electromagnetic problems. *International Journal of RF and Microwave Computer-Aided Engineering, 20*(5), 492–501.

Couckuyt, I., Forrester, A., Gorissen, D., De Turck, F., & Dhaene, T. (2012). Blind Kriging: Implementation and performance analysis. *Advances in Engineering Software, 49*, 1–13.

Crevecoeur, G., Sergeant, P., Van Dupre, L., & de Walle, R. (2010). A two-level genetic algorithm for electromagnetic optimization. *IEEE Transactions on Magnetics, 46*(7), 2585–2595.

CST Microwave Studio. (2022). Dassault Systemes, France.

Davidson, D. B. (2010). *Computational electromagnetics for RF and microwave engineering* (2nd ed.). Cambridge University Press.

de la Rubia, V., Razafison, U., & Maday, Y. (2009). Reliable fast frequency sweep for microwave devices via the reduced-basis method. *IEEE Transactions on Microwave Theory and Techniques, 57*(12), 2923–2937.

Declercq, F., Couckuyt, I., Rogier, H., & Dhaene, T. (2013). Environmental high frequency characterization of fabrics based on a novel surrogate modelling antenna technique. *IEEE Transactions on Antennas and Propagation, 61*(10), 5200–5213.

Dennis, J. M., Vertenstein, M., Worley, P. H., Mirin, A. A., Craig, A. P., & Jacob, R. (2012). Computational performance of ultra-high-resolution capability in the Community Earth System Model. *The International Journal of High Performance Computing Applications, 26*(1), 5–16.

Deschrijver, D., Haegeman, B., & Dhaene, T. (2007). Orthonormal vector fitting: A robust macromodeling tool for rational approximation of frequency domain responses. *IEEE Transactions on Advanced Packaging, 30*(2), 216–225.

Director, S. W., & Rohrer, R. A. (1969). The generalized adjoint network and network sensitivities. *IEEE Transactions on Circuit Theory, 16*(3), 318–323.

Dray, S. (2008). On the number of principal components: A test of dimensionality based on measurements of similarity between matrices. *Computational Statistics & Data Analysis, 52*(4), 2228–2237.

Du, J., & Roblin, C. (2017). Statistical modeling of disturbed antennas based on the polynomial chaos expansion. *IEEE Antennas and Wireless Propagation Letters, 16*, 1843–1846.

Du, J., & Roblin, C. (2018). Stochastic surrogate models of deformable antennas based on vector spherical harmonics and polynomial chaos expansions: Application to textile antennas. *IEEE Transactions on Antennas and Propagation, 66*(7), 3610–3622.

Easum, J. A., Nagar, J., Werner, P. L., & Werner, D. H. (2018). Efficient multiobjective antenna optimization with tolerance analysis through the use of surrogate models. *IEEE Transactions on Antennas and Propagation, 66*(12), 6706–6715.

Echeverria, D., & Hemker, P. W. (2005). Space mapping and defect correction. *Computational Methods in Applied Mathematics, 5*(2), 107–136.

Echeverria, D., Lahaye, D., Encica, L., Lomonova, E. A., Hemker, P. W., & Vandenput, A. J. A. (2006). Manifold-mapping optimization applied to linear actuator design. *IEEE Transactions on Magnetics, 42*(4), 1183–1186.

El Sabbagh, M. A., Bakr, M. H., & Nikolova, N. K. (2006). Sensitivity analysis of the scattering parameters of microwave filters using the adjoint network method. *International Journal of RF and Microwave Computer-Aided Engineering, 16*, 596–606.

emTM Version 16.56 (2020). Sonnet Software, Inc., Sonnet Software, Inc., 126 N. Salina Street, Syracuse, NY 13202, USA

Fakhfakh, M., Tlelo-Cuautle, E., & Siarry, P. (Eds.). (2015). *Computational intelligence in analog and mixed-signal (AMS) and radio-frequency (RF) circuit design*. Springer.

Fang, W.-T., & Lin, Y.-S. (2016). Highly integrated switched beamformer module for 2.4-GHz wireless transceiver application. *IEEE Transactions on Microwave Theory and Techniques, 64*(9), 2933–2942.

Feng, F., Zhang, C., Na, W., Zhang, J., Zhang, W., & Zhang, Q. (2019). Adaptive feature zero assisted surrogate-based EM optimization for microwave filter design. *IEEE Microwave and Wireless Components Letters, 29*(1), 2–4.

FLUENT, ver. 15.0, ANSYS Inc. (2015). Southpointe, 275 Technology Drive, Canonsburg, PA 15317, USA.

Foo, J., & Karniadakis, G. E. (2010). Multi-element probabilistic collocation method in high dimensions. *Journal of Computational Physics, 229*(5), 1536–1557.

Forrester, A. I. J., & Keane, A. J. (2009). Recent advances in surrogate-based optimization. *Progress in Aerospace Sciences, 45*(1), 50–79.

Forrester, A. I. J., Sóbester, A., & Keane, A. J. (2007). Multi-fidelity optimization via surrogate modelling. *Proceedings of Royal Society A: Mathematical, Physical and Engineering Sciences, 463*(2088).

Fusion 360. (2022). Autodesk, 111 McInnis Parkway, San Rafael, 94903 California, USA.

Gibson, W. C. (2007). *The method of moments in electromagnetics*. Chapman and Hall/CRC.

Giles, M., & Pierce, N. (2000). An introduction to the adjoint approach to design. *Flow, Turbulence and Combustion, 65*(3–4), 393–415.

Gorissen, D., Dhaene, T., & De Turck, F. (2009). Evolutionary model type selection for global surrogate modeling. *Journal of Machine Learning Research, 10*, 2039–2078.

Gorissen, D., Crombecq, K., Couckuyt, I., Dhaene, T., & Demeester, P. (2010). A surrogate modeling and adaptive sampling toolbox for computer based design. *Journal of Machine Learning Research, 11*, 2051–2055.

Griewank, A. (2000). *Evaluating derivatives: Principles and techniques of algorithmic differentiation*. Society for Industrial and Applied Mathematics (SIAM).

Gutiérrez-Ayala, V., & Rayas-Sánchez, J. E. (2010). Neural input space mapping optimization based on nonlinear two-layer perceptrons with optimized nonlinearity. *International Journal of RF and Microwave Computer-Aided Engineering, 20*, 512–526.

Hajjaj, A. Z., Hafiz, M. A., & Younis, M. I. (2017). TI - mode coupling and nonlinear resonances of MEMS arch resonators for bandpass filters. *Scientific Reports, 7*, 41820.

Han, Z.-H., Görtz, S., & Zimmermann, R. (2013). Improving variable-fidelity surrogate modeling via gradient-enhanced kriging and a generalized hybrid bridge function. *Aerospace Science and Technology, 25*(1), 177–189.

Haykin, S. (1998). *Neural networks: A comprehensive foundation* (2nd ed.). Prentice Hall.

Hazdra, P., Polivka, M., & Sokol, V. (2005). Microwave antennas and circuits modeling using electromagnetic field simulator. *Radioengineering, 14*(4), 2–10.

Henneron, T., & Clénet, S. (2014). Model order reduction of non-linear magnetostatic problems based on POD and DEI methods. *IEEE Transactions on Magnetics, 50*(2), 33–36.

HFSS. (2021). Release 19.0, ANSYS, http://www.ansoft.com/products/hf/hfss/, 2600 Ansys Dr., Canonsburg, PA 15317, USA.

Hosder, S. (2012). Stochastic response surfaces based on non-intrusive polynomial chaos for uncertainty quantification. *International Journal of Mathematical Modelling and Numerical Optimisation, 3*(1/2), 117–139.

Hu, X., Chen, X., Parks, G. T., & Yao, W. (2016). Review of improved Monte Carlo methods in uncertainty-based design optimization for aerospace vehicles. *Progress in Aerospace Sciences, 86*, 20–27.

Inventor. (2019). Autodesk, 111 McInnis Parkway San Rafael, 94903 California, USA.

Ittipiboon, A., Petosa, A., & Cuhaci, M. (1998). Effects of inter-deed coupling on the axial ratio performance of circular polarized microstrip antennas. In *Symposium on antenna technology and applied electromagnetics, Ottawa* (pp. 209–212).

Jacobs, J. P. (2012). Bayesian support vector regression with automatic relevance determination kernel for modeling of antenna input characteristics. *IEEE Transactions on Antennas and Propagation, 60*(4), 2114–2118.

Jameson, A. (1988). Aerodynamic design via control theory. *Journal of Scientific Computing, 3*, 233–260.

Jia, X., & Lu, G. (2019). A hybrid Taguchi binary particle swarm optimization for antenna designs. *IEEE Antennas and Wireless Propagation Letters, 18*(8), 1581–1585.
Jiang, C., Cai, X., Qiu, H., Gao, L., & Li, P. (2018). A two-stage support vector regression assisted sequential sampling approach for global metamodeling. *Structural and Multidisciplinary Optimization, 58*(4), 1657–1672.
Jin, J. (2002). *The finite element method in electromagnetics* (2nd ed.). Wiley.
Jin, Y. (2011). Surrogate-assisted evolutionary computation: Recent advances and future challenges. *Swarm and Evolutionary Computation, 1*(2), 61–70.
Jin, R., Chen, W., & Simpson, T. (2001). Comparative studies of metamodelling techniques under multiple modelling criteria. *Structural and Multidisciplinary Optimization, 23*(1), 1–3.
Jones, D. R. (2001). A taxonomy of global optimization methods based on response surfaces. *Journal of Global Optimization, 21*, 345–383.
Jones, D., Schonlau, M., & Welch, W. (1998). Efficient global optimization of expensive black-box functions. *Journal of Global Optimization, 13*, 455–492.
Kabir, H., Wang, Y., Yu, M., & Zhang, Q. J. (2008). Neural network inverse modeling and applications to microwave filter design. *IEEE Transactions on Microwave Theory and Techniques, 56*(4), 867–879.
Khalatpour, A., Amineh, R. K., Cheng, Q. S., Bakr, M. H., Nikolova, N. K., & Bandler, J. W. (2011). Accelerating input space mapping optimization with adjoint sensitivities. *IEEE Microwave and Wireless Components Letters, 21*(6), 280–282.
Khuri, A. I., & Mukhopadhyay, S. (2010). Response surface methodology: Advanced review. *Computational Statistics, 2*(2), 128–149.
Kleijnen, J. P. C. (2009). Kriging metamodeling in simulation: A review. *European Journal of Operational Research, 192*(3), 707–716.
Ko, J., Byun, J., Park, J., & Kim, H. (2011). Robust design of dual band/polarization patch antenna using sensitivity analysis and Taguchi's method. *IEEE Transactions on Magnetics, 47*(5), 1258–1261.
Kolda, T. G., Lewis, R. M., & Torczon, V. (2003). Optimization by direct search: New perspectives on some classical and modern methods. *SIAM Review, 45*, 385–482.
Kouassi, A., Nguyen-Trong, N., Kaufmann, T., Lallechere, S., Bonnet, P., & Fumeaux, C. (2016). Reliability-aware optimization of a wideband antenna. *IEEE Transactions on Antennas and Propagation, 64*(2), 450–460.
Koziel, S. (2010). Shape-preserving response prediction for microwave design optimization. *IEEE Transactions on Microwave Theory and Techniques, 58*(11), 2829–2837.
Koziel, S. (2015). Fast simulation-driven antenna design using response-feature surrogates. *International Journal of RF and Microwave Computer-Aided Engineering, 25*(5), 394–402.
Koziel, S., & Abdullah, M. (2021). Machine-learning-powered EM-based framework for efficient and reliable design of low scattering metasurfaces. *IEEE Transactions on Microwave Theory and Techniques, 69*(4), 2028–2041.
Koziel, S., & Bandler, J. W. (2015a). Rapid yield estimation and optimization of microwave structures exploiting feature-based statistical analysis. *IEEE Transactions on Microwave Theory and Techniques, 63*(1), 107–114.
Koziel, S., & Bandler, J. W. (2015b). Reliable microwave modeling by means of variable-fidelity response features. *IEEE Transactions on Microwave Theory and Techniques, 63*(12), 4247–4254.
Koziel, S., & Leifsson, L. (Eds.). (2013a). *Surrogate-based modeling and optimization. Applications in engineering*. Springer.
Koziel, S., & Leifsson, L. (2013b). Surrogate-based aerodynamic shape optimization by variable-resolution models. *AIAA Journal, 51*(1), 94–106.
Koziel, S., & Leifsson, L. (2016). *Simulation-driven design by knowledge-based response correction techniques*. Springer.

Koziel, S., & Ogurtsov, S. (2012). Model management for cost-efficient surrogate-based optimization of antennas using variable-fidelity electromagnetic simulations. *IET Microwaves, Antennas & Propagation, 6*, 1643–1650.
Koziel, S., & Ogurtsov, S. (2014). *Antenna design by simulation-driven optimization. Surrogate-based approach*. Springer.
Koziel, S., & Ogurtsov, S. (2019a). *Simulation-based optimization of antenna arrays*. World Scientific.
Koziel, S., & Ogurtsov, S. (2019b). Surrogate-assisted tolerance analysis of low sidelobe linear arrays with microstrip corporate feeds. *International Journal of Numerical Modelling: Electronic Networks, Devices and Fields, 32*(2), e2533.
Koziel, S., & Pietrenko-Dabrowska, A. (2019a). Efficient gradient-based algorithm with numerical derivatives for expedited optimization of multi-parameter miniaturized impedance matching transformers. *Radioengineering, 28*(3), 572–578.
Koziel, S., & Pietrenko-Dabrowska, A. (2019b). Reduced-cost electromagnetic-driven optimization of antenna structures by means of trust-region gradient-search with sparse Jacobian updates. *IET Microwaves, Antennas & Propagation, 13*(10), 1646–1652.
Koziel, S., & Pietrenko-Dabrowska, A. (2020a). *Performance-driven surrogate modeling of high-frequency structures*. Springer.
Koziel, S., & Pietrenko-Dabrowska, A. (2020b). Expedited feature-based quasi-global optimization of multi-band antennas with Jacobian variability tracking. *IEEE Access, 8*, 83907–83915.
Koziel, S., & Pietrenko-Dabrowska, A. (2020c). Recent advances in high-frequency modeling by means of domain confinement and nested kriging. *IEEE Access, 8*, 189326–189342.
Koziel, S., & Pietrenko-Dabrowska, A. (2021a). Recent advances in accelerated multi-objective design of high-frequency structures using knowledge-based constrained modeling approach. *Knowledge Based Systems, 214*, 106726.
Koziel, S., & Pietrenko-Dabrowska, A. (2021b). Global EM-driven optimization of multi-band antennas using knowledge-based inverse response-feature surrogates. *Knowledge Based Systems, 227*, 107189.
Koziel, S., & Pietrenko-Dabrowska, A. (2021c). Knowledge-based performance-driven modeling of antenna structures. *Knowledge Based Systems, 237*, 107698.
Koziel, S., & Pietrenko-Dabrowska, A. (2021d). Robust parameter tuning of antenna structures by means of design specification adaptation. *IEEE Transactions on Antennas and Propagation, 69*(12), 8790–8798.
Koziel, S., & Pietrenko-Dabrowska, A. (2022). Constrained multi-objective optimization of compact microwave circuits by design triangulation and Pareto front interpolation. *European Journal of Operational Research, 299*(1), 302–312.
Koziel, S., & Unnsteinsson, S. D. (2018). Expedited design closure of antennas by means of trust-region-based adaptive response scaling. *IEEE Transactions on Antennas and Propagation, 17*(6), 1099–1103.
Koziel, S., Cheng, Q. S., & Bandler, J. W. (2008). Space mapping. *IEEE Microwave Magazine, 9*(6), 105–122.
Koziel, S., Bandler, J. W., & Madsen, K. (2009). Space mapping with adaptive response correction for microwave design optimization. *IEEE Transactions on Microwave Theory and Techniques, 57*, 478–486.
Koziel, S., Bandler, J. W., & Cheng, Q. S. (2011). Constrained parameter extraction for microwave design optimisation using implicit space mapping. *IET Microwaves, Antennas & Propagation, 5*, 1156–1163.
Koziel, S., Yang, X. S., & Zhang, Q. J. (Eds.). (2013). *Simulation-driven design optimization and modeling for microwave engineering*. Imperial College Press.
Koziel, S., Bekasiewicz, A., & Leifsson, L. (2016). Cost-efficient modeling of input characteristics of narrow-band antennas using response features. *10th European Conference on Antennas and Propagation (EuCAP), Davos, 2016*, 1–4.

Koziel, S., Pietrenko-Dabrowska, A., & Ullah, U. (2021a). Low-cost modeling of microwave components by means of two-stage inverse/forward surrogates and domain confinement. *IEEE Transactions on Microwave Theory and Techniques, 69*(12), 5189–5202.

Koziel, S., Calik, N., Mahouti, P., & Belen, M. A. (2021b). Accurate modeling of antenna structures by means of domain confinement and pyramidal deep neural networks. *IEEE Transactions on Antennas and Propagation, 70*(3), 2174–2188.

Krause, E., & Jäger, W. (Eds.). (2005). *High performance computing in science and engineering*. Transactions of the High Performance Computing Center.

Kuo, C.-N., Houshmand, B., & Itoh, T. (1997). Full-wave analysis of packaged microwave circuits with active and nonlinear devices: An FDTD approach. *IEEE Transactions on Microwave Theory and Techniques, 45*(5), 819–826.

Kurgan, P., & Koziel, S. (2019). Selection of circuit geometry for miniaturized microwave components based on concurrent optimization of performance and layout area. *AEU-International Journal of Electronics and Communications, 108*, 287–294.

Leary, S., Bhaskar, A., & Keane, A. (2003). Optimal orthogonal-array-based latin hypercubes. *Journal of Applied Statistics, 30*, 585–598.

Lee, S. H., & Jin, J. M. (2007). Adaptive solution space projection for fast and robust wideband finite-element simulation of microwave components. *IEEE Microwave and Wireless Components Letters, 17*(7), 474–476.

Levy, H., & Lessman, F. (1992). *Finite difference equations*. Dover Publications.

Li, X. (2010). Finding deterministic solution from underdetermined equation: Largescale performance modeling of analog/RF circuits. *IEEE Transactions on Computer-Aided Design of Integrated Circuits and Systems, 29*(11) 1661–1668.

Li, X., Zhang, W., Wang, F., Sun, S., & Gu, C. (2012). Efficient parametric yield estimation of analog/mixed-signal circuits via Bayesian model fusion. In *2012 IEEE/ACM International Conference on Computer-Aided Design (ICCAD), San Jose, CA, USA* (pp. 627–634).

Li, W., Zhang, Y., & Shi, X. (2019). Advanced fruit fly optimization algorithm and its application to irregular subarray phased array antenna synthesis. *IEEE Access, 7*, 165583–165596.

Li, S., Fan, X., Laforge, P. D., & Cheng, Q. S. (2020). Surrogate model-based space mapping in postfabrication bandpass filters' tuning. *IEEE Transactions on Microwave Theory and Techniques, 68*(6), 2172–2182.

Liang, Y., Lu, L., Xiao, Q., & Yan, S. (2020). Evaluating fast algorithms for convolutional neural networks on FPGAs. *IEEE Transactions on Computer-Aided Design of Integrated Circuits and Systems, 39*(4), 857–870.

Liersch, C. M., & Hepperle, M. (2011). A distributed toolbox for multidisciplinary preliminary aircraft design. *CEAS Aeronautical Journal, 2*(1–4), 57–68.

Lim, D., Jin, Y., Ong, Y., & Sendhoff, B. (2010). Generalizing surrogate-assisted evolutionary computation. *IEEE Transactions on Evolutionary Computation, 14*(3), 329–355.

Lin, Y., Bao, L., & Wei, Y. (2007). A model-order reduction method based on Krylov subspaces for mimo bilinear dynamical systems. *Journal of Applied Mathematics and Computing, 25*(1–2), 293.

Liu, B. (2017). Posterior exploration based sequential Monte Carlo for global optimization. *Journal of Global Optimization, 69*(4), 847–868.

Liu, B., Aliakbarian, H., Ma, Z., Vandenbosch, G. A. E., Gielen, G., & Excell, P. (2014). An efficient method for antenna design optimization based on evolutionary computation and machine learning techniques. *IEEE Transactions on Antennas and Propagation, 62*(1), 7–18.

Liu, Y., Shi, Y., Zhou, Q., & Xiu, R. (2016). A sequential sampling strategy to improve the global fidelity of metamodels in multi-level system design. *Structural and Multidisciplinary Optimization, 53*(6), 1295–1313.

Liu, H., Ong, Y. S., & Cai, J. (2018a). A survey of adaptive sampling for global metamodeling in support of simulation-based complex engineering design. *Journal of Structural and Multidisciplinary Optimization, 57*(1), 416.

Liu, H., Hervas, J. R., Ong, Y. S., Cai, J., & Wang, Y. (2018b). An adaptive RBF-HDMR modeling approach under limited computational budget. *Structural and Multidisciplinary Optimization, 57*(3), 1–18.

Lv, Z., Wang, L., Han, Z., Zhao, J., & Wang, W. (2019). Surrogate-assisted particle swarm optimization algorithm with Pareto active learning for expensive multi-objective optimization. *IEEE/CAA Journal of Automatica Sinica, 6*(3), 838–849.

Ma, X., & Zabaras, N. (2010). An adaptive high-dimensional stochastic model representation technique for the solution of stochastic partial differential equations. *Journal of Computational Physics, 229*, 3884–3915.

Ma, J., Ren, Z., Zhao, G., Zhang, Y., & Koh, C. (2018). A new reliability analysis method combining adaptive kriging with weight index Monte Carlo simulation. *IEEE Transactions on Magnetics, 54*(3), 1–4. Art no. 7001904.

Mackman, T. J., Allen, C. B., Ghoreyshi, M., & Nadcock, K. J. (2013). Comparison of adaptive sampling methods for generation of surrogate aerodynamic models. *AIAA Journal, 51*(4), 797–808.

Mahrokh, M., & Koziel, S. (2021). Improved-efficacy EM-based antenna miniaturization by multi-fidelity simulations and objective function adaptation. *Energies, 15*(2), 403.

Mandal, A., Zafar, H., Das, S., & Vasilakos, A. V. (2012). A modified differential evolution algorithm for shaped beam linear array antenna design. *Progress in Electromagnetics Research, 125*, 439–457.

Mandal, J. K., Mukhopadhyay, S., & Dutta, P. (Eds.). (2018). *Multi-objective optimization: Evolutionary to hybrid framework*. Springer.

Manfredi, P., Ginste, D. V., Stievano, I. S., De Zutter, D., & Canavero, F. G. (2017). Stochastic transmission line analysis via polynomial chaos methods: An overview. *IEEE Electromagnetic Compatibility Magazine, 6*(3), 77–84. Third Quarter 2017.

Mehmani, A., Chowdhury, S., Tong, W., & Messac, A. (2015). Adaptive switching of variable-fidelity models in population-based optimization. In N. Lagaros & M. Papadrakakis (Eds.), *Engineering and applied sciences optimization. Computational methods in applied sciences* (Vol. 38). Springer.

Mendes, M. H. S., Soares, G. L., Coulomb, J., & Vasconcelos, J. A. (2013). Appraisal of surrogate modeling techniques: A case study of electromagnetic device. *IEEE Transactions on Magnetics, 49*(5), 1993–1996.

Mirjalili, S., & Dong, J. S. (2019). *Multi-objective optimization using artificial intelligence techniques*. Springer Briefs in Applied Sciences and Technology.

Momentum. (2020) Keysight Technologies, Fountaingrove Parkway 1400, Santa Rosa, CA 95403–1799.

Mukhopadhyay, N. (2011). Sequential sampling. In M. Lovric (Ed.), *International Encyclopedia of statistical science*. Springer.

Muller, J., & Shoemaker, C. A. (2014). Influence of ensemble surrogate models and sampling strategy on the solution quality of algorithms for computationally expensive black-box global optimization problems. *Journal of Global Optimization, 60*(2), 123–144.

Multiphysics Simulation. (2022). ANSYS Inc., Southpointe, 275 Technology Drive, Canonsburg, PA 15317, USA.

Needell, D., & Tropp, J. A. (2009). CoSaMP: Iterative signal recovery from incomplete and inaccurate samples. *Applied and Computational Harmonic Analysis, 26*(3), 301–321.

NI AWR Design Environment. (2019). National Instruments, 11500 N Mopac Expy, Austin, TX 78759, USA.

NI Microwave Office. (2019). National Instruments, 11500 N Mopac Expy, Austin, TX 78759, USA.

Nikolova, N. K., Li, Y., Li, Y., & Bakr, M. H. (2006). Sensitivity analysis of scattering parameters with electromagnetic time-domain simulators. *IEEE Transactions on Microwave Theory and Techniques, 54*(4), 1598–1610.

Nocedal, J., & Wright, S. J. (2000). *Numerical optimization* (Springer series in operations research). Springer.

Palacios, F., Colonno, M. R., Aranake, A. C., Campos, A., Copeland, S. R., Economon, T. D., Lonkar, A. K., Lukaczyk, T. W., Taylor, T. W. R., & Alonso, J. J. (2013). Stanford University unstructures (SU^2): An open-source integrated computational environment for multi-physics simulation and design. In *AIAA Aerospace Sciences Meeting, Grapevine, TX, USA*.

Pantoja, M. F., Meincke, P., & Bretones, A. R. (2007). A hybrid genetic-algorithm space-mapping tool for the optimization of antennas. *IEEE Transactions on Antennas and Propagation, 55*(3), 777–781.

Papadimitriou, D. I., & Giannakoglou, K. C. (2008). Aerodynamic shape optimization using first and second order adjoint and direct approaches. *Archives of Computational Methods in Engineering, 15*, 447–488.

Park, D., Chung, I. B., & Choi, D. H. (2018). Surrogate based global optimization using adaptive switching infill sampling criterion. In A. Schumacher, T. Vietor, S. Fiebig, K. U. Bletzinger, & K. Maute (Eds.), *Advances in structural and multidisciplinary optimization. WCSMO 2017* (pp. 692–699). Springer.

Paulotto, S., Baccarelli, P., Frezza, F., & Jackson, D. R. (2008). Full-wave modal dispersion analysis and broadside optimization for a class of microstrip CRLH leaky-wave antennas. *IEEE Transactions on Microwave Theory and Techniques, 56*(12), 2826–2837.

Pietrenko-Dabrowska, A., & Koziel, S. (2019). Expedited antenna optimization with numerical derivatives and gradient change tracking. *Engineering Computations, 37*(4), 1179–1193.

Pietrenko-Dabrowska, A., & Koziel, S. (2020a). Computationally-efficient design optimization of antennas by accelerated gradient search with sensitivity and design change monitoring. *IET Microwaves, Antennas & Propagation, 14*(2), 165–170.

Pietrenko-Dabrowska, A., & Koziel, S. (2020b). Cost-efficient EM-driven size reduction of antenna structures by multi-fidelity simulation models. *Electronics, 10*, 1536.

Pietrenko-Dabrowska, A., & Koziel, S. (2021). Design centering of compact microwave components using response features and trust regions. *Energies, 14*, 1–15. paper no. 8550.

Pietrenko-Dabrowska, A., Koziel, S., & Al-Hasan, M. (2020). Expedited yield optimization of narrow- and multi-band antennas using performance-driven surrogates. *IEEE Access, 8*, 143104–143113.

Pietrenko-Dabrowska, A., Koziel, S., & Al-Hasan, M. (2021). Accelerated parameter tuning of antenna structures using inverse and feature-based forward kriging surrogates. *International Journal of Numerical Modelling: Electronic Networks, Devices and Fields, 34*(5), e2880.

Pietrenko-Dabrowska, A., Koziel, S., & Ullah, U. (2022). Reduced-cost two-level surrogate antenna modeling using domain confinement and response features. *Scientific Reports, 12*, 4667.

Pironneau, O. (1984). *Optimal shape design for elliptic systems*. Springer-Verlag.

Pistikopoulos, E. N., Georgiadis, M. C., & Dua, V. (2007). *Multi-parametric programming*. Wiley VCH.

Pozar, D. M. (2012). *Microwave engineering* (4th ed.). Wiley.

Qin, W., & Xue, Q. (2013). Elliptic response bandpass filter based on complementary CMRC. *Electronics Letters, 49*(15), 945–947.

Queipo, N. V., Haftka, R. T., Shyy, W., Goel, T., Vaidynathan, R., & Tucker, P. K. (2005). Surrogate-based analysis and optimization. *Progress in Aerospace Sciences, 41*(1), 1–28.

Rajab, K. Z., Fuh, K. F., Mittra, R., & Lanagan, M. (2005). Dielectric property measurement using a resonant nonradiative dielectric waveguide structure. *IEEE Microwave and Wireless Components Letters, 15*(2), 104–106.

Rayas-Sanchez, J. E. (2016). Power in simplicity with ASM: Tracing the aggressive space mapping algorithm over two decades of development and engineering applications. *IEEE Microwave Magazine, 17*(4), 64–76.

Rayas-Sanchez, J. E., Koziel, S., & Bandler, J. W. (2021). Advanced RF and microwave design optimization: A journey and a vision of future trends. *IEEE Journal of Microwaves, 1*(1), 481–493.

Robinson, T. D., Eldred, M. S., Willcox, K. E., & Haimes, R. (2008). Surrogate-based optimization using multifidelity models with variable parameterization and corrected space mapping. *AIAA Journal, 46*(11), 2814–2822.

Rosenberg, A., Selvaraj, S., & Sharma, A. (2014). A novel dual-rotor turbine for increased wind energy capture. *Journal of Physics: Conference Series, 524.*

Rossi, M., Dierck, A., Rogier, H., & Vande Ginste, D. (2014). A stochastic framework for the variability analysis of textile antennas. *IEEE Transactions on Antennas and Propagation, 62*(12), 6510–6514.

Rozhenko, A. I. (2018). Comparison of radial basis functions. *Numerical Analysis and Applications, 11*(3), 220–235.

Sadrossadat, S. A., Cao, Y., & Zhang, Q. (2013). Parametric modeling of microwave passive components using sensitivity-analysis-based adjoint neural-network technique. *IEEE Transactions on Microwave Theory and Techniques, 61*(5), 1733–1747.

Salleh, M. K. M., Prigent, G., Pigaglio, O., & Crampagne, R. (2008). Quarter-wavelength side-coupled ring resonator for bandpass filters. *IEEE Transactions on Microwave Theory and Techniques, 56*(1), 156–162.

San, H., Kobayashi, H., Kawakami, S., & Kuroiwa, N. (2004). A noise-shaping algorithm of multi-bit DAC nonlinearities in complex bandpass ΔΣAD modulators. *IEICE Transactions on Fundamentals of Electronics, Communications and Computer Sciences, E87-A*(4), 792–800.

Sarkar, T. K., Chen, H., Palma, M. S., & Zhu, M. (2019). Lessons learned using a physics based macro model for analysis of radio wave propagation in wireless transmission. *IEEE Transactions on Antennas and Propagation, 67*(4), 2150–2157.

Sato, Y., Campelo, F., & Igarashi, H. (2015). Fast shape optimization of antennas using model order reduction. *IEEE Transactions on Magnetics, 51*(3), 1–4.

Semcad, X. (2022). Schmid & Partner Engineering AG, Switzerland.

Sen, S., & Moyra, T. (2019). Compact microstrip low-pass filtering power divider with wide harmonic suppression. *IET Microwaves, Antennas & Propagation, 13*(12), 2026–2031.

Sevgi, L. (2014). *Electromagnetic modeling and simulation.* IEEE Press Series on Electromagnetic Wave Theory.

Shaker, G. S. A., Bakr, M. H., Sangary, N., & Safavi-Naeini, S. (2009). Accelerated antenna design methodology exploiting parameterized Cauchy models. *Journal of Progress in Electromagnetic Research (PIER B), 18*, 279–309.

Shan, S., & Wang, G. (2011). Survey of modeling and optimization strategies to solve high-dimensional design problems with computationally-expensive black-box functions. *Structural and Multidisciplinary Optimization, 41*(219), 219–241.

Shen, Y., & Wang, J. (2008). An improved algebraic criterion for global exponential stability of recurrent neural networks with time-varying delays. *IEEE Transactions on Neural Networks, 19*(3), 528–531.

Shitvov, A., Schuchinsky, A. G., Steer, M. B., & Wetherington, J. M. (2014). Characterisation of nonlinear distortion and intermodulation in passive devices and antennas. In *8th European Conference on Antennas and Propagation (EuCAP 2014)* (pp. 1454–1458). The Hague.

Siegler, J., Ren, J., Leifsson, L., Koziel, S., & Bekasiewicz, A. (2016). Supersonic airfoil shape optimization by variable-fidelity models and manifold mapping. *Procedia Computer Science, 80*, 1103–1113.

Simpson, T. W., Peplinski, J., Koch, P. N., & Allen, J. K. (2001). Metamodels for computer-based engineering design: Survey and recommendations. *Engineering Computers, 17*, 129–150.

Smith, G. C., & Fortin, A. S. (2022). Verification of eddy properties in operational oceanographic analysis systems. *Ocean Modelling, 72*, 101982.

Smola, A. J., & Schölkopf, B. (2004). A tutorial on support vector regression. *Statistics and Computing, 14*, 199–222.

Sobester, A., & Forrester, A. I. J. (2015). *Aircraft aerodynamic design: Geometry and optimization*. John Wiley & Sons.

Star-CCM+. (2015). CD-adapco Group, 60 Broadhollow Road, Melville, NY 11747, USA.

Styblinski, M. A., & Opalski, L. J. (1986). Algorithms and software tools for IC yield optimization based on fundamental fabrication parameters. *IEEE Transactions on Computer-Aided Design of Integrated Circuits and Systems, 5*(1), 79–89.

Sudret, B. (2008). Global sensitivity analysis using polynomial chaos expansions. *Reliability Engineering & System Safety, 93*(7), 964–979.

Sullivan, D. M. (2013). *Electromagnetic simulation using the FDTD method* (2nd ed.). John Wiley & Sons.

Swanson, D. G., & Hoefer, W. J. R. (2003). *Microwave circuit modeling using electromagnetic field simulation*. Artech House Microwave Library.

Taflove, A., & Hagness, S. C. (Eds.). (2005). *Computational electrodynamics. The finite-difference time-domain approach* (3rd ed.). Artech House.

Tang, D., & Luo, X. (2021). Compact filtering balun with wide stopband and low radiation loss using hybrid microstrip and substrate-integrated defected ground structure. *IEEE Microwave and Wireless Components Letters, 31*(6), 549–552.

Tang, M., Chen, X., Li, M., & Ziolkowski, R. W. (2018). Particle swarm optimized, 3-D-printed, wideband, compact hemispherical antenna. *IEEE Antennas and Wireless Propagation Letters, 17*(11), 2031–2035.

Tao, J., Liao, C., Zeng, X., & Li, X. (2016). Harvesting design knowledge from internet: High-dimensional performance trade-off modeling for large-scale analog circuits. *IEEE Transactions on Computer-Aided Design of Integrated Circuits and Systems, 35*(1), 23–36.

Tao, J., Wang, F., Cachecho, P., Zhang, W., Sun, S., Li, X., Kanj, R., Gu, C., & Zeng, X. (2019). Large-scale circuit performance modeling by bayesian model fusion. In I. Elfadel, D. Boning, & X. Li (Eds.), *Machine learning in VLSI computer-aided design*. Springer.

Taran, N., Ionel, D. M., & Dorrell, D. G. (2018). Two-level surrogate-assisted differential evolution multi-objective optimization of electric machines using 3-D FEA. *IEEE Transactions on Magnetics, 54, no. 11*, paper 8107605.

Toivanen, J. I., Makinen, R. A. E., Jarvenpaa, S., Yla-Oijala, P., & Rahola, J. (2009). Electromagnetic sensitivity analysis and shape optimization using method of moments and automatic differentiation. *IEEE Transactions on Antennas and Propagation, 57*(1), 168–175.

Toktas, A., Ustun, D., & Tekbas, M. (2019). Multi-objective design of multi-layer radar absorber using surrogate-based optimization. *IEEE Transactions on Microwave Theory and Techniques, 67*(8), 3318–3329.

Toropov, V. V. (1989). Simulation approach to structural optimization. *Structural Optimization, 1*, 37–46.

Tropp, J. A. (2004). Greed is good: Algorithmic results for sparse approximation. *IEEE Transactions on Information Theory, 50*(10), 2231–2242.

Ullah, U., & Koziel, S. (2018). A broadband circularly polarized wide-slot antenna with a miniaturized footprint. *IEEE Antennas and Wireless Propagation Letters, 17*(12), 2454–2458.

Upreti, S. R. (2017). *Process modeling and simulation for chemical engineers: Theory and practice*. John Wiley & Sons.

Van Nechel, E., Ferranti, F., Rolain, Y., & Lataire, J. (2018). Model-driven design of microwave filters based on scalable circuit models. *IEEE Transactions on Microwave Theory and Techniques, 66*(10), 4390–4396.

Wang, F., Zhang, W., Sun, S., Li, X., & Gu, C. (2013). Bayesian model fusion: largescale performance modeling of analog and mixed-signal circuits by reusing early-stage data. In *Design Automation Conference (DAC), Austin, Texas, USA*.

Wang, F., Cachecho, P., Zhang, W., Sun, S., Li, X., Kanj, R., & Gu, C. (2016). Bayesian model fusion: Large-scale performance modeling of analog and mixed-signal circuits by reusing early-stage data. *IEEE Transactions on Computer-Aided Design of Integrated Circuits and Systems, 35*(8), 1255–1268.

Wang, K., Ding, D., & Chen, R. (2018). A surrogate modeling technique for electromagnetic scattering analysis of 3-d objects with varying shape. *IEEE Antennas and Wireless Propagation Letters, 17*(8), 1524–1527.

Webb, J. P. (2004). Matching a given field using hierarchal vector basis functions. *Electromagnetics, 24*, 113–122.

Wehner, M. F., Bala, G., Duffy, P., Mirin, A. A., & Romano, R. (2010). Towards direct simulation of future tropical cyclone statistics in a high-resolution global atmospheric model. *Advances in Meteorology, 2010*, 1–13.

Wei, X., Wu, Y.-Z., & Chen, L.-P. (2012). A new sequential optimal sampling method for radial basis functions. *Applied Mathematics and Computation, 218*(19), 9635–9646.

Wei, F., Guo, Y. J., Qin, P., & Wei Shi, X. (2015). Compact balanced dual- and tri-band bandpass filters based on stub loaded resonators. *IEEE Microwave and Wireless Components Letters, 25*(2), 76–78.

White, J. F. (2004). *High frequency techniques: An introduction to RF and microwave design and computer simulation*. Wiley-IEEE Press.

Wild, S. M., Regis, R. G., & Shoemaker, C. A. (2008). ORBIT: Optimization by radial basis function interpolation in trust-regions. *SIAM Journal on Scientific Computing, 30*, 3197–3219.

Willcox, K., & Peraire, J. (2002). Balanced model reduction via the proper orthogonal decomposition. *AIAA Journal, 40*(11), 2323–2330.

Wu, X., Peng, X., Chen, W., & Zhang, W. (2019). A developed surrogate-based optimization framework combining HDMR-based modeling technique and TLBO algorithm for high-dimensional engineering problems. *Structural and Multidisciplinary Optimization, 60*(2), 663–680.

Wu, Q., Wang, H., & Hong, W. (2020). Multistage collaborative machine learning and its application to antenna modeling and optimization. *IEEE Transactions on Antennas and Propagation, 68*(5), 3397–3409.

Wu, Q., Chen, W., Yu, C., Wang, H., & Hong, W. (2021). Multilayer machine learning-assisted optimization-based robust design and its applications to antennas and arrays. *IEEE Transactions on Antennas and Propagation, 69*(9), 6052–6057.

XFDTD. (2021). Remcom, Inc., South Allen 315, Suite 416, State College, PA 16801.

Xiao, L., Shao, W., Ding, X., & Wang, B. (2018). Dynamic adjustment kernel extreme learning machine for microwave component design. *IEEE Transactions on Microwave Theory and Techniques, 66*(10), 4452–4461.

Xiong, F., Xiong, Y., Chen, W., & Yang, S. (2009). Optimizing Latin hypercube design for sequential sampling of computer experiments. *Engineering Optimization, 41*(8), 793–810.

Xiu, D. (2009). Fast numerical methods for stochastic computations: A review. *Communications in Computational Physics, 5*(2–4), 242–272.

Xu, S., Liu, H., Wang, X., & Jiang, X. (2014). A robust error-pursuing sequential sampling approach for global metamodeling based on Voronoi diagram and cross validation. *ASME Journal of Mechanical Design, 136*(7).

Yang, X. S. (2010). *Engineering optimization: An introduction with metaheuristic applications*. Wiley.

Yang, S. H., & Kiang, J. F. (2015). Optimization of sparse linear arrays using harmony search algorithms. *IEEE Transactions on Antennas and Propagation, 63*(11), 4732–4738.

Yang, Z., Qiu, H., Gao, L., Jiang, C., & Zhang, J. (2019). Two-layer adaptive surrogate-assisted evolutionary algorithm for high-dimensional computationally expensive problems. *Journal of Global Optimization, 74*(2), 327–359.

Yelten, M. B., Zhu, T., Koziel, S., Franzon, P. D., & Steer, M. B. (2012). Demystifying surrogate modeling for circuits and systems. *IEEE Circuits and Systems Magazine, 12*(1), 45–63.

Yondo, R., Andrés, E., & Eusebio, V. (2018). A review on design of experiments and surrogate models in aircraft real-time and many-query aerodynamic analyses. *Progress in Aerospace Sciences, 96*, 23–61.

Yondo, R., Bobrowski, K., Andrés, E., & Valero, E. (2019). A review of surrogate modeling techniques for aerodynamic analysis and optimization: Current limitations and future challenges in industry. In E. Minisci, M. Vasile, J. Periaux, N. Gauger, K. Giannakoglou, & D. Quagliarella (Eds.), *Advances in evolutionary and deterministic methods for design, optimization and control in engineering and sciences, computational methods in applied sciences* (Vol. 48). Springer.

Zaslavski, A. J. (2010). Parametric optimization. In *Optimization on metric and normed spaces, springer optimization and its applications* (Vol. 44). Springer.

Zhang, C., Feng, F., Gongal-Reddy, V., Zhang, Q. J., & Bandler, J. W. (2015). Cognition-driven formulation of space mapping for equal-ripple optimization of microwave filters. *IEEE Transactions on Microwave Theory and Techniques, 63*(7), 2154–2165.

Zhang, J., Zhang, C., Feng, F., Zhang, W., Ma, J., & Zhang, Q. J. (2018a). Polynomial chaos-based approach to yield-driven EM optimization. *IEEE Transactions on Microwave Theory and Techniques, 66*(7), 3186–3199.

Zhang, C., Jin, J., Na, W., Zhang, Q., & Yu, M. (2018b). Multivalued neural network inverse modeling and applications to microwave filters. *IEEE Transactions on Microwave Theory and Techniques, 66*(8), 3781–3797.

Zhang, W., Shen, Z., Xu, K., & Shi, J. (2019). A compact wideband phase shifter using slotted substrate integrated waveguide. *IEEE Microwave and Wireless Components Letters, 29*(12), 767–770.

Zhang, Z., Cheng, Q. S., Chen, H., & Jiang, F. (2020a). An efficient hybrid sampling method for neural network-based microwave component modeling and optimization. *IEEE Microwave and Wireless Components Letters, 30*(7), 625–628.

Zhang, W., Feng, F., Yan, S., Na, W., Ma, J., & Zhang, Q. (2020b). EM-centric multiphysics optimization of microwave components using parallel computational approach. *IEEE Transactions on Microwave Theory and Techniques, 68*(2), 479–489.

Zhang, Z., Chen, H., Yu, Y., Jiang, F., & Cheng, Q. S. (2021). Yield-constrained optimization design using polynomial chaos for microwave filters. *IEEE Access, 9*, 22408–22416.

Zhao, P., & Wu, K. (2020). Homotopy optimization of microwave and millimeter-wave filters based on neural network model. *IEEE Transactions on Microwave Theory and Techniques, 68*(4), 1390–1400.

Zhou, Z., Ong, Y. S., Nair, P. B., Keane, A. J., & Lum, K. Y. (2007). Combining global and local surrogate models to accelerate evolutionary optimization. *IEEE Transactions on Systems, Man, and Cybernetics Part C (Applications and Reviews), 37*(1), 66–76.

Zhu, L. (2002). Realistic equivalent circuit model of coplanar waveguide open circuit: Lossy shunt resonator network. *IEEE Microwave and Wireless Components Letters, 12*(5), 175–177.

Zhu, Y., & Cangellaris, A. C. (2001). A new finite element model for reduced order electromagnetic modeling. *IEEE Microwave and Wireless Components Letters, 11*(5), 211–213.

Chapter 2
Simulation-Driven Design of High-Frequency Structures

The primary area of interest of this book is simulation-driven design of high-frequency structures using the response feature technology. For the convenience of the reader, this chapter formulates some of the common design tasks, including local and global optimization, multi-criterial design, surrogate modeling, as well as uncertainty quantification. It also introduces the notation and terminology utilized in this context, such as the system responses, parameter space, objective function, constraints, etc. The discussion is accompanied by representative examples of design problems in the field of antenna and microwave engineering.

2.1 Overview. Notation and Terminology

Here, we introduce the basic notation utilized throughout the book. One of the central concepts is design parameters x_k, $k = 1, \ldots, n$, which represent the degrees of freedom of the system at hand and can be manipulated while solving a particular design task. The goal of this manipulation is to improve the appropriately defined performance metrics (cf. Sect. 2.2). Design variables are most often geometry parameters (system dimensions, Koziel & Ogurtsov, 2014), but they can also be material parameters. The parameters are aggregated into a vector $\boldsymbol{x} = [x_1 \ldots x_n]^T$, where the superscript T represents transposition.

Simulation-driven design tasks are solved over a parameter space X, which is most often an interval $X = [\boldsymbol{l}\ \boldsymbol{u}]$, where $\boldsymbol{l} = [l_1 \ldots l_n]^T$ and $\boldsymbol{u} = [u_1 \ldots u_n]^T$ are lower and upper bounds on the system parameters, i.e., we have $l_k \leq x_k \leq u_k$, $k = 1, \ldots, n$. Here, n stands for the parameter space dimensionality. In practice, we might have additional constraints, which are related directly or indirectly to the parameter vector $\boldsymbol{x}$ (cf. Sect. 2.2).

The system at hand is evaluated using a computational model, which, for the purpose of this book, is normally based on full-wave electromagnetic (EM) analysis.

A. Pietrenko-Dabrowska, S. Koziel, *Response Feature Technology for High-Frequency Electronics. Optimization, Modeling, and Design Automation*,
https://doi.org/10.1007/978-3-031-43845-5_2

On a generic level, the relevant responses of the system will be denoted as $\boldsymbol{R}(\boldsymbol{x})$. Occasionally, when using EM simulation of different fidelities a subscript f will be used to represent a high-fidelity (fine) model $\boldsymbol{R}_f(\boldsymbol{x})$ and a subscript c to represent a low-fidelity (coarse) model $\boldsymbol{R}_c(\boldsymbol{x})$ of the system. While $\boldsymbol{R}(\boldsymbol{x})$ stand for the aggregated outputs, which are typically frequency characteristics, we might also use notation $\boldsymbol{R}(\boldsymbol{x},f)$, which represent the responses at the design $\boldsymbol{x}$ and frequency f. Furthermore, we may distinguish specific types of system characteristics, which are listed below:

- $S_{11}(\boldsymbol{x},f)$ – reflection response at the design $\boldsymbol{x}$ and frequency f
- $S_{kj}(\boldsymbol{x},f)$ – scattering parameters (k and j representing the system ports)
- $G(\boldsymbol{x},f)$ – realized gain at the design $\boldsymbol{x}$ and frequency f
- $AR(\boldsymbol{x},f)$ – axial ratio at the design $\boldsymbol{x}$ and frequency f
- $A(\boldsymbol{x})$ – footprint area at the design $\boldsymbol{x}$

Other types of responses will be introduced in a particular context as necessary.

2.2 Local and Global Parameter Tuning

Parameter tuning is perhaps the most common type of simulation-based design task undertaken in practice. It is employed to improve the performance of the system at hand by adjusting its geometry and/or material parameters. Often, it is carried out in a local sense, especially when a decent initial design is available, typically obtained at the initial stages of the design process (selection of the component topology, parametric studies; Bashir et al., 2017; Palazzi et al., 2018). In many cases, global optimization is necessary, which might be due to unavailability of a sufficiently good starting point, multimodality of the problem, or simply a large parameter space (Liu et al., 2020; Pietrenko-Dabrowska & Koziel, 2021; Zhu et al., 2017).

Parameter tuning is formulated as a nonlinear minimization problem of the form (Koziel & Ogurtsov, 2019)

$$\boldsymbol{x}^* = \arg\min_{\boldsymbol{x}\in X_f} U(\boldsymbol{x}) \tag{2.1}$$

where $U(\boldsymbol{x})$ is a scalar objective function, whereas $\boldsymbol{x} \in X_f \subseteq R^n$. In (2.1), X_f is the feasible space, which is defined as a set of parameter vectors that satisfy the assumed design constraints; $\boldsymbol{x}^*$ is the design optimum to be found. The problem (2.1) can also be formulated with explicit constraints as

$$\boldsymbol{x}^* = \arg\min U(\boldsymbol{x}) \tag{2.2}$$

subject to

$$g_k(\boldsymbol{x}) \le 0, k = 1, \ldots, n_g \tag{2.3}$$

$$h_k(\boldsymbol{x}) = 0, k = 1, \ldots, n_h \tag{2.4}$$

Most practical problems utilize box constraints, i.e., the lower and upper bounds for design variables, $\boldsymbol{l} \le \boldsymbol{x} \le \boldsymbol{u}$, where the inequalities are understood component-wise (cf. Sect. 2.1). Other constraints may be related to either geometry conditions (e.g., maximum component footprint area) or electrical/field properties of the system at hand, e.g., maximum in-band reflection over a target frequency range, maximum axial ratio, tolerance on the target power split ratio, maximum side lobe level, etc.

Majority of constraints are expensive to evaluate (require EM analysis of the system); therefore their explicit handling is normally inconvenient. An alternative is implicit treatment, where the problem is turned into a formally unconstrained one, whereas possible constraint violations are added to the primary objective in the form of penalty factors. The reformulated problem is stated as

$$\boldsymbol{x}^* = \arg\min_{\boldsymbol{x}} U_P(\boldsymbol{x}) \tag{2.5}$$

where the cost function U_P is defined as

$$U_P(\boldsymbol{x}) = U(\boldsymbol{x}) + \sum\nolimits_{k=1}^{n_g+n_h} \beta_k c_k(\boldsymbol{x}) \tag{2.6}$$

The penalty functions $c_k(\boldsymbol{x})$ measure violations of the requirements imposed on particular circuit characteristics, whereas β_k are the penalty coefficients. Typically, the penalty functions are defined to quantify relative violation of the constraints, e.g., with respect to the assumed acceptance threshold for the inequality constraints.

Tables 2.1 and 2.2 provide examples of EM-driven design tasks and the associated objective functions for antenna and microwave components, respectively. In both cases, implicit constraint handling is assumed. Figures 2.1 and 2.2 show the examples of initial and optimized designs for a quasi-Yagi antenna and a dual-band power divider, respectively.

Local parameter tuning tasks are often solved using gradient-based methods (Haq & Koziel, 2018; Hassan et al. 2015a, b), whereas popular global optimization techniques often involve nature-inspired algorithms (Jia & Lu, 2019; Li et al., 2019; Tang et al., 2018). In both cases, especially if the system at hand is evaluated using full-wave EM simulation, the primary bottleneck is high cost of the optimization process. This issue led to the development of accelerated methods, largely based on surrogate modeling techniques. Interested reader can find introduction to numerical optimization in Chap. 3.

Table 2.1 Selected EM-driven antenna design tasks

Task description	Objective function
Improve in-band matching $\|S_{11}\|$ at target frequencies $f_{0.j}$, $j = 1, \ldots, N$	$U(\boldsymbol{x}) = \max_{j \in \{1,\ldots,N\}} \|S_{11}(\boldsymbol{x}, f_{0.j})\|$
Improve in-band matching $\|S_{11}\|$ over target (relative) bandwidths B_k, centered at $f_{0.j}$, $j = 1, \ldots, N$	$U(\boldsymbol{x}) = \max_{j \in \{1,\ldots,N\}} \left\{ f_{0.j}\left(1 - \frac{B_j}{2}\right) \le f \le f_{0.j}\left(1 + \frac{B_j}{2}\right) : \|S_{11}(\boldsymbol{x}, f)\| \right\}$
Maximize average in-band gain while ensuring reflection level of -10 dB or better. Absolute operating bands are B_k, centered at $f_{0.k}$, $k = 1, \ldots, N$	$U(\boldsymbol{x}) = -\frac{1}{\sum_{k=1}^{N} B_k} \sum_{k=1}^{N} \int_{f_{0.k} - B_k/2}^{f_{0.k} + B_k/2} G(\boldsymbol{x}, f)df + \beta c(\boldsymbol{x})^2$ where $c(\boldsymbol{x}) = \max_{k \in \{1,\ldots,N\}} \left\{ f_{0.k} - \frac{B_k}{2} \le f \le f_{0.k} - \frac{B_k}{2} : \max\left\{ \frac{\|S_{11}(\boldsymbol{x}, f)\| + 10}{10}, 0 \right\} \right\}$
Minimize in-band axial ratio while ensuring reflection level of -10 dB or better. Relative operating bands are B_k, centered at $f_{0.k}$, $k = 1, \ldots, N$	$U(\boldsymbol{x}) = \max_{k \in \{1,\ldots,N\}} \left\{ \begin{array}{c} f_{0.k}\left(1 - \frac{B_k}{2}\right) \le f \le f_{0.k}\left(1 + \frac{B_k}{2}\right) : \\ AR(\boldsymbol{x}, f_{0.k}) \end{array} \right\} + \beta c(\boldsymbol{x})^2$ where $c(\boldsymbol{x}) = \max_{k \in \{1,\ldots,N\}} \left\{ f_{0.k} - \frac{B_k}{2} \le f \le f_{0.k} - \frac{B_k}{2} : \max\left\{ \frac{\|S_{11}(\boldsymbol{x}, f)\| + 10}{10}, 0 \right\} \right\}$
Reduce antenna footprint $A(\boldsymbol{x})$ while ensuring reflection level of -10 dB or better at the center frequency f_0, and broadside gain of 5 dB or better at all operating frequencies $f_{0.k}$, $k = 1, \ldots, N$	$U(\boldsymbol{x}) = A(\boldsymbol{x}) + \beta_1 c_1(\boldsymbol{x})^2 + \beta_2 c_2(\boldsymbol{x})^2$ where $c_1(\boldsymbol{x}) = \max_{k \in \{1,\ldots,N\}} \left\{ \frac{\|S_{11}(\boldsymbol{x}, f_{0.k})\| + 10}{10}, 0 \right\}$ $c_2(\mathbf{x}) = \max_{k \in \{1,\ldots,N\}} \left\{ \frac{5 - G(\mathbf{x}, f_{0.k})}{5}, 0 \right\}$

2.3 Multi-objective Optimization

Majority of engineering design problems are of multi-objective nature, i.e., there are several performance figures that have to be controlled at the same time. In high-frequency electronics, typical figures include bandwidth, gain, radiation pattern, axial ratio, power split ratio, sidelobe level, as well as the structure size (for compact structures) (Azaro et al., 2008; Ding et al., 2008; Gautam et al., 2013; Kuwahara, 2005; Smolders & Johannsen, 2011; Viani et al., 2008). In most cases, either because the designer's priorities are known beforehand or just to simplify the design task, the problem can be reformulated into a single-objective one, which can be achieved by selecting the primary goal and handling the remaining objectives as constraints (Koziel & Pietrenko-Dabrowska, 2020). Another popular technique is aggregation of the objectives into a scalar cost function using a weighted sum approach (Kuwahara, 2005; Lizzi et al., 2009) or penalty functions (Koziel, 2017). Nevertheless, in some situations, it might be important to acquire more comprehensive information about the system at hand, in particular, to identify the best possible trade-offs between the conflicting criteria. If this is the case, defaulting to genuine

Table 2.2 Selected EM-driven microwave design tasks

Task description	Objective function
Improve matching $\lvert S_{11}\rvert$ of an impedance transformer at target frequencies $f_{0.j}$, $j = 1, \ldots, N$	$U(\boldsymbol{x}) = \max_{k\in\{1,\ldots,N\}} \lvert S_{11}(\boldsymbol{x}, f_{0.k}) \rvert$
Improve matching $\lvert S_{11}\rvert$ of a bandpass filter over the frequency range $F = [f_1\ f_2]$; maintain in-band transmission ripple lower than 0.5 dB within F	$U(\boldsymbol{x}) = \max_{f_1\le f\le f_2} \lvert S_{11}(\boldsymbol{x}, f) \rvert + \beta c(\boldsymbol{x})$ where $c(\boldsymbol{x}) = \max_{f_1\le f\le f_2} \left\{\frac{\lvert S_{21}(\boldsymbol{x},f)\rvert - 0.5}{0.5}, 0\right\}$
Improve matching $\lvert S_{11}\rvert$ and isolation $\lvert S_{41}\rvert$ of a microwave coupler, and ensure power split $d_S(\boldsymbol{x},f) = \lvert\, \lvert S_{21}(\boldsymbol{x},f)\rvert - \lvert S_{31}(\boldsymbol{x},f)\rvert \,\rvert = K_j$, both at target frequencies $f_{0.j}$, $j = 1, \ldots, N$	$U(\boldsymbol{x}) = \max_{k\in\{1,\ldots,N\}} \{\lvert S_{11}(\boldsymbol{x}, f_{0.k})\rvert, \lvert S_{41}(\boldsymbol{x}, f_{0.k})\rvert\} + \beta c(\boldsymbol{x})^2$ where $c(\boldsymbol{x}) = \left\Vert \begin{bmatrix} \lvert S_{31}(\boldsymbol{x}, f_{0.1})\rvert - \lvert S_{21}(\boldsymbol{x}, f_{0.1})\rvert \\ \vdots \\ \lvert S_{31}(\boldsymbol{x}, f_{0.N})\rvert - \lvert S_{21}(\boldsymbol{x}, f_{0.N})\rvert \end{bmatrix} - \begin{bmatrix} K_1 \\ \vdots \\ K_N \end{bmatrix} \right\Vert$
Reduce footprint $A(\boldsymbol{x})$ of a microstrip coupler while maintaining matching and isolation at -20 dB or better, and equal power split ratio, both over operating bands B_k, centered at $f_{0.k}$, $k = 1, \ldots, N$	$U(\boldsymbol{x}) = A(\boldsymbol{x}) + \beta_1 c_1(\boldsymbol{x})^2 + \beta_2 c_2(\boldsymbol{x})^2$ where $c_1(\boldsymbol{x}) = \max_{k\in\{1,\ldots,N\}} \left\{ \begin{array}{c} f_{0.k} - \frac{B_k}{2} \le f \le f_{0.k} - \frac{B_k}{2} : \\ \max\left\{ \frac{\max\{\lvert S_{11}(\boldsymbol{x},f)\rvert, \lvert S_{41}(\boldsymbol{x},f)\rvert\} + 20}{20} \right\} \end{array} \right\}$ $c_2(\boldsymbol{x}) = \max_{k\in\{1,\ldots,N\}} \{\lvert\lvert S_{31}(\boldsymbol{x}, f_{0.k})\rvert - \lvert S_{21}(\boldsymbol{x}, f_{0.k})\rvert\rvert\}$

multi-objective optimization becomes necessary. Multi-objective optimization leads to a set of alternative designs representing a so-called Pareto front (Fonseca, 1995). Subsequently, a decision-making process allows for selecting one of these designs as a final solution.

Perhaps the most popular multi-objective optimization approaches are population-based metaheuristics (Deb, 2001; Talbi, 2009). These algorithms are capable of yielding the entire Pareto set in one algorithm run; however, the computational cost of evolutionary optimization may be extraordinary: thousands or even tens of thousands of objective function evaluations. In particular, direct multi-objective optimization of expensive EM simulation models of high-frequency structures is normally prohibitive.

For the purposes of considerations of this chapter, let $\boldsymbol{F}(\boldsymbol{x}) = [F_1(\boldsymbol{x}) \ldots F_{Nobj}(\boldsymbol{x})]^T$ be a vector of design objectives, all to be minimized. As before, $\boldsymbol{x} \in X$ is a vector of designable parameters, whereas X is the parameter space. Formally speaking, the goal is to solve argmin$\{\boldsymbol{x}: \boldsymbol{F}(\boldsymbol{x})\}$. In most practical cases, the objectives are conflicting with each other, so that improvement of one leads to degradation of the others. Thus, one of the fundamental issues is how to compare the designs against each other.

Assuming that $N_{obj} > 1$, any two designs $\boldsymbol{x}^{(1)}$ and $\boldsymbol{x}^{(2)}$ for which $F_k(\boldsymbol{x}^{(1)}) < F_k(\boldsymbol{x}^{(2)})$ and $F_l(\boldsymbol{x}^{(2)}) < F_l(\boldsymbol{x}^{(1)})$ for at least one pair $k \neq l$ are not commensurable, i.e., none is

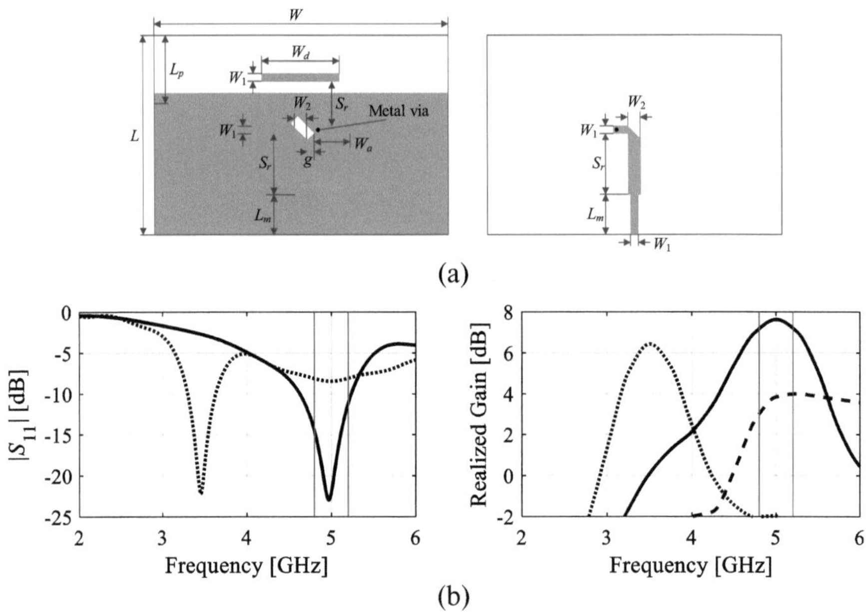

Fig. 2.1 Optimization of a quasi-Yagi antenna (Koziel et al., 2020): (**a**) antenna geometry (front and back sides), (**b**) initial (····) and optimized responses (—); design specifications are to maximize antenna gain at 5.0 GHz while maintaining $|S_{11}| \leq -10$ dB in the frequency range from 4.8 GHz to 5.2 GHz

better than the other in the multi-objective sense. For the sake of comparing the designs, a Pareto dominance relation $\prec$ can be used, which is defined as follows (Fonseca, 1995): for the two designs $\boldsymbol{x}$ and $\boldsymbol{y}$, we have $\boldsymbol{x} \prec \boldsymbol{y}$ ($\boldsymbol{x}$ dominates $\boldsymbol{y}$) if $F_k(\boldsymbol{x}) \leq F_k(\boldsymbol{y})$ for all $k = 1, \ldots, N_{\text{obj}}$, and $F_k(\boldsymbol{x}) < F_k(\boldsymbol{y})$ for at least one k. The goal of multi-objective optimization is to find a representation of a so-called Pareto front (or Pareto optimal set) $\boldsymbol{F}(X_P)$ of the design space image through $\boldsymbol{F}(X)$, such that for any $\boldsymbol{x} \in X_P$, there is no $\boldsymbol{y} \in X$ for which $\boldsymbol{y} \prec \boldsymbol{x}$ (Fonseca, 1995). The concept of the Pareto dominance relation is explained in Fig. 2.3.

Figure 2.4 explains the concept of the Pareto front for the two-objective design problem. The part of the boundary of the design space image through $\boldsymbol{F}$, $\boldsymbol{F}(X)$, marked with the thick line, is the Pareto front, i.e., there are no designs in X that would dominate any design in X_P. In general, the Pareto front is at most an $N_{\text{obj}} - 1$ dimensional manifold in the feature space (e.g., a curve for $N_{\text{obj}} = 2$ and a surface for $N_{\text{obj}} = 3$, etc.).

Figure 2.5 illustrates a Pareto set for a UWB dipole antenna (Koziel & Bekasiewicz, 2016) optimized for two design objectives: reduction of the structure size and minimization of the maximum in-band reflection level. It can be observed that the antenna size changes from slightly over 700 mm^2 to around 1200 mm^2, whereas the corresponding reflection level changes from -9 dB to -14 dB.

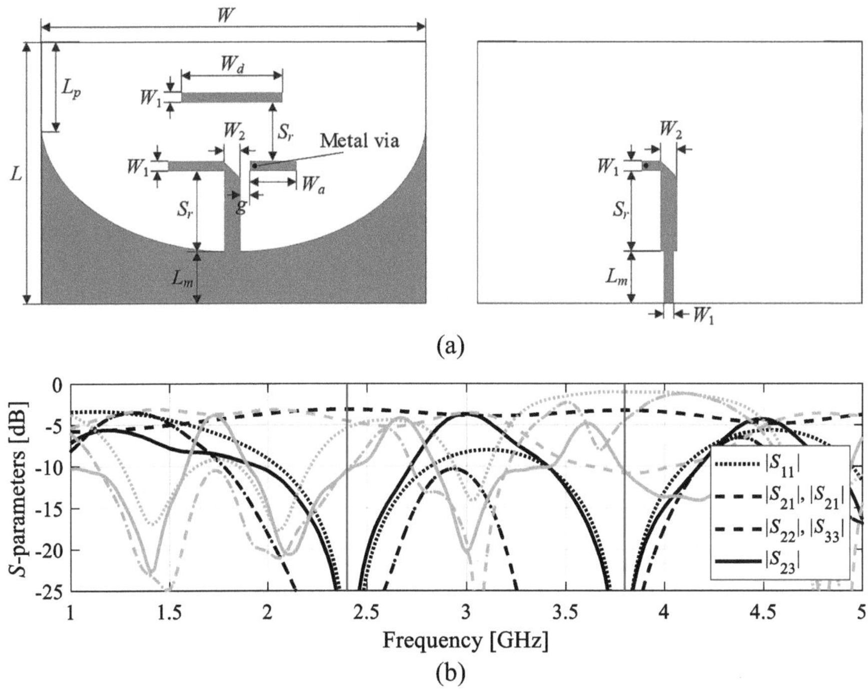

Fig. 2.2 Optimization of a dual-band power divider (Koziel et al., 2022): (**a**) circuit geometry and (**b**) initial (gray) and optimized responses (black); design specifications are to minimize the input matching $|S_{11}|$, output matching $|S_{22}|$, $|S_{33}|$, and isolation $|S_{23}|$, while maintaining equal power division ratio, all at 2.4 GHz and 3.8 GHz

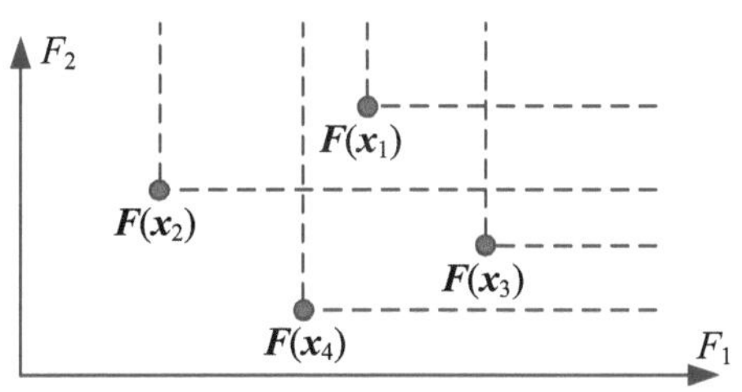

Fig. 2.3 Illustration of the Pareto dominance relation for a two-objective feature space. For the four designs $\boldsymbol{x}_k$, $k = 1, 2, 3, 4$, we have the following relations: $\boldsymbol{x}_2 \prec \boldsymbol{x}_1$, $\boldsymbol{x}_4 \prec \boldsymbol{x}_3$, and $\boldsymbol{x}_4 \prec \boldsymbol{x}_1$; the remaining pairs of designs are not in a dominance relation with each other

The relationship between design parameters and objectives and, consequently, the geometry of the Pareto front depend on various factors, including complexity of the system under design. In general, the following three situations are possible (Tan et al., 2005):

- Totally conflicting objectives, i.e., there are no two designs $\boldsymbol{x}, \boldsymbol{y} \in X$ such that $\boldsymbol{x} \prec \boldsymbol{y}$ or $\boldsymbol{y} \prec \boldsymbol{x}$. In particular, no optimization is necessary for this class of problem

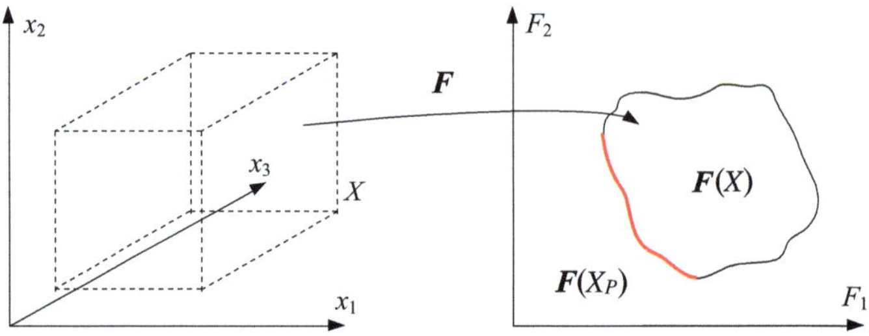

Fig. 2.4 Conceptual illustration of the Pareto front for a two-objective design problem and a three-dimensional design space: design space X (left), its image $\boldsymbol{F}(X)$ (right), and the image of the Pareto front $\boldsymbol{F}(X_P)$ marked using a thick line

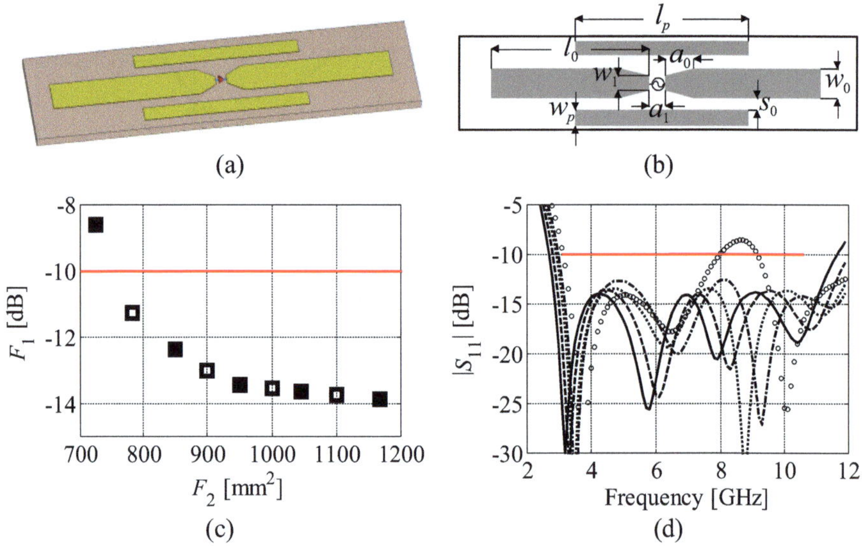

Fig. 2.5 Pareto front illustration for a UWB dipole antenna designed for two objectives (Koziel & Bekasiewicz, 2016), F_1 – size reduction and F_2 – minimization of the maximum in-band reflection: (**a**) visualization, (**b**) geometrical details, (**c**) Pareto set (designs 1, 3, 5, 7, and 9 marked black), and (**d**) reflection characteristics of designs: 1 (°°°), 3 (–·), 5 (····), 7 (– –), 9 (—).

because the feasible set X already represents the global Pareto optimal solutions, i.e., we have $X_P = X$.

- Nonconflicting objectives. In this case, for any two designs $\boldsymbol{x}, \boldsymbol{y} \in X$ we either have $\boldsymbol{x} \prec \boldsymbol{y}$ or $\boldsymbol{y} \prec \boldsymbol{x}$. This class of problems can be converted into a single-objective task by considering only one objective (as the improvement of any of the objectives leads to the improvement of all the others). For this class of problems, the Pareto set is usually a single point.
- Partially conflicting objectives. This case covers all other situations, i.e., when the objectives are neither totally conflicting nor nonconflicting.

For partially conflicting objectives, the image of the problem domain $\boldsymbol{F}(X)$ is a "thick" set, and, consequently, the Pareto front is a proper subset of the domain, i.e., $X_P \subset X$ but $X_P \neq X$. Needless to say, the last situation is the most interesting and the most common one for majority of real-world engineering design problems, including those related to antenna and microwave design.

In the course of multi-objective optimization, a decision has to be made toward selecting a single final design. A decision-making process is based upon the set of trade-off solutions identified in the process, but it can also be interleaved with the optimization itself. The following approaches can be distinguished (Coello Coello et al., 2007):

- A priori preference articulation, where the multi-objective problem is transformed into a single-objective one (e.g., by objective aggregation) prior to optimization.
- Progressive preference articulation, where decision and optimization are intertwined; in particular, partial preference information is provided upon which optimization proceeds.
- Posteriori preference articulation, where a set of efficient candidate solutions is found using a specific algorithm and then a decision is made on the choice of the best solution.

If the relative importance of various objectives is clearly defined and known beforehand, a priori preference articulation seems to be a proper approach. If the knowledge of the problem is limited, the initial goals may be set for the objectives and subsequently modified during the optimization process, especially if it turns out that some objectives are difficult to be met (progressive preference articulation). In general, however, it might be preferable to acquire as much information about the Pareto optimal designs as possible. This would give an idea about available trade-offs between design objectives and allow for selecting the most suitable solution (posteriori preference articulation). As mentioned before, this approach might be very expensive in computational terms. For the purpose of this book, multi-objective optimization is merely understood as a procedure of Pareto set identification.

In practice, multi-objective problems are often treated in a simplistic manner, because of the abundance of numerical algorithms for solving single-objective problems. As mentioned earlier, perhaps the most popular approach is objective aggregation, using, e.g., a weighted sum approach (Marler & Arora, 2010), where a scalar cost function is composed as a linear combination of objectives with the proportionality factors selected to express the designer's priorities. Genuine multi-objective design is most often carried out using population-based nature-inspired algorithms (Goudos et al., 2009; Zhang et al., 2019a), which is because algorithms of this class offer both global search capability and the capability of rendering the entire Pareto set in a single run (Baumgartner et al., 2018). In the case of computationally expensive objectives, such as those considered in this book, straightforward application of nature-inspired methods is not possible. However, their surrogate-assisted versions (Lim et al., 2015; Xia et al., 2014; Xiao et al., 2018) may mitigate the high-cost issues to a certain extent. A conceptual illustration of one of the possible methodologies of this sort has been shown in Fig. 2.6.

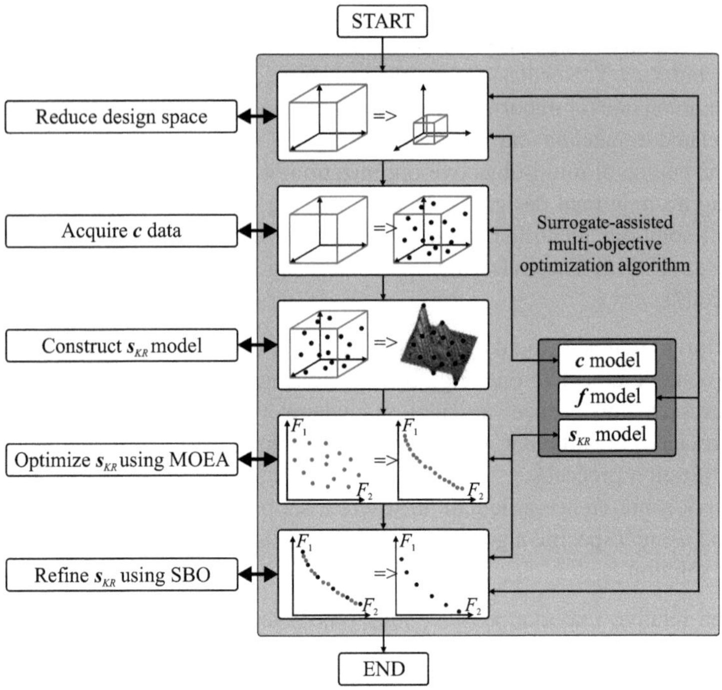

Fig. 2.6 Flowchart of the surrogate-assisted procedure for computationally efficient multi-objective design (Koziel & Bekasiewicz, 2016). The framework employs initial parameter space reduction, variable-fidelity simulation models ($\boldsymbol{c}$ and $\boldsymbol{f}$ refer to the low- and high-fidelity simulation models, respectively; $\boldsymbol{s}_{KR}$ stands for the kriging-based surrogate model), and design refinement scheme required to bring the Pareto optimal designs to the high-fidelity level of accuracy. MOEA stands for multi-objective evolutionary algorithm

2.4 Surrogate Modeling

Surrogate modeling has been attracting considerable attention in the field of high-frequency electronics over the last two decades or so. Widespread utilization of full-wave EM simulation models created a number of practical problems related to high computational cost of massive system evaluations, entailed by numerical design procedures, such as parametric optimization or uncertainty quantification. Fast replacement models are instrumental in accelerating EM-driven procedures. Chapter 4 of this book provides a brief introduction to the subject. Here, we only briefly outline the major components of the modeling process.

The purpose of surrogate modeling has been shown in Fig. 2.7. Given the high-fidelity EM model $\boldsymbol{R}_f(\boldsymbol{x})$, the goal is to construct its reliable replacement, the surrogate $\boldsymbol{R}_s(\boldsymbol{x})$, so that the domain of the latter coincides with that of the high-fidelity model, and $\boldsymbol{R}_s(\boldsymbol{x})$ is possibly close to $\boldsymbol{R}_f(\boldsymbol{x})$ in the respective region of validity. The two major classes of surrogate modeling techniques include data-driven (or approximation) models and physics-based ones (Bandler et al., 2004;

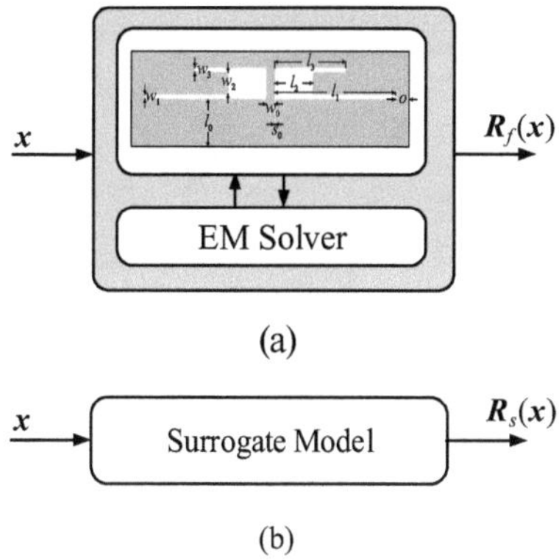

Fig. 2.7 Surrogate modeling concept: (**a**) antenna structure evaluated using full-wave EM analysis; antenna outputs $\boldsymbol{R}_f(\boldsymbol{x})$ are functions of design parameters (vector $\boldsymbol{x}$); (**b**) surrogate model $\boldsymbol{R}_s(\boldsymbol{x})$ constructed to produce antenna outputs as a function of $\boldsymbol{x}$, however, at considerably lower computational cost. Ideally, the model should be constructed so that the domain of $\boldsymbol{R}_s$ coincides with that of $\boldsymbol{R}_f$, and $\boldsymbol{R}_s(\boldsymbol{x}) \approx \boldsymbol{R}_f(\boldsymbol{x})$ over the region of validity

Forrester & Keane, 2009; Koziel & Pietrenko-Dabrowska, 2022a). Data-driven surrogates are constructed by approximating sampled EM simulation data, whereas physics-based ones are typically obtained through appropriate correction of the underlying low-fidelity representation (e.g., equivalent circuit or coarse-discretization EM analysis; Koziel & Ogurtsov, 2012).

The modeling process consists of several steps, which are briefly outlined below:

- *Design of experiments*. This is a common term for strategies for allocating samples to become the training/testing data set for surrogate model construction. Nowadays, as most of the data comes from computer simulation, space-filling designs are recommended, which aim at allocating the samples in a possibly uniform manner (Giunta et al., 2003; Santner et al., 2003; Simpson et al., 2001). If a feedback from the system at hand is utilized, sequential sampling methods might be more efficient (Lehmensiek et al., 2002), where new samples are allocated based on, e.g., estimated levels of model error throughout the parameter space.
- *Training data acquisition*. This step is associated with evaluating the system outputs at the locations assigned in the previous stage. This is a single most expensive part of the modeling process. The number of samples is often limited by available computational budget.
- *Model identification*. This is the actual model construction stage, where the simulation data acquired earlier is approximated (or, in the case of physics-based surrogates, where the low-fidelity model is corrected) to produce the surrogate (Queipo et al., 2005). There are numerous modeling methods available in the literature, some of which are outlined in Chap. 4 of this book.

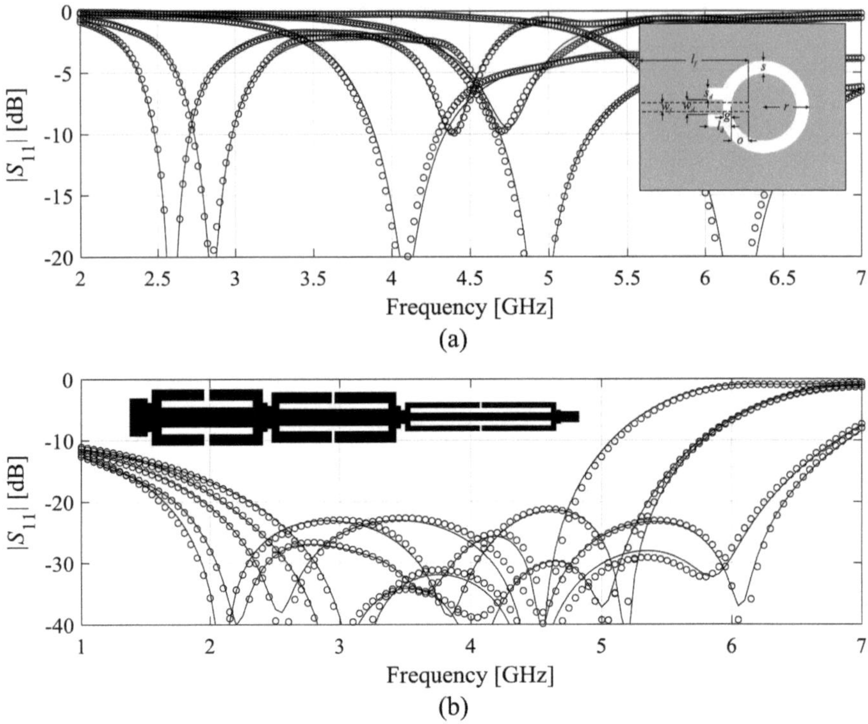

Fig. 2.8 Examples of data-driven surrogate models for (**a**) a microstrip antenna and (**b**) a microwave coupler: (—) EM simulation, (o) surrogate model. The insets show the structure geometries

- *Model validation*. Upon constructing the model, its predictive power has to be verified. Normally, we are interested in the approximation capability (the ability to approximate the training data set) and generalization capability (the ability to predict the system outputs at the locations not used in the model identification process). There are various validation methods, including a split-sample approach (Queipo et al., 2005) and cross-validation (Geisser, 1993). If insufficient accuracy is detected, the model can be further refined by allocating new data samples (cf. sequential sampling; Crombecq et al., 2011; Liu et al., 2016; Wang, 2003; Tong, 2006; Davis & Ierapetritou, 2010).

Figure 2.8 shows the examples of high-frequency structures and the comparison between high-fidelity and surrogate model outputs. The visual alignment shown in the picture corresponds to a relative modeling error at the level of a few percent, which is normally the accuracy required to ensure design utility of the surrogate.

2.5 Uncertainty Quantification

High-frequency systems are normally designed in the nominal sense, with possible deviations of geometry and material parameters (e.g., due to fabrication inaccuracies) being generally neglected. Yet, uncertainties may have detrimental effects on the system operation and performance; therefore, the ability to quantify them is of utmost importance. Among possible types of uncertainties, those pertinent to imperfect manufacturing (Biernacki et al., 2012; Liang et al., 1993) play the most important role in practice. They are stochastic, which makes statistical analysis imperative for their evaluation (Hassan et al., 2015a, b; Koziel et al., 2005). Diminishing the impact of parameter deviations requires handling of appropriately defined figures of merit such as minimization of the response variance or maximization of the yield (Leifsson et al., 2020; Prasad et al., 2016). In the case of high-frequency systems, yield seems to be more suitable because performance specifications are often expressed in a minimax form, in particular, lower/upper acceptance thresholds for *S*-parameters, etc. (Koziel & Pietrenko-Dabrowska, 2021, 2022b).

Evaluation of the effects of parameter tolerances involves statistical analysis (Li et al., 2015). For reliability, it is normally carried out using full-wave electromagnetic (EM) analysis, which is associated with significant computational costs. Nevertheless, it is mandatory whenever simpler models (e.g., equivalent circuits) are insufficient to account for cross-coupling and similar effects (Jin et al., 2017). Improving computational efficiency of statistical analysis can be achieved using simplistic yet inaccurate methods (e.g., worst-case analysis; Zhang & Rahmat-Samii, 2018), or surrogate-assisted techniques (Ma et al., 2018), where repetitive EM simulations are replaced by evaluation of a fast replacement model, usually prepared beforehand. Popular modeling methods include polynomial approximation (Matoglu et al., 2004), neural networks (NNs) (Kim et al., 2020), or polynomial chaos expansion (PCE) (Du & Roblin, 2017; Rossi et al., 2014). Despite their advantages, surrogate-based methods are affected by the curse of dimensionality leading to excessive costs of model rendition for more complex systems. Alleviation of these difficulties is partially possible by the employment of hybrid techniques (e.g., PC kriging; Leifsson et al., 2020), dimensionality reduction (Ochoa & Cangellaris, 2013), multi-resolution methods (space mapping, Zhang et al., 2019b; co-kriging, Lefebvre et al., 2015), or model order reduction (Spina et al., 2014).

Reducing the effects of manufacturing tolerances is even more essential than their evaluation. The relevant procedures are often referred to as robust design (also, yield-driven design, tolerance-aware optimization, Kouassi et al., 2016; Abdel-Malek et al., 2006). In practical terms, this can be accomplished by improving the statistical merit functions of choice, particularly the yield. Unfortunately, yield maximization is a CPU-heavy task, to the extent of being prohibitive when directly executed at the level of EM simulation models. Surrogate-assisted procedures offer viable workarounds (Du & Roblin, 2017; Kim et al., 2020). Some of the most popular modeling methods utilized in this context are polynomial approximations (Matoglu et al., 2004), space mapping (Abdel-Malek et al., 2006), NNs (Rayas-Sanchez & Gutierrez-Ayala, 2006), and PCE (Zhang et al., 2018). As for statistical analysis, the bottleneck is a potentially high cost of the surrogate model setup, related to the dimensionality of the parameter space and parameter ranges. A partial mitigation has been offered by sequential approximation optimization (SAO) (Koziel & Bekasiewicz, 2018), where the metamodel is rendered along the optimization path, within the limited-volume domains centered at the current design produced by the robust design procedure. Another option is to employ the response feature technology (Koziel & Bandler, 2015). This method capitalizes on reduced nonlinearity of the relationships between the appropriately selected characteristic points of the system responses and geometry parameters of the structure under design. The latter enables a construction of accurate metamodels using small training data sets (Pietrenko-Dabrowska et al., 2022).

As mentioned earlier, in antenna and microwave design, performance requirements are frequently formulated in a minimax form, i.e., by appointing upper/lower acceptance levels on the electrical characteristics of the device at hand. These may include maximum in-band reflection (antennas) or maximum power split error within the operating band (couplers) (Budimir & Goussetis, 2003). As a consequence, a commonly utilized figure of merit is the yield (Koziel et al., 2005), i.e., the percentage of designs for which the target specifications are met given the assumed deviations of parameters.

Let $d\boldsymbol{x}$ denote a vector of deviations of the circuit parameters from their nominal values. The deviations originate from manufacturing inaccuracy and are quantified using probability distributions that are specific to the fabrication process, for example, joint Gaussian distribution with zero mean and variance σ (common for all parameters). In a more generic setup, the distribution can be determined by a covariance matrix that accounts for the circuit topology (e.g., correlations between deviations for certain parameters).

The fabrication yield Y is computed by integrating the probability density function $p(\boldsymbol{x},\boldsymbol{x}^{(0)})$ that describes deviations of $\boldsymbol{x}$ from the nominal design $\boldsymbol{x}^{(0)}$. The integration is executed over the feasible set X_f, which contains all designs satisfying the performance specifications. We have (Kim et al., 2020)

$$Y\left(\boldsymbol{x}^{(0)}\right) = \int_{X_f} p\left(\boldsymbol{x},\boldsymbol{x}^{(0)}\right) d\boldsymbol{x} \tag{2.7}$$

As the geometry of the feasible set is not known explicitly, in practice, the yield is estimated through numerical integration, most often by means of Monte Carlo

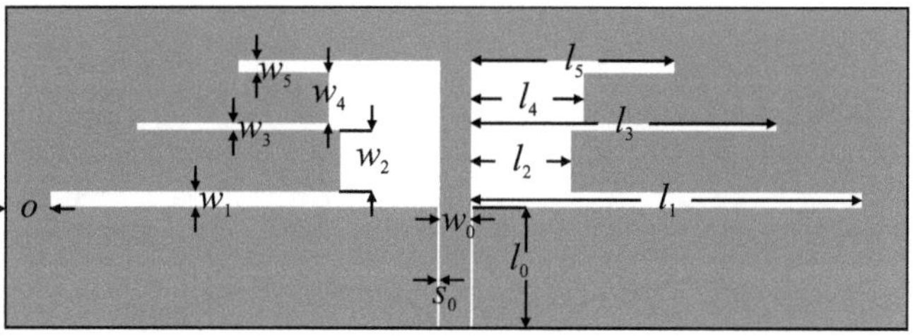

Fig. 2.9 Geometry of a triple-band uniplanar dipole antenna

simulation. Given a set of random observables $d\boldsymbol{x}^{(k)}$, $k = 1, \ldots, N_r$, allocated according to the density function p, we get

$$Y\left(\boldsymbol{x}^{(0)}\right) = N_r^{-1} \sum\nolimits_{k=1}^{N_r} H\left(\boldsymbol{x}^{(0)} + d\boldsymbol{x}^{(k)}\right) \tag{2.8}$$

where $H(\boldsymbol{x}) = 1$ if the performance specifications are satisfied for the design $\boldsymbol{x}$, $H(\boldsymbol{x}) = 0$ otherwise.

The yield optimization problem is formulated as

$$\boldsymbol{x}^* = \arg\min_{\boldsymbol{x}} \{-Y(\boldsymbol{x})\} \tag{2.9}$$

Solving (2.9) entails repetitive yield estimations, which is associated with a prohibitive computational cost when executed directly at the level of EM analysis. The majority of practical approaches employ fast surrogate models for the purpose of evaluating (2.8).

For the sake of illustration, consider a triple-band antenna shown in Fig. 2.9. The structure is described using ten geometry parameters, $\boldsymbol{x} = [l_1\ l_2\ l_3\ l_4\ l_5\ w_1\ w_2\ w_3\ w_4\ w_5]^T$. The design objective is to minimize the in-band reflection within 4-percent symmetric bandwidths centered at the frequencies $f_{01} = 2.45$ GHz, $f_{02} = 3.6$ GHz, and $f_{03} = 5.3$ GHz. The nominal design is $\boldsymbol{x}^{(0)} = [35.42\ 11.54\ 26.07\ 8.09\ 17.14\ 0.60\ 0.99\ 1.44\ 0.78\ 1.17]^T$. Deviations of geometry parameters are described by the independent uniform probability distributions with the maximum deviation $d_{\max} = 0.05$ mm. The estimated yield at $\boldsymbol{x}^{(0)}$ is 58%. The surrogate-assisted methodology using domain-confined surrogates (Pietrenko-Dabrowska et al., 2020) produces the design $\boldsymbol{x}^* = [35.37\ 11.52\ 26.0\ 8.09\ 17.14\ 0.59\ 1.01\ 1.43\ 0.80\ 1.22]^T$ featuring the yield of 69%. Figure 2.10 shows visualization of the EM-based Monte Carlo analysis at the nominal design and at $\boldsymbol{x}^*$.

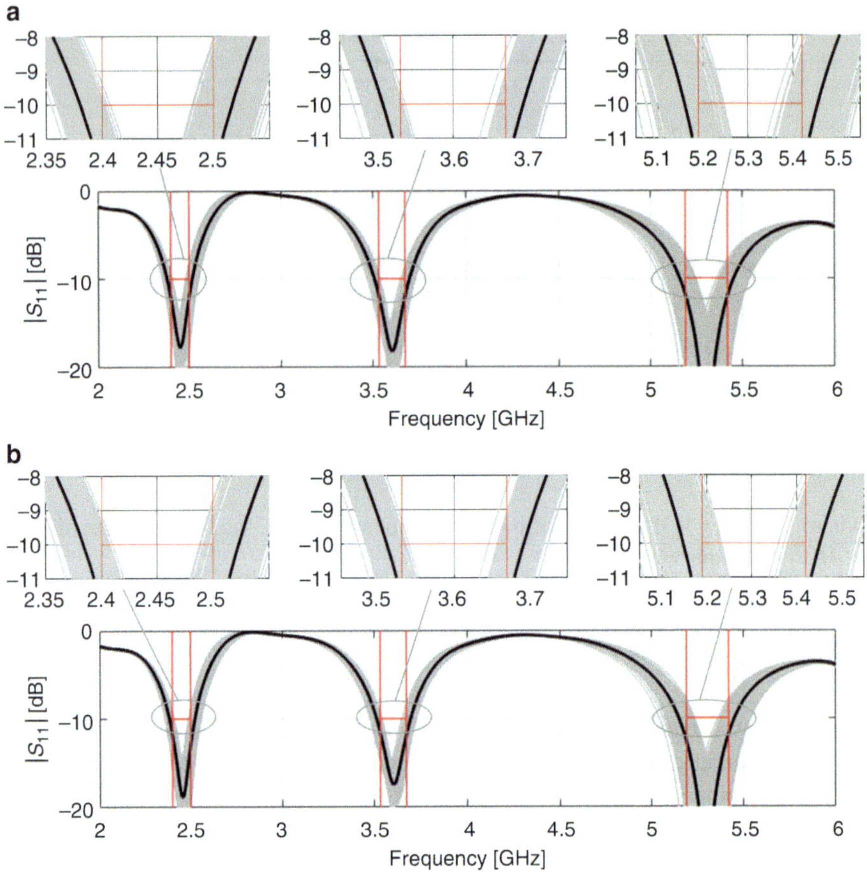

Fig. 2.10 Monte Carlo analysis of antenna of Fig. 2.8 using EM simulations (gray plots): (**a**) nominal design, (**b**) yield-optimized design obtained using the surrogate-assisted procedure (Pietrenko-Dabrowska et al., 2020). Black plots show the antenna response at the nominal and optimized designs, respectively. Vertical lines indicate target operating bandwidths

References

Abdel-Malek, H. L., Hassan, A. S. O., Soliman, E. A., & Dakroury, S. A. (2006). The ellipsoidal technique for design centering of microwave circuits exploiting space-mapping interpolating surrogates. *IEEE Transactions on Microwave Theory and Techniques, 54*, 3731–3738.

Azaro, R., Zeni, E., Rocca, P., & Massa, A. (2008). Innovative design of a planar fractal-shaped GPS/GSM/Wi-Fi antenna. *Microwave and Optical Technology Letters, 50*, 825–829.

Bandler, J. W., Cheng, Q. S., Dakroury, S. A., Mohamed, A. S., Bakr, M. H., Madsen, K., & Søndergaard, J. (2004). Space mapping: The state of the art. *IEEE Transactions on Microwave Theory and Techniques, 52*, 337–361.

Bashir, U., Jha, K. R., Mishra, G., Singh, G., & Sharma, S. K. (2017). Octahedron-shaped linearly polarized antenna for multistandard services including RFID and IoT. *IEEE Transactions on Antennas and Propagation, 65*, 3364–3373.

Baumgartner, P., Baurnfeind, T., Biro, O., Hackl, A., Magele, C., Renhart, W., & Torchio, R. (2018). Multi-objective optimization of Yagi-Uda antenna applying enhanced firefly algorithm with adaptive cost function. *IEEE Transactions on Magnetics, 54*(3), article no. 8000504.

Biernacki, R., Chen, S., Estep, G., Rousset, J., & Sifri, J. (2012). Statistical analysis and yield optimization in practical RF and microwave systems. In *IEEE MTT-S international microwave symposium digest Montreal, Canada* (pp. 1–3).

Budimir, D., & Goussetis, G. (2003). Design of asymmetrical RF and microwave bandpass filters by computer optimization. *IEEE Transactions on Microwave Theory and Techniques, 51*, 1174–1178.

Coello Coello, C. A., Lamont, G. B., & Van Veldhuizen, D. (2007). *Evolutionary algorithms for solving multi-objective problems* (2nd ed.). Springer.

Crombecq, K., Laermans, E., & Dhaene, T. (2011). Efficient space-filling and non-collapsing sequential design strategies for simulation-based modeling. *European Journal of Operational Research, 214*, 683–696.

Davis, E., & Ierapetritou, M. (2010). A centroid-based sampling strategy for kriging global modeling and optimization. *AICHE Journal, 56*, 220–240.

Deb, K. (2001). *Multi-objective optimization using evolutionary algorithms*. Wiley.

Ding, M., Jin, R., Geng, J., Wu, Q., & Yang, G. (2008). Auto-design of band-notched UWB antennas using mixed model of 2D GA and FDTD. *Electronics Letters, 44*, 257–258.

Du, J., & Roblin, C. (2017). Statistical modeling of disturbed antennas based on the polynomial chaos expansion. *IEEE Antennas and Wireless Propagation Letters, 16*, 1843–1847.

Fonseca, C. M. (1995). *Multiobjective genetic algorithms with application to control engineering problems*. PhD thesis, Department of Automatic Control and Systems Engineering, University of Sheffield, UK.

Forrester, A. I. J., & Keane, A. J. (2009). Recent advances in surrogate-based optimization. *Progress in Aerospace Sciences, 45*, 50–79.

Gautam, A. K., Yadav, S., & Kanaujia, B. K. (2013). A CPW-fed compact UWB microstrip antenna. *IEEE Antennas and Wireless Propagation Letters, 12*, 151–154.

Geisser, S. (1993). *Predictive inference*. Chapman and Hall.

Giunta, A. A., Wojtkiewicz, S. F., & Eldred, M. S. (2003). Overview of modern design of experiments methods for computational simulations. *Aerospace Sciences Meeting and Exhibit*, paper AIAA 2003-0649.

Goudos, K., Zaharis, Z. D., Kampitaki, D. G., Rekanos, I. T., & Hilas, C. S. (2009). Pareto optimal design of dualband base station antenna arrays using multi-objective particle swarm optimization with fitness sharing. *IEEE Transactions on Magnetics, 45*(3), 1522–1525.

Haq, M. A., & Koziel, S. (2018). Ground plane alterations for design of high-isolation compact wideband MIMO antenna. *IEEE Access, 6*, 48978–48983.

Hassan, A. S. O., Abdel-Malek, H. L., Mohamed, A. S. A., Abuelfadl, T. M., & Elqenawy, A. E. (2015a). Statistical design centering of RF cavity linear accelerator via non-derivative trust region optimization. In *IEEE international conference numerical EM multiphysics modeling optimization (NEMO)* (pp. 1–3).

Hassan, E., Noreland, D., Augustine, R., Wadbro, E., & Berggren, M. (2015b). Topology optimization of planar antennas for wideband near-field coupling. *IEEE Transactions on Antennas and Propagation, 63*, 4208–4213.

Jia, X., & Lu, G. (2019). A hybrid Taguchi binary particle swarm optimization for antenna designs. *IEEE Antennas and Wireless Propagation Letters, 18*, 1581–1585.

Jin, H., Zhou, Y., Huang, Y. M., Ding, S., & Wu, K. (2017). Miniaturized broadband coupler made of slow-wave half-mode substrate integrated waveguide. *IEEE Microwave and Wireless Components Letters, 27*, 132–134.

Kim, D., Kim, M., & Kim, W. (2020). Wafer edge yield prediction using a combined long short-term memory and feed- forward neural network model for semiconductor manufacturing. *IEEE Access, 8*, 215125–215132.

Kouassi, A., Nguyen-Trong, N., Kaufmann, T., Lallechere, S., Bonnet, P., & Fumeaux, C. (2016). Reliability-aware optimization of a wideband antenna. *IEEE Transactions on Antennas and Propagation, 64*, 450–460.

Koziel, S. (2017). Objective relaxation algorithm for reliable simulation-driven size reduction of antenna structure. *IEEE Antennas and Wireless Propagation Letters, 16*, 1949–1952.

Koziel, S., & Bandler, J. W. (2015). Rapid yield estimation and optimization of microwave structures exploiting feature-based statistical analysis. *IEEE Transactions on Microwave Theory and Techniques, 63*, 107–114.

Koziel, S., & Bekasiewicz, A. (2016). *Multi-objective design of antennas using surrogate models*. World Scientific.

Koziel, S., & Bekasiewicz, A. (2018). Sequential approximate optimization for statistical analysis and yield optimization of circularly polarized antennas. *IET Microwaves, Antennas & Propagation, 12*, 2060–2064.

Koziel, S., & Ogurtsov, S. (2012). Model management for cost-efficient surrogate-based optimization of antennas using variable-fidelity electromagnetic simulations. *IET Microwaves, Antennas & Propagation, 6*, 1643–1650.

Koziel, S., & Ogurtsov, S. (2014). *Antenna design by simulation-driven optimization. Surrogate-based approach*. Springer.

Koziel, S., & Ogurtsov, S. (2019). *Simulation-based optimization of antenna arrays*. World Scientific.

Koziel, S., & Pietrenko-Dabrowska, A. (2020). On computationally-efficient reference design acquisition for reduced-cost constrained modeling and re-design of compact microwave passives. *IEEE Access, 8*, 203317–203330.

Koziel, S., & Pietrenko-Dabrowska, A. (2021). Global EM-driven optimization of multi-band antennas using knowledge-based inverse response-feature surrogates. *Knowledge-Based Systems, 227*, 107189.

Koziel, S., & Pietrenko-Dabrowska, A. (Eds.). (2022a). *Surrogate modeling for high-frequency design: Recent advances*. World Scientific.

Koziel, S., & Pietrenko-Dabrowska, A. (2022b). Tolerance-aware multi-objective optimization of antennas by means of feature-based regression surrogates. *IEEE Transactions on Antennas and Propagation*. Early Access.

Koziel, S., Bandler, J., Mohamed, A., & Madsen, K. (2005). Enhanced surrogate models for statistical design exploiting space mapping technology. In *IEEE MTT-S international microwave symposium digest Long Beach, CA* (pp. 1–4).

Koziel, S., Pietrenko-Dabrowska, A., & Al-Hasan, M. (2020). Frequency-based regularization for improved reliability optimization of antenna structures. *IEEE Transactions on Antennas and Propagation, 69*, 4246–4251.

Koziel, S., Pietrenko-Dabrowska, A., & Plotka, P. (2022). Design specification management with automated decision-making for reliable optimization of miniaturized microwave components. *Scientific Reports, 12*, 829.

Kuwahara, Y. (2005). Multiobjective optimization design of Yagi–Uda antenna. *IEEE Transactions on Antennas and Propagation, 53*, 1984–1992.

Lefebvre, J.-P., Dompierre, B., Robert, A., Le Bihan, M., Wyart, E., & Sainvitu, C. (2015). Failure probability assessment using co-kriging surrogate models. *Procedia Engineering, 133*, 622–630.

Lehmensiek, R., Meyer, P., & Muller, M. (2002). Adaptive sampling applied to multivariate, multiple output rational interpolation models with application to microwave circuits. *International Journal of RF Microwave CAE, 12*(4), 332–340.

Leifsson, L., Du, X., & Koziel, S. (2020). Efficient yield estimation of multi-band patch antennas by polynomial chaos-based kriging. *International Journal of Numerical Modelling, 66*, e2722.

Li, X., Zhou, J., Duan, B., Yang, Y., Zhang, Y., & Fang, J. (2015). Performance of planar arrays for microwave power transmission with position errors. *IEEE Antennas and Wireless Propagation Letters, 14*, 1794–1797.

Li, W., Zhang, Y., & Shi, X. (2019). Advanced fruit fly optimization algorithm and its application to irregular subarray phased array antenna synthesis. *IEEE Access, 7*, 165583–165596.

Liang, J. F., Chang, H. C., & Zaki, K. A. (1993). Design and tolerance analysis of thick iris waveguide bandpass filters. *IEEE Transactions on Magnetics, 29*, 1605–1609.

Lim, D. K., Woo, D. K., Yeo, H. K., Jung, S. Y., Ro, J. S., & Jung, H. K. (2015). A novel surrogate-assisted multi-objective optimization algorithm for an electromagnetic machine design. *IEEE Transactions on Magnetics, 51*(3), paper 8200804.

Liu, Z., Yang, M., & Li, W. (2016). A sequential Latin hypercube sampling method for metamodeling. In L. Zhang, X. Song, & Y. Wu (Eds.), *Theory, methodology, tools and applications for Modeling and simulation of complex systems* (Vol. 643, pp. 176–185). AsiaSim 2016, Comm. Computer Information Science.

Liu, Y., Li, M., Haupt, R. L., & Guo, Y. J. (2020). Synthesizing shaped power patterns for linear and planar antenna arrays including mutual coupling by refined joint rotation/phase optimization. *IEEE Transactions on Antennas and Propagation, 68*, 4648–4657.

Lizzi, L., Viani, F., & Massa, A. (2009). Dual-band spline-shaped PCB antenna for Wi-Fi applications. *IEEE Antennas and Wireless Propagation Letters, 8*, 616–619.

Ma, B., Lei, G., Liu, C., Zhu, J., & Guo, Y. (2018). Robust tolerance design optimization of a PM claw pole motor with soft magnetic composite cores. *IEEE Transactions on Magnetics, 54*, 8102404.

Marler, R. T., & Arora, J. S. (2010). The weighted sum method for multi-objective optimization: New insights. *Structural and Multidisciplinary Optimization, 41*, 853–862.

Matoglu, E., Pham, N., De Araujo, D., Cases, M., & Swaminathan, M. (2004). Statistical signal integrity analysis and diagnosis methodology for high-speed systems. *IEEE Transactions on Advanced Packaging, 27*, 611–629.

Ochoa, J. S., & Cangellaris, A. C. (2013). Random-space dimensionality reduction for expedient yield estimation of passive microwave structures. *IEEE Transactions on Microwave Theory and Techniques, 61*, 4313–4321.

Palazzi, V., Hester, J., Bito, J., Alimenti, F., Kalialakis, C., Collado, A., Mezzanotte, P., Georgiadis, A., Roselli, L., & Tentzeris, M. (2018). A novel ultra-lightweight multiband rectenna on paper for RF energy harvesting in the next generation LTE bands. *IEEE Transactions on Antennas and Propagation, 66*, 366–379.

Pietrenko-Dabrowska, A., & Koziel, S. (2021). Globalized parametric optimization of microwave components by means of response features and inverse metamodels. *Scientific Reports, 11*, paper No. 23718.

Pietrenko-Dabrowska, A., Koziel, S., & Al-Hasan, M. (2020). Expedited yield optimization of narrow- and multi-band antennas using performance-driven surrogates. *IEEE Access*, 143104–143113.

Pietrenko-Dabrowska, A., Koziel, S., & Ullah, U. (2022). Reduced-cost two-level surrogate antenna modeling using domain confinement and response features. *Scientific Reports, 12*, Art. No. 4667.

Prasad, A. K., Ahadi, M., & Roy, S. (2016). Multidimensional uncertainty quantification of microwave/RF networks using linear regression and optimal design of experiments. *IEEE Transactions on Microwave Theory and Techniques, 64*, 2433–2446.

Queipo, N. V., Haftka, R. T., Shyy, W., Goel, T., Vaidynathan, R., & Tucker, P. K. (2005). Surrogate based analysis and optimization. *Progress in Aerospace Sciences, 41*, 1–28.

Rayas-Sanchez, J. E., & Gutierrez-Ayala, V. (2006). EM-based statistical analysis and yield estimation using linear-input and neural-output space mapping. In *IEEE MTT-S international microwave symposium digest (IMS)* (pp. 1597–1600).

Rossi, M., Dierck, A., Rogier, H., & Vande Ginste, D. (2014). A stochastic framework for the variability analysis of textile antennas. *IEEE Transactions on Antennas and Propagation, 62*, 6510–6514.

Santner, T. J., Williams, B., & Notz, W. (2003). *The design and analysis of computer experiments*. Springer-Verlag.

Simpson, T. W., Pelplinski, J. D., Koch, P. N., & Allen, J. K. (2001). Metamodels for computer-based engineering design: Survey and recommendations. *Engineering with Computers, 17*, 129–150.

Smolders, A. B., & Johannsen, U. (2011). Axial ratio enhancement for circularly-polarized millimeter-wave phased-arrays using a sequential rotation technique. *IEEE Transactions on Antennas and Propagation, 59*, 3465–3469.

Spina, D., Ferranti, F., Antonini, G., Dhaene, T., & Knockaert, L. (2014). Efficient variability analysis of electromagnetic systems via polynomial chaos and model order reduction. *IEEE Transactions on Components, Packaging and Manufacturing Technology, 4*, 1038–1051.

Talbi, E.-G. (2009). *Metaheuristics: From design to implementation*. Wiley.

Tan, K., Khor, E., & Lee, T. (2005). *Multiobjective evolutionary algorithms and applications*. Springer-Verlag.

Tang, M., Chen, X., Li, M., & Ziolkowski, R. W. (2018). Particle swarm optimized, 3-D-printed, wideband, compact hemispherical antenna. *IEEE Antennas and Wireless Propagation Letters, 17*, 2031–2035.

Tong, C. (2006). Refinement strategies for stratified sampling algorithms. *Reliability Engineering and System Safety, 91*, 1257–1265.

Viani, F., Lizzi, L., Azaro, R., & Massa, A. (2008). A miniaturized UWB antenna for wireless dongle devices. *IEEE Antennas and Wireless Propagation Letters, 7*, 714–717.

Wang, G. G. (2003). Adaptive response surface algorithm using inherited Latin hypercube design points. *Journal of Mechanical Design, 125*, 210–220.

Xia, B., Ren, Z., & Koh, C. S. (2014). Utilizing kriging surrogate models for multi-objective robust optimization of electromagnetic devices. *IEEE Transactions on Magnetics, 50*(2), paper 7017104.

Xiao, S., Liu, G. Q., Zhang, K. L., Jing, Y. Z., Duan, J. H., Di Barba, P., & Sykulski, J. K. (2018). Multi-objective Pareto optimization of electromagnetic devices exploiting kriging with Lipschitzian optimized expected improvement. *IEEE Transactions on Magnetics, 54*(3), 7001704.

Zhang, B., & Rahmat-Samii, Y. (2018). Robust optimization with worst case sensitivity analysis applied to array synthesis and antenna designs. *IEEE Transactions on Antennas and Propagation, 66*, 160–171.

Zhang, J., Zhang, C., Feng, F., Zhang, W., Ma, J., & Zhang, Q. J. (2018). Polynomial chaos-based approach to yield-driven EM optimization. *IEEE Transactions on Microwave Theory and Techniques, 66*, 3186–3199.

Zhang, C., Fu, X., Chen, X., Peng, S., & Min, X. (2019a). Synthesis of uniformly excited sparse rectangular planar array for sidelobe suppression using multi-objective optimization algorithm. *Journal of Engineering, 2019*(19), 6278–6281.

Zhang, J., Feng, F., Na, W., Yan, S., & Zhang, Q. (2019b). Parallel space-mapping based yield-driven EM optimization incorporating trust region algorithm and polynomial chaos expansion. *IEEE Access, 7*, 143673–143683.

Zhu, D. Z., Werner, P. L., & Werner, D. H. (2017). Design and optimization of 3-D frequency-selective surfaces based on a multiobjective lazy ant colony optimization algorithm. *IEEE Transactions on Antennas and Propagation, 65*, 7137–7149.

Chapter 3
Fundamentals of Numerical Optimization

The main topic of this book is the response feature technology for simulation-driven design of high-frequency components, including optimization and surrogate modeling. For the convenience of the reader, this chapter summarizes the fundamentals of numerical optimization. The material covered here is not supposed to be a systematic and exhaustive presentation of the subject (interested readers may refer to numerous textbooks such as Nocedal & Wright, 2006; Yang, 2010a; and Conn et al., 2009), but it should give the reader sufficient background information necessary to understand the remaining parts of the book. Here, we cover conventional methods, including gradient-based algorithms for both unconstrained and constrained optimization, derivative-free techniques such as pattern search, as well as global optimization by means of nature-inspired metaheuristics. An introduction to surrogate modeling and optimization will be provided in Chap. 4.

3.1 Optimization Problem Formulation

The optimization problem to be solved is formulated as

$$\boldsymbol{x}^* = \arg\min_{\boldsymbol{x}\in X} f(\boldsymbol{x}) \tag{3.1}$$

where $f(\boldsymbol{x})$ is a scalar objective function, whereas $\boldsymbol{x} \in X \subseteq R^n$; X is the objective function domain (also referred to as a design/search space); $\boldsymbol{x}^*$ is the design optimum to be found. The problem (3.1) can also be formulated with explicit constraints as

$$\boldsymbol{x}^* = \arg\min f(\boldsymbol{x}) \tag{3.2}$$

A. Pietrenko-Dabrowska, S. Koziel, *Response Feature Technology for High-Frequency Electronics. Optimization, Modeling, and Design Automation*,
https://doi.org/10.1007/978-3-031-43845-5_3

subject to

$$g_k(\boldsymbol{x}) \leq 0, k = 1, \ldots, N \tag{3.3}$$

$$h_k(\boldsymbol{x}) = 0, k = 1, \ldots, M. \tag{3.4}$$

Most practical problems utilize box constraints, i.e., the lower and upper bounds for design variables, $\boldsymbol{l} \leq \boldsymbol{x} \leq \boldsymbol{u}$, where the inequalities are understood as component-wise.

For simplicity, the algorithms described in this chapter are formulated for scalar objective functions. Yet, for practical design tasks, the objective function is often of the form $U(\boldsymbol{f}(\boldsymbol{x}))$, i.e., it is a composition of the merit function U and a vector-valued system response $\boldsymbol{f}(\boldsymbol{x})$, cf. Chap. 2. Figure 3.1 shows the generic flowchart of simulation-driven optimization of high-frequency components.

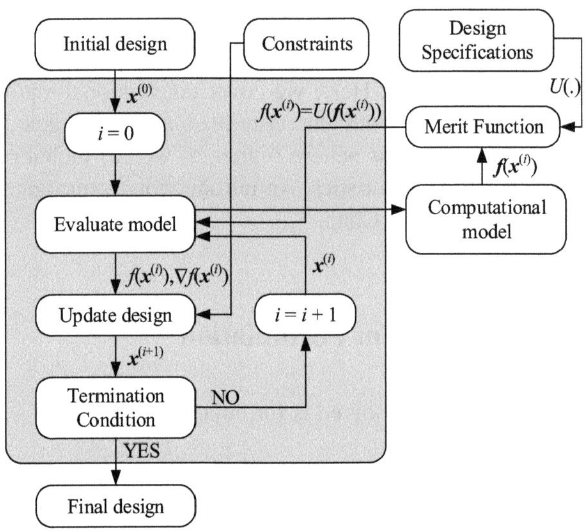

Fig. 3.1 Direct simulation-driven optimization flowchart. The candidate designs generated by the algorithm are evaluated using computational model pertinent to the problem at hand. The objective function is calculated based on given performance requirements. The evaluation is for verification purposes but also to provide the optimizer with information to search for better designs (subjected to constraints, if any). The search process may be guided by the model response only, or the response and its derivatives (if available)

3.2 Gradient-Based Optimization

Gradient-based techniques belong to the most popular optimization methods (Nocedal & Wright, 2006). The search process is based on the gradient $\nabla f(\boldsymbol{x})$ of the objective function $f(\boldsymbol{x})$

$$\nabla f(\boldsymbol{x}) = \left[\frac{\partial f}{\partial x_1}(\boldsymbol{x}) \frac{\partial f}{\partial x_2}(\boldsymbol{x}) \ \ldots \ \frac{\partial f}{\partial x_n}(\boldsymbol{x})\right]^T \tag{3.5}$$

Assuming that the function $f(\boldsymbol{x})$ is sufficiently smooth (i.e., at least first-order continuously differentiable), the gradient $\nabla f(\boldsymbol{x})$ determines the local behavior of the function in the vicinity of $\boldsymbol{x}$. Consider the first-order Taylor expansion of f around $\boldsymbol{x}$. We have

$$f(\boldsymbol{x} + \boldsymbol{h}) \cong f(\boldsymbol{x}) + \nabla f(\boldsymbol{x})^T \cdot \boldsymbol{h} < f(\boldsymbol{x}) \tag{3.6}$$

for sufficiently small $\boldsymbol{h}$ and assuming that $\boldsymbol{h}$ is a descent direction, i.e., $\nabla f(\boldsymbol{x})^T \cdot \boldsymbol{h} < 0$. Note that the largest local reduction of the function value is obtained for $\boldsymbol{h} = -\nabla f(\boldsymbol{x})$, which is referred to as the steepest-descent direction.

There are two basic ways of utilizing the gradient in the optimization process: (i) by moving along a descent direction, or (ii) by exploiting a local approximation model of the objective function constructed with ∇f. As explained later, moving along the steepest-descent direction may not be the best strategy.

Although it is possible to conduct the optimization process utilizing only the gradient information, more comprehensive information about the local behavior of the function can be obtained from the second-order derivatives, specifically, the Hessian $\boldsymbol{H}(\boldsymbol{x})$

$$\boldsymbol{H}(\boldsymbol{x}) = \begin{bmatrix} \frac{\partial^2 f}{\partial x_1 \partial x_1}(\boldsymbol{x}) & \cdots & \frac{\partial^2 f}{\partial x_1 \partial x_n}(\boldsymbol{x}) \\ \vdots & \ddots & \vdots \\ \frac{\partial^2 f}{\partial x_n \partial x_1}(\boldsymbol{x}) & \cdots & \frac{\partial^2 f}{\partial x_n \partial x_n}(\boldsymbol{x}) \end{bmatrix} \tag{3.7}$$

In particular, the gradient and the Hessian together allow us to identify the local minimizer of a function. The points for which the gradient ∇f vanishes are referred to as the *stationary points* of f. Vanishing of the gradient is a necessary condition for the minimum: if $\boldsymbol{x}^*$ is a local minimizer of f then $\nabla f(\boldsymbol{x}^*) = 0$. Unfortunately, this is not a sufficient condition: the stationary point may or may not be a local minimizer as shown in Fig. 3.2. This is determined by the properties of the Hessian. In particular, if $\boldsymbol{x}^*$ is a stationary point and $\boldsymbol{H}(\boldsymbol{x}^*)$ is positive definite (Nocedal & Wright, 2006), then $\boldsymbol{x}^*$ is a local minimizer of f. On the other hand, if $\boldsymbol{x}^*$ is a local minimizer of f, then $\boldsymbol{H}(\boldsymbol{x}^*)$ is positive semidefinite.

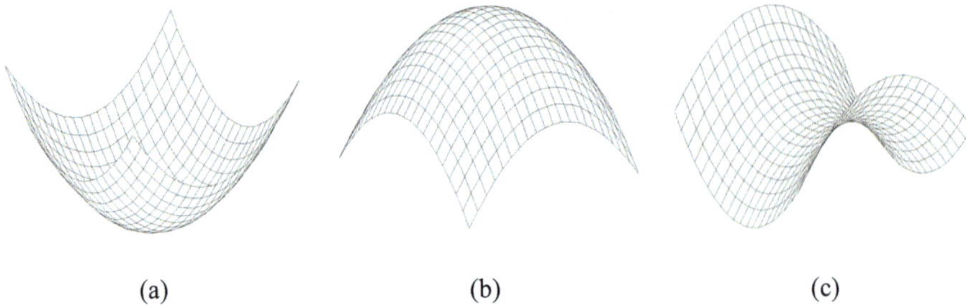

Fig. 3.2 Graphical illustration of stationary points in a two-dimensional search space: (**a**) local minimizer, (**b**) local maximizer, and (**c**) a saddle point. The Hessian is positive/negative definite in the local minimum/maximum, and it is neither positive nor negative definite in the saddle point

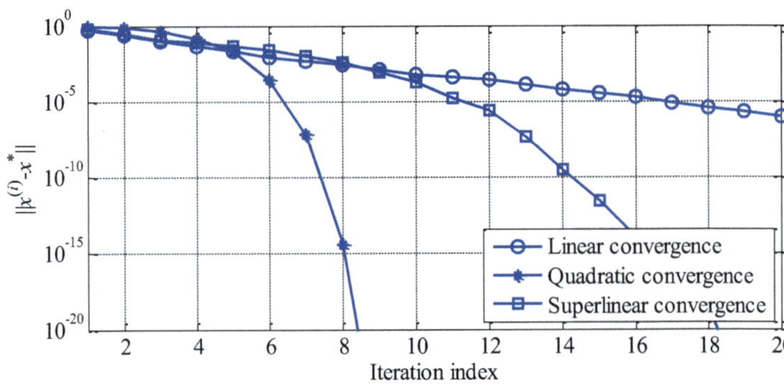

Fig. 3.3 Conceptual illustration of various convergence rates of optimization algorithms: linear (o), quadratic (*), and superlinear (□)

3.2.1 Optimization by Descent Methods

Here, we consider iterative methods generating a sequence of vectors $\boldsymbol{x}^{(i)}$, $i = 0, 1, 2, \ldots$, that are supposed to approximate the solution of (3.1). The fundamental requirement is that $f(\boldsymbol{x}^{(i+1)}) < f(\boldsymbol{x}^{(i)})$, i.e., an improvement with respect to the cost function value is enforced between iterations. The new vector $\boldsymbol{x}^{(i+1)}$ is found starting from $\boldsymbol{x}^{(i)}$ by moving along a descent direction, where f decreases, at least locally.

An important performance measure of an iterative method is the convergence rate, which describes how quickly we approach the minimizer $\boldsymbol{x}^*$. Linear convergence is defined by a condition $\|\boldsymbol{x}^{(i+1)} - \boldsymbol{x}^*\| \leq c_1\|\boldsymbol{x}^{(i)} - \boldsymbol{x}^*\|$ with $0 < c_1 < 1$. Quadratic convergence is defined by $\|\boldsymbol{x}^{(i+1)} - \boldsymbol{x}^*\| \leq c_2\|\boldsymbol{x}^{(i)} - \boldsymbol{x}^*\|^2$ with $0 < c_2 < 1$. Finally, superlinear convergence, defined as $\|\boldsymbol{x}^{(i+1)} - \boldsymbol{x}^*\|/\|\boldsymbol{x}^{(i)} - \boldsymbol{x}^*\| \rightarrow 0$ for $i \rightarrow \infty$, is faster than a linear one but not as good as a quadratic. Note that the aforementioned conditions are supposed to be satisfied when we are close to the optimum (see Fig. 3.3).

Let $\boldsymbol{x}^{(0)}$ be the initial solution. A generic descent algorithm works as follows (Nocedal & Wright, 2006):

1. Set $i = 0$;
2. Find a search direction $\boldsymbol{h}_d$;
3. Find a step length α;
4. Set $\boldsymbol{x}^{(i)} = \boldsymbol{x}^{(i)} + \alpha \boldsymbol{h}_d$;
5. Set $i = i + 1$;
6. If the termination condition is not satisfied, go to 2;
7. END.

The two critical stages of the iteration include finding a descent direction $\boldsymbol{h}_d$ (Step 2), and finding an appropriate length of the step taken along this direction (Step 3). As mentioned before, a descent direction is a vector $\boldsymbol{h}_d$ for which $\nabla f(\boldsymbol{x})^T \cdot \boldsymbol{h}_d < 0$ (cf. (3.6)). The step length $\alpha \boldsymbol{h}_d$ has to be determined to ensure that $f(\boldsymbol{x}^{(i+1)}) < f(\boldsymbol{x}^{(i)})$. This is realized using a separate algorithm as explained later. There are various termination conditions that can be utilized in practical implementations such as convergence in argument, i.e., $\|\boldsymbol{x}^{(i+1)} - \boldsymbol{x}^{(i)}\| \leq \varepsilon_1$, vanishing of the gradient, i.e., $\|\nabla f(\boldsymbol{x}^{(i)})\| \leq \varepsilon_2$, convergence in the function value, i.e., $f(\boldsymbol{x}^{(i)}) - f(\boldsymbol{x}^{(i+1)}) \leq \varepsilon_3$, or a combination of the above.

A procedure of finding the appropriate length of the step along the descent direction is referred to as a line search (Antoniou & Lu, 2007). More specifically, one considers a one-dimensional function of a length parameter α.

$$\varphi(\alpha) = f(\boldsymbol{x} + \alpha \boldsymbol{h}_d) \tag{3.8}$$

There are two types of line search that are utilized in practice. In a traditional (and more intuitive) *exact line search*, the step length α_e is found as $\alpha_e = \text{argmin}\{\alpha > 0: f(\boldsymbol{x} + \alpha \boldsymbol{h}_d)\}$. On the other hand, *soft line search* only requires satisfaction of certain (and usually rather loose) conditions such as the Wolfe conditions (Fletcher, 1987):

$$\varphi(\alpha_s) \leq \lambda(\alpha_s) \quad \text{where} \quad \lambda(\alpha) = \varphi(0) + \rho \varphi'(0) \cdot \alpha \quad \text{with} \quad 0 < \rho < 0.5 \tag{3.9}$$

$$\varphi'(\alpha_s) \geq \beta \cdot \varphi'(0) \quad \text{with} \quad \rho < \beta < 1 \tag{3.10}$$

The first condition ensures sufficient decrease of the cost function f. The second condition allows for avoiding steps that are too short. Both conditions are illustrated in Fig. 3.4.

It can be shown that the descent methods with line search satisfying (3.9), (3.10) are convergent to at least a first-order stationary point of f (i.e., where the function gradient vanishes), assuming that the function f is sufficiently smooth (here, uniformly continuous). It is also worth mentioning that soft line search is, in practice, more efficient than exact line search: although the latter normally requires a smaller number of iterations to converge, the cost per iteration is considerably higher than for soft line search, and so is the overall computational cost of the optimization process.

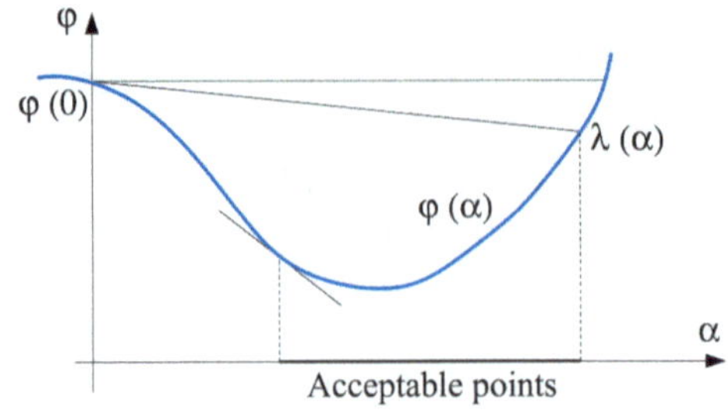

Fig. 3.4 Illustration of the Wolfe conditions. The lower bound of the acceptable point interval is determined by (3.10), whereas the upper bound is obtained from (3.9)

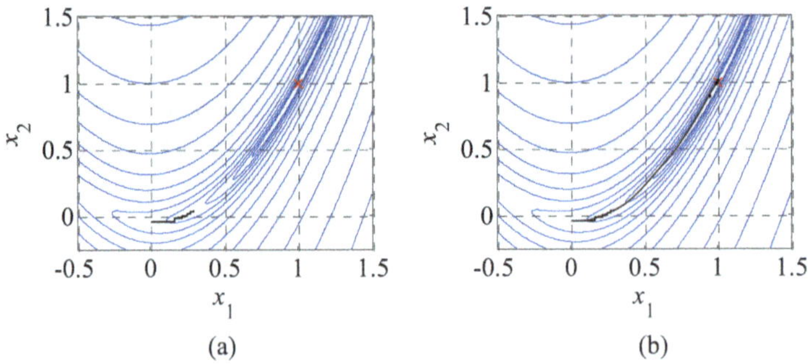

Fig. 3.5 Optimization of the Rosenbrock function (Rosenbrock, 1960): (**a**) steepest-descent methods where exact line search fails to find the optimum at $\boldsymbol{x}^* = [1\ 1]^T$ (marked ×): the narrow and curved valley containing the minimum slows down the algorithm convergence preventing the method from reaching $\boldsymbol{x}^*$); (**b**) conjugate-gradient algorithm with the Fletcher-Reeves update method and soft line search: the optimum is found in less than 30 iterations

At first glance, selecting the steepest-descent direction $\boldsymbol{h} = -\nabla f(\boldsymbol{x})$ as the search direction seems to be a good idea. Unfortunately, a steepest-descent method (where, $\boldsymbol{h} = -\nabla f(\boldsymbol{x})$ at each iteration) performs poorly when close to the optimum. This can be illustrated using a famous example of the Rosenbrock function $f(\boldsymbol{x}) = f([x_1\ x_2]^T) = (1 - x_1)^2 + 100(x_2 - x_1^2)^2$ (Rosenbrock, 1960). Figure 3.5a shows how the steepest-descent method fails to find the optimum of this function: due to a narrow and curved valley, the steps taken in subsequent iterations of the steepest-descent method (with exact line search) become shorter and shorter as they are perpendicular to each other. On the other hand, the steepest-descent method may be useful in the initial stages of the optimization process (away from the optimum), which leads to various hybrid methods.

Improved efficiency of the optimization process can be achieved by choosing—as the search direction—a linear combination of the previous direction $\boldsymbol{h}_{\text{prev}}$ and the current gradient, i.e.,

$$\boldsymbol{h} = -\nabla f\left(\boldsymbol{x}^{(i)}\right) + \gamma \boldsymbol{h}_{\text{prev}} \tag{3.11}$$

This idea is exploited by the so-called conjugate-gradient methods (Yang, 2010a). The difference between various conjugate-gradient methods is in the strategy of updating the search direction. An important example is the Fletcher-Reeves method (Fletcher & Reeves, 1964) with

$$\gamma = \frac{\nabla f(\boldsymbol{x})^T \nabla f(\boldsymbol{x})}{\nabla f(\boldsymbol{x}_{\text{prev}})^T \nabla f(\boldsymbol{x}_{\text{prev}})} \tag{3.12}$$

Another example is a Polak-Ribiére method (Polak & Ribiére, 1969) with the updating formula

$$\gamma = \frac{\left(\nabla f(\boldsymbol{x}) - \nabla f(\boldsymbol{x}_{\text{prev}})\right)^T \nabla f(\boldsymbol{x})}{\nabla f(\boldsymbol{x}_{\text{prev}})^T \nabla f(\boldsymbol{x}_{\text{prev}})} \tag{3.13}$$

It should be noted that the direction $\boldsymbol{h}$ obtained from (3.11) is not guaranteed to be downhill (e.g., when the soft line search is used to find the iteration step). A safeguard is to default to $\boldsymbol{h} = -\nabla f(\boldsymbol{x})$ in case $\nabla f(\boldsymbol{x})^T \boldsymbol{h} \geq 0$. Figure 3.5b shows the Fletcher-Reeves conjugate-gradient method (with soft line search) optimizing the Rosenbrock function. Conjugate-gradient methods exhibit the linear convergence rate (Crowder & Wolfe, 1972).

An alternative implementation of descent methods is by utilizing trust-region framework (Conn et al., 2000) rather than line search. Here, the iteration step is determined by means of the local expansion model of f created using first-order or, optionally, higher-order derivatives. The first-order Taylor expansion of f is defined as

$$f(\boldsymbol{x} + \boldsymbol{h}) \approx q(\boldsymbol{h}) \;\; \text{with} \;\; q(\boldsymbol{h}) = f(\boldsymbol{x}) + \nabla f(\boldsymbol{x}) \boldsymbol{h}^T \tag{3.14}$$

The second-order Taylor expansion (Bischof et al., 1993) is given by

$$f(\boldsymbol{x} + \boldsymbol{h}) \approx q(\boldsymbol{h}) \;\; \text{with} \;\; q(\boldsymbol{h}) = f(\boldsymbol{x}) + \nabla f(\boldsymbol{x}) \cdot \boldsymbol{h}^T + \frac{1}{2} \boldsymbol{h}^T \boldsymbol{H}(\boldsymbol{x}) \boldsymbol{h} \tag{3.15}$$

The model $q(\boldsymbol{h})$ is a good approximation of the function f in the vicinity of $\boldsymbol{x}$, i.e., for sufficiently small vectors $\boldsymbol{h}$ (assuming that the function f is sufficiently smooth). In the trust-region framework, the candidate iteration step size $\boldsymbol{h}_{tr}$ is determined as

$$\boldsymbol{h}_{tr} = \arg \min_{\boldsymbol{h}, \, \|\boldsymbol{h}\| \leq \delta} q(\boldsymbol{h}) \tag{3.16}$$

where $\delta > 0$ is the trust-region radius.

The step $\boldsymbol{h}_{tr}$ obtained using (3.16) is only accepted if it leads to the improvement of the cost function, i.e., $f(\boldsymbol{x} + \boldsymbol{h}_{tr}) < f(\boldsymbol{x})$. Otherwise it is rejected and the search continues from $\boldsymbol{x}$ using a reduced trust-region radius δ. The value of δ is adjusted using a so-called *gain ratio* defined as

$$r = \frac{f(\boldsymbol{x}) - f(\boldsymbol{x} + \boldsymbol{h})}{q(0) - q(\boldsymbol{h})} \tag{3.17}$$

A typical adjustment scheme would be $\delta \leftarrow 2\delta$ if $r > 0.75$ and $\delta \leftarrow \delta/3$ if $r < 0.25$. Because both models (3.14) and (3.15) are first-order consistent with f at $\boldsymbol{x}$ (i.e., $q(0) = f(\boldsymbol{x})$ and $\nabla q(0) = \nabla f(\boldsymbol{x})$), the condition $f(\boldsymbol{x} + \boldsymbol{h}_{tr}) < f(\boldsymbol{x})$ would be satisfied for a sufficiently small δ.

3.2.2 Newton and Quasi-Newton Methods

Performance of gradient-based search can be improved by exploiting higher-order derivatives. In particular, utilization of the Hessian, gives rise to a family of Newton and quasi-Newton methods outlined in this section.

Assuming that the cost function f is at least twice continuously differentiable, it can be locally represented using its second-order Taylor approximation (3.15). If the Hessian $\boldsymbol{H}(\boldsymbol{x})$ of f at $\boldsymbol{x}$ is positive definite, then the model $q(\boldsymbol{h})$ has a unique minimizer for $\boldsymbol{h}$ such that $\nabla q(\boldsymbol{h}) = 0$, i.e.,

$$\nabla f(\boldsymbol{x}) + \boldsymbol{H}(\boldsymbol{x})\boldsymbol{h} = 0 \tag{3.18}$$

This observation is a basis of the Newton's method (Antoniou & Lu, 2007) where the next iteration point is obtained by solving (3.18) as follows:

$$\boldsymbol{x}^{(i+1)} = \boldsymbol{x}^{(i)} - \left[\boldsymbol{H}\left(\boldsymbol{x}^{(i)}\right)\right]^{-1} \nabla f\left(\boldsymbol{x}^{(i)}\right) \tag{3.19}$$

In order for the algorithm (3.19) to be well defined, the Hessian $\boldsymbol{H}(\boldsymbol{x})$ has to be non-singular. Moreover, if the Hessian is positive definite for all iterations and the starting point is sufficiently close to the optimum, the method usually converges very quickly (quadratically) to a minimum.

The basic Newton algorithm of (3.19) does not work well in practice. It is not globally convergent for many problems, it may converge to a maximum or a saddle point, and—most importantly—the problem (3.19) may be ill-conditioned. Furthermore, the Newton method requires analytical second-order derivatives.

The fundamental difficulty, i.e., the Hessian not being positive definite, can be alleviated by using the so-called damped Newton method, where (3.19) is replaced by

$$\boldsymbol{x}^{(i+1)} = \boldsymbol{x}^{(i)} - \left[\boldsymbol{H}\left(\boldsymbol{x}^{(i)}\right) + \mu \boldsymbol{I}\right]^{-1} \nabla f\left(\boldsymbol{x}^{(i)}\right) \tag{3.20}$$

where $\boldsymbol{I}$ is the identity matrix. By choosing sufficiently large $\mu > 0$ the matrix $\boldsymbol{H}(\boldsymbol{x}) + \mu\boldsymbol{I}$ can be made positive definite because of being dominated by the term $\mu\boldsymbol{I}$. The step $\boldsymbol{h}_\mu$ found as a solution to the problem $[\boldsymbol{H}(\boldsymbol{x}) + \mu\boldsymbol{I}]\boldsymbol{h}_\mu = -\nabla f(\boldsymbol{x})$ is a minimizer of the following model

$$q_\mu(\boldsymbol{h}) = q(\boldsymbol{h}) + \frac{1}{2}\mu\boldsymbol{h}^T\boldsymbol{h} = f(\boldsymbol{x}) + \nabla f(\boldsymbol{x}) \cdot \boldsymbol{h}^T + \frac{1}{2}\boldsymbol{h}^T[\boldsymbol{H}(\boldsymbol{x}) + \mu\boldsymbol{I}]\boldsymbol{h} \tag{3.21}$$

Additionally, $\boldsymbol{h}_\mu$ is a descent direction for f at $\boldsymbol{x}$. Moreover, large steps are penalized in (3.20) because of the term $\mu\boldsymbol{h}^T\boldsymbol{h}/2$. It is also interesting to note that for very large μ, the algorithm defaults to the steepest-descent method because $\boldsymbol{h}_\mu \approx -\nabla f(\boldsymbol{x})/\mu$.

In practical implementations of the damped Newton's method, large values of μ are utilized at the early stages of the iteration process (3.20), whereas μ is reduced as the algorithm converges. When close to the optimum (and with small value of μ), the algorithm gets back to the original Newton's method, which allows for finding the minimizer of f. One of the most popular implementations of this concept is the Levenberg-Marquardt (LM) algorithm (Bates & Watts, 1988), where the value of μ is increased at the beginning of each iteration until $\boldsymbol{H}(\boldsymbol{x}) + \mu\boldsymbol{I}$ becomes positive definite, and then μ is updated based on the value of the gain ratio r (3.17). The typical updating formula is $\mu \leftarrow \mu \cdot \max\{1/3, 1 - (2r - 1)^3\}$ if $r > 0$, and $\mu = 2\mu$ otherwise (Nielsen, 1999). The flow of the LM algorithm is shown in Fig. 3.6.

Damped Newton methods resolve all of the issues of the basic Newton's algorithm but one: a potentially high-cost of obtaining second-order derivatives of the objective function f. This issue can be addressed by the so-called quasi-Newton methods, where the exact Hessian (or, even better, its inverse) is approximated using appropriate updating formulas. Until now, the best and the most popular updating formula is the Broyden-Fletcher-Goldfarb-Shanno (BFGS) one (Nocedal & Wright, 2006). BFGS preserves positive symmetry and definiteness of the approximation. It is defined as follows for a Hessian approximation $\boldsymbol{B}$:

```
x = x^(0); μ = μ^(0); found = false; k = 0;
while ~found & k ≤ k_max
  while f''(x)+μI not positive definite
    μ = 2*μ;
  end
  Solve (f''(x)+μI)h_dn = –f'(x); Compute gain factor r;
  if r > δ
    x = x + h_dn; μ = μ*max{1/3,1-2(2r-1)^3};
  else
    μ = μ*2;
  end
  k = k+1; Update found;
end
```

Fig. 3.6 Pseudocode of the Levenberg-Marquardt algorithm

```
x = x^(0); D = D^(0); k = 0; n_f = 0;
while ||f'(x)|| > ε & k < k_max & n_f < n_f.max
  h_qn = -Df'(x);
  [t,k_f] = soft_line_search(x,h_qn);
  n_f = n_f + k_f;
  x_new = x + t·h_qn; k = k + 1;
  if h_qn^T f'(x_new) > h_qn^T f'(x)
    Update D;
  end
  x = x_new;
end
```

Fig. 3.7 Pseudocode of the Quasi-Newton algorithm with BFGS update. Initialization of the inverse Hessian by the identify matrix is recommended. Updating of the matrix $\boldsymbol{D}$ is realized using (3.23)

$$\begin{gathered}\boldsymbol{B}_{\text{new}} = \boldsymbol{B} + \frac{1}{\boldsymbol{h}^T\boldsymbol{y}}\boldsymbol{y}\boldsymbol{y}^T - \frac{1}{\boldsymbol{h}^T\boldsymbol{u}}\boldsymbol{u}\boldsymbol{u}^T, \\ \text{where} \quad \boldsymbol{h} = \boldsymbol{x}_{\text{new}} - \boldsymbol{x}, \ \boldsymbol{y} = \nabla f(\boldsymbol{x}_{\text{new}}) - \nabla f(\boldsymbol{x}), \ \boldsymbol{u} = \boldsymbol{B}\boldsymbol{h}\end{gathered} \tag{3.22}$$

Approximation of the inverse of Hessian $\boldsymbol{D}$ is defined as

$$\begin{gathered}\boldsymbol{D}_{\text{new}} = \boldsymbol{D} + \kappa_1 \boldsymbol{h}\boldsymbol{h}^T - \kappa_2\left(\boldsymbol{h}\boldsymbol{v}^T + \boldsymbol{v}\boldsymbol{h}^T\right), \\ \text{where} \ \boldsymbol{h} = \boldsymbol{x}_{\text{new}} - \boldsymbol{x}, \ \boldsymbol{y} = \nabla f(\boldsymbol{x}_{\text{new}}) - \nabla f(\boldsymbol{x}), \ \boldsymbol{v} = \boldsymbol{B}\boldsymbol{y} \\ \kappa_2 = \frac{1}{\boldsymbol{h}^T\boldsymbol{y}}, \ \kappa_1 = \kappa_2\left(1 + \kappa_2\left(\boldsymbol{y}^T\boldsymbol{v}\right)\right)\end{gathered} \tag{3.23}$$

The quasi-Newton algorithm with the BFGS formula uses soft line search with loose conditions (e.g., $\rho = 0.0001$, and $\beta = 0.9$, cf. (3.9), (3.10)). The algorithm flow is shown in Fig. 3.7.

Quasi-Newton algorithms belong to the most efficient methods for unconstrained optimization. On the other hand, conjugate-gradient methods may be better when the number of design variables is large: Newton-like methods rely on matrix operations while conjugate-gradient methods only use vector operations, which are cheaper; also, Newton-type methods require more storage capacity ($n \times n$ matrix vs. a few vectors for conjugate gradients).

3.2.3 Qualitative Comparison of Descent Methods

Table 3.1 shows a qualitative comparison of the descent methods discussed in this chapter. It contains the most important characteristics of algorithms such as the computational complexity and convergence rates, but also some practical remarks and observations.

Table 3.1 Qualitative comparison of gradient-based descent methods

Method	Complexity	Convergence	Remarks
Steepest descent	$O(n)$	Linear	Not used in practice; fails for many problems; slow convergence
Conjugate gradient	$O(n)$	Linear	Reliable and computationally cheap but relatively slow convergence
Newton's	$O(n^3)$	Quadratic	Best convergence rate but requires analytical Hessian; not convergent for many problems; may converge to a maximum or saddle point; may be ill-conditioned
Damped Newton's	$O(n^3)$	Superlinear	Alleviates many of Newton's method drawbacks but still requires Hessian data
Quasi-Newton (BFGS updating)	$O(n^2)$	Superlinear	One of the best methods for unconstrained optimization to date

3.2.4 Remarks on Constrained Optimization

The algorithms discussed so far in this chapter allow unconstrained optimization. On the other hand, vast majority of real-world problems are constrained. Typical constraints include lower/upper bounds for design parameters but also linear and nonlinear equality and inequality constraints. The purpose of this section is to provide the reader with some remarks concerning constrained optimization methods. More detailed exposition of the subject can be found in numerical optimization textbooks, e.g., (Bertsekas, 1982; Nocedal & Wright, 2006).

The formulation of a constrained optimization problem has been presented in Sect. 3.1, see (3.2) through (3.4). The set of points satisfying the constraints (3.3) and (3.4) is referred to as a feasible region (cf. Fig. 3.8). Constrained optimization seeks for the minimum of the function f within its feasible domain.

The concepts of the optimum and optimality conditions are more complex than for the unconstrained case. The first-order necessary conditions, the so-called Karush-Kuhn-Tucker or KKT conditions (Kuhn & Tucker, 1951), for an optimum of a constrained problem (3.2–3.4), state that—at the local optimum $\boldsymbol{x}^*$ of f—there exist constants $\mu_1, \ldots, \mu_N$, and $\lambda_1, \lambda_2, \ldots, \lambda_M$, such that

$$\nabla f(\boldsymbol{x}^*) + \sum_{k=1}^{N} \mu_k \nabla g_k(\boldsymbol{x}^*) + \sum_{k=1}^{M} \lambda_k \nabla h_k(\boldsymbol{x}^*) = 0 \tag{3.24}$$

and

$$g_k(\boldsymbol{x}^*) \leq 0, \qquad \mu_k g_k(\boldsymbol{x}^*) = 0, \qquad k = 1, \ldots, N \tag{3.25}$$

where

$$\mu_k \geq 0, \qquad k = 1, \ldots, N \tag{3.26}$$

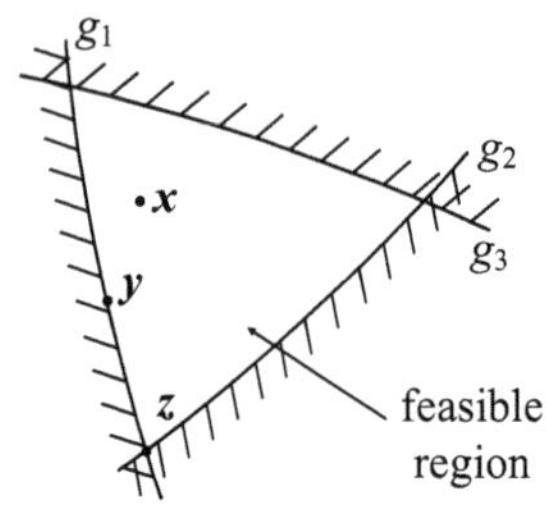

Fig. 3.8 Illustration of the feasible region for an example problem with three inequality constraints g_k, $k = 1, 2, 3$. None of the constraints is active at point $\boldsymbol{x}$, constraint g_1 is active at point $\boldsymbol{y}$, and two constraints g_1 and g_3 are active at point $\boldsymbol{z}$

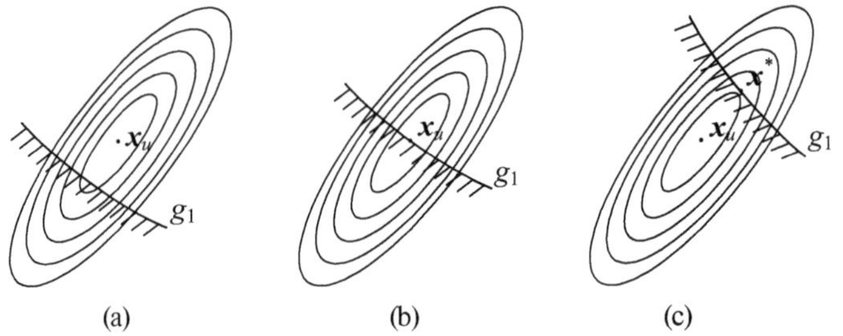

Fig. 3.9 Strongly and weakly active constraints ($\boldsymbol{x}_u$ is the unconstrained minimum, $\boldsymbol{x}^*$ is the constrained minimum): (**a**) constraint g_1 is inactive at $\boldsymbol{x}_u$ because $g_1(\boldsymbol{x}_u) > 0$ (therefore, $\boldsymbol{x}_u$ is both unconstrained and constrained minimizer of f); (**b**) g_1 is weakly active at $\boldsymbol{x}_u$ as $g_1(\boldsymbol{x}_u) = 0$ and $f'(\boldsymbol{x}_u) = \lambda g_1(\boldsymbol{x}_u)$ with $\lambda = 0$ ($\boldsymbol{x}_u$ is both unconstrained and constrained minimizer of f); (**c**) g_1 is strongly active: $g_1(\boldsymbol{x}_u) < 0$ ($\boldsymbol{x}_u$ is infeasible); $f'(\boldsymbol{x}^*) = \lambda g_1(\boldsymbol{x}^*)$ with $\lambda > 0$ (i.e., $\boldsymbol{x}^*$ is a constrained minimizer of f)

and provided that all the functions (both objective and constraints) are continuously differentiable. The function at the left-hand side of (3.26) is referred to as the Lagrangian function, whereas coefficients μ and λ are called Lagrange multipliers.

Figure 3.9 provides examples of unconstrained and constrained minima of the function f for a case with a single inequality constraint. Depending on the location of the unconstrained minimum and the constraint, it may be inactive, weakly active, or strongly active at the minimum. In the last case, the constrained minimum of the function is different from the unconstrained one.

One can distinguish two general approaches to handling constrained optimization. Assuming that the unconstrained minimum of f is not in the interior of the feasible region, the constrained minimum must be on its boundary. The so-called active set methods search for the optimum along the boundary. Inequality constraints are handled by keeping track of the set of active constraints while moving downhill along the boundary.

There is another group of methods that is based on approaching the optimal solution in an iterative way, either from within the feasible region (so-called interior point methods) or by possible use of infeasible points (however, not by moving along the feasible region boundary). In the latter approaches, the objective function is modified and corresponding unconstrained optimization problems are solved at each iteration.

For the rest of this section, we will discuss a few specific techniques for constraint handling. One of the simplest methods is the penalty function approach, where the original problem (3.2) with constraints (3.3), (3.4) is replaced by

$$\arg\min_{\boldsymbol{x}} \phi(\boldsymbol{x},\boldsymbol{\alpha},\boldsymbol{\beta}) = \arg\min_{\boldsymbol{x}} \left\{ f(\boldsymbol{x}) + \sum_{k=1}^{N} \alpha_k \overline{g}_k^2(\boldsymbol{x}) + \sum_{k=1}^{M} \beta_k h_k^2(\boldsymbol{x}) \right\} \tag{3.27}$$

where $\boldsymbol{\alpha} = [\alpha_1 \ldots \alpha_N]$, $\boldsymbol{\beta} = [\beta_1 \ldots \beta_M]$, α_k, $\beta_k >> 1$, are the penalty factors, and $\overline{g}_k(\boldsymbol{x}) = \max\{0, g_k(\boldsymbol{x})\}$ (inequalities only contribute to ϕ if they are active). In practice, one solves a sequence of problems (3.27) for increasing values of penalty factors using the solution of the previous iteration as the starting point for the next one. An example illustrating the penalty function approach is shown in Fig. 3.10. We attempt to minimize the function $f(x,y) = x^2 + 4y^2$ with an equality constraint $h_1(x,y) = (x + 2)^2 + (y + 2)^2 – 2$. The unconstrained minimum is $[0\ 0]^T$, whereas the constrained minimum is $[1.389 – 0.725]^T$. The sequence of Fig. 3.10a through f shows the contour plots of the penalized cost function (3.27) for the increasing value of the penalty factor β_1 (by one order of magnitude in each iteration).

Another technique for constraint handling is the barrier method. It replaces the original constrained problem (3.2) by the following formulation (for simplicity, only inequality-type constraints are assumed here):

$$\arg\min_{\boldsymbol{x}} \phi(\boldsymbol{x},\mu) = \arg\min_{\boldsymbol{x}} \left\{ f(\boldsymbol{x}) - \mu \sum_{k=1}^{N} \ln(-g_k(\boldsymbol{x})) \right\} \tag{3.28}$$

where $\mu > 0$ is a barrier parameter. The search starts from a feasible point. A sequence of unconstrained problems is solved with μ decreasing to zero so that the minimum of $\phi(\boldsymbol{x},\mu)$ converges to a solution of f.

Other approaches to constrained optimization are the augmented Lagrangian methods (Tapia, 1978). These are a class of algorithms that are similar to penalty methods in the sense that the original constrained problem is solved as a sequence of suitably formulated unconstrained tasks. The unconstrained objective is the Lagrangian of the constrained problem with an additional penalty term (so-called augmentation). Assuming—for the sake of simplicity—equality-only constraints $h_k(\boldsymbol{x})$, $k = 1, \ldots, M$, the augmented Lagrangian algorithm generates a new approximation of the constrained solution to (3.2) as

$$\boldsymbol{x}^{(i+1)} = \arg\min_{\boldsymbol{x}} \phi^{(i)}(\boldsymbol{x}) = \arg\min_{\boldsymbol{x}} \left\{ f(\boldsymbol{x}) + \frac{\mu^{(i)}}{2} \sum_{k=1}^{M} h_k^2(\boldsymbol{x}) - \sum_{k=1}^{M} \lambda_k h_k(\boldsymbol{x}) \right\} \tag{3.29}$$

The starting point is the previous approximation $\boldsymbol{x}^{(i)}$. In each iteration, the coefficients μ_k are updated as follows: $\lambda_k \leftarrow \lambda_k - \mu^{(i)} h_k(\boldsymbol{x}^{(i+1)})$. The coefficients λ_k are the estimates of the Lagrange multipliers (cf. (3.24)), and their accuracy increases

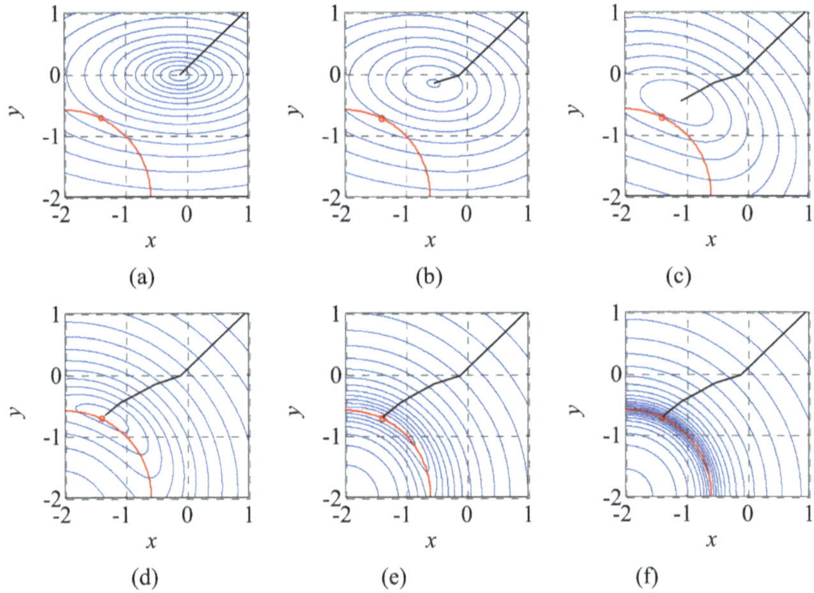

Fig. 3.10 Constrained optimization of the function $f(x,y) = x^2 + 4y^2$; the equality constraint is $h_1(x, y) = (x + 2)^2 + (y + 2)^2 - 2$. The unconstrained minimum is $[0\ 0]^T$, the constrained minimum is $[1.389 - 0.725]^T$. Plots (a) through (f) show the contour plots of the penalized cost function (3.27) for the increasing values of the penalty factor β_1. Constrained minimum marked using a circle

as the optimization process progresses. The value of μ is increased in each iteration, however, unlike in the penalty method, it is not necessary to ensure $\mu \to \infty$.

One of the most popular techniques for solving constrained optimization problems is nowadays sequential quadratic programming (SQP; Han, 1977; Bertsekas, 1982). At each iteration of SQP, the following quadratic programming subproblem is utilized to compute a search direction $\boldsymbol{h}^{(i)}$:

$$\boldsymbol{h}^{(i)} = \arg\min_{\boldsymbol{h}} \left\{ f\left(\boldsymbol{x}^{(i)}\right) - \nabla f\left(\boldsymbol{x}^{(i)}\right)^T \boldsymbol{h} + \frac{1}{2}\boldsymbol{h}^T \boldsymbol{H}\left(\boldsymbol{x}^{(i)}\right)\boldsymbol{h} \right\} \tag{3.30}$$

so that

$$g_k\left(\boldsymbol{x}^{(i)}\right) + \nabla g_k\left(\boldsymbol{x}^{(i)}\right)^T \boldsymbol{h} \le 0, \quad k = 1, \ldots, N \tag{3.31}$$

and

$$h_k\left(\boldsymbol{x}^{(i)}\right) + \nabla h_k\left(\boldsymbol{x}^{(i)}\right)^T \boldsymbol{h} = 0, \quad k = 1, \ldots, M \tag{3.32}$$

Here, $\boldsymbol{x}^{(i)}$ is a current iteration point, whereas $\boldsymbol{H}$ is a symmetric, positive definite matrix (preferably an approximation of the Hessian of the Lagrangian of f, cf. (3.24)). The new approximation $\boldsymbol{x}^{(i+1)}$ is subsequently obtained using line search.

3.3 Derivative-Free Optimization

Derivatives of the objective function provide a lot of information that can be utilized in the optimization process as elaborated earlier in this chapter. Unfortunately, despite the strong theoretical foundations of the gradient-based techniques and their good performance on smooth objective functions, they cannot be applied to a large number of practical optimization problems. An obvious reason for this fact is that derivative information may not be available or it might be too expensive to compute (e.g., through finite differentiation of an expensive objective function evaluated with simulations). Nowadays, another and even more important issue arises which is that gradient-based search does not perform well then the objective function is noisy. This is normally the case for objective functions evaluated using computer simulation such as finite-element analysis. One of the sources of the numerical noise is the utilization of adaptive meshing techniques in many commercial solvers. Adaptive meshing may lead to considerable changes of the mesh topology even for very slight changes of the geometry parameters of the structure under analysis. Another issue is termination of the simulation process upon achieving certain (required) resolution of the solution in terms of convergence, e.g., based on the residual energy for the finite-volume transient solvers.

It should also be emphasized that majority of gradient-based optimization algorithms (in particular those discussed in Sect. 3.2) are local methods. At the same time, increased complexity of contemporary engineering systems, including microwave components or antenna structures (and, to even higher extent, antenna arrays), often results in non-convex, multi-modal, and rugged functional landscapes. For such cases, global optimization may be necessary to find a satisfactory design.

The aforementioned difficulties have significantly increased popularity of derivative-free optimization techniques where the search process does not rely on derivative data (Conn et al., 2009). These techniques include local methods, i.e., mostly pattern search algorithms (Kolda et al., 2003), trust-region methods based on interpolation and approximation models (Conn et al., 2000), global optimization methods such as nature-inspired population-based metaheuristics (Yang, 2010a), as well as surrogate-based optimization (SBO) algorithms (Queipo et al., 2005; Koziel et al., 2011). Section 3.4 provides an overview of population-based methods, whereas surrogate-assisted techniques are outlined in Chap. 4. In the remaining parts of this chapter, we briefly present the concept of the pattern search (Sect. 3.3.1) as well as two specific algorithms: the Hooke-Jeeves direct search (Sect. 3.3.2) and the Nelder-Mead algorithm (Sect. 3.3.3).

3.3.1 Pattern Search Methods

Pattern search is a popular class of optimization algorithms in which the search is restricted to a predefined grid. The objective function is evaluated on a stencil determined by a set of directions that are suitable from either geometric or algebraic viewpoints. The initial grid is subsequently modified (in particular, refined) during the optimization run. In the simplest case, the grid may be rectangular, i.e., spanned by the vectors aligned with the axes of the coordinate system. Computationally more efficient versions of the pattern search methods may rely on the grids generated using so-called positive spanning directions (Davis, 1954) ensuring local improvement of the objective function (assuming its continuity) for a sufficiently refined grid while using only $n + 1$ neighboring points (versus $2n + 1$ points for n-dimensional rectangular grids).

A generic illustration of the pattern search is conceptually illustrated in Fig. 3.11 (Kolda et al., 2003). For the sake of simplicity, the optimization process in Fig. 3.11 is restricted to the rectangular grid and thus explores a grid-restricted vicinity of the current design. If the operations performed on the current grid fail to improve the objective function, the grid is refined to allow smaller steps. A typical termination criterion involves reaching a required resolution (e.g., a user-defined minimum stencil size). There are numerous variations of the pattern search algorithms. For a review see, e.g., (Kolda et al., 2003).

Pattern search methods are known to be relatively robust; however, they usually exhibit slow convergence compared to the gradient-based algorithms. On the other hand, they do not need derivative information; more importantly, they are relatively immune to numerical noise. The interest in pattern search methods has recently been revived not only due to the development of rigorous convergence theory (e.g., Conn et al., 2009) but also because these methods easily benefit from parallel computing.

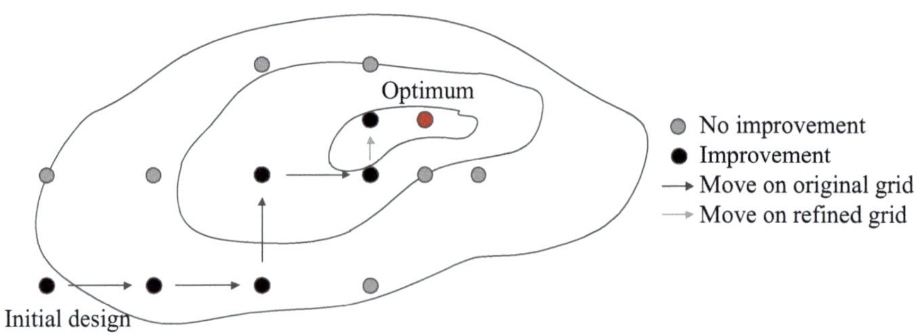

Fig. 3.11 Conceptual illustration of the pattern search process, here using a rectangular grid. The exploratory movements are restricted to the grid around the initial design. In the case of failure of making a successful move, the grid is refined to allow smaller steps (Koziel & Ogurtsov, 2014). Practical implementations of pattern search routines typically use more sophisticated strategies (e.g., grid-restricted line search)

3.3.2 Hooke-Jeeves Direct Search

The Hooke-Jeeves direct search (Hooke & Jeeves, 1961) is a pattern search method. Actually, it is the first method to use the term of "direct search" and to exploit the idea of a pattern. The method is based on two types of steps, exploratory and pattern. The algorithm flow is as follows:

1. Set $j = 0$; select a base point $\boldsymbol{x}^j$ and the search directions $\boldsymbol{d}_k$, $k = 1, \ldots, n$.
2. (Exploratory move) Perform local search by sequential exploration of all search directions in an opportunistic manner, i.e., $\boldsymbol{x}^j + \boldsymbol{d}_1$, then $\boldsymbol{x}^j + \boldsymbol{d}_1 + \boldsymbol{d}_2$ (if $\boldsymbol{x}^j + \boldsymbol{d}_1$ resulted in a cost function improvement) or $\boldsymbol{x}^j - \boldsymbol{d}_1$ otherwise, etc. The search is continued until all search directions have been used.
3. In case there is no cost function improvement in Step 2, the step size is reduced and the search restarts from the previous best point.
4. In case there was an improvement in Step 2, a pattern move is executed as follows. Let $\boldsymbol{x}_{\text{tmp}}^{j+1} = \boldsymbol{x}^{j+1} + \alpha(\boldsymbol{x}^{j+1} - \boldsymbol{x}^j)$ be a temporary head, where $\boldsymbol{x}_{j+1}$ is the best point found in Step 2, and $\alpha \geq 1$ is an acceleration factor (typically, $\alpha = 2$).
5. If $f(\boldsymbol{x}_{\text{tmp}}^{j+1}) < f(\boldsymbol{x}^{j+1})$, set $\boldsymbol{x}^{j+1} = \boldsymbol{x}_{\text{tmp}}^{j+1}$. Continue expansion as in Step 4 as long as the cost function value is decreased.
6. If $f(\boldsymbol{x}_{\text{tmp}}^{j+1}) \geq f(\boldsymbol{x}^{j+1})$ and the termination condition is not satisfied, set $j = j + 1$ and go to 2.

The termination condition may be based on reducing the search vectors below a user-defined threshold. It should be noted that the Hooke-Jeeves procedure is intrinsically serial. It might be beneficial in situations where an optimum is far from the initial guess. In case the initial search vectors are large, the procedure may be able to avoid some local minima. To some extent, it is also robust against noisy cost functions.

3.3.3 Nelder-Mead Algorithm

The Nelder-Mead algorithm is a very popular derivative-free downhill simplex procedure for unconstrained optimization (Nelder & Mead, 1965). The search process is based on manipulating a finite set of candidate solutions forming a simplex. A simplex or n-simplex is a convex hull of a set of $n + 1$ affinely independent points $\boldsymbol{x}^{(i)}$ in a vector space (normally, we consider n-simplex in an n-dimensional space). The set $X = \{\boldsymbol{x}^{(i)}\}_{i = 1,\ldots,k}$ is affinely independent if the vectors $\boldsymbol{v}^{(i)} = \boldsymbol{x}^{(i+1)} - \boldsymbol{x}^{(1)}$, $i = 1,\ldots,k - 1$ are linearly independent. A convex hull $H(X)$ of X is defined as a set of all convex combinations of the points from X, i.e., $H(X) = \{\sum_{i = 1,\ldots,k} a_i \boldsymbol{x}^{(i)}: \boldsymbol{x}^{(i)} \in X, a_i \geq 0, \sum_{i = 1,\ldots,k} a_i = 1\}$. Examples of 1-, 2-, and 3-simplex are, respectively, a line segment, a triangle, and a tetrahedron (see Fig. 3.12).

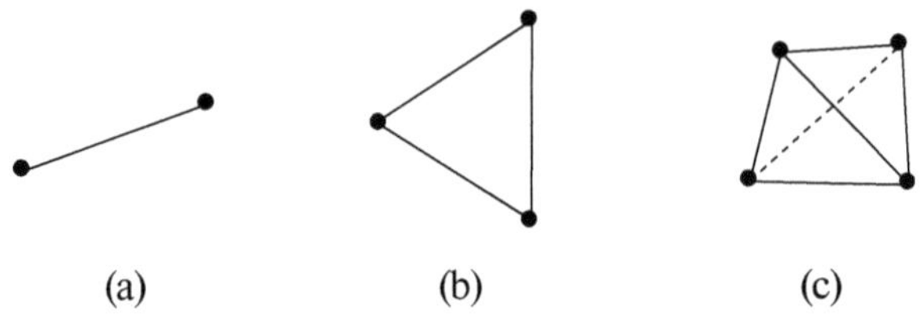

Fig. 3.12 The concept of a simplex: (**a**) 1-simplex, (**b**) 2-simplex, (**c**) 3-simplex

The Nelder-Mead algorithm follows the procedure that consists of processing and updating the simplex vertices based on the corresponding objective function values (see also Fig. 3.13):

- Order the vertices so that $f(\boldsymbol{x}^{(1)}) \leq f(\boldsymbol{x}^{(2)}) \leq \ldots f(\boldsymbol{x}^{(n+1)})$. An auxiliary point $\boldsymbol{x}^{(0)}$ is then defined to be center of gravity of all vertices but $\boldsymbol{x}^{(n+1)}$.
- Compute a *reflection* $\boldsymbol{x}^r = (1+\alpha)\boldsymbol{x}^{(0)} - \alpha\boldsymbol{x}^{(n+1)}$ (a typical value of α is 1); If $f(\boldsymbol{x}^{(1)}) \leq f(\boldsymbol{x}^r) < f(\boldsymbol{x}^{(n)})$ reject $\boldsymbol{x}^{(n+1)}$ and update the simplex using $\boldsymbol{x}^r$.
- If $f(\boldsymbol{x}^r) < f(\boldsymbol{x}^{(1)})$ compute an *expansion* $\boldsymbol{x}^e = \rho\boldsymbol{x}^r + (1-\rho)\boldsymbol{x}^{(0)}$ (a typical value of ρ is 2); If $f(\boldsymbol{x}^e) < f(\boldsymbol{x}^r)$ reject $\boldsymbol{x}^{(n+1)}$ and update the simplex using $\boldsymbol{x}^e$.
- If $f(\boldsymbol{x}^{(n)}) \leq f(\boldsymbol{x}^r) < f(\boldsymbol{x}^{(n+1)})$ compute a *contraction* $\boldsymbol{x}^c = (1+\gamma)\boldsymbol{x}^{(0)} - \gamma\boldsymbol{x}^{(n+1)}$ (a typical value of γ is 0.5); If $f(\boldsymbol{x}^c) \leq f(\boldsymbol{x}^r)$ reject $\boldsymbol{x}^{(n+1)}$ and update the simplex using $\boldsymbol{x}^c$.
- If $f(\boldsymbol{x}^c) > f(\boldsymbol{x}^r)$ or $f(\boldsymbol{x}^r) \geq f(\boldsymbol{x}^{(n+1)})$ shrink the simplex: $\boldsymbol{x}^{(i)} = \boldsymbol{x}^{(1)} + \sigma(\boldsymbol{x}^{(i)} - \boldsymbol{x}^{(1)})$, $i = 1, \ldots, n+1$ (a typical value of σ is 0.5).

The outlined procedure describes one iteration of the algorithm which is continued until the termination condition is met. Typically, the termination condition is set as a reduction of the simplex size below a user-defined threshold. The Nelder-Mead algorithm is characterized by a relatively low computational cost of only a few objective function evaluations per iteration; however, its convergence is rather slow.

3.4 Global Optimization: Nature-Inspired Population-Based Algorithms

From practical perspective (with our main focus being design of high-frequency structures), one of the most important features of the optimization methods presented so far in this chapter is their locality. Gradient-based algorithms as well as pattern-search derivative-free techniques normally allow for finding an optimum that is located in the vicinity of the initial solution, or, more generally, in the region of attraction of the nearest local optimum. Unfortunately, the objective functions that one needs to deal with in a large number of real-world engineering problems are characterized by multiple local optima. Moreover, the functional landscape of the problem at hand is often unknown in terms of the nonlinearity of the objective function, importance of particular variables, but also the number and the location of

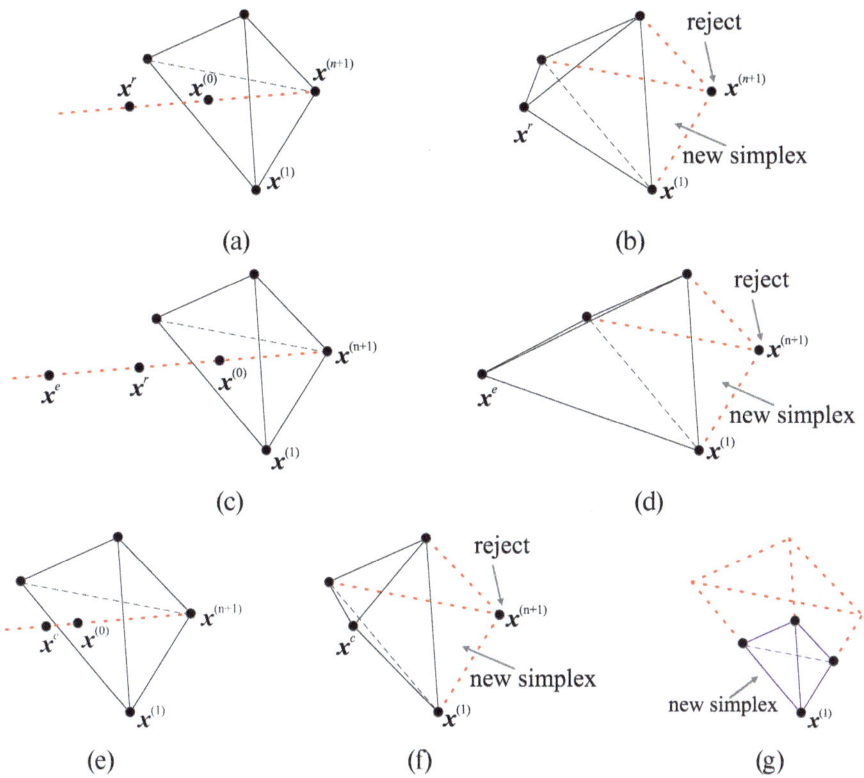

Fig. 3.13 Operations of the Nelder-Mead algorithm: (**a**) reflection: $\boldsymbol{x}^r = (1+\alpha)\boldsymbol{x}^{(0)}-\alpha\boldsymbol{x}^{(n+1)}$, (**b**) acceptance of the reflection point $\boldsymbol{x}^r$ if $f(\boldsymbol{x}^{(1)}) \leq f(\boldsymbol{x}^r) < f(\boldsymbol{x}^{(n)})$, (**c**) expansion: $\boldsymbol{x}^e = \rho\boldsymbol{x}^r + (1-\rho)\boldsymbol{x}^{(0)}$, (**d**) acceptance of the expansion point $\boldsymbol{x}^e$ if $f(\boldsymbol{x}^e) < f(\boldsymbol{x}^r)$, (**e**) contraction: $\boldsymbol{x}^c = (1+\gamma)\boldsymbol{x}^{(0)}-\gamma\boldsymbol{x}^{(n+1)}$, (**f**) acceptance of the contraction point $\boldsymbol{x}^c$ if $f(\boldsymbol{x}^c) \leq f(\boldsymbol{x}^r)$, (**g**) shrinking of the simplex if neither of the above operations led to the objective function improvement

the optima. In many cases, finding a reasonably good starting point is a non-trivial task by itself. In such situations, straightforward application of local algorithms often leads to unsatisfactory results; therefore, global search may be necessary.

Global optimization is a broad subject with numerous methods and algorithms developed over the last few decades (Goldberg, 1989; Horst et al., 2000; Horst & Tuy, 1996; Hansen & Walster, 2003). The algorithms for global optimization can be roughly categorized into deterministic (Horst & Tuy, 1996; Zhigljavsky & Žilinskas, 2008) and stochastic ones (Goldberg, 1989; Talbi, 2009). The most successful deterministic methods include cutting plane methods, branch and bound methods, and interval methods (Luenberger, 2003; Jaulin et al., 2001; Hansen & Walster, 2003). Stochastic methods include simulated annealing (van Laarhoven & Aarts, 1987), Monte-Carlo methods (Kroese et al., 2011), and population-based approaches (Goldberg, 1989; Talbi, 2009).

This section does not attempt to give a comprehensive introduction to global optimization. Instead, we focus on nature-inspired population-based metaheuristics as the most popular approaches to global optimization problems nowadays (Talbi, 2009; Yang, 2008), including optimization of antennas (e.g., Deb et al. 2014; Kerkhoff & Ling, 2007; Kibria et al., 2014), antenna arrays (Karamalis et al., 2009; Jang et al., 2016; Elragal et al., 2011; Datta & Misra, 2009), and microwave components (Matekovits et al., 2004; Wang et al., 2005; Ülker, 2008; Luo et al., 2019). There are several reasons for such popularity. The most important reasons include simple implementation, versatility, ability to handle "difficult" functions (discontinuous, noisy), as well as suitability for parallelization.

3.4.1 Fundamentals of Population-Based Metaheuristics

Metaheuristics are derivative-free global optimization methods. They have been developed out of practical necessity of handling optimization problems that were challenging (or even unmanageable) for conventional algorithms, including gradient-based routines. The most important challenges included non-differentiable, discontinuous, or noisy objective functions, as well as the presence of multiple local optima. The last issue is particularly troublesome because it makes the optimization outcome very much dependent on the initial solution when local methods are used.

The majority of metaheuristic algorithms are based on mimicking of natural processes (e.g., biological or social systems). It should be emphasized right away that there is usually little (or none) rigorous explanation why a specific scheme should work and why it should be better than other schemes. Available performance comparisons are exclusively based on numerical experiments (e.g., Datta & Misra, 2009; Deb et al., 2014) and may be contradictory with each other depending on the inclination of a particular author toward a particular technique.

Most metaheuristics process sets (or populations) of potential solutions to the optimization problem at hand are often referred to as individuals or agents. In the course of the optimization process, the method-specific interactions between individuals and selection methods are applied in order to identify and explore promising regions of the design space. Some older metaheuristic algorithms (e.g., simulated annealing; Kirkpatrick et al., 1983; van Laarhoven & Aarts, 1987) process a single solution rather than a set of individuals.

The first population-based optimization methods were developed in late 1960s (evolution strategies, ES, Beyer, 2001). A sudden increase of their popularity begun in late 1980s and in 1990s (genetic algorithms, GAs, Goldberg, 1989; evolutionary algorithms, EAs, Back et al., 2000). Today, metaheuristics are standard techniques widely used and, sometimes, overused in almost every engineering discipline. Apart from the already mentioned algorithms, other popular methods include particle swarm optimizers (PSO) (Kennedy et al., 2001), differential evolution (DE) (Price et al., 2005), and, more recently, firefly algorithm, cuckoo search, bat algorithm, and

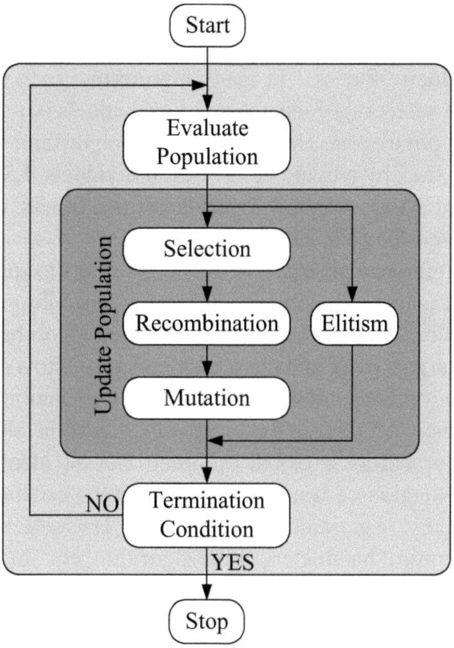

Fig. 3.14 Generic flow of a population-based metaheuristic algorithm. Evaluation of the individuals is followed by a selection process where parent individuals are chosen to create an updated population. The selected individuals undergo the recombination and/or mutation processes. In some implementations, elitism is also utilized to prevent the best individual (or individuals) from extinction by identifying and inserting that individual directly into the new population with bypassing the recombination procedures. Termination condition may be based on the maximum number of iterations or certain convergence criteria (e.g., reduction of the population diversity below a user-defined threshold)

others (Yang, 2010a, b, 2014). There are many new (at least by the name) techniques proposed every year (Duman et al., 2012; Cheng & Prayogo, 2014; Tamura & Yasuda, 2011 etc.); however, it seems that these are mostly minor variations of the existing methods.

Below, we outline and discuss a generic flow of a typical population-based metaheuristic algorithm (see Fig. 3.14 for a flow diagram). We will use the symbol *P* to denote a set of individuals (population) processed by the algorithm:

1. Initialize population *P*;
2. Evaluate population *P*;
3. Select parent individuals *S* from *P*;
4. Create a new population *P* from the parent individuals *S* by applying recombination operators;
5. Introduce local perturbations in individuals of *P* by applying mutation operators;
6. If termination condition is not satisfied go to 2;
7. END.

In most algorithms, the population is initialized randomly. In Step 2, the individuals in the population are evaluated, and the corresponding values of the objective function determine their "fitness." In many algorithms, such as GAs and EAs, an important step is the selection of the subset of individuals (so-called parent individuals) to form a new population. Depending on the algorithm, the selection process can be deterministic (i.e., by picking up the best individuals, ES) or partially random where probability of being selected depends on the fitness value, but there is a chance to survive even for poor individuals, EAs.

The amount of preference given to the best individuals to survive is referred to as selection pressure. It is one of the important factors controlling the performance and convergence properties of the population-based search algorithm. If it is too low, the algorithm may converge slowly and the search may turn into a random search; if it is too high, the best individuals quickly take over the population and one may face a premature convergence, which usually negatively affects the algorithm performance (the search concludes before a global optimum can be identified). The opposite mechanism is a production of a new information (or "genetic material" in GAs), which is controlled by, e.g., mutation operators. The selection pressure can be adjusted by certain control parameters of the algorithm; e.g., in tournament selection it is adjusted by the candidate set size (Tan et al., 2005).

In certain metaheuristic algorithms, e.g., PSO or DE, no selection is used, i.e., individuals are modified from iteration to iteration but never die out. This is the case when the algorithm is based on mimicking certain social behavior (e.g., a flock of birds in PSO, or a swarm of fireflies) so that individuals in the population are relocated according to certain rules (e.g., being attracted to the best solution found so far) rather than replacing the old population by the new one according to a certain generational model and recombination operators.

In general, there are two fundamental types of operators that can be used to modify individuals in the population: exploratory ones (e.g., crossover in EAs or ES) and exploitative ones (e.g., mutation in GAs). The purpose of exploratory operators is to combine information contained in the parent individuals to create an offspring. An example of such an operator used by evolutionary algorithms with natural (floating point) representation is an arithmetic crossover, where a new individual $\boldsymbol{c}$ can be created as a convex combination of the parents $\boldsymbol{p}_1$ and $\boldsymbol{p}_2$, i.e., $\boldsymbol{c} = \alpha\boldsymbol{p}_1 + (1\text{–}\alpha)\boldsymbol{p}_2$, where $0 < \alpha < 1$ is a random number (Eiben & Smith, 2003). The purpose of this type of operators is to allow making large "steps" in the design space and, consequently, explore the new and promising regions (Michalewicz, 1996).

Exploitative operators introduce small perturbations (e.g., $\boldsymbol{p} \leftarrow \boldsymbol{p} + \boldsymbol{r}$, where $\boldsymbol{r}$ is a random vector selected according to a normal probability distribution with zero mean and a certain problem dependent variance). These operators allow exploitation of a given region of the design space improving local search properties of the algorithm (Deb, 2001; Coello Coello et al., 2007). In some algorithms, e.g., PSO, the difference between both types of operators is not as clear: modification of the individual may be based on the best solution found so far by that given individual as well as the best solution found by the entire population (Clerc, 2006; Kennedy et al., 2001). On the other hand, the operators used by algorithms such as PSO should be interpreted as exchange of information between individuals that affects their behavior rather than the individuals themselves.

3.4.2 Evolution Strategies

Evolution strategies (ES) are probably the first population-based search methods ever conceived. They were developed in mid 1960s in Germany by I. Rechenberg and H.P. Schwefel and applied for solving continuous optimization problems, specifically, aerodynamic shape optimization (Rechenberg, 1965; Schwefel, 1968). ES comes in several variations (Hansen et al., 2015). The initial versions were only using mutation operators. Later on, recombination operators were introduced to allow the exchange of information between individuals (Schwefel & Rudolph, 1995). More recent versions of ES utilize covariance matrix adaptation (CMA-ES) to shape the mutation distribution so as to adjust it to a particular landscape of the optimized cost function (Hansen & Kern, 2004).

Evolution strategies use a natural representation (i.e., floating point numbers) and process a population of μ individuals so that, in each iteration, λ offsprings are created from the set of parent individuals using problem-specific recombination operators. The new set of parent individuals is chosen either from offsprings (so-called (μ,λ)-ES) or from offsprings and parents (so-called (μ + λ)-ES). Each individual contains both the solution data (design parameters) and the set of strategy parameters that determine its further mutation process.

The basic flow of the evolution strategies algorithm is shown in Fig. 3.15. Evolutionary strategies use random selection. The new population is chosen using the deterministic truncation selection based on objective function values of individuals. The mutation operators for unconstrained real-valued search space are defined as

$$(\boldsymbol{x},\boldsymbol{s}) \leftarrow \begin{cases} \boldsymbol{s} \leftarrow \boldsymbol{s} \cdot e^{\tau N(0,1)} \\ \boldsymbol{x} \leftarrow \boldsymbol{x} + \boldsymbol{s} \cdot N(0,\boldsymbol{I}) \end{cases} \tag{3.33}$$

where $N(0,1)$ and $N(0,\boldsymbol{I})$ are (0,1) normally distributed random scalars and vectors ($\boldsymbol{I}$ stands for the identity matrix of the size equal to the number of design variables n), and τ is the rate of self-adaptation (learning parameters). Parameter $\boldsymbol{s}$ controls the

```
initialize population Pμ = {(x(1),s(1)),...,(x(μ),s(μ))};
while ~termination_condition
  select a parent population Pλ = {(x(1),s(1)),...,(x(λ),s(λ))} from Pμ;
  for j = 1:λ
    mutate s(j);
    mutate x(j) using updated value of s(j);
  end
  select a new population Pμ from Pλ ((μ,λ)-selection) or from
    Pλ ∪ Pλ ((μ+λ)-selection);
end
```

Fig. 3.15 Pseudocode of the basic version of the evolution strategies algorithm

```
initialize population Pμ = {(x(1),s(1)),...,(x(μ),s(μ))};
while ~termination_condition
  for j = 1:λ
    select a parent set Pρ = {(y(1),t(1)),...,(y(ρ),r(ρ))} from Pμ;
    create (x(j),s(j)) by recombining (y(1),t(1)) to (y(ρ),r(ρ));
    mutate s(j);
    mutate x(j) using updated value of s(j);
  end
  select a new population Pμ from Pλ ((μ,λ)-selection) or from
    Pλ ∪ Pλ ((μ+λ)-selection);
end
```

Fig. 3.16 Pseudocode of the ES algorithm with recombination

strength of the design parameter mutation. It is suggested (Schwefel, 1988) that τ is proportional to $n^{-1/2}$. In more advanced implementations of ES, the identity matrix $\boldsymbol{I}$ in $N(0,\boldsymbol{I})$ may be replaced by a general covariance matrix with evolving entries (Hansen et al., 2015) that allows for taking into account correlations between variables, thus leading to more efficient search for the optimum.

Later versions of ES introduced recombination to allow exchange of information between individuals (Beyer, 2001), cf. Figure 3.16. The two most typical recombination operators used by ES are the arithmetical averaging

$$\boldsymbol{x} = \frac{1}{\rho}\sum_{j=1}^{\rho} \boldsymbol{y}^{(j)} \tag{3.34}$$

where $\boldsymbol{y}^{(j)}$ are parent individuals, and coordinate-wise random transfer

$$\boldsymbol{x} = [x_1 \ \ldots \ x_n]^T = \left[y_1^{(r_1)} \ \ldots \ y_n^{(r_n)}\right]^T \tag{3.35}$$

where r_1 to r_n are randomly selected from the set $\{1, 2, \ldots, \rho\}$.

Recommendations and rules of thumb concerning practical usage of ES can be found in the literature, e.g., Beyer and Schwefel (2002).

3.4.3 Genetic Algorithms

Genetic algorithms (GAs) are probably the most famous class of population-based metaheuristics around. They have been developed in late 1970s, mostly by J. Holland, K. DeJong and D. Goldberg (Holland, 1975; De Jong, 1975; Goldberg, 1989). Because the initial idea was based on mimicking biological evolution, GA processed populations of individuals on a genotypic level of binary strings while evaluating solutions on a phenotypic level pertinent to a problem at hand (e.g., floating point representation for continuous optimization). Nowadays, integer or floating point representation is preferred.

3.4.3.1 Algorithm Flow and Representation

The overall GA flow is similar to a generic structure shown in Fig. 3.14. Typically, GAs process individuals that are represented in certain internal format (referred to as *genotype*). Recombination and mutation operations are performed on a genotype; however, assessment of individual's fitness is performed on decoded individuals (referred to as *fenotype*). Traditional GAs utilize bit-string representation (Goldberg, 1989). Gray coding is often utilized instead of standard binary coding because small changes of the genotype imply small changes of the fenotype with the Gray coding.

Nowadays, a natural representation is preferred, e.g., floating point representation is used for continuous optimization problems, integer representation is used for combinatorial optimization (e.g., traveling salesman problem), etc. It is recommended that a representation is tailored to the problem in order to facilitate genetic operations or ensure feasibility of individuals (Michalewicz, 1996; Caorsi et al., 2000; Lai & Jeng, 2006; Ding et al., 2008).

3.4.3.2 Crossover

Crossover is a fundamental operator in GAs. It is designed to exchange information between individuals. Figure 3.17 shows several variations of two-parent crossover for bit-string representation. Crossover is explorative, that is, it allows for making large steps and thus helps discovering the promising area in the search space. Typical crossover rate, p_c, is high (e.g., 0.8), so that most of the parents exchange their data. Crossover may be generalized to involve more parents; however, no significant advantages of this approach have been reported (Coello Coello et al., 2007). It should be noticed that crossover does not produce new genetic information; therefore, it cannot be used as the only operator in a genetic algorithm.

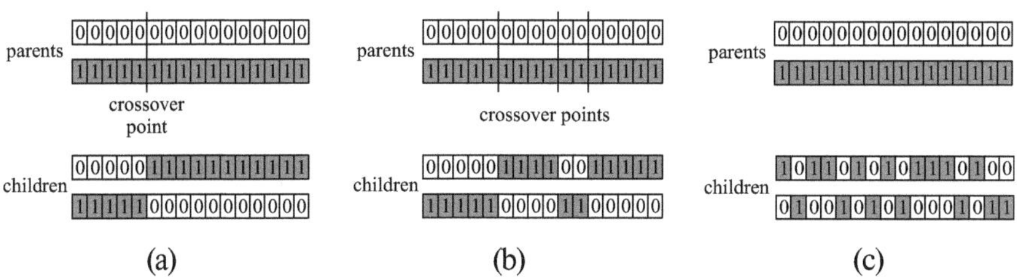

Fig. 3.17 Crossover operators for bit-string GAs: (**a**) one-point crossover, (**b**) multi-point crossover, and (**c**) uniform crossover (bits are changed randomly)

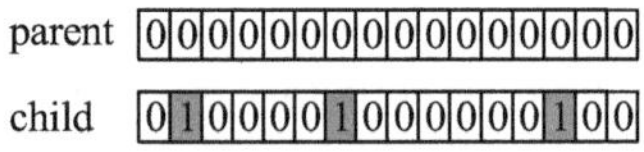

Fig. 3.18 Mutation operator for bit-string GAs. Each gene is flipped independently with a probability p_m

Crossover operators for floating point representation are arranged differently. For the sake of presentation, we will denote the parent vectors as $\boldsymbol{x} = [x_1 \ldots x_n]^T$ and $\boldsymbol{y} = [y_1 \ldots y_n]^T$, whereas the child (offspring) will be denoted as $\boldsymbol{z} = [z_1 \ldots z_n]^T$. The following three basic types of crossover operators can be distinguished (Eiben & Smith, 2003):

- Discrete: $z_i = x_i$ or y_i (random selection).
- Intermediate: $z_i = ax_i + (1\text{-}a)y_i$ with $0 \leq a \leq 1$ (a selected randomly).
- Arithmetic: $\boldsymbol{z} = a\boldsymbol{x} + (1 - a)\boldsymbol{y}$ with $0 \leq a \leq 1$ (a selected randomly).

Variations and mixtures of the above operators are also used in practical implementations of GAs.

3.4.3.3 Mutation

The purpose of mutation is to introduce small, random changes into the genotype. The mutation rate p_m is normally small, typically between $1/N$ and $1/L$ (where N and L are the population size and the chromosome length, respectively). Bit-flip mutation operator for traditional GAs is shown in Fig. 3.18. Mutation is an exploitative (local) operator; in other words, it facilitates optimization within a promising area using available information encoded in the chromosome. Mutation produces new genetic information and therefore, in principle, may be used as the only operator in a genetic algorithm. However, it is recommended to use both mutation and crossover in general (Goldberg, 1989).

The mutation operator for floating point representation (we assume that the individual is denoted as $\boldsymbol{x} = [x_1 \ldots x_n]^T$) can be written as $x_i \rightarrow x_i' = x_i + \Delta x_i$, where Δx_i is a random deviation that may be drawn with uniform or non-uniform (e.g., normal) distribution; in both cases, x_i' is eventually curtailed to the range if x_i (Eiben & Smith, 2003). Non-uniform distributions that promote smaller changes are typically preferred, e.g., (Michalewicz, 1996),

$$\Delta x_i = \begin{cases} (x_{i.\max} - x_i) \cdot (2(r - 0.5))^\beta & \text{if} \quad r > 0.5 \\ (x_{i.\min} - x_i) \cdot (2(0.5 - r))^\beta & \text{otherwise} \end{cases} \tag{3.36}$$

where $r \in [0,1]$ is a random number; β is normally larger than 1 (e.g., 3) and may increase in later stages of the algorithm run. In a particular implementation of (3.36), mutation is arranged to ensure satisfaction of the box constraints, i.e., $x_{i.\min} \leq x_i + \Delta x_i \leq x_{i.\max}$ for $i = 1, \ldots, n$.

3.4.3.4 Selection

Selection is another important GA component. It is a process of choosing individuals that will become parents of a new population. Selection is based on the quality of individuals, traditionally referred to as fitness. Normally, better individuals are given higher value of fitness; for minimization of the objective function f, the fitness of an individual $\boldsymbol{x}$ could be $-f(\boldsymbol{x})$. Genetic algorithms utilize partially stochastic selection, i.e., better individuals are favored, but there is an element of randomness that allows weak individuals to be selected as well (although with lower probability). On the one hand, this sort of strategy comes from the overall background of GAs (mimicking of natural evolution); on the other hand, it helps avoiding premature convergence of the algorithm.

Chronologically, one of the first selection schemes used in GAs was the roulette wheel selection (Coello Coello et al., 2007), where the probability of an individual to be selected was proportional to its fitness value. An obvious disadvantage of this scheme is a risk of a premature convergence: a single highly fit individual may quickly take over the population. Another disadvantage is considerable reduction of the selection pressure when all fitness values are on a similar level (this can be controlled by fitness scaling; Goldberg, 1989).

One of the most popular selection schemes today, free from the aforementioned drawbacks, is a tournament selection which works as follows (Bäck, 1996):

1. Randomly pick up a subset of k individuals from the current population (k is a tournament size).
2. Select an individual with the highest fitness value within the subset.
3. Repeat Steps 1 and 2 N times in order to create the parent set.

The selection pressure can be adjusted by changing the tournament size, from pure random selection for $k = 1$, to fully deterministic selection for $k = N$ (the latter would lead to instant convergence of the algorithm).

3.4.3.5 Elitism

An obvious issue related to stochastic selection is that the best individual (or individuals) is not guaranteed to survive. Even if it does, it may be altered by recombination and/or mutation operators. Elitism is a technique frequently used in GAs and other types metaheuristics (Zitzler & Thiele, 1999; Lim & Isa, 2014; Yang, 2014) to avoid losing the best individual(s) found so far: the set of best individuals (typically one) is stored and deterministically inserted into the new population (cf. Fig. 3.19). Various versions of this technique have been considered; for example, there are algorithms where individuals may survive within the population up to a certain number of iterations (Tan et al., 2005).

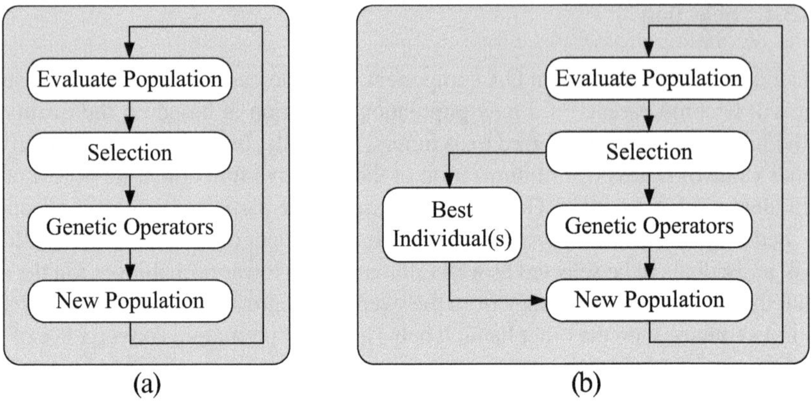

Fig. 3.19 Flowchart of a GA without (**a**) and with elitism (**b**). In the elitist GA, the best individual (or individuals) is found during the selection process and directly inserted into the new population bypassing the genetic operations

3.4.3.6 Selected Topics

In this section, we focus on two practical issues pertinent to GAs, namely, various population models and controlling the selection pressure.

There are three basic approaches to replacing the old population by a new one (Coello Coello et al., 2007):

1. Generational models: each individual survives for exactly one generation, and the entire set of parents is replaced by the offspring.
2. Steady-state models: one offspring is generated per iteration, and one member of the population is replaced.
3. Generation gap: the fraction of the population is replaced varying from 1.0 (as in generational models) to $1/N$ (as in steady-state models).

The most popular model is a generational one; however, other models may exhibit certain advantages. One of these is lower computational costs of the algorithm iteration for the generation gap models and, obviously, the steady-state one.

Selection pressure has been already mentioned on a few occasions in this chapter. It is basically the amount of preference given to the better individuals to survive. It is mainly controlled by the selection scheme utilized in the algorithm. Selection pressure is critical for controlling the operation and performance of GAs. In particular, if it is too high, it may lead to premature convergence, i.e., the situation when all individuals become (almost) identical before the search space has been properly explored. Production of a new genetic material is a mechanism opposite to selection pressure, and it is mainly dependent on the mutation rate. Normally, we want to keep a balance between the selection pressure and the production of a new genetic material so that the premature convergence is avoided, but the evolution is not a random search either. In practice, such a balance can be achieved by monitoring diversity of the population and adjusting the mutation rate accordingly (self-adaptation) (Hesser & Männer, 1991). Diversity of the population can be measured using standard deviation of individuals (details depend on a particular representation used in the algorithm). The mutation rate can be reduced at the final stages of the algorithm run in order to improve exploitation at the expense of exploration and, therefore, obtain more precise localization of the optimal solution.

3.4.4 Evolutionary Algorithms

The term of evolutionary algorithms has been conceived in early 1990s (see, e.g., Michalewicz, 1996). It refers to the entire class of population-based search methods that are distinguished from other types of optimization algorithms by the following features:

- The use of population
- The use of stochastic search operators
- The use of stochastic selection

Genetic algorithms are often considered to be special case of evolutionary algorithms. As a matter of fact, genetic algorithms using floating point representation are considered by some to be evolutionary algorithms but not genetic algorithms.

There is a little of theory explaining why evolutionary algorithms work. Early, semi-heuristic results for bit-string-based genetic algorithms (so-called schemata theorem by J. Holland; Holland, 1975) suggesting that genetic algorithms work by discovering, exploiting, and combining "building blocks" (groups of closely interacting genes) have been later criticized partially because of the success of algorithms using natural representations (Eiben & Smith, 2003; Liu et al., 2014; Ding & Wang, 2013; Alander et al., 2004; Yeung et al., 2008). The results obtained using Markov Chain analysis, where GAs are treated as stochastic processes (Horn, 1993; David, 1994; Suzuki, 1995), consider algorithm convergence to a global solution in a probabilistic sense; however, practical significance of such results is minor.

Evolutionary algorithms (EAs) can be considered as robust problem solvers in the following sense (cf. Fig. 3.20):

- EAs do not perform as good as problem-specific methods for problems for which the latter methods were designed.
- EAs perform better than a random search "on average."
- EAs perform better than problem-specific methods "on average."

These features are generally common for other types of nature-inspired metaheuristics as well (cf. Sects. 3.4.5, 3.4.6, 3.4.7, and 3.4.8).

On a practical note, it should be emphasized that "standard" off-the-shelf implementation of an evolutionary algorithm may not perform well when applied to a particular engineering problem. This is in line with so-called no free lunch (NFL) theorems (Wolpert & Macready, 1997, 2005) that basically say—when referred to

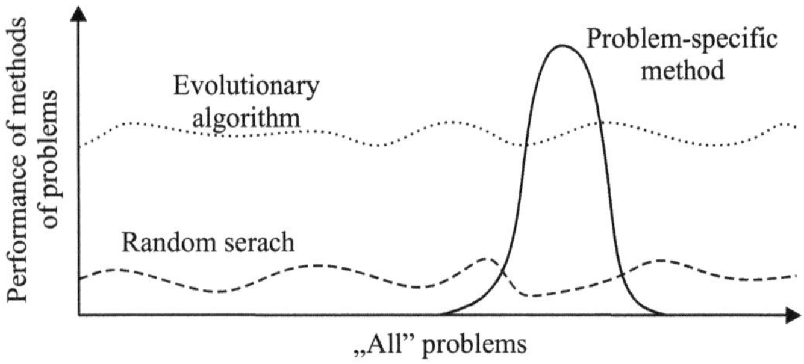

Fig. 3.20 Robustness of evolutionary algorithms versus pure random search and problem-specific methods

evolutionary methods—that performance of all algorithms is the same when averaged over all problems. Consequently, a general recommendation is that EAs should be individually designed for solving any given task in order to maximize performance and reliability. There are various (and problem-specific) ways of doing so, including development of appropriate representation, tailored recombination, and mutation operators.

It is normally better to preserve feasibility of solutions by suitably defined genetic operations rather than to enforce extinction of infeasible individuals by assigning them low fitness values. A famous example is a partially mapped crossover operator for the TSP problem as described by Goldberg (Goldberg, 1989).

3.4.5 Particle Swarm Optimization

Nowadays, particle swarm optimization (PSO) belongs to the most popular nature-inspired algorithms. PSO is based on mimicking the swarm behavior such as fish or bird schooling in nature (Kennedy, 1997; Kennedy et al., 2001). The first version of PSO was developed in mid 1990s. Since then, many variations of the algorithm have been applied to virtually every area of optimization, computational intelligence, and engineering design including high-frequency engineering (Lizzi et al., 2008; Poli et al., 2014; Pavlidis et al., 2005; Wang et al., 2005; Rocca et al., 2012; Luo et al., 2019; Bi et al., 2021).

A PSO algorithm processes a set of individuals (traditionally referred to as swarm) that are iteratively relocated within the design space and interact with other individuals by means of social influence and social learning. Individuals (or particles) are vectors in a multidimensional space and are represented by their positions and velocities. Particles are relocated based on their velocities but also information about the best positions found so far (both individual and overall for the swarm). The movement of a swarming particle consists of two major components: a stochastic component and a deterministic component. Each particle is attracted toward the position of the current global best $\boldsymbol{g}$ and its own best location $\boldsymbol{x}_i^*$ found during the optimization run. Additionally, there is a stochastic component controlled by suitable algorithm parameters.

Let $\boldsymbol{x}_i$ and $\boldsymbol{v}_i$ denote the position and the velocity vectors of the ith particle, respectively. The velocity and position are updated as follows:

$$\boldsymbol{v}_i \leftarrow \chi\left[\boldsymbol{v}_i + c_1\boldsymbol{r}_1 \bullet \left(\boldsymbol{x}_i^* - \boldsymbol{x}_i\right) + c_2\boldsymbol{r}_2 \bullet (\boldsymbol{g} - \boldsymbol{x}_i)\right] \tag{3.37}$$

$$\boldsymbol{x}_i \leftarrow \boldsymbol{x}_i + \boldsymbol{v}_i \tag{3.38}$$

Here, $\boldsymbol{r}_1$ and $\boldsymbol{r}_2$ are vectors whose components are uniformly distributed random numbers between 0 and 1; • denotes component-wise multiplication. The control parameter χ is typically equal to 0.7298 (Kennedy et al., 2001); c_1 and c_2 are acceleration constants determining how much the particle is directed toward good positions (they represent a "cognitive" and a "social" components, respectively); usually, $c_1, c_2 \approx 2$. The sum of $c_1 + c_2$ should not be smaller than 4.0; as the values of these parameters increase, χ gets smaller as does the damping effect.

There are many different variants and extensions of the standard PSO algorithm (Clerc & Kennedy, 2002; Kennedy et al., 2001; Fong et al., 2016; Bansal et al., 2011; Nickabadi et al., 2011; Kao & Zahara, 2008).

3.4.6 Differential Evolution

Differential evolution (DE) is currently one of the most popular population-based metaheuristic algorithms for continuous global optimization. DE was developed in late 1990s by R. Storn and K. Price (Storn & Price, 1997). It is a vector-based evolutionary algorithm and can be considered as a further development to genetic algorithms. It is a derivative-free stochastic search algorithm with self-organizing tendency. The design parameters are represented as n-dimensional vectors (also called agents). The new agents are allocated in the search space by combining the positions of other existing agents. More specifically, an intermediate agent is generated from two agents randomly chosen from the current population. This temporary agent is then mixed with a predetermined target agent. The new agent is accepted for the next generation if and only if it yields reduction in objective function. In other words, DE is using a sort of deterministic selection.

The basic DE algorithm uses a random initialization. A new agent $\boldsymbol{y} = [y_1 \ldots y_n]$ is created from the exisiting one $\boldsymbol{x} = [x_1 \ldots x_n]$ as follows:

1. Three agents $\boldsymbol{a} = [a_1 \ldots a_n]$, $\boldsymbol{b} = [b_1 \ldots b_n]$ and $\boldsymbol{c} = [c_1 \ldots c_n]$ are randomly extracted from the population ($\boldsymbol{a} \neq \boldsymbol{b} \neq \boldsymbol{c} \neq \boldsymbol{x}$).

2. A position index $p \in \{1, ..., N\}$ is determined randomly (N is the population size).
3. The new agent $\boldsymbol{y}$ is determined as follows (separately for each of its components $i = 1, ..., n$):

   (a) Select a random number $r_i \in (0,1)$ with uniform probability distribution.
   (b) If $i = p$ or $r_i < CR$ let $y_i = a_i + F(b_i - c_i)$, otherwise let $y_i = x_i$; here, $0 \leq F \leq 2$ is the differential weight and $0 \leq CR \leq 1$ is the crossover probability, both defined by the user.
   (c) If $f(\boldsymbol{y}) < f(\boldsymbol{x})$ then replace $\boldsymbol{x}$ by $\boldsymbol{y}$; otherwise reject $\boldsymbol{y}$ and keep $\boldsymbol{x}$.

The above DE scheme is often referred to as *DE/rand/1/bin*, meaning that the donor vector $\boldsymbol{a}$ is selected randomly ("rand"), there is one pair of vectors selected to calculated the mutation differential ("1"), and that the recombination is binomial ("bin"). Some of DE variations, using the terminology explained above are as follows (Mezura-Montes et al., 2006a, b, 2010):

- *DE/rand/p/bin*—explained above; p denotes the number of pairs of vectors used to calculate the mutation differential (p can be equal to 1).
- *DE/rand/p/exp*—as above but with exponential crossover (Noman & Iba, 2005).
- *DE/best/p/bin*, *DE/best/p/exp*—a donor vector is the best solution found so far in the population.
- *DE/current-to-rand/p/bin*, *DE/current-to-best/p/exp*—variants with arithmetic recombination.
- *DE/current-to-rand/p/bin*—a variant with combined arithmetic-discrete recombination.

It should be emphasized that although DE resembles some other stochastic optimization techniques, unlike traditional EAs, DE perturbs the solutions in the current generation vectors with scaled differences of two randomly selected agents. As a consequence, no separate probability distribution has to be used, and thus the scheme presents some degree of self-organization. Additionally, DE is simple to implement, uses very few control parameters, and has been observed to perform satisfactorily in multi-modal optimization problems (Chakraborty, 2008; Rocca et al., 2011).

3.4.7 *Firefly Algorithm*

Firefly algorithm (FA) is one of the most recent yet already popular metaheuristics. It was developed in 2007 by X.S. Yang (Yang, 2008) by modeling behavior of firefly using a few idealized rules, namely: (i) fireflies are attracted to each other and (ii) attractiveness is proportional to their brightness (i.e., the less bright one moves toward the brighter one); (iii) the brightness of a firefly depends on the objective function landscape (e.g., for maximization problem, the brightness may be proportional to the objective function value). Figure 3.21 shows the algorithmic flow of the firefly algorithm.

```
objective function f(x), x = [x1 ... xd]^T;
initialize population x^(i), i = 1, ..., N;
define light absorption coefficient γ;
while ~termination_condition
  for i = 1:N
    for j = 1:N
      if I(x^(i)) < I(x^(j))
        move firefly i towards j
      end
      vary attractiveness with distance r proportionally to exp(−γr);
      evaluate new solutions and update light intensity;
    end
  end
  rank the fireflies and find the global best g*;
end
```

Fig. 3.21 Flow of the firefly algorithm. Light intensity $I(\boldsymbol{x}^{(i)})$ is determined by $f(\boldsymbol{x}^{(i)})$ (e.g., $I(\boldsymbol{x}^{(i)}) = -f(\boldsymbol{x}^{(i)})$ for a minimization problem). Termination condition is typically based on the maximum number of iterations

Formally, the updating formula for the ith firefly is as follows:

$$\boldsymbol{x}_i \leftarrow \boldsymbol{x}_i + \beta_0 e^{-\gamma r_{ij}^2} \left(\boldsymbol{x}_j - \boldsymbol{x}_i\right) + \alpha \boldsymbol{\varepsilon}_i \tag{3.39}$$

where $r_{ij} = \|\boldsymbol{x}_i - \boldsymbol{x}_j\|$ is the Euclidean distance between the ith and jth firefly (this can be replaced by other type of norm), β_0 is the attractiveness parameter at a zero distance, $\alpha \in (0,1]$ is a randomization parameter, whereas $\boldsymbol{\varepsilon}_i$ is a random vector drawn with Gaussian or uniform probability distribution. Note that (3.39) is essentially a random walk toward brighter fireflies (Yang, 2008). The convergence rate of the algorithm critically depends on the parameter γ (taken in the range 0.1 to 10 in most applications).

It should be noted that FA is, as a matter of fact, quite closely related to PSO. In particular, by setting $\gamma \to 0$, FA becomes a special case of accelerated PSO (Yang, 2008). Also, by removing the inner loop (cf. Fig. 3.21), and replacing the light intensity by the global best g^*, one essentially gets the standard PSO (Weyland, 2015). On the other hand, for $\gamma \to \propto$, FA turns into random search because the influence of other fireflies on relocation of any given firefly becomes negligible. In practice, FA is normally operating between these two extreme cases thus arguably better than both random search and PSO (Yang, 2008).

3.4.8 Other Methods

Metaheuristic optimization has been constantly growing in popularity over the last three decades. With the major methods and algorithms developed before 1980s (e.g., ES, GAs, EAs) and 1990s (PSO, DE, ant colony optimization), the number of new algorithms as well as improvement of the existing ones proposed in the literature has been accelerating.

Examples include bee algorithm (Yang, 2005), bacteria foraging algorithm (Gazi & Passino, 2004), harmony search (Geem et al., 2001), bat algorithm (Yang, 2010a), gravitational search algorithm (Rashedi et al., 2009), wind-driven optimization (Bayraktar et al., 2013), spider monkey optimization (Al-Azza et al., 2016), or cat swarm optimization (Ram et al., 2015). Most of these techniques are based on observations of specific natural or social phenomena; however, the general concepts are usually similar (cf. Sect. 3.4.1). A good exposition of some of the recent methods can be found in (Yang, 2010a). At the same time, it can be argued that the differences between many of the recently proposed techniques are quite negligible. In particular, the majority of them are actually small variations of the PSO algorithm (Weyland, 2015). This, in conjunction with the very limited (or non-existing) theoretical justification, makes development of the new algorithmic variations rather questionable.

3.5 Summary

In this chapter, we outlined several types of conventional optimization techniques including gradient-based and derivative-free methods, along with nature-inspired metaheuristics. Gradient-based techniques (Sect. 3.2) are primarily local ones so that their performance depends on the starting point. Also, they require sensitivity data, often acquired through finite differentiation, which makes the optimization process computationally expensive. On the other hand, derivative-free methods (Sect. 3.3) are suitable for handling analytically intractable (noisy, discontinuous) functions. Global and multi-objective optimization is—in a large part—dominated by nature-inspired algorithms outlined in Sect. 3.4. Unfortunately, global search capability comes at a price of tremendous computational costs. For typical population-based procedure, the required number of objective function evaluations is many hundreds or even thousands. Also a majority of the methods described here are suitable for unconstrained optimization (apart from box constraints that can be easily handled by appropriate definitions of the recombination operators). Constrained metaheuristic optimization is a research area in its own (Mezura-Montes & Coello Coello, 2011; Koziel & Michalewicz, 1998) and exceeds the scope of this chapter.

Poor computational efficiency of conventional optimization methods hinders their applicability for EM-driven design, which is especially pertinent to population-based techniques. From this perspective, a promising way of handling computationally expensive models in high-frequency design is the surrogate-assisted optimization. Introduction to surrogate-based methods has been provided in Chap. 4. Another alternative is a response feature technology, which is the major topic of this book. Some of the optimization and modeling methods outlined in this chapter and Chap. 4 will be employed as the underlying algorithms in the feature-based design frameworks (Chaps. 5, 6, 7, 8, 9, 10, 11, 12, 13, 14, and 15).

References

Alander, J. T., Zinchenko, L. A., & Sorokin, S. N. (2004). Comparison of fitness landscapes for evolutionary design of dipole antennas. *IEEE Transactions on Antennas and Propagation, 52*, 2932–2940.

Al-Azza, A. A., Al-Jodah, A. A., & Harackiewicz, F. J. (2016). Spider monkey optimization: A novel technique for antenna optimization. *IEEE Antennas and Wireless Propagation Letters, 15*, 1016–1019.

Antoniou, A., & Lu, W.-S. (2007). *Practical optimization: Algorithms and engineering applications*. Springer US.

Bäck, T. (1996). *Evolutionary algorithms in theory and practice*. New York: Oxford University Press.

Back, T., Fogel, D. B., & Michalewicz, Z. (Eds.). (2000). *Evolutionary computation 1: Basic algorithms and operators*. Institute of Physics Publishing.

Bansal, J. C., Singh, P. K., Saraswat, M., Verma, A., Jadon, S. S., & Abraham, A. (2011). Inertia weight strategies in particle swarm optimization. *World Congress Nature Biologically Inspired Comp*, 633–640.

Bates, D. M., & Watts, D. G. (1988). *Nonlinear regression and its applications*. John Wiley & Sons.

Bayraktar, Z., Komurcu, M., Bossard, J. A., & Werner, D. H. (2013). The wind driven optimization technique and its application in electromagnetics. *IEEE Transactions on Antennas and Propagation, 61*(5), 2745–2757.

Bertsekas, D. P. (1982). *Constrained optimization and laguange multiplier methods*. Academic Press.

Beyer, H.-G. (2001). *The theory of evolution strategies*. Springer-Verlag.

Beyer, H.-G., & Schwefel, H.-P. (2002). Evolution strategies: A comprehensive introduction. *Natural Computing, 1*, 3–52.

Bi, L., Cao, W., Hu, W., & Wu, M. (2021). Intelligent tuning of microwave cavity filters using granular multi-swarm particle swarm optimization. *IEEE Transactions on Industrial Electronics, 68*(12), 12901–12,911.

Bischof, C., Corliss, G., & Griewank, A. (1993). Structured second- and higher-order derivatives through univariate Taylor series. *Optimization Methods and Software, 2*, 211–232.

Caorsi, S., Massa, A., & Pastorino, M. (2000). A computational technique based on a real-coded genetic algorithm for microwave imaging purposes. *IEEE Transactions in Geoscience Remote Sensing, 38*, 1697–1708.

Chakraborty, U. (2008). *Advances in differential evolution*. Springer.

Cheng, M. Y., & Prayogo, D. (2014). Symbiotic organism search: A new metaheuristic optimization algorithm. *Computers & Structures, 139*, 98–112.

Clerc, M. (2006). *Particle swarm optimization*. ISTE.

Clerc, M., & Kennedy, J. (2002). The particle swarm – explosion, stability, and convergence in a multidimensional complex space. *IEEE Transactions on Evolutionary Computation, 6*, 58–73.

Coello Coello, C. A., Lamont, G. B., & Van Veldhuizen, D. (2007). *Evolutionary algorithms for solving multi-objective problems* (2nd ed.). Springer.

Conn, A. R., Gould, N. I. M., & Toint, P. L. (2000). *Trust region methods*. In *MPS-SIAM series on optimization*. MPS-SIAM.

Conn, A. R., Scheinberg, K., & Vincente, L. N. (2009). *Introduction to derivative-free optimization, MPS-SIAM series on optimization*. MPS-SIAM.

Crowder, H. P., & Wolfe, P. (1972). Linear convergence of the conjugate gradient method. *IBM J. Res. Development*, 16, 431–433.

Datta, T., & Misra, I. S. (2009). A comparative study of optimization techniques in adaptive antenna array processing: The bacteria-foraging algorithm and particle-swarm optimization. *IEEE Antennas and Propagation Magazine, 51*(6), 69–81.

David, H. (1994). A Markov Chain analysis of genetic algorithms with a state dependent fitness function. *Complex Systems, 8*, 407–417.

Davis, C. (1954). Theory of positive linear dependence. *American Journal of Mathematics, 76*, 733–746.

De Jong, A. K. (1975). *An Analysis of the Behavior of a Class of Genetic Adaptive Systems*. PhD thesis, University of Michigan.

Deb, K. (2001). *Multi-objective optimization using evolutionary algorithms*. John Wiley & Sons.

Deb, A., Roy, J. S., & Gupta, B. (2014). Performance comparison of differential evolution, particle swarm optimization and genetic algorithm in the design of circularly polarized microstrip antennas. *IEEE Transactions on Antennas and Propagation, 62*(8), 3920–3928.

Ding, D., & Wang, G. (2013). Modified multiobjective evolutionary algorithm based on decomposition for antenna design. *IEEE Transactions on Antennas and Propagation, 61*, 5301–5307.

Ding, M., Jin, R., Geng, J., Wu, Q., & Yang, G. (2008). Auto-design of band-notched UWB antennas using mixed model of 2D GA and FDTD. *Electronics Letters, 44*, 257–258.

Duman, E., Uysal, M., & Alkaya, A. F. (2012). Migrating birds optimization: A new metaheuristic approach and its performance on quadratic assignment problem. *Information Sciences, 217*, 65–77.

Eiben, A. E., & Smith, J. E. (2003). *Introduction to evolutionary computing*. Springer-Verlag.

Elragal, H. M., Mangoud, M. A., & Alsharaa, M. T. (2011). Hybrid differential evolution and enhanced particle swarm optimization technique for design of reconfigurable phased antenna arrays. *IET Microwaves, Antennas & Propagation, 5*(11), 1280–1287.

Fletcher, R. (1987). *Practical methods of optimization* (2nd ed.). JohnWiley & Sons.

Fletcher, R., & Reeves, C. M. (1964). Function minimization by conjugate gradients. *The Computer Journal, 7*, 149–154.

Fong, S., Wong, R., & Vasilakos, A. V. (2016). Accelerated PSO swarm search feature selection for data stream mining big data. *IEEE Transactions on Services Computing, 9*, 33–45.

Gazi, V., & Passino, K. M. (2004). Stability analysis of social foraging swarms. *IEEE Transactions on Systems, Man, and Cybernetics—Part B: Cybernetics, 34*, 539–557.

Geem, Z. W., Kim, J. H., & Loganathan, G. V. (2001). A new heuristic optimization algorithm: Harmony search. *Simulation, 76*, 60–68.

Goldberg, D. E. (1989). *Genetic algorithms in search, optimization and machine learning*. Addison-Wesley Longman Publishing.

Han, S. P. (1977). A globally convergent method for nonlinear programming. *Journal of Optimization Theory and Applications, 22*, 297–309.

Hansen, N., & Kern, S. (2004). Evaluating the CMA evolution strategy on multimodal test functions. In X. Yao et al. (Eds.), *Parallel problem solving from nature* (pp. 282–291). Springer-Verlag.

Hansen, E., & Walster, G. W. (2003). *Global optimization using interval analysis* (2nd ed.). Marcel Dekker.

Hansen, N., Arnold, D. V., & Auger, A. (2015). Evolution strategies. In J. Kacprzyk & W. Pedrycz (Eds.), *Springer handbook of computational intelligence* (pp. 871–898). Springer-Verlag.

Hesser, J., & Männer, R. (1991). Towards an optimal mutation probability for genetic algorithms. In H.-P. Schwefel & R. Männer (Eds.), *Parallel problem solving from nature* (pp. 23–32). Springer-Verlag.

Holland, J. H. (1975). *Adaptation in natural and artificial systems: An introductory analysis with applications to biology, control, and artificial intelligence*. The MIT Press.

Hooke, R., & Jeeves, T. A. (1961). Direct search solution of numerical and statistical problems. *Journal of the ACM, 8*(2), 212–229.

Horn, J. (1993). Finite Markov chain analysis of genetic algorithms with niching. *Proceedings of the Second International Conference on Genetic Algorithms*, 110–117.

Horst, R., & Tuy, H. (1996). *Global optimization: Deterministic approaches* (3rd ed., p. 1996). Springer-Verlag.

Horst, R., Pardalos, P. M., & Thoai, N. V. (2000). *Introduction to global optimization* (2nd ed.). Kluwer Academic Publishers.

Jang, C. H., Hu, F., Li, J., & Zhu, D. (2016). Low-redundancy large linear arrays synthesis for aperture synthesis radiometers using particle swarm optimization. *IEEE Transactions on Antennas and Propagation, 64*(6), 2179–2188.

Jaulin, L., Kieffer, M., Didrit, O., & Walter, E. (2001). *Applied interval analysis*. Springer-Verlag.

Kao, Y.-T., & Zahara, E. (2008). A hybrid genetic algorithm and particle swarm optimization for multimodal functions. *Applied Soft Computing, 8*, 849–857.

Karamalis, P. D., Kanatas, A. G., & Constantinou, P. (2009). A genetic algorithm applied for optimization of antenna arrays used in mobile ratio channel characterization devices. *IEEE Transactions on Instrumentation and Measurement, 58*(8), 2475–2487.

Kennedy, J. (1997). The particle swarm: Social adaptation of knowledge. *Proceedings of International Conference on Evolutionary Computation*, 303–308.

Kennedy, J., Eberhart, R. C., & Shi, Y. (2001). *Swarm intelligence*. Academic Press.

Kerkhoff, A. J., & Ling, H. (2007). Design of a band-notched planar monopole antenna using genetic algorithm optimization. *IEEE Transactions on Antennas and Propagation, 55*, 604–610.

Kibria, S., Islam, M. T., & Yatim, B. (2014). New compact dual-band circularly polarized universal RFID reader antenna using ramped convergence particle swarm optimization. *IEEE Transactions on Antennas and Propagation, 62*(5), 2795–2801.

Kirkpatrick, S., Gelatt, C. D., & Vecchi, M. P. (1983). Optimization by simulated annealing. *Science, New Series, 220*, 671–680.

Kolda, T. G., Lewis, R. M., & Torczon, V. (2003). Optimization by direct search: New perspectives on some classical and modern methods. *SIAM Review, 45*, 385–482.

Koziel, S., & Michalewicz, Z. (1998). A decoder-based evolutionary algorithm for constrained parameter optimization problems. In T. Bäck, A. E. Eiben, M. Schoenauer, & H.-P. Schwefel (Eds.), *Proceedings of. parallel problem solving from nature, Lecture notes in computer Science* (Vol. 1498, pp. 231–240). Springer-Verlag.

Koziel, S., & Ogurtsov, S. (2014). *Antenna design by simulation-driven optimization*. Springer.

Koziel, S., Echeverría Ciaurri, D., & Leifsson, L. (2011). Surrogate-based methods. In S. Koziel & X.-S. Yang (Eds.), *Computational optimization, methods and algorithms* (pp. 33–59). Springer-Verlag.

Kroese, D. P., Taimre, T., & Botev, Z. I. (2011). *Handbook of Monte Carlo methods*. John Wiley & Sons.

Kuhn, H. W., & Tucker, A. W. (1951). Nonlinear programming. In J. Neyman (Ed.), *Proceedings of Berkeley symposium on mathematical statistics probability* (pp. 481–492). University of California Press.

Lai, M.-I., & Jeng, S.-K. (2006). Compact microstrip dual-band bandpass filters design using genetic-algorithm techniques. *IEEE Transactions on Microwave Theory and Techniques, 54*, 160–168.

Lim, W. H., & Isa, N. A. M. (2014). An adaptive two-layer particle swarm optimization with elitist learning strategy. *Information Sciences, 273*, 49–72.

Liu, B., Aliakbarian, H., Zhongkun, M., Vandenbosch, G. A. E., Gielen, G., & Excell, P. (2014). An efficient method for antenna design optimization based on evolutionary computation and machine learning techniques. *IEEE Transactions on Antennas and Propagation, 62*, 7–18.

Lizzi, L., Viani, F., Azaro, R., & Massa, A. (2008). A PSO-driven spline-based shaping approach for ultra-wideband (UWB) antenna synthesis. *IEEE Transactions on Antennas and Propagation, 56*, 2613–2621.

Luenberger, D. G. (2003). *Linear and nonlinear programming* (2nd ed.). Kluwer Academic Publishers.

Luo, X., Yang, B., & Qian, H. J. (2019). Adaptive synthesis for resonator-coupled filters based on particle swarm optimization. *IEEE Transactions on Microwave Theory and Techniques, 67*(2), 712–725.

Matekovits, L., Mussetta, M., Pirinoli, P., Selleri, S., & Zich, R. E. (2004). Particle swarm optimization of microwave microstrip filters. *IEEE Antennas and Propagation Society Symposium, 3*, 2731–2734.

Mezura-Montes E., & Coello-Coello, C. A. (2011). Constraint-handling in nature-inspired numerical optimization: Past, present and future. *Swarm and Evolutionary Computation, 1*, 173–194.

Mezura-Montes, E., Velázquez-Reyes, J., & Coello Coello, C. A. (2006a). A comparative study of differential evolution variants for global optimization. *Conference on Genetic and Evolutionary Computation.*

Mezura-Montes, E., Reyes-Sierra, M., & Coello Coello, C. A. (2006b). Multi-objective optimization using differential evolution: A survey of the state-of-the-art. In U. K. Chakraborty (Ed.), *Advances in differential evolution* (pp. 173–196).

Mezura-Montes, E., Miranda-Varela, M. E., & Gómez-Ramón, R. C. (2010). Differential evolution in constrained numerical optimization: An empirical study. *Information Sciences, 180*, 4223–4262.

Michalewicz, Z. (1996). *Genetic algorithms + data structures = Evolution programs*. Springer-Verlag.

Nelder, J. A., & Mead, R. (1965). A simplex method for function minimization. *The Computer Journal, 7*, 308–313.

Nickabadi, A., Ebadzadeh, M. M., & Safabakhsh, R. (2011). A novel particle swarm optimization algorithm with adaptive inertia weight. *Applied Soft Computing, 11*, 3658–3670.

Nielsen, H. B. (1999). Damping parameter in Marquardt's method. IMM DTU. Report IMM-REP-1999-05.

Nocedal, J., & Wright, S. (2006). *Numerical optimization* (2nd ed.). Springer.

Noman, N., & Iba, H. (2005). Enhancing differential evolution performance with local search for high dimensional function optimization. In H.-G. Beyer (Ed.), *Proceedings of the 7th annual conference on Genetic and evolutionary computation* (pp. 967–974). New York: ACM.

Pavlidis, N. G., Parsopoulos, K. E., & Vrahatis, M. N. (2005). Computing Nash equilibria through computational intelligence methods. *Journal of Computational and Applied Mathematics, 175*, 113–136.

Polak, E., & Ribiére, G. (1969). Note sur la convergence de méthodes de directions eonjuguées. *Revue Francaise d'lnformatique et de Recherche Opirationnelle, 16*, 35–43.

Poli, L., Rocca, P., Oliveri, G., & Massa, A. (2014). Failure correction in time-modulated linear arrays. *IET Radar, Sonar and Navigation, 8*, 195–201.

Price, K., Storn, R. M., & Lampinen, J. A. (2005). *Differential evolution: A practical approach to global optimization*. Springer-Verlag.

Queipo, N. V., Haftka, R. T., Shyy, W., Goel, T., Vaidynathan, R., & Tucker, P. K. (2005). Surrogate based analysis and optimization. *Progress in Aerospace Science, 41*, 1–28.

Ram, G., Mandal, D., Kar, R., & Ghoshal, S. P. (2015). Cat swarm optimization as applied to time-modulated concentric circular antenna array: Analysis and comparison with other stochastic optimization methods. *IEEE Transactions on Antennas and Propagation, 63*(9), 4180–4183.

Rashedi, E., Nezamabadi-pour, H., & Saryazdi, S. (2009). GSA: A gravitational search algorithm. *Information Sciences, 179*, 2232–2248.

Rechenberg, I. (1965). Cybernetic solution path of an experimental problem. Royal Aircraft Establishment, Library Translation Number 1122, Farnborough, UK.

Rocca, P., Oliveri, G., & Massa, A. (2011). Differential Evolution as applied to electromagnetics. *IEEE Antennas and Propagation Magazine, 53*, 38–49.

Rocca, P., Poli, L., Oliveri, G., & Massa, A. (2012). Adaptive nulling in time-varying scenarios through time-modulated linear arrays. *IEEE Antennas and Wireless Propagation Letters, 11*, 101–104.

Rosenbrock, H. H. (1960). An automatic method for finding the greatest or least value of a function. *The Computer Journal, 3*, 175–184.

Schwefel, H.-P. (1968). Experimentelle optimierung einer zweiphasendüse tell 1. AEG Research Institute Project MHD-Staustrahlrohr 11034/68, Technical report 35.

Schwefel, H.-P. (1988). Collective intelligence in evolving systems. In W. Wolff, C.-J. Soeder, & F. R. Drepper (Eds.), *Ecodynamics* (pp. 95–100). Springer-Verlag.

Schwefel, H.-P., & Rudolph, G. (1995). Contemporary evolution strategies. In F. Moran, A. Moreno, J. J. Merelo, & P. Chacon (Eds.), *Advances in artificial life* (pp. 891–907). Springer-Verlag.

Storn, R., & Price, K. (1997). Differential evolution – a simple and efficient heuristic for global optimization over continuous spaces. *Journal of Global Optimization, 11*, 341–359.

Suzuki, J. (1995). Finite Markov chain analysis of genetic algorithms. *IEEE Transactions on Systems, Man, and Cybernetics, 25*.

Talbi, E.-G. (2009). *Metaheuristics: From design to implementation*. John Wiley & Sons.

Tamura, K., & Yasuda, K. (2011). Spiral dynamics inspired optimization. *Journal of Advanced Computational Intelligence and Intelligent Informatics, 15*(8), 1116–1122.

Tan, K., Khor, E., & Lee, T. (2005). *Multiobjective evolutionary algorithms and applications*. Springer-Verlag.

Tapia, R. A. (1978). Quasi-Newton methods for equality constrained optimization: Equivalence of existing methods and a new implementation. In O. Mangasarian, R. Meyer, & S. Robinson (Eds.), *Nonlinear programming* (Vol. 3, pp. 125–164). Academic Press.

Ülker, S. (2008). Particle swarm optimization application to microwave circuits. *Microwave and Optical Technology Letters, 50*, 1333–1336.

van Laarhoven, P. J., & Aarts, E. H. (1987). *Simulated annealing: Theory and applications*. Springer Netherlands.

Wang, W., Lu, Y., Fu, J. S., & Xiong, Y. Z. (2005). Particle swarm optimization and finite-element based approach for microwave filter design. *IEEE Transactions on Magnetics, 41*(5), 1800–1803.

Weyland, D. (2015). A critical analysis of the harmony search algorithm – how not to solve sudoku. *Operations Research Perspectives, 2*, 97–105.

Wolpert, D. H., & Macready, W. G. (1997). No free lunch theorems for optimization. *IEEE Transactions on Evolutionary Computation, 1*, 67.

Wolpert, D. H., & Macready, W. G. (2005). Coevolutionary free lunches. *IEEE Transactions on Evolutionary Computation, 9*(6), 721–735.

Yang, X. S. (2005). Engineering optimization via nature-inspired virtual bee algorithms. *Lecture Notes in Computer Science*, 317–323.

Yang, X. S. (2008). *Nature-inspired metaheuristic algorithms*. Luniver Press.

Yang, X.-S. (2010a). *Engineering optimization: An introduction with metaheuristic applications*. John Wiley & Sons.

Yang, X.-S. (2010b). A New Metaheuristic Bat-Inspired algorithm. In J. R. González, D. A. Pelta, C. Cruz, G. Terrazas, & N. Krasnogor (Eds.), *Nature inspired cooperative strategies for optimization* (pp. 65–74). Springer.

Yang, X.-S. (Ed.). (2014). *Cuckoo search and firefly algorithm: Theory and applications*. Springer International Publishing.

Yeung, S. H., Man, K. F., Luk, K. M., & Chan, C. H. (2008). A trapeizform U-slot folded patch feed antenna design optimized with jumping genes evolutionary algorithm. *IEEE Transactions on Antennas and Propagation, 56*, 571–577.

Zhigljavsky, A., & Žilinskas, A. (2008). *Stochastic global optimization*. Springer US.

Zitzler, E., & Thiele, L. (1999). Multiobjective evolutionary algorithms: A comparative case study and the strength Pareto approach. *IEEE Transactions on Evolutionary Computation, 3*, 257–271.

Chapter 4
Fundamentals of Surrogate Modeling and Surrogate-Assisted Optimization

Over the last two decades or so, utilization of surrogate modeling tools has increased tremendously in the field of high-frequency engineering. As mentioned earlier, the feature-based technology can be incorporated into surrogate-assisted procedures. For the convenience of the reader, this chapter summarizes the fundamentals of surrogate modeling and surrogate-based optimization. Similarly as in Chap. 3, this material is not a systematic and exhaustive presentation of the subject, yet it should give the reader sufficient background information necessary to understand the remaining parts of the book. More comprehensive treatment of the subject can be found in the vast literature, e.g., Forrester et al., 2008; Gramacy, 2021; Koziel & Ogurtsov, 2019; Koziel & Pietrenko-Dabrowska, 2020; Koziel & Pietrenko-Dabrowska, 2022. Here, we cover the concept of surrogate-based optimization, discuss surrogate modeling flow and basic methods (both data-driven and physics-based), as well as outline selected surrogate-assisted design procedures employed in high-frequency engineering.

4.1 Surrogate-Assisted Optimization

Development of optimization methods has always been driven by practical necessity oriented toward handling specific classes of objective functions. For example, gradient-based algorithms have been created for handling smooth functions, e.g., those that are at least continuously differentiable up to a certain order (Nocedal & Wright, 2000). Population-based metaheuristics were designed to handle multimodal and even discontinuous functions; however, the function evaluation cost is supposed to be low due to a large number of objective functions required for metaheuristics algorithms to converge.

As mentioned in Chap. 1, contemporary engineering design is heavily based on computational models. Their advantages include a potential for design automation

A. Pietrenko-Dabrowska, S. Koziel, *Response Feature Technology for High-Frequency Electronics. Optimization, Modeling, and Design Automation*,
https://doi.org/10.1007/978-3-031-43845-5_4

and a reduced need for costly prototyping. The downside—from the perspective of simulation-driven design—is the high cost. In the case of high-frequency structures, typical simulation times are from a few minutes for models of simple microstrip devices to many hours for finely discretized models of complex antenna arrays (Afshinmanesh et al., 2008; Chamaani et al., 2010; Hanninen, 2012; Koziel & Ogurtsov, 2015). Obviously, hundreds and thousands of simulations required by conventional optimization routines add up to impractically high overall computational costs.

Surrogate-based optimization (SBO) (Queipo et al., 2005; Koziel et al., 2011a; Koziel & Leifsson, 2013a; Forrester & Keane, 2009; Koziel & Ogurtsov, 2014; Leifsson & Koziel, 2015a; Yelten et al., 2012) seems to be a promising approach to solving design optimization problems that involve expensive simulation models. The key concept behind surrogate-assisted methods is to replace direct optimization of the high-fidelity simulation model by an iterative process yielding a sequence $\boldsymbol{x}^{(i)}$, $i = 0, 1, \ldots$, of approximate solutions to the original optimization problem (3.1). The designs $\boldsymbol{x}^{(i)}$ are obtained using a fast (yet reasonably accurate) representation of the high-fidelity model, referred to as a surrogate. Normally, the surrogate model is updated at each iteration using the available high-fidelity model data. In the case of local methods, this could be a high-fidelity model response evaluated at the most recent design. Whereas in the case of a global search, there might be several designs selected using an appropriate infill strategy (Forrester et al., 2008; Koziel et al., 2011a; Koziel & Ogurtsov, 2014).

A popular way to generate a new design is optimization of the current surrogate model. In this case, the surrogate-based optimization process can be written as (Koziel et al., 2011a, see also Fig. 4.1)

$$\boldsymbol{x}^{(i+1)} = \arg\min_{\boldsymbol{x}} U\left(s^{(i)}(\boldsymbol{x})\right) \tag{4.1}$$

where $s^{(i)}$ is the surrogate model at the ith iteration. The initial design $\boldsymbol{x}^{(0)}$ is often obtained using engineering insight or by optimizing an available lower-fidelity model.

An important advantage of SBO is its computational efficiency. Assuming that the surrogate model is fast, the cost of solving (4.1) is typically negligible so that the overall optimization cost is merely determined by the number of the high-fidelity model evaluations. On the other hand, the number of iterations required by a typical SBO algorithm is substantially smaller than those of the majority of conventional methods when solving (3.1) (Koziel et al., 2006).

There are various ways to build a surrogate model, some of them being problem-dependent. In certain (and rather simple) cases, the surrogates can be obtained from analytical models. Another option is utilization of simplified physics models such as equivalent circuit representations in microwave engineering (Bandler et al., 2004a). Finally, the surrogates can be constructed as data-driven models also referred to as response surface approximation (RSA) models (Liu et al., 2014b). The RSA models can be constructed beforehand using either extensive sampling of the design space (Koziel & Ogurtsov, 2011), or sequential sampling techniques (Forrester & Keane, 2009).

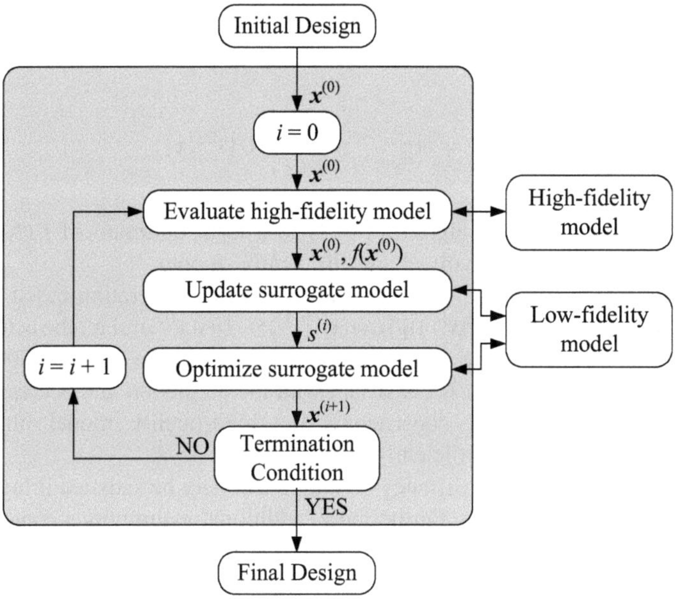

Fig. 4.1 Surrogate-based optimization (SBO) process. The surrogate model s is iteratively updated and optimized to yield a sequence $\boldsymbol{x}^{(i)}$, $i = 0, 1, \ldots$, of approximate solutions to the original problem $\boldsymbol{x}^* = \mathrm{argmin}\{\boldsymbol{x}: U(f(\boldsymbol{x}))\}$. The high-fidelity model f is evaluated for verification and to provide data needed to update the surrogate. The computational benefits of SBO are due to (i) typically small number of iterations and (ii) a low cost of each iteration (as the search is executed at the level of cheap surrogate)

In the case of high-frequency components, the range of possible surrogate models is rather limited. As reliable equivalent network models are hardly available, which is especially the case for antenna structures, the surrogate model is often constructed using coarse discretization simulations (Koziel & Ogurtsov, 2012a; Koziel et al., 2016). An important practical challenge here is finding a proper balance between the model accuracy and speed (Koziel & Ogurtsov, 2012a; Leifsson & Koziel, 2015b). In practical design cases, the achievable time evaluation ratio between the high- and low-fidelity models is between 5 and 50 or so. One of the consequences is that the computational cost of the surrogate model evaluation when solving (4.1) cannot be neglected. As a matter of fact, it may become a significant contributor to the overall optimization cost, which leads to additional challenges concerning the development and implementation of the SBO algorithms (Koziel et al., 2011a; Koziel & Ogurtsov, 2014; Forrester & Keane, 2009).

Reliability of the SBO process depends, among others, on the surrogate model accuracy. From the viewpoint of local optimization, it is important that the surrogate is consistent with the high-fidelity model at the current design $\boldsymbol{x}^{(i)}$, i.e., $s^{(i)}(\boldsymbol{x}^{(i)}) = f(\boldsymbol{x}^{(i)})$ (zero-order consistency (Alexandrov & Lewis, 2001)), and $J[s^{(i)}(\boldsymbol{x}^{(i)})] = J[f(\boldsymbol{x}^{(i)})]$ (first-order consistency), where $f(\boldsymbol{x})$ represents the response of high-fidelity model and $J[\cdot]$ stands for the model Jacobian. Assuming first-order consistency and embedding the surrogate-based algorithm in the trust region framework (Conn et al., 2000)

$$x^{(i+1)} = \arg \min_{x,\ \|x - x^{(i)}\| \le \delta^{(i)}} U(s^{(i)}(x)) \tag{4.2}$$

ensures convergence of the algorithm (4.1) to a local optimum of f (Alexandrov et al., 1998) if the models involved are sufficiently smooth.

The trust region radius $\delta^{(i)}$ in (4.2) is updated in each iteration based on a gain ratio $\rho = [U(f(\mathbf{x}^{(i+1)})) - U(f(\mathbf{x}^{(i)}))]/[U(s(\mathbf{x}^{(i+1)})) - U(s(\mathbf{x}^{(i)}))]$, i.e., the actual versus surrogate-predicted objective function improvement (Conn et al., 2000); $\delta^{(i)}$ is increased if the improvement is consistent with the prediction and decreased otherwise. Given the first-order consistency, the high-fidelity model objective is guaranteed to improve for sufficiently small values of δ.

Clearly, the first-order consistency condition can only be satisfied if high-fidelity model gradients are available. Furthermore, additional assumptions concerning the smoothness of the functions involved are necessary to warrant convergence (Echeverría & Hemker, 2008). Convergence of various SBO algorithms can be ensured under other scenarios of, e.g., space mapping (Koziel et al., 2006, 2008b), manifold mapping (Echeverría & Hemker 2008), and surrogate management framework (Booker et al., 1999).

SBO can be implemented either as a local or as a global search algorithm. In local optimization, trust region-like convergence safeguards are often used (cf. (4.2)), and at least zero-order consistency conditions are ensured by appropriate construction of the surrogate model (Koziel et al., 2010a). For global search, the surrogate is constructed in a larger portion of the design space. The new candidate solutions are typically found using global methods such as evolutionary algorithms. The surrogate model itself is updated using appropriate statistical infill criteria based either on the expected improvement of the objective function or on minimization of the (global) modeling error (Forrester & Keane, 2009; Liu et al., 2014b).

4.2 Surrogate Modeling: Data-Driven Surrogates

Data-driven models (also referred to as approximation models) belong to the most popular class of surrogates. A large variety of modeling methods have been developed (Simpson et al., 2001, Søndergaard, 2003; Forrester & Keane, 2009, Couckuyt, 2013, Lophaven et al., 2002, Gorissen et al., 2010a); yet there are certain features common to all types of approximation models. These include:

- They are data-driven, i.e., the model is constructed merely from sampled training data so that no problem-specific knowledge is required.

- They are generic, therefore, applicable to a wide range of problems and easily transferrable between various areas.
- They are typically based on explicit analytical formulas.
- They are cheap to evaluate.

A disadvantage of data-driven surrogates is a large number of training data points necessary to ensure good predictive power of the model. The necessary training data set rapidly grows with the number of the parameters, which is termed as the curse of dimensionality (Forrester & Keane, 2009; Gorissen et al., 2010b). Furthermore, the number of necessary data points quickly grows with the parameter ranges. In the case of high-frequency components, data-driven modeling is essentially limited to a few (< 5–6) dimensions unless the ranges of geometry parameters are considerably restricted. On the other hand, local approximation models can be quite useful as auxiliary optimization tools.

In this section, we outline the surrogate modeling flow as well as discuss popular approximation techniques and model validation methods.

4.2.1 Modeling Flow for Data-Driven Surrogates

A surrogate is constructed by approximating sampled data of the high-fidelity model. There are four steps involved in the modeling process:

- Design of experiments (DOE), which is allocation of the training points in the design space using a strategy of choice. Due to computational budget constraints as well as potential numerical problems of handling large amounts of training data, the number of samples is normally limited in practice. Nowadays, space-filling DOEs are preferred (Simpson et al., 2001; Forrester & Keane, 2009), cf. Sect. 4.2.2.
- Training data acquisition at the locations assigned at the first stage.
- Identification of the surrogate model. This step aims at finding the model parameters which are either determined by solving a suitably defined minimization problem, e.g., using kriging (Kleijnen, 2009), neural networks (Rayas-Sanchez, 2004), or by solving an appropriate regression problem analytically, e.g., with polynomial approximation (Queipo et al., 2005).
- Validation of the model. This step estimates the generalization error of the surrogate, i.e., the predictive power at the designs not seen during the identification stage.

In practice, surrogate model construction is carried out iteratively, by repeating the above steps until the model accuracy reaches the required level. At each iteration, a new set of so-called infill samples is added using a given strategy referred to as infill criteria (Forrester & Keane, 2009), and the model is reidentified. Figure 4.2 shows the block diagram of the surrogate modeling process.

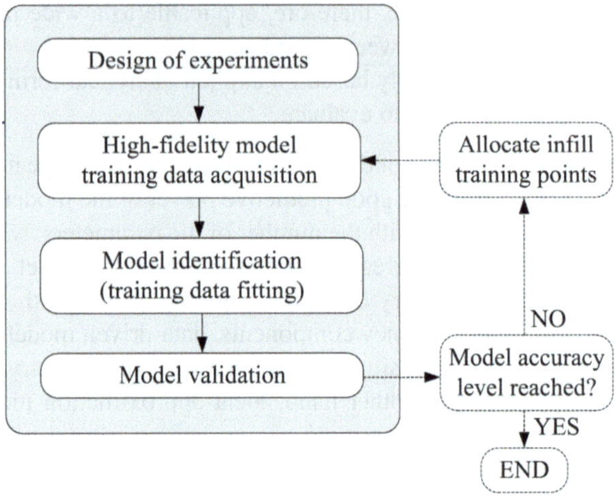

Fig. 4.2 A flowchart of data-driven surrogate model construction. Optional iterative model enhancement involving infill training points is indicated using dashed lines

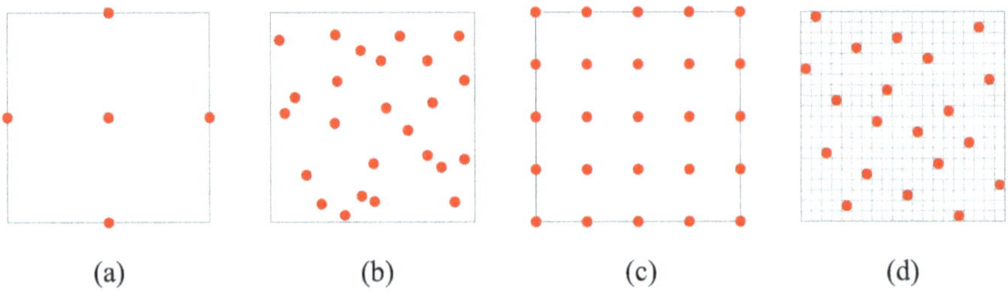

Fig. 4.3 Popular DOE techniques: (**a**) factorial designs (here, star distribution); (**b**) random sampling; (**c**) uniform grid sampling; (**d**) Latin hypercube sampling (LHS)

4.2.2 Design of Experiments

Design of experiments (DOE) is a strategy for allocating the training data points in the design space (Giunta et al., 2003; Santner et al., 2003; Koehler & Owen, 1996). The high-fidelity model data are subsequently acquired at these locations to construct the surrogate. As mentioned before, the training points may be selected beforehand or using the feedback about the model performance, e.g., by allocating more points in the regions where the system exhibits more linear behavior (Couckuyt, 2013; Devabhaktuni et al., 2001).

Traditional DOE strategies are factorial designs with the samples allocated in the corners, edges, or faces of the design space. An example of factorial DOE is shown in Fig. 4.3a. Factorial designs allow for estimating the main effect and interactions between design variables without excessive number of samples. The most "economical" factorial DOE is a so-called star distribution (Fig. 4.3a) which is often used in conjunction with space mapping (Cheng et al., 2006). In factorial DOEs, the samples are normally spread as far from each other as possible. This is quite helpful to reduce the effects of uncertainties pertinent to evaluation of the system responses (e.g., through physical measurements).

Nowadays most of the data comes from deterministic computer simulations. Thus a majority of modern DOE approaches attempt to allocate the training points uniformly within the design space (Queipo et al., 2005). Such approaches are useful for constructing surrogates when the knowledge about the system is limited. Popular space-filing DOEs include pseudorandom sampling (Giunta et al., 2003; Fig. 4.3b), uniform grid sampling (Fig. 4.3c), and Latin hypercube sampling (LHS, McKay et al., 1979; Fig. 4.3d). Although simple, pseudo-random sampling does not ensure good uniformity. Grid sampling is only applicable for low-dimensional spaces. LHS and its improvements (e.g., Beachkofski & Grandhi, 2002; Leary et al., 2003; Ye, 1998; Palmer & Tsui, 2001) offer good uniformity and flexibility. In order to allocate p samples with LHS, the range for each parameter is divided into p bins which yields a total number of p^n bins in the design space for n design variables. The samples are randomly selected in the design space so that (i) each sample is randomly placed inside a bin and (ii) there is exactly one sample in each bin for all one-dimensional projections of the p samples and bins.

An example realization of 20-sample LHS in two-dimensional space has been shown in Fig. 4.3d. Other uniform sampling techniques that are commonly used include orthogonal array sampling (Queipo et al., 2005), quasi-Monte Carlo sampling (Giunta et al., 2003), or Hammersley sampling (Hammersley, 1960). Uniform sample allocation can also be obtained through optimization (Leary et al., 2003).

An important improvement over conventional space-filling design is sequential sampling (Lehmensiek et al., 2002). It is an iterative process, where the data obtained from the previous iterations (both the surrogate model and samples) are analyzed in order to allocate the new (infill) samples, primarily in the regions that are more difficult to approximate. As a result, a more efficient distribution of samples is obtained compared to traditional DOEs (Sasena et al., 2002). An important consideration for sequential DOE is the balance between exploration and exploitation (Crombecq et al., 2011). In the context of DOE, exploration aims at identifying the new (not explored before) regions of the design space that contain discontinuities, optima, plateaus, etc. In practice, it amounts to filling up the domain as uniformly as possible; therefore, the system response is not involved in the process. However, the locations of the infill samples in each iteration depend on the samples already distributed in the design space, which is the major difference between the sequential and one-shot DOEs. Exploitation, on the other hand, focuses on the regions of interest that have been already identified and tries to allocate infill samples therein. This is in order to obtain a better representation of the specific parts of the space, e.g., vicinities of the optima. As opposed to exploration, exploitation involves the system outputs, evaluated at the previously allocated points, to guide the sampling process (Crombecq et al., 2009).

There has been a considerable variety of sequential DOE strategies developed. Probably the most popular ones are sequential LHS (Liu et al., 2016; Wang, 2003; Tong, 2006). Other techniques include low-discrepancy sequences such as Halton sequence or Sobol sequence (Giunta et al., 2003; Chi et al., 2005), techniques based

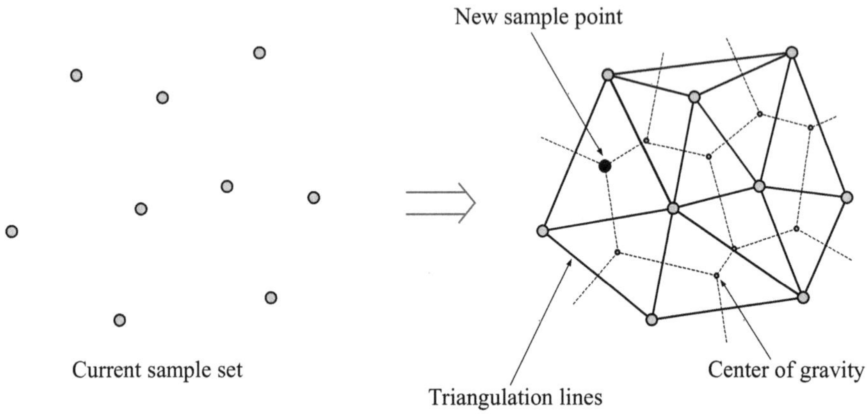

Fig. 4.4 Sequential sampling using Delaunay triangulation in two-dimensional space. The current experimental design marked using large gray circles. Solid lines represent its Delaunay triangulation, whereas dashed lines determine the gravity centers of the corresponding simplexes. The new sample (marked using a large black circle) is allocated in the center of the largest-volume simplex. The procedure (triangulation and allocation of the new sample) is iterated until allocating a required number of points

on Delaunay triangulation (Davis & Ierapetritou, 2010), as well as techniques involving a Voronoi tessellation (Crombecq et al., 2011). All of the aforementioned methods are exploratory ones. For illustration purposes, Fig. 4.4 shows the concept of Delaunay triangulation-based sampling.

4.2.3 Data-Driven Modeling Techniques

Data-driven modeling includes numerous techniques such as polynomial regression (Queipo et al., 2005), radial basis functions (Wild et al., 2008), kriging (Forrester & Keane, 2009), neural networks (Haykin, 1998), support vector regression (Gunn, 1998), or Gaussian Process Regression (Rasmussen & Williams, 2006). Selected methods are briefly outlined below. More detailed information can be found in the references listed above, as well as in review works, e.g., in (Simpson et al., 2001; Koziel & Pietrenko-Dabrowska, 2020).

Throughout this section, we denote the training data set as $\{\boldsymbol{x}^{(i)}, f(\boldsymbol{x}^{(i)})\}$, $i = 1, \ldots, p$, where $\boldsymbol{x}^{(i)}$ are data points, $f(\boldsymbol{x}^{(i)})$ are corresponding high-fidelity number evaluations, and p is the number of the training points.

4.2.3.1 Polynomial Regression

Perhaps the most popular yet simple approximation technique is polynomial regression (Queipo et al., 2005). Low-order polynomials are quite useful particularly for local modeling or to construct rough initial surrogates indicating the trends of the function of interest. The surrogate model is of the form

$$s(\boldsymbol{x}) = \sum_{j=1}^{K} \beta_j v_j(\boldsymbol{x}) \tag{4.3}$$

where v_j are the (polynomial) basis functions and β_j are the model parameters that can be found as a least-square solution to the linear system $\boldsymbol{F} = \boldsymbol{X\beta}$. $\boldsymbol{F} = [f(\boldsymbol{x}^{(1)})\ f(\boldsymbol{x}^{(2)}) \ldots f(\boldsymbol{x}^{(p)})]^T$, $\boldsymbol{X}$ is a $p \times K$ matrix containing the basis function values at the sample points, and $\boldsymbol{\beta} = [\beta_1\ \beta_2 \ldots \beta_K]^T$. If $p \geq K$ and rank($\boldsymbol{X}$) $= K$, the model parameter can be determined in the least-squares sense through the pseudoinverse $\boldsymbol{X}^+$ of $\boldsymbol{X}$ as $\boldsymbol{\beta} = \boldsymbol{X}^+\boldsymbol{f} = (\boldsymbol{X}^T\boldsymbol{X})^{-1}\boldsymbol{X}^T\boldsymbol{f}$ (Golub & Van Loan, 1996).

One of the simplest examples of a regression model is a second-order polynomial one defined as

$$s(\boldsymbol{x}) = s\left([x_1\ x_2\ \ldots\ x_n]^T\right) = \beta_0 + \sum_{j=1}^{n} \beta_j x_j + \sum_{i=1}^{n} \sum_{j \leq i}^{n} \beta_{ij} x_i x_j \tag{4.4}$$

with the basis functions being monomials: 1, x_j, and $x_i x_j$. Note that the basis functions do not need to be polynomials. In particular applications, these could be, e.g., exponential or sinusoidal.

4.2.3.2 Radial Basis Functions

Another popular technique for constructing data-driven surrogate models is radial basis function (RBF) interpolation (Forrester & Keane, 2009; Wild et al., 2008). The RBF model is defined as a linear combination of radially symmetric functions ϕ

$$s(\boldsymbol{x}) = \sum_{j=1}^{K} \lambda_j \phi\left(\left\|\boldsymbol{x} - \boldsymbol{c}^{(j)}\right\|\right) \tag{4.5}$$

where $\boldsymbol{\lambda} = [\lambda_1\ \lambda_2 \ldots \lambda_K]^T$ is the vector of model parameters and $\boldsymbol{c}^{(j)}, j = 1, \ldots, K$, are the known basis function centers. The parameters can be calculated as $\boldsymbol{\lambda} = \boldsymbol{\Phi}^+\boldsymbol{F} = (\boldsymbol{\Phi}^T\boldsymbol{\Phi})^{-1}\boldsymbol{\Phi}^T\boldsymbol{F}$, where $\boldsymbol{F} = [f(\boldsymbol{x}^{(1)})\ f(\boldsymbol{x}^{(2)}) \ldots f(\boldsymbol{x}^{(p)})]^T$, and the $p \times K$ matrix $\boldsymbol{\Phi} = [\Phi_{kl}]_{k = 1,\ldots,p;\ l = 1,\ldots,K}$, with the entries defined as

$$\Phi_{kl} = \phi\left(\left\|\boldsymbol{x}^{(k)} - \boldsymbol{c}^{(l)}\right\|\right) \tag{4.6}$$

If $p = K$ and the centers of the basis functions coincide with the data points and are all different, then $\boldsymbol{\Phi}$ is a regular square matrix, and the model coefficients can be found as $\boldsymbol{\lambda} = \boldsymbol{\Phi}^{-1}\boldsymbol{F}$. A popular basis function is a Gaussian one, $\phi(r) = \exp.(-r^2/2\sigma^2)$, where σ is the scaling parameter. Other basis functions include multiquadric $\phi(r) = (1 + \sigma^2 r^2)^{1/2}$, inverse quadratic $\phi(r) = (1 + \sigma^2 r^2)^{-1}$, inverse multiquadric $\phi(r) = (1 + \sigma^2 r^2)^{-1/2}$, and thin plate spline $\phi(r) = r^2\ln(r)$.

4.2.3.3 Kriging

Kriging belongs to the most popular surrogate modeling techniques today (Matheron, 1963; Journel & Huijbregts, 1981; Simpson et al., 2001; Kleijnen, 2009; O'Hagan, 1978). A basic formulation of kriging assumes that the function of interest is of the form

$$f(\boldsymbol{x}) = \boldsymbol{g}(\boldsymbol{x})^T \boldsymbol{\beta} + Z(\boldsymbol{x}) \tag{4.7}$$

where $\boldsymbol{g}(\boldsymbol{x}) = [g_1(\boldsymbol{x})\ g_2(\boldsymbol{x}) \ldots g_K(\boldsymbol{x})]^T$ are known (e.g., constant) functions, $\boldsymbol{\beta} = [\beta_1\ \beta_2 \ldots \beta_K]^T$ are the unknown model parameters (hyperparameters), and $Z(\boldsymbol{x})$ is a realization of a normally distributed Gaussian random process with zero mean and variance σ^2. The regression part $\boldsymbol{g}(\boldsymbol{x})^T\boldsymbol{\beta}$ is a trend function for f, and $Z(\boldsymbol{x})$ takes into account localized variations. The covariance matrix of $Z(\boldsymbol{x})$ is given as

$$Cov\left[Z\left(\boldsymbol{x}^{(i)}\right)Z\left(\boldsymbol{x}^{(j)}\right)\right] = \sigma^2 \boldsymbol{R}\left(\left[R\left(\boldsymbol{x}^{(i)}, \boldsymbol{x}^{(j)}\right)\right]\right) \tag{4.8}$$

where $\boldsymbol{R}$ is a $p \times p$ correlation matrix with $R_{ij} = R(\boldsymbol{x}^{(i)},\boldsymbol{x}^{(j)})$. $R(\boldsymbol{x}^{(i)}, \boldsymbol{x}^{(j)})$ is the correlation function between the data points $\boldsymbol{x}^{(i)}$ and $\boldsymbol{x}^{(j)}$.

The most popular correlation function is a Gaussian one

$$R(\boldsymbol{x}, \boldsymbol{y}) = \exp\left[-\sum\nolimits_{k=1}^{n} \theta_k |x_k - y_k|^2\right] \tag{4.9}$$

where θ_k are the correlation parameters to be determined; x_k and y_k are the kth components of the vectors $\boldsymbol{x}$ and $\boldsymbol{y}$, respectively. The kriging predictor (Simpson et al., 2001; Journel & Huijbregts, 1981) is defined as

$$s(\boldsymbol{x}) = \boldsymbol{g}(\boldsymbol{x})^T \boldsymbol{\beta} + \boldsymbol{r}^T(\boldsymbol{x}) \boldsymbol{R}^{-1} (\boldsymbol{F} - \boldsymbol{G}\boldsymbol{\beta}) \tag{4.10}$$

where $\boldsymbol{r}(\boldsymbol{x}) = [R(\boldsymbol{x}, \boldsymbol{x}^{(1)}) \ldots R(\boldsymbol{x}, \boldsymbol{x}^{(p)})]^T$, $\boldsymbol{F} = [f(\boldsymbol{x}^{(1)})\ f(\boldsymbol{x}^{(2)}) \ldots f(\boldsymbol{x}^{(p)})]^T$ and $\boldsymbol{G}$ is a $p \times K$ matrix with $G_{ij} = g_j(\boldsymbol{x}^{(i)})$. We also have $\boldsymbol{\beta} = (\boldsymbol{G}^T\boldsymbol{R}^{-1}\boldsymbol{G})^{-1}\boldsymbol{G}^T\boldsymbol{R}^{-1}\boldsymbol{F}$. Model fitting is accomplished by maximum likelihood for θ_k (Journel & Huijbregts, 1981).

4.2.3.4 Artificial Neural Networks

Artificial neural networks (ANNs) (Haykin, 1998) are a large subject and an extensive research area. One of the important applications of ANNs is classification. However, from the point of view of this chapter, ANNs can be considered as rather complex nonlinear regression models.

The key component of ANN is the neuron (or single-unit perceptron) (Haykin, 1998; Minsky & Papert, 1969). A neuron realizes a nonlinear operation (cf. Fig. 4.5a), where w_1 through w_n are regression coefficients, β is the bias value of the neuron, and T is a user-defined slope parameter. The most common neural network architecture is the multilayer feed-forward network shown in Fig. 4.5b.

There are two fundamental stages of ANN construction: selection of the network architecture and the network training. For example, for a multilayer ANN (cf. Fig. 4.5b), one needs to select the number of hidden layers as well as the number of neurons in these layers. Both affect the network flexibility, i.e., its ability to approximate the training data. Training of the network, i.e., assignment of the values to the neuron weights, is realized as a nonlinear least-squares regression problem. A popular technique for solving this regression problem is the error back-propagation algorithm (Simpson et al., 2001; Haykin, 1998). For large networks with complex architectures, the use of global optimization methods might be necessary (Alba & Marti, 2006). It should be mentioned that finding the optimal ANN architecture is a nontrivial problem yet of primary importance: the network that is too simple may not be able to approximate the training data well; on the other hand, if it has too many degrees of freedom, it may exhibit poor generalization capability.

Nowadays, ANN models evolved into many variations, some of which exhibit a high level of complexity (e.g., convolutional neural networks (Liang et al., 2020), recurrent neural networks (Shen & Wang, 2008), deep neural networks (Koziel et al., 2021), or multipath networks (Szegedy et al., 2015)).

4.2.3.5 Support Vector Regression

Support vector regression (SVR) (Gunn, 1998) features good generalization capability (Angiulli, et al., 2007) and easy training by means of quadratic programming (Smola & Schölkopf, 2004). SVR has gained popularity in various areas including electrical engineering and aerodynamic design (Pattipati et al., 2011; Rojo-Alvarez et al., 2005; Meng & Xia, 2007; Xia et al., 2007; Andrés et al., 2012; Zhang & Han, 2013).

In this section, we formulate SVR for a vector-valued function $\boldsymbol{f}$. We denote the training samples by $r^k = [r_1{}^k\ r_2{}^k \ldots r_m{}^k]^T = \boldsymbol{f}(\boldsymbol{x}^k)$, $k = 1, 2, \ldots, N$. SVR is supposed to approximate r^k at all points $\boldsymbol{x}^k$. For linear regression, we aim at approximating a training data set; here, the pairs $D_j = \{(\boldsymbol{x}^1, r_j{}^1), \ldots, (\boldsymbol{x}^N, r_j{}^N)\}$, $j = 1, 2, \ldots, m$, by a

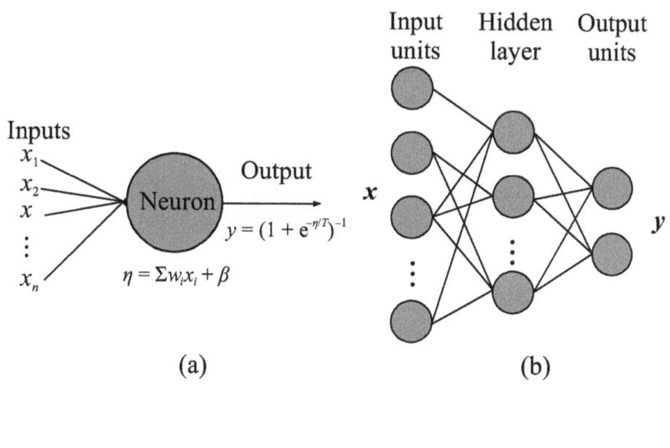

Fig. 4.5 Basic concepts of artificial neural networks: (**a**) structure of a neuron; (**b**) two-layer feed-forward neural network architecture

linear function $f_j(\boldsymbol{x}) = \boldsymbol{w}_j^T\boldsymbol{x} + b_j$. The optimal regression function is given by the minimum of the functional (Smola & Schölkopf, 2004)

$$\Phi_j(\boldsymbol{w}, \xi) = \frac{1}{2}\left\|\boldsymbol{w}_j\right\|^2 + C_j\sum_{i=1}^{N}\left(\xi_{j.i}^{+} + \xi_{j.i}^{-}\right) \tag{4.11}$$

In (4.11), C_j is a user-defined value, and $\xi_{j.i}^{+}$ and $\xi_{j.i}^{-}$ are the slack variables representing upper and lower constraints on the system output. The typical cost function used in SVR is an ε-insensitive loss function defined as $L_\varepsilon(y) = 0$ for $|f_j(\boldsymbol{x}) - y| < \varepsilon$ and $L_\varepsilon(y) = |f_j(\boldsymbol{x}) - y|$ otherwise. The value of C_j determines the trade-off between the flatness of f_j and the amount up to which deviations larger than ε are tolerated (Gunn, 1998).

Here, we describe nonlinear regression employing the kernel approach in which the linear function $\boldsymbol{w}_j^T\boldsymbol{x} + b_j$ is replaced by the $\Sigma_i\gamma_{j.i}K(\boldsymbol{x}^k,\boldsymbol{x}) + b_j$, where K is a kernel function. The SVR model is defined as

$$s(\boldsymbol{x}) = \left[\sum\nolimits_{i=1}^{N}\gamma_{1.i}K\left(\boldsymbol{x}^i,\boldsymbol{x}\right) + b_1 \quad \ldots. \quad \sum\nolimits_{i=1}^{N}\gamma_{m.i}K\left(\boldsymbol{x}^i,\boldsymbol{x}\right) + b_m\right]^T \tag{4.12}$$

with the parameters $\gamma_{j.i}$ and b_j, $j = 1, \ldots, m$, $i = 1, \ldots, N$ obtained according to a general SVR methodology. In particular, Gaussian kernels of the form $K(\boldsymbol{x},\boldsymbol{y}) = \exp.(-0.5\cdot||\boldsymbol{x} - \boldsymbol{y}||^2/c^2)$ with $c > 0$ can be used, where c is the scaling parameter. Both c and parameters C_j and ε can be adjusted to minimize the generalization error calculated using cross-validation (Queipo et al., 2005).

4.2.3.6 Other Approximation Methods

The approximation techniques outlined in Sects. 4.2.3.1, 4.2.3.2, 4.2.3.3, 4.2.3.4, and 4.2.3.5 include some of the most popular methods. Others were mentioned in Chap. 1. One of these is moving least-squares (MLS) (Levin, 1998). In MLS, the error contribution from each training point $\boldsymbol{x}^{(i)}$ is multiplied by a weight ω_i that depends on the distance between $\boldsymbol{x}$ and $\boldsymbol{x}^{(i)}$. This improves the model flexibility but also increases computational complexity as the model has to be identified for each $\boldsymbol{x}$ separately. A typical choice for the weights is

$$\omega_i\left(\left\|\boldsymbol{x} - \boldsymbol{x}^{(i)}\right\|\right) = \exp\left(-\left\|\boldsymbol{x} - \boldsymbol{x}^{(i)}\right\|^2\right) \tag{4.13}$$

Gaussian process regression (GPR) (Rasmussen & Williams, 2006) is a surrogate modeling technique that addresses the approximation problem from a stochastic point of view. From this perspective and since Gaussian processes are mathematically tractable, it is relatively easy to compute error estimations for GPR-based surrogates in the form of uncertainty distributions. Under certain conditions, GPR models can be shown to be equivalent to large neural networks while requiring much less regression parameters (Rasmussen & Williams, 2006). In recent years, GPR has been successfully applied for modeling of antenna structures (Jacobs, 2012, 2016; Jacobs & Koziel, 2014).

Another technique to mention is co-kriging (Forrester et al., 2007; Toal & Keane, 2011). Co-kriging is an extension of kriging interpolation (cf. Sect. 4.2.3.3) that allows for combining information from computational models of various fidelity levels. The fundamental benefit of such combination is that—by exploiting knowledge embedded in the lower-fidelity model—the surrogate can be created at much lower computational cost than for the models exclusively based on high-fidelity data. Co-kriging is a rather recent method with relatively few applications in engineering (Toal & Keane, 2011; Huang & Gao, 2012; Laurenceau & Sagaut, 2008; Koziel et al., 2013a; Pietrenko-Dabrowska & Koziel, 2020).

4.2.4 Model Validation

Upon identification, the surrogate model has to be validated in order to assess its predictive power also referred to as generalization capability, i.e., an ability to predict the response of the high-fidelity model at the locations other than those utilized for training. The most common validation approach is a split-sample method (Queipo et al., 2005), where a part of the available data set (the training set) is used to construct the surrogate, whereas another part (the test set) serves purely for model validation. The method is straightforward; however, the error estimation depends strongly on how the data set has been partitioned.

A more reliable estimation of the generalization error of the surrogate can be obtained by means of cross-validation (Queipo et al., 2005; Geisser, 1993). In this approach, the available data set is split into K subsets, and each of these subsets is sequentially used as a testing set for a surrogate constructed on the other $K-1$ subsets. The prediction error can be estimated with all the K error measures obtained in this process. The primary advantage of cross-validation is that the error estimation is less biased than with the split-sample method.

As mentioned before, the entire cycle of constructing the surrogate and its validation may be iterated. The process of adding new (infill) samples based on infill criteria (Forrester & Keane, 2009) is often referred to as sequential sampling. It is normally continued until the prescribed surrogate accuracy level is reached. In the modeling context (as opposed to optimization), the objective is the improvement of the global accuracy of the surrogate, and the new samples are inserted at the locations where the estimated modeling error is the highest. It should also be mentioned that finding infill point locations normally requires solving global optimization problems (Couckuyt, 2013); such problems are often handled with population-based metaheuristics.

4.3 Surrogate Modeling: Physics-Based Surrogates

Data-driven surrogates are perfect solutions for many practical modeling problems. Their major advantages are versatility and low evaluation cost. Unfortunately, application of data-driven surrogates in high-frequency design is rather limited. Their fundamental issue is high cost of training data acquisition. Other issues include high dimensionality of design spaces for real-world antenna or microwave structures as well as wide parameter ranges that are necessary for the models configured over reasonable ranges of operating conditions (e.g., resonant frequencies, bandwidth). Some of these issues may be addressed using physics-based surrogates, which are briefly outlined in this section.

Physics-based surrogates are constructed by suitable correction or enhancement of an underlying low-fidelity model. In the case of high-frequency structures, the low-fidelity model is often obtained from EM analysis performed with coarse discretization simulations. In some cases, it could also be an equivalent network or even an analytical model. The correction process aims at improving alignment between the low-fidelity model and the high-fidelity model, either locally or across the entire design space.

As the low-fidelity model embeds certain knowledge about the system of interest, the models are normally well correlated. Consequently, only a limited amount of high-fidelity data is normally sufficient to ensure reasonable accuracy of the surrogate model. For the same reason, feature-based surrogates typically feature good generalization. The latter translates into improved computational efficiency of the SBO process. Here, we outline the main concepts of physics-based surrogate modeling. More detailed treatment of the subject can be found in the literature (Koziel et al., 2011a; Koziel & Ogurtsov, 2014; Koziel et al. 2010c).

The low-fidelity (coarse) model of the system of interest will be denoted as $c(\boldsymbol{x})$. In order to explain the physics-based surrogate modeling concept, we discuss a simple case of multiplicative response correction in the context of the SBO algorithm (4.1). For ensuring convergence of a sequence $\{\boldsymbol{x}^{(i)}\}$ of approximate solutions to the original problem (3.1), produced by (4.1), it is imperative to have the surrogate and the high-fidelity model well (at least locally) aligned. The surrogate $s^{(i)}(\boldsymbol{x})$ at iteration i can be constructed with response correction:

$$s^{(i)}(\boldsymbol{x}) = \beta_k(\boldsymbol{x})c(\boldsymbol{x}) \tag{4.14}$$

where $\beta_k(\boldsymbol{x}) = \beta_k(\boldsymbol{x}^{(i)}) + \nabla\beta(\boldsymbol{x}^{(i)})^T(\boldsymbol{x} - \boldsymbol{x}^{(i)})$, and $\beta(\boldsymbol{x}) = f(\boldsymbol{x})/c(\boldsymbol{x})$. This construction ensures both the zero-order consistency, $s^{(i)}(\boldsymbol{x}^{(i)}) = f(\boldsymbol{x}^{(i)})$, and the first-order consistency, $\nabla s^{(i)}(\boldsymbol{x}^{(i)}) = \nabla f(\boldsymbol{x}^{(i)})$ (Alexandrov & Lewis, 2001). Quality of a surrogate constructed with (4.14) is illustrated in Fig. 4.6.

Response correction can also be realized in a multipoint fashion where high-fidelity data at several designs are utilized for model alignment. For example, the surrogate can be defined for vector-valued models as (Koziel & Leifsson, 2013b)

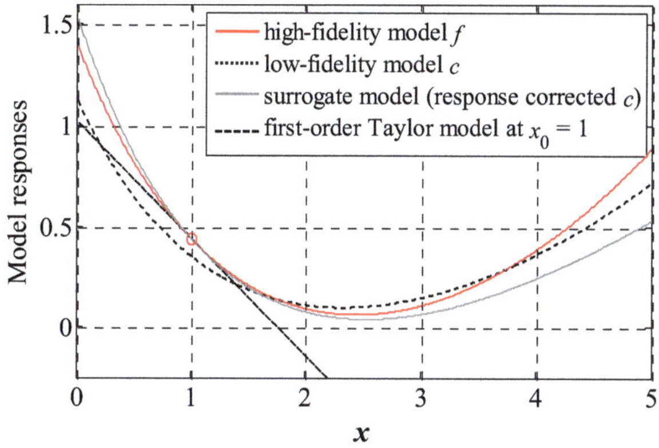

Fig. 4.6 Visualization of the response correction (4.14) for the example analytical functions c (low-fidelity model) and f (high-fidelity model). The correction is established at $\boldsymbol{x}_0 = 1$. It should be emphasized that the surrogate exhibits good alignment with the high-fidelity model in a relatively wide vicinity of $\boldsymbol{x}_0$, especially compared to the first-order Taylor model set up using the same data from f (the value and the gradient at $\boldsymbol{x}_0$)

$$\boldsymbol{s}^{(i)}(\boldsymbol{x}) = \boldsymbol{\Lambda}^{(i)} \circ \boldsymbol{c}(\boldsymbol{x}) + \boldsymbol{\Delta}_r^{(i)} + \boldsymbol{\delta}^{(i)} \tag{4.15}$$

with column vectors $\boldsymbol{\Lambda}^{(i)}$, $\boldsymbol{\Delta}_r^{(i)}$, and $\boldsymbol{\delta}^{(i)}$ and $\circ$ denoting component-wise multiplication. The global response correction parameters $\boldsymbol{\Lambda}^{(i)}$ and $\boldsymbol{\Delta}_r^{(i)}$ are obtained as

$$\left[\boldsymbol{\Lambda}^{(i)}, \boldsymbol{\Delta}_r^{(i)}\right] = \arg\min_{[\boldsymbol{\Lambda}, \boldsymbol{\Delta}_r]} \sum_{k=0}^{i} \left\| \boldsymbol{f}\left(\boldsymbol{x}^{(k)}\right) - \boldsymbol{\Lambda} \circ \boldsymbol{c}\left(\boldsymbol{x}^{(k)}\right) + \boldsymbol{\Delta}_r \right\|^2 \tag{4.16}$$

The response scaling (4.16) is supposed to improve alignments of the models for all previous iterations. The (local) additive term $\boldsymbol{\delta}^{(i)}$ is then defined as

$$\boldsymbol{\delta}^{(i)} = \boldsymbol{f}\left(\boldsymbol{x}^{(i)}\right) - \left[\boldsymbol{\Lambda}^{(i)} \circ \boldsymbol{c}\left(\boldsymbol{x}^{(i)}\right) + \boldsymbol{\Delta}_r^{(i)}\right] \tag{4.17}$$

(4.17) ensures a perfect match between the surrogate and the high-fidelity model at the current design $\boldsymbol{x}^{(i)}$, $\boldsymbol{s}^{(i)}(\boldsymbol{x}^{(i)}) = \boldsymbol{f}(\boldsymbol{x}^{(i)})$.

All correction terms $\boldsymbol{\Lambda}^{(i)}$, $\boldsymbol{\Delta}_r^{(i)}$, and $\boldsymbol{\delta}^{(i)}$ can be obtained analytically by solving appropriate linear regression problems (Koziel & Leifsson, 2013b).

As an example, let us consider a dielectric resonator antenna (DRA) shown in Fig. 4.7. The DRA is suspended above the ground plane in order to enhance its impedance bandwidth. The high-fidelity model $\boldsymbol{f}$ is simulated using the CST MWS transient solver (CST, 2022) (~800,000 mesh cells, evaluation time 20 min). The low-fidelity model $\boldsymbol{c}$ is also evaluated in CST (~30,000 mesh cells, evaluation time 40 seconds). Figure 4.8a shows the responses of the low- and high-fidelity DRA model at several designs. As illustrated in Fig. 4.8b, conventional one-point response correction perfectly aligns $\boldsymbol{c}$ and $\boldsymbol{f}$ at the design where it is established, but the

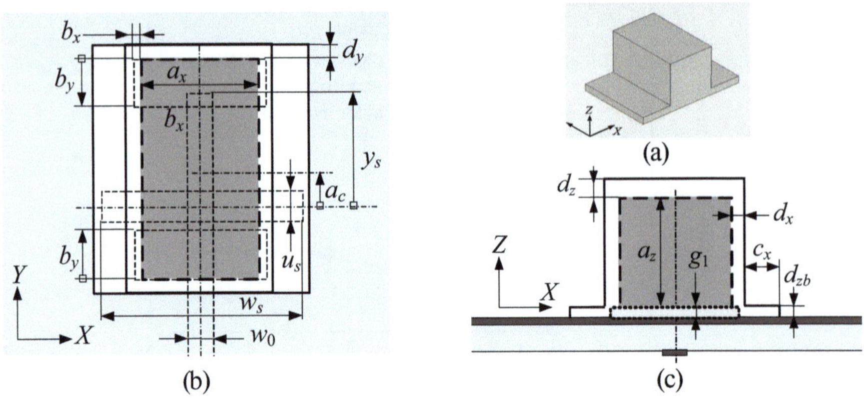

Fig. 4.7 Suspended DRA (**a**) 3D view of its housing, top (**b**) and front (**c**) views

alignment is not as good for other designs. On the other hand, as shown in Fig. 4.8c, the multipoint response correction improves model alignment at all designs involved in the model construction (note that Fig. 4.8c only shows the global part $\boldsymbol{\Lambda}^{(i)} \circ \boldsymbol{c}(\boldsymbol{x}) + \boldsymbol{\Delta}_r^{(i)}$ without $\boldsymbol{\delta}^{(i)}$ which would give a perfect alignment at $\boldsymbol{x}^{(i)}$).

Another approach to low-fidelity model correction is to apply the correction at the level of the model domain. Perhaps the most popular example of such a procedure is input space mapping (ISM) (Bandler et al., 2004a), where the surrogate is created as

$$\boldsymbol{s}^{(i)}(\boldsymbol{x}) = \boldsymbol{c}\left(\boldsymbol{x} + \boldsymbol{q}^{(i)}\right) \tag{4.18}$$

where the low-fidelity model and the surrogate are vector-valued.

The model parameters $\boldsymbol{q}^{(i)}$ are obtained by minimizing the misalignment between the surrogate and the high-fidelity model, $||\boldsymbol{f}(\boldsymbol{x}^{(i)}) - \boldsymbol{c}(\boldsymbol{x}^{(i)} + \boldsymbol{q}^{(i)})||$; $\boldsymbol{x}^{(i)}$ is a reference design (e.g., the most recent design encountered during the optimization run) at which the surrogate model is established. The advantages of input space mapping are demonstrated in Fig. 4.9 for microwave filter design (Hong & Lancaster, 2001). The high-fidelity model is evaluated using EM simulation, whereas the low-fidelity model is an equivalent circuit. The response of interest is the reflection coefficient $|S_{11}|$ as a function of frequency. For this particular example, ISM offers both good approximation and generalization capability (cf. Fig. 4.9d).

An alternative way of low-fidelity model correction is to exploit parameters that are normally fixed in the high-fidelity model such as microstrip substrate height and/or dielectric permittivity (Koziel et al., 2008a; Bandler et al., 2004b). Because the surrogate model is just an auxiliary tool and it is not supposed to be built or measured, such parameters can be freely adjusted in the low-fidelity model to improve its alignment with the high-fidelity one. Implicit space mapping (Bandler et al., 2003, 2004b; Koziel et al., 2010b) is a technique that utilizes this concept. More specifically, the surrogate is obtained as

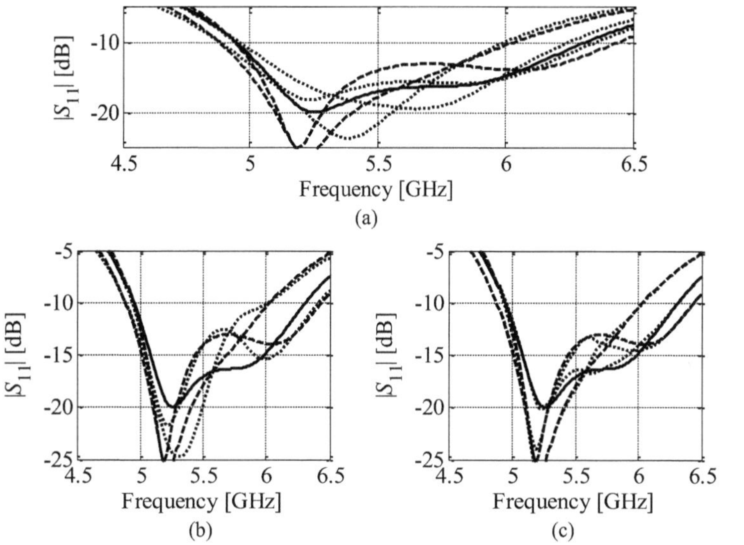

Fig. 4.8 Suspended DRA of Fig. 4.7: (**a**) low- (····) and high- (- - - and —) fidelity model responses at three designs; (**b**) ~~OSM~~-corrected low- (····) and high- (- - - and —) fidelity model responses at the same designs (~~OSM~~ correction at the design marked —); (**c**) multipoint-corrected low- (····) and high- (- - - and —) fidelity model responses

$$\boldsymbol{s}^{(i)}(\boldsymbol{x}) = \boldsymbol{c}_I\left(\boldsymbol{x}, \boldsymbol{p}^{(i)}\right) \tag{4.19}$$

The vector $\boldsymbol{p}$ in (4.19) denotes a set of implicit space mapping (preassigned) parameters, and $\boldsymbol{c}_I$ is the low-fidelity model with explicit dependence on these parameters. When the surrogate is established at the current design $\boldsymbol{x}^{(i)}$, the vector $\boldsymbol{p}^{(i)}$ is typically obtained by minimizing the norm-wise discrepancy between the models, $||\boldsymbol{f}(\boldsymbol{x}^{(i)}) - \boldsymbol{c}_I(\boldsymbol{x}^{(i)},\boldsymbol{p})||$. Illustration of implicit space mapping for the filter of Fig. 4.9a, b is shown in Fig. 4.10. Here, the implicit space mapping parameters are substrate thicknesses of the microstrip-line components (rectangle elements in Fig. 4.9b).

Another type of low-fidelity model enhancement is derived from observation of the vector-valued responses of the system which are evaluations of the same design but at different values of certain parameters such as the time, frequency, or a specific geometry parameter. Subsequently and based on such observations, the low-fidelity model is enhanced by applying a linear or nonlinear scaling to these parameters.

A representative example of such a correction method is frequency scaling. It is popular in electrical engineering where the figures of interest are often responses over frequency, e.g., S-parameters or gain (Koziel et al., 2006; Koziel & Ogurtsov, 2014).

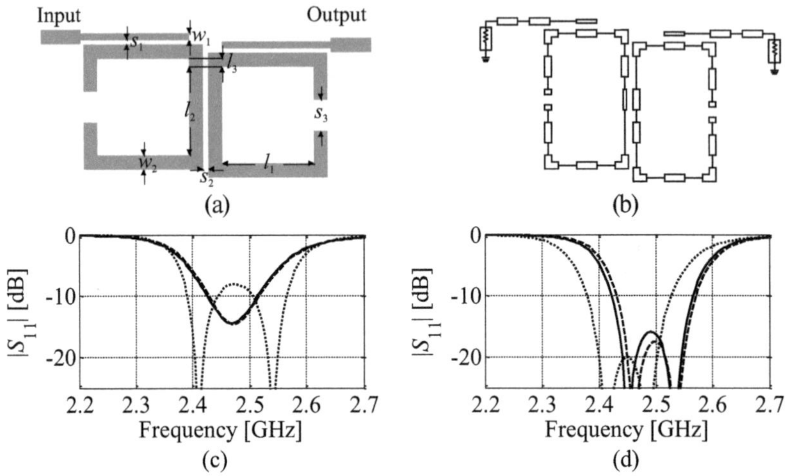

Fig. 4.9 Low-fidelity model correction through parameter shift (input space mapping): (**a**) microstrip filter geometry (high-fidelity model ***f*** evaluated using EM simulation); (**b**) low-fidelity model ***c*** (equivalent circuit); (**c**) response of ***f*** (—) and ***c*** (····), as well as response of the surrogate model ***s*** (– –) created using input space mapping; (**d**) surrogate model verification at a different design (other than that at which the model was created) ***f*** (—), ***c*** (····), and ***s*** (– –). Good alignment indicates excellent generalization of the model

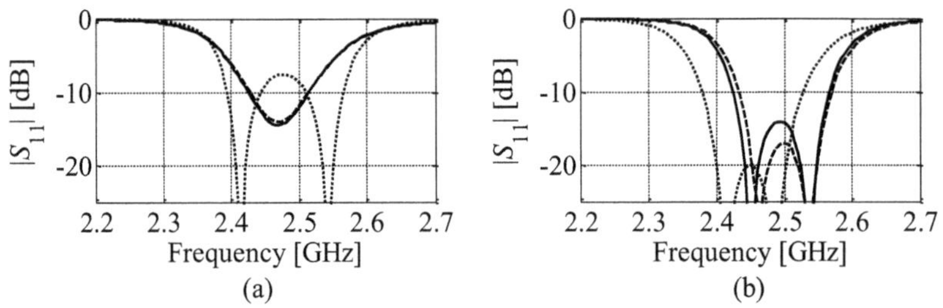

Fig. 4.10 Low-fidelity model correction through implicit space mapping applied to a microstrip filter of Fig. 4.9: (**a**) response of ***f*** (—) and ***c*** (····), as well as response of the surrogate model ***s*** (– –) created using implicit space mapping, (**b**) surrogate model verification at a different design (other than that at which the model was created) ***f*** (—), ***c*** (····), and ***s*** (– –)

Assume that $\boldsymbol{f}(\boldsymbol{x}) = [f(\boldsymbol{x},\omega_1)\, f(\boldsymbol{x},\omega_2) \ldots f(\boldsymbol{x},\omega_m)]^T$ where $f(\boldsymbol{x},\omega_k)$ is the evaluation of the high-fidelity model at a frequency ω_k. Similarly, we have $\boldsymbol{c}(\boldsymbol{x}) = [c(\boldsymbol{x},\omega_1)\; c(\boldsymbol{x},\omega_2) \ldots c(\boldsymbol{x},\omega_m)]^T$. The frequency-scaled surrogate model $\boldsymbol{s}_F(\boldsymbol{x})$ is then defined as

$$\boldsymbol{s}_F(\boldsymbol{x},[F_0\;\; F_1]) = [c(\boldsymbol{x}, F_0 + F_1\omega_1) \qquad \ldots \qquad c(\boldsymbol{x}, F_0 + F_1\omega_m)]^T \tag{4.20}$$

Here, F_0 and F_1 are scaling parameters obtained as $[F_0, F_1] = \mathrm{argmin}\{[F_0, F_1]: \| \boldsymbol{f}(\boldsymbol{x}^{(j)}) - \boldsymbol{s}_F(\boldsymbol{x}^{(j)},[F_0\;\; F_1])\|\}$. Figure 4.11 shows an example of frequency scaling

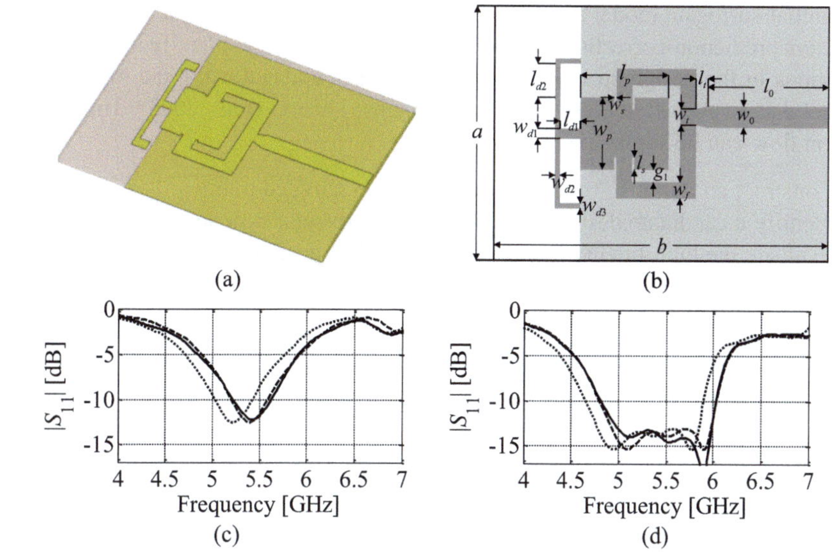

Fig. 4.11 Low-fidelity model correction with frequency scaling: (**a**) antenna view; (**b**) antenna geometry (both ***f*** and ***c*** evaluated using EM simulation, coarse discretization used for ***c***). Response of ***f*** (—) and ***c*** (····), as well as response of the surrogate model ***s*** (- - -) created using frequency scaling at (**c**) a certain reference design, and (d) another (test) design. Note that the surrogate properly accounts for the frequency shifts between ***c*** and ***f***

applied to the low-fidelity model of a substrate-integrated cavity antenna. In this example, both the low- and high-fidelity models are evaluated using coarse- and fine-discretization EM simulations. More extensive discussion on physics-based surrogates can be found in the literature, e.g., Koziel & Ogurtsov, 2014; Bandler et al., 2004a; Koziel et al., 2013b; Leifsson & Koziel, 2015a.

4.4 Optimization with Data-Driven Surrogates

This section briefly outlines the concepts of optimization exploiting data-driven surrogate models. In particular, we discuss a response surface methodology as well as global optimization using kriging surrogates. Furthermore, we highlight a practically important issue of balancing design space exploration and exploitation.

4.4.1 Optimization by Means of Response Surfaces

Optimization involving data-driven surrogates, sometimes referred to as response surface approximations (RSAs), consists of two major stages: (i) a construction of the initial surrogate model and (ii) an iterative prediction-correction scheme. In the iterative prediction-correction scheme, the surrogate is used to identify the promising locations in the design space; then the surrogate is updated using the high-fidelity model data evaluated at these promising designs (Queipo et al., 2005). The optimization flow can be summarized as follows:

1. Construct an initial surrogate model using pre-sampled training data.
2. Identify a candidate design by optimizing the surrogate model.
3. Evaluate the high-fidelity model at the candidate design.
4. Update the surrogate by incorporating new high-fidelity data.
5. If the termination condition is not satisfied, go to 2.

See also Fig. 4.12 for an illustration of this concept.

It should be emphasized that generating the new sample points by optimizing the surrogate is only one of the possible infill criteria. Other criteria may aim at improving global accuracy of the surrogate (see Sect. 4.4.3 for more details). In the case of local optimization, the optimization algorithm can be embedded in a trust region-like framework (Forrester & Keane, 2009) for improved convergence.

A simple yet quite common variation of the above scheme is when the surrogate model is constructed only once and optimized without further refinement. This so-called one-shot approach normally requires a considerable amount of training samples to ensure sufficient accuracy, and it is preferred for constructing multiple-use models (Kabir et al., 2008).

4.4.2 Sequential Approximate Optimization

One of the simplest realizations of surrogate-assisted optimization involving data-driven surrogates is sequential approximate optimization (SAO) (Jacobs et al., 2004; Giunta & Eldred, 2000; Pérez et al., 2002; Hosder et al., 2001; Roux et al., 1998; Sobieszczanski-Sobieski & Haftka, 1997; Giunta, 1997). The basic idea of SAO is to restrict the search to a small region of the design and utilize a local approximation model (e.g., low-order polynomial) as a prediction tool. Once prediction is rendered, the search region is modified—upon optimizing the surrogate—using a chosen relocation strategy.

The SAO flow is as follows (cf. Fig. 4.13):

1. Set $k = 0$ (iteration count).
2. Evaluate the high-fidelity model f at the design $\boldsymbol{x}^{(k)}$.
3. Construct a search region $r^{(k)}$ based on a relocation strategy.
4. Sample $r^{(k)}$ using a selected design-of-experiment plan.
5. Acquire high-fidelity data at the designs selected in Step 4.
6. Construct the surrogate $s^{(k)}$ using the data set of Step 5.
7. Find a new design $\boldsymbol{x}^{(k+1)}$ by optimizing the surrogate $s^{(k)}$ within $r^{(k)}$.
8. Terminate if the stopping criteria are met; otherwise set $k = k + 1$ and go to 2.

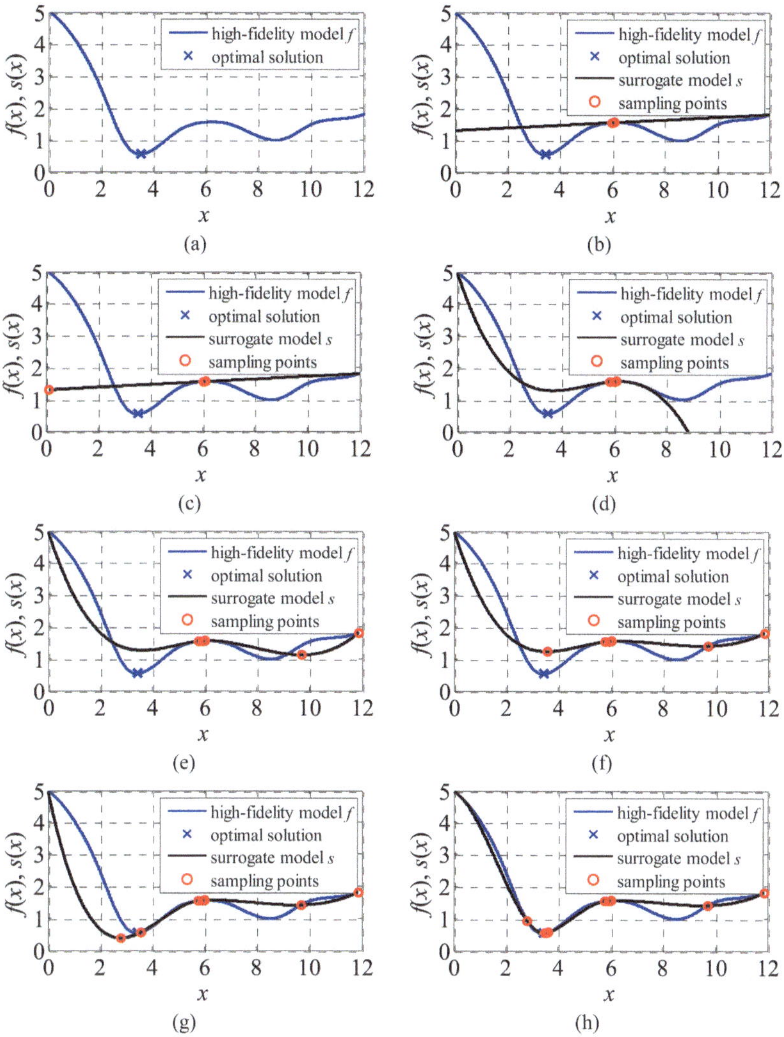

Fig. 4.12 RSA-based SBO illustration: (**a**) high-fidelity model f (····) to be minimized and the minimum of f (×); (**b**) initial surrogate model (here, cubic splines) constructed using two data points; (**c**) a new data point obtained by optimizing the surrogate (evaluation of f at this point is used to update the surrogate); (**d**)–(**h**) after a few iterations, the surrogate model optimum adequately predicts the high-fidelity minimizer; global accuracy of the surrogate is not of primary concern

The number of training samples utilized at each iteration is small because of using simple surrogates. This is an advantage of SAO. On the other hand, only local search is possible with SAO.

An important part of the SAO process is the relocation strategy, that is, the adjustment of size and location of the new region before executing the next iteration

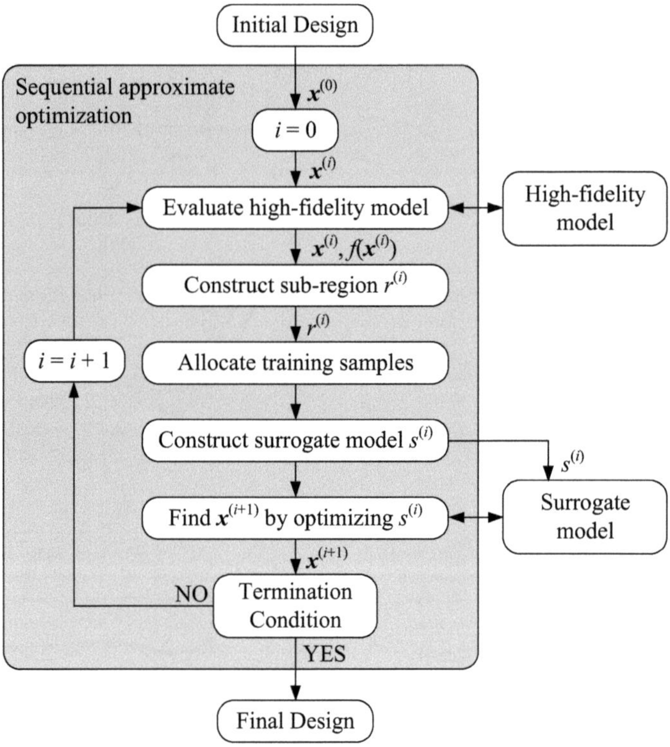

Fig. 4.13 Flowchart of sequential approximate optimization

(Step 3). A typical strategy is to allocate the center of the new region at the most current optimum (Jacobs et al., 2004). In multipoint strategies, the last iteration optimum may become the corner point for the next search subregion (Toropov et al., 1993). Adjustment of the subregion size may involve trust region-like frameworks (Giunta & Eldred, 2000; Koziel & Ogurtsov, 2012b).

4.4.3 *SBO with Kriging Surrogates: Exploration Versus Exploitation*

Choosing appropriate infill criteria is an important aspect of surrogate-assisted optimization (Forrester & Keane, 2009). The two main goals for selecting the infill points are (i) a reduction of the objective function value and (ii) an improvement of the global accuracy of the surrogate (Couckuyt, 2013).

The simplest infill strategy, suitable for local search, is to allocate a single sample at the surrogate model optimum. This strategy is capable of converging at least to a local minimum of the high-fidelity model assuming that the optimization algorithm is embedded in the trust region framework (Forrester & Keane, 2009) and the first-order consistency surrogate model (Alexandrov et al., 1998). In more general terms, allocation of new training points may be oriented toward global search or constructing a globally accurate surrogate. Clearly, the latter requires a feedback regarding the prediction power of the surrogate model and its generalization capability. In this context, kriging seems to be the most advantageous type of data-driven surrogate because it provides information about the expected model error (Kleijnen, 2009; Jones et al., 1998; Gorissen et al., 2010a).

The following infill criteria based on this feature of kriging models are commonly used:

1. Maximization of the expected improvement, that is, the improvement one expects at an untried point $\boldsymbol{x}$ (Jones et al., 1998).
2. Minimization of the predicted objective function $\widehat{y}(\boldsymbol{x})$. A reasonable global accuracy of the surrogate has to be assumed (Liu et al., 2012).
3. Minimization of the statistical lower bound, i.e., $LB(\boldsymbol{x}) = \widehat{y}(\boldsymbol{x}) - As(\boldsymbol{x})$ (Forrester & Keane, 2009), where $\widehat{y}(\boldsymbol{x})$ is the surrogate model prediction and $s^2(\boldsymbol{x})$ is the variance; A is a user-defined coefficient.
4. Maximization of the probability of improvement, i.e., identifying locations that give the highest chance of improving the objective function value (Forrester & Keane, 2009).
5. Maximization of the mean square error, i.e., finding locations where the mean square error predicted by the surrogate is the highest (Liu et al., 2012).

It should be emphasized that identifying the new samples according to the above infill criteria requires global optimization, typically realized using population-based metaheuristic algorithms (Couckuyt, 2013). Some comments on constrained SBO processes can be found in (Forrester & Keane, 2009; Queipo et al., 2005; Liu et al., 2012).

The aforementioned methods are also referred to as adaptive sampling techniques or sequential sampling. They may either aim at constructing globally accurate surrogates or carrying out global optimization. Putting more focus on design space exploitation usually leads to reduced computational costs. Design space exploration normally results in higher costs and global search capability (Forrester & Keane, 2009). On the other hand, global exploration is often impractical, especially for expensive functions with a medium/large number of optimization variables (more than a few tens). In the optimization context, there should be a balance between exploitation and exploration. Minimization of the statistical lower bound is an example of achieving such a balance controlled by the constant A (from pure exploitation, i.e., $LB(\boldsymbol{x}) \to \widehat{y}(\boldsymbol{x})$, for $A \to 0$ to pure exploration for $A \to \infty$). However, choosing a good value of A is a nontrivial task (Forrester & Keane, 2009).

Surrogate-assisted optimization methods with adaptive sampling are often referred to as efficient global optimization (EGO) techniques. A detailed discussion of EGO is given in Jones et al., (1998) and Forrester and Keane (2009). Kriging-based SBO can also be combined with pattern search methods, e.g., the surrogate management framework (SMF) (Booker et al., 1999). Several EGO-like methods, e.g., the SADEA framework enhanced by memetic operations such as solution pre-screening, have been proposed recently (Liu et al., 2014a).

4.4.4 *Summary*

Surrogate-assisted algorithms utilizing data-driven models work by iterative exploration of the design space so as to improve the model accuracy while searching for the optimum design. One of the important problems for this class of methods is finding a balance between exploration of the design space and exploitation of the promising regions. In a majority of high-frequency design problems, the computational cost of high-fidelity model evaluation is rather high; thus, the cost of training data acquisition is typically high. Therefore, approximation-based surrogates may be appropriate for problems with small or medium numbers of designable parameters. For more complex problems, the computational cost of creating the approximation-based surrogate may be prohibitive.

4.5 SBO Using Physics-Based Surrogates

Good generalization capability belongs to the most attractive features of physics-based surrogates. It is a result of the knowledge about the system of interest embedded in the underlying low-fidelity models. As a consequence, a reliable physics-based surrogate can be constructed with a significantly smaller number of high-fidelity training points than approximation-based models (Koziel et al., 2011a). In this section, we outline selected physics-based SBO methods, including some variations of space mapping (Bandler et al., 2004a; Koziel et al., 2006, 2008a), approximation model management optimization (AMMO, Alexandrov & Lewis, 2001), manifold mapping (Echeverría & Hemker, 2005), shape-preserving response prediction (Koziel, 2010a), as well as adaptive response scaling (Koziel & Unnsteinsson, 2018).

4.5.1 *Space Mapping*

Space mapping (SM) (Bandler et al., 1994; Bandler et al., 2004a; Koziel et al., 2006) refers to a family of design optimization techniques originally developed for solving expensive problems in computational electromagnetics, in particular, microwave engineering. The popularity of SM has spread across various engineering disciplines (Redhe & Nilsson, 2004; Bandler et al., 2004a; Priess et al., 2011; Marheineke et al., 2012; Koziel & Leifsson, 2012a; Tu et al., 2013).

In its early versions, SM was exclusively based on transformation of the low-fidelity model domain (aggressive SM, input SM; Bandler et al., 1995; Bandler et al., 1994), which was sufficient for handling many engineering problems, especially in electrical engineering, where the ranges of both the low- and high-fidelity model responses are similar. Other SM variations were developed to handle situations where the low- and high-fidelity models are severely misaligned in terms of the response levels (output SM; Bandler et al., 2003, 2004b; Koziel et al., 2006, 2008a; Robinson et al., 2008).

Original SM assumes existence of a mapping $\boldsymbol{P}$ between the high- and low-fidelity model domains (Bandler et al., 1994), such that $\boldsymbol{x}_c = \boldsymbol{P}(\boldsymbol{x}_f)$ and $\boldsymbol{c}(\boldsymbol{P}(\boldsymbol{x}_f)) \approx \boldsymbol{f}(\boldsymbol{x}_f)$. Given $\boldsymbol{P}$, the direct solution of the original problem (3.1) can be replaced by finding $\boldsymbol{x}_f^{\#} = \boldsymbol{P}^{-1}(\boldsymbol{x}_c^*)$. Here, $\boldsymbol{x}_c^*$ is the optimal design of $\boldsymbol{c}$ defined as $\boldsymbol{x}_c^* = \arg\min\{\boldsymbol{x}_c\text{: } U(\boldsymbol{c}(\boldsymbol{x}_c))\}$; $\boldsymbol{x}_f^{\#}$ can be considered as a reasonable estimate of $\boldsymbol{x}_f^*$. In other words, the problem (3.1) can be reformulated as

$$\boldsymbol{x}_f^{\#} = \arg\min_{\boldsymbol{x}_f} U(\boldsymbol{c}(\boldsymbol{P}(\boldsymbol{x}_f))) \tag{4.21}$$

where $\boldsymbol{c}(\boldsymbol{P}(\boldsymbol{x}_f))$ is a surrogate model. However, $\boldsymbol{P}$ is not given explicitly: it can only be evaluated at any $\boldsymbol{x}_f$ by means of the parameter extraction (PE) procedure, $\boldsymbol{P}(\boldsymbol{x}_f) = \arg\min\{\boldsymbol{x}_c\text{: } \|\boldsymbol{f}(\boldsymbol{x}_f) - \boldsymbol{c}(\boldsymbol{x}_c)\|$. One of the practical issues is possible nonuniqueness of the solution to (4.21) (Bandler et al., 1995). Another issue is the assumption on the similarity of high- and low-fidelity model ranges, which is a very strong assumption (Alexandrov & Lewis, 2001). These and other issues led to improvements, including parametric SM which is outlined later in this section.

Aggressive SM (ASM) (Bandler et al., 1995) is one of the first versions of SM, and it is still popular in microwave engineering (Sans et al., 2014, Rayas-Sanchez, 2016). Assuming uniqueness of $\boldsymbol{x}_c^*$, the solution to (4.21) is equivalent to reducing the residual vector $\boldsymbol{f} = \boldsymbol{f}(\boldsymbol{x}_f) = \boldsymbol{P}(\boldsymbol{x}_f) - \boldsymbol{x}_c^*$ to zero. The first step of the ASM algorithm is to find $\boldsymbol{x}_c^*$. Next, ASM iteratively solves the nonlinear system $\boldsymbol{f}(\boldsymbol{x}_f) = 0$ for $\boldsymbol{x}_f$. At the jth iteration, the calculation of the error vector $\boldsymbol{f}^{(j)}$ requires an evaluation of $\boldsymbol{P}^{(j)}(\boldsymbol{x}_f^{(j)}) = \boldsymbol{P}(\boldsymbol{x}_f^{(j)}) = \arg\min\{\boldsymbol{x}_c\text{: } \boldsymbol{f}(\boldsymbol{x}_f^{(j)}) - \boldsymbol{c}(\boldsymbol{x}_c)\|\}$. The quasi-Newton step in the high-fidelity model space is given by

$$\boldsymbol{B}^{(j)}\boldsymbol{h}^{(j)} = -\boldsymbol{f}^{(j)} \tag{4.22}$$

where $\boldsymbol{B}^{(j)}$ is the approximation of the space mapping Jacobian $\boldsymbol{J}_P = \boldsymbol{J}_P(\boldsymbol{x}_f) = [\partial \boldsymbol{P}^T/\partial \boldsymbol{x}_f]^T = [\partial(\boldsymbol{x}_c^T)/\partial \boldsymbol{x}_f]^T$. Solving (4.22) for $\boldsymbol{h}^{(j)}$ gives the next iterate $\boldsymbol{x}_f^{(j+1)} = \boldsymbol{x}_f^{(j)} + \boldsymbol{h}^{(j)}$. The algorithm terminates if $\|\boldsymbol{f}^{(j)}\|$ is sufficiently small. The output of the algorithm is an approximation to $\boldsymbol{x}_f^{\#} = \boldsymbol{P}^{-1}(\boldsymbol{x}_c^*)$. A popular way of obtaining the matrix $\boldsymbol{B}$ is through a rank one Broyden update (Broyden, 1965) of the form

$$\boldsymbol{B}^{(j+1)} = \boldsymbol{B}^{(j)} + \frac{\left[\boldsymbol{f}^{(j+1)} - \boldsymbol{f}^{(j)} - \boldsymbol{B}^{(j)}\boldsymbol{h}^{(j)}\right]\boldsymbol{h}^{(j)T}}{\left\|\boldsymbol{h}^{(j)}\right\|^2} \tag{4.23}$$

Several improvements of the ASM algorithm have been proposed (Bakr et al., 1998, 1999). ASM has been the most popular SM approach in microwave engineering till now (Rayas-Sanchez, 2016).

Parametric SM is another type of space mapping. It is more generic and characterized by explicit analytical form of the surrogate model. In parametric SM, the optimization algorithm is an iterative process (4.1). The two simple examples of input and implicit SM were shown in Sect. 4.3. In general, the input SM surrogate model can take the form (Koziel et al., 2006)

$$\boldsymbol{s}^{(i)}(\boldsymbol{x}) = \boldsymbol{c}\left(\boldsymbol{B}^{(i)} \cdot \boldsymbol{x} + \boldsymbol{q}^{(i)}\right) \tag{4.24}$$

Here, $\boldsymbol{B}^{(i)}$ and $\boldsymbol{q}^{(i)}$ are matrices obtained by minimizing misalignment between the surrogate and the high-fidelity model as

$$\left[\boldsymbol{B}^{(i)}, \boldsymbol{q}^{(i)}\right] = \arg\min_{[\boldsymbol{B},\boldsymbol{q}]} \sum\nolimits_{k=0}^{i} w_{i.k} \left\|\boldsymbol{f}\left(\boldsymbol{x}^{(k)}\right) - \boldsymbol{c}\left(\boldsymbol{B} \cdot \boldsymbol{x}^{(k)} + \boldsymbol{q}\right)\right\| \tag{4.25}$$

(4.25) is, in fact, a nonlinear regression problem with $w_{i.k}$ being the weighting factors. A common choice of $w_{i.k}$ is $w_{i.k} = 1$ for all i and all k (all previous designs contribute equally) or $w_{i.1} = 1$ and $w_{i.k} = 0$ for $k < i$ (the surrogate depends on the most recent design only).

In general, the SM surrogate model is constructed as follows:

$$\boldsymbol{s}^{(i)}(\boldsymbol{x}) = \bar{\boldsymbol{s}}\left(\boldsymbol{x}, \boldsymbol{p}^{(i)}\right) \tag{4.26}$$

where $\bar{\boldsymbol{s}}$ is a generic SM surrogate model, i.e., the low-fidelity model $\boldsymbol{c}$ composed with suitable (usually linear) transformations. The parameters $\boldsymbol{p}$ are obtained in the extraction process similar to (4.25). More information about specific SM surrogates can be found in the literature (Bandler et al., 2004a; Koziel et al., 2006).

Despite its simplicity and demonstrated efficiency, there are some practical issues that are still preventing SM from being widely accepted by industry designers. These issues include (i) finding an appropriate balance between the quality and the speed of the underlying low-fidelity model, as well as (ii) selection of appropriate SM transformations (Koziel & Bandler, 2007a, b). Resolving the above listed-issues is not a trivial task (Koziel et al., 2008b). As a matter of fact, similar issues are actually common to the majority of physics-based SBO methods. On the other hand, SM has been found to be an efficient tool for yielding satisfactory designs in various engineering disciplines (Redhe & Nilsson, 2004; Bandler et al., 2004a; Marheineke et al., 2012). A number of enhancements of SM algorithms have been suggested to alleviate some of the difficulties such as potential convergence problems (Koziel et al., 2010a; Koziel et al., 2010b).

4.5.2 Approximation Model Management Optimization (AMMO)

AMMO (Alexandrov & Lewis, 2001) is a simple algorithmic framework with the surrogate model constructed through response correction. It exploits sensitivity data to ensure first-order consistency conditions, in particular, $s^{(i)}(\boldsymbol{x}^{(i)}) = f(\boldsymbol{x}^{(i)})$, and $\nabla s^{(i)}(\boldsymbol{x}^{(i)}) = \nabla f(\boldsymbol{x}^{(i)})$. Additionally, AMMO utilizes the trust region methodology (Conn et al., 2000) to guarantee convergence of the optimization process to the high-fidelity model optimum. Assuming that $\beta(\boldsymbol{x}) = f(\boldsymbol{x})/c(\boldsymbol{x})$ is the correction function and

$$\beta_i(\boldsymbol{x}) = \beta\left(\boldsymbol{x}^{(i)}\right) + \nabla\beta\left(\boldsymbol{x}^{(i)}\right)^T\left(\boldsymbol{x} - \boldsymbol{x}^{(i)}\right) \quad (4.27)$$

the surrogate model is defined as

$$s^{(i)}(\boldsymbol{x}) = \beta_i(\boldsymbol{x})c(\boldsymbol{x}) \quad (4.28)$$

(4.28) can be shown to satisfy the aforementioned consistency conditions. Clearly, (4.28) requires both the low- and high-fidelity derivatives.

4.5.3 Manifold Mapping (MM)

MM can be considered as a variation of output space mapping. It is essentially the response correction technique (Echeverría & Hemker, 2005; Echeverría & Hemker, 2008) with the surrogate model set up using multiple high-fidelity data points as follows:

$$\boldsymbol{s}^{(i)}(\boldsymbol{x}) = \boldsymbol{f}\left(\boldsymbol{x}^{(i)}\right) + \boldsymbol{S}^{(i)}\left(\boldsymbol{c}(\boldsymbol{x}) - \boldsymbol{c}\left(\boldsymbol{x}^{(i)}\right)\right) \quad (4.29)$$

Here, $\boldsymbol{S}^{(i)} = \boldsymbol{\Delta F} \cdot \boldsymbol{\Delta C}^{\dagger}$ is the $m \times m$ correction matrix, where

$$\boldsymbol{\Delta F} = \left[\boldsymbol{f}\left(\boldsymbol{x}^{(i)}\right) - \boldsymbol{f}\left(\boldsymbol{x}^{(i-1)}\right) \quad \ldots \quad \boldsymbol{f}\left(\boldsymbol{x}^{(i)}\right) - \boldsymbol{f}\left(\boldsymbol{x}^{(\max\{i-n,0\})}\right)\right] \quad (4.30)$$

$$\boldsymbol{\Delta C} = \left[\boldsymbol{c}\left(\boldsymbol{x}^{(i)}\right) - \boldsymbol{c}\left(\boldsymbol{x}^{(i-1)}\right) \quad \ldots \quad \boldsymbol{c}\left(\boldsymbol{x}^{(i)}\right) - \boldsymbol{c}\left(\boldsymbol{x}^{(\max\{i-n,0\})}\right) \right] \tag{4.31}$$

The pseudoinverse, denoted by †, is defined as $\boldsymbol{\Delta C}^{\dagger} = \boldsymbol{V}_{\boldsymbol{\Delta C}} \boldsymbol{\Sigma}_{\boldsymbol{\Delta C}}^{\dagger} \boldsymbol{U}_{\boldsymbol{\Delta C}}^{T}$, where $\boldsymbol{U}_{\boldsymbol{\Delta C}}$, $\Sigma_{\boldsymbol{\Delta C}}$, and $\boldsymbol{V}_{\boldsymbol{\Delta C}}$ are the factors in the singular value decomposition of the matrix $\boldsymbol{\Delta C}$. The matrix $\Sigma_{\boldsymbol{\Delta C}}{}^{\dagger}$ is the result of inverting the nonzero entries in $\Sigma_{\boldsymbol{\Delta C}}$, leaving the zeroes invariant. Upon convergence, the linear correction $\boldsymbol{S}^{*}$ (being the limit of $\boldsymbol{S}^{(i)}$ with $i \rightarrow \infty$) maps the point $\boldsymbol{c}(\boldsymbol{x}^{*})$ to $\boldsymbol{f}(\boldsymbol{x}^{*})$ and the tangent plane for $\boldsymbol{c}(\boldsymbol{x})$ at $\boldsymbol{c}(\boldsymbol{x}^{*})$ to the tangent plane for $\boldsymbol{f}(\boldsymbol{x})$ at $\boldsymbol{f}(\boldsymbol{x}^{*})$ (Echeverría & Hemker, 2008). If the sensitivity data is available for both the low- and high-fidelity models, the correction matrix $\boldsymbol{S}^{(i)}$ can be defined using the exact Jacobians of the models.

4.5.4 *Shape-Preserving Response Prediction (SPRP)*

Shape-preserving response prediction (SPRP) is a parameter-less approach (Koziel, 2010a, 2012) where the surrogate model is constructed assuming that the change of the high-fidelity model response due to the adjustment of the design variables can be predicted using the actual changes of the low-fidelity model response. It is critically important for SPRP that the low-fidelity model is physics-based, which ensures that the effect of the design parameter variations on the model response is similar for both models. The change of the low-fidelity model response is described by the translation vectors corresponding to a certain (finite) number of characteristic points of the model's response. These translation vectors are subsequently used to predict the change of the high-fidelity model response with the actual response $\boldsymbol{f}(\boldsymbol{x}^{(i)})$ of the model $\boldsymbol{f}$ at the current iteration point, treated as a reference.

Figure 4.14 explains the concept of SPRP using an example of a microstrip bandstop filter (Koziel, 2010b). Figure 4.14a shows the example of the low-fidelity model response $|S_{21}|$ (transmission coefficient) in the frequency range from 8 GHz to 18 GHz, at the design $\boldsymbol{x}^{(i)}$, as well as the low-fidelity model response at some other design $\boldsymbol{x}$. Circles denote five characteristic points of $\boldsymbol{c}(\boldsymbol{x}^{(i)})$, here, selected to represent $|S_{21}| = -3$ dB, $|S_{21}| = -20$ dB and the local $|S_{21}|$ maximum (at about 13 GHz). Squares denote corresponding characteristic points for $\boldsymbol{c}(\boldsymbol{x})$, while small line segments represent the translation vectors that determine the "shift" of the characteristic points of $\boldsymbol{c}$ when changing the design variables from $\boldsymbol{x}^{(i)}$ to $\boldsymbol{x}$. Because the low-fidelity model is physics-based, the high-fidelity model response at the given design, here, $\boldsymbol{x}$, can be predicted using the same translation vectors applied to the corresponding characteristic points of the fine model response at $\boldsymbol{x}^{(i)}$, $\boldsymbol{f}(\boldsymbol{x}^{(i)})$, cf. Fig. 4.14b.

As indicated in Fig. 4.14c, the predicted high-fidelity model response at the design $\boldsymbol{x}$ is in a very good agreement with the actual response, $\boldsymbol{f}(\boldsymbol{x})$.

A rigorous formulation of SPRP is available in the literature (Koziel, 2010a, b). It should be mentioned that an important assumption of SPRP is that the shapes of the high- and low-fidelity model responses are similar in overall. This assumption means

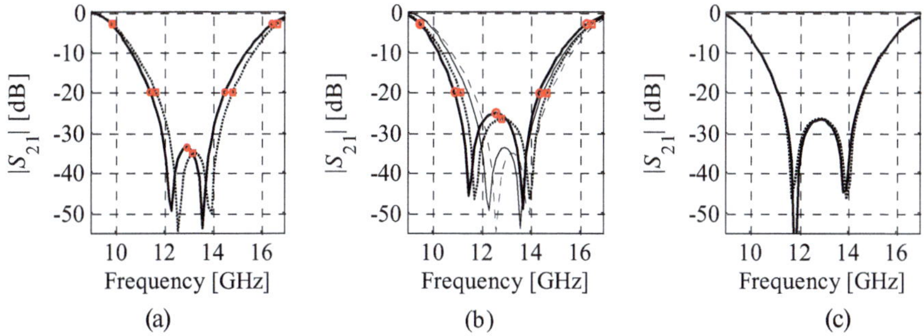

Fig. 4.14 SPRP concept (Koziel, 2010a): (**a**) example low-fidelity model response at the design $\boldsymbol{x}^{(i)}$, $\boldsymbol{c}(\boldsymbol{x}^{(i)})$ (—) and at another design $\boldsymbol{x}$, $\boldsymbol{c}(\boldsymbol{x})$ (····), characteristic points of $\boldsymbol{c}(\boldsymbol{x}^{(i)})$ (○) and $\boldsymbol{c}(\boldsymbol{x})$ (□), and the translation vectors (short lines); (**b**) high-fidelity model response at $\boldsymbol{x}^{(i)}$, $\boldsymbol{f}(\boldsymbol{x}^{(i)})$ (—) and the predicted fine model response at $\boldsymbol{x}$ (····) obtained using SPRP based on characteristic points of Fig. 4.14a; characteristic points of $\boldsymbol{f}(\boldsymbol{x}^{(i)})$ (○) and the translation vectors (short lines) were used to find the characteristic points (□) of the predicted high-fidelity model response; low-fidelity model responses $\boldsymbol{c}(\boldsymbol{x}^{(i)})$ and $\boldsymbol{c}(\boldsymbol{x})$ are plotted using thin solid and dotted line, respectively, and (**c**) predicted (····) and actual (—) high-fidelity model response at $\boldsymbol{x}$

that the characteristic points of the coarse model $\boldsymbol{c}$ and the fine model $\boldsymbol{f}$ responses are in one-to-one correspondence. If this assumption is not satisfied, the SPRP surrogate cannot be evaluated because the translation vectors are not well defined. Generalizations of SPRP that relax the above-listed assumption can be found in some cases (Koziel, 2010b). In the context of antenna optimization, handling reflection characteristics of narrow-band and multi-band structures as well as array radiation pattern optimization (e.g., for sidelobe reduction) are the examples of problems SPRP is well suited for.

4.5.5 Adaptive Response Scaling

The foundation of adaptive response scaling (Koziel & Unnsteinsson, 2018) is to construct a mapping between the low- ($\boldsymbol{c}$) and high-fidelity ($\boldsymbol{f}$) models of the structure under design. As opposed to many other physics-based surrogate-assisted optimization methods (e.g., space mapping), ARS can be considered as nonparametric, therefore, more flexible and more generic. Here, the ARS concept is explained using a narrow-band antenna example. Figure 4.15 shows the high- and low-fidelity reflection responses at a certain reference design $\boldsymbol{x}^{(i)}$ and a test design $\boldsymbol{x}$. ARS exploits the correlation between the models, and, in the case of complex characteristics, the scaling process is conducted independently for the real and imaginary parts of the response(s). ARS modeling is realized in several stages. At the first stage, a frequency relationship between $\boldsymbol{c}$ and $\boldsymbol{f}$ is established at the design $\boldsymbol{x}^{(i)}$ (in the optimization context, this would be the current iteration point found by the

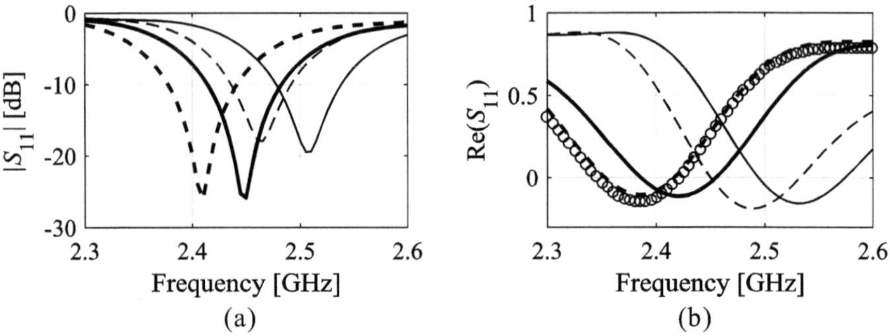

Fig. 4.15 Example narrow-band antenna responses: (**a**) $|S_{11}|$ at a reference design $\boldsymbol{x}^{(i)}$ (—) and at evaluation design $\boldsymbol{x}$ (- - -); high- and low-fidelity models shown using thick and thin lines, respectively. Designs are taken arbitrarily for illustration purposes; (**b**) responses at $\boldsymbol{x}^{(i)}$ (—) and $\boldsymbol{x}$ (- - -): $\boldsymbol{f}$ (thick lines) and $\boldsymbol{c}$ (thin lines); ARS surrogate model response shown using circles

algorithm). The frequency scaling function F is obtained by solving a nonlinear regression problem of the form

$$F^{(i)}(\omega) = \arg\min_F \int_{\omega_{\min}}^{\omega_{\max}} \left| r_f\left(\boldsymbol{x}^{(i)}, \omega\right) - r_c\left(\boldsymbol{x}^{(i)}, F(\omega)\right) \right| d\omega \tag{4.32}$$

The function F can be realized by means of cubic splines (Koziel & Unnsteinsson, 2018); r_f/r_c are the high- and low-fidelity responses of interest (e.g., Re(S_{11})); the purpose of $F^{(i)}$ is to reduce the ("horizontal") discrepancies between the responses within the frequency range of interest $\omega_{\min}$ to $\omega_{\max}$.

In the next stage, a frequency relationship between the low-fidelity model responses at the design $\boldsymbol{x}^{(i)}$ and at the (evaluation) design $\boldsymbol{x}$ is established. This is done similarly as in (4.32), i.e.,

$$F(\boldsymbol{x}, \omega) = \arg\min_F \int_{\omega_{\min}}^{\omega_{\max}} \left| r_c(\boldsymbol{x}, \omega) - r_c\left(\boldsymbol{x}^{(i)}, F(\omega)\right) \right| d\omega \tag{4.33}$$

The third stage is to quantify the level (amplitude) relationship between the frequency-scaled low-fidelity model at the reference design and the low-fidelity model at the evaluation design (here, ÷ denotes component-wise division)

$$A(\boldsymbol{x}, \omega) = [r_c(\boldsymbol{x}, \omega) + M] \div \left[r_c\left(\boldsymbol{x}^{(i)}, F(\boldsymbol{x}, \omega)\right) + M \right] \tag{4.34}$$

The shift by M (we use $M = 1$) is necessary to avoid division by zero (for frequencies for which $r_c(\boldsymbol{x}^{(i)}, F(\boldsymbol{x},\omega)) = 0$).

At the prediction stage, ARS surrogate model r_s estimates the high-fidelity model response at the evaluation design $\boldsymbol{x}$. We have

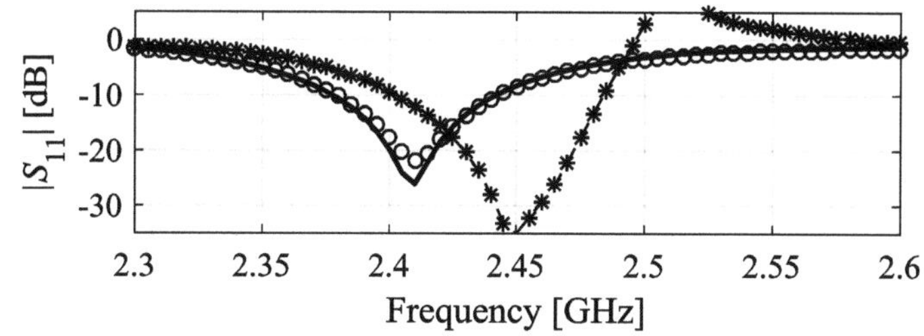

Fig. 4.16 High-fidelity model response at the evaluation design $\boldsymbol{x}$ (thick line), ARS surrogate model response (o), and conventional response correction (output space mapping) (*). Note severe distortion of the latter due to inability to account for frequency changes of the low-fidelity model response

$$r_s(\boldsymbol{x}, \omega) = A\left(\boldsymbol{x}, F^{(i)}(\omega)\right) \circ \left[r_f\left(\boldsymbol{x}^{(i)}, F(\boldsymbol{x}, \omega)\right) + M\right] - M \tag{4.35}$$

where ° denotes component-wise multiplication.

The above concepts are explained in Figs. 4.15b and 4.16. Figure 4.15b shows Re (S_{11}) of the responses shown in Fig. 4.15a as well as the response of the ARS surrogate. The prediction power of ARS is good also at the level of $|S_{11}|$ (cf. Fig. 4.16). This is not the case for conventional correction (output space mapping) which fails to account for frequency changes of the response. For practical applications, ARS is realized at the level of real and imaginary parts of the reflection response which makes it more accurate and further recalculated to whatever form is required by calculation of the design objectives (here, the absolute value).

In the original implementation of ARS, a construction of the surrogate model $\boldsymbol{s}^{(i)}(\boldsymbol{x})$ at the current design $\boldsymbol{x}^{(i)}$ is followed by its optimization to yield the next approximation

$$\boldsymbol{x}^{(i+1)} = \arg\min\left\{\boldsymbol{x} : U\left(\boldsymbol{s}^{(i)}(\boldsymbol{x})\right)\right\} \tag{4.36}$$

of the optimum design $\boldsymbol{x}^* = \arg\min\{\boldsymbol{x}: U(\boldsymbol{f}(\boldsymbol{x}))\}$ (U is the objective function). Subsequently, the high-fidelity model is evaluated, and the surrogate is reset (both at $\boldsymbol{x}^{(i+1)}$). This procedure works fine assuming that the low-fidelity model is very fast (e.g., an equivalent circuit). In the case of coarse discretization EM antenna models, iterative solving of (4.36) is just too expensive.

In order to make ARS feasible for a more generic case, in particular, coarse discretization low-fidelity model (as in antenna design), a modified procedure should be used, which generates a sequence of designs $\boldsymbol{x}^{(i)}$, $i = 0, 1, \ldots$, (approximations to $\boldsymbol{x}^*$) as

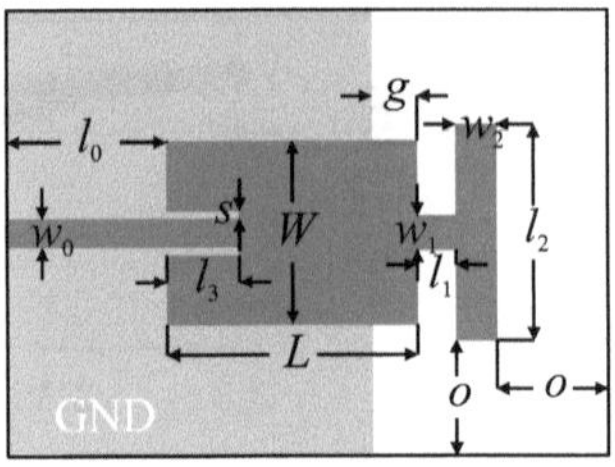

Fig. 4.17 Dual-band patch antenna; ground plane marked using the light-shade gray

$$x^{(i+1)} = \arg \min_{\mathbf{x},\ \left\|\mathbf{x}-\mathbf{x}^{(i)}\right\| \le \delta^{(i)}} U\left(s\left(x^{(i)}\right) + J_{s^{(i)}}\left(x^{(i)}\right) \cdot \left(x - x^{(i)}\right)\right) \tag{4.37}$$

where $J_s^{(i)}$ is a Jacobian of the ARS surrogate estimated using finite differentiation. The cost of each iteration of (4.37) is one evaluation of the high-fidelity model and n evaluations of $\boldsymbol{c}$ (n being the number of design variables).

The entire process is embedded in the trust region (TR) framework (Conn et al., 2000), in which $\delta^{(i)}$ is the TR radius, updated in each iteration. The design $x^{(1)}$ is obtained through direct optimization of the low-fidelity model.

For illustration purposes, consider a dual-band planar antenna shown in Fig. 4.17. The structure is implemented on RF-35 substrate (h = 0.762 mm, ε_r = 3.5). The design variables are $x^{\mathrm{I}} = [L\ l_1\ l_2\ l_3\ W\ w_1\ w_2\ g]^T$; $o = 7$, $l_0 = 10$, $s = 0.5$, and $w_0 = 1.7$, are fixed (all dimensions in mm). The EM models are implemented in CST Microwave Studio: $\boldsymbol{f}$ (~3,300,000 cells, 90 minutes) and $\boldsymbol{c}$ (~82,000, ~138,000 and ~ 154,000 mesh cells, simulation time 88, 139, and 169 seconds, for the three mesh densities corresponding to the global mesh parameter LPW (lines per wavelength) of 12, 14, and 16. Some discussion concerning the low-fidelity model selection can be found in Koziel & Ogurtsov, 2014.

The dual-band patch has been optimized using ARS to maximize the symmetric bandwidth at the operating frequencies of 2.45 GHz and 5.3 GHz. The objective function was defined as follows. Let f_{k1} and f_{k2}, k = 1, 2, denote the frequencies corresponding to the beginning and end of the operating band for the lower and upper band, respectively. This information can be extracted from the simulated antenna response. We have $U(\boldsymbol{f}(\boldsymbol{x})) = -\min\{B_1,B_2\}$, where $B_k = 2\cdot\min\{f_{k0} - f_{k1}, f_{k2} - f_{k0}\}$, k = 1, 2, is a symmetric part of the antenna bandwidth. In case the resonance is not sufficiently deep (i.e., above −10 dB), a penalty term is used instead proportional to the difference between −10 dB and the resonance level at its minimum.

The ARS-optimized design for the low-fidelity models with LPW = 16 is $x^* = [15.83\ 3.57\ 12.63\ 5.54\ 16.43\ 1.57\ 9.00\ 6.70]^T$; see also Fig. 4.18.

ARS has been compared to build-in CST optimizers (a trust region algorithm, a Nelder-Mead algorithm, genetic algorithm, and particle swarm optimizer, cf. Table 4.1) as well as direct optimization of the high-fidelity model. It can be observed that ARS offers significant cost savings without compromising accuracy.

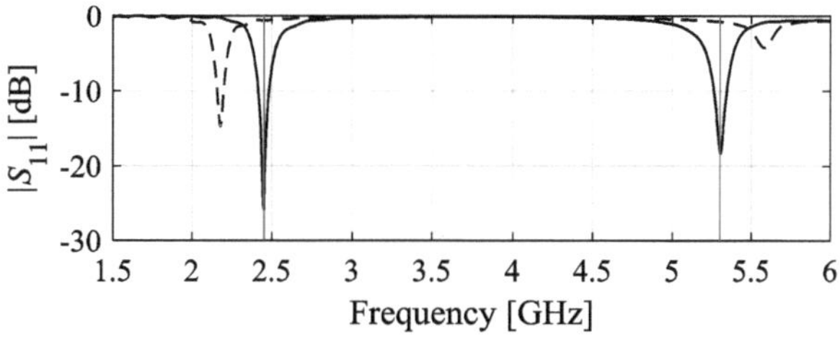

Fig. 4.18 Initial (- - -) and optimized (—) responses of the dual-band antenna of Fig. 4.17, obtained using ARS

Table 4.1 Dual-band antenna: Optimization results

Optimization algorithm	Optimization cost: Number of low-fidelity model evaluations c	Number of high-fidelity model evaluations f	Total[a]	Objective function[b]
CST built-in (trust region)	–	22	22	Failed
CST built-in (Nelder-Mead)	–	488	488	0.071
CST built-in (GA)	–	500	500	0.027
CST built-in (PSO)	–	500	500	0.006
Direct optimization	–	75	75	0.068
ARS (c with LPW = 12)	112	8	9.7	0.050
ARS (c with LPW = 14)	103	8	10.5	0.036
ARS (c with LPW = 16)	117	7	10.4	0.065

[a]Optimization cost expressed in the equivalent number of high-fidelity number evaluations.
[b]Minimum of the symmetric bandwidth for the two operating frequencies at the final design

This is despite the fact that the time evaluation of the high- and low-fidelity model is quite low. It should also be noted that in some cases, CST optimizers fail or provide relatively poor results, especially the stochastic methods, which were terminated upon reaching 500 EM simulations of the antenna at hand. On the other hand, the number of high-fidelity model evaluations during ARS optimization is low (typically a few) which translates into low overall cost. Finally, ARS performance for various low-fidelity model discretization densities has been investigated. It can be observed that the results are consistent although slight degradation is observed when discretization is reduced, which can be attributed to the increased level of numerical noise pertinent to low-fidelity EM analysis.

The optimized antenna has been manufactured and measured for the sake of additional validation. Figure 4.19 shows the photograph of the fabricated prototype, whereas Figs. 4.20 and 4.21 show the reflection characteristic and the E- and H-plane patterns, respectively. The agreement between the simulations and measurements is

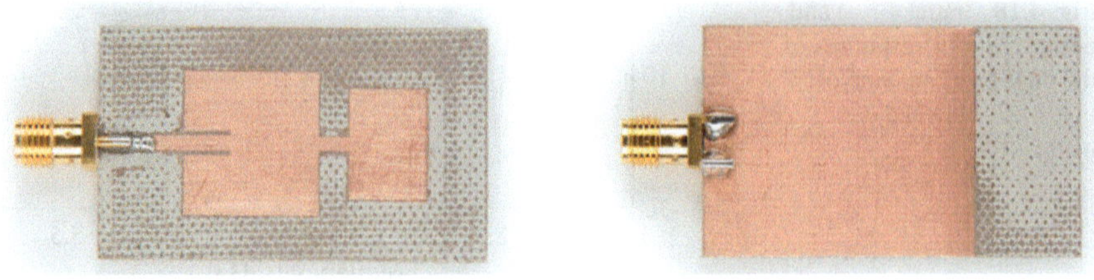

Fig. 4.19 Photographs of the fabricated prototype of the ARS-optimized dual-band antenna

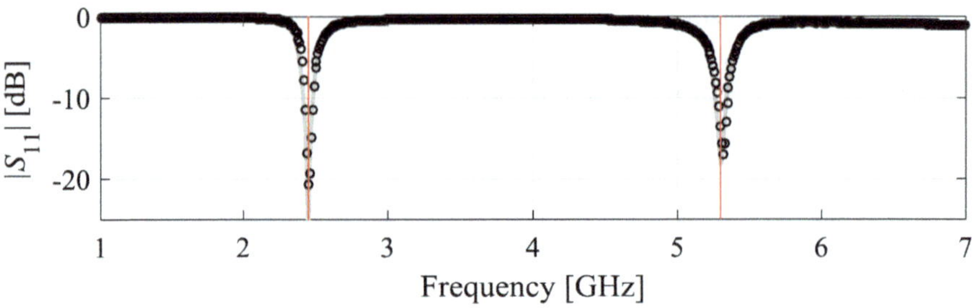

Fig. 4.20 Simulated (gray) and measured (o) reflection characteristics of the dual-band antenna of Fig. 4.19

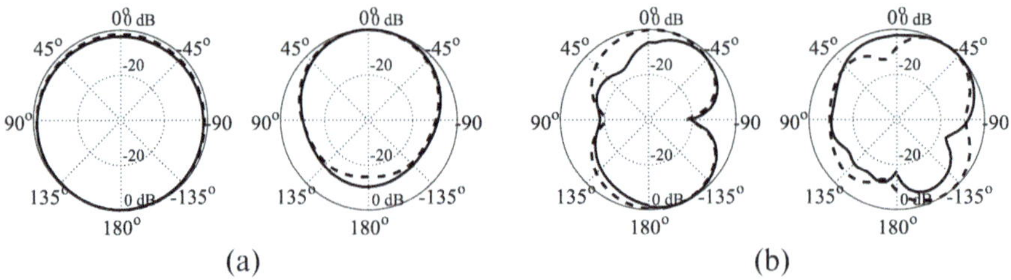

Fig. 4.21 Simulated (- - -) and measured (—) radiation patterns of the dual-band antenna of Fig. 4.19: (**a**) H-plane at 2.45 GHz and 5.3 GHz, (**b**) E-plane at 2.45 GHz and 5.3 GHz

acceptable which additionally validates the reliability of the optimization procedure. Discrepancies for E-plane radiation pattern result from the measurement setup (shadowing effect of the 90-degree bend used to mount the structure as well as a piece of cable extending from the measurement tower; neither of which has been included in the computational model of the antenna).

4.5.6 Summary

Physics-based SBO algorithms offer improved computational efficiency due to exploiting the problem-specific knowledge embedded in the low-fidelity model. At the same time, they are more difficult in implementation and less transferrable between various classes of problems. Nevertheless, the physics-based SBO methods seem to be the most promising approaches for handling complex design problems in high-frequency engineering, particularly, problems that involve expensive EM-simulations. It should be mentioned that this chapter only presented selected methods. Discussion of other techniques can be found in the literature (Koziel et al., 2009, 2011b; Koziel & Leifsson, 2012b; Robinson et al., 2008; Koziel & Pietrenko-Dabrowska, 2020, 2022).

References

Afshinmanesh, F., Marandi, A., & Shahabadi, M. (2008). Design of a single-feed dual-band dual-polarized printed microstrip antenna using a Boolean particle swarm optimization. *IEEE Transactions on Antennas and Propagation, 56*, 1845–1852.

Alba, E., & Marti, R. (Eds.). (2006). *Metaheuristic procedures for training neural networks*. Springer.

Alexandrov, N. M., & Lewis, R. M. (2001). An overview of first-order model management for engineering optimization. *Optical Engineering, 2*, 413–430.

Alexandrov, N. M., Dennis, J. E., Lewis, R. M., & Torczon, V. (1998). A trust region framework for managing use of approximation models in optimization. *Structural and Multidisciplinary Optimization, 15*, 16–23.

Andrés, E., Salcedo-Sanz, S., Monge, F., & Pérez-Bellido, A. M. (2012). Efficient aerodynamic design through evolutionary programming and support vector regression algorithms. *Expert Systems with Applications: An International Journal, 39*, 10700–10708.

Angiulli, G., Cacciola, M., & Versaci, M. (2007). Microwave devices and antennas modelling by support vector regression machines. *IEEE Transactions on Magnetics, 43*, 1589–1592.

Bakr, M. H., Bandler, J. W., Biernacki, R. M., Chen, S. H., & Madsen, K. (1998). A trust region aggressive space mapping algorithm for EM optimization. *IEEE Transactions on Microwave Theory and Techniques, 46*, 2412–2425.

Bakr, M. H., Bandler, J. W., Georgieva, N. K., & Madsen, K. (1999). A hybrid aggressive space-mapping algorithm for EM optimization. *IEEE Transactions on Microwave Theory and Techniques, 47*, 2440–2449.

Bandler, J. W., Biernacki, R. M., Chen, S. H., Grobelny, P. A., & Hemmers, R. H. (1994). Space mapping technique for electromagnetic optimization. *IEEE Transactions on Microwave Theory and Techniques, 42*, 2536–2544.

Bandler, J. W., Biernacki, R. M., Chen, S. H., Hemmers, R. H., & Madsen, K. (1995). Electromagnetic optimization exploiting aggressive space mapping. *IEEE Transactions on Microwave Theory and Techniques, 43*, 2874–2882.

Bandler, J. W., Cheng, Q. S., Gebre-Mariam, D. H., Madsen, K., Pedersen, F., & Søndergaard, J. (2003). EM-based surrogate modeling and design exploiting implicit, frequency and output space mappings. In *IEEE International Microwave Symposium Digest*, Philadelphia, PA (pp. 1003–1006).

Bandler, J. W., Cheng, Q. S., Dakroury, S. A., Mohamed, A. S., Bakr, M. H., Madsen, K., & Søndergaard, J. (2004a). Space mapping: The state of the art. *IEEE Transactions on Microwave Theory and Techniques, 52*, 337–361.

Bandler, J. W., Cheng, Q. S., Nikolova, N. K., & Ismail, M. A. (2004b). Implicit space mapping optimization exploiting preassigned parameters. *IEEE Transactions on Microwave Theory and Techniques, 52*, 378–385.

Beachkofski, B., & Grandhi, R. (2002). Improved distributed hypercube sampling. In *Structures, structural dynamics, materials conference*, Paper AIAA 2002-1274.
Booker, A. J., Dennis, J. E., Frank, P. D., Serafini, D. B., Torczon, V., & Trosset, M. W. (1999). A rigorous framework for optimization of expensive functions by surrogates. *Structural Optimization, 17*, 1–13.
Broyden, C. G. (1965). A class of methods for solving nonlinear simultaneous equations. *Mathematics of Computation, 19*, 577–593.
Chamaani, S., Abrishamian, M. S., & Mirtaheri, S. A. (2010). Time-domain design of UWB Vivaldi antenna array using multiobjective particle swarm optimization. *IEEE Antennas and Wireless Propagation Letters, 9*, 666–669.
Cheng, Q. S., Koziel, S., & Bandler, J. W. (2006). Simplified space mapping approach to enhancement of microwave device models. *International Journal of RF and Microwave Computer-Aided Engineering, 16*, 518–535.
Chi, H., Mascagni, M., & Warnock, T. (2005). On the optimal Halton sequence. *Mathematics and Computers in Simulation, 70*, 9–21.
Conn, A. R., Gould, N. I. M., & Toint, P. L. (2000). *Trust region methods* (MPS-SIAM series on optimization). MPS-SIAM.
Couckuyt, I., (2013). *Forward and inverse surrogate modeling of computationally expensive problems*. PhD thesis. Ghent University.
Crombecq, K., Gorissen, D., Tommasi, L. D., & Dhaene, T. (2009). A novel sequential design strategy for global surrogate modeling, In *Proceedings of the 41st Winter Simulation Conference* (pp. 731–742).
Crombecq, K., Laermans, E., & Dhaene, T. (2011). Efficient space-filling and non-collapsing sequential design strategies for simulation-based modeling. *European Journal of Operational Research, 214*(3), 683–696.
CST Microwave Studio. (2022). Dassault Systemes, France.
Davis, E., & Ierapetritou, M. (2010). A centroid-based sampling strategy for kriging global modeling and optimization. *AICHE Journal, 56*(1), 220–240.
Devabhaktuni, V. K., Yagoub, M. C. E., & Zhang, Q. J. (2001). A robust algorithm for automatic development of neural-network models for microwave applications. *IEEE Transactions on Microwave Theory and Techniques, 49*, 2282–2291.
Echeverría, D., & Hemker, P. W. (2005). Space mapping and defect correction. *Computational Methods in Applied Mathematics, 5*, 107–136.
Echeverría, D., & Hemker, P. W. (2008). Manifold mapping: A two-level optimization technique. *Computing and Visualization in Science, 11*, 193–206.
Forrester, A. I. J., & Keane, A. J. (2009). Recent advances in surrogate-based optimization. *Progress in Aerospace Sciences, 45*, 50–79.
Forrester, A. I. J., Sobester, A., & Keane, A. J. (2007). Multi-fidelity optimization via surrogate modeling. *Proceedings of the Royal Society A, 463*, 3251–3269.
Forrester, A., Sobester, A., & Keane, A. (2008). *Engineering design via surrogate modeling: A practical guide*. Wiley.
Geisser, S. (1993). *Predictive inference*. Chapman and Hall.
Giunta, A. A. (1997). *Aircraft multidisciplinary design optimization using design of experiments theory and response surface modeling methods*. PhD thesis, Virginia Polytechnic Institute and State University.
Giunta, A. A., & Eldred, M. S. (2000). Implementation of a trust region model management strategy in the DAKOTA optimization toolkit. In *Proceedings of the symposium multidisciplinary analysis Optimization*, Long Beach, CA, AIAA-2000-4935.
Giunta, A. A., Wojtkiewicz, S. F., & Eldred, M. S. (2003). Overview of modern design of experiments methods for computational simulations. In *Aerospace sciences meeting exhibit*, paper AIAA 2003–0649.
Golub, G. H., & Van Loan, C. F. (1996). *Matrix computations* (3rd ed.). Johns Hopkins University Press.

Gorissen, D., Crombecq, K., Couckuyt, I., Dhaene, T., & Demeester, P. (2010a). A surrogate modeling and adaptive sampling toolbox for computer based design. *Journal of Machine Learning Research, 11*, 2051–2055.

Gorissen, D., Couckuyt, I., Laermans, E., & Dhaene, T. (2010b). Multiobjective global surrogate modeling, dealing with the 5-percent problem. *Engineering with Computers, 26*, 81–98.

Gramacy, R. B. (2021). *Surrogates. Gauusian process modeling, design, and optimization for the applied sciences* (Texts in statistical science). Chapman & Hall/CRC.

Gunn, S. R. (1998). *Support vector machines for classification and regression* (Technical report). School of Electronics and Computer Science, University of Southampton.

Hammersley, J. M. (1960). Monte-Carlo methods for solving multivariable problems. *Annals New York Academy of Sciences, 86*, 844–874.

Hanninen, I. (2012). *Optimization of a reflector antenna system, Whitepaper*. Computer Simulation Technology AG.

Haykin, S. (1998). *Neural networks: A comprehensive foundation* (2nd ed.). Prentice Hall.

Hong, J.-S., & Lancaster, M. (2001). *Microstrip filters for RF/Microwave applications*. Wiley.

Hosder, S., Watson, L. T., Grossman, B., Mason, W. H., & Kim, H. (2001). Polynomial response surface approximations for the multidisciplinary design optimization of a high speed civil transport. *Optical Engineering, 2*, 431–452.

Huang, L., & Gao, Z. (2012). Wing-body optimization based on multi-fidelity surrogate model. In *International Congress of the Aeronautical Sciences*, Brisbane, Australia.

Jacobs, J. P. (2012). Bayesian support vector regression with automatic relevance determination kernel for modeling of antenna input characteristics. *IEEE Transactions on Antennas and Propagation, 60*, 2114–2118.

Jacobs, J. P. (2016). Characterization by Gaussian processes of finite substrate size effects on gain patterns of microstrip antennas. *IET Microwaves, Antennas & Propagation, 10*(11), 1189–1195.

Jacobs, J. P., & Koziel, S. (2014). Cost-effective global surrogate modeling of planar microwave filters using multi-fidelity Bayesian support vector regression. *International Journal of RF and Microwave CAE, 24*, 11–17.

Jacobs, J. H., Etman, L. F. P., van Keulen, F., & Rooda, J. E. (2004). Framework for sequential approximate optimization. *Structural and Multidisciplinary Optimization, 27*, 384–400.

Jones, D., Schonlau, M., & Welch, W. (1998). Efficient global optimization of expensive black-box functions. *Journal of Global Optimization, 13*, 455–492.

Journel, A. G., & Huijbregts, C. J. (1981). *Mining geostatistics*. Academic.

Kabir, H., Wang, Y., Yu, M., & Zhang, Q. J. (2008). Neural network inverse modeling and applications to microwave filter design. *IEEE Transactions on Microwave Theory and Techniques, 56*, 867–879.

Kleijnen, J. P. C. (2009). Kriging metamodeling in simulation: A review. *European Journal of Operational Research, 192*(3), 707–716.

Koehler, J. R., & Owen, A. B. (1996). Computer experiments. In S. Ghosh & C. R. Rao (Eds.), *Handbook of statistics* (Vol. 13, pp. 261–308). Elsevier Science B.V.

Koziel, S. (2010a). Shape-preserving response prediction for microwave design optimization. *IEEE Transactions on Microwave Theory and Techniques, 58*, 2829–2837.

Koziel, S. (2010b). Shape-preserving response prediction for microwave circuit modeling. In *IEEE MTT-S International Microwave Symposium Digest*, Anaheim, CA, (pp. 1660–1663).

Koziel, S. (2012). Accurate low-cost microwave component models using shape-preserving response prediction. *International Journal of Numerical Modelling, 25*, 152–162.

Koziel, S., & Bandler, J. W. (2007a). Coarse and surrogate model assessment for engineering design optimization with space mapping, In *IEEE MTT-S International Microwave Symposium Digest*, Honolulu, HI, (pp. 107–110).

Koziel, S., & Bandler, J. W. (2007b). Space-mapping optimization with adaptive surrogate model. *IEEE Transactions on Microwave Theory and Techniques, 55*, 541–547.

Koziel, S., & Leifsson, L. (2012a). Knowledge-based airfoil shape optimization using space mapping. In *AIAA applied aerodynamics conference*.

Koziel, S., & Leifsson, L. (2012b). Multi-fidelity airfoil shape optimization with adaptive response prediction. In *AIAA/ISSMO multidisciplinary analysis and optimization conference*.

Koziel, S., & Leifsson, L. (Eds.). (2013a). *Surrogate-based modeling and optimization. Applications in engineering*. Springer.

Koziel, S., & Leifsson, L. (2013b). Multi-point response correction for reduced-cost EM-simulation-driven design of antenna structures. *Microwave and Optical Technology Letters, 55*(9), 2070–2074.

Koziel, S., & Ogurtsov, S. (2011). Rapid design optimization of antennas using space mapping and response surface approximation models. *International Journal of RF and Microwave CAE, 21*, 611–621.

Koziel, S., & Ogurtsov, S. (2012a). Model management for cost-efficient surrogate-based optimization of antennas using variable-fidelity electromagnetic simulations. *IET Microwaves, Antennas & Propagation, 6*, 1643–1650.

Koziel, S., & Ogurtsov, S. (2012b). Fast simulation-driven design of microwave structures using improved variable-fidelity optimization technique. *Engineering Optimization, 44*, 1007–1019.

Koziel, S., & Ogurtsov, S. (2014). *Antenna design by simulation-driven optimization*. Springer.

Koziel, S., & Ogurtsov, S. (2015). Phase-spacing optimization of linear microstrip antenna arrays using simulation-based surrogate superposition models. *International Journal of RF and Microwave Computer-Aided Engineering, 25*(6), 536–547.

Koziel, S., & Ogurtsov, S. (2019). *Simulation-based optimization of antenna arrays*. World Scientific.

Koziel, S., & Pietrenko-Dabrowska, A. (2020). *Performance-driven surrogate modeling of high-frequency structures*. Springer.

Koziel, S., & Pietrenko-Dabrowska, A. (Eds.). (2022). *Surrogate modeling for high-frequency design: Recent advances*. World Scientific.

Koziel, S., & Unnsteinsson, S. D. (2018). Expedited design closure of antennas by means of trust-region-based adaptive response scaling. *IEEE Antennas and Wireless Propagation Letters, 17*(6), 1099–1103.

Koziel, S., Bandler, J. W., & Madsen, K. (2006). A space mapping framework for engineering optimization: Theory and implementation. *IEEE Transactions on Microwave Theory and Techniques, 54*, 3721–3730.

Koziel, S., Cheng, Q. S., & Bandler, J. W. (2008a). Space mapping. *IEEE Microwave Magazine, 9*, 105–122.

Koziel, S., Bandler, J. W., & Madsen, K. (2008b). Quality assessment of coarse models and surrogates for space mapping optimization. *Optical Engineering, 9*, 375–391.

Koziel, S., Bandler, J. W., & Madsen, K. (2009). Space mapping with adaptive response correction for microwave design optimization. *IEEE Transactions on Microwave Theory and Techniques, 57*, 478–486.

Koziel, S., Bandler, J. W., & Cheng, Q. S. (2010a). Robust trust-region space-mapping algorithms for microwave design optimization. *IEEE Transactions on Microwave Theory and Techniques, 58*, 2166–2174.

Koziel, S., Cheng, Q. S., & Bandler, J. W. (2010b). Implicit space mapping with adaptive selection of preassigned parameters. *IET Microwaves, Antennas & Propagation, 4*, 361–373.

Koziel, S., Bandler, J. W., & Cheng, Q. S. (2010c). Adaptively constrained parameter extraction for robust space mapping optimization of microwave circuits. In *IEEE MTT-S International Microwave Symposium Digest* (pp. 205–208).

Koziel, S., Echeverría Ciaurri, D., & Leifsson, L. (2011a). Surrogate-based methods. In S. Koziel & X.-S. Yang (Eds.), *Computational optimization, methods and algorithms* (pp. 33–59). Springer-Verlag.

Koziel, S., Bandler, J. W., & Cheng, Q. S. (2011b). Tuning space mapping design framework exploiting reduced EM models. *IET Microwaves, Antennas & Propagation, 5*, 1219–1226.

Koziel, S., Ogurtsov, S., Couckuyt, I., & Dhaene, T. (2013a). Variable-fidelity electromagnetic simulations and co-kriging for accurate modeling of antennas. *IEEE Transactions on Antennas and Propagation, 61*, 1301–1308.

Koziel, S., Yang, X. S., & Zhang, Q. J. (Eds.). (2013b). *Simulation-driven design optimization and modeling for microwave engineering*. Imperial College Press.

Koziel, S., Bekasiewicz, A., Kurgan P., & Bandler, J. W. (2016). Rapid multi-objective de-sign optimisation of compact microwave couplers by means of physics-based surrogates. *IET Microwaves, Antennas & Propagation, 10*(5), 479-486.

Koziel, S., Calik, N., Mahouti, P., & Belen, M. A. (2021). Accurate modeling of antenna structures by means of domain confinement and pyramidal deep neural networks. *IEEE Transactions on Antennas and Propagation, 70*(3), 2174–2188.

Laurenceau, J., & Sagaut, P. (2008). Building efficient response surfaces of aerodynamic functions with kriging and cokriging. *AIAA Journal, 46*, 498–507.

Leary, S., Bhaskar, A., & Keane, A. (2003). Optimal orthogonal-array-based Latin hypercubes. *Journal of Applied Statistics, 30*, 585–598.

Lehmensiek, R., Meyer, P., & Muller, M. (2002). Adaptive sampling applied to multivariate, multiple output rational interpolation models with application to microwave circuits. *International Journal of RF and Microwave CAE, 12*(4), 332–340.

Leifsson, L., & Koziel, S. (2015a). *Simulation-driven aerodynamic design using variable-fidelity models*. Imperial College Press.

Leifsson, L., & Koziel, S. (2015b). Variable-resolution shape optimization: Low-fidelity model selection and scalability. *International Journal of Mathematical Modelling and Numerical Optimisation, 6*, 1–21.

Levin, D. (1998). The approximation power of moving least-squares. *Mathematics of Computation, 67*, 1517–1531.

Liang, Y., Lu, L., Xiao, Q., & Yan, S. (2020). Evaluating fast algorithms for convolutional neural networks on FPGAs. *IEEE Transactions on Computer-Aided Design of Integrated Circuits and Systems, 39*(4), 857–870.

Liu, J., Han, Z., & Song, W. (2012). *Comparison of infill sampling criteria in kriging-based aerodynamic optimization*. International Council of the Aeronautical Sciences.

Liu, B., Aliakbarian, H., Zhongkun, M., Vandenbosch, G. A. E., Gielen, G., & Excell, P. (2014a). An efficient method for antenna design optimization based on evolutionary computation and machine learning techniques. *IEEE Transactions on Antennas and Propagation, 62*, 7–18.

Liu, B., Zhang, Q., & Gielen, G. (2014b). A Gaussian process surrogate model assisted evolutionary algorithm for medium scale expensive black box optimization problems. *IEEE Transactions on Evolutionary Computation, 18*, 180–192.

Liu, Z., Yang, M., & Li, W. (2016). A sequential Latin hypercube sampling method for metamodeling. In L. Zhang, X. Song, & Y. Wu (Eds.), *Theory, methodology, tools and applications for modeling and simulation of complex systems* (Vol. 643, pp. 176–185). AsiaSim 2016, Comm. Computer Information Science.

Lophaven, S. N., Nielsen, H. B., & Søndergaard, J. (2002). *DACE: a Matlab kriging toolbox*. Technical University of Denmark.

Marheineke, N., Pinnau, R., & Reséndiz, E. (2012). Space mapping-focused control techniques for particle dispersions in fluids. *Optical Engineering, 13*, 101–120.

Matheron, G. (1963). Principles of geostatistics. *Economic Geology, 58*, 1246–1266.

McKay, M., Conover, W., & Beckman, R. (1979). A comparison of three methods for selecting values of input variables in the analysis of output from a computer code. *Technometrics, 21*, 239–245.

Meng, J., & Xia, L. (2007). Support-vector regression model for millimeter wave transition. *International Journal of Infrared and Millimeter Waves, 28*, 413–421.

Minsky, M. I., & Papert, S. A. (1969). *Perceptrons: An introduction to computational geometry*. The MIT Press.

Nocedal, J., & Wright, S. J. (2000). *Numerical optimization, Springer series in operations research*. Springer.

O'Hagan, A. (1978). Curve fitting and optimal design for predictions. *Journal of the Royal Statistical Society Series B, 40*, 1–42.

Palmer, K., & Tsui, K.-L. (2001). A minimum bias Latin hypercube design. *IIE Transactions, 33*, 793–808.

Pattipati, B., Sankavaram, C., & Pattipati, K. R. (2011). System identification and estimation framework for pivotal automotive battery management system characteristics. *IEEE Transactions on Systems, Man, and Cybernetics, Part C, 41*, 869–884.

Pérez, V. M., Renaud, J. E., & Watson, L. T. (2002). Interior point sequential approximate optimization methodology. *Proceedings of the Symposium multidisciplinary analysis optimization*, Atlanta, GA, AIAA-2002-5505.

Pietrenko-Dabrowska, A., & Koziel, S. (2020). Antenna modeling using variable-fidelity EM simulations and constrained co-kriging. *IEEE Access, 8*(1), 91048–91056.

Priess, M., Koziel, S., & Slawig, T. (2011). Surrogate-based optimization of climate model parameters using response correction. *Journal of Computer Science, 2*, 335–344.

Queipo, N. V., Haftka, R. T., Shyy, W., Goel, T., Vaidynathan, R., & Tucker, P. K. (2005). Surrogate based analysis and optimization. *Progress in Aerospace Sciences, 41*, 1–28.

Rasmussen, C. E., & Williams, C. K. I. (2006). *Gaussian processes for machine learning*. MIT Press.

Rayas-Sanchez, J. E. (2004). EM-based optimization of microwave circuits using artificial neural networks: The state-of-the-art. *IEEE Transactions on Microwave Theory and Techniques, 52*, 420–435.

Rayas-Sanchez, J. E. (2016). Power in simplicity with ASM: Tracing the aggressive space mapping algorithm over two decades of development and engineering applications. *IEEE Microwave Magazine, 17*(4), 64–76.

Redhe, M., & Nilsson, L. (2004). Optimization of the new Saab 9-3 exposed to impact load using a space mapping technique. *Structural and Multidisciplinary Optimization, 27*, 411–420.

Robinson, T. D., Eldred, M. S., Willcox, K. E., & Haimes, R. (2008). Surrogate-based optimization using multifidelity models with variable parameterization and corrected space mapping. *AIAA Journal, 46*, 2316–2326.

Rojo-Alvarez, J. L., Camps-Valls, G., Martinez-Ramon, M., Soria-Olivas, E., Navia-Vazquez, A., & Figueiras-Vidal, A. R. (2005). Support vector machines framework for linear signal processing. *Signal Processing, 85*, 2316–2326.

Roux, W. J., Stander, N., & Haftka, R. T. (1998). Response surface approximations for structural optimization. *International Journal for Numerical Methods in Engineering, 42*, 517–534.

Sans, M., Selga, J., Rodriguez, A., Bonache, J., Boria, V. E., & Martin, F. (2014). Design of planar wideband bandpass filters from specifications using a two-step aggressive space mapping (ASM) optimization algorithm. *IEEE Transactions on Microwave Theory and Techniques, 62*, 3341–3350.

Santner, T. J., Williams, B., & Notz, W. (2003). *The design and analysis of computer experiments*. Springer-Verlag.

Sasena, M., Parkinson, M., Goovaerts, P. (2002). Adaptive experimental design applied to an ergonomics testing procedure. In *ASME 2002 design Engineering technical conferences and computer and information in engineering conference*, Montreal, Canada.

Shen, Y., & Wang, J. (2008). An improved algebraic criterion for global exponential stability of recurrent neural networks with time-varying delays. *IEEE Transactions on Neural Networks, 19*(3), 528–531.

Simpson, T. W., Pelplinski, J. D., Koch, P. N., & Allen, J. K. (2001). Metamodels for computer-based engineering design: Survey and recommendations. *Engineering with Computers, 17*, 129–150.

Smola, A. J., & Schölkopf, B. (2004). A tutorial on support vector regression. *Statistics and Computing, 14*, 199–222.

Sobieszczanski-Sobieski, J., & Haftka, R. T. (1997). Multidisciplinary aerospace design optimization: Survey of recent developments. *Structural Optimization, 14*, 1–23.
Søndergaard, J. (2003). *Optimization using surrogate models – By the space mapping technique*. PhD thesis, Informatics and Mathematical Modelling, Technical University of Denmark, Lyngby.
Szegedy, C., Liu, W., Jia, Y., Sermanet, P., Reed, S. E., Anguelov, D., Erhan, D., Vanhoucke, V., & Rabinovich, A. (2015). Going deeper with convolutions. In *2015 IEEE Conference on Computer Vision and Pattern Recognition (CVPR)* (pp. 1–9).
Toal, D. J. J., & Keane, A. J. (2011). Efficient multipoint aerodynamic design optimization via Cokriging. *Journal of Aircraft, 48*, 1685–1695.
Tong, C. (2006). Refinement strategies for stratified sampling algorithms. *Reliability Engineering and System Safety, 91*(10–11), 1257–1265.
Toropov, V. V., Filatov, A. A., & Polynkin, A. A. (1993). Multiparameter structural optimization using FEM and multipoint explicit approximations. *Structural Optimization, 6*, 7–14.
Tu, S., Cheng, Q. S., Zhang, Y., Bandler, J. W., & Nikolova, N. K. (2013). Space mapping optimization of handset antennas exploiting thin-wire models. *IEEE Transactions on Antennas and Propagation, 61*, 3797–3807.
Wang, G. G. (2003). Adaptive response surface algorithm using inherited Latin hypercube design points. *Journal of Mechanical Design, 125*(2), 210–220.
Wild, S. M., Regis, R. G., & Shoemaker, C. A. (2008). ORBIT: Optimization by radial basis function interpolation in trust-regions. *SIAM Journal on Scientific Computing, 30*, 3197–3219.
Xia, L., Xu, R. M., & Yan, B. (2007). LTCC interconnect modeling by support vector regression. *Progress in Electromagnetics Research, 69*, 67–75.
Ye, K. Q. (1998). Orthogonal column Latin hypercubes and their application in computer experiments. *Journal of the American Statistical Association, 93*, 1430–1439.
Yelten, M. B., Zhu, T., Koziel, S., Franzon, P. D., & Steer, M. B. (2012). Demystifying surrogate modeling for circuits and systems. *IEEE Circuits and Systems Magazine, 12*, 45–63.
Zhang, K., & Han, Z. (2013). Support vector regression-based multidisciplinary design optimization in aircraft conceptual design. *AIAA Aerospace Sciences Meeting*, AIAA paper 2013-1160.

Chapter 5
Basics of Response Feature Technology

Chapters 2, 3, and 4 provided fundamentals of simulation-driven high-frequency design with the emphasis on numerical modeling and optimization. In particular, in Chap. 2, a number of EM-driven tasks have been formulated, such as parametric optimization, multi-objective design, uncertainty quantification, as well as general-purpose surrogate modeling. Chapter 3 outlined a range of optimization methodologies, including gradient-based and derivative-free algorithms, as well as nature-inspired metaheuristic procedures. The basics of surrogate modeling and surrogate-assisted optimization have been covered in Chap. 4. Along with the discussion of the aforementioned methods, the major challenges of simulation-driven design have been outlined, the most important of which is a high computational cost associated with massive EM simulations of the system at hand (Koziel & Ogurtsov, 2019; Haq & Koziel, 2018; Hassan et al., 2015; Li et al., 2019; Jia & Lu, 2019; Tang et al., 2018). Consequently, the majority of the conventional optimization and modeling procedures as well as statistical design methods cannot be directly applied to handle full-wave models. Surrogate-assisted techniques can be employed to mitigate these problems to a certain extent, as shown in Chap. 4 (Koziel et al., 2006; Toropov et al., 1993; Queipo et al., 2005). Notwithstanding, a typically high nonlinearity of frequency characteristics of antennas and microwave components creates additional problems, e.g., the necessity of using global optimization strategies for parameter tuning (Koziel & Pietrenko-Dabrowska, 2021) or insufficiency of standard modeling methods when constructing fast replacement models (Koziel et al., 2021). The response feature technology, the main topic of this book, offers some unique alternatives when it comes to handling EM-driven design tasks. The benefits of this technology come in different forms, either by expediting the optimization procedures (Koziel, 2015), by simplifying the problem at hand (Pietrenko-Dabrowska & Koziel, 2020), or by regularizing the functional landscape to be handled by numerical algorithms (Koziel & Bandler, 2015). This chapter gives the first exposition of the response feature approach. More specifically, it introduces the concept of characteristic points, discusses their properties, and demonstrates the benefits of

A. Pietrenko-Dabrowska, S. Koziel, *Response Feature Technology for High-Frequency Electronics. Optimization, Modeling, and Design Automation*,
https://doi.org/10.1007/978-3-031-43845-5_5

applying them for design purposes, here, local gradient-based parameter tuning. The remaining parts of the book will cover utilization of response features in different contexts, including accelerated local and global search, multi-objective design, as well as low-cost statistical analysis and surrogate model rendition.

5.1 Case Study: Antenna Parameter Tuning

In this section, we introduce the concept of response features in the context of parameter tuning of antenna structures (Koziel, 2015). The focus is on feature point definition and selection, their relations to designable parameters of the problem at hand, as well as potential advantages of using this approach, especially in terms of accelerating design optimization procedures.

5.1.1 Problem Statement

Here, we consider the antenna parameter tuning problem, formulated as in Chap. 2 (Sect. 2.2). We have (Koziel & Ogurtsov, 2019)

$$\boldsymbol{x}^* = \arg\min_{\boldsymbol{x}\in X_f} U(\boldsymbol{x}) \tag{5.1}$$

where $U(\boldsymbol{x})$ is a scalar objective function, whereas $\boldsymbol{x} \in X_f \subseteq R^n$. In (2.1), X_f is the feasible space, which is defined as a set of parameter vectors that satisfy the assumed design constraints; $\boldsymbol{x}^*$ is the design optimum to be found.

The objective function U is evaluated based on the antenna responses obtained through high-fidelity EM analysis, here, represented as $\boldsymbol{R}_f(\boldsymbol{x})$. The analytical formulation of U is problem-dependent (cf. Table 5.1). The fundamental issue is the high computational cost of solving (5.1). One of the purposes of the response feature technology is to reduce this cost, as elaborated on in the remaining parts of this chapter.

Table 5.1 Planar inverted-F antenna—design optimization cost for Case 1

Algorithm	Algorithm component	Number of model evaluations	CPU Cost Absolute	CPU Cost Relative to $\boldsymbol{R}_f$
Direct $\boldsymbol{R}_f$ optimization (pattern search)	Evaluating $\boldsymbol{R}_f$	$205 \times \boldsymbol{R}_f$	2665 min	205.0
Space mapping	Evaluating $\boldsymbol{R}_c$	$345 \times \boldsymbol{R}_c$	288 min	22.2
	Evaluating $\boldsymbol{R}_f$	$5 \times \boldsymbol{R}_f$	65 min	5.0
	Total cost	N/A	353 min	27.2
Feature-based optimization	Evaluating $\boldsymbol{R}_c$	$50 \times \boldsymbol{R}_c$	42 min	4.0
	Evaluating $\boldsymbol{R}_f$	$8 \times \boldsymbol{R}_f$	104 min	9.0
	Total cost	N/A	146 min	13.0

5.1.2 Response Features

The overall goal is to use the response feature technology to reduce the computational cost of the optimization process (Koziel, 2015). The technique is founded on the idea that instead of handling complete (often highly nonlinear) antenna responses, only selected response features (e.g., frequencies corresponding to $|S_{11}|$ at specific levels, e.g., -10 dB) are processed. The dependence of those features on the designable parameters is usually less nonlinear than for the entire responses, which speeds up the optimization process. The main prerequisite is that the response features (also referred to as characteristic points) allow us to account for the assumed design objectives, i.e., they can be used to quantify the quality of the design. In other words, the design problem formulated using response features is equivalent to that formulated in a conventional manner.

Figure 5.1 shows a family of reflection responses of the antenna considered in Sect. 5.1.4, evaluated along a certain line segment in the design space, $\boldsymbol{x} = t\boldsymbol{x}_a + (1 - t)\boldsymbol{x}_b$, $0 \leq t \leq 1$, where $\boldsymbol{x}_a$ and $\boldsymbol{x}_b$ are arbitrarily selected vectors, as well as the plots of the corresponding features. Figure 5.1b shows $|S_{11}|$ versus the line segment parameter t for various frequencies, here, 1.9 GHz, 2.0 GHz, and 2.1 GHz. When creating a conventional response surface approximation model, the response at each frequency is modelled separately. As indicated in the plots, the responses are highly nonlinear.

At the same time, the behavior of the response features (here, the three points corresponding to the -15 dB levels of $|S_{11}|$ and $|S_{11}|$ at the center frequency) is less nonlinear and, consequently, easier to handle. In the optimization context, the estimation of the next candidate design is usually obtained from a local model established in the vicinity of the current design (e.g., first-order Taylor expansion, etc.) so that more linear behavior of the optimized function results in faster convergence of the optimization algorithm.

We will use the symbols $f_k(\boldsymbol{x})$ and $l_k(\boldsymbol{x})$ to denote the frequency and level of the kth feature, $k = 1, \ldots, p$, where p is the number of features for a given problem. The response features should be selected in accordance to given performance specifications so that the original formulation of the design problem (5.1) can be replaced by an equivalent problem formulated for the features

$$\boldsymbol{x}^* = \arg\min_{\boldsymbol{x}} U_F\left(\boldsymbol{R}_{F.f}(\boldsymbol{x})\right) \tag{5.2}$$

where $\boldsymbol{R}_{F.f} \in R^{2p}$ denotes the feature-based model (as a function of the design variables $\boldsymbol{x}$), whereas U_F is the objective function at the feature level. Using the previously introduced notation, we have $\boldsymbol{R}_{F.f}(\boldsymbol{x}) = [f_1(\boldsymbol{x})\, f_2(\boldsymbol{x}) \ldots f_p(\boldsymbol{x})\; l_1(\boldsymbol{x})\; l_2(\boldsymbol{x}) \ldots l_p(\boldsymbol{x})]^T$.

Figure 5.2 shows a typical situation for a narrow-band antenna. In this specific example, the goal is maximization of the fractional bandwidth at -15 dB level. For illustration, the three response features are utilized corresponding to the above response level and to the response minimum (although only the first two features are used in the optimization process).

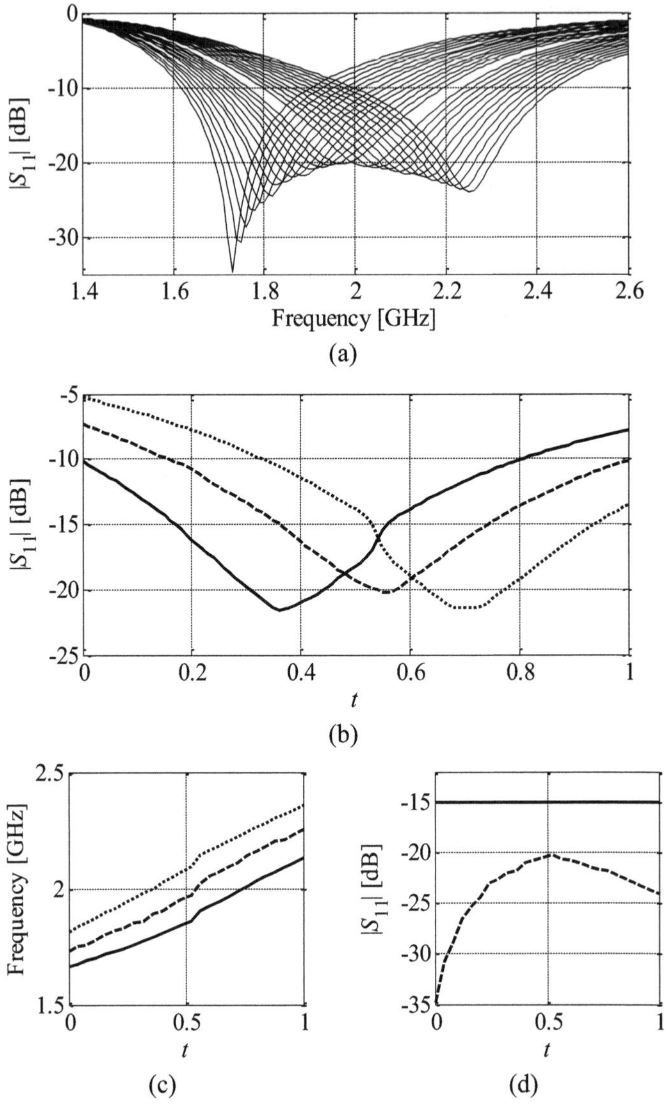

Fig. 5.1 Response features: (**a**) family of reflection responses of the antenna structure considered in Sect. 5.1.4, evaluated along a certain line segment in the design space, example features corresponding to −15 dB levels (o) and center frequency (□); (**b**) response variability at selected frequencies, 1.9 GHz (—), 2.0 GHz (- - -), and 2.1 GHz (. . .), indicating their high nonlinearity; (**c**) variability of the response features (frequency components), here, corresponding to −15 dB levels as well as the center frequency; (**d**) variability of the level components of the response features of (**c**) (Koziel, 2015)

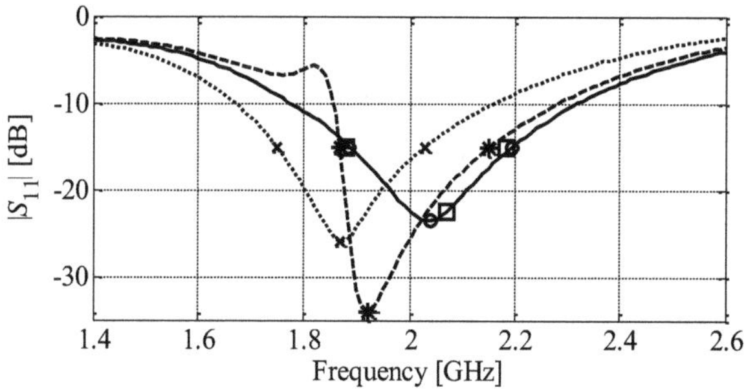

Fig. 5.2 Feature prediction using estimated feature gradients: high-fidelity model at the reference design $\boldsymbol{x}$ (. . . .) and corresponding feature points (×), high-fidelity model at the perturbed design $\boldsymbol{x} + d\boldsymbol{x}$ (—) and the corresponding feature points (o), linear expansion model constructed at the frequency-based response level and evaluated at $\boldsymbol{x} + d\boldsymbol{x}$ (- - -) together with the corresponding feature points (*), and the high-fidelity feature points predicted using feature gradients (□). The feature points predicted using conventional linear expansion (at the frequency-based response level) are inaccurate, whereas the prediction using feature gradients is more reliable (Koziel, 2015)

It can be observed that the prediction of the feature location obtained using the feature model gradients $\nabla f_k(\boldsymbol{x})$ and $\nabla l_k(\boldsymbol{x})$ (i.e., $f_k(\boldsymbol{x} + d\boldsymbol{x}) \approx f_k(\boldsymbol{x}) + \nabla f_k(\boldsymbol{x})d\boldsymbol{x}$, and $l_k(\boldsymbol{x} + d\boldsymbol{x}) \approx l_k(\boldsymbol{x}) + \nabla l_k(\boldsymbol{x})d\boldsymbol{x}$) for a specific search step size $d\boldsymbol{x}$, versus the actual feature location verified by EM simulation, is good. This is not the case when the same search step is utilized for the first-order Taylor model constructed for the frequency-based response, i.e., $\boldsymbol{R}_f(\boldsymbol{x} + d\boldsymbol{x}) \approx \boldsymbol{R}_f(\boldsymbol{x}) + \boldsymbol{J}_{\boldsymbol{R}f}(\boldsymbol{x})d\boldsymbol{x}$.

The reason, again, is a less nonlinear behavior of the response features with respect to the location in the design space. Perhaps an even more important reason is that modeling of frequency coordinates of the features, i.e., $f_k(\boldsymbol{x})$, directly describes relocation of important parameters (such as bandwidth, center frequency), whereas conventional description does this indirectly through the modeling of the entire responses.

5.1.3 Feature-Based Antenna Optimization

The algorithm considered here (Koziel, 2015) is formulated as an iterative process generating a series $\boldsymbol{x}^{(i)}$ of approximations to $\boldsymbol{x}^*$

$$\boldsymbol{x}^{(i+1)} = \arg\min_{\boldsymbol{x}} U_F(\boldsymbol{R}_l^{(i)}(\boldsymbol{x})) \tag{5.3}$$

where $\boldsymbol{R}_l^{(i)}$ is a linear model of the response features set up at the current design $\boldsymbol{x}^{(i)}$ using finite differentiation; recall (cf. (5.2) and the text below it) that U_F is the objective function encoding the design specifications at the feature level.

The model $\boldsymbol{R}_l^{(i)}$ is defined as

$$\boldsymbol{R}_l^{(i)}(\boldsymbol{x}) = \begin{bmatrix} \begin{bmatrix} f_1(\boldsymbol{x}^{(i)}) \\ \cdots \\ f_p(\boldsymbol{x}^{(i)}) \end{bmatrix} + \begin{bmatrix} \dfrac{f_1(\boldsymbol{x}_{d.1}^{(i)}) - f_1(\boldsymbol{x}^{(i)})}{d_1^{(i)}} \\ \cdots \\ \dfrac{f_p(\boldsymbol{x}_{d.p}^{(i)}) - f_p(\boldsymbol{x}^{(i)})}{d_p^{(i)}} \end{bmatrix} \circ (\boldsymbol{x} - \boldsymbol{x}^{(i)}) \\ \begin{bmatrix} l_1(\boldsymbol{x}^{(i)}) \\ \cdots \\ l_p(\boldsymbol{x}^{(i)}) \end{bmatrix} + \begin{bmatrix} \dfrac{l_1(\boldsymbol{x}_{d.1}^{(i)}) - l_1(\boldsymbol{x}^{(i)})}{d_1^{(i)}} \\ \cdots \\ \dfrac{l_p(\boldsymbol{x}_{d.p}^{(i)}) - l_p(\boldsymbol{x}^{(i)})}{d_p^{(i)}} \end{bmatrix} \circ (\boldsymbol{x} - \boldsymbol{x}^{(i)}) \end{bmatrix} \tag{5.4}$$

where $\boldsymbol{x}^{(i)} = [x_1^{(i)}\ x_2^{(i)}\ \ldots\ x_n^{(i)}]^T$ is a current design, whereas $\boldsymbol{x}_{d.k}^{(i)} = [x_1^{(i)} \ldots x_k^{(i)} + d_k^{(i)} \ldots x_n^{(i)}]^T$ are the perturbed designs. Here, $\boldsymbol{d}^{(i)} = [d_1^{(i)}\ d_2^{(i)} \ldots d_n^{(i)}]^T$ is the perturbation size which may, in general, be iteration dependent. The symbol $\circ$ denotes component-wise multiplication. Note that frequency and level coordinates of the feature points are modeled independently. To ensure convergence, the algorithm is embedded in the trust region framework (Conn et al., 2000), i.e., (5.3) is replaced by

$$\boldsymbol{x}^{(i+1)} = \arg \min_{\left\|\boldsymbol{x} - \boldsymbol{x}^{(i)}\right\| \le \delta^{(i)}} U_F\left(\boldsymbol{R}_l^{(i)}(\boldsymbol{x})\right) \tag{5.5}$$

where $\boldsymbol{R}_l^{(i)}$ is optimized within the trust region defined as $||\boldsymbol{x} - \boldsymbol{x}^{(i)}|| \le \delta^{(i)}$; $\delta^{(i)}$ is the trust region size updated using conventional rules (Conn et al., 2000). The new design $\boldsymbol{x}^{(i+1)}$ produced by (5.5) is only accepted if it leads to improvement of $\boldsymbol{R}_{F.f}(\boldsymbol{x})$, i.e., $\boldsymbol{R}_{F.f}(\boldsymbol{x}^{(i+1)}) < \boldsymbol{R}_{F.f}(\boldsymbol{x}^{(i)})$.

In order to speed up the optimization process, the gradients of the features are estimated (cf. (5.4)) using low-fidelity model evaluations rather than the high-fidelity ones. In particular, the left-hand side terms $f_k(\boldsymbol{x}^{(i)})$ and $l_k(\boldsymbol{x}^{(i)})$ come from the high-fidelity model responses, whereas the finite differentiation terms $[f_k(\boldsymbol{x}_{d.k}^{(i)}) - f_k(\boldsymbol{x}^{(i)})]$ and $[l_k(\boldsymbol{x}_{d.k}^{(i)}) - l_k(\boldsymbol{x}^{(i)})]$ come from the low-fidelity model. This is crucial to obtain considerable computational savings: gradient estimation using a high-fidelity model requires n evaluations of $\boldsymbol{R}_f$, whereas the same estimation realized with the low-fidelity model requires n evaluations of $\boldsymbol{R}_c$, which normally corresponds (CPU time-wise) to much less than a single evaluation of $\boldsymbol{R}_f$. Although the low- and high-fidelity models are normally misaligned (see Fig. 5.3), the gradient estimations of the

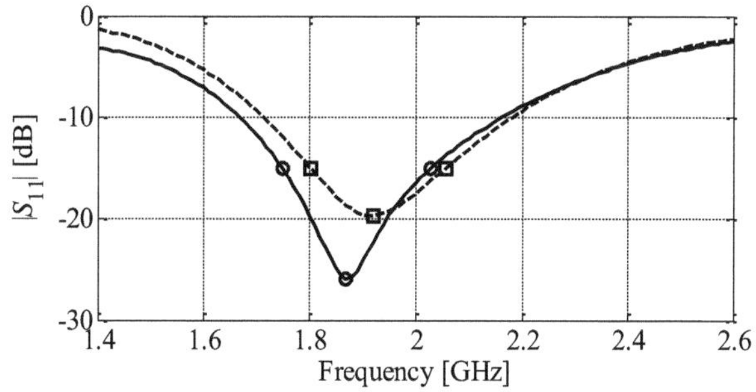

Fig. 5.3 High- (—) and low-fidelity (- - -) model responses at a selected design, as well as the corresponding feature points (−15 dB levels and center frequency), (o) and (□), respectively (Koziel, 2015)

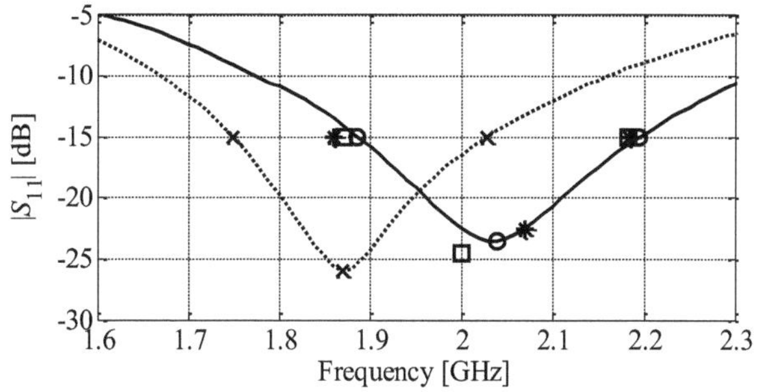

Fig. 5.4 Feature prediction using estimated low-fidelity feature gradients: high-fidelity model at the reference design $\boldsymbol{x}$ (. . . .) and corresponding feature points (×), high-fidelity model at the perturbed design $\boldsymbol{x} + d\boldsymbol{x}$ (—) and the corresponding feature points (o), feature points predicted using feature gradients estimated using high-fidelity model responses (□), feature points predicted using feature gradients estimated using low-fidelity model responses (*). It can be observed that the feature point prediction obtained using feature gradients estimated with a low-fidelity model is quite reliable and comparable to that obtained using a high-fidelity model, particularly for −15 dB levels (Koziel, 2015)

features are quite similar as illustrated in Fig. 5.4. In particular, the feature point prediction obtained using the low-fidelity model is comparable to that obtained using the high-fidelity one.

5.1.4 Illustration Example 1: Planar Inverted-F Antenna (PIFA)

Consider a PIFA antenna (Volakis, 2007) shown in Fig. 5.5. The design variables are $\boldsymbol{x} = [v_0\ v_1\ v_2\ v_3\ v_4\ v_5\ v_6]^T$. The initial design is $\boldsymbol{x}^{(0)} = [2\ 8\ 5\ 2\ 10\ 25\ 45]^T$ mm. Fixed parameters are $[u_0\ u_1\ u_2\ u_3\ u_4\ u_5\ u_6\ u_7\ w_0\ r_0]^T = [6.15\ 50.0\ 15.0\ 10.5\ 29.35\ 11.65\ 5.0\ 1.0\ 0.5]^T$ mm; 0.508 mm substrate (Fig. 5.5a on the right) and the box (on the left) are of Rogers TMM4 and TMM 6. The high-fidelity model $\boldsymbol{R}_f$ is evaluated in CST Microwave Studio (CST, 2013) (~650,000 mesh cells, simulation time 13 min). The low-fidelity model $\boldsymbol{R}_c$ is also evaluated in CST (~22,000 mesh cells, simulation time 50 s).

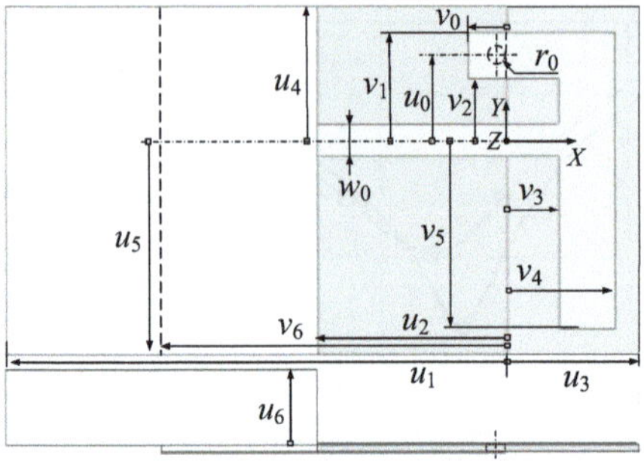

(a)

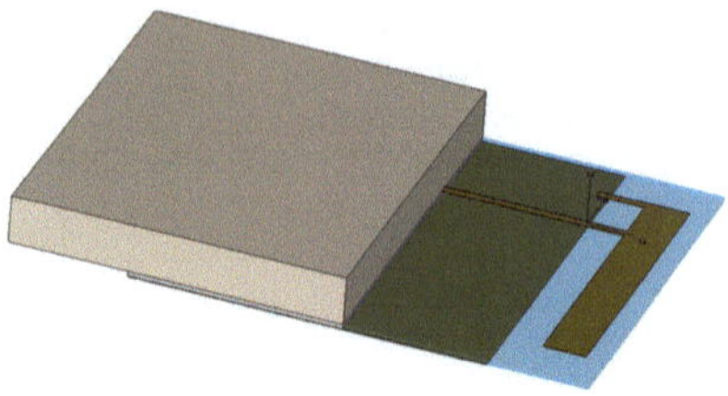

(b)

Fig. 5.5 PIFA geometry: (**a**) top and side view, substrate shown transparent; (**b**) perspective view

The antenna was optimized using the methodology of Sect. 5.1.3, assuming two scenarios: (i) obtaining as wide fractional bandwidth as possible at $|S_{11}| = -15$ dB (symmetrically) around 2.0 GHz (Case 1) and (ii) minimizing $|S_{11}|$ at the center frequency of 2.0 GHz (Case 2).

For the first scenario, the final design $\boldsymbol{x}^* = [2.028\ 8.226\ 6.587\ 2.618\ 9.134\ 22.87\ 45.24]^T$ mm was obtained in 6 iterations of the algorithm (5.2), with the total cost corresponding to about 11 evaluations of the high-fidelity model (8 × $\boldsymbol{R}_f$ and 50 × $\boldsymbol{R}_c$). Figure 5.6 shows the high-fidelity model responses at the initial and the final designs, as well as the evolution of the fractional bandwidth and the convergence plot. It should be noted that the considered approach is also efficient in terms of low-fidelity model evaluations, which is important when the time evaluation ratio between the $\boldsymbol{R}_f$ and $\boldsymbol{R}_c$ is low as in the present example.

For the second design scenario, the final design $\boldsymbol{x}^* = [2.116\ 8.186\ 6.454\ 2.341\ 8.8123\ 22.72\ 45.27]^T$ mm (Fig. 5.7) was obtained in seven iterations at the total cost of about 13 evaluations of $\boldsymbol{R}_f$ (9 × $\boldsymbol{R}_f$ and 58 × $\boldsymbol{R}_c$).

For the sake of comparison, the antenna was also optimized using the two benchmark methods: (i) space mapping (SM) (Koziel et al., 2008) and (ii) the pattern search algorithm (Kolda et al., 2003). The SM algorithm utilized here exploits the low-fidelity model $\boldsymbol{R}_c$ as the underlying coarse model and two types of model correction, specifically, frequency scaling (Koziel & Ogurtsov, 2014) and additive response correction (Koziel & Leifsson, 2012). The SM surrogate is reset at every iteration using the high-fidelity model data from the most current design; it is subsequently re-optimized using pattern search. The pattern search algorithm used in both direct search and space mapping procedure is based on the implementation described in (Koziel, 2010).

The final designs obtained using all three methods are comparable; however, the design costs differ considerably. The design cost breakdown for both cases is shown in Tables 5.1 and 5.2, respectively. It can be observed that direct optimization of the high-fidelity model is clearly the most expensive approach (205 × $\boldsymbol{R}_f$ and 141 × $\boldsymbol{R}_f$ for Case 1 and 2, respectively). Space mapping offers significant reduction of the design cost (to less than 30 equivalent evaluations of the high-fidelity model). However, due

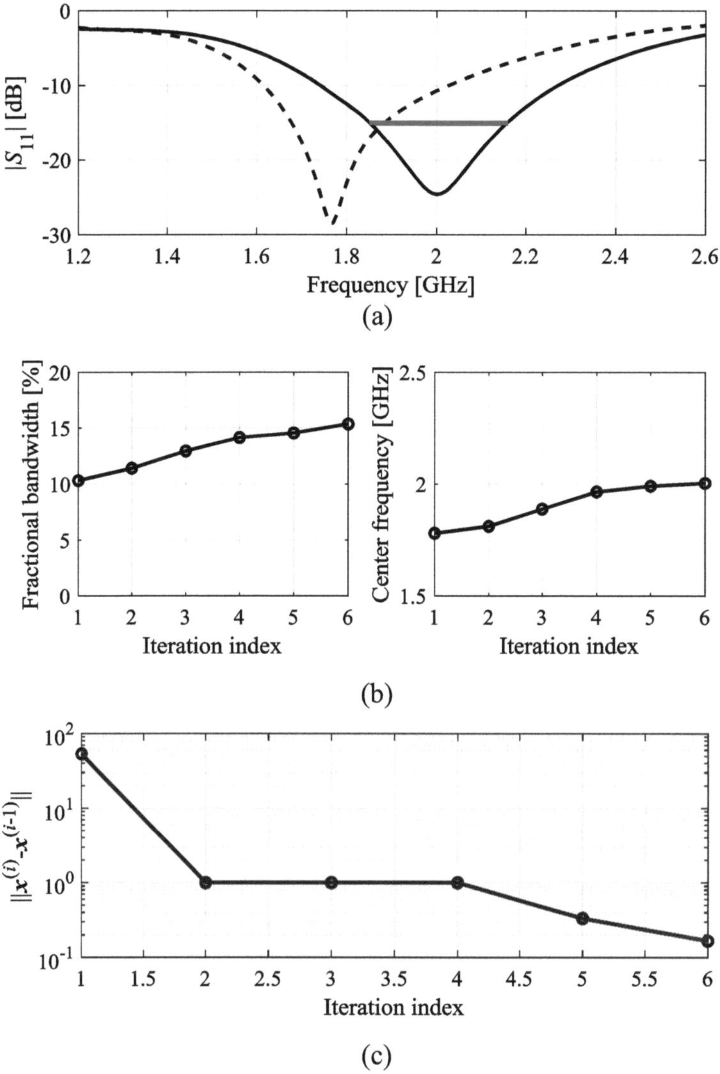

Fig. 5.6 PIFA optimization (Case 1): (**a**) initial (- - -) and optimized (—) response (fractional bandwidth around 2 GHz of 15 percent); (**b**) evolution of the fractional bandwidth and the center frequency vs. iteration index; (**c**) convergence plot

to the time evaluation ratio between the high- and low-fidelity model (about 16:1), the computational overhead related to surrogate model optimization (around 350 × $\boldsymbol{R}_c$) becomes the major contributor (~80 percent) to the total cost.

The feature-based optimization algorithm is more efficient with this respect (the low-fidelity model evaluation corresponds to only about 30 percent of the total cost), which results in substantial savings compared to SM even though the number of algorithm iterations is larger.

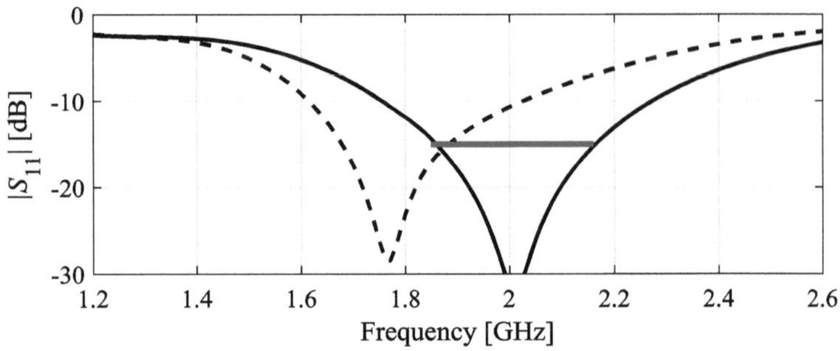

Fig. 5.7 PIFA optimization (Case 2): initial (- - -) and optimized (—) response ($|S_{11}| = -32$ dB at the center frequency). The horizontal line marks the design specifications

Table 5.2 Planar inverted-F antenna—design optimization cost for Case 2

Algorithm	Algorithm component	Number of model evaluations	CPU Cost	
			Absolute	Relative to $\boldsymbol{R}_f$
Direct $\boldsymbol{R}_f$ optimization (pattern search)	Evaluating $\boldsymbol{R}_f$	141 × $\boldsymbol{R}_f$	1833 min	141.0
Space mapping	Evaluating $\boldsymbol{R}_c$	350 × $\boldsymbol{R}_c$	292 min	22.4
	Evaluating $\boldsymbol{R}_f$	6 × $\boldsymbol{R}_f$	78 min	6.0
	Total cost	N/A	370 min	28.4
Feature-based optimization	Evaluating $\boldsymbol{R}_c$	58 × $\boldsymbol{R}_c$	48 min	3.7
	Evaluating $\boldsymbol{R}_f$	9 × $\boldsymbol{R}_f$	117 min	9.0
	Total cost	N/A	165 min	12.7

5.1.5 Single-Brick Dielectric Resonator Antenna

Consider a single-brick dielectric resonator antenna (DRA) shown in Fig. 5.8 (Petosa, 2007). The rectangular dielectric resonator (DR) is excited at the $TE_{\delta 11}$ mode with a 50 ohm microstrip through a slot made in the metal ground plane. Substrate is 0.5-mm-thick RO4003C ($\varepsilon_r = 3.38$) (RO4000, 2010) of infinite lateral extends. Metallization of the ground and the microstrip trace (the width w_0 of 1.15 mm) is with 0.05-mm-thick copper. DR-relative permittivity and loss tangent are 10 and 1e-4, respectively. The design variables are $\boldsymbol{x} = [a_x\ a_y\ a_z\ a_c\ u_s\ w_s\ y_s]^T$. The high-fidelity model $\boldsymbol{R}_f$ is simulated using the CST MWS transient solver (~600,000 mesh cells, evaluation time ~ 15 min). The low-fidelity model $\boldsymbol{R}_c$ is also evaluated in CST MWS (~37,000 mesh cells, evaluation time ~ 30 s).

The design objective is to obtain $|S_{11}| \le -10$ dB at least over 8% of the bandwidth (5.28 to 5.72 GHz) centered at 5.5 GHz. Design constraints imposed at the DRA radiation are the following: (i) realized gain is to be not less than 3 dB for the zero zenith angle, and (ii) realized gain of back radiation is to be less than −10 dB. For the sake of simplicity, the gain constraints are handled using penalty functions.

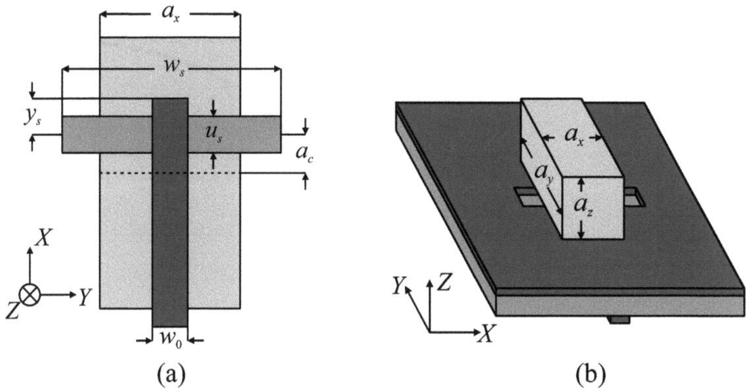

Fig. 5.8 Single-brick DRA: (**a**) bottom view, (**b**) 3D view

The initial design is $\boldsymbol{x}^{(0)} = [8.0\ 14.0\ 6.0\ 1.0\ 2.0\ 9.0\ 8.0]^T$ mm. The final design obtained using the feature-based approach is $\boldsymbol{x}^* = [9.57\ 14.56\ 7.15\ 0.59\ 2.04\ 8.33\ 7.90]^T$ mm. Figure 5.9 shows the low- and high-fidelity model $|S_{11}|$ versus frequency at the initial design and the high-fidelity model response at the final design. Figure 5.10 shows the realized gain response at the final design for the selected zenith angle in the E-plane (YOZ) versus frequency.

Similarly, as for the previous example, the DRA antenna was optimized directly using pattern search (Koziel, 2010), as well as space mapping involving frequency scaling and additive response correction (Koziel & Ogurtsov, 2011) (cf. Table 5.3).

The results (cf. Table 5.3) are in good agreement with those provided in Sect. 5.1.4. In particular, although the discussed approach requires more algorithm iterations than space mapping, its overall cost is considerably lower because of much smaller number of low-fidelity model evaluations.

5.1.6 Discussion

Numerical results provided in Sects. 5.1.4 and 5.1.5 demonstrate the ability of the response feature technique to realize low-cost design optimization of antenna structures. It should be emphasized that considerable savings are obtained not only in terms of the number of high-fidelity model evaluations but also the low-fidelity ones. Utilization of variable-resolution models for sensitivity estimation, as well as other methods of expediting feature-based optimization procedures, will be discussed in the later chapters of this book.

A comparison with a representative physics-based surrogate-assisted optimization approach (here, space mapping) indicates that even though the number of algorithm iterations (and, consequently, high-fidelity model simulations) is larger for feature-based technique than for SM, the number of low-fidelity model

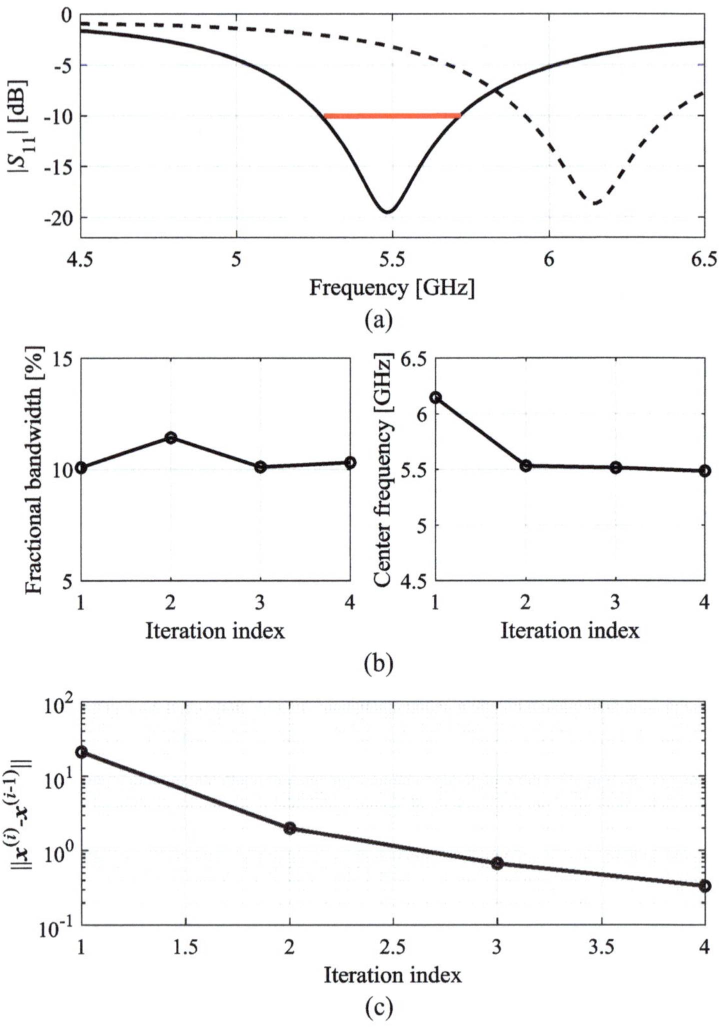

Fig. 5.9 DRA optimization: (**a**) initial (- - -) and optimized (—) response (fractional bandwidth around 2 GHz of 15 percent); the horizontal line marks the design specifications; (**b**) evolution of the fractional bandwidth and the center frequency vs. iteration index; (**c**) convergence plot

evaluations is considerably larger for the latter (contributing to around 80% of the total design cost). Due to being efficient in this respect, feature-based optimization allows for using finer discretization for low-fidelity models, which would result in better accuracy and, consequently, improved robustness of the design process.

Another aspect which is worth mentioning is a good scalability of the feature-based optimization: the cost of each algorithm iteration grows linearly with the number n of design variables, whereas conventional optimization of the surrogate model utilized in other algorithms (typically, pattern search) scale, more or less, as

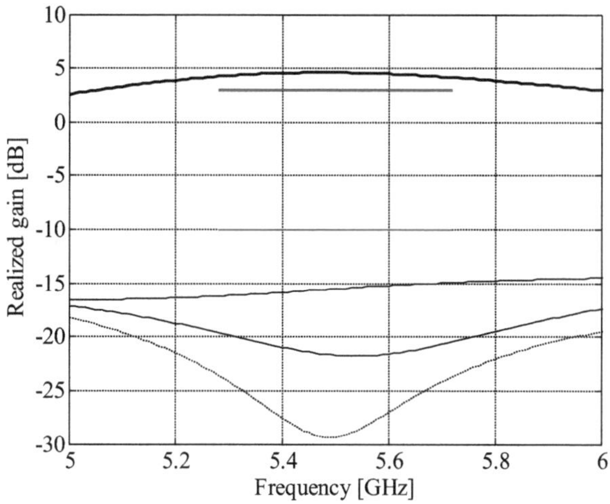

Fig. 5.10 Realized gain of the DRA at the final design in the E-plane (YOZ): for zenith angle of 0° () and back radiation, zenith angles of 135° (positive *Y*-direction, —), 180° (- - -), and 135° (negative *Y*-direction, ····). Design constraints over the impedance bandwidth are shown with the upper horizontal line of 3 dB level and lower horizontal line of −10 dB level

Table 5.3 Dielectric resonator antenna—design optimization cost

Algorithm	Algorithm component	Number of model evaluations	CPU Cost Absolute	CPU Cost Relative to $\boldsymbol{R}_f$
Direct $\boldsymbol{R}_f$ optimization (pattern search)	Evaluating $\boldsymbol{R}_f$	$172 \times \boldsymbol{R}_f$	2580 min	172.0
Space mapping	Evaluating $\boldsymbol{R}_c$	$180 \times \boldsymbol{R}_c$	90 min	6.0
	Evaluating $\boldsymbol{R}_f$	$3 \times \boldsymbol{R}_f$	45 min	3.0
	Total cost	N/A	135 min	9.0
Feature-based optimization	Evaluating $\boldsymbol{R}_c$	$26 \times \boldsymbol{R}_c$	13 min	0.9
	Evaluating $\boldsymbol{R}_f$	$5 \times \boldsymbol{R}_f$	75 min	5.0
	Total cost	N/A	88 min	5.9

n^2. This makes the feature-based approach potentially attractive for higher-dimensional problems.

The specific design cases considered in this section concern narrow-band antennas; however, the technique is general. Section 5.2, as well as the remaining parts of this book, will provide an exposition of the response feature technology for handling EM-driven tasks pertinent to a variety of antenna and microwave components.

5.2 Case Study: Feature-Based Optimization of Dual-Band Coupler

This section illustrates utilization of response features in parameter tuning of a dual-band branch-line coupler (Koziel & Bekasiewicz, 2015). Our main purpose at this point is to demonstrate again the benefits of reformulating the design task in terms of suitably defined characteristic points of the system outputs, especially when it comes to "flattening" the functional landscape of the objective function being optimized.

5.2.1 Compact Dual-Band Microstrip Coupler: Geometry and Design Task

Consider a compact dual-band branch-line coupler shown in Fig. 5.11. The circuit is implemented on Taconic RF-35 dielectric substrate (h = 0.762 mm; ε_r = 3.5; tanδ = 0.0018). The structure is constructed using transmission lines of three different characteristic impedances with equal electrical lengths (Park & Lee, 2005). The compact design is based on conventional dual-band BLC realization and modified so that the interior of a coupler is filled as much as possible. The area reduction is obtained by means of meandering of the coupler transmission lines in the following manner. In the first step, the internal BLC sections are folded. This step is crucial as the central location of the lines determines the size of the final circuit. Subsequently, the remaining sections are meandered into the coupler interior.

The BLC geometry is described by seven independent adjustable parameters $\boldsymbol{x} = [w_1\ w_2\ w_3\ l_1\ l_2\ l_3\ l_{31}]^T$. The dependent variables are $l_{1m} = 0.2\cdot\text{abs}(l_3 + 3w_3 + 3l_{3m} + 4w_1)$, $l_{2m} = 0.2\cdot(l_3 + w_3 + l_{3m} + 0.5w_1–4.5w_2)$ and $l_{32} = w_1 + l_1 - w_3$, whereas $l_{3m} = 0.2$ and $w_0 = 1.7$. The area occupied by the coupler is defined as $A \times B$ rectangle with $A = 2\cdot(l_1 + 2w_1) + 5w_3 + 6l_{3m}$ and $B = 2\cdot(5w_1 + 5l_{1m}) + w_3$. The unit of all dimensions is mm. Due to considerable cross-couplings between the adjacent sections, accurate responses of the miniaturized circuit can be obtained only by means of EM simulations. The EM model of the structure is implemented in CST Microwave studio using about 300,000 mesh cells. Its average simulation time is 20 min.

The following design objectives are considered in the optimization process:

- Achieving equal power split between the transmission and the coupling ports (i.e., $|S_{21}| = |S_{31}|$) at the assumed operating frequencies f_1 = 0.92 GHz and f_2 = 1.9 GHz
- Minimizing the return loss $|S_{11}|$ at f_1 and f_2
- Minimization of isolation $|S_{41}|$ at f_1 and f_2

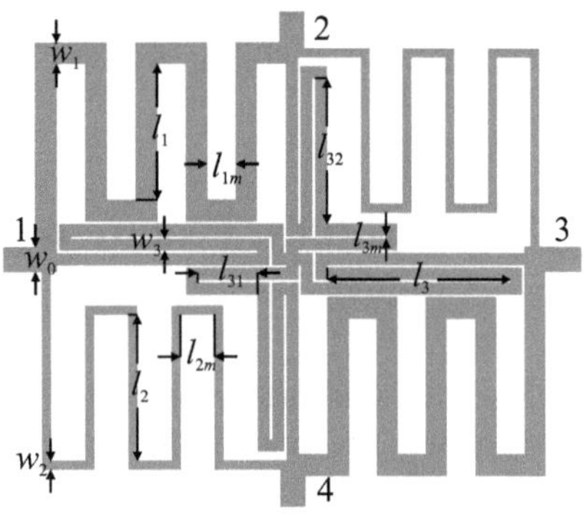

Fig. 5.11 Geometry of a compact dual-band branch-line coupler

The physical consistency of the circuit geometry is ensured by imposing the following lower and upper bounds for design variables: $\boldsymbol{l} = [0.7\ 0.3\ 0.5\ 6\ 6\ 4\ 10]^T$ and $\boldsymbol{u} = [1.7\ 1.3\ 1.5\ 10\ 10\ 8\ 14]^T$.

5.2.2 Response Features

As mentioned earlier, the response features for a given design problem and a specific circuit should be selected so that they are sufficient to account for the design objectives. The particular set of response features selected for the circuit in Fig. 5.11 has been shown in Fig. 5.12. Because we are interested in ensuring equal power split as well as good matching and isolation at both operating frequencies f_1 and f_2, the features of interest are local maxima of $|S_{21}|$ and $|S_{31}|$ and minima of $|S_{11}|$ and $|S_{41}|$ around f_1 and f_2.

These points are marked in Fig. 5.12 using circles. Each feature point is characterized by the two coordinates: the frequency (F) and level (L). Figure 5.13 shows a family of coupler responses corresponding to designs along a certain line segment $t\boldsymbol{x}_a + (1 - t)\boldsymbol{x}_b$ in the design space, parameterized by a parameter t ranging from 0 to 1, where $\boldsymbol{x}_a$ and $\boldsymbol{x}_b$ are the two arbitrarily selected points in the design space. Note that despite the highly nonlinear dependence of S-parameters on the coupler geometry, the behavior of the feature points (cf. Fig. 5.14) is much less nonlinear.

Consequently, expressing the design objectives in terms of the feature points and using feature-based surrogates is expected to speed up the optimization process considerably. The reason is that generalization capability of the surrogate model formulated at the level of the feature points is expected to be much better than for the surrogate constructed directly at the level of S-parameters.

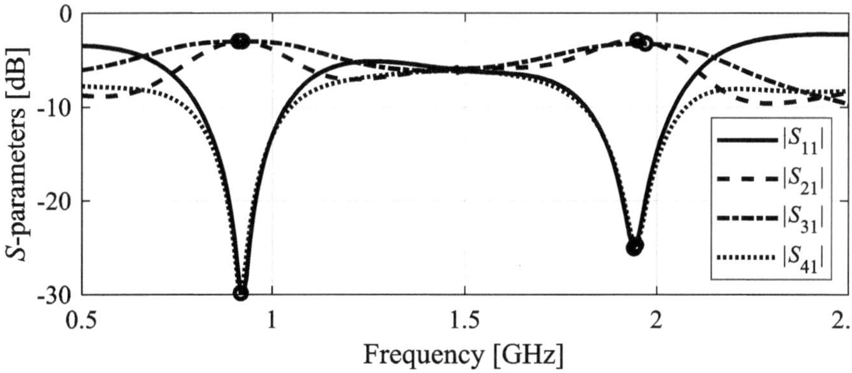

Fig. 5.12 Typical responses of a dual-band coupler. Feature points are marked with the circles, here, corresponding to the maxima of $|S_{21}|$ and $|S_{31}|$ around the operating frequencies f_1 and f_2, as well as the minima of $|S_{11}|$ and $|S_{41}|$ around f_1 and f_2

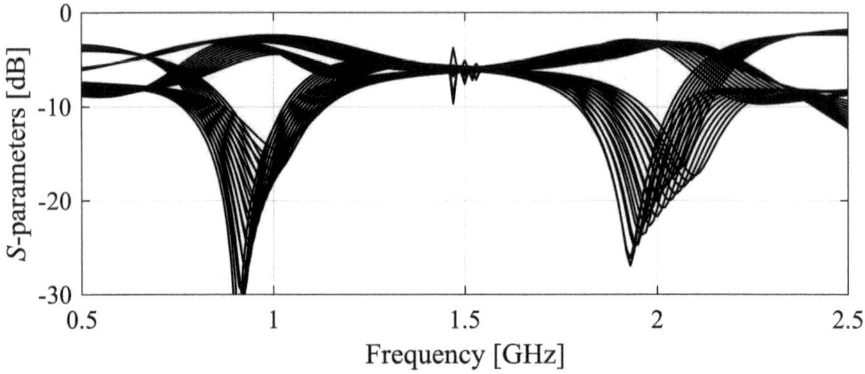

Fig. 5.13 Family of coupler responses corresponding to its geometries along a certain line segment in the design space showing highly nonlinear behavior of S-parameters on geometry variables of the circuit

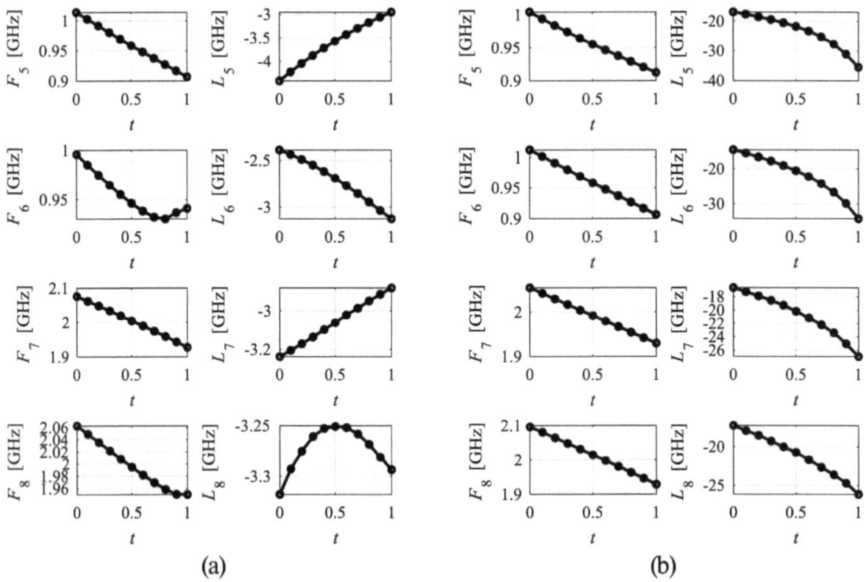

Fig. 5.14 Feature points (as marked in Fig. 5.12) corresponding to the coupler geometries along the same line segment parameterized by $0 \leq t \leq 1$ as in Fig. 5.13: (**a**) points corresponding to the maxima of $|S_{21}|$ and $|S_{31}|$ around the operating frequencies f_1 and f_2; (**b**) points corresponding to the minima of $|S_{11}|$ and $|S_{41}|$ around f_1, and f_2. Note that dependence of the feature coordinates on the coupler geometry is much less nonlinear than the dependence of S-parameters versus frequency

5.2.3 *Objective Function and Constraints: Optimization Algorithm*

Our goal is to formulate the coupler design problem through its response features. In particular, we want to solve the following minimization task:

$$\boldsymbol{x}^* = \arg\min_{\boldsymbol{x}} U_F(\boldsymbol{F}(\boldsymbol{x}), \boldsymbol{L}(\boldsymbol{x})) \tag{5.6}$$

Here, U_F is the objective function formulated in terms of the feature vectors $\boldsymbol{F}(\boldsymbol{x}) = [f_{11.1}\ f_{21.1}\ f_{31.1}\ f_{41.1}\ f_{11.2}\ f_{21.2}\ f_{31.2}\ f_{41.2}]^T$, and $\boldsymbol{L}(\boldsymbol{x}) = [l_{11.1}\ l_{21.1}\ l_{31.1}\ l_{41.1}\ l_{11.2}\ l_{21.2}\ l_{31.2}\ l_{41.2}]^T$, where $f_{j1.k}$ is the frequency corresponding to the minimum (for $j = 1, 4$) or maximum (for $j = 2, 3$) of the respective response $|S_{j1}|$ at the kth operating band, $k = 1, 2$, whereas $l_{j1.k}$ denotes the corresponding response levels; $\boldsymbol{x}$ is a vector of geometry parameters to be adjusted. Note that, in this case, a slightly different notation is used than in Sect. 5.1, where all feature point coordinates were aggregated into a single vector. The notation above can be viewed as an alternative one, whenever it is more convenient to separate the frequency and level coordinates of the characteristic points.

The feature-based objective function is formulated to minimize the level coordinates $l_{11.k}$ and $l_{41.k}$ for $k = 1, 2$, as well as to align the frequency coordinates $f_{21.k}$ and $f_{31.k}$ with f_1 (for $k = 1$) and f_2 (for $k = 2$), respectively. Furthermore, we consider the following equality constraints:

$$c_k(\boldsymbol{L}(\boldsymbol{x})) = 0, \qquad k = 1, 2 \tag{5.7}$$

which are formulated in terms of the feature point level coordinates, i.e., $c_k = l_{21.k} - l_{31.k}$. The constraints (5.7) allow for enforcing equal power split at both frequencies. Given the above, the analytical formulation of the objective function takes the form of

$$\begin{aligned} U_F(\boldsymbol{F}(\boldsymbol{x}), \boldsymbol{L}(\boldsymbol{x})) &= \max\{l_{11.1}, l_{41.1}, l_{11.2}, l_{41.2}\} \\ &+ \beta_F \left(\begin{bmatrix} f_{21.1} \\ f_{31.1} \\ f_{21.2} \\ f_{31.2} \end{bmatrix} - \begin{bmatrix} f_1 \\ f_1 \\ f_2 \\ f_2 \end{bmatrix} \right)^2 + \sum_{k=1}^{2} \beta_k c_k(\boldsymbol{L}(\boldsymbol{x}))^2 \end{aligned} \tag{5.8}$$

where β_F and β_k, $k = 1, 2$, are the penalty coefficients (Koziel, 2017). By setting these to sufficiently large values (e.g., 10^3), it is possible to enforce approximate satisfaction of the constraints (5.7) as well as the alignment of the operating frequencies with their target values.

The problem (5.6), (5.7), and (5.8) has been solved using the trust region algorithm (Conn et al., 2000) with numerical derivatives (Levy & Lessman, 1992).

5.2.4 Numerical Results and Experimental Validation

The coupler structure of Fig. 5.11 has been optimized using the feature-based approach of Sect. 5.2.1. The initial design $\boldsymbol{x}^{(0)} = [1.39\ 0.58\ 0.99\ 8.66\ 9.19\ 5.02\ 12.73]^T$ mm has been obtained by optimizing the equivalent circuit model of the circuit. Figure 5.15 shows the responses at $\boldsymbol{x}^{(0)}$ indicating non-equal power split at both operating frequencies as well as the frequency shift for matching and isolation characteristics for f_2.

Owing to close-to-linear dependence of the feature points on $\boldsymbol{x}$, only three iterations of the trust region algorithm, i.e., 24 EM simulations of the coupler, were sufficient to yield the final design $\boldsymbol{x}^* = [1.61\ 0.66\ 0.91\ 8.66\ 9.55\ 5.47\ 12.24]^T$ mm. The footprint of the miniaturized coupler is $29.5 \times 35.3 = 1041$ mm^2. It can be observed (cf. Fig. 5.16) that the responses are now very close to being centered around the operating frequencies and almost an equal power split has been achieved ($|S_{21}| - |S_{31}| < 0.02$ dB for f_1, and $|S_{21}| - |S_{31}| < 0.05$ dB for f_2). Small deviations from the "ideal" equal power split and from perfect centering of the matching and isolation responses around the operating frequencies are mostly due to the fact that the number of objectives and constraints is larger than the number of degrees of freedom in the circuit.

Consequently, the "ideal" dual-band coupler response may not be achievable in this case. For the sake of comparison, direct optimization of the coupler structure at the level of S-parameters vs. frequency has been conducted using a pattern search algorithm (Koziel, 2010). A design of a similar quality has been obtained, however, at a much higher computational cost of 122 evaluations of the EM coupler model.

The optimized coupler has been fabricated and measured. A photograph of the manufactured circuit is shown in Fig. 5.17. A comparison of the simulated and the measured responses in terms of its S-parameters is provided in Fig. 5.18. Although the obtained results are generally in good agreement, a slight frequency shift is observed for the lower operating frequency. This is mostly because the EM model

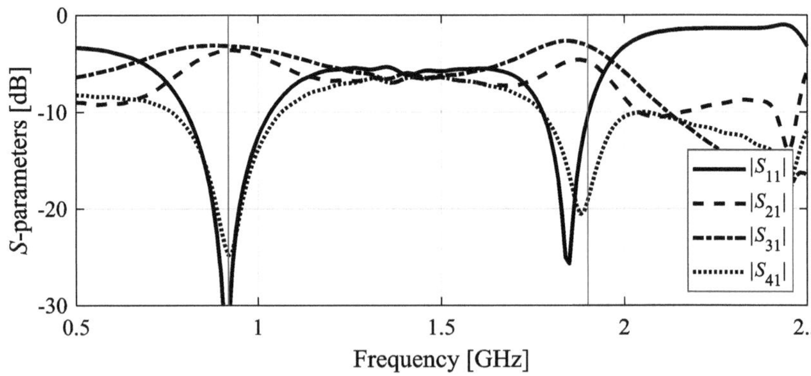

Fig. 5.15 Responses of the compact dual-band coupler at the initial design. Operating frequencies $f_1 = 0.92$ GHz and $f_2 = 1.90$ GHz are marked using vertical lines

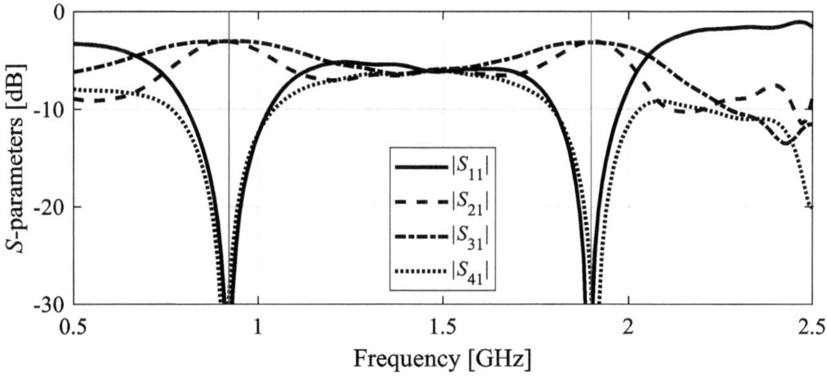

Fig. 5.16 Responses of the compact dual-band coupler at the design obtained by feature-based optimization. Operating frequencies $f_1 = 0.92$ GHz and $f_2 = 1.90$ GHz are marked using vertical lines

Fig. 5.17 Photograph of the fabricated prototype of the dual-band branch-line coupler of Fig. 5.11

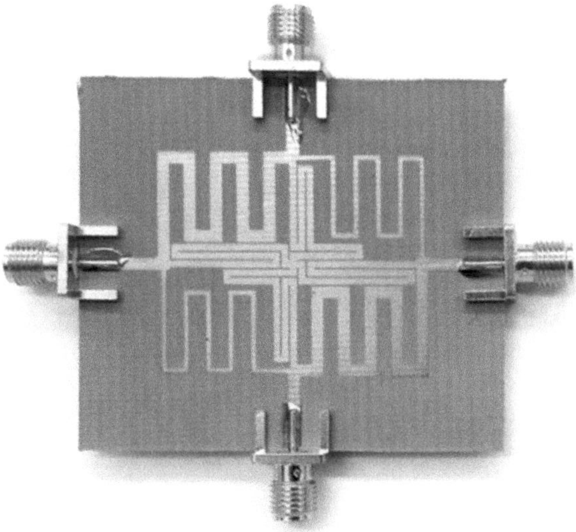

involved in the optimization process lacks connectors. The assumed value of phase difference for the branch-line couplers should be ±90° for both operating frequencies. However, it can be observed that both the simulated and measured phase differences slightly deviate from the required values. This is because the phase was not considered during the optimization process: the number of design objectives would be too large compared to the number of degrees of freedom in the coupler.

The detailed data on BLC properties is gathered in Table 5.4. It should be noted that the power split level is slightly lower for the measured coupler responses, which is due to higher losses introduced by the connectors. On the other hand, the power split imbalance is similar to the simulated results. The bandwidth BW_1, defined as the

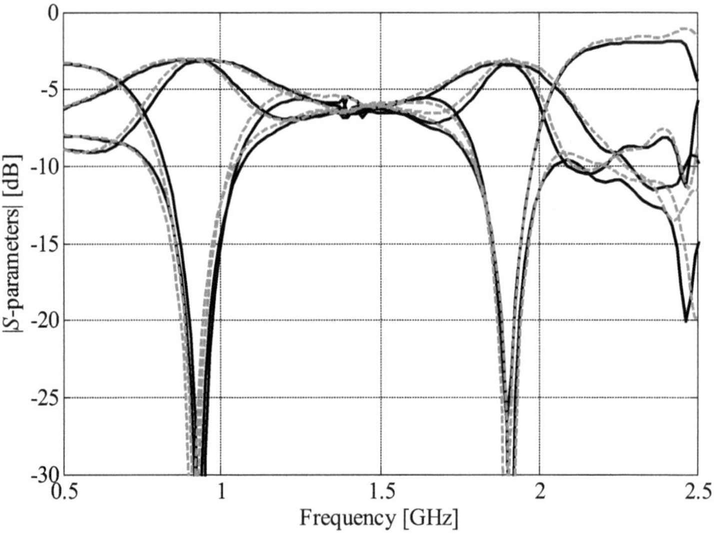

Fig. 5.18 The measured (black line) and simulated (gray dashed line) scattering parameters of the compact dual-band branch-line coupler

Table 5.4 A comparison of simulated and measured coupler responses

	Simulation		Measurement	
	f_1	f_2	f_1	f_2
S_{11} [dB]	−38.8	−33.7	−26.4	−23.5
S_{21} [dB]	−3.06	−3.06	−3.18	−3.37
S_{31} [dB]	−3.04	−3.11	−3.14	−3.46
S_{41} [dB]	−29.3	−27.9	−33.8	−39.8
BW_1 [GHz]	0.88 to 0.94	1.88 to 1.92	0.90 to 0.97	1.88 to 1.92
BW_2 [GHz]	0.88 to 0.94	1.87 to 1.94	0.91 to 0.97	1.88 to 1.93
$\arg(S_{21}/S_{31})$ [°]	−91.8	95.9	−91.4	95.5

frequency range where both return loss $|S_{11}|$ and isolation $|S_{41}|$ are below −20 dB level, is 7% and 2% for f_1 and f_2, respectively. The bandwidth BW_2 (defined as $||S_{21}| - |S_{31}|| \leq 0.2$ dB) is 7% for 0.92 GHz and 4% for 1.9 GHz.

5.3 Summary

This chapter can be considered an introduction to the response feature technique, which was explained and demonstrated using specific design examples of antenna and microwave component parameter tuning. The next chapters will discuss various applications of the response feature approaches, not only in the context of local optimization but also globalized search, multi-criterial design, statistical analysis, design centering, and general-purpose surrogate modeling. The case studies described in Sects. 5.1 and 5.2 illustrate potential advantages of the employment of characteristic points but also indicate possible difficulties related to the necessity of defining and extracting the response features from EM simulated system outputs, which may not always be straightforward or even possible. These and other issues will be discussed throughout the remaining parts of this book.

References

Conn, A. R., Gould, N. I. M., & Toint, P. L. (2000) *Trust region methods*. MPS-SIAM Series on Optimization, Philadelphia.

CST Microwave Studio, ver. 2013. (2013). CST AG, Bad Nauheimer Str. 19, D-64289 Darmstadt, Germany.

Haq, M. A., & Koziel, S. (2018). Ground plane alterations for design of high-isolation compact wideband MIMO antenna. *IEEE Access, 6*, 48978–48983.

Hassan, E., Noreland, D., Augustine, R., Wadbro, E., & Berggren, M. (2015). Topology optimization of planar antennas for wideband near-field coupling. *IEEE Transactions on Antennas and Propagation, 63*, 4208–4213.

Jia, X., & Lu, G. (2019). A hybrid Taguchi binary particle swarm optimization for antenna designs. *IEEE Antennas and Wireless Propagation Letters, 18*, 1581–1585.

Kolda, T. G., Lewis, R. M., & Torczon, V. (2003). Optimization by direct search: New perspectives on some classical and modern methods. *SIAM Review, 45*, 385–482.

Koziel, S. (2010). Computationally efficient multi-fidelity multi-grid design optimization of microwave structures. *Applied Computational Electromagnetics Society Journal, 25*(7), 578–586.

Koziel, S. (2015). Fast simulation-driven antenna design using response-feature surrogates. *International Journal of RF and Microwave CAE, 25*(5), 394–402.

Koziel, S. (2017). Objective relaxation algorithm for reliable simulation-driven size reduction of antenna structure. *IEEE Antennas and Wireless Propagation Letters, 16*(1), 1949–1952.

Koziel, S., & Bandler, J. W. (2015). Reliable microwave modeling by means of variable-fidelity response features. *IEEE Transactions on Microwave Theory and Techniques, 63*(12), 4247–4254.

Koziel, S., & Bekasiewicz, A. (2015). Fast simulation-driven feature-based design optimization of compact dual-band microstrip branch-line coupler. *International Journal of RF and Microwave CAE, 26*(1), 13–20.

Koziel, S., & Leifsson, L. (2012). Response correction techniques for surrogate-based design optimization of microwave structures. *International Journal of RF and Microwave Microwave CAE, 22*(2), 211–223.

Koziel, S., & Ogurtsov, S. (2011). Bandwidth enhanced design of dielectric resonator antennas using surrogate-based optimization. In *IEEE International Symposium on Antennas and Propagation*, Spokane, WA, July 3–8.

Koziel, S., & Ogurtsov, S. (2014). *Antenna design by simulation-driven optimization. Surrogate-based approach*. Springer.

Koziel, S., & Ogurtsov, S. (2019). *Simulation-based optimization of antenna arrays*. World Scientific.

Koziel, S., & Pietrenko-Dabrowska, A. (2021). Global EM-driven optimization of multi-band antennas using knowledge-based inverse response-feature surrogates. *Knowledge Based Systems, 227*, paper 107189.

Koziel, S., Bandler, J. W., & Madsen, K. (2006). A space mapping framework for engineering optimization: Theory and implementation. *IEEE Transactions on Microwave Theory and Techniques, 54*, 3721–3730.

Koziel, S., Cheng, Q. S., & Bandler, J. W. (2008). Space mapping. *IEEE Microwave Magazine, 9*(6), 105–122.

Koziel, S., Calik, N., Mahouti, P., & Belen, M. A. (2021). Accurate modeling of antenna structures by means of domain confinement and pyramidal deep neural networks. *IEEE Transactions on Antennas and Propagation, 70*(3), 2174–2188.

Levy, H., & Lessman, F. (1992). *Finite difference equations*. Dover Publications.

Li, W., Zhang, Y., & Shi, X. (2019). Advanced fruit fly optimization algorithm and its application to irregular subarray phased array antenna synthesis. *IEEE Access, 7*, 165583–165596.

Park, M.-J., & Lee, B. (2005). Dual-band, cross coupled branch line coupler. *IEEE Microwave and Wireless Components Letters, 15*(10), 655–657.

Petosa, A. (2007). *Dielectric resonator antenna handbook*. Artech House.

Pietrenko-Dabrowska, A., & Koziel, S. (2020). Simulation-driven antenna modeling by means of response features and confined domains of reduced dimensionality. *IEEE Access, 8*, 228942–228954.

Queipo, N. V., Haftka, R. T., Shyy, W., Goel, T., Vaidynathan, R., & Tucker, P. K. (2005). Surrogate based analysis and optimization. *Progress in Aerospace Sciences, 41*, 1–28.

RO4000® Series High Frequency Circuit Materials. (2010). Data Sheet, Rogers Corporation, Publication #92-004.

Tang, M., Chen, X., Li, M., & Ziolkowski, R. W. (2018). Particle swarm optimized, 3-D-Printed, wideband, compact hemispherical antenna. *IEEE Antennas and Wireless Propagation Letters, 17*, 2031–2035.

Toropov, V. V., Filatov, A. A., & Polynkin, A. A. (1993). Multiparameter structural optimization using FEM and multipoint explicit approximations. *Structural Optimization, 6*, 7–14.

Volakis, J. L. (Ed.). (2007). *Antenna engineering handbook*. McGraw-Hill.

Chapter 6
Response Features for Local Optimization

As elaborated on in Chap. 5, response feature technology exploits a particular structure of the system responses and capitalizes on the dependence of characteristic points of the system outputs on the design variables being considerably less nonlinear than of the entire characteristics. Reformulating design optimization tasks in terms of response features allows for a notable reduction of their computational overhead and improves their reliability. In order to employ the response feature technology, the designer needs to tailor it to a design task at hand. In particular, this requires a cautious selection of the characteristic points of the system response (or responses) so that they reflect the design specifications. A proper definition of response features necessitates a thorough scrutiny of the outputs of a structure under design. In high-frequency electronics, the designers encounter a large variety of the types and shapes of system responses, e.g., narrow-band, multi-band, and broadband. The main goal of this chapter is to present how to select and define response features for structures of different kinds of outputs and for various design problems. The employment of the response feature technology for solving design optimization tasks necessitates reformulation of the objective function in terms of response features, which is also an important topic of this chapter. The notion of response features has been introduced in the Chap. 5; hence, the reader should refer to it while analyzing the examples presented herein. Some of the application case studies considered in this chapter utilize certain acceleration mechanisms that will be a main topic of Chap. 7, which covers expedited feature-based optimization. The rationale behind including them here is to present the reader with a large variety of system outputs and diverse design tasks.

A. Pietrenko-Dabrowska, S. Koziel, *Response Feature Technology for High-Frequency Electronics. Optimization, Modeling, and Design Automation*,
https://doi.org/10.1007/978-3-031-43845-5_6

6.1 Conventional Feature-Based Optimization

This section introduces the conventional approach to feature-based optimization (FBO), including response feature selection for solving local design optimization problems. We present several high-frequency structures with different definitions of characteristic points for the respective optimization tasks. The provided examples serve to illustrate how to reformulate the objective function with the use of response features. Section 6.1.1 explains the aforementioned concepts using the examples of a dual-band patch antenna and a triple band antenna, both optimized for matching enhancement at their operating frequencies. Whereas Sect. 6.1.2 presents a bandwidth maximization of a three-section impedance matching transformer, along with its size reduction. The example of a SIW filter optimized for insertion and return loss improvement is provided in Sect. 6.1.3. The delineation of a side-lobe reduction of a linear antenna array of Sect. 6.1.4, being the last example of the conventional FBO, concludes the section.

6.1.1 Matching Enhancement of Multi-band Antennas

Two following examples are considered in this chapter: a dual-band patch antenna and a triple-band dipole antenna optimized for best matching. In both cases, the response features are selected to be the antenna resonant frequencies, as well as the reflection levels at these resonances. The section begins with a general formulation of an antenna optimization task, followed by a definition of the characteristic points of the antenna response, along with a feature-based formulation of the antenna matching enhancement problem. The last part of the section is devoted to the presentation of the results which have been obtained using the methodology proposed in (Koziel & Bekasiewicz, 2017a).

6.1.1.1 Design Optimization of High-Frequency Structures: Problem Statement

In order to make the chapter self-contained, let us recall the conventional (i.e., not feature-based) formulation of the design optimization problem, which has been addressed in Chap. 3 (cf. (3.1))

$$\boldsymbol{x}^* = \arg\min_{\boldsymbol{x}} U(\boldsymbol{x}, f) \tag{6.1}$$

where U is the scalar objective function expressed in terms of the EM-simulated model response, e.g., antenna reflection S_{11} versus frequency f. Other possible antenna characteristics include the realized gain $G(\boldsymbol{x}, f)$ or the axial ratio AR

$(\boldsymbol{x}, f)$. In the following parts of this section, the responses of other high-frequency devices, such as transformers or filters, will also be considered.

In the example considered in this section, minimization of the reflection of a dual-band antenna is of interest. Let us first provide a formulation of a classical design optimization task for a general multi-band antenna optimization case. In this case, the goal is to minimize $\max\{|S_{11}(\boldsymbol{x}, f_{0.1})|, \ldots, |S_{11}(\boldsymbol{x}i)f_{0.\mathrm{NK}}|\}$ with respect to $\boldsymbol{x}$, in particular, to ensure that the reflection is less than -10 dB at the antenna operating frequencies $f_{0.k}$, $k = 1, \ldots, N_K$. The minimax merit function may be then defined as

$$U(\boldsymbol{x},f) = \max\{|S_{11}(\boldsymbol{x},f_{0.1})|, \ldots, |S_{11}(\boldsymbol{x},f_{0.N_K})|\} \tag{6.2}$$

Observe, that the formulation (6.2) requires handling antenna reflection, which is highly nonlinear as a function of geometry parameters and frequency.

6.1.1.2 Feature Points Definition and Feature-Based Optimization Task

In the following, the feature vectors $\boldsymbol{F}(\boldsymbol{x}) = [f_1(\boldsymbol{x}), \ldots, f_p\, \boldsymbol{x})]^T$ and $\boldsymbol{L}(\boldsymbol{x}) = [l_1(\boldsymbol{x}), \ldots, l_p\, \boldsymbol{x})]^T$ will be used, whose entries are the frequency and level coordinates of the feature points, respectively, where p stands for the number of features. The general feature-based formulation of the optimization task is given as follows

$$\boldsymbol{x}^* = \arg\min_{\boldsymbol{x}} U_F(\boldsymbol{F}(\boldsymbol{x}), \boldsymbol{L}(\boldsymbol{x})) \tag{6.3}$$

where U_F is an objective function expressed in terms of the response features. As mentioned on several occasions, the major advantage of the formulation (6.3) is that the respective functional landscape is much less nonlinear than that of the original problem (6.1). This is particularly visible for functional landscapes of sharp resonant-type of responses that are highly nonlinear when being considered from the point of view of individual frequencies. The structures that exhibit such responses are much easier to handle when considered from the point of view of the frequency location of the resonance as well the corresponding response level. This has been illustrated in Fig. 6.1 for an exemplary narrow-band antenna. Another important advantage is that, because of directly handling the frequencies of the feature points, it is possible to easily control the frequency location of the resonances, which may prove difficult when solving (6.1).

In the case of antenna reflection enhancement at the resonant frequencies $f_{0.k}$ of a multi-band antenna, the function U_F utilizes p feature points $[f_1(\boldsymbol{x})\; l_1(\boldsymbol{x})], \ldots, [f_p(\boldsymbol{x})\; l_p(\boldsymbol{x})]$ corresponding to the reflection response minima, and is defined as follows:

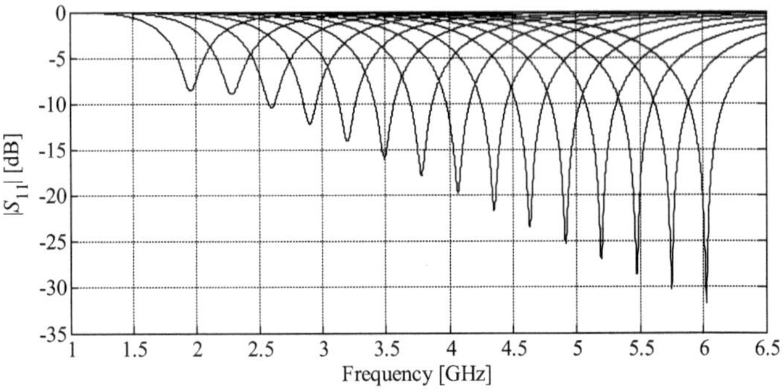

(a)

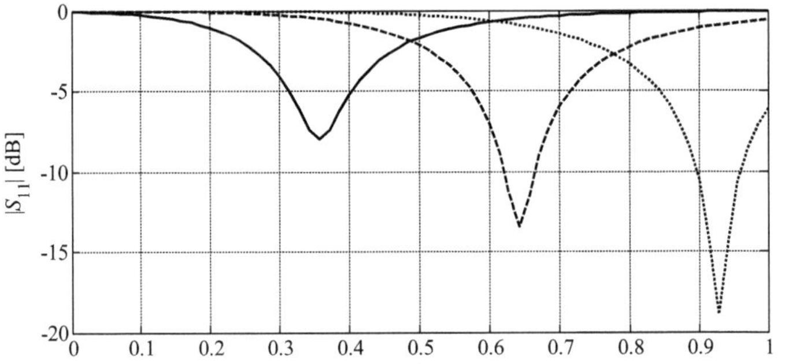

(b)

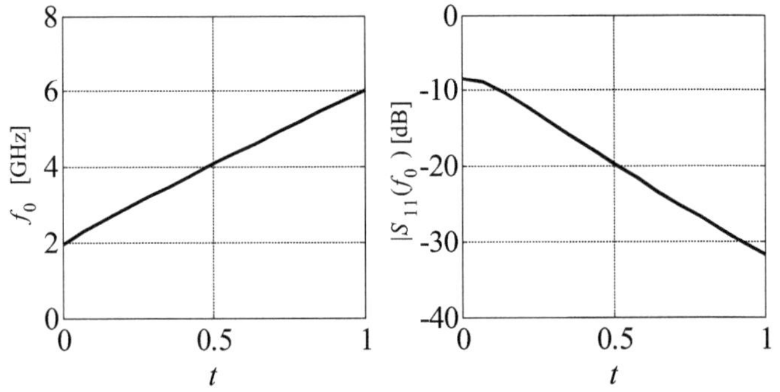

(c)

Fig. 6.1 S-parameter versus feature-based representation: (**a**) responses of the example narrow-band antenna evaluated along a line segment in the design space parameterized by t, in the range $0 \leq t \leq 1$; (**b**) reflection variability along the same line segment for the frequencies 3.0 GHz (—), 4.0 GHz (- - -), and 5.0 GHz (····); (**c**) behavior of the resonant frequency f_0 and the reflection level at f_0 corresponding to the curves of figure (**a**) are much less nonlinear than for the entire responses of figure (**b**)

$$U_F(\boldsymbol{x}) = \max\{l_1(\boldsymbol{x}), \ldots, l_p(\boldsymbol{x})\} + \beta \left\| \begin{bmatrix} f_1(\boldsymbol{x}) \\ \vdots \\ f_p(\boldsymbol{x}) \end{bmatrix} - \begin{bmatrix} f_{0.1} \\ \vdots \\ f_{0.p} \end{bmatrix} \right\|^2 \tag{6.4}$$

Observe that in (6.4), we use p response features, where p is equal to the number of antenna resonant frequencies N_K. In (6.4), the main objective is to minimize antenna reflection, whereas the penalty term (i.e., the second term of (6.4)) permits control over the operating frequencies of the structure. The value of the coefficient β should be sufficiently large to ensure that the contribution of the penalty term is comparable to the primary objective for unacceptably large frequency allocation error. Unlike the highly nonlinear objective function (6.2), the function (6.4) is a weakly nonlinear function of the design variables and the frequency, which, as will be shown later, translates into the increased reliability of the optimization process and sizeable cost savings. The problem (6.3) is solved using the trust region (TR) gradient-based algorithm (Conn et al., 2000). The details on adopting the TR routine to solving optimization problems formulated in terms of response features are provided in the next section.

6.1.1.3 Trust Region Gradient-Based Search

The examples of this chapter utilize a trust region (TR) gradient-based optimization algorithm (Conn et al., 2000) as a core optimization procedure. This section briefly recapitulates the TR routine and delineates its adaptation for solving feature-based optimization tasks. Still, it should be noted, that, in principle, the optimization problem such as (6.3) can be solved using any available algorithm of the designer's choice.

The trust region algorithm solves the optimization task in an iterative manner, by yielding a sequence of designs $\boldsymbol{x}^{(i)}$, $i = 0, 1, \ldots$, which approximate the optimal solution $\boldsymbol{x}^*$ as follows

$$\boldsymbol{x}^{(i+1)} = \arg \min_{\left\|\boldsymbol{x} - \boldsymbol{x}^{(i)}\right\| \leq \delta^{(i)}} U_F\left(\boldsymbol{F}_S^{(i)}(\boldsymbol{x}), \boldsymbol{L}_S^{(i)}(\boldsymbol{x})\right) \tag{6.5}$$

where the feature vectors $\boldsymbol{F}$ and $\boldsymbol{L}$ in (6.3) are replaced by their linear approximation models $\boldsymbol{F}_S^{(i)}$ and $\boldsymbol{L}_S^{(i)}$, determined at the current design $\boldsymbol{x}^{(i)}$, whereas $\delta^{(i)}$ denotes the search radius, which is updated using the standard TR rules based on the gain ratio (Conn et al., 2000).

The linear models $\boldsymbol{F}_S^{(i)}$ and $\boldsymbol{L}_S^{(i)}$ are constructed through finite differentiation, based on n perturbed designs around $\boldsymbol{x}^{(i)}$, i.e., $\boldsymbol{x}^{(i)} + \boldsymbol{h}_k$, $k = 1, \ldots, n$, where $\boldsymbol{h}_k = [0 \ldots 0\ d_k\ 0 \ldots 0]^T$, $d_k > 0$, and the corresponding feature points $\boldsymbol{F}(\boldsymbol{x}^{(i)} + \boldsymbol{h}_k)$ and $\boldsymbol{L}(\boldsymbol{x}^{(i)} + \boldsymbol{h}_k)$ are extracted from the respective antenna responses. For $\boldsymbol{F}_S^{(i)}$ we have (the formula for $\boldsymbol{L}_S^{(i)}$ is analogous):

$$\boldsymbol{F}_S^{(i)}(\boldsymbol{x}) = \boldsymbol{F}\left(\boldsymbol{x}^{(i)}\right) + \begin{bmatrix} \left[\boldsymbol{F}\left(\boldsymbol{x}^{(i)} + \boldsymbol{h}_1\right) - \boldsymbol{F}\left(\boldsymbol{x}^{(i)}\right)\right]^T / d_1 \\ \cdots \\ \left[\boldsymbol{F}\left(\boldsymbol{x}^{(i)} + \boldsymbol{h}_n\right) - \boldsymbol{F}\left(\boldsymbol{x}^{(i)}\right)\right]^T / d_n \end{bmatrix}^T \cdot \left(\boldsymbol{x} - \boldsymbol{x}^{(i)}\right) \tag{6.6}$$

6.1.1.4 Dual-Band Planar Antenna: Results

Consider a dual-band planar antenna shown in Fig. 6.2 (Koziel et al., 2016). The structure consists of two radiating elements in the form of a quasi-microstrip patch with an inset feed and a monopole radiator, both connected in a cascade. The antenna is excited through a 50 Ohm microstrip line. The lower and upper resonant frequencies are introduced by the monopole and patch, respectively. The ground plane is trimmed below the patch component to increase the number of degrees of freedom (hence, it is referred to as quasi-microstrip).

The antenna is implemented on a 0.762-mm-thick Taconic RF-35 dielectric substrate with relative permittivity of 3.5 and loss tangent of 0.0018. The design variables are $\boldsymbol{x} = [L\ l_1\ l_2\ l_3\ W\ w_1\ w_2\ g]^T$. The parameters $o = 7$, $w_0 = 1.7$, $l_0 = 10$, and $s = 0.5$ are fixed. The unit for all parameters is mm. The lower and upper bounds for the parameters are $\boldsymbol{l} = [10\ 1\ 5\ 1\ 4\ 0.2\ 1\ {-4}]^T$ and $\boldsymbol{u} = [20\ 6\ 17\ 7\ 16\ 4\ 6\ 6]^T$, respectively. The EM antenna model $\boldsymbol{R}$ is implemented in CST Microwave Studio and contains ~1,900,000 hexahedral mesh cells.

Two cases have been considered in terms of design requirements (Koziel & Bekasiewicz, 2017a):

- Case I: operating frequencies 3.5 GHz and 5.3 GHz
- Case II: operating frequencies 2.4 GHz and 4.8 GHz

In both cases, the goal is to allocate the resonances at the respective operating frequencies and to minimize reflection at these frequencies. The initial design is $\boldsymbol{x}^{\text{init}} = [17.0\ 1.8\ 11.7\ 4.1\ 12.9\ 0.7\ 1.3\ 0.2]^T$ mm. As shown in Fig. 6.3, this design is

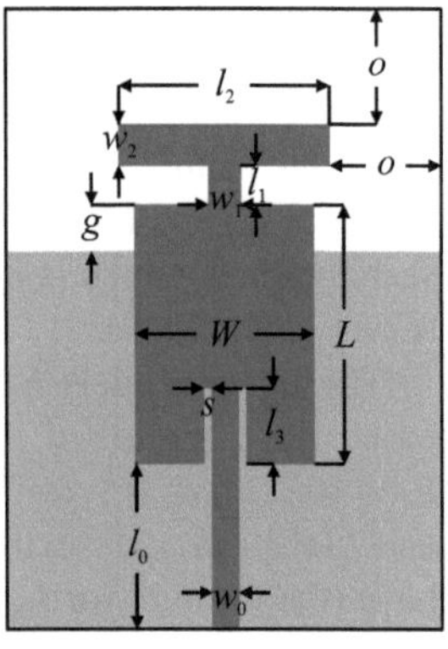

Fig. 6.2 Geometry of the dual-band patch antenna (Koziel et al., 2016). Ground plane marked using the lighter shade of gray

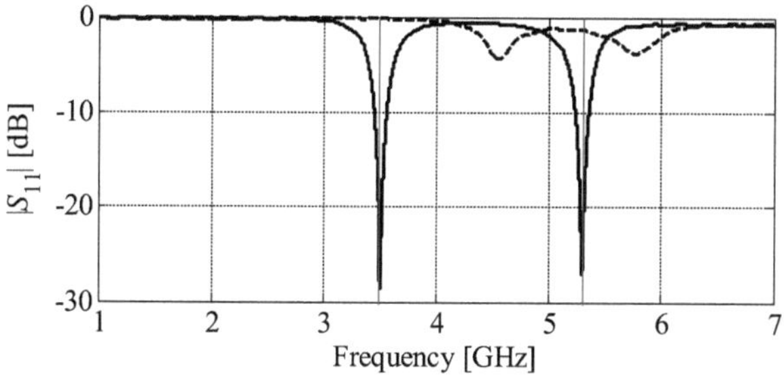

Fig. 6.3 Optimization results, Case I. Antenna responses at the initial design (- - -) and at the design found by feature-based optimization algorithm (—)

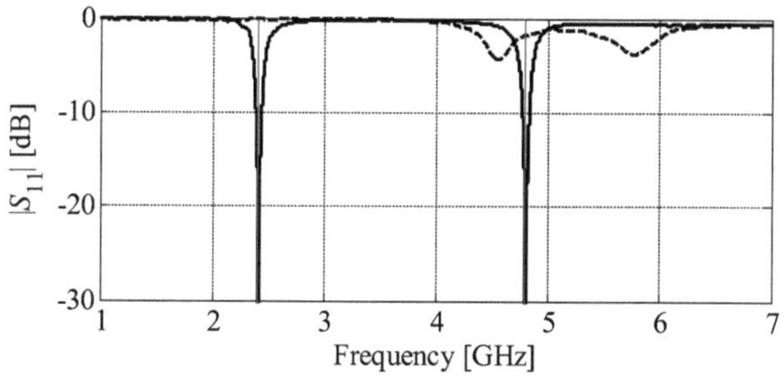

Fig. 6.4 Optimization results, Case II. Antenna responses at the initial design (- - -) and at the design found by feature-based optimization algorithm (—)

poor, i.e., the resonant frequencies at this design are distant from the intended operating frequencies for both considered design scenarios. More importantly, as the reflection response is flat and close to zero at the initial design in the vicinity of the target operating frequencies, local optimization would get stuck at the initial design when the original formulation of the problem is considered (cf. (6.1)). This has been verified using the pattern-search algorithm (Kolda et al., 2003). Consequently, in this case, global optimization methods would have to be used, which are associated with high computational expenses.

Figure 6.3 shows the antenna responses at the initial and at the final design optimized to operate at the frequencies 3.5 GHz and 5.3 GHz (Case I): $\boldsymbol{x}^{*\mathrm{I}} = [17.17\ 2.19\ 13.20\ 5.00\ 12.78\ 0.94\ 4.00\ 2.62]^T$ mm found by FBO. Note that the design quality is very good with the resonances well centered at the target operating frequencies, and the reflection levels at these frequencies are below –25 dB. The optimization cost has been equal to 72 evaluations of the EM model (8 algorithm iterations), which is low given the complexity of the problem (8 design variables). Local optimization working for the original formulation of the problem failed to find the acceptable design.

Figure 6.4 shows the results for Case II. The optimum design is $\boldsymbol{x}^{*\mathrm{II}} = [17.64\ 2.89\ 13.84\ 6.90\ 13.55\ 0.20\ 5.03\ 4.09]^T$ mm, and it has been obtained at the cost of 90 evaluations of the EM antenna model (10 algorithm iterations). Again, the design quality is very good with resonances centered at the operating frequencies and the reflection minima below −30 dB. Note that the optimum design is at a considerable distance from the initial one (response-wise). The optimization launched from $\boldsymbol{x}^{*\mathrm{I}}$ has been capable of rendering a solution comparable to $\boldsymbol{x}^{*\mathrm{II}}$ in merely four iterations.

The example of this section shows that reformulation of the design problem in a feature space of suitably defined characteristic points is the key component of the FBO procedure that ensures its reliability and computational efficiency. Using the considered approach, the optimized designs have been yielded at a low cost corresponding to just a few dozen of EM simulations.

6.1.1.5 Triple-Band Antenna: Results

The second example of this section is a triple-band uniplanar dipole antenna shown in Fig. 6.5. Here, also the response features are the antenna resonant frequencies, along with the corresponding reflection levels. The structure is based on geometry introduced in (Chen et al., 2006). The utilized substrate is a 0.762-mm-thick Taconic RF-35. The antenna is constructed as a stack of three narrow, long ground plane slits separated by two thicker, short slots. The radiator is fed through a 50 Ohm coplanar waveguide (CPW). The vector of adjustable parameters is $\boldsymbol{x} = [l_1\ l_2\ l_3\ l_4\ l_5\ w_1\ w_2\ w_3\ w_4\ w_5]^T$, whereas dimensions $l_0 = 30$, $w_0 = 3$, $s_0 = 0.15$, and $o = 5$ remain fixed. The unit for all geometric parameters is mm. The search space is defined using the following lower and upper bounds: $\boldsymbol{l} = [30\ 5\ 22\ 5\ 15\ 0.2\ 1.2\ 0.2\ 1.2\ 0.2]^T$ and $\boldsymbol{u} = [50\ 15\ 29\ 15\ 21\ 2.2\ 4.2\ 2.2\ 4.2\ 2.2]^T$. The EM antenna model $\boldsymbol{R}$ (~100,000 cells; simulation: 6 min) is prepared in CST Microwave Studio and simulated using its time domain solver.

Similarly as in Sect. 6.1.1.4, the goal of the optimization has been to minimize the antenna reflection for the given operating frequencies. The initial design is $\boldsymbol{x}^{\text{init}} = [36\ 14\ 26\ 12\ 20\ 1\ 3\ 1\ 2\ 1]^T$ mm. As it can be seen from Fig. 6.6, the initial design is closer to the desired specifications compared to the example discussed in previous section. Consequently, for the Case I, original formulation of the design problem given by (6.1) could be utilized to find the desired design solution using local search algorithm (here, pattern search). However, for the Case II, the said local algorithm has been stuck at the local minimum which has been away from the desired solution.

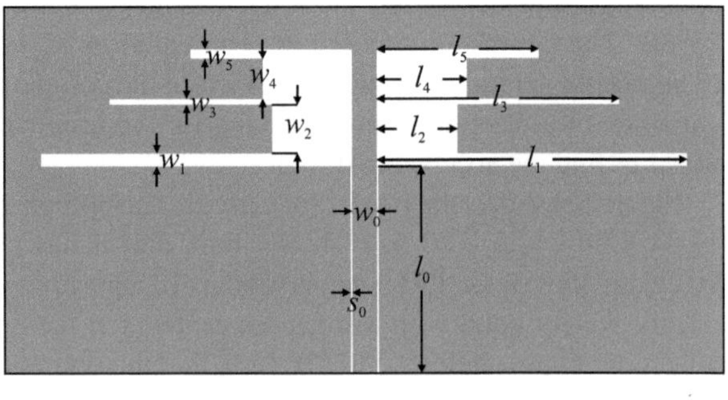

Fig. 6.5 Geometry of the triple-band dipole antenna (Chen et al., 2006)

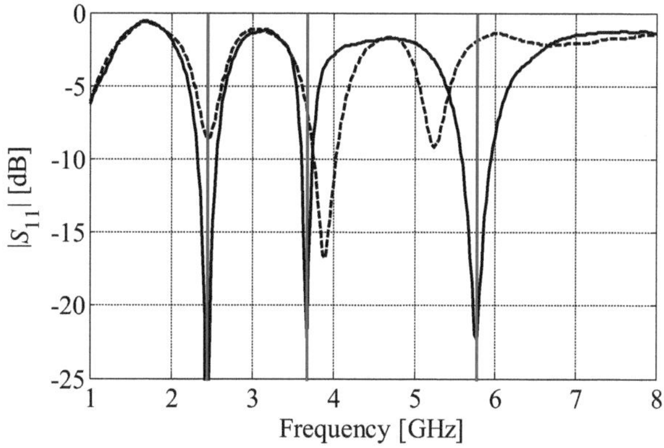

Fig. 6.6 Triple-band dipole antenna optimization results: Case I. Antenna responses at the initial design (- - -) and at the design found by feature-based optimization algorithm (—) (Koziel & Bekasiewicz, 2017a)

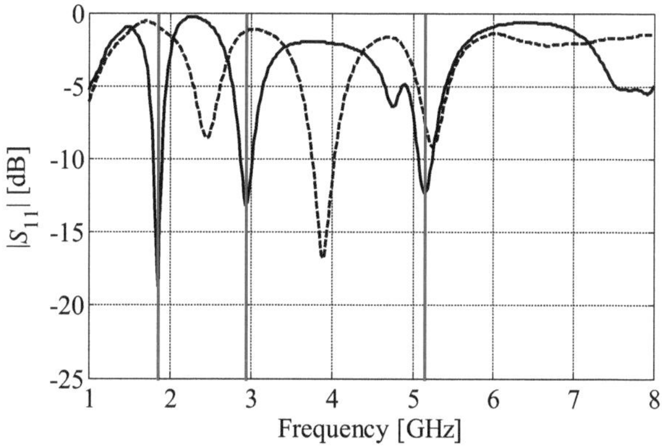

Fig. 6.7 Triple-band dipole antenna optimization results: Case II. Antenna responses at the initial design (- - -) and at the design found by feature-based optimization algorithm (—) (Koziel & Bekasiewicz, 2017a)

For this antenna, two sets of design specifications are considered:

- Case I: operating frequencies 2.45 GHz 3.65 GHz, and 5.77 GHz
- Case II: operating frequencies 1.85 GHz 2.95 GHz, and 5.15 GHz

For Case I, the comparison of the antenna responses at the initial and the design optimized using feature-based method is shown in Fig. 6.6. The final vector of parameters is $\boldsymbol{x}^{*\mathrm{I}} = [39.04\ 14.99\ 28.26\ 12.26\ 18.91\ 1.20\ 1.2\ 0.2\ 1.35\ 0.77]^T$. It should be noted that the structure response is well centered at the given operating frequencies with reflection levels below -22 dB. The cost of the algorithm operation has been 130 evaluations of the antenna model (12 iterations). At the same time, the pattern-search algorithm required 278 EM simulations to obtain similar antenna solution. Thus, for the considered case, numerical cost of the pattern-search optimization is almost 50 percent higher than FBO.

For Case II, the vector of optimal parameters is: $\boldsymbol{x}^{*\mathrm{II}} = [45.28\ 9.48\ 28.99\ 10.63\ 19.29\ 1.45\ 3.55\ 0.62\ 1.68\ 0.43]^T$. Frequency characteristics of the structure at the initial and optimized design are compared in Fig. 6.7. Again, the antenna features good electrical performance at the target operating frequencies with reflection below the level of -12 dB. The numerical cost of the design process is 100 evaluations of

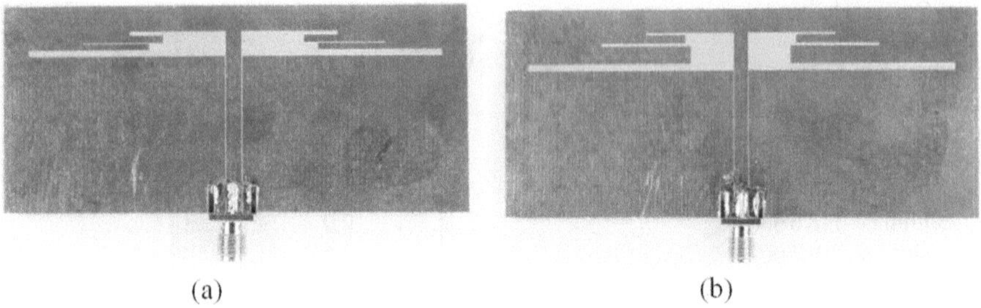

Fig. 6.8 Photographs of fabricated dipole antenna prototypes: (**a**) Case I and (**b**) Case II

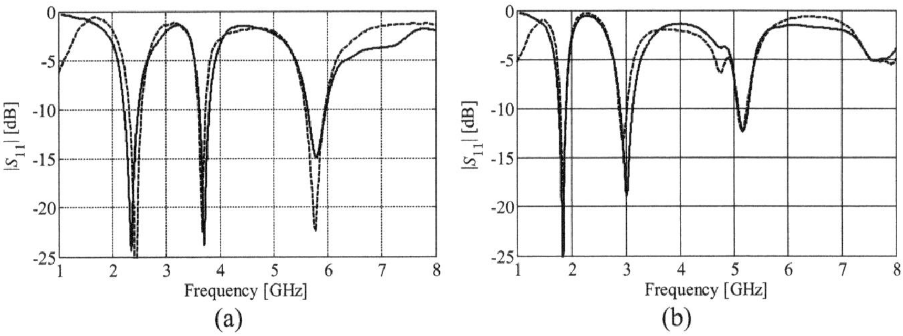

Fig. 6.9 Comparison of simulated (– –) and measured (—) reflection characteristics obtained for the triple-band dipole antenna: (**a**) Case I and (**b**) Case II

the EM antenna model (9 iterations). Having in mind noticeable distance of the first two operating frequencies from the initial design, the cost is very low.

The results obtained for both antenna designs have been experimentally validated. The photographs of the fabricated structures are shown in Fig. 6.8, whereas Fig. 6.9 presents comparison of the simulated and measured reflection characteristics. For Case I, the measured lowest operating frequency is slightly shifted down by about 50 MHz. Nonetheless, the remaining bands are well aligned.

For Case II (Fig. 6.9b), the measured bandwidth is slightly broader around the second operating frequency compared to the simulated one. The difference, however, is only about 60 MHz. Overall, the results are in a good agreement. Small discrepancies between the simulated and measured responses are due to utilization of the simplified antenna EM model without EM connector. The latter has been excluded from simulations in order to reduce the computational cost of the optimization process. As in the previous section, remarkable efficiency of the considered procedure has been a result of reformulating the design task in terms of response features. This allowed for obtaining high-quality optimal designs of satisfactorily allocated operating frequencies at the cost of around hundred EM simulations.

6.1.2 *Feature-Based Optimization of an Impedance Matching Transformer*

This section delineates the feature-based bandwidth maximization on the example of a three-section impedance matching transformer (Koziel & Bekasiewicz, 2016a). As in Sect. 6.1.1, the feature point vectors of the transformer under design will be denoted as $\boldsymbol{F}(\boldsymbol{x}) = [f_1(\boldsymbol{x})\, f_2(\boldsymbol{x}) \ldots f_p(\boldsymbol{x})]^T$, and $\boldsymbol{L}(\boldsymbol{x}) = [l_1(\boldsymbol{x})\; l_2(\boldsymbol{x}) \ldots l_p(\boldsymbol{x})]^T$. For the exemplary three-section transformer, we have $p = 4$, with the feature points corresponding to local response maxima, as well as points defining the -20-dB bandwidth, as shown in Fig. 6.10. Observe that the number and allocation of the feature points is such that the information they carry is sufficient to determine satisfaction/violation of given design specifications.

Here, the goal is to maximize the transformer bandwidth, with the function U defined as

$$U(\boldsymbol{x}) = -B(\boldsymbol{x}) = -2\min\{f_0 - f_{\min}(\boldsymbol{x}), f_{\max}(\boldsymbol{x}) - f_0\} \tag{6.7}$$

where $f_{\min}(\boldsymbol{x})$ and $f_{\max}(\boldsymbol{x})$ are the frequencies defining the left- and right-hand side edges of the -20-dB bandwidth B of the circuit (here, the bandwidth is defined symmetrically with respect to the center frequency f_0).

As in the previous section, in order to utilize feature-based optimization concept, the original design optimization problem (6.1) has to be reformulated in terms of the

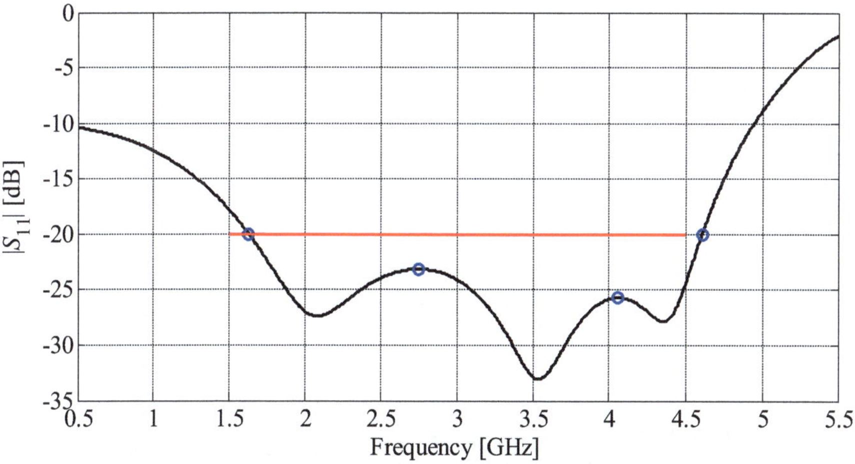

Fig. 6.10 Typical reflection characteristic of an impedance matching transformer (—) at a reasonably good initial design. There are $p = 4$ feature points, which are marked using (o) and correspond to local response maxima, as well as points defining the -20-dB bandwidth. These points are sufficient to handle transformer design optimization-oriented toward maximizing the bandwidth or minimizing the circuit size for a given (minimum) bandwidth. Design specs marked with thick horizontal line correspond to the latter case

feature point vectors $\boldsymbol{F}(\boldsymbol{x})$ and $\boldsymbol{L}(\boldsymbol{x})$. Thus, we aim at minimizing the merit function $U_F(\boldsymbol{F}(\boldsymbol{x}), \boldsymbol{L}(\boldsymbol{x}))$, as in (6.3). Again, the advantage of the formulation (6.3) over (6.1) consists in U_F being less nonlinear function of $\boldsymbol{x}$ than U. Reformulation of the merit function (6.7) in the feature space can be the following:

$$U_F(\boldsymbol{x}) = -2\min\{f_0 - f_1(\boldsymbol{x}), f_p(\boldsymbol{x}) - f_0\} + \\ +\beta\left[\frac{\max\{\max\{l_2(\boldsymbol{x}), \ldots, l_{p-1}(\boldsymbol{x})\} + 20, 0\}}{20}\right]^2 \tag{6.8}$$

Here, the penalty factor (i.e., the second factor of (6.8)) is added to ensure that the maximum level of the reflection response does not exceed −20 dB within the bandwidth determined by f_1 and f_p.

6.1.2.1 Bandwidth Maximization

We consider a three-section 50 to 100 Ohm transformer shown in Fig. 6.11. The design variables are $\boldsymbol{x} = [l_{1.1}\ l_{1.2}\ w_{1.1}\ w_{1.2}\ w_{1.0}\ l_{2.1}\ l_{2.2}\ w_{2.1}\ w_{2.2}\ w_{2.0}\ l_{3.1}\ l_{3.2}\ w_{3.1}\ w_{3.2}\ w_{3.0}]^T$. In the example of this section, further reduction of the optimization cost is obtained by utilization of variable-fidelity electromagnetic (EM) simulations. Nevertheless, the very formulation of the design optimization task is generic and pertains also to a single-fidelity regime. The following EM simulation models are employed, both implemented in CST Microwave Studio: the high-fidelity model $\boldsymbol{R}_f$ (~870,000 cells, simulation time 10 minutes) and the low-fidelity model $\boldsymbol{R}_c$ (~240,000 cells, simulation time 2 minutes).

Let us consider a feature-based bandwidth maximization around the center frequency $f_0 = 3$ GHz. Here, the final design $\boldsymbol{x}^{*\mathrm{I}} = [3.47\ 0.25\ 0.72\ 0.49\ 1.34\ 3.78\ 0.23\ 0.49\ 0.45\ 0.92\ 4.52\ 0.20\ 0.39\ 0.21\ 0.57]^T$ mm, has been obtained at the total

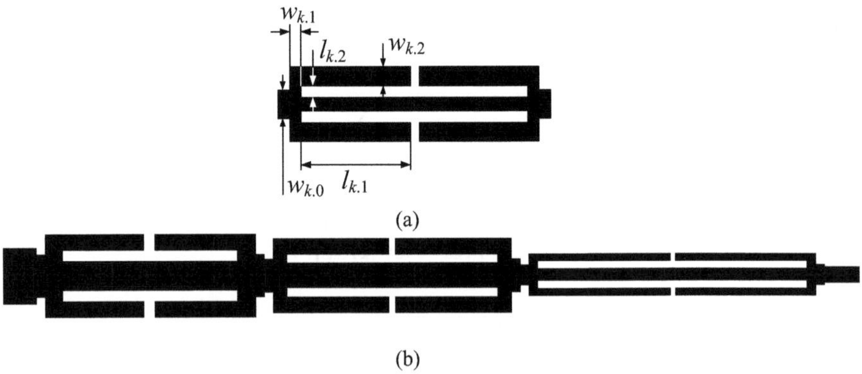

Fig. 6.11 Compact impedance matching transformer: (**a**) CMRC dimensions with k being the cell index, (**b**) transformer structure

Table 6.1 Cost breakdown of matching enhancement of impedance matching transformer

	Number of model evaluations		Total cost	
Optimization method	Low-fidelity	High-fidelity	Absolute [hours]	Relative to $\boldsymbol{R}_f$
FBO	$75 \times \boldsymbol{R}_c$	$6 \times \boldsymbol{R}_f$	3.5	21.0
Pattern search	–	$560 \times \boldsymbol{R}_f$	93.3	560

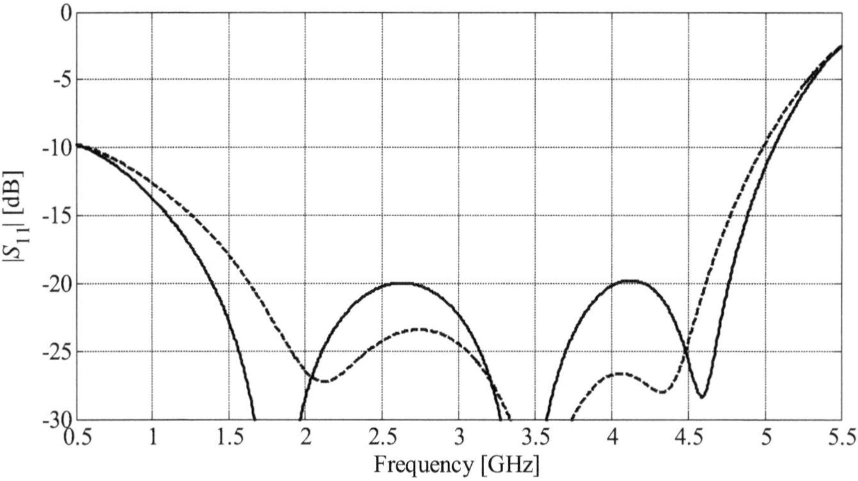

Fig. 6.12 Feature-based matching enhancement of the transformer of Fig. 6.11 with the use of the algorithm of Sect. 6.1.2: initial design (- - -), final design (—)

cost corresponding to about 21 evaluations of the high-fidelity model $\boldsymbol{R}_f$. The detailed cost breakdown has been shown in Table 6.1. Note that the number of both the low- and high-fidelity model evaluations is small, especially considering a large number of geometry parameters.

The reflection responses of the transformer are shown in Fig. 6.12. The bandwidth of the circuit is 3.23 GHz, whereas its footprint is 64 mm^2. Note that the local maxima of the response are at the level of −20 dB which indicates that the design is essentially at the global optimum for given design specifications. For comparison, the high-fidelity transformer model has been also optimized using a pattern-search algorithm (Conn et al., 2009). The final design obtained this way is similar in terms of the quality, yet, the CPU cost is much higher (cf. Table 6.1).

6.1.2.2 Size Reduction

The second example of this section is a feature-based miniaturization of the impedance matching transformer of the previous section. As the objective is to minimize the circuit size $A(\boldsymbol{x})$, the merit function may be defined as

Table 6.2 Cost breakdown of size reduction of impedance matching transformer

	Number of model evaluations		Total cost	
Optimization method	Low-fidelity	High-fidelity	Absolute [hours]	Relative to $\boldsymbol{R}_f$
FBO	$91 \times \boldsymbol{R}_c$	$8 \times \boldsymbol{R}_f$	4.4	26.2
Pattern search	–	$613 \times \boldsymbol{R}_f$	102.2	613

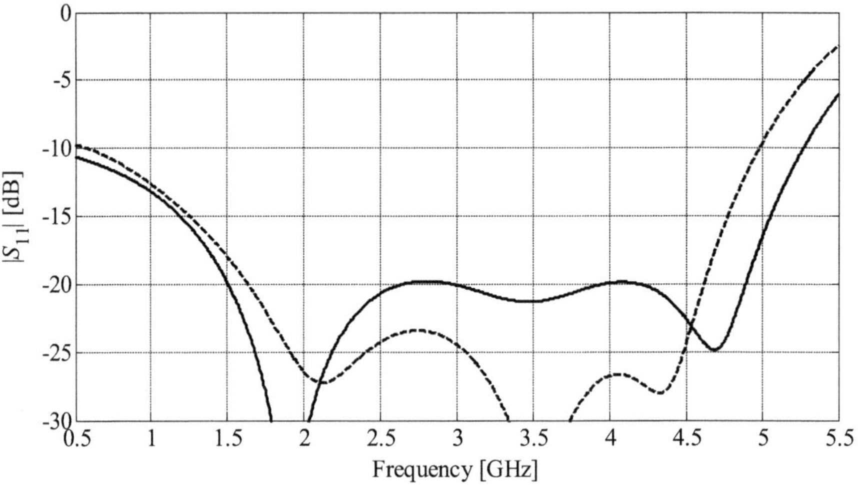

Fig. 6.13 Feature-based size reduction of the transformer of Fig. 6.11 with the use of the algorithm of Sect. 6.1.2: initial design (- - -), final design (—)

$$U(\boldsymbol{R}_f(\boldsymbol{x})) = A(\boldsymbol{x}) + \beta[\min\{B(\boldsymbol{x}) - B_{\min}, 0\}/B_{\min}]^2 \quad (6.9)$$

where $B_{\min}$ is the minimum acceptable bandwidth. In the formulation (6.9), a penalty function approach is utilized where a relative violation of the above requirement is added to the primary objective after being multiplied by the penalty factor β. The value of the penalty factor should be sufficiently large so that the penalty term contributes to the overall objective function value in a considerable manner in case of violating the bandwidth requirement above an acceptable level.

Design problem of this section is formulated as a feature-based size reduction while ensuring a minimum bandwidth from 1.5 GHz to 4.5. The initial design $\boldsymbol{x}^{(0)}$ = [3.46 0.28 0.79 0.53 1.35 3.79 0.25 0.48 0.47 0.91 4.53 0.20 0.39 0.20 0.58]^T has been obtained by optimizing the equivalent circuit model of the transformer. The final design $\boldsymbol{x}^{*\mathrm{II}}$ = [3.43 0.15 0.74 0.27 1.34 3.77 0.21 0.39 0.38 0.92 4.51 0.17 0.31 0.17 0.57]^T mm has been obtained at the total cost corresponding to about 26 evaluations of the high-fidelity model $\boldsymbol{R}_f$ (see Table 6.2 for the cost breakdown). The reflection responses of the transformer are shown in Fig. 6.13. The bandwidth is 3 GHz as required, and the circuit footprint is only 44 mm^2. The high-fidelity transformer model has been also optimized directly using a pattern-search algorithm (Kolda et al., 2003). Similarly as for the example of Sect. 6.1.2.1., the final design is similar to that found by FBO, but the CPU cost is dramatically higher (cf. Table 6.2).

Numerical results obtained for an example of a three-section 50–100 Ohm transformer indicate that the discussed approach can be utilized both for bandwidth enhancement, as well as size reduction (while maintaining the bandwidth at a prescribed minimum level). It should be emphasized that in both cases, the computational cost of the optimization process corresponds to less than 30 high-fidelity EM simulations of the transformer structure, which is a very small number considering dimensionality of the problem at hand (15 geometry parameters). The cost of the benchmark method (pattern-search algorithm) is over 200 times higher.

For both examples of this section, i.e., feature-based bandwidth enhancement and size reduction of a three-section impedance transformer, the optimal designs satisfying the design specifications have been rendered using the discussed FBO approach at the cost of merely around 20 high-fidelity EM simulations, despite the fact that the number of designable parameters of the transformer is as high as 15.

6.1.3 Feature-Based Optimization of a SIW Filter

This section describes a feature-based optimization of a substrate-integrated waveguide (SIW) filter (Glubokov & Koziel, 2014). Filter design aims at attaining given performance specifications, usually set as maximum levels of the return loss $|S_{11}|$ and transmission $|S_{21}|$ responses for specific frequency ranges. Let us consider a typical filter response that slightly violates given specifications ($|S_{11}| \leq -20$ dB, $f = 9.9\ldots10.1$ GHz) shown in Fig. 6.14. In order to satisfy the specifications, it is sufficient to bring the local maxima of the return loss to −20 dB and ensure that the slopes of the response reach this level at 9.9 and 10.1 GHz.

Generally, a response satisfies the specifications if all the points of the set (i.e., the local maxima of $|S_{11}|$ characteristics, as well as the crossing points corresponding to the bandwidth-determining frequencies) are located at the assumed positions. Thus, depending on the type of the characteristic point, only one of the feature point coordinates is useful for practical purposes. For the local maxima, the level value l_k is considered, whereas for the crossing points, the frequency f_k is taken into account. Therefore, instead of employing the previously defined vectors $\boldsymbol{F}(\boldsymbol{x}) = [f_1(\boldsymbol{x})\ f_2(\boldsymbol{x})\ \ldots\ f_p(\boldsymbol{x})]^T$ and $\boldsymbol{L}(\boldsymbol{x}) = [l_1(\boldsymbol{x})\ l_2(\boldsymbol{x})\ \ldots\ l_p(\boldsymbol{x})]^T$, here, we use the following p-element vector of the aggregated features $\boldsymbol{P}_f(\boldsymbol{x}) = [f_1, f_2, \ldots, f_M, l_{M+1}, \ldots, l_p]^T$, where M refers to the number of the filter local maxima. We also describe a kth characteristic point as $\boldsymbol{C}_k(\boldsymbol{x}) = \{f_k, l_k\}$, $k = 1, \ldots, p$.

The microwave filter optimization problem is formulated similarly as in Sect. 6.1.1.2. (cf. (6.3)), yet, using the vector $\boldsymbol{P}_f(\boldsymbol{x})$, i.e., we have

$$\boldsymbol{x}^* = \arg\min_{\boldsymbol{x}} U_F\big(\boldsymbol{P}_f(\boldsymbol{x})\big), \tag{6.10}$$

with

$$U_F = \left\|\boldsymbol{P}_f(\boldsymbol{x}) - \boldsymbol{P}_{fobj}\right\| \tag{6.11}$$

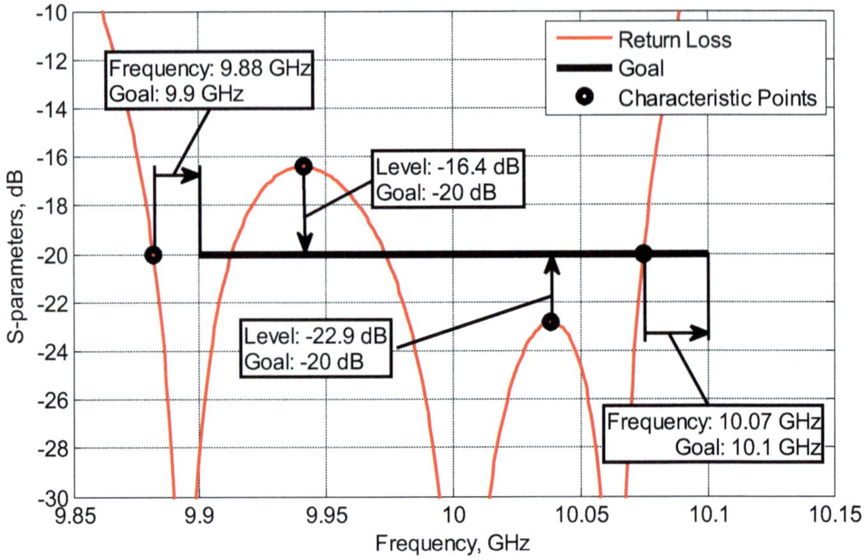

Fig. 6.14 Frequency responses of a typical microwave filter with marked design specifications and characteristic points

where $\boldsymbol{P}_{f\,\mathrm{obj}}$ refers to the target vector comprising the design specifications, i.e., the target values of the frequencies and levels of interest.

The optimization problem (6.10) and (6.11) is solved by a standard gradient-based approach (Conn et al., 2000), exploiting a linear model of the feature vector at a current design at the ith iteration $\boldsymbol{x}^{(i)}$:

$$\boldsymbol{P}_{L.f}(\boldsymbol{x}) = \boldsymbol{P}_f\left(\boldsymbol{x}^{(i)}\right) + \boldsymbol{J}_{\boldsymbol{P}_f}\left(\boldsymbol{x}^{(i)}\right) \cdot \left(\boldsymbol{x} - \boldsymbol{x}^{(i)}\right) \tag{6.12}$$

where the Jacobian $\boldsymbol{J}_{\boldsymbol{P}f}$ of the feature vector $\boldsymbol{P}_f$ is estimated using finite differentiation. For solving the task (6.10) and (6.11), we employ the TR routine approximating the solution $\boldsymbol{x}^*$ by

$$\boldsymbol{x}^{(i+1)} = \arg \min_{\left\|\boldsymbol{x} - \boldsymbol{x}^{(i)}\right\| \le \delta^{(i)}} \left\|\boldsymbol{P}_{L_f}(\boldsymbol{x}) - \boldsymbol{P}_{fobj}\right\| \tag{6.13}$$

where (cf. (6.5)): $\boldsymbol{x}^{(i)}$, $i = 1, 2, \ldots$, is a sequence of designs approximating $\boldsymbol{x}^*$, whereas $\delta^{(i)}$ is the trust region radius at iteration i.

As in the case of any other feature-based optimization algorithm, the computational efficiency of the considered procedure originates from the fact that while typical filter responses are highly nonlinear functions of frequency (with sharp minima and maxima), the dependence of the response features on the design variables is much more linear as illustrated in Fig. 6.15. Thus, the function $\boldsymbol{P}_f(\boldsymbol{x})$ is much more tractable than the original response $\boldsymbol{R}_f(\boldsymbol{x})$.

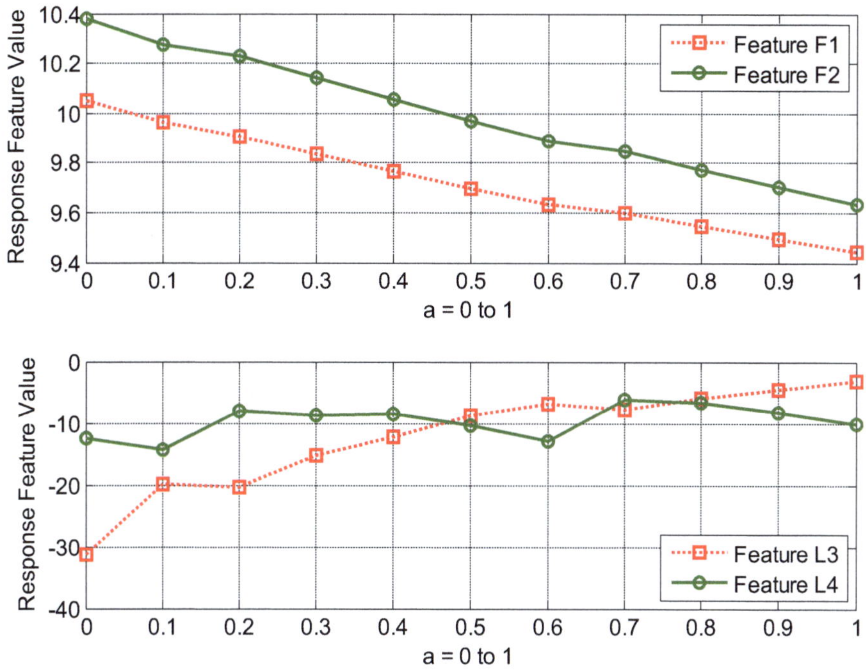

Fig. 6.15 Typical behavior of the response features $\boldsymbol{P}_f(\boldsymbol{x})$ corresponding to the high-fidelity model $\boldsymbol{R}_f(\boldsymbol{x})$ with respect to linear variation of the design parameters: $\boldsymbol{x} = a \cdot \boldsymbol{x}_A + (1 - a) \cdot \boldsymbol{x}_B$, $0 \le a \le 1$; $\boldsymbol{x}_A$ and $\boldsymbol{x}_B$ – arbitrary (preselected) points in the design space

In the presented example, additional cost reduction is achieved by employing coarse discretization EM model. Nevertheless, the SIW filter design optimization problem may be also solved using only fine (high-fidelity) EM model of the filter under design. Let us denote by $\boldsymbol{R}_c(\boldsymbol{x})$ the response vector of the low-fidelity model, along with the corresponding feature vector $\boldsymbol{P}_c(\boldsymbol{x})$.

Figure 6.16 illustrates the correlation between low- and high-fidelity model responses (and the corresponding response features) for a 3rd-order filter. Both responses have the same feature structure and show similar behavior. One can notice the frequency shift between the two responses and a small discrepancy between the values of the local maxima characteristic points. These issues are minor, as we are mainly interested in their changes with respect to small perturbations of the design parameters, which are essentially the same for both considered models.

Here, we exploit a coarse discretization EM-simulation to build low-fidelity models for SIW structures. This type of model can be obtained from the respective fine model by changing standard mesh properties, such as number of lines per wavelength (LPW) of the CST Microwave Studio solver. Assuming that the low- and high-fidelity models are physically related, which is the case when both models are evaluated using the same EM solver at different discretization levels, the

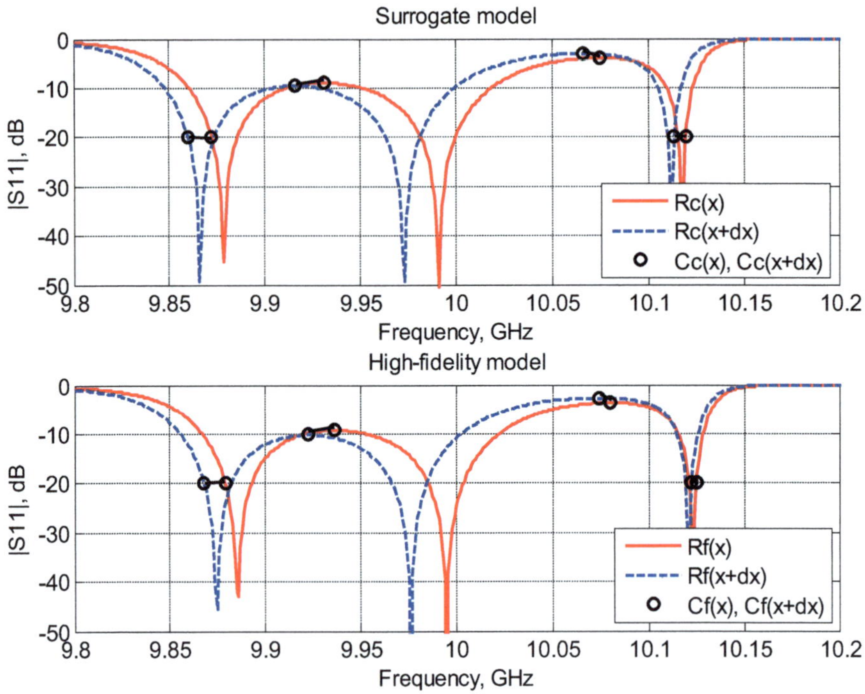

Fig. 6.16 Responses $\boldsymbol{R}_c(\boldsymbol{x})$ and $\boldsymbol{R}_f(\boldsymbol{x})$ of the coarse and fine model of the 3rd-order SIW filter, respectively, along with the corresponding characteristic points $\boldsymbol{C}_c(\boldsymbol{x})$ and $\mathbf{C}_f(\boldsymbol{x})$) and their behavior with respect to small perturbation of the design variable $\boldsymbol{dx}$

dependence of the response features on the design parameters of the filter will be similar for both models. Consequently, in (6.12), the high-fidelity model Jacobian can be replaced by that of the low-fidelity model, thereby increasing efficiency of the optimization procedure.

6.1.3.1 FBO of SIW Filter: Results

For the sake of verification, the discussed optimization procedure is applied to a 3rd-order cross-coupled folded SIW bandpass filter shown in Fig. 6.17. The design parameters are $\boldsymbol{x} = [L_{\text{res1}}\ L_{\text{res2}}\ W_{12}\ W_{13}\ W_{\text{in}}]^T$. The width of all the resonators and substrate parameters are fixed to $W_{\text{res}} = 8$ mm, $H_{\text{sub}} = 1$ mm, and $\varepsilon_r = 2.2$ and remain constant throughout the optimization. The design specifications for the filter are given as follows: $|S_{11}| \leq -20$ dB for 9.9 GHz $\leq f \leq$ 10.1 GHz and $|S_{21}| \leq -30$ dB for $f \geq$ 10.25 GHz. Here, the considered concept is also applied to the transmission losses response where an additional characteristic point represents the local maximum of the $|S_{21}|$ in the upper stopband.

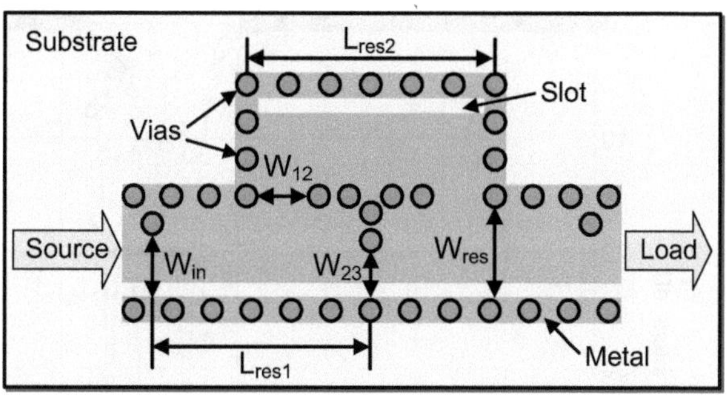

Fig. 6.17 Configuration of the 3rd-order folded SIW filter

The initial design, $\boldsymbol{x}_0 = [14\ 15.5\ 5\ 3\ 4.5]^T$ mm, has been found by manual tuning of the filter surrogate model. The following values of LPW parameter have been chosen: LPW = 7 (coarse model) and LPW = 25 (fine model).

Figure 6.18 shows the optimal response $\boldsymbol{R}_f(\boldsymbol{x}_{\mathrm{opt}})$ compared to the $\boldsymbol{R}_f(\boldsymbol{x}_0)$, where $\boldsymbol{x}_{\mathbf{opt}} = [13.51\ 15.52\ 5.24\ 2.96\ 4.82]^T$ mm, along with the filter specifications. The optimum has been obtained within six algorithm iterations at the cost of 7 evaluations of the fine model and 30 evaluations of the coarse model (less than $9 \times \boldsymbol{R}_f$ in total).

This section described FBO of SIW filter aimed at maintaining the assumed maximum levels of the return loss and transmission responses for the required frequency ranges. As in previous sections, by exploiting close-to-linear dependence of the response features on designable parameters, the optimum designs have been obtained at the cost corresponding to merely few evaluations of the high-fidelity EM-model of the structure under design.

6.1.4 FBO for the Design of Linear Microstrip Antenna Array Apertures

This section presents the last example of the conventional approach to feature-based optimization applied to the design of antenna array apertures (Koziel & Ogurtsov, 2018). Here, the feature points are defined as the angle and level locations of the first null as well as the side lobes. The focus is on optimization of the radiation pattern for side-lobe level (SLL) reduction.

We use the following notation. Let $\boldsymbol{R}_f(\boldsymbol{x})$ be an EM-simulated pattern of the array at hand and $\boldsymbol{R}_c(\boldsymbol{x})$ be a pattern obtained with the array factor (AF) model which includes the simulated element pattern using pattern multiplication (Mailloux, 2005).

From the point of view of SLL minimization as well as controlling the main beam width, one does not need to consider the entire pattern but only certain number of characteristic points, here, defined as angles and level locations of the first null as well as local side-lobe maxima. Figure 6.19a shows exemplary EM-simulated and AF patterns of an equally spaced 12-element microstrip array for various excitation

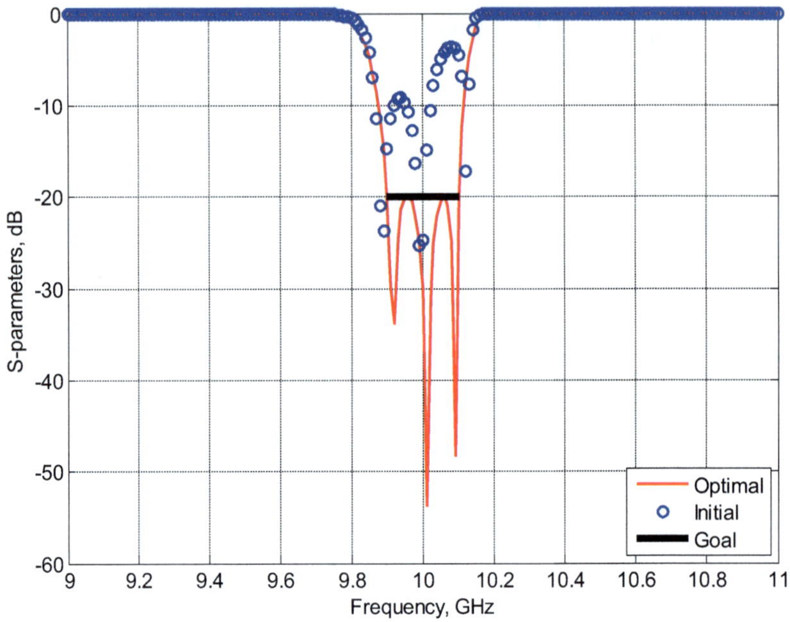

(a)

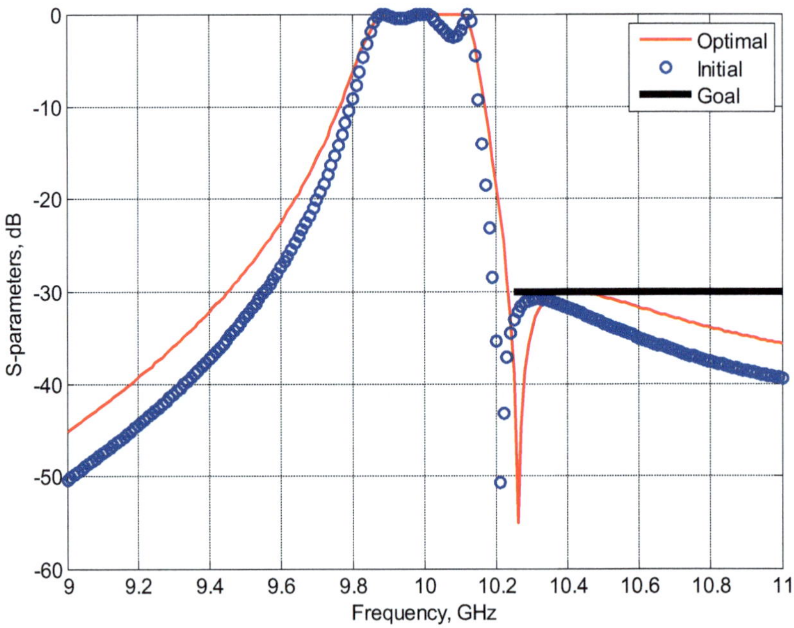

(b)

Fig. 6.18 Characteristics of the 3rd order SIW filter: initial $\boldsymbol{R}_f(\boldsymbol{x}_0)$ and optimal $\boldsymbol{R}_f(\boldsymbol{x}_{\mathbf{opt}})$ response: (**a**) the return loss $|S_{11}|$; (**b**) the transmission loss $|S_{21}|$

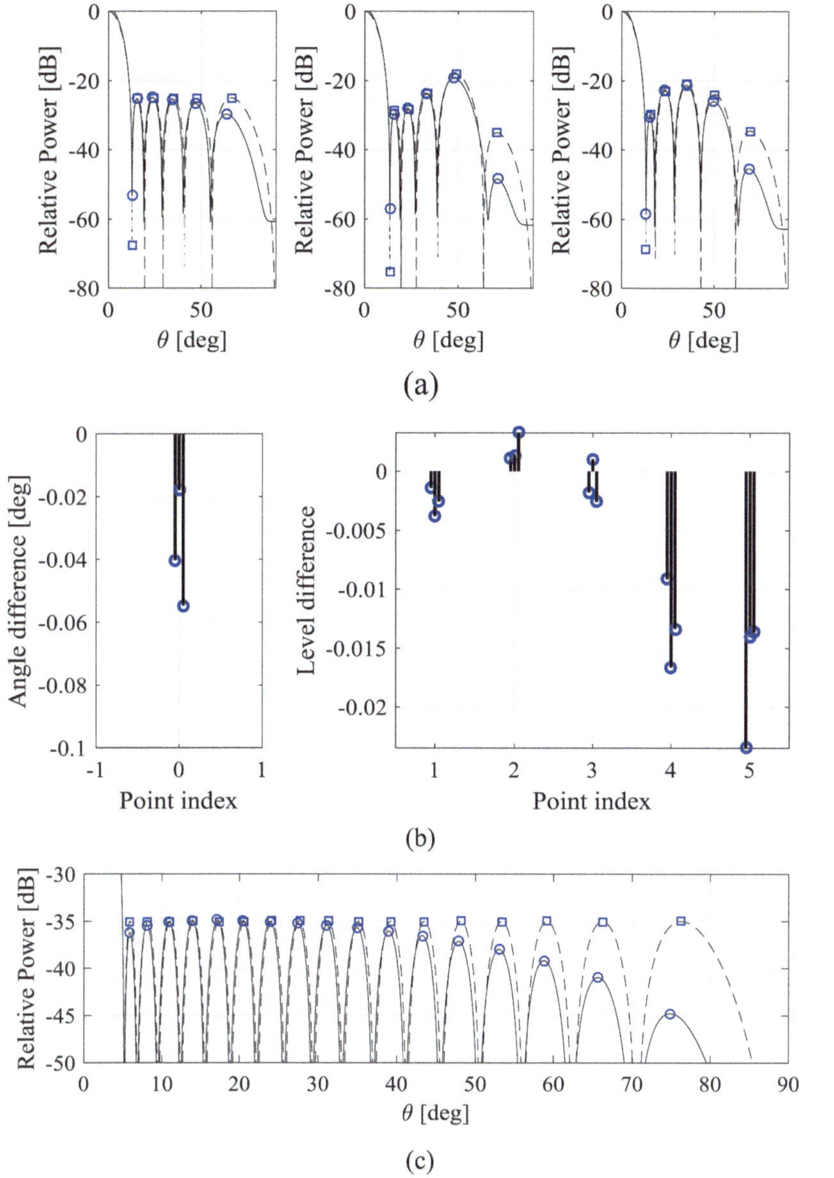

Fig. 6.19 Feature-based design concept illustrated for a 12-element microstrip array (AF model marked with a dashed line, EM model marked using a solid line): (**a**) power patterns and feature points for various excitation tapers, (**b**) differences of angle and level components of the feature points (first null angle on the left; local maxima levels in a linear scale on the right) showing good correlation between the models despite considerable absolute discrepancies, (**c**) discrepancies between AF and EM models can be significant (here, shown for 36-element array)

amplitudes along with the aforementioned characteristic points. Figure 6.19b shows the angle and level components of the points for the corresponding excitation tapers. It can be observed that good correlation is maintained so that feature-based correction of AF model can be utilized for design purposes. At the same time, absolute discrepancies between the models can be significant as illustrated in Fig. 6.19c.

Let $\boldsymbol{P}_f(\boldsymbol{x})$ and $\boldsymbol{P}_c(\boldsymbol{x})$ denote the vectors of the feature points of the EM-simulated and AF model patterns of the array, respectively. We have $\boldsymbol{P}_f(\boldsymbol{x}) = [p_0\, p_1 \ldots p_K\, l_0\, l_1 \ldots l_K]^T$, where p_j and l_j, $j = 0, 1, \ldots, K$ are angle and level components of the respective points ($j = 0$ corresponds to the first null, $j > 0$ are for pattern maxima). The vector $\boldsymbol{P}_c(\boldsymbol{x})$ has the same structure, i.e., it comprises angle and level components of the characteristic points of the AF model patterns of the array.

The design problem can be formulated as

$$\boldsymbol{x}^* = \arg\min_{\boldsymbol{x}} U\left(\boldsymbol{P}_f(\boldsymbol{x})\right) \tag{6.14}$$

where

$$U\left(\boldsymbol{P}_f\right) = \max_{j=1,\ldots,K} l_j + \beta[\max(p - p_{\max}, 0)/p_{\max}]^2 \tag{6.15}$$

The first term in (6.15) is basically SLL, whereas the second term is a penalty term in which a relative violation of the maximum main beam width $p_{\max}$ is added to the primary objective upon violation of the condition $p \le p_{\max}$. The coefficient β is set to 10^3 to ensure a sufficient contribution of the penalty term to the primary objective. Because the objective function (6.15) is smooth (the model optimized is an analytical AF model), the penalty coefficient can be set at a large value for a precise control of the main beam width.

Solving (6.14) directly is too expensive in practice. The approach discussed here is an iterative TR-based procedure that yields a sequence of approximations $\boldsymbol{x}^{(i)}$, $i = 0, 1, \ldots$, to the solution $\boldsymbol{x}^*$ as

$$\boldsymbol{x}^{(i+1)} = \arg \min_{\boldsymbol{x};\ \left\|\boldsymbol{x}-\boldsymbol{x}^{(i)}\right\| \le \delta^{(i)}} U\left(\boldsymbol{P}_c(\boldsymbol{x}) + \Delta\left(\boldsymbol{x}^{(i)}\right)\right) \tag{6.16}$$

where $\delta^{(i)}$ is a trust region radius adjusted adaptively (Conn et al., 2000) and

$$\Delta\left(\boldsymbol{x}^{(i)}\right) = \boldsymbol{P}_f\left(\boldsymbol{x}^{(i)}\right) - \boldsymbol{P}_c\left(\boldsymbol{x}^{(i)}\right) \tag{6.17}$$

The correction term (6.17) "translates" the feature points of the AF model into the points of the EM model. Iterating of (6.16) is necessary because the correlation between patterns of the low- and high-fidelity models is not perfect. Note that each iteration of (6.16) requires only one evaluation of the high-fidelity model $\boldsymbol{R}_f$. In practice, the number of iterations is small so that the overall cost of the optimization process is low as demonstrated through the examples of Sect. 6.1.4.2.

6.1.4.1 Setup of the Studies

Here we consider linear array apertures comprising 12, 24, and 36 linearly polarized microstrip patch antennas (MPAs), as outlined in Fig. 6.20. The operating frequency, i.e., the frequency of H-plane radiation pattern evaluation and optimization, is 5.8 GHz. The array elements are half wavelength center-to-center spaced at 5.8 GHz. The MPAs are energized at the dominant TM_{010} mode (Balanis, 2005) through the ground plane slots. The MPAs reside on the 1.524-mm-thick Arlon AD 250 substrate. MPAs inputs are 50 Ohm microstrips residing on the 0.762-mm-thick Arlon AD 250 substrate. The two dielectric substrates share the same ground plane. Metallization is 17.5 μm copper.

Prior to optimization of the array aperture, the single element has been optimized (Koziel & Ogurtsov, 2014) so that at 5.8 GHz its broadside radiation power density is maximum, and the reflection coefficient is about −25 dB. The EM model of the single MPA element includes periodic boundary conditions of the CST Microwave Studio solver that allows for a better consistency between full-wave simulations and the AF model. The lateral extensions of the array aperture ground plane and the substrate of the microstrip inputs are modelled as infinite. The lateral extensions of the MPAs' substrate are finite.

The problem of SLL minimization is solved for the H-plane patterns at 5.8 GHz. The design variables for the array apertures are the MPAs' excitation amplitudes. In the EM models, the amplitudes are realized as the in-phase incident waves at 5.8 GHz stimulating the microstrip inputs of the MPAs.

The initial excitation amplitudes of the Tschebyscheff tapers correspond to a particular array size and the specified SLL (Stegen, 1953). The null-to-null beam widths of the major lobe of the initial designs are used to set up design constraints so that no beam broadening occurs due to optimization.

Because of the shape of the element patterns, as well as EM interactions of the MPAs within the apertures, the EM-simulated H-plane patterns of the arrays are not equally rippled. At the same time, the simulated side lobes are under the specified SLLs. Still, by reducing the level differences between the side lobes, it is possible to

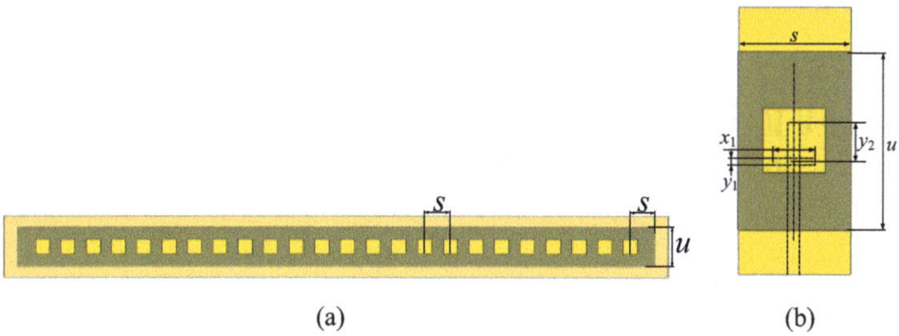

Fig. 6.20 Linear array of microstrip patch antennas (MPAs), top views: (**a**) 24 element aperture; (**b**) MPA. MPA patch is 14.35 mm × 14.35 mm. Ground plane slot dimensions are $x_1 = 10.0$ mm and $y_1 = 0.75$ mm; the slot is 5.0 mm down off the patch center. Open-end microstrip stub length $y_2 = 6.0$ mm. Spacing $s = 25.9$ mm, $u = 40.0$ mm

yield certain SLL reduction. On the other hand, it has not been possible to obtain perfect (equally rippled) patterns due to the aforementioned constraint imposed on the major lobe beam width and anisotropy of the element pattern.

It is worth noticing that if geometry of an antenna array comprising N elements is kept fixed, simulation-based optimization of the array excitation requires N full-wave simulations (without utilization of symmetry) (Koziel & Ogurtsov, 2015). This computational cost is spent on setting a reusable EM model of the aperture. Such model can be utilized in studying how a particular aperture performs in terms of radiation patterns, peak realized gain, and active reflection coefficients for different scan angles in phased array applications. However, such costs might not only be quite high for electromagnetically large apertures of graded composition but also they may be excessive for apertures producing either the fixed beam patterns or just few scanned patterns.

6.1.4.2 Results

Optimization results for the array apertures with 12, 24, and 36 MPAs for different SLLs are shown in Figs. 6.21, 6.22, and 6.23, respectively. The effect of cross-pol component on the patterns of Figs. 6.21, 6.22, and 6.23 is under the shown dB ranges. Summary of the results in terms of the SLLs with the related optimization costs is given in Table 6.3.

In all cases, simulation-based optimization has resulted in tuning of the initial Tschebyscheff tapers. For the aperture with 12 MPAs, the tapers have been tuned from [1.0000 0.9307 0.8031 0.6372 0.4572 0.4225] to [1.0000 0.9395 0.7868 0.6585 0.4086 0.4127] for the –25 dB SLL case, from [1.0000 0.9153 0.7629

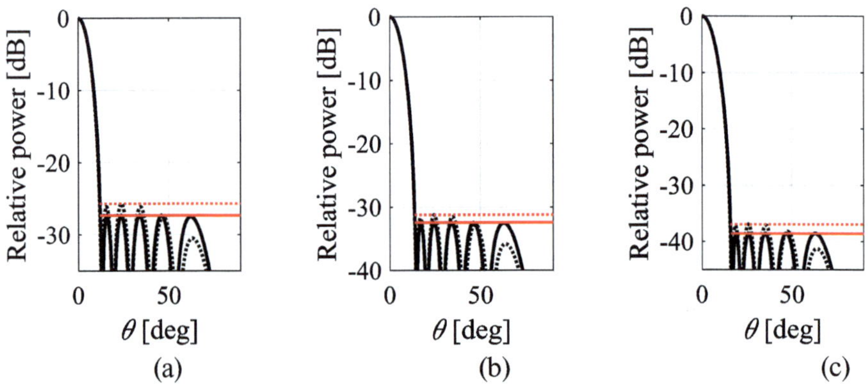

Fig. 6.21 Feature-based design results for the array aperture with 12 MPAs: (**a**) –25 dB, (**b**) –30 dB, and (**c**) –35 dB SLL cases. EM-simulated H-plane patterns at 5.8 GHz, shown for the [0, 90] degree sector, with the original Tschebyscheff tapers (···) at the initial designs and the optimized Tschebyscheff tapers (—). SLLs with the original Tschebyscheff tapers (···) and with the optimized tapers (—)

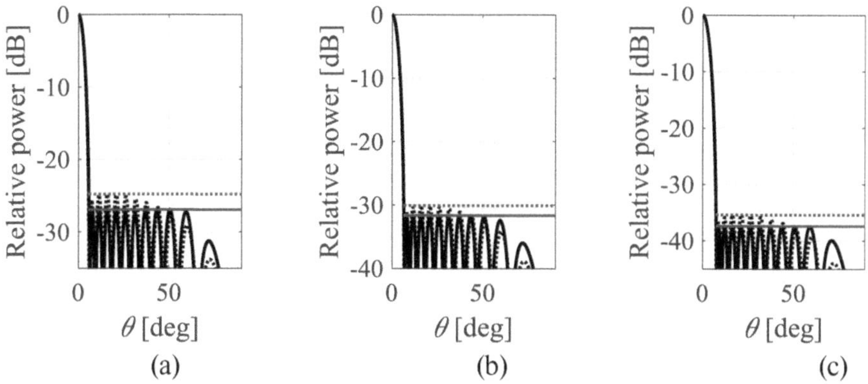

Fig. 6.22 Feature-based design results for the array aperture with 24 MPAs: (**a**) –25 dB, (**b**) –30 dB, and (**c**) –35 dB SLL cases. EM-simulated H-plane patterns at 5.8 GHz, shown for the [0, 90] degree sector, with the original Tschebyscheff tapers (···) at the initial designs and the optimized Tschebyscheff tapers (—). SLLs with the original Tschebyscheff tapers (···) and with the optimized tapers (—)

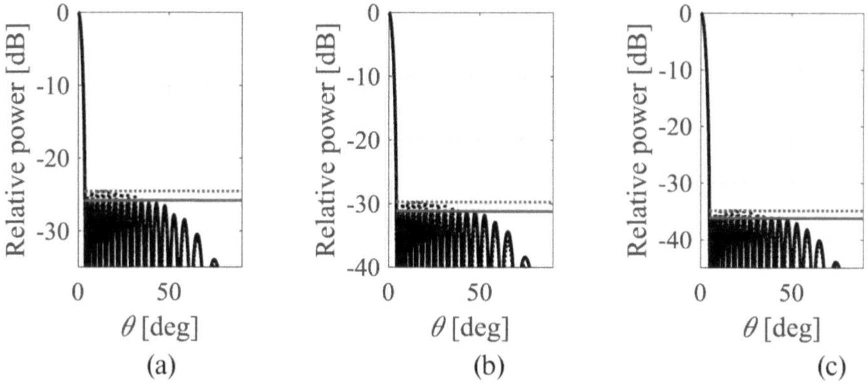

Fig. 6.23 Feature-based design results for the array aperture with 36 MPAs: (**a**) –25 dB, (**b**) –30 dB, and (**c**) –35 dB SLL cases. EM-simulated H-plane patterns at 5.8 GHz, shown for the [0, 90] degree sector, with the original Tschebyscheff tapers (···) at the initial designs and the optimized Tschebyscheff tapers (—). SLLs with the original Tschebyscheff tapers (···) and with the optimized tapers (—)

Table 6.3 Uniplanar UWB antenna: optimization cost breakdown

		Number of model evaluations		Total cost	
Optimization method	Design objective	Low-fidelity	High-fidelity	Absolute [hours]	Relative to $\boldsymbol{R}_f$
FBO	$\|S_{11}\|$	$134 \times \boldsymbol{R}_c$	$13 \times \boldsymbol{R}_f$	15.2	25.3
	Size	$167 \times \boldsymbol{R}_c$	$16 \times \boldsymbol{R}_f$	18.8	31.3
Pattern search	$\|S_{11}\|$	–	$321 \times \boldsymbol{R}_f$	192.6	321
	Size	–	$478 \times \boldsymbol{R}_f$	286.8	478

0.5720 0.3767 0.2641] to [1.0000 0.9200 0.7547 0.5840 0.3544 0.2666] for the −30 dB SLL case, and from [1.0000 0.9004 0.7251 0.5141 0.3106 0.1723] to [1.0000 0.9004 0.7165 0.5123 0.2943 0.1675] for the −35 dB SLL case.

In all cases, optimization of the aperture excitation tapers has been performed at a relatively low cost, from 4 to 6 simulations of the corresponding EM models of the array apertures. The SLLs have been lowered from 1.2 to 2.1 dB in comparison to those with the initial aperture excitation tapers. At the same time, the major lobe is left essentially intact in all cases.

For printed circuit linear arrays, high evaluation cost of their EM models makes utilization of conventional optimization approaches impractical. At the same time, the presented feature-based approach enables lowering SLLs of printed circuit linear arrays at the level of full-wave simulations at a very low cost of a handful of EM analyses.

6.1.5 *Conventional Feature-Based Optimization: Summary*

This section discussed the conventional approach to feature-based optimization. Several high-frequency structures, two multi-band antennas, an antenna array, an impedance matching transformer, and a SIW filter, have been utilized to show a variety of possibilities of a problem-specific selection of response features. In addition, the reformulation of objective functions in terms of response features for various design tasks has been described as well. In all presented cases, feature-based optimization of the respective structures allowed for obtaining optimal designs meeting the specifications at a cost significantly lower than the cost of the design optimization task formulated conventionally at the level of the entire characteristics. At the same time, as indicated in many cases, conventional optimization problems have been unable to find the satisfactory solutions.

The remainder of this chapter provides three examples of a more intricate approach to feature-based optimization. Section 6.2 outlines FBO with variable number of features on the example of a matching enhancement and size reduction of a UWB antenna. FBO with circuit decomposition of an impedance matching transformer is recalled in Sect. 6.3. The last technique considered in this chapter is a feature-based miniaturization of multi-band antennas discussed in Sect. 6.4.

6.2 FBO with Variable Number of Features

This section delineates feature-based optimization of ultrawide-band antennas with a variable number of features (Koziel & Bekasiewicz, 2017b) aimed at matching enhancement and size reduction. For wideband structures, the definition and extraction of the feature coordinates is more challenging than for narrow-band ones, FBO design of which has been described in Sect. 6.1. This is because both the lower edge of the antenna bandwidth has to be maintained at acceptable level, and the reflection has to be controlled in the entire UWB band. FBO of UWB antennas is aggravated even further by the fact that the number of feature points (specifically, the number of local maxima) may change depending on the particular antenna geometry and its dimensions. This section outlines a FBO framework developed to handle this potential variability.

6.2.1 Response Features for UWB Antennas

Frequency responses of an exemplary UWB antenna (here, the uniplanar structure considered as one of the examples) are shown in Fig. 6.24. Normally, the fulfillment of the design specifications can be assessed based on the following properties of the response: (i) reflection levels for the in-band maxima and at the upper corner frequency and (ii) allocation of the frequency point at which $|S_{11}| \leq -10$ dB right before the first resonance of the response. It should be noted, however, that the reference level for (ii) can be set to a lower value in order to ensure a safety margin. Also, as shown in Fig. 6.24a, such a point may not exist. Moreover, the number of the reflection maxima may change with the antenna geometry (cf. Fig. 6.24c). The algorithm discussed in this section allows for handling such situations.

The conventional formulation of the design problem is given by (6.1) (see Sect. 6.1). In the case of design optimization of UWB antennas, the specific formulation of the objective function U of (6.1) may be as follows:

$$U(\boldsymbol{x},f) = \max\left\{ |S_{11}(\boldsymbol{x})|_{3.1\ \mathrm{GHz} \leq f \leq 10.6\ \mathrm{GHz}} \right\} \tag{6.18}$$

The design task (6.1) and (6.18) may be reformulated in terms of the response features according to (6.3).

As mentioned before, the first feature point for UWB structures $[f_1, l_1]$ corresponds to the lowest frequency f_1 at which the reflection response attains the reference level l_1 (here, -15 dB). If the reflection level at the first resonance is above l_1, the first feature point is defined as the reflection level and frequency at the first resonance.

The remaining feature points are defined as described above. The in-band reflection level and allocation of the f_1 point are controlled using the following objective function:

$$U_F(\boldsymbol{x}) = \max\{l_2(\boldsymbol{x}), \ldots, l_K(\boldsymbol{x})\} + \beta\left[\left(\max\{f_1 - f_{1.\mathrm{target}}, 0\}\right)/f_{1.\mathrm{target}}\right]^2 \tag{6.19}$$

where $f_{1.\mathrm{target}} = 3.1$ GHz stands for the lower edge of the UWB band and β is a penalty coefficient (here, we use $\beta = 1000$).

If the desired reflection level for the first resonance is over l_1, the objective function (6.19) is automatically switched to

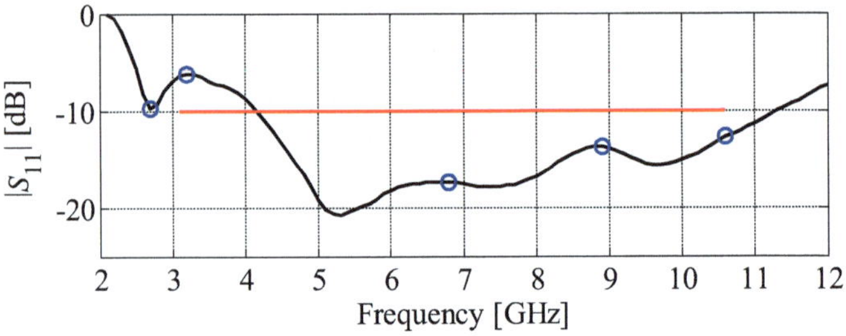

(a)

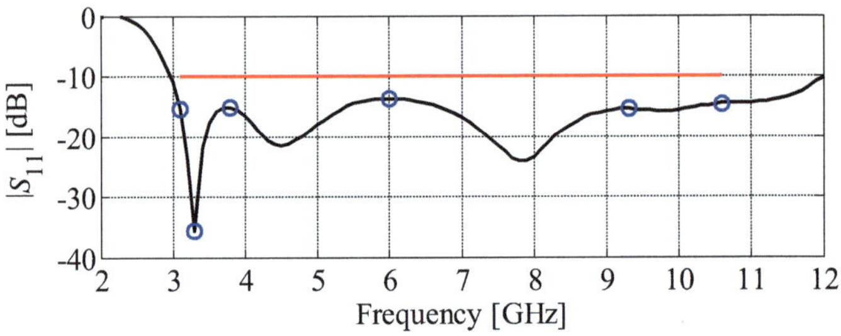

(b)

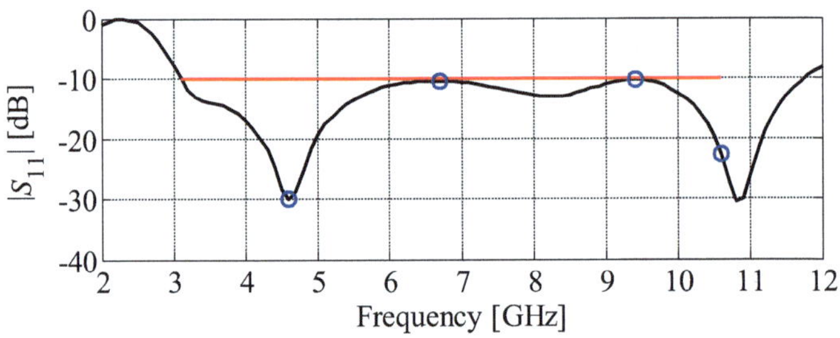

(c)

Fig. 6.24 Reflection characteristics of the uniplanar UWB antenna (—) and the characteristic points (○) utilized by feature-based optimization algorithm: (**a**) poor design which lacks the −15 dB characteristic point, (**b**) good design with the existing −15 dB point which can be exploited to control the lower corner frequency of the antenna frequency band, (**c**) design with two local maxima in the frequency band of interest. Note that the thick horizontal line defines design specifications, i.e., $|S_{11}| \leq -10$ dB within 3.1 GHz to 10.6 GHz band

$$U_F(\boldsymbol{x}) = \max\{l_1(\boldsymbol{x}) + 5, l_2(\boldsymbol{x}), \ldots, l_K(\boldsymbol{x})\} \tag{6.20}$$

The term $l_1(\boldsymbol{x})$ + 5 in (6.20) is used to enforce the −15 dB level of the first resonance as opposed to −10 dB required for remaining feature coordinates.

The design problem (6.18) is solved iteratively similarly as in Sect. 6.1.1.3 (cf. (6.5)). Low cost of the process is ensured by performing finite differentiation at the low-fidelity model level $\boldsymbol{R}_c$ of the antenna at hand.

There are two practical issues concerning the implementation of the FBO algorithm described in this section. The first one is the extraction of the feature points from the $|S_{11}|$ response of the antenna of interest. This can be done easily by scanning the response from the lowest toward the highest frequency and detecting the location of the first resonance, as well as all subsequent local maxima.

The second issue is the fact that the number of relevant feature points may change in the course of optimization run. Yet, in practice, this does not pose a problem because the linear approximation models $\boldsymbol{F}_S^{(i)}$ and $\boldsymbol{L}_S^{(i)}$ used while solving (6.5) are determined at the current point $\boldsymbol{x}^{(i)}$ using small perturbations with respect to $\boldsymbol{x}^{(i)}$. This virtually guarantees feature point consistency for the purpose of an approximation model construction. If the number of feature points changes at the new design $\boldsymbol{x}^{(i+1)}$, the subsequent approximation models are determined using the "new" feature set.

6.2.2 Case Studies

The discussed feature-based design optimization task is illustrated using two design examples: an 11-parameter uniplanar UWB antenna fed through a coplanar waveguide (CPW) (Qing & Chen, 2009) and a seven-variable rectangular-shaped UWB monopole (Koziel & Bekasiewicz, 2016b). Both antenna structures are validated through the measurements of their optimized designs.

6.2.2.1 Case I: Uniplanar UWB Antenna with CPW Feed

Consider the uniplanar UWB antenna shown in Fig. 6.25. The structure consists of a fork-shaped driven element and a ground plane with open slot. It is fed through the CPW line. The antenna is designed on a 0.762-mm-thick Taconic RF-35. The adjustable parameters are $\boldsymbol{x} = [l_0\ l_1\ l_{2r}\ l_{3r}\ l_4\ l_5\ w_1\ w_2\ w_3\ w_4\ g]^T$. The dimensions $w_f = 3.5$ and $s = 0.16$ are constant to ensure 50 Ohm input impedance of the feed

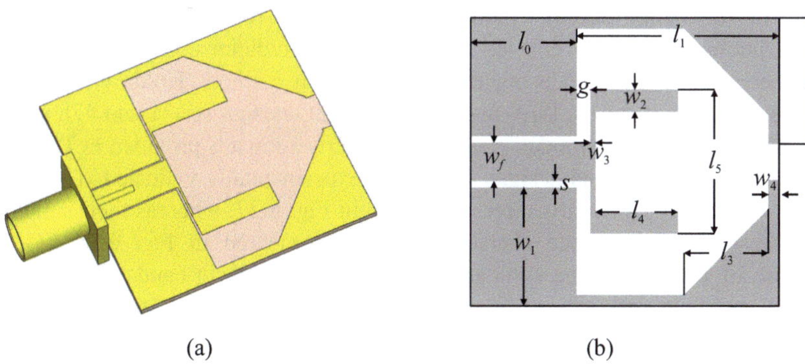

Fig. 6.25 Uniplanar UWB antenna: (**a**) 3D view; (**b**) geometrical details

line. Relative variables are $l_2 = (0.5w_f + s + w_1)\cdot\max\{l_{2r}, l_{3r}\}$ and $l_3 = (0.5w_f + s + w_1)\cdot l_{3r}$. All dimensions except l_{2r} and l_{3r} are in mm. The ranges of design variables are $\boldsymbol{l} = [5\ 15\ 0.2\ 0.2\ 4\ 8\ 7\ 0.5\ 0.2\ 0.2\ 0.2]^T$ and $\boldsymbol{u} = [15\ 25\ 1\ 0.8\ 11\ 16\ 15\ 3.5\ 2.5\ 2\ 2]^T$.

Two models of the structure have been implemented in the CST Microwave Studio and simulated using its transient solver. The high-fidelity model $\boldsymbol{R}_f$ includes 6,800,000 mesh cells (simulation time: 36 minutes), whereas its low-fidelity counterpart contains ~1,200,000 cells (CPU-time: 198 seconds). For the sake of a reliable comparison of the simulation and measurement results, the antenna model is supplemented with the SMA connector.

The considered antenna structure has been optimized with respect to two objectives: (i) minimization of reflection (see Sect. 6.2.1) and (ii) size reduction while ensuring acceptable electrical performance (i.e., $|S_{11}| \leq -10$ dB in 3.1 GHz to 10.6 GHz band). The latter is controlled using the following objective function:

$$U_F(\boldsymbol{x}) = A(\boldsymbol{x}) + \beta_1[(\max\{l_2(\boldsymbol{x}), \ldots l_K(\boldsymbol{x})\} + 10)/10]^2 + + \beta_2\left[\left(\max\{f_1 - f_{1.\text{target}} 0\}\right)/f_{1.\text{target}}\right]^2 \tag{6.21}$$

where $A(\boldsymbol{x})$ is the antenna size, whereas the penalty terms are enforcing the antenna reflection response to stay below -10 dB level within the UWB band, and β_1 and β_2 refer to the respective penalty factors.

The initial design for FBO is $\boldsymbol{x}^0 = [10\ 19\ 0.8\ 0.6\ 7.7\ 12\ 11.3\ 2\ 0.5\ 1\ 0.8]^T$. The final design optimized with respect to reflection is $\boldsymbol{x}^* = [9.29\ 18.36\ 0.76\ 0.75\ 8.93\ 16\ 11.41\ 2.64\ 0.22\ 1.09\ 0.26]^T$. The dimensions of the optimized structure are 26.6 mm × 27.7 mm, and its footprint is 737 mm^2. For the area-wise optimization, the initial design is set to $\boldsymbol{x}^0 = \boldsymbol{x}^*$, and the final design is $\boldsymbol{x}^{**} = [4.86\ 14.86\ 0.92\ 0.59\ 8.57\ 14.72\ 6.68\ 2.8\ 0.57\ 0.2\ 0.31]^T$. The antenna size at $\boldsymbol{x}^{**}$ is 17.2 mm × 19.7 mm = 339 mm^2, and it is 54% smaller compared to $\boldsymbol{x}^*$. The reflection responses of the antenna structure at the initial and optimized designs are shown in Fig. 6.26.

The computational cost of reflection enhancement corresponds to 25.3 simulations of the high-fidelity model $\boldsymbol{R}_f$, whereas—for the footprint-oriented optimization—the design cost is about 31.3 $\boldsymbol{R}_f$ simulations. Both optimizations have been also performed using the pattern-search method (Kolda et al., 2003). The numerical cost for the $|S_{11}|$- and area-wise FBO design is 321 and 478 simulations of the high-fidelity model. A detailed cost breakdown is provided in Table 6.3.

The optimized antenna designs have been experimentally validated. The photographs of the fabricated structures are shown in Fig. 6.27, whereas their simulated and measured performance characteristics are compared in Fig. 6.28. For the response of Fig. 6.28a, the simulated and measured peak in-band reflection is –10.2 dB and -10 dB, respectively. The measured bandwidth is 100 MHz broader than the simulated one. For the antenna response shown in Fig. 6.28b, the peak reflection is -11.4 dB for simulation and -10.7 dB for measurement. Additionally, the measured antenna response is characterized by slightly narrower -10 dB

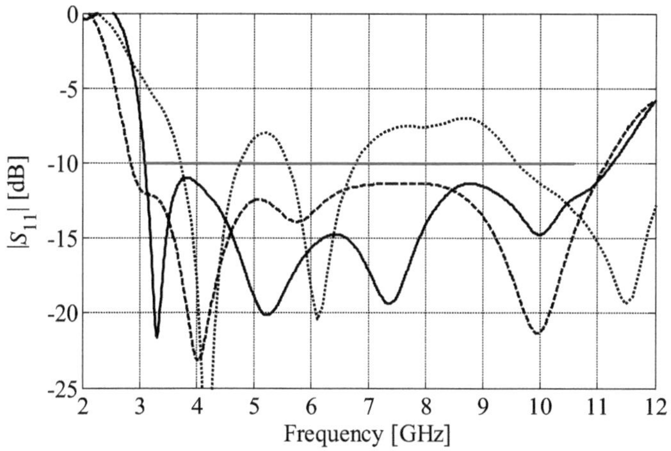

Fig. 6.26 Uniplanar UWB antenna characteristics at: the initial design (····), reflection-wise optimized design (- - -), and the design optimized with respect to size (—)

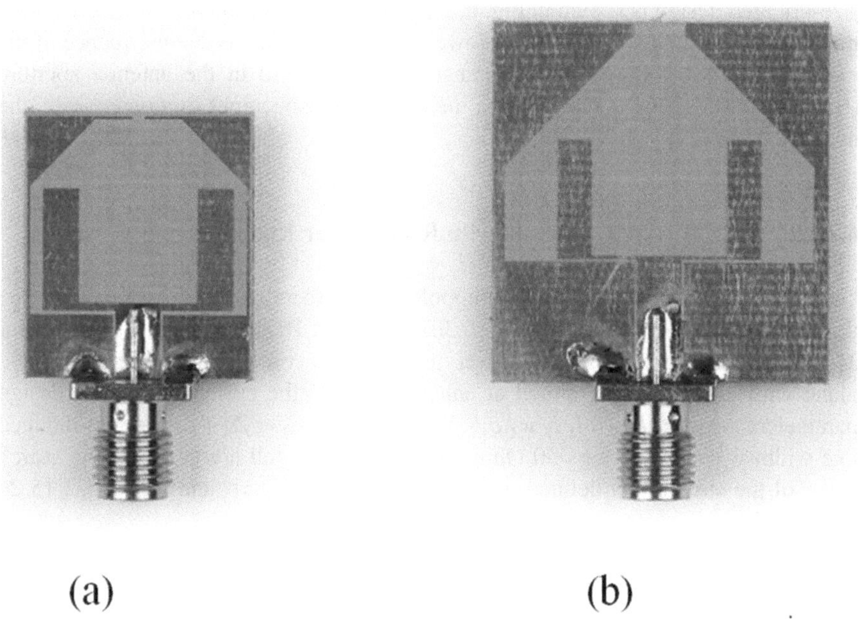

Fig. 6.27 Uniplanar UWB antenna: photographs of the designs optimized w.r.t: (**a**) size and (**b**) in-band reflection

bandwidth (7.8 GHz) with respect to the results obtained in the course of simulation (8.3 GHz). The discrepancies between the characteristics are mostly due to fabrication tolerances as well as the influence of electrically large measurement setup and lack of cable current suppressor (Hertel, 2005). Overall, the agreement between the simulated and the measured antenna responses is acceptable.

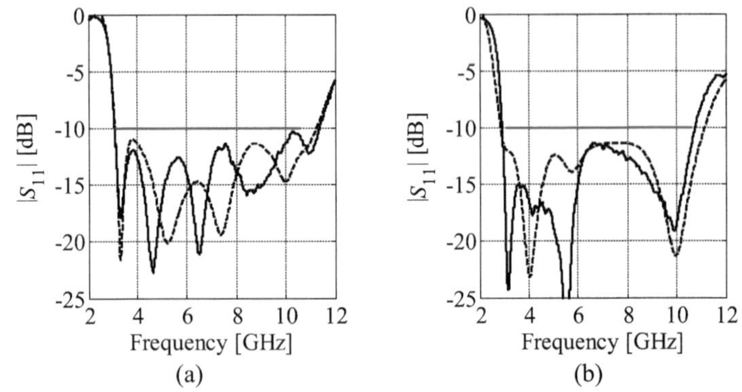

Fig. 6.28 Comparison of simulated (- - -) and measured (—) reflection characteristics of the uniplanar antenna at the designs optimized w.r.t.: (**a**) size, (**b**) in-band reflection

A comparison of the E-field radiation patterns (E-plane) for the reflection- and size-oriented design is shown in Fig. 6.29. As expected, the smaller antenna exhibits more omnidirectional properties for lower frequencies. This is a consequence of the radiation pattern being a Fourier transform of the field in the antenna aperture (Balanis, 2005). Thus, the antenna field response becomes narrower with the increasing size.

6.2.2.2 Case II: UWB Antenna with Rectangular Radiator

Our second example is a UWB monopole with a rectangular-shaped radiator fed through the microstrip line (see Fig. 6.30) (Koziel & Bekasiewicz, 2016b). Similar as the antenna of the previous section, the structure contains the modified ground plane with L-shaped stub aimed at enhancement of the current path. The design parameters are $\boldsymbol{x} = [l_0\ g\ a\ l_1\ l_2\ w_1\ o]^T$. We also have $w_0 = 2o + a$, whereas the feed line width $w_f = 1.7$ ensures 50 Ohm input impedance (all are in mm). The search space of the structure is defined by $\boldsymbol{l} = [4\ {-2}\ 4\ 5\ 1\ 0.5\ 0.5]^T$ and $\boldsymbol{u} = [15\ 2\ 15\ 20\ 10\ 3.5\ 5.5]^T$.

The high- and low-fidelity EM antenna models are both implemented in CST Microwave Studio ($\boldsymbol{R}_f$: ~4,600,000 mesh cells, simulation time 40 minutes, and $\boldsymbol{R}_c$: ~850,000 cells, 2 minutes). As in Sect. 6.2.2.1, the models are supplemented with SMA connectors to ensure reliable simulation results.

The antenna has been optimized with respect to reflection and size. The design objectives for both cases are defined in Sects. 6.2.1 and 6.2.2.1, respectively. For the reflection-wise optimization, the initial design is $\boldsymbol{x}^0 = [7\ 1\ 10\ 11\ 3\ 1\ 1]^T$, and the optimal parameter set is $\boldsymbol{x}^* = [4.73\ 1.15\ 9.74\ 13.76\ 4.88\ 2.96\ 3.38]^T$. The antenna dimensions are 16.5 mm × 21.4 mm = 354 mm^2. For the size-oriented design, the initial geometry is set to $\boldsymbol{x}^0 = \boldsymbol{x}^*$, whereas the optimal parameter set is $\boldsymbol{x}^{**} = [4.57$

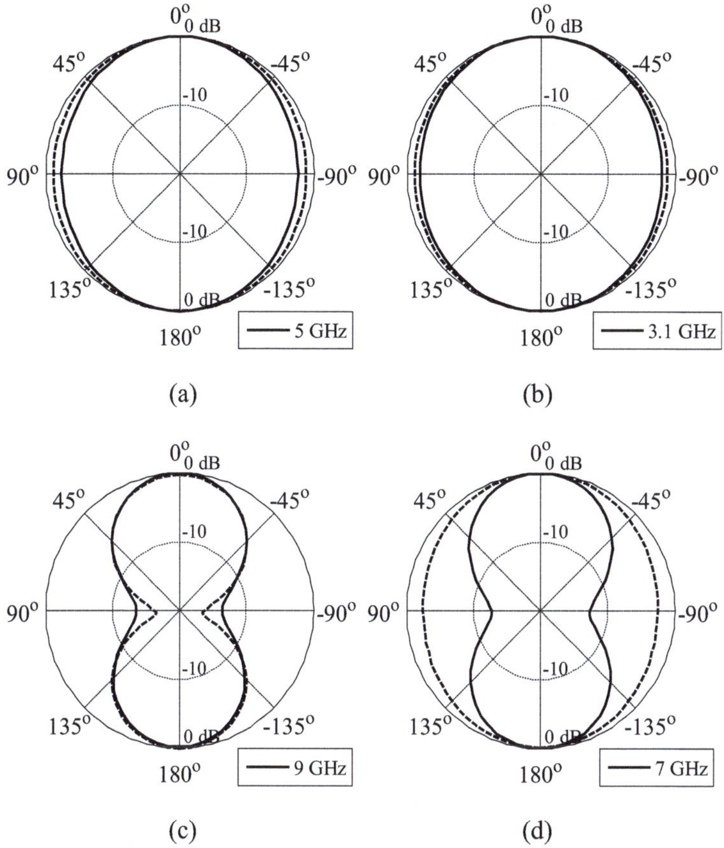

Fig. 6.29 E-field radiation patterns of the uniplanar UWB antenna obtained for the size- (– –) and reflection-wise (—) optimized designs

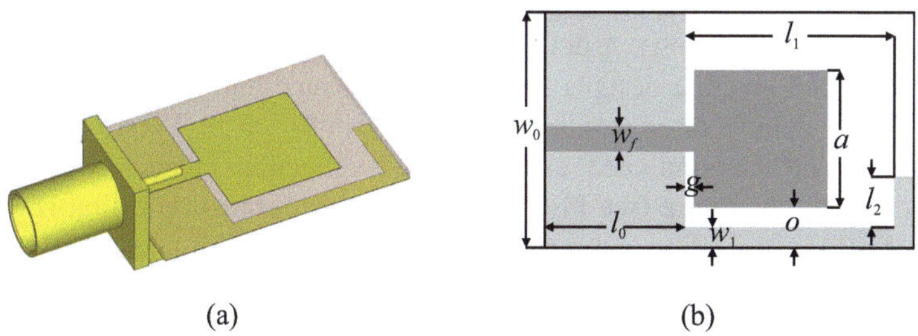

Fig. 6.30 UWB monopole with rectangular radiator: (**a**) 3D view (**b**) geometrical details

1.29 9 13.18 6.32 1.35 1.96]^T with corresponding antenna size of 12.9 mm × 19 mm = 247 mm^2 (over 30% smaller compared to $\boldsymbol{x}^*$). Figure 6.31 provides reflection characteristics at the initial and the optimized designs.

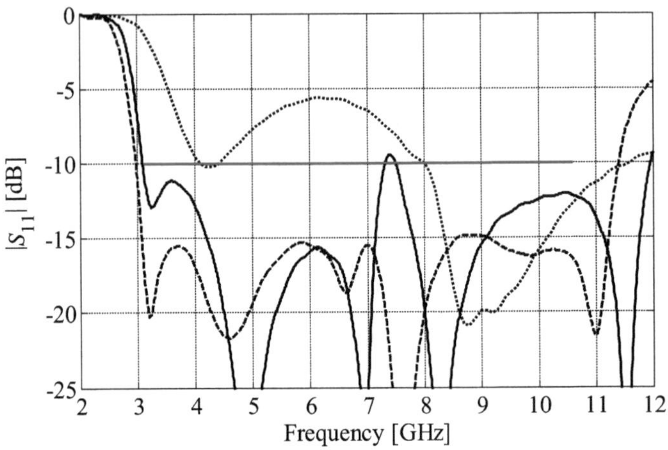

Fig. 6.31 Rectangular UWB monopole characteristics at the initial design (····), reflection-wise optimized design (- - -), and the design optimized with respect to size (—)

Table 6.4 UWB monopole with rectangular radiator: optimization cost breakdown

Optimization method	Design objective	Number of model evaluations		Total cost	
		Low-fidelity	High-fidelity	Absolute [hours]	Relative to $\boldsymbol{R}_f$
FBO	$\|S_{11}\|$	$72 \times \boldsymbol{R}_c$	$11 \times \boldsymbol{R}_f$	9.7	14.6
	Size	$93 \times \boldsymbol{R}_c$	$14 \times \boldsymbol{R}_f$	12.4	18.6
Pattern search	$\|S_{11}\|$	–	$231 \times \boldsymbol{R}_f$	154	231
	Size	–	$324 \times \boldsymbol{R}_f$	216	324

The numerical cost of rendering monopole design corresponds to 14.6 and 18.6 evaluations of the high-fidelity model for the reflection- and size-oriented optimization, respectively. At the same time, the cost of pattern-search-based design is 231 and 324 $\boldsymbol{R}_f$ simulations. A detailed cost breakdown is gathered in Table 6.4.

The obtained antenna designs have been fabricated and measured. The photographs of the manufactured structures are shown in Fig. 6.32. Figure 6.33 provides comparisons of the simulated and measured reflection characteristics of both designs. For the first case (see Fig. 6.33a), the peak in-band reflection is −9.2 dB and −9.4 dB for simulation and measurement, respectively. Simulated and measured responses of the second design (cf. Fig. 6.33b) are −14.8 dB and −14.1 dB. Overall, the agreement between the simulation and measurement results is good.

Figure 6.34 shows radiation patterns of the antenna. Similarly as in Sect. 6.2.2.1, the antenna design optimized for size reduction features slightly more omnidirectional characteristics. However, at 7 GHz frequency, this effect is reduced due to increased shielding of the ground plane stub. For the 9 GHz frequency, the characteristics are similar.

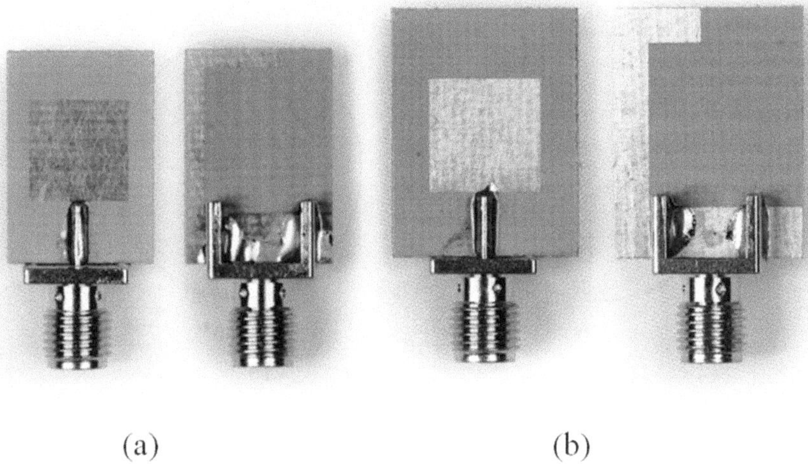

Fig. 6.32 Photographs of rectangular monopole designs optimized w.r.t: (a) size; and (b) in-band reflection

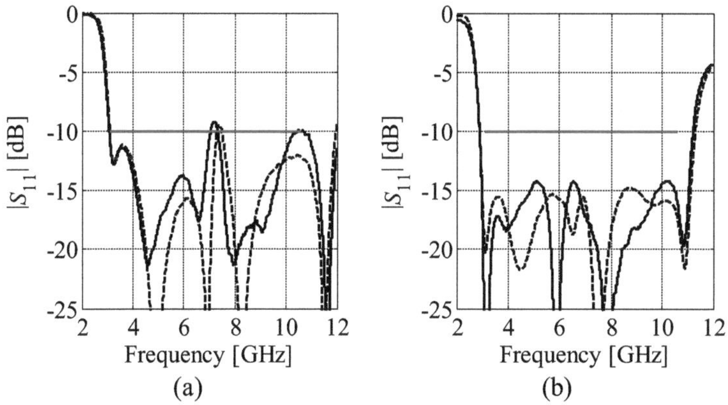

Fig. 6.33 Comparison of simulated (- - -) and measured (—) reflection characteristics of the rectangular UWB monopole at the designs optimized w.r.t.: (a) size and (b) in-band reflection

This section considered rapid design optimization of UWB antennas by means of response features. Appropriate definition of the objective function allows for simultaneous control of the lower frequency of the antenna response as well as the maximum reflection level within the entire UWB band. As demonstrated through the design of a uniplanar UWB structure and a rectangular-shaped UWB monopole, utilization of response features and variable-fidelity EM simulations ensures low cost of the optimization process. Numerical results indicated superiority of the FBO algorithm over conventional techniques, here, pattern search. The considered two antenna examples—four design cases—have been positively verified by means of reflection measurements.

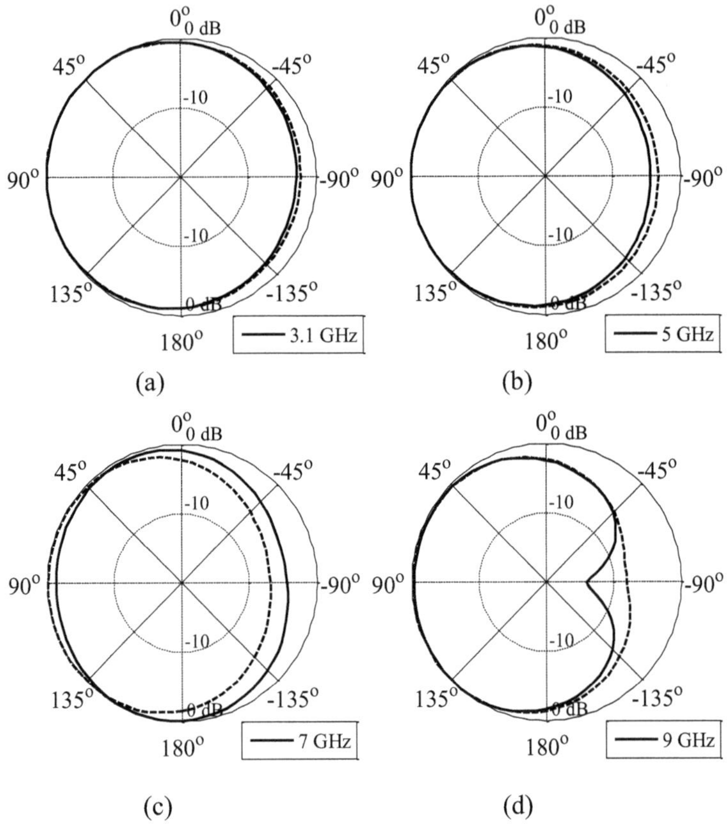

Fig. 6.34 E-plane radiation patterns of the rectangular UWB monopole obtained for the size- (– –) and reflection-wise (—) optimized designs

6.3 FBO with Circuit Decomposition

This section discusses accelerated design optimization of miniaturized microwave structures with structure decomposition (Bekasiewicz & Koziel, 2016) on the example of a compact impedance matching transformer. The first stage of the design process involves feature-based optimization of the elementary blocks of the structure (i.e., CMRCs). The entire structure composed of the optimized cells is then fine-tuned using local response surface approximation surrogates and response correction techniques.

6.3.1 Conventional and Compact Circuit. Structure Decomposition

Conceptual illustration of a miniaturization of the impedance matching transformer using CMRCs is shown in Fig. 6.35. A conventional structure is composed of transmission lines, which are replaced by CMRCs in its compact version. Here, we utilize structure decomposition to speed up the design process. The design specifications are transformed to the level of individual CMRCs (using the optimized dimensions of the conventional transformer). The CMRCs are optimized using feature-based surrogate-based technique. Finally, a fast tuning procedure is carried out for the entire circuit. It is necessary because cross-coupling effects between the cells are not taken into account when optimizing CMRCs separately. The tuning process exploits variable-fidelity EM simulation models and response correction techniques.

In the first step of the process, the conventional transformer circuit is optimized. This gives us the initial values for the input/output line widths w_k of the CMRCs and determine their target electrical lengths θ_k. Conventional transformer is designed by (i) optimization of its equivalent network model and (ii) design closure by means of iterative construction and optimization of its local approximation models.

6.3.2 Cell Optimization Using Response Features

Even though the EM model of CMRC is much faster than the model of the entire transformer, cell optimization is one of the major contributors to the overall design cost. This is because CMRC optimization has to be performed several times (depending on the number of transformer sections), and the design problem is more complex due to necessity of controlling both the return loss and electrical length of the cell. As in other examples of this chapter, in order to reduce that cost,

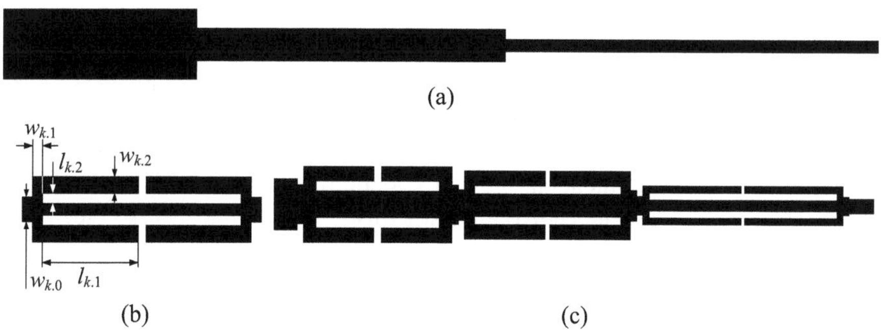

Fig. 6.35 Impedance matching transformer miniaturization using CMRCs: (**a**) conventional structure, (**b**) example CMRC, (**c**) compact transformer structure

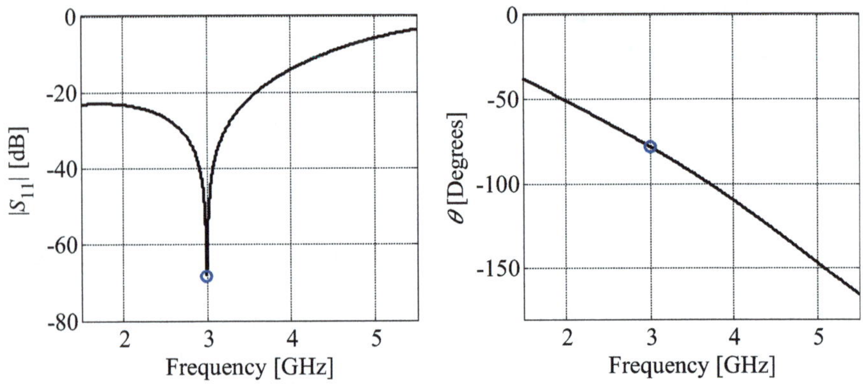

Fig. 6.36 Typical response of CMRC (—) and the feature points (o): reflection (left) and electrical length (right)

feature-based optimization (FBO) paradigm is utilized. Figure 6.36 shows a typical response of CMRC with the two feature points of interest marked (minimum of the return loss characteristic and the electrical length of the cell at the frequency of interest f_0). Despite high nonlinearity of the return loss characteristic, the frequency location of its minimum depends almost linearly on the geometry parameters of the cell as shown in Fig. 6.37. Consequently, optimization of the cell is much faster (compared to handling *S*-parameter response vs. frequency) when reformulated in terms of the features. More specifically, we minimize the following objective function

$$U(\boldsymbol{x}) = l(\boldsymbol{x}) + \beta_1 \left[\frac{f(\boldsymbol{x}) - f_0}{f_0} \right]^2 + \beta_2 \left[\frac{\theta(\boldsymbol{x}) - \theta_0}{\theta_0} \right]^2 \tag{6.22}$$

where l is the return loss value at its minimum, f is the frequency location of the minimum and θ is the electrical length of the cell (all being function of geometry vector $\boldsymbol{x}$) and f_0 and θ_0 are the target values. It should be noted that the objective function (6.22) permits simultaneous control of the CMRC resonance depth (which is the primary objective here), allocation of the resonance at the required frequency f_0, and obtaining the required electrical length of the cell. By increasing the values of the penalty factors β_1 and β_2, one can enforce smaller errors $|f(\boldsymbol{x}) - f_0|$ and $|\theta(\boldsymbol{x}) - \theta_0|$. With the above formulation, the cell is optimized using gradient search with finite differentiation, embedded in the trust region framework (see also Sect. 6.1.1.3).

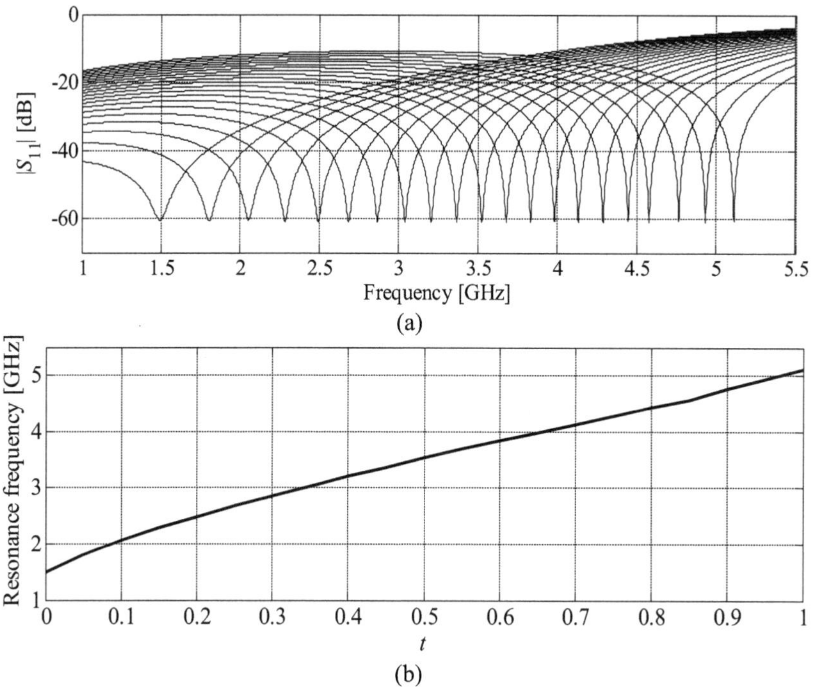

Fig. 6.37 Variability of CMRC responses in the *S*-parameter and feature response spaces: (**a**) family of reflection responses of CMRC evaluated along a line segment $\boldsymbol{x} = t{\cdot}\boldsymbol{x}_A + (1 - t){\cdot}\boldsymbol{x}_B$ parameterized by $0 \le t \le 1$ (with $\boldsymbol{x}_A$ and $\boldsymbol{x}_B$ being some arbitrarily selected designs); (**b**) frequency location of the feature point corresponding to the CMRC resonance. It can be observed that despite the fact that $|S_{11}|$ response is highly nonlinear, the frequency coordinate of the response feature point changes in a close-to-linear fashion

6.3.3 Simulation-Based Transformer Tuning

The entire transformer has to be further tuned because the CMRCs have been optimized at the coarse discretization EM simulation level, and cross-couplings between the cells have not been taken into account.

We denote by $\boldsymbol{x}$ the aggregated geometry parameter vector of the entire circuit; $\boldsymbol{x}^{(0)}$ is the design representing the optimized CMRCs, whereas, as in previous sections, $\boldsymbol{R}_c$ and $\boldsymbol{R}_f$ denote the coarse- and fine-discretization EM model of the transformer structure. The problem at hand can be formulated as

$$\boldsymbol{x}^* = \arg\min_{\boldsymbol{x}} U_T\left(\boldsymbol{R}_f(\mathbf{x})\right) \tag{6.23}$$

where U_T is the objective function encoding the specifications at the transformer level. Here, $U_T(\boldsymbol{R}_f(\boldsymbol{x}))$ is the maximum $|S_{11}|$ over the frequency range of interest. The tuned design is found iteratively as

$$\boldsymbol{x}^{(i+1)} = \arg \min_{\boldsymbol{x}^{(i)} - \boldsymbol{d} \leq \boldsymbol{x} \leq \boldsymbol{x}^{(i)} + \boldsymbol{d}} U_T\left(\boldsymbol{R}_q(\boldsymbol{x}) + \left[\boldsymbol{R}_f\left(\boldsymbol{x}^{(i)}\right) - \boldsymbol{R}_c\left(\boldsymbol{x}^{(i)}\right)\right]\right) \tag{6.24}$$

where $\boldsymbol{R}_q$ is the second-order polynomial (without mixed terms) approximating $\boldsymbol{R}_c$ in the vicinity of the current design $\boldsymbol{x}^{(i)}$. It is defined as

$$\boldsymbol{R}_q(\boldsymbol{x}) = \boldsymbol{R}_q\left([x_1 \;\; x_2 \;\; \ldots \;\; x_n]^T\right) = q_0 + \sum_{k=1}^{n} q_k x_k + \sum_{k=1}^{n} q_{n+k} x_k^2 \tag{6.25}$$

The model coefficients are found by solving linear regression problems $\boldsymbol{R}_q(\boldsymbol{x}_B{}^{(j)}) = \boldsymbol{R}_c(\boldsymbol{x}_B{}^{(j)})$, $j = 0, 1, \ldots, 2n$, where $\boldsymbol{x}_B{}^{(j)}$ are the base designs with $\boldsymbol{x}_B{}^{(0)} = \boldsymbol{x}^{(i)}$ and the rest of the points being $2n$ perturbations around $\boldsymbol{x}^{(i)}$ (Koziel, 2010). Note that the model (6.25) is interpolative; in particular, we have $\boldsymbol{R}_q(\boldsymbol{x}^{(i)}) = \boldsymbol{R}_c(\boldsymbol{x}^{(i)})$. The correction term $[\boldsymbol{R}_f(\boldsymbol{x}^{(i)}) - \boldsymbol{R}_c(\boldsymbol{x}^{(i)})]$ accounts for the misalignment between $\boldsymbol{R}_c$ and $\boldsymbol{R}_f$ at $\boldsymbol{x}^{(i)}$. In practice, up to three iterations of (6.24) are sufficient for the tuning process to converge. At the tuning stage, all CMRC parameters, including the line widths w_k are adjusted.

6.3.4 *Optimization Algorithm*

The overall optimization flow can be summarized as follows:

1. Optimize conventional transformer to obtain the target electrical lengths θ_k and the input/output line widths w_k of CMRCs.
2. Using FBO, optimize CMRCs for the target lengths θ_k and assuming line widths w_k obtained in Step 1 (cf. Sect. 6.3.2).
3. Fine-tune the entire transformer (cf. Sect. 6.3.3) using the CMRC designs obtained in Step 2 as a starting point.

Note that the high-fidelity model $\boldsymbol{R}_f$ is only utilized at the tuning stage (Step 3). As mentioned before, because the pre-tuning design is already reasonably good, only two or three iterations of (6.24) are sufficient to converge to the final transformer dimensions.

6.3.5 *Results*

Our verification case is a three-section 50 to 100 Ohm transformer shown in Fig. 6.35c. The structure is implemented on a 0.762-mm-thick Taconic RF-35 substrate ($\varepsilon_r = 3.5$, $\tan\delta = 0.0018$). The design variables are $\boldsymbol{x} = [l_{1.1} \; l_{1.2} \; w_{1.1} \; w_{1.2} \; w_{1.0} \; l_{2.1} \; l_{2.2} \; w_{2.1} \; w_{2.2} \; w_{2.0} \; l_{3.1} \; l_{3.2} \; w_{3.1} \; w_{3.2} \; w_{3.0}]^T$. We use three EM simulation models, all implemented in CST [13]: (i) CMRC model $\boldsymbol{R}_C$ (~180,000 mesh cells,

Table 6.5 Optimized dimensions of CMRCs

Cell	Geometry parameter					Required length θ	Obtained length θ
	$l_{k.1}$	$l_{k.2}$	$w_{k.1}$	$w_{k.2}$	$w_{k.0}$		
1	3.18	0.50	0.71	0.55	1.35	−73.3°	−73.3°
2	3.31	0.50	0.39	0.53	0.91	−75.8°	−75.8°
3	3.58	0.32	0.20	0.52	0.58	−78.1°	−78.0°

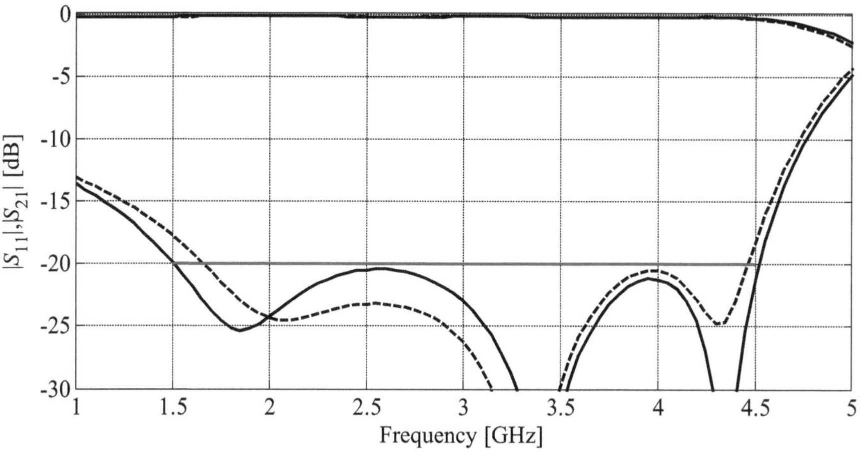

Fig. 6.38 Three-section impedance transformer of Fig. 6.35(a): pre- (- - -) and post-tuning (—) responses

simulation time 50 seconds), (ii) low-fidelity transformer model $\boldsymbol{R}_c$ (~290,000 cells, 2.5 minutes), and (iii) high-fidelity transformer model $\boldsymbol{R}_f$ (~870,000 cells, 10 minutes). Design specifications are $|S_{11}| \leq -20$ dB for 1.5 GHz to 4.5 GHz.

The electrical lengths of the transformer sections and the widths of the input/output lines (parameters $w_{k.0}$, $k = 1, 2, 3$, in the vector $\boldsymbol{x}$) have been obtained by optimizing the conventional circuit. Each CMRC has been optimized using FBO algorithm (cf. Sect. 6.3.2). The design specifications for the cells and their optimized dimensions are gathered in Table 6.5. The final design has been obtained by executing three iterations of the tuning procedure (Sect. 6.3.3). The optimized parameters are $\boldsymbol{x}^* = [3.20\ 0.46\ 0.65\ 0.52\ 1.29\ 3.38\ 0.48\ 0.37\ 0.49\ 0.86\ 3.51\ 0.29\ 0.19\ 0.52\ 0.53]^T$. The pre- and post-tuning responses of the transformer are shown in Fig. 6.38.

The overall optimization cost corresponds to only 31 evaluations of the high-fidelity transformer model (~5 hours of CPU time), which is low concerning a large number of geometry parameters involved in the process. The cost includes 56 evaluations of the CMRC model (for cell optimization), 93 evaluations of the low-fidelity model $\boldsymbol{R}_c$ (for construction of the approximation model $\boldsymbol{R}_q$, and 4 evaluations of the high-fidelity model $\boldsymbol{R}_f$. For the sake of comparison, conventional optimization of CMRCs using pattern search involves 210 model evaluations, which

Table 6.6 Three-section impedance matching transformer design. Optimization cost

Algorithm	Design stage	Computational cost		
		Number of model evaluations	Absolute time [min]	Relative to transformer model $\boldsymbol{R}_T$
Decomposition-based optimization with response features	CMRC 1	$21 \times \boldsymbol{R}_C$	17.5	1.8
	CMRC 2	$17 \times \boldsymbol{R}_C$	14.2	1.4
	CMRC 3	$18 \times \boldsymbol{R}_C$	15.0	1.5
	Design tuning	$93 \times \boldsymbol{R}_c + 4 \times \boldsymbol{R}_f$	272.5	27.2
	Total	N/A	319.2	**31.9**
Decomposition-based optimization[a]	CMRC 1	$71 \times \boldsymbol{R}_C$	59.2	5.9
	CMRC 2	$72 \times \boldsymbol{R}_C$	60.0	6.0
	CMRC 3	$67 \times \boldsymbol{R}_C$	55.8	5.6
	Design tuning	$93 \times \boldsymbol{R}_c + 4 \times \boldsymbol{R}_f$	272.5	27.2
	Total	N/A	447.5	**44.7**

[a]CMRCs optimized using pattern search

translates into the overall optimization cost of about 45 evaluations of $\boldsymbol{R}_f$ (~8 h, which is about 60 percent higher than when using FBO). The detailed cost breakdown has been shown in Table 6.6.

This section presented feature-based optimization with circuit decomposition. The discussed approach combines several techniques, including (i) structure decomposition, (ii) feature-based optimization of elementary building blocks of the circuit (here, CMRCs), and (iii) simulation-driven tuning exploiting local approximation models and variable-fidelity EM simulations. As demonstrated, feature-based design of CMRC leads to considerable reduction of the design cost as compared to the benchmark method even though the latter utilizes both decomposition and variable-fidelity tuning. The overall optimization cost for a verification example of a 3-section is only about 30 evaluations of the high-fidelity EM model.

6.4 Feature-Based Size Reduction

The last example of this section is miniaturization of multi-band antennas by means of response features (Koziel & Bekasiewicz, 2018). The fundamental difficulty of the considered design task is the necessity of controlling several antenna resonances, both in terms of their frequency allocations and depths, along with attempting to reduce physical dimensions (here, footprint areas) of the structure. This has been made possible by employing response features technology.

6.4.1 Problem Formulation

In this section, we use the following notation. Let $\boldsymbol{R}(\boldsymbol{x})$ denote the computational (EM-simulated) model of the antenna structure under design, where $\boldsymbol{x}$ is a vector of adjustable geometry parameters. We also denote by $A(\boldsymbol{x})$ the antenna size and by $E(\boldsymbol{x})$ a merit function that represents the antenna performance with respect to electrical specifications (here, the reflection level for the operating frequencies of interest). The design problem can be formulated as

$$\boldsymbol{x}^* = \arg\min_{\boldsymbol{x}} U(\boldsymbol{R}(\boldsymbol{x})) = \arg\min_{\boldsymbol{x}} U(A(\boldsymbol{x}), E(\boldsymbol{x})) \tag{6.26}$$

where $U(\boldsymbol{R}(\boldsymbol{x}))$ is a scalar objective function calculated based on the response $\boldsymbol{R}(\boldsymbol{x})$. A particular definition of U depends on the design goals. Here, the primary objective is size reduction and the objective function is defined as

$$U(A(\boldsymbol{x}), E(\boldsymbol{x})) = A(\boldsymbol{x}) + \beta \cdot c(E(\boldsymbol{x}))^2 \tag{6.27}$$

In (6.27), c is a penalty function that evaluates possible violation of the electrical specifications, and β is a penalty factor, the value of which is set to ensure considerable contribution of the penalty term in case violation of the performance specifications above acceptable level. Here, the problem (1) is solved using a trust region gradient search (Conn et al., 2000), which is delineated in Sect. 6.1.1.3.

6.4.2 Controlling Reflection Response Using Response Features

An important issue in the design of multi-band antennas is that their reflection response is a highly nonlinear function of frequency. This often results in inferior performance of the optimization algorithm and makes it difficult to handle other performance figures, here, the antenna size. Thus, the optimization task (6.26) is solved using FBO as the way of improving the optimization process efficiency. Typical choice of response features for multi-band antennas has been illustrated in Fig. 6.39.

The feature points corresponding to antenna resonances are useful to control antenna operating frequencies, whereas the −10 dB points are convenient to control the antenna bandwidth. In this section, we are mostly concerned with an appropriate allocation of the operating frequencies and the resonances depth; therefore the first type of characteristic points is utilized. As stated before, the major benefit of using features is that their dependence on antenna geometry parameters is much less nonlinear than for original responses as shown in Fig. 6.35c.

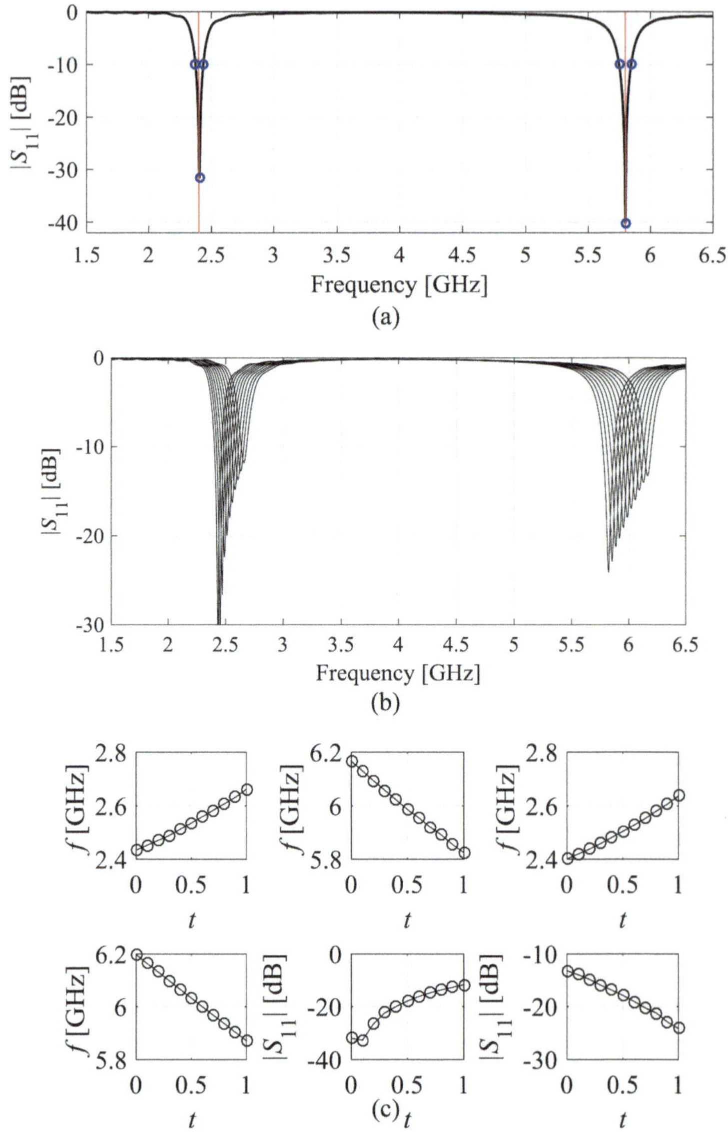

Fig. 6.39 Design using response features: (**a**) typical response of a multi-band antenna (here, dual-band) with several response features marked (circles), corresponding to the frequency and level locations of the antenna resonances, as well as frequencies corresponding to −10 dB reflection levels; (**b**) reflection responses of the antenna structure evaluated along an arbitrary line segment in the design space (parameterized using $0 \le t \le 1$); (**c**) coordinates of selected feature points: dependence of feature point coordinates on geometry parameters is much less nonlinear compared to original frequency responses, thus easier to model

6.4.3 Objective Function for FBO Design

In this section, we provide the formulation of the objective function for the FBO algorithm. As in previous sections, $\boldsymbol{F}(\boldsymbol{x}) = [f_1(\boldsymbol{x})\ f_2(\boldsymbol{x})\ \ldots\ f_N(\boldsymbol{x})]^T$ is the vector of resonant frequencies for the antenna at hand, whereas $\boldsymbol{L}(\boldsymbol{x}) = [l_1(\boldsymbol{x})\ l_2(\boldsymbol{x})\ \ldots\ l_N(\boldsymbol{x})]^T$ stands for the reflection coefficient values at the resonant frequencies $F(\boldsymbol{x})$, respectively, where N represents the number of antenna bands.

At the level of feature points, the objective function can be reformulated as follows:

$$U_F(A(\boldsymbol{x}), F(\boldsymbol{x}), L(\boldsymbol{x})) = A(\boldsymbol{x}) + \beta_1 \cdot \left\| F(\boldsymbol{x}) - F_{\text{target}} \right\|^2 + \beta_2 \cdot c(L(\boldsymbol{x}))^2 \tag{6.28}$$

In (6.28), the antenna size $A(\boldsymbol{x})$ is the primary objective, whereas the remaining two terms are penalty functions introduced to control the frequency allocation of the antenna resonances and the resonance depths. Here, F_{target} is a vector of desired operating frequencies, whereas c is a penalty function defined as $c(L(\boldsymbol{x})) = \max\{\max\{j = 0, \ldots, N: l_j(\boldsymbol{x}) + S_{\max}, 0\}/S_{\max}$, and β_1 and β_2 are penalty coefficients. The second term of (6.28) is a norm-wise difference between the target and the actual resonant frequencies which allows us to enforce the antenna operating frequencies to assume the required values. The last term penalizes the design in case the reflection level at any of the resonances exceeds $S_{\max}$ (user-defined). The latter allows for an approximate control of the resonance depth at the optimized design.

The design problem according to FBO is the same as (6.26) except it utilizes the objective function (6.28). The feature-based optimization algorithm takes the form of

$$\boldsymbol{x}^{(i+1)} = \arg \min_{\boldsymbol{x};\ \left\| \boldsymbol{x} - \boldsymbol{x}^{(i)} \right\| \le \boldsymbol{d}^{(i)}} U_F\left(A(\boldsymbol{x}), \boldsymbol{F}_S^{(i)}(\boldsymbol{x}), \boldsymbol{L}_S^{(i)}(\boldsymbol{x})\right) \tag{6.29}$$

where $\boldsymbol{F}_S^{(i)}(\boldsymbol{x})$ and $\boldsymbol{L}_S^{(i)}(\boldsymbol{x})$ are linear expansion models of $\boldsymbol{F}(\boldsymbol{x})$ and $\boldsymbol{L}(\boldsymbol{x})$, respectively. The necessary Jacobians are estimated using finite differentiation.

6.4.4 Results

The design optimization approach of this section is demonstrated using two antennas: a dual-band quasi-patch antenna (Koziel & Bekasiewicz, 2017c) and a triple-band uniplanar dipole antenna (Chen et al., 2006). The numerical results are validated experimentally.

6.4.4.1 Dual-Band Quasi-Patch Antenna

Consider a planar dual-band antenna shown in Fig. 6.40. The structure is a slightly modified version of the design of the design proposed by (Koziel & Bekasiewicz, 2017c). Introduced changes ensure consistency of the geometry for the selected ranges of design parameters. The antenna is composed of two radiators connected in series, i.e., microstrip patch with inset feed and a rectangular monopole, and fed through a 50 Ohm microstrip line. Moreover, its ground plane is trimmed to increase the number of degrees of freedom for better control of the reflection characteristic. The vector of design parameters is $\boldsymbol{x} = [L\ l_1\ l_2\ l_{\mathrm{sr}}\ W\ w_1\ w_2\ o_r\ o_0\ l_0]^T$, with $o = (w_2 + o_0 + l_1)\cdot o_r$ and $l_s = W\cdot l_{\mathrm{sr}}$. All variables except o_r and l_{sr} (which are unit-less) are in mm. Dimension w_0 is fixed to 1.7 mm to ensure 50 Ohm input impedance, whereas w_s is set to 0.5 mm. The latter parameter is set constant due to the lack of its effect on antenna size.

The structure is implemented on a 0.762-mm-thick Rogers RO4350B dielectric substrate with permittivity of 3.48 and loss tangent of 0.0031. The electromagnetic model is prepared in CST Microwave Studio and evaluated using its time domain solver. It consists of ~350,000 mesh cells, and its average simulation time on a dual Intel Xeon E5540 machine with 64 GB RAM is 5 min.

The structure footprint for the design objective (6.28) is defined as $A(\boldsymbol{x}) = (\max\{W, l_2\} + 2o_0)\cdot(l_0 + L + l_1 + w_2 + o_0)$. The antenna reflection should be maintained at the level of $S_{\max} = -15$ dB for the following vector of operating frequencies $F_{\mathrm{target}} = [2.4\ 5.8]$ GHz. The feature points w.r.t. structure resonant frequencies and reflection levels for the given input parameters are obtained through evaluation of the linear models of $\boldsymbol{F}(\boldsymbol{x})$ and $\boldsymbol{L}(\boldsymbol{x})$.

The search space bounds $\boldsymbol{l}/\boldsymbol{u}$ are $\boldsymbol{l} = [10\ 0.5\ 10\ 0.1\ 10\ 0.2\ 1.0\ {-1}\ 0.5\ 5]^T$ and $\boldsymbol{u} = [20\ 5\ 20\ 0.5\ 20\ 2.0\ 11.0\ 1\ 10\ 15]^T$. It should be noted that positive and negative values of the parameter o_r correspond to trimming the ground plane below the monopole and the patch radiating elements, respectively (see also Fig. 6.40).

The antenna has been miniaturized using procedure of Sect. 6.4.3. The initial design is set to $\boldsymbol{x}_{\mathrm{init}} = [14\ 3.5\ 12.5\ 0.3\ 15.5\ 0.6\ 7\ 0.3\ 7\ 10]^T$. In the first step, the structure has been optimized to minimize the reflection level at the F_{target} frequencies. The obtained design—utilized as a starting point for size reduction—is $\boldsymbol{x}^{(0)} = [14.18\ 3.03\ 12.5\ 0.37\ 14.77\ 0.45\ 7.78\ 0.33\ 5.12\ 9.53]^T$. It should be noted that $\boldsymbol{x}^{(0)}$ has been found only for the purpose of estimating the antenna

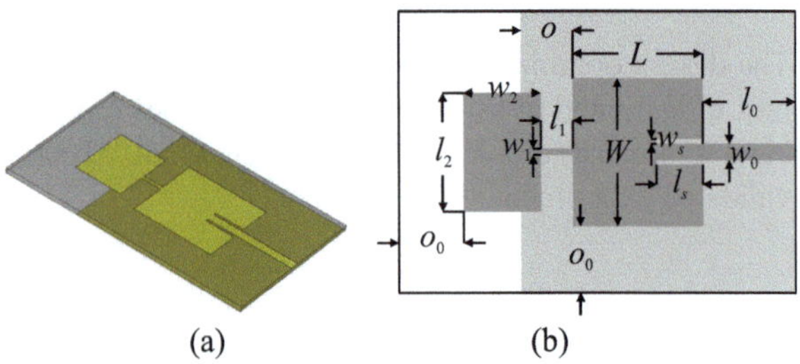

Fig. 6.40 Planar dual-band antenna (Koziel & Bekasiewicz, 2017c): (**a**) visualization, (**b**) geometrical details.

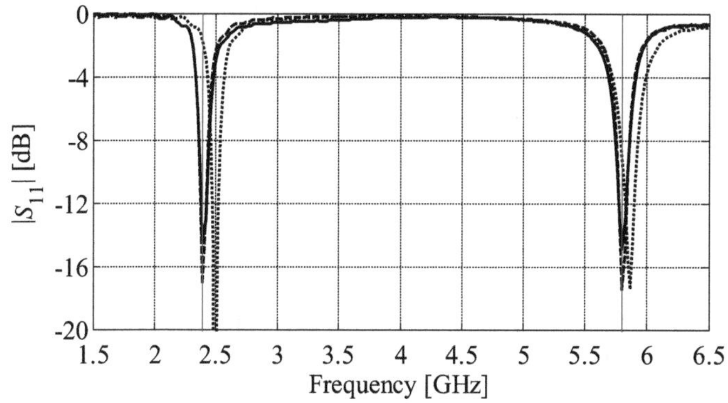

Fig. 6.41 Reflection characteristics of considered dual-band antenna at $\boldsymbol{x}_{\text{init}}$ (···), $\boldsymbol{x}^{(0)}$ (– –) and $\boldsymbol{x}^*$ (—)

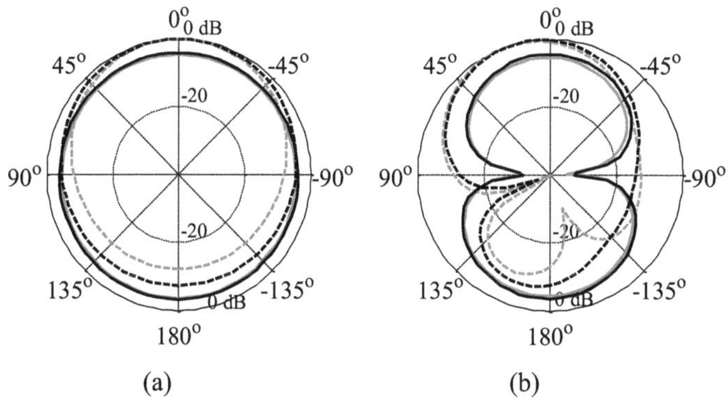

Fig. 6.42 E-field radiation patterns of the antenna for $\boldsymbol{x}^{(0)}$ (gray) and $\boldsymbol{x}^*$ (black) at 2.4 GHz (—) and 5.8 GHz (– –): (**a**) H-plane, (**b**) E-plane

miniaturization rate. Finally, the design $\boldsymbol{x}^* = [14.36\ 3.22\ 12.03\ 0.48\ 12.03\ 0.2\ 7\ 0.44\ 1.14\ 7.66]^T$ has been obtained. The antenna footprints for $\boldsymbol{x}^{(0)}$ and $\boldsymbol{x}^*$ are 991 mm^2 and 478 mm^2, respectively. The size reduction for the miniaturized design—calculated as $(1 - A(\boldsymbol{x}^*)/A(\boldsymbol{x}^{(0)}))\cdot 100\%$—is 52 percent.

Figure 6.41 shows reflection characteristics of the considered antenna at all three designs. The obtained results indicate that both $\boldsymbol{x}^{(0)}$ and $\boldsymbol{x}^*$ match the requirements concerning the center frequencies. However, the miniaturized design violates the –15 dB requirement by 0.5 dB and 0.2 dB at 2.4 GHz and 5.8 GHz, respectively. This is because the penalty function approach utilized in (6.28) only allows for an approximate satisfaction of this sort of constraint. Furthermore, the focus has been on appropriate allocation of the operating frequencies, which is quite precise as indicated in Fig. 6.41. E-field radiation patterns obtained for designs $\boldsymbol{x}^{(0)}$ and $\boldsymbol{x}^*$ in E- and H-plane are compared in Fig. 6.42. The obtained results indicate that the miniaturized structure features more omnidirectional characteristics. Particularly, the effect is noticeable for the higher operating frequency.

It should be reiterated that reduction of the antenna footprints has been achieved merely by adjustment of their geometry parameters without any topological modifications of the respective structures. This demonstrates the importance and benefits of rigorous numerical optimization when it comes to achieving the best possible antenna performance.

6.4.4.2 Triple-Band Uniplanar Dipole Antenna

The second example is a tri-band dipole shown in Fig. 6.43. The antenna is also implemented on a 0.762-mm-thick Rogers RO4350B substrate. The structure is based on (Chen et al., 2006) and modified in order to achieve a more compact geometry. It consists of a three narrow slots separated by two thicker ones. The radiator is fed through a 50 Ohm coplanar waveguide. The size of the antenna is reduced by folding narrow slots so that the available surface is better utilized. The design variables are $\boldsymbol{x} = [l_{0r}\ l_{1r}\ l_2\ l_{3r}\ l_4\ l_5\ w_1\ w_2\ w_3\ w_4\ w_5\ o]^T$. Consistency of the geometry is ensured using the following set of relative parameters: $l_{10} = \min\{\max\{0.5{\cdot}l_{1r}, l_2\}, l_5\}$, $d_3 = \max\{l_{10} + w_2 + w_3, l_2 + w_2 + w_3\}$, $d_1 = \min\{\max\{l_{3r} - d_3, 0.01\}, l_{1r} - l_{10}\}$, $l_{30} = \min\{d_3, l_{3r}\}$, $l_{31} = \max\{l_{3r} - d_3, 0.01\}$, $l_0 = l_{0r} + \max\{o + d_1, l_{00}\}$, $l_{11} = \max\{\min\{l_{1r} - l_{10} - l_{31}, 2{\cdot}w_1 + w_3 + 2{\cdot}w_2\}, 0.01\}$, and $l_{12} = \max\{l_{1r} - l_{10} - l_{31} - l_{11}, 0.01\}$. The dimensions $w_0 = 3$ and $s_0 = 0.15$ are fixed to ensure 50 Ohm input impedance. Moreover, l_{00} is set to 10. The unit for all parameters is mm.

The antenna model is implemented in CST Studio (~400,000 cells, 7 min simulation). The design objective is to minimize antenna geometry defined as a rectangle $A(\boldsymbol{x}) = X{\cdot}Y$, where $X = (2\ (o + \max\{l_{10} + (l_{11} - w_1), l_5\}) + w_0)$ and $Y = \max\{o + l_0 + w_1 + w_2 + w_3 + w_4 + w_5, l_0 - d_1 + l_{12} + o\}$. Reduction of antenna size is to be achieved together with maintaining the reflection below -15 dB for the following frequencies $F_{\text{target}} = [2.45\ 3.5\ 5.35]$ GHz. The design bounds are $\boldsymbol{l} = [0\ 30\ 5\ 22\ 5\ 15\ 0.2\ 1.2\ 0.2\ 1.2\ 0.2\ 1]^T$ and $\boldsymbol{u} = [30\ 50\ 15\ 29\ 15\ 21\ 2.2\ 4.2\ 2.2\ 4.2\ 2.2\ 10]^T$.

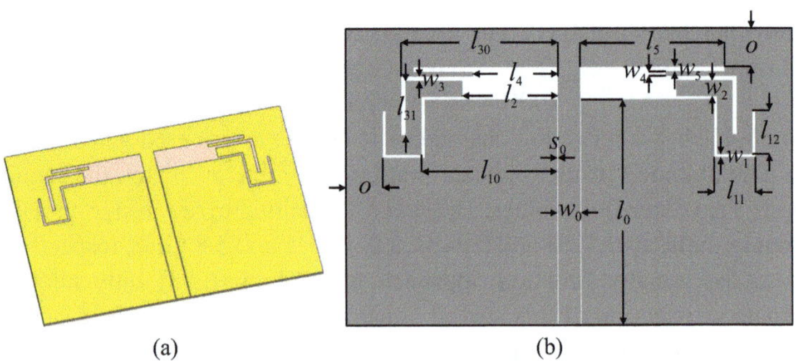

Fig. 6.43. A uniplanar dipole antenna: (**a**) visualization, (**b**) geometrical details.

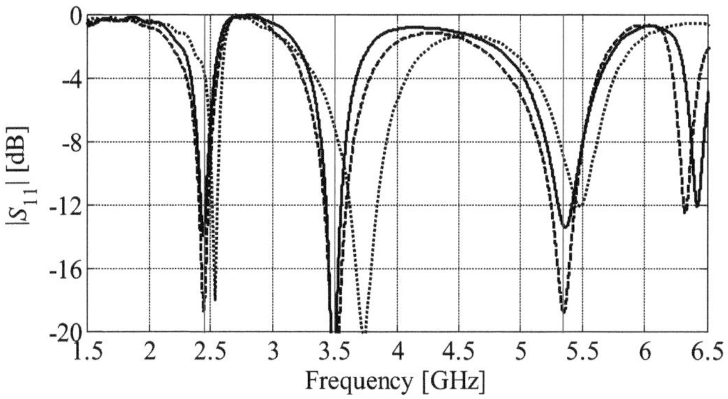

Fig. 6.44 Reflection characteristics of tri-band uniplanar dipole at $\boldsymbol{x}_{\rm init}$ (···), $\boldsymbol{x}^{(0)}$ (– –) and $\boldsymbol{x}^*$ (—)

The initial design is $\boldsymbol{x}_{\rm init} = [16\ 35\ 11\ 27\ 12\ 20\ 0.5\ 2\ 0.6\ 0.5\ 0.7\ 5]^T$. The starting point for size reduction $\boldsymbol{x}^{(0)} = [18.44\ 39.12\ 13.74\ 28.99\ 11.65\ 20.25\ 0.57\ 2.55\ 0.34\ 1.2\ 0.2\ 3.95]^T$ has been obtained by optimizing antenna for the best matching at the $F_{\rm target}$ frequencies. The antenna size for $\boldsymbol{x}^{(0)}$ is 2340 mm^2.

The final design rendered using the technique of Sect. 6.4.3 is $\boldsymbol{x}^* = [17.81\ 38.53\ 13.82\ 28.75\ 11.62\ 20.11\ 0.95\ 1.2\ 0.41\ 1.85\ 0.2\ 1]^T$. The footprint of the miniaturized antenna is 1706 mm^2, i.e., over 27 percent smaller compared to the design $\boldsymbol{x}^{(0)}$. Reflection responses of the structure for the considered designs are shown in Fig. 6.44. Similarly as in the first example, the operating frequencies for $\boldsymbol{x}^{(0)}$ and $\boldsymbol{x}^*$ are well centered, whereas the miniaturized design slightly violates requirement upon -15 dB reflection (1 dB at 2.45 GHz and 1.5 dB at 5.35 GHz). Overall, the characteristics can be considered satisfactory. It should be noted that lower threshold reflection could be maintained by setting higher value of the penalty factor in (6.28), yet this might lead to premature convergence of the optimization algorithm due to the increased steepness of the functional landscape. Radiation characteristics for $\boldsymbol{x}^{(0)}$ and $\boldsymbol{x}^*$ designs are shown in Fig. 6.45. As for the previous example, size reduction resulted in slightly more omnidirectional patterns.

6.4.4.3 Experimental Validation

The miniaturized antenna designs obtained in Sects. 6.4.4.1 and 6.4.4.2 have been fabricated and measured. Figure 6.46 shows the photographs of the antenna prototypes. Comparisons of the obtained simulated and measured reflection characteristics are given in Fig. 6.47 The responses are in acceptable agreement. For the quasi-patch antenna, a slight misalignment between the measured and simulated resonances for the lower operating frequency can be observed. For the dipole antenna, misalignment can be observed for two lower frequencies. Radiation pattern characteristics of both structures obtained in H-plane and E-plane are shown in Figs. 6.48 and 6.49.

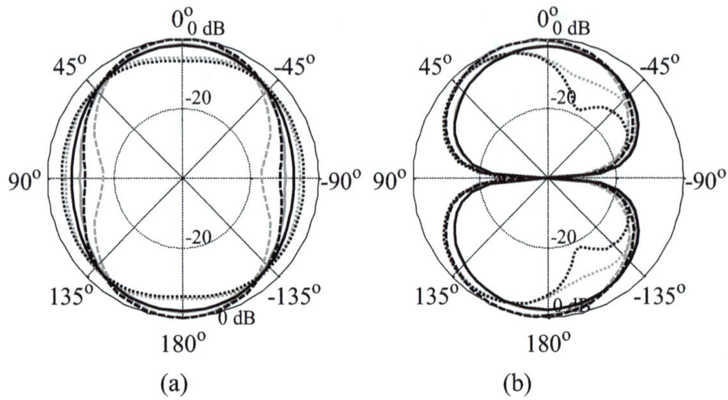

Fig. 6.45 E-field radiation patterns of the antenna for $\boldsymbol{x}^{(0)}$ (gray) and $\boldsymbol{x}^*$ (black) at 2.45 GHz (—), 3.5 GHz (– –), and 5.35 GHz (···): (**a**) H-plane, (**b**) E-plane

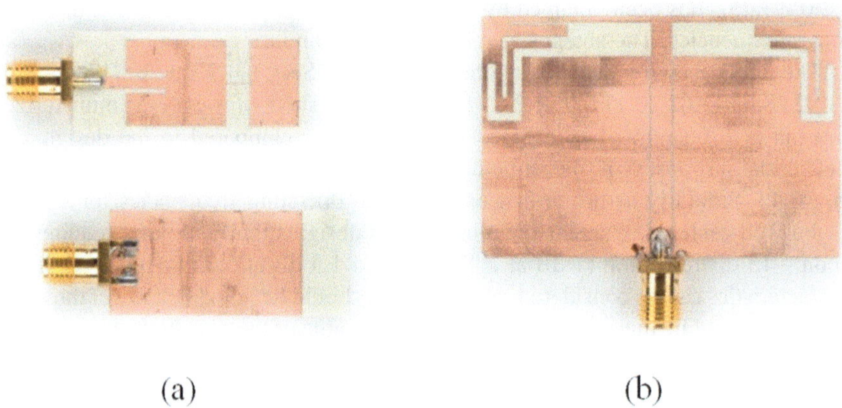

Fig. 6.46 Photographs of the fabricated antenna prototypes: (**a**) quasi-patch antenna optimized for minimum size, (**b**) uniplanar dipole optimized for minimum size

The simulated and measured responses are in good agreement. It should be noted that discrepancies between the results are due to utilization of simplified simulation models that lack SMA connectors. To some extent, they are also derived from fabrication tolerances, as well as electrically large measurement setup which has not been accounted for in the simulation models.

This section discussed simulation-driven feature-based miniaturization of multi-band antennas. The presented approach is based on appropriate formulation of the design problem with implicit handling of antenna reflection specifications (including allocation of the operating frequencies and reflection levels at the resonances). The algorithm has been demonstrated using two antenna examples, a dual-band patch antenna, and a triple-band uniplanar dipole. In both cases, significant reduction of the footprints has been achieved, over 60 percent for the first antenna and almost

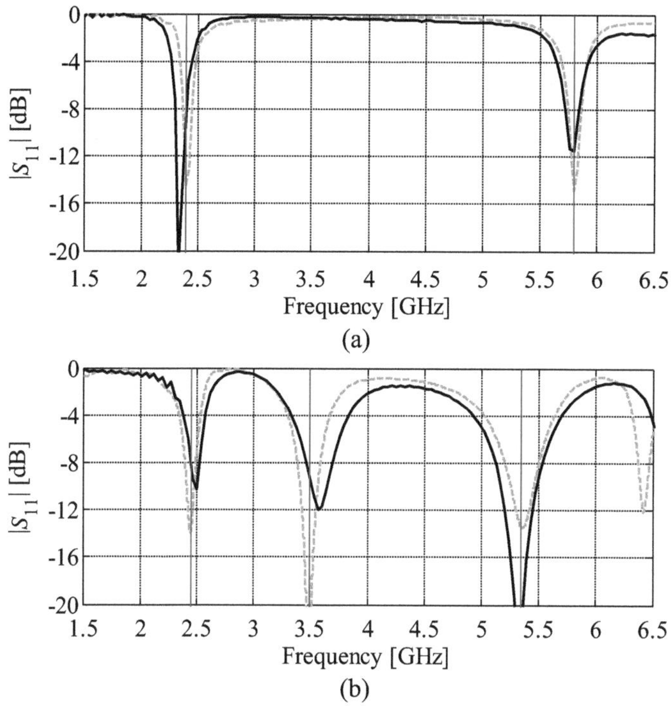

Fig. 6.47 Comparison of simulated (gray) and measured (black) reflection characteristics obtained for (**a**) dual-band antenna, (**b**) tri-band dipole

30 percent for the second antenna. These results demonstrate the importance of simultaneous adjustment of all relevant geometry parameters in order to obtain truly optimum designs that still satisfy prescribed performance specifications.

6.5 Summary

This chapter discussed the conventional approach to feature-based optimization with the emphasis on an appropriate selection of response features for solving local design optimization tasks. The benchmark high-frequency structures of this chapter varied in type and shapes of their respective characteristics in order to present the reader with various possible definitions of the characteristic points. Furthermore, the examples of this chapter were aimed at illustrating the reformulation of the objective function with the use of response features. Among others, we considered matching enhancement of multi-band antennas (Sect. 6.1.1), bandwidth maximization, as well as size reduction of an impedance matching transformer (Sect. 6.1.2). Section 6.1.3 discussed optimization of a SIW filter for insertion and return loss enhancement, whereas Sect. 6.1.4 covered side-lobe reduction of a linear antenna array. The

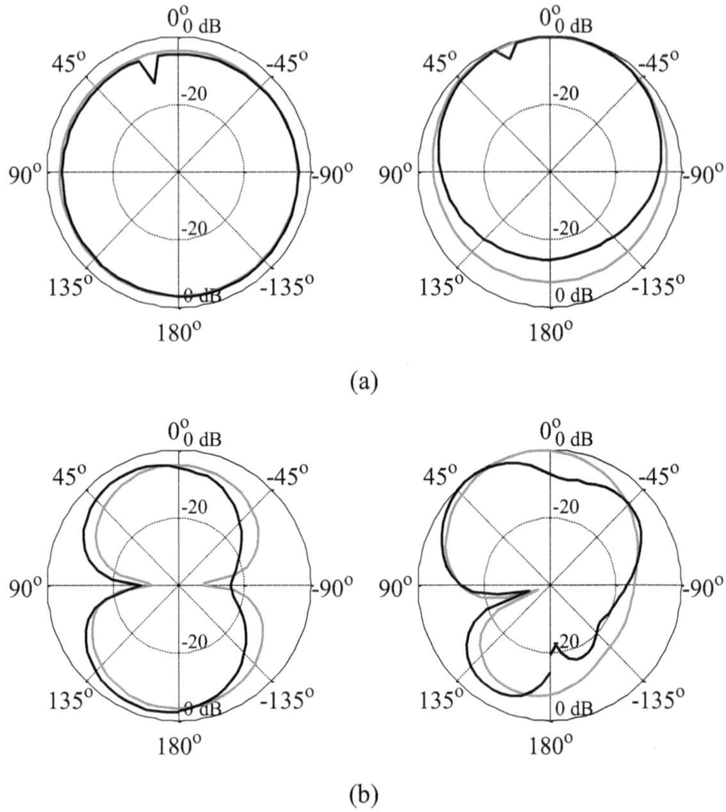

Fig. 6.48 Quasi-patch antenna: a comparison of simulated (gray) and measured (black) E-field radiation patterns in H-plane (left) and E-plane (right) for (**a**) 2.4 GHz frequency, (**b**) 5.8 GHz frequency

subsequent sections showcased slightly more involved examples of feature-based optimization. Section 6.2 outlined matching enhancement and size reduction of a UWB antenna with variable number of response features. Subsequently, FBO with circuit decomposition of an impedance matching transformer (Sect. 6.3) was discussed. Finally, the feature-based miniaturization of multi-band antennas was presented in Sect. 6.4. In all examples of this chapter, considerable cost savings were obtained owing to the employment of the feature-based technology. Furthermore, in many of the considered cases, optimization using standard formulation of the design task (i.e., at the level of the entire frequency characteristics) simply failed to yield satisfactory results. This indicates that apart from significant cost reduction, feature-based local optimization also enables the improvement of the optimization process reliability.

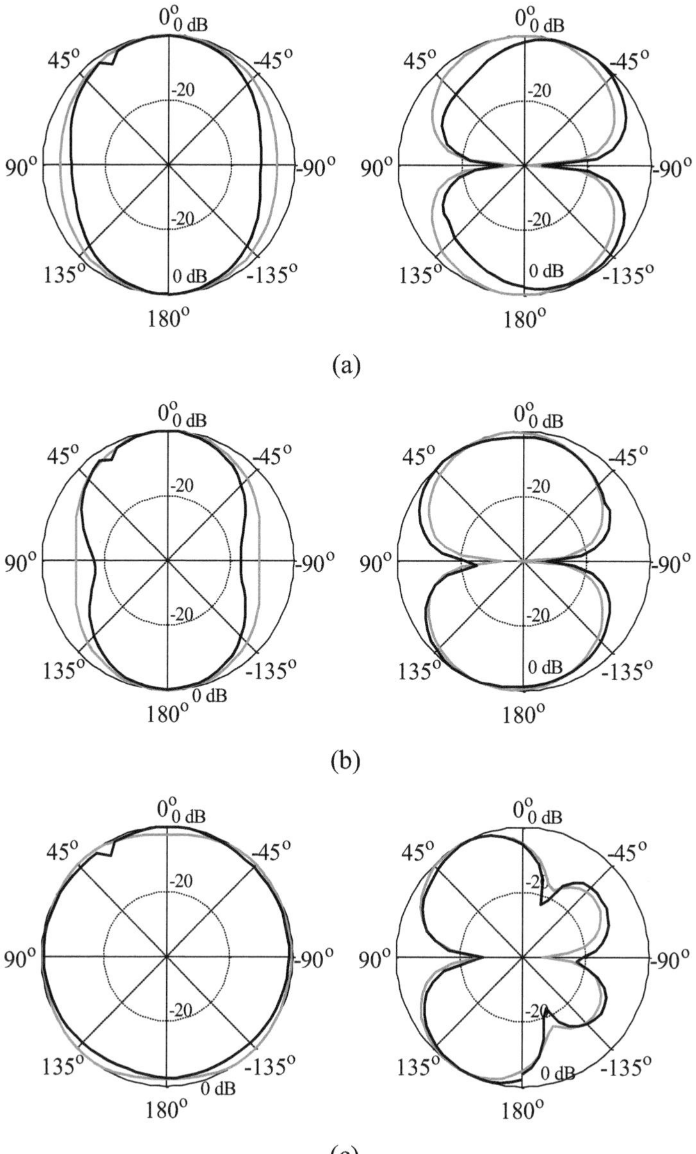

Fig. 6.49 Uniplanar dipole antenna: a comparison of simulated (gray) and measured (black) E-field radiation patterns in H-plane (left) and E-plane (right) for frequencies: (**a**) 2.45 GHz, (**b**) 3.5 GHz, (**c**) 5.35 GHz

References

Balanis, C. A. (2005). *Antenna theory: Analysis and design* (3rd ed.). Wiley-Interscience.

Bekasiewicz, A., & Koziel, S. (2016). Response features and circuit decomposition for accelerated EM-driven design of compact impedance matching transformers. *Microwave and Optical Technology Letters, 58*(9), 2130–2133. [p29].

Chen, Y. -C., Chen, S. -Y., & Hsu, P. (2006). *Dual-band slot dipole antenna fed by a coplanar waveguide*. IEEE international symposium on antennas and propagation, pp. 3589–3592.

Conn, A. R., Gould, N. I. M., & Toint, P. L. (2000). *Trust region methods*. MPS-SIAM Series on Optimization, Philadelphia, PA, USA: MPS-SIAM.

Conn, A. R., Scheinberg, K., & Vincente, L. N. (2009). *Introduction to derivative-free optimization*, MPS-SIAM Series on Optimization, Philadelphia, PA, USA: MPS-SIAM.

Glubokov, O., & Koziel, S. (2014). *EM-driven tuning of substrate integrated waveguide filters exploiting feature-space surrogates*. The IEEE MTT-S international microwave symposium (IMS2014), 2014, pp. 1–3. [e03].

Hertel, T. W. (2005). *Cable-current effects of miniature UWB antennas*. IEEE antennas and propagation society international symposium, vol. 3A, pp. 524–527.

Kolda, T. G., Lewis, R. M., & Torczon, V. (2003). Optimization by direct search: New perspectives on some classical and modern methods. *SIAM Review, 45*(3), 385–482.

Koziel, S. (2010). Computationally efficient multi-fidelity multi-grid design optimization of microwave structures. *The Applied Computational Electromagnetics Society Journal, 25*(7), 578–586.

Koziel, S., & Bekasiewicz, A. (2016a). Computationally-efficient design closure of miniaturized impedance matching transformers using response features. *The International Journal of RF and Microwave Computer-Aided Engineering, 26*(5), 396–401. [p30].

Koziel, S., & Bekasiewicz, A. (2016b). *Low-cost multi-objective optimization of antennas using Pareto front exploration and response features*. International symposium on antennas and propagation (APSURSI), pp. 571–572.

Koziel, S., & Bekasiewicz, A. (2017a). Rapid design optimization of multi-band antennas by means of response features. *Metrology and Measuring Systems, 24*(2), 337–346. [e02].

Koziel, S., & Bekasiewicz, A. (2017b). Expedited simulation-driven design optimization of UWB antennas by means of response features. *The International Journal of RF and Microwave Computer-Aided Engineering, 27*(6), paper no. e21102. [p27].

Koziel, S., & Bekasiewicz, A. (2017c). Rapid dimension scaling of dual-band antennas using variable-fidelity EM models and inverse surrogates. *Journal of Electromagnetic Waves and Applications, 31*, 297–308.

Koziel, S., & Bekasiewicz, A. (2018). Simulation-driven size-reduction-oriented design of multi-band antennas by means of response features. *IET Microwaves, Antennas and Propagation, 12*(7), 1093–1098. [p23].

Koziel, S., & Ogurtsov, S. (2014). Simulation-based design of microstrip linear antenna arrays using fast radiation response surrogates. *IEEE Antennas and Wireless Propagation Letters, 14*, 759–762.

Koziel, S., & Ogurtsov, S. (2015). Fast simulation-driven optimization of planar microstrip antenna arrays using surrogate superposition models. *The International Journal of RF and Microwave Computer-Aided Engineering, 25*(5), 371–381.

Koziel, S., & Ogurtsov, S. (2018). Rapid design closure of linear microstrip antenna array apertures using response features. *IEEE Antennas and Wireless Propagation Letters, 17*(4), 645–648. [p22].

Koziel, S., Bekasiewicz, A., & Leifsson, L. (2016). Rapid EM-driven antenna dimension scaling through inverse modeling. *IEEE Antennas and Wireless Propagation Letters, 15*, 714–717.

Mailloux, R. J. (2005). *Phased array antenna handbook* (2nd ed.). Artech House.
Qing, X., & Chen, Z. N. (2009). Compact coplanar waveguide-fed ultra-wideband monopole-like slot antenna. *IET Microwaves, Antennas & Propagation, 3*(5), 889–898.
Stegen, R. J. (1953). Excitation coefficients and beamwidths of Tschebyscheff arrays. *Proceedings of IRE, 41*(11), 1671–1674.

Chapter 7
Accelerated Feature-Based Local Optimization with Variable-Fidelity EM Simulations

This chapter discusses an approach to accelerated feature-based local optimization with the use of variable-fidelity EM simulation models. The considered acceleration mechanism constitutes one of the available methods of expediting local FBO procedures. Other options include the employment of adjoint sensitivities as well as sparse sensitivity updates, which will be considered in Chap. 8. Utilization of variable-fidelity EM simulations is directly aimed at lowering the computational cost of the optimization process. This is achieved by delegating some of the computational effort, normally executed using high-fidelity EM analysis, into a cheaper low-fidelity model. In the majority of variable-fidelity optimization frameworks, only two levels of fidelities are exploited: coarse/fine and low-/high-fidelity models. The low-fidelity model may be based on coarse-discretization EM analysis (typically in antenna design), but also equivalent networks (more often to be utilized in microwave design). The examples of this chapter utilize the former type. In a majority of frameworks, the discretization density of the coarse model is adjusted using engineering experience and visual inspection of the family of EM-simulated responses of the structure under design, evaluated at different levels. Appropriate adjustment of the model fidelity is not a trivial task: setting discretization too low may lead to a failure of the optimization process. Yet, increasing the discretization overly may compromise the cost efficacy of the process. The FBO frameworks of this chapter utilize dedicated procedures for an automated determination of the discretization level of a low-fidelity model to ensure that the trade-off between reliability and computational cost is maintained. Specifically, in the considered algorithms, low-fidelity EM simulations are employed for the evaluation of the system gradients, thereby allowing us to achieve a considerable speedup of the optimization process. The discussed frameworks are demonstrated using several examples of antennas and microwave structures. Comprehensive benchmarking and application case studies, along with experimental verification, are also provided.

A. Pietrenko-Dabrowska, S. Koziel, *Response Feature Technology for High-Frequency Electronics. Optimization, Modeling, and Design Automation*,
https://doi.org/10.1007/978-3-031-43845-5_7

7.1 Variable-Fidelity Feature-Based Optimization of Antennas

This section delineates feature-based variable-fidelity optimization of two narrow-band antenna structures: a dielectric resonator antenna (DRA) and a slot-ring coupled patch antenna (Koziel & Bekasiewicz, 2018). Formulation of a general optimization problem, both the conventional and feature-based ones, is recapitulated in Sect. 7.1.1. Section 7.1.2 describes the employed method for an automated determination of the mesh density of the low-fidelity model. The results of a bandwidth enhancement of a DRA antenna are provided in Sect. 7.1.3, whereas the results of a matching improvement of a slot-ring antenna are given in Sect. 7.1.4. Section 7.2 outlines feature-based minimization of the return loss and isolation of two branch-line couplers, also in a variable-fidelity regime. In the examples of this chapter, the main acceleration mechanism, apart from employing feature-based formulation of the respective optimization tasks, consists in employing variable-fidelity EM simulations, specifically, for calculating the response gradients using a less expensive low-fidelity model.

7.1.1 Problem Formulation

Let us recall the conventional formulation of the design task (see (6.1)). However, here, we will express the objective function U in terms of the EM-simulated model response $\boldsymbol{R}(\boldsymbol{x})$ in order to allow for differentiating between the model of various fidelities later on

$$\boldsymbol{x}^* = \arg\min_{\boldsymbol{x}} U\left(\boldsymbol{R}_f(\boldsymbol{x})\right) \tag{7.1}$$

In this chapter, the following notation is used. $\boldsymbol{R}_f(\boldsymbol{x})$ represents the high-fidelity EM model of the antenna or other high-frequency structure under design. As in the previous chapters, $\boldsymbol{x}$ is a vector of adjustable parameters to be optimized, and $\boldsymbol{x}^*$ is the optimum design to be found.

Here, minimization of the structure reflection in a frequency range of interest $[f_0 - d_f, f_0 + d_f]$ is considered. The objective function for such optimization problem is often given as $U = \max\{|S_{11}|(\boldsymbol{x}, f) \text{ for } f_0 - d_f \leq f \leq f_0 + d_f\}$, where the reflection coefficient $|S_{11}|(\boldsymbol{x}, f)$ is the device response of interest.

In this chapter, cost efficacy of FBO procedure is improved by employing variable-fidelity EM simulations. More specifically, the optimization burden is shifted from the high-fidelity model $\boldsymbol{R}_f$ to the low-fidelity model $\boldsymbol{R}_c$ that is computationally cheaper. The latter is constructed by appropriate enhancement (or correction) of the underlying low-fidelity model. In the case of antennas, the low-fidelity model is most often obtained from coarse-discretization EM simulations of the structure of interest.

The main topic of this book is the employment of the response features technology for solving various design tasks. Thus, the design problem is formulated in terms of appropriately defined feature points and their coordinates, which allows for reducing the computational cost of solving the problem (7.1).

Similarly as in Chap. 6, the coordinates of the kth feature point are denoted as $f_k(\boldsymbol{x})$ and $l_k(\boldsymbol{x})$ for $k = 1, \ldots, p$, where p is the total number of points defined for the problem at hand. In the example of this section, we will use the following (overall) feature vector $\boldsymbol{R}_{F.f}$. This is in contrast to Chap. 6 (see (6.3)), where separate feature vectors $\boldsymbol{F}(\boldsymbol{x})$ and $\boldsymbol{L}(\boldsymbol{x})$ comprising frequency and level coordinates of the feature points, respectively, were employed. Thus, we reformulate the design task (7.1) in terms of the response features as follows

$$\boldsymbol{x}^* = \arg\min_{\boldsymbol{x}} U_F\left(\boldsymbol{R}_{F.f}(\boldsymbol{x})\right) \tag{7.2}$$

where U_F denotes the objective function defined at the level of feature coordinates and $\boldsymbol{R}_{F.f} \in R^{2p}$ is the feature-based model of the structure at hand, i.e., $\boldsymbol{R}_{F.f}(\boldsymbol{x}) = [f_1(\boldsymbol{x})\ f_2(\boldsymbol{x}) \ldots f_p(\boldsymbol{x})\ l_1(\boldsymbol{x})\ l_2(\boldsymbol{x}) \ldots l_p(\boldsymbol{x})]^T$.

FBO generates a series of approximations $\boldsymbol{x}^{(i)}$ to $\boldsymbol{x}^*$ given by

$$\boldsymbol{x}^{(i+1)} = \arg\min_{\|\boldsymbol{x}-\boldsymbol{x}^{(i)}\| \le \delta^{(i)}} U_F\left(\boldsymbol{R}_l^{(i)}(\boldsymbol{x})\right) \tag{7.3}$$

In (7.3), $\boldsymbol{R}_l^{(i)}$ denotes a finite-differentiation-based linear model of the feature points obtained at the current design $\boldsymbol{x}^{(i)}$. The algorithm (7.3) is embedded in the trust-region framework to ensure convergence (Conn et al., 2000).

The linear model $\boldsymbol{R}_l^{(i)}$ is defined as

$$\boldsymbol{R}_l^{(i)}(\boldsymbol{x}) = \begin{bmatrix} [f_1(\boldsymbol{x}^{(i)}) & \ldots & f_p(\boldsymbol{x}^{(i)})]^T + \boldsymbol{Df}(\boldsymbol{x}^{(i)}) \circ (\boldsymbol{x} - \boldsymbol{x}^{(i)}) \\ [l_1(\boldsymbol{x}^{(i)}) & \ldots & l_p(\boldsymbol{x}^{(i)})]^T + \boldsymbol{Dl}(\boldsymbol{x}^{(i)}) \circ (\boldsymbol{x} - \boldsymbol{x}^{(i)}) \end{bmatrix} \tag{7.4}$$

where

$$\boldsymbol{Df}\left(\boldsymbol{x}^{(i)}\right) = \begin{bmatrix} \left(f_1\left(\boldsymbol{x}_{d.1}^{(i)}\right) - f_1\left(\boldsymbol{x}^{(i)}\right)\right)/d_1^{(i)} \\ \cdots \\ \left(f_p\left(\boldsymbol{x}_{d.1}^{(i)}\right) - f_p\left(\boldsymbol{x}^{(i)}\right)\right)/d_p^{(i)} \end{bmatrix}, \quad \boldsymbol{Dl}\left(\boldsymbol{x}^{(i)}\right) = \begin{bmatrix} \left(l_1\left(\boldsymbol{x}_{d.1}^{(i)}\right) - l_1\left(\boldsymbol{x}^{(i)}\right)\right)/d_1^{(i)} \\ \cdots \\ \left(l_p\left(\boldsymbol{x}_{d.1}^{(i)}\right) - l_p\left(\boldsymbol{x}^{(i)}\right)\right)/d_p^{(i)} \end{bmatrix} \tag{7.5}$$

Here, $\boldsymbol{x}^{(i)} = \left[x_1^{(i)} x_2^{(i)} \ldots x_n^{(i)}\right]^T$ is the design at iteration i, $\boldsymbol{x}_{d.k}^{(i)} = \left[x_1^{(i)} \ldots x_k^{(i)} + d_k^{(i)} \ldots x_n^{(i)}\right]^T$ are the perturbations with respect to $\boldsymbol{x}^{(i)}$, whereas $\circ$ denotes component-wise multiplication. It should be noted that the perturbation size vector $\boldsymbol{d}^{(i)} = \left[d_1^{(i)} d_2^{(i)} \ldots d_n^{(i)}\right]^T$ is set based on sensitivity analysis so that the norm-wise response change is similar for perturbations with respect to all parameters.

Although reformulating the design problem in terms of response features already leads to reduction in the cost of the optimization process, its further acceleration is possible by calculating the gradients (7.5) using the low-fidelity model $\boldsymbol{R}_c$. It should be emphasized that the low-fidelity model may be 10–50 times faster than the high-fidelity one; thus, utilization of $\boldsymbol{R}_c$ instead of $\boldsymbol{R}_f$ in (7.5) may significantly reduce the cost of the gradient estimation.

7.1.2 Automated Low-Fidelity Model Selection

In this section, the method for automated determination of the mesh density of the low-fidelity model is considered (Koziel & Bekasiewicz, 2018). The approach discussed below is based on a statistical analysis of the correlations between perturbations of the feature coordinates for models of various fidelities. The procedure allows for determination of the cheapest low-fidelity model that is sufficiently accurate to ensure reliability of the optimization process. The presented technique is embedded into the FBO framework of Sect. 7.1.1.

Correlation between the high- and low-fidelity EM models is an important factor from the point of view of reliability of the surrogate-assisted optimization process (7.3)–(7.5). The level of absolute model misalignment (e.g., frequency shift) is not of primary importance. This is because the low-fidelity model is only utilized in (7.5) to perform finite differentiation for the construction of the linear expansion model. In other words, it is only necessary that the changes of the low-fidelity model responses (at the level of feature coordinates) follow the changes of the high-fidelity one. There are two scenarios that may lead to practical problems, i.e., the low-fidelity model being too coarse or too fine. In the first case, the gradient estimation based on the low-fidelity model evaluations may be inaccurate, which could cause divergence of the optimization algorithm. In the second case, the cost of each iteration (7.3) increases without any further gains in terms of reliability.

In most practical cases, the model assessment is realized based on engineering experience, i.e., by visual inspection of the misalignment between responses of models with different mesh densities (Koziel & Ogurtsov, 2014). Nevertheless, unattended selection of the model fidelity is important from the point of view of automating the design optimization process.

The discussed model selection procedure is based on correlation analysis between the models of various fidelities. The correlation is examined based on the reference designs $\boldsymbol{x}^{(1)}$ and $\boldsymbol{x}^{(2)}$. Once the fine- and coarse-discretized models are evaluated at $\boldsymbol{x}^{(1)}$ and $\boldsymbol{x}^{(2)}$, the following vectors are defined

$$\Delta\boldsymbol{F} = \boldsymbol{F}\left(\boldsymbol{x}^{(2)}\right) - \boldsymbol{F}\left(\boldsymbol{x}^{(1)}\right), \quad \Delta\boldsymbol{F}_c = \boldsymbol{F}_c\left(\boldsymbol{x}^{(2)}\right) - \boldsymbol{F}_c\left(\boldsymbol{x}^{(1)}\right) \tag{7.6}$$

Here, (7.6) represents the variations of the frequency components of the feature coordinates for the high- and low-fidelity model of the structure at hand, i.e., $\boldsymbol{F}(\boldsymbol{x}^{(k)}) = [f_1(\boldsymbol{x}^{(k)}) \ldots f_p(\boldsymbol{x}^{(k)})]^T$, $k = 1, 2$. The definition for $\boldsymbol{F}_c$ is analogous. We also have $\boldsymbol{L}(\boldsymbol{x}^{(k)}) = [l_1(\boldsymbol{x}^{(k)}) \ldots l_p(\boldsymbol{x}^{(k)})]^T$, $k = 1, 2$; with $\boldsymbol{L}_c$ defined in a similar manner. For the sake of illustration, we consider a design example with five feature points that correspond to -10 dB and -15 dB levels of reflection (on the left- and right-hand side slopes), as well as the peak minimum value of the reflection response. In the analysis below, for the sake of brevity, we only consider the frequency coordinates of the feature points. After defining the feature points, the data sets $\{\Delta\boldsymbol{F}, \Delta\boldsymbol{F}_c\}$ and $\{\Delta\boldsymbol{L}, \Delta\boldsymbol{L}_c\}$ (the latter defined for the vectors of the level coordinates similarly as in (7.6)), are utilized for a construction of linear regression models. Then, the determination coefficient r^2 (Draper & Smith, 1998) is computed. The factor r^2 reflects the level of correlation between the fine- and coarse-discretization data sets. In other words, it represents the reliability of gradient estimation (cf. (7.4)) when the low-fidelity model is utilized for sensitivity analysis instead of high-fidelity one. In particular, when $r^2 = 1.0$, the correlation between the models is ideal. Here, it is assumed that the low-fidelity model is sufficiently accurate if the corresponding determination coefficient is $r^2 = 0.9$ or higher.

In order to find the appropriate discretization level, the correlation analysis described in the previous paragraph has to be performed for a family of the low-fidelity models of increasing mesh density. Here, the computational models of the antenna structures are implemented in CST Microwave Studio and simulated using its time domain solver. The mesh density is adjusted by setting the value of the length per wavelength (LPW) coefficient. The LPW factor for the fine model is set to 40. The analysis of correlation is performed for LPW coefficient values of 8 and higher.

Figure 7.1 shows the selected scatter plots $\Delta\boldsymbol{F}_c$ versus $\Delta\boldsymbol{F}$ and their corresponding regression lines obtained for the DRA structure discussed in Sect. 7.1.4. The LPW coefficient values for which the plots have been obtained was set to 10, 16, 24, and 40, respectively. Based on these results, one can conclude that the mesh density for LPW = 10 is too low, whereas the model alignment for LPW = 16 is significantly improved. At the same time, the model with LPW = 24 can be considered sufficient to be utilized for estimating gradients in (7.5). The correlations for LPW = 40 correspond to the high-fidelity model; thus, they are ideal.

From a practical point of view, it should be emphasized that the specific values of determination coefficient depend on a particular design for which they are calculated. Also, some fluctuations may be present for low LPW values that are the result of numerical noise. Here, this problem is mitigated through analysis of correlation at K reference designs so that $(K - 1)K/2$ independent vectors $\Delta\boldsymbol{F}$ and $\Delta\boldsymbol{F}_c$ are obtained for each selected LPW. The r^2 plots for LPW factors in a range from 8 to 40 and r^2 averaged for all available reference designs are shown in Fig. 7.2. The obtained results indicate that for LPW smaller than 16, the coefficient of determination is too low for reliable gradient estimation in (7.5), yet it becomes rather stable for LPW greater than 16. Obviously, for LPW = 40 the perfect alignment is observed; thus, $r^2 = 1$.

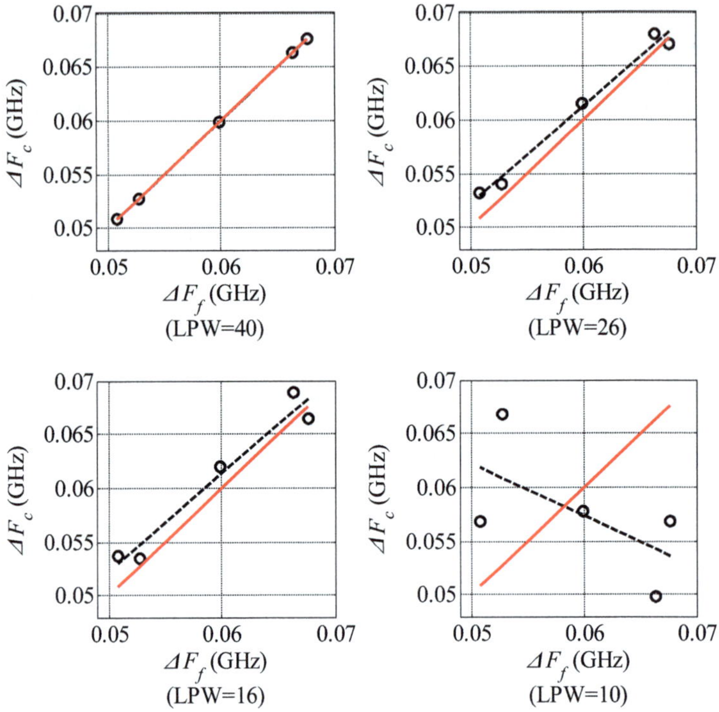

Fig. 7.1 Scatter plots $\Delta \boldsymbol{F}_c$ versus $\Delta \boldsymbol{F}$ obtained for two designs $\boldsymbol{x}^{(1)}$ and $\boldsymbol{x}^{(2)}$ for the LPW coefficients set to 40 (fine model), 24, 16, and 10. Solid lines represent regression functions for the reference data sets $\{\Delta \boldsymbol{F}, \Delta \boldsymbol{F}\}$ and $\{\Delta \boldsymbol{L}, \Delta \boldsymbol{L}\}$, whereas the linear regression function is marked using dashed lines

The above considerations lead to the coarse-discretization model selection procedure that consists of the following steps:

1. Perform evaluation of the EM model of the structure at hand for K designs $\boldsymbol{x}^{(k)}$, $k = 1, \ldots, K$, for the selected range of LPW coefficients (the reference designs can be allocated randomly within the design space).
2. Determine the average coefficient of determination r^2 for the data sets $\{\Delta \boldsymbol{F}, \Delta \boldsymbol{F}_c\}$ obtained for all LPW values.
3. Select the coarsest low-fidelity model for which the average of r^2 factors is above 0.9.

The lowest acceptable LPW factor of the low-fidelity model for the design example discussed above is 16.

Note that the computational cost of the model selection procedure is low due to the fact that coarse-discretization EM simulations are significantly faster than the high-fidelity ones. In practice, the overall cost is equivalent to a few evaluations of the fine model. More importantly, the procedure is fully automated.

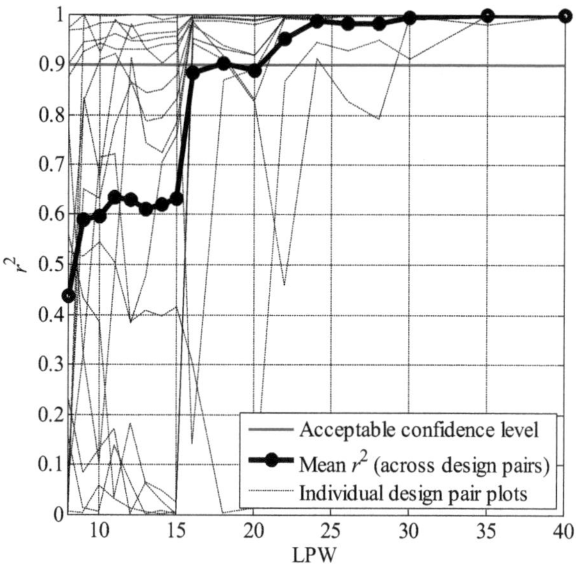

Fig. 7.2 Compact DRA: r^2 plots for the $\{\Delta\boldsymbol{F}, \Delta\boldsymbol{F}_c\}$ data sets versus values of LPW coefficient. Dashed lines represent data obtained for all design pairs $\{\boldsymbol{x}^{(k)}, \boldsymbol{x}^{(l)}\}$, $k \neq l$, whereas thick sold line is average of r^2. Horizontal line denotes acceptable confidence level (here set to 0.9)

7.1.3 Bandwidth Enhancement of a Quasi-Isotropic DRA

In this section, variable-fidelity FBO optimization algorithm is demonstrated using the example of a quasi-isotropic dielectric resonator antenna (DRA) presented in Fig. 7.3 (Pan et al., 2014). The structure consists of a cuboid dielectric resonator implemented on a Taconic CER-10 dielectric substrate ($\varepsilon_r = 10$, $\tan\delta = 0.0035$). The antenna is fed from the bottom through a cylindrical probe allocated in the resonator and excited through a 50-Ohm coaxial input. The vector of antenna parameters is $\boldsymbol{x} = [a\ b\ c\ o_1\ o_2\ l]^T$. Dimensions $d = 1.26$ and $g = 0.82$ remain fixed in the optimization process. The unit for all dimensions is mm. The high-fidelity EM model of the structure contains about 1,000,000 mesh cells, and its simulation time on a dual Intel Xeon E5540 machine is about 8 min. The minimum lines per wavelength (LPW) factor that defines the mesh density of the low-fidelity antenna model is set to 8 (the lower the factor the coarser the mesh).

The considered design objective is a maximization of -10 dB bandwidth of the antenna as well as ensuring its symmetry around the center frequency of 2.4 GHz. The scatter and the r^2 plots of the structure are shown in Figs. 7.1 and 7.2, respectively. The LPW factor of 16 for the low-fidelity model was determined based on the considered the approach. Figure 7.4 shows a comparison of the antenna responses at the initial and the optimized designs $\boldsymbol{x}^{\text{init}} = [25\ 25\ 15\ 0.1\ 5\ 5]^T$ and $\boldsymbol{x}^* = [25.18\ 26.26\ 15.97\ 0.15\ 4.42\ 5.52]^T$, respectively. The latter was obtained by the discussed feature-based optimization algorithm with variable-fidelity models.

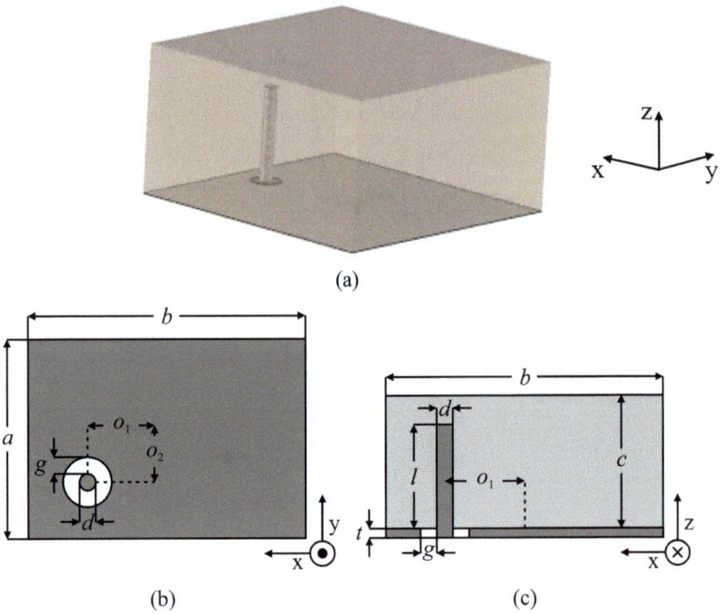

Fig. 7.3 Compact DRA structure: (**a**) 3D view; (**b**) bottom view; (**c**) cross-section view. Dielectric resonator and metallization are represented by light and dark gray

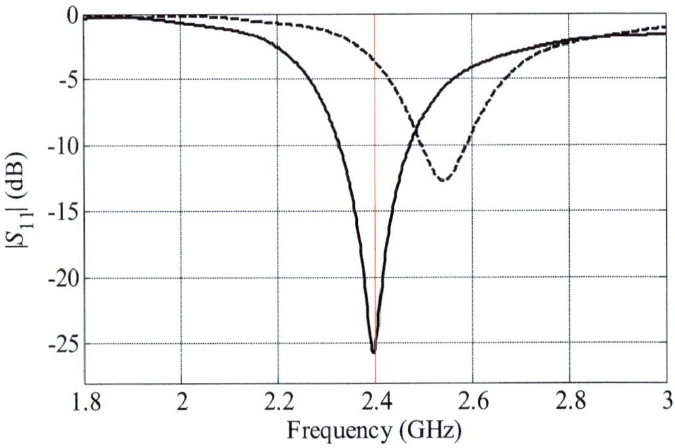

Fig. 7.4 Compact DRA: reflection characteristics at the initial design (- - -) and the design obtained by means of FBO exploiting the low-fidelity model with LPW coefficient set to 16 (—). The vertical line indicates the intended operating frequency

Table 7.1 Optimization cost of compact DRA

Algorithm	Number of model evaluations		Optimization cost	
	Low-fidelity	High-fidelity	Absolute [hours]	Relative to $\boldsymbol{R}_f$
Single-fidelity feature-based ($\boldsymbol{R}_c$ with LPW = 40 = $\boldsymbol{R}_f$)	35	6	5.5	41.0
Variable-fidelity feature-based ($\boldsymbol{R}_c$ with LPW = 24)	41	7	2.0	15.1
Variable-fidelity feature-based ($\boldsymbol{R}_c$ with LPW = 16)	41	8	1.4	10.2
Variable-fidelity feature-based ($\boldsymbol{R}_c$ with LPW = 10)	N/A (algorithm is divergent)			
Pattern search	–	161	21.5	161

For the sake of comparison, the antenna was also optimized using the coarse models with LPW factors set to 10 and 24. For the first of these two factors, the structure model was too coarse. Consequently, the algorithm failed to find a satisfactory design solution as expected.

Although the optimization using the model with LPW = 24 leads to a similar final design, the computational cost of the process is higher than for a models with LPW = 16 (cf. Table 7.1). At the same time, FBO with the LPW = 40 (the high-fidelity antenna model) yields a similar design at a much higher cost. For the sake of comparison, the antenna optimization has been also performed using the pattern search algorithm (Koziel, 2010). The results gathered in Table 7.1 indicate superiority in terms of the cost of FBO over direct optimization of the high-fidelity model using pattern search. Moreover, the computational savings for the variable-fidelity FBO procedure with respect to the single-fidelity routine are considerable. Depending on the setup, the employment of variable-fidelity simulations allows for speeding up the optimization process three (for LPW = 24) or even four times (LPW = 16).

7.1.4 Matching Improvement of a Slot-Ring Coupled Patch Antenna

This section presents feature-based variable-fidelity optimization of a slot-ring coupled patch antenna for matching improvement. In this case, simulation results are supported by physical measurement of the fabricated antenna prototype

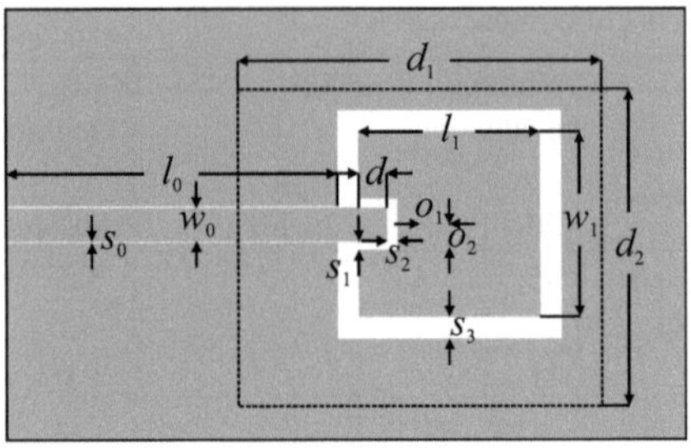

Fig. 7.5 CPW-fed slot-ring coupled patch antenna: geometry with highlighted design parameters. Note that dashed line represents the patch located on the opposite side of the ground plane with respect to the ground plane (gray). White color represents ground plane slots

Our second design example is a patch antenna shown in Fig. 7.5 (Chen, 2007). The structure consists of a driven element in the form of a rectangular patch coupled to the rectangle-shaped ground plane slot localized on the other side of the substrate. The slot is coupled to a coplanar waveguide (CPW) line, which is utilized as the antenna feed. The allocation of the patch can be adjusted with respect to the center of the ground plane slot.

The structure is implemented on a 0.762 mm thick Taconic RF-35 substrate ($\varepsilon_r = 3.5$, $\tan\delta = 0.0018$). The vector of design variables is $\boldsymbol{x} = [d\ d_1\ d_2\ l_1\ s_1\ s_2\ s_3\ w_1\ o_1\ o_2]^T$, whereas dimensions $w_0 = 3$, $l_0 = 30$ and $s_0 = 0.15$ remain fixed (all parameters are in mm).

For this example, we are interested in minimizing the antenna reflection at the operating frequency f_0 (and, at the same time, allocating therein the minimum at the operating frequency). We consider two cases: (i) $f_0 = 2.45$ GHz (case I), and (ii) $f_0 = 1.6$ GHz (case II). As before, the EM models are implemented in CST Microwave Studio. The high-fidelity model (LPW = 40) consists of 1,200,000 mesh cells, and its simulation time is about 12 min. The minimum LPW for low-fidelity model is 8.

Figure 7.6 shows the r^2 plots of the antenna of Fig. 7.5. Here, we select the low-fidelity model with LPW = 20 (simulation time: 140 seconds). Figure 7.7 shows the antenna responses at the initial design $\boldsymbol{x}^{\text{init}} = [4\ 5\ 5\ 16\ 0.3\ 1\ 1\ 16\ 0\ 0]^T$ and the optimized designs $\boldsymbol{x}^{*\text{I}} = [2.1\ 3.67\ 4.94\ 15.95\ 0.81\ 1.48\ 0.20\ 16.38\ -0.29\ -0.11]^T$ (case I) as well as $\boldsymbol{x}^{*\text{II}} = [2.39\ 6.73\ 4.77\ 16.80\ 1.61\ 1.05\ 1.90\ 16.65\ -0.30\ -0.13]^T$ (case II) obtained using the discussed algorithm. Similarly as for the first example, the antenna was also optimized with the low-fidelity models of different LPW, here, 10 (evaluation time: 30 seconds) and 28 (evaluation time: 5 min). As expected, the algorithm failed to find a satisfactory design when working with too coarse low-fidelity model. Optimization using LPW = 28 yields similar designs; however, the computational cost of the process is higher as shown in Tables 7.2 and 7.3. Clearly, direct optimization of the high-fidelity model leads to even higher costs.

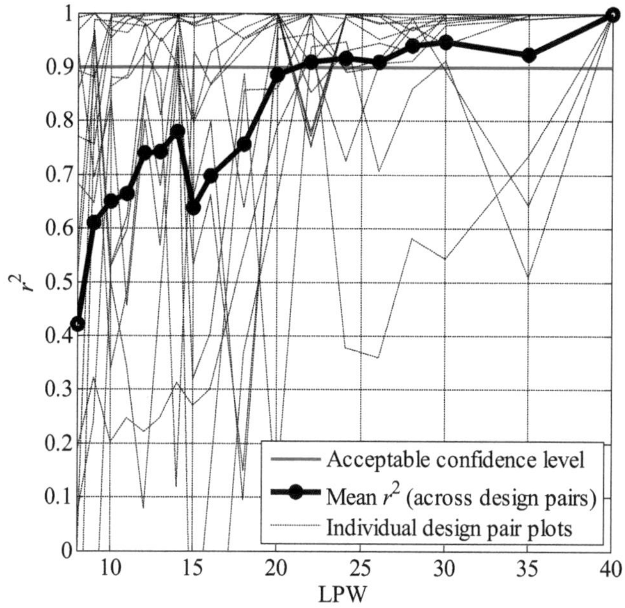

Fig. 7.6 Slot-ring coupled patch antenna: r^2 plots for the $\{\Delta \boldsymbol{F}, \Delta \boldsymbol{F}_c\}$ data sets versus model discretization level (here, determined by the LPW parameter). Original data are plotted using dotted lines for all design pairs $\{\boldsymbol{x}^{(k)}, \boldsymbol{x}^{(l)}\}$, $k \neq l$. The mean r^2 value is plotted with continuous line. Acceptable confidence level (here set to 0.9) is marked using a horizontal line

For additional verification, the designs $\boldsymbol{x}^{*\mathrm{I}}$ and $\boldsymbol{x}^{*\mathrm{II}}$ have been fabricated and measured. Figure 7.8 shows the photographs of the antenna prototypes, whereas Fig. 7.9 shows a comparison of the simulated and measured antenna responses. It can be observed that the measured responses agree well with the simulated ones. Small frequency shift toward lower frequencies is due to SMA connectors that were not included in the computational model of the antenna.

The variable-fidelity method of this section is based on correlation analysis between the coarse and fine models of the structure at hand. The analysis is performed at the level of appropriately defined feature points of the antenna frequency response. The discussed approach allows for obtaining the low-fidelity model featuring the most suitable mesh density of the antenna and, consequently, to determine the fastest coarse model that ensures sufficient level of correlation with the fine model. The procedure has been numerically verified based on two case studies: a dielectric resonator antenna and a patch antenna coupled through ground plane slot. Further validation has been provided by physical measurements of the fabricated coupled patch antenna prototype.

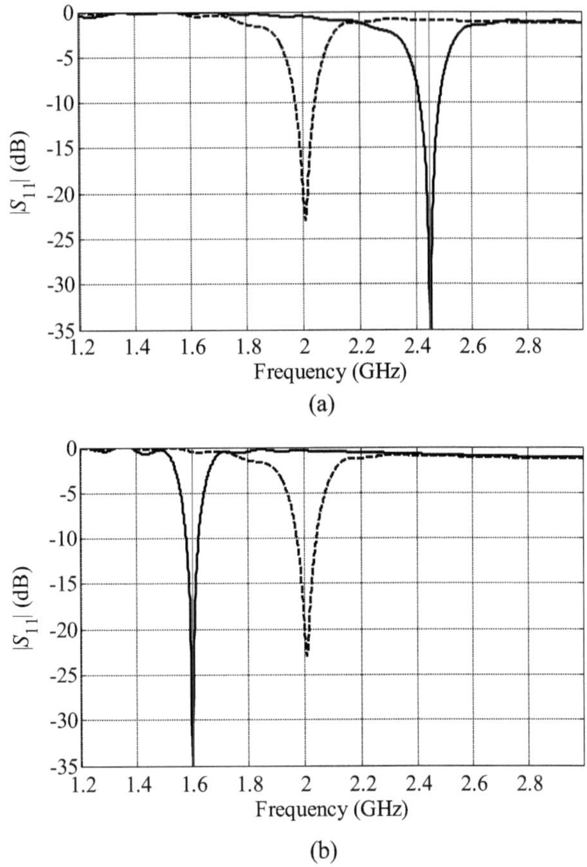

Fig. 7.7 Slot-ring coupled patch antenna: return loss responses at the initial design (- - -) and the design optimized using the feature-based algorithm exploiting the low-fidelity model with LPW = 20 (—): (**a**) case I (operating frequency 2.45 GHz), (**b**) case II (operating frequency 1.6 GHz). The vertical lines indicate the intended operating frequencies

Table 7.2 Optimization cost of slot-ring coupled patch antenna (case I)

Algorithm	Number of model evaluations		Optimization cost	
	Low-fidelity	High-fidelity	Absolute [hours]	Relative to $\boldsymbol{R}_f$
Single-fidelity feature-based ($\boldsymbol{R}_c$ with LPW = 40 = $\boldsymbol{R}_f$)	80	10	18	90
Variable-fidelity feature-based ($\boldsymbol{R}_c$ with LPW = 24)	80	11	8.9	44.3
Variable-fidelity feature-based ($\boldsymbol{R}_c$ with LPW = 16)	60	11	4.5	22.4
Variable-fidelity feature-based ($\boldsymbol{R}_c$ with LPW = 10)	N/A (algorithm is divergent)			
Pattern search	–	356	71.2	356

Table 7.3 Optimization cost of slot-ring coupled patch antenna (case II)

Algorithm	Number of model evaluations		Optimization cost	
	Low-fidelity	High-fidelity	Absolute [hours]	Relative to $\boldsymbol{R}_f$
Single-fidelity feature-based ($\boldsymbol{R}_c$ with LPW = 40 = $\boldsymbol{R}_f$)	70	11	16.2	81
Variable-fidelity feature-based ($\boldsymbol{R}_c$ with LPW = 24)	60	10	7	35
Variable-fidelity feature-based ($\boldsymbol{R}_c$ with LPW = 16)	50	9	3.7	18.5
Variable-fidelity feature-based ($\boldsymbol{R}_c$ with LPW = 10)	N/A (algorithm is divergent)			
Pattern search	–	329	65.8	329

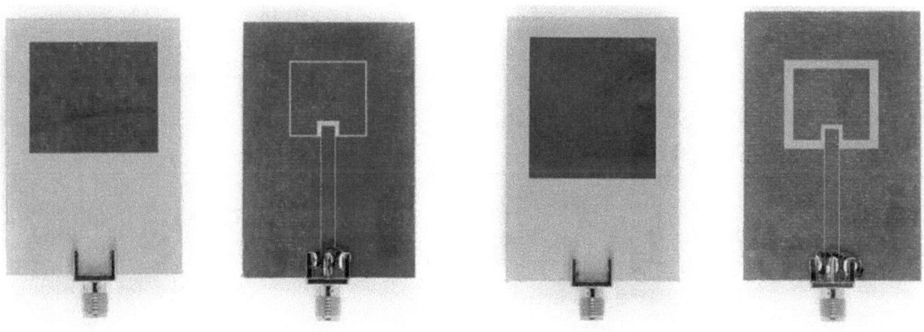

(a) (b)

Fig. 7.8 Slot-ring coupled patch antenna: fabricated prototypes for design $\boldsymbol{x}^{*\mathrm{I}}$ (**a**) and design $\boldsymbol{x}^{*\mathrm{II}}$ (**b**): front (left) and back (right)

7.2 Variable-Fidelity Feature-Based Optimization of Microwave Structures

This section discusses FBO of branch-line couplers. The structures are optimized for minimization of the return loss and isolation, as well for maintaining the equal power split. As in the previous section, a feature-based optimization algorithm is accelerated using variable-fidelity EM models, which are employed for calculating the response gradients in a cost-efficient manner.

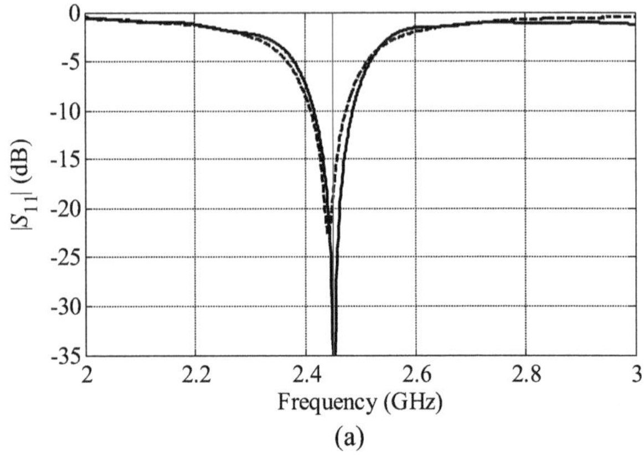

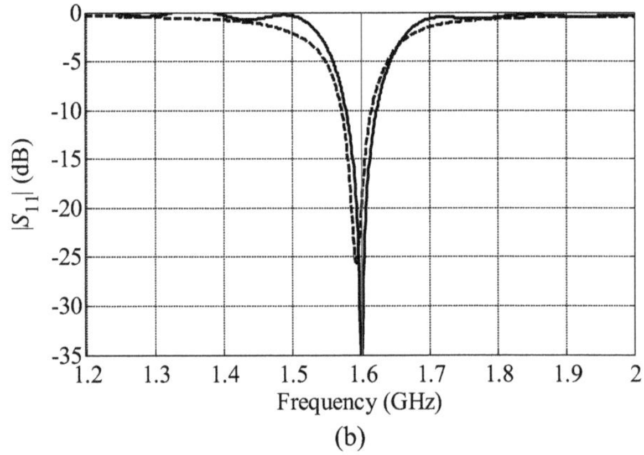

Fig. 7.9 Slot-ring coupled patch antenna: simulated (—) and measured (- - -) reflection responses for the design $\boldsymbol{x}^{*\mathrm{I}}$ (**a**) and design $\boldsymbol{x}^{*\mathrm{II}}$ (**b**)

7.2.1 Response Features for Multi-Band Couplers

In the design of coupler structures, the design objectives are normally formulated in terms of the scattering parameters and include (i) achieving equal power split (i.e., $|S_{21}| = |S_{31}|$), as well as (ii) minimization of $|S_{11}|$ and $|S_{41}|$. Both objectives are to be simultaneously fulfilled for one (for single-band couplers) or several operating frequencies (for multi-band couplers), denoted here as f_k, $k = 1, \ldots, N_f$. The second objective is often formulated in terms of obtaining sufficient −20 dB bandwidth for both $|S_{11}|$ and $|S_{41}|$.

The typical responses for a compact dual-band coupler are shown in Fig. 7.10 (Koziel & Bekasiewicz, 2015). These are the actual characteristics of the structure considered in Sect. 7.2.3. One can notice that the responses are highly nonlinear functions of frequency, therefore, difficult to optimize. Because we are interested in ensuring equal power split as well as good matching and isolation at all operating frequencies f_k, the response features of interest are selected as local maxima of $|S_{21}|$

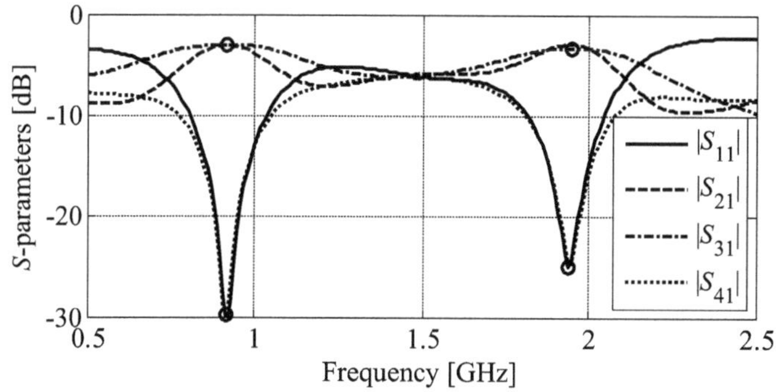

Fig. 7.10 Typical responses of a dual-band coupler (Koziel & Bekasiewicz, 2015). Selected feature points are marked with the circles (here, corresponding to the maximum of $|S_{21}|$ around the lower operating frequency f_1, the maximum of $|S_{31}|$ around the higher operating frequency f_2, as well as the two minima: of $|S_{41}|$ around f_1, and of $|S_{11}|$ around f_2)

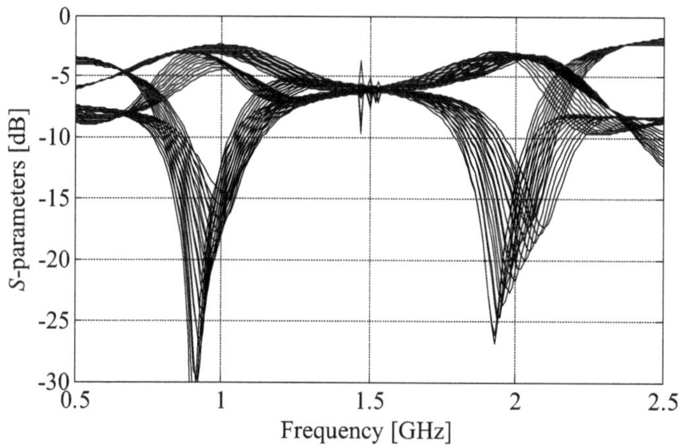

Fig. 7.11 Family of coupler responses corresponding to its geometries along certain line segment in the design space showing highly nonlinear behavior of S-parameters on geometry variables of the circuit

and $|S_{31}|$ and minima of $|S_{11}|$ and $|S_{41}|$ around f_1 and f_2. Some of these points are marked in Fig. 7.10. Each feature point is characterized by the two coordinates: frequency and level. Figure 7.11 shows a family of coupler responses of Fig. 7.10 corresponding to designs along certain line segment in the design space. Note that despite highly nonlinear dependence of S-parameters on the coupler geometry parameters, behavior of the feature points (cf. Fig. 7.12) is close to linear. Consequently, expressing the design goals in terms of the feature points and using feature-based formulation of the design task is expected to speed up the optimization process considerably.

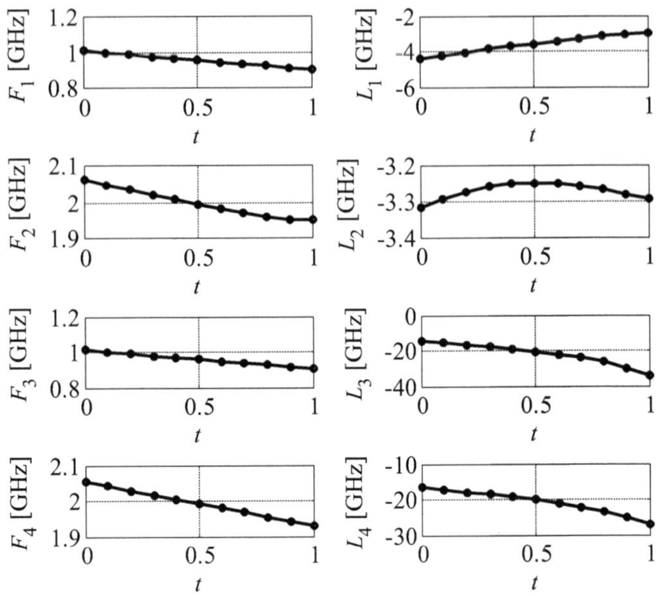

Fig. 7.12 Selected feature points (as marked in Fig. 7.10) corresponding to coupler geometries along the same line segment parameterized by t as in Fig. 7.11. Note that dependence of feature coordinates on coupler geometry is close to linear

7.2.2 Design Objectives

The coupler design problem is formulated in the feature space as in (7.2). In the examples of this section, the objective function is defined as the maximum of the feature point levels corresponding to the minima of $|S_{11}|$ and $|S_{41}|$ at all operating frequencies with added penalty terms that are proportional to the deviations between the frequency coordinates of all feature points and the respective operating frequencies (Koziel & Bekasiewicz, 2018). Thus, the primary objective is to improve matching and isolation at all the operating frequencies f_k, $k = 1, \ldots, N_f$, and move the points of equal power split as close to the respective frequencies f_k as possible.

We also consider the following equality constraints (to enforce equal power split at all frequencies). The constraints $c_k(\boldsymbol{x}) = 0$, $k = 1, \ldots, N_f$, are formulated using the feature point level coordinates. We have $c_k = l_{21.k} - l_{31.k}$, where $l_{21.k}$ and $l_{31.k}$ are the levels corresponding to the maxima of $|S_{21}|$ and $|S_{31}|$ around the operating frequency f_k. The discussed design optimization task is solved according to the methodology described in Sect. 7.1.1.

7.2.3 Automated Low-Fidelity Model Selection

The utilized procedure for an automated selection of the fidelity of the low-fidelity model is similar to that delineated in Sect. 7.1.2. Here, also, we consider a family of low-fidelity models with various (increasing) discretization levels. As in Sect. 7.1.1, CST Microwave Studio is utilized for EM analysis, with LPW parameter defining the discretization level of the model. The high-fidelity model corresponds to LPW = 40. We use LPW = 8 and up for the low-fidelity models.

Figure 7.13 shows the scatter plots $\boldsymbol{\Delta F_c}$ versus $\boldsymbol{\Delta F}$ and $\boldsymbol{\Delta L_c}$ versus $\boldsymbol{\Delta L}$, along with the regression lines, for the coupler of Sect. 7.2.4 for the three values of LPW: 10, 16, and 26. It can be observed that the model with LPW = 10 is clearly too coarse, especially in terms of the level components. Alignment is greatly improved for LPW = 16; however, the deviations from the regression line seem still too large for the level components. The model with LPW = 26 exhibits very good correlations with the high-fidelity model.

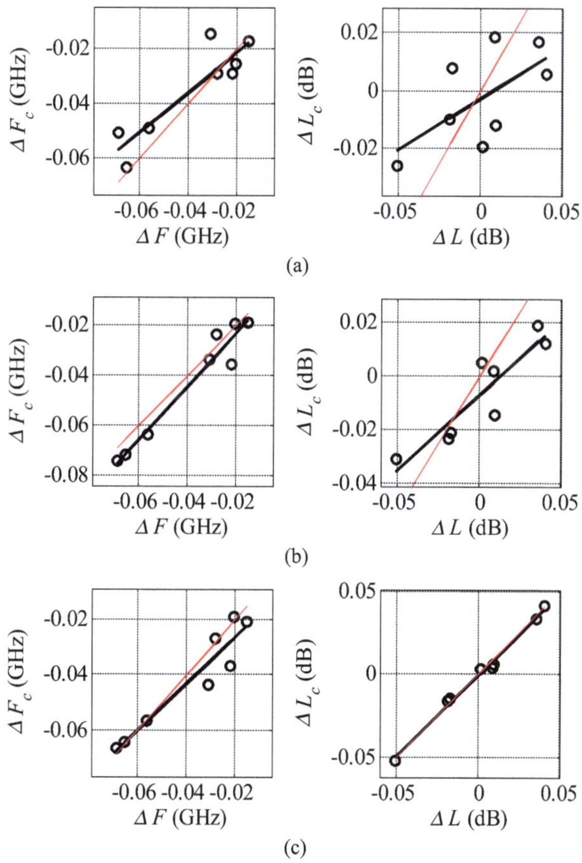

Fig. 7.13 Scatter plots $\boldsymbol{\Delta F_c}$ versus $\boldsymbol{\Delta F}$ (left) and $\boldsymbol{\Delta L_c}$ versus $\boldsymbol{\Delta L}$ (right) for the three low-fidelity models defined with LPW = 10 (**a**), 16 (**b**), and 26 (**c**). The black solid lines denote the linear regression function; the red lines are regression functions for the reference data sets $\{\boldsymbol{\Delta F}, \boldsymbol{\Delta F}\}$ and $\{\boldsymbol{\Delta L}, \boldsymbol{\Delta L}\}$

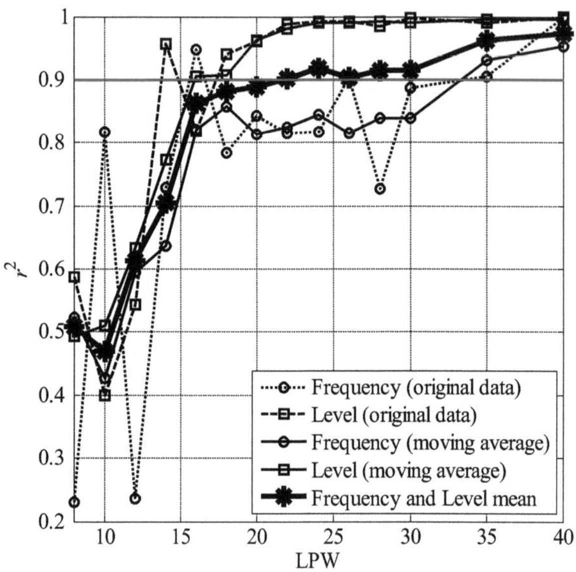

Fig. 7.14 Plots of r^2 for the $\{\Delta \boldsymbol{F}, \Delta \boldsymbol{F}_c\}$ and $\{\Delta \boldsymbol{L}, \Delta \boldsymbol{L}_c\}$ data sets versus model discretization level (here, determined by the LPW parameter). Original data are plotted using dotted (frequency components) and dashed lines (level component). Moving averages are plotted with continuous line, whereas the frequency/level mean is plotted using thick line with star markers. Acceptable confidence level (0.9) is marked using a horizontal line

Figure 7.14 shows the r^2 plots for the range of LPW values from 8 to 40. The value of r^2 is very low for LPW smaller than 15 and stabilizes for LPW > 20. Obviously, $r^2 = 1$ for LPW $= 40$. It can also be seen that—especially for higher values of LPW—correlation is better for the level components than for the frequency components. An important aspect is that the values of r^2 exhibit large fluctuations, particularly for small values of LPW. This is partially the effect of the numerical noise (decreasing with the increase of the discretization density) but also an arbitrary choice of the reference design. Therefore, for the sake of the model selection, we calculate the moving average of the r^2 values (calculated for the three consecutive points along the characteristic) and then consider the mean of the frequency and the level characteristics. As indicated in Fig. 7.14, the mean is much smoother than the "raw" characteristics. For the acceptance level for r^2 equal to 0.9, the threshold value of LPW $= 22$.

7.2.4 *Results for Compact Dual-Band Coupler I*

In this section, we discuss a compact dual-band branch-line coupler structure designed for the two operating frequencies $f_1 = 0.92$ GHz and $f_2 = 1.9$ GHz. Description of the coupler structure is followed by the numerical results and comparison with the benchmark methods, including single-fidelity feature-based optimization. Experimental validation is also provided.

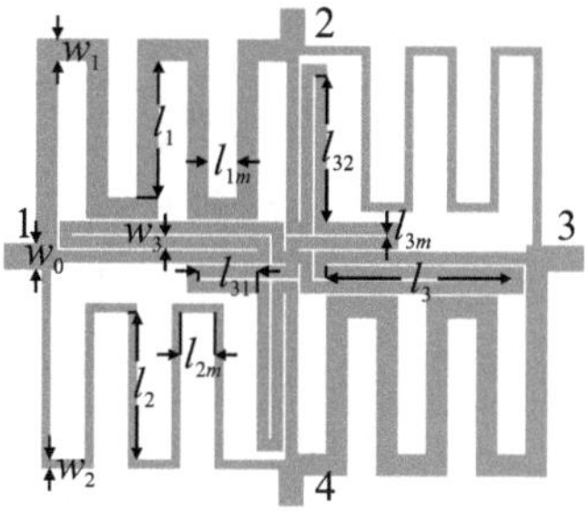

Fig. 7.15 Geometry of a compact dual-band branch-line coupler (BLC I)

Consider a compact dual-band branch-line coupler shown in Fig. 7.15 (Koziel & Bekasiewicz, 2015). The circuit is implemented on Taconic RF-35 dielectric substrate ($h = 0.762$ mm; $\varepsilon_r = 3.5$; $\tan\delta = 0.0018$). The structure is constructed using the transmission lines of three different characteristic impedances with equal electrical lengths (Park & Lee, 2005). The compact design is based on a conventional dual-band BLC realization and modified so that the interior of a coupler is filled as much as possible. The BLC geometry is described by seven independent adjustable parameters $\boldsymbol{x} = [w_1\ w_2\ w_3\ l_1\ l_2\ l_3\ l_{31}]^T$. The dependent variables are $l_{1m} = 0.2\cdot\text{abs}(l_3 + 3w_3 + 3l_{3m} + 4w_1)$, $l_{2m} = 0.2\cdot(l_3 + w_3 + l_{3m} + 0.5w_1 - 4.5w_2)$ and $l_{32} = w_1 + l_1 - w_3$, whereas $l_{3m} = 0.2$ and $w_0 = 1.7$ (the unit is mm). The area occupied by the coupler is defined as $A \times B$ rectangle with $A = 2\cdot(l_1 + 2w_1) + 5w_3 + 6l_{3m}$ and $B = 2\cdot(5w_1 + 5l_{1m}) + w_3$. The high-fidelity EM model of the structure is implemented in CST Microwave Studio (~500,000 mesh cells; LPW = 40). Its average simulation time is 15 min.

The following design objectives are considered in the optimization process: (i) equal power split between the transmission and coupling ports (i.e., $|S_{21}| = |S_{31}|$), (ii) minimization of the return loss $|S_{11}|$, and (iii) minimization of isolation $|S_{41}|$. The assumed operating frequencies are $f_1 = 0.92$ GHz and $f_2 = 1.9$ GHz. The physical consistency of the circuit geometry is ensured by imposing the following lower and upper bounds for design variables: $\boldsymbol{l} = [0.7\ 0.3\ 0.5\ 6\ 6\ 4\ 10]^T$ and $\boldsymbol{u} = [1.7\ 1.3\ 1.5\ 10\ 10\ 8\ 14]^T$.

The low-fidelity model has been selected by considering various values of LPW in the range from 8 to 40. The selected scatter plots as well as the r^2 plots have been presented in Figs. 7.13 and 7.14, respectively. Based on this analysis, the model with LPW = 22 has been selected as the low-fidelity model for the variable-fidelity feature-based optimization.

The initial design $\boldsymbol{x}^{(0)} = [1.39\ 0.58\ 0.99\ 8.66\ 9.19\ 5.02\ 12.73]^T$ mm has been obtained by optimizing the equivalent circuit model of the circuit. Figure 7.16 shows the responses at $\boldsymbol{x}^{(0)}$ indicating non-equal power split at both operating frequencies and the frequency shift for matching and isolation characteristics for f_2. The final design $\boldsymbol{x}^* = [1.451\ 0.554\ 0.865\ 9.008\ 9.821\ 5.159\ 12.043]^T$ mm has been obtained after three iterations of the algorithm (7.3). The footprint of the coupler is $29.3 \times 34.2 = 1002$ mm^2. The optimized circuit responses are shown in Fig. 7.17.

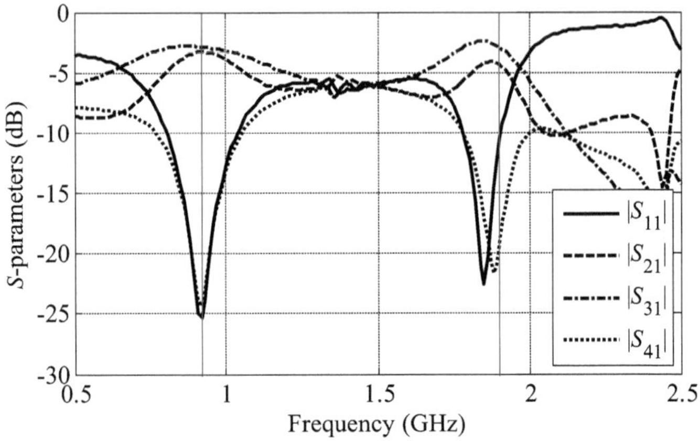

Fig. 7.16 Responses of the compact dual-band coupler (BLC I) at the initial design. Operating frequencies $f_1 = 0.92$ GHz and $f_2 = 1.90$ GHz are marked using vertical lines

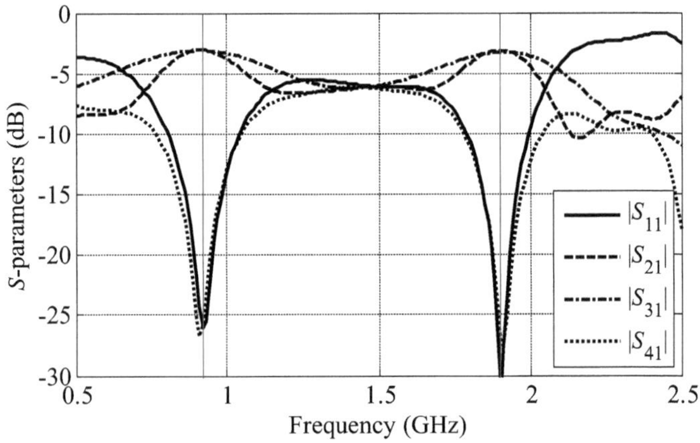

Fig. 7.17 Responses of the compact dual-band coupler (BLC I) at the design obtained by feature-based optimization using the low-fidelity model with LPW = 22. Operating frequencies $f_1 = 0.92$ GHz and $f_2 = 1.90$ GHz are marked using vertical lines

It can be observed that they are well centered around the operating frequencies, and that almost equal power split has been achieved ($|S_{21}| - |S_{31}| < 0.03$ dB for f_1, and $|S_{21}| - |S_{31}| < 0.1$ dB for f_2).

For the sake of comparison, direct optimization of the coupler structure at the level of *S*-parameters versus frequency has been conducted using a pattern search algorithm (Kolda et al., 2003). A design of a similar quality has been obtained, however, at a much higher computational cost of 122 evaluations of the EM coupler model.

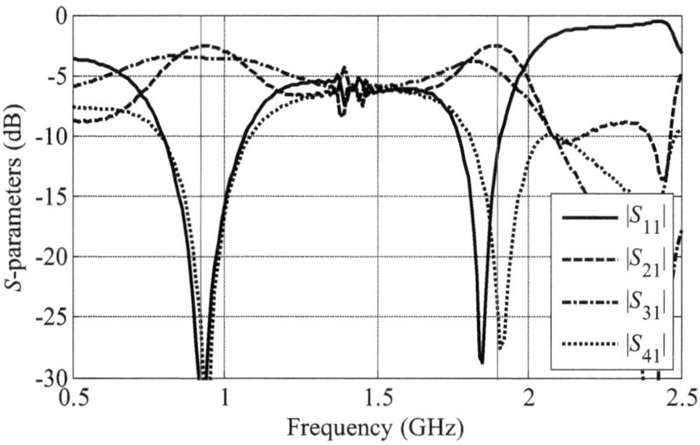

Fig. 7.18 Responses of the dual-band coupler (BLC I) after the first iteration of the feature-based optimization using the low-fidelity model with LPW = 10. Divergence toward a poor design can be observed indicating that the low-fidelity model is too coarse to provide reliable estimation of the feature point gradients

Table 7.4 Optimization of a dual-band coupler (BLC I). Design cost

Algorithm	Number of model evaluations		Optimization cost	
	Low-fidelity	High-fidelity	Absolute [hours]	Relative to high-fidelity model
Variable-fidelity feature-based (low-fidelity model with LPW = 22)	24	4	2.3	9.3
Single-fidelity feature-based (low-fidelity model with LPW = 40)	–	25	6.25	25.0
Variable-fidelity feature-based (low-fidelity model with LPW = 10)	N/A (algorithm is divergent)			
Pattern search	–	122	30.5	122.0

The importance of the low-fidelity model selection has been illustrated by executing a single-fidelity feature-based optimization (i.e., the simulation model of LPW = 40, i.e., identical to the high-fidelity one), as well as variable-fidelity optimization with the low-fidelity model of LPW = 10 (i.e., too coarse according to the correlation analysis of Sect. 7.2.2). In the first case, a similar final design was obtained, however, at a much higher computational cost. In the second case, the algorithm started diverging already after the first iteration (cf. Fig. 7.18), indicating that the low-fidelity model with LPW = 10 does not provide reliable estimation of the feature point gradients. The cost breakdown for all considered optimization approaches has been gathered in Table 7.4.

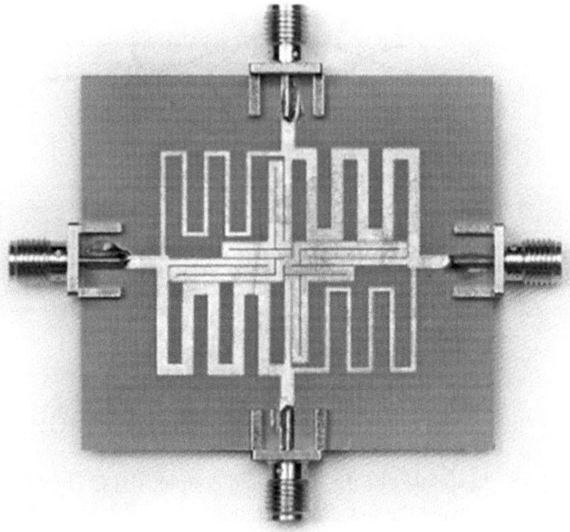

Fig. 7.19 Photograph of the fabricated prototype of the dual-band branch-line coupler of Fig. 7.15 (BLC I)

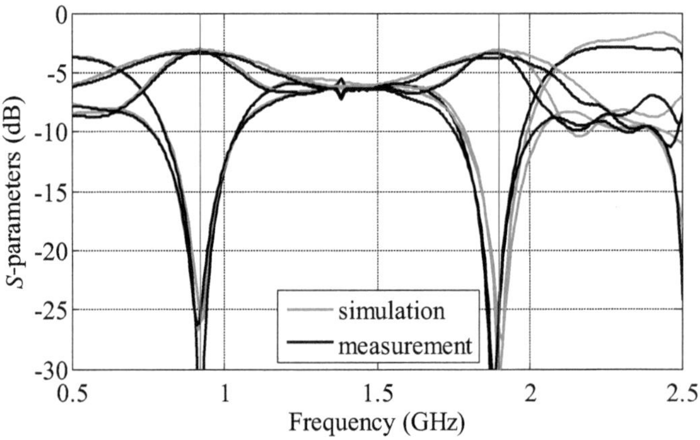

Fig. 7.20 The measured (black line) and simulated (gray dashed line) scattering parameters of the compact dual-band branch-line coupler (BLC I). The vertical lines indicate the intended operating frequencies

The optimized coupler structure has been fabricated and measured. A photograph of the manufactured circuit is shown in Fig. 7.19. Comparison of the simulated and the measured responses in terms of its *S*-parameters is provided in Fig. 7.20. Although the obtained results are generally in a good agreement, a slight frequency shift is observed for the upper operating frequency. This is mostly because the EM model involved in the optimization process lacks connectors.

The detailed data on BLC properties is gathered in Table 7.5. It should be noted that the power split level is slightly lower for the measured coupler responses, which is due to higher losses introduced by the connectors, as well as tolerance of the

Table 7.5 Simulated versus measured BLC I responses

	Simulation		Measurement	
	f_1	f_2	f_1	f_2
S_{11} [dB]	−26.1	−31.1	−38.0	−37.1
S_{21} [dB]	−3.03	−3.08	−3.24	−3.42
S_{31} [dB]	−3.07	−3.23	−3.34	−3.76
S_{41} [dB]	−25.8	−27.0	−25.2	−24.5
BW_1 [GHz]	0.89 to 0.94	1.88 to 1.93	0.87 to 0.93	1.84 to 1.91
BW_2 [GHz]	0.86 to 0.97	1.83 to 1.96	0.84 to 0.95	1.89 to 1.95

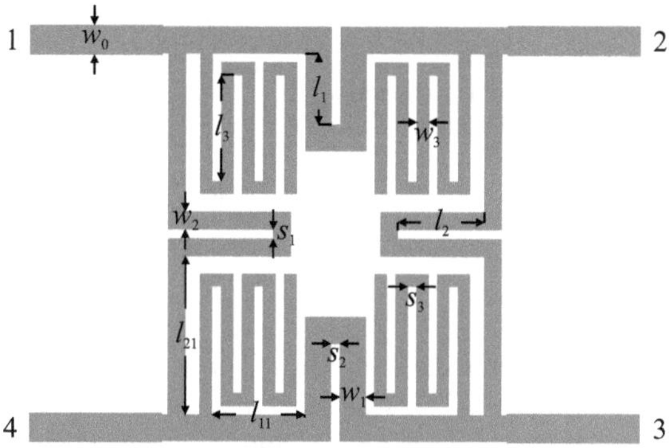

Fig. 7.21 Geometry of a compact dual-band branch-line coupler (BLC II)

fabrication process. On the other hand, the power split imbalance is similar to the simulated results. For the measured structure, the bandwidth BW_1, defined as the frequency range where both return loss $|S_{11}|$ and isolation $|S_{41}|$ are below −20 dB level, is 6.5% and 3.5% for f_1 and f_2, respectively. The bandwidth BW_2 (defined as $||S_{21}| - |S_{31}|| \leq 0.2$ dB) is 12% for 0.92 GHz and 3% for 1.9 GHz.

7.2.5 Results for Compact Dual-Band Coupler II

This section presents the second example, which is also a compact dual-band branch-line coupler, designed for the two operating frequencies $f_1 = 0.9$ GHz and $f_2 = 2.0$ GHz (Koziel & Bekasiewicz, 2018). Numerical results are accompanied by experimental validation of the fabricated coupler prototype.

Consider a compact dual-band branch-line coupler shown in Fig. 7.21. The circuit is implemented on Taconic RF-35 dielectric substrate ($h = 0.762$ mm; $\varepsilon_r = 3.5$; $\tan\delta = 0.0018$). The structure is constructed using three types of transmission lines. The miniaturized design is based on conventional realization of a dual-band branch-line coupler with quarter-wave open circuit stubs (Cheng & Wong, 2004). The introduced geometry modifications include folding sections and meandering of the stubs, both to the interior of the circuit. The structure is described by a vector of nine

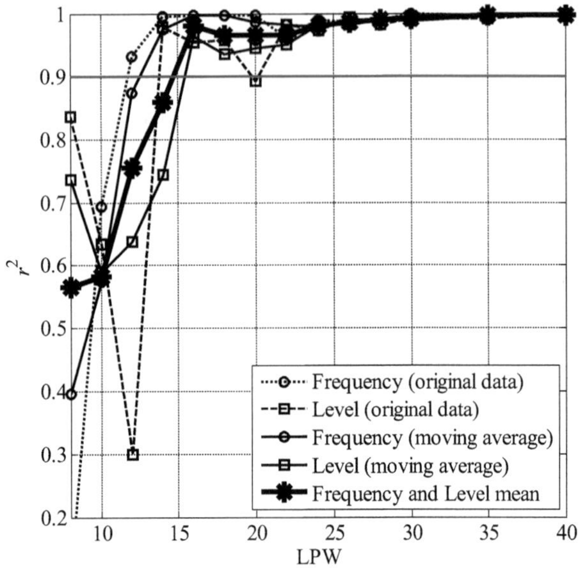

Fig. 7.22 Plots of r^2 for the $\{\Delta \boldsymbol{F}, \Delta \boldsymbol{F}_c\}$ and $\{\Delta \boldsymbol{L}, \Delta \boldsymbol{L}_c\}$ data sets versus model discretization level (here, determined by the LPW parameter) for BLC II. Original data are plotted using dotted (frequency components) and dashed lines (level component). Moving averages are plotted with continuous line, whereas the frequency/level mean is plotted using thick line with star markers. Acceptable confidence level (0.9) is marked using a horizontal line

adjustable parameters: $\boldsymbol{x} = [l_1\ l_2\ l_3\ s_1\ s_2\ s_3\ w_1\ w_2\ w_3]^T$. The dependent parameters are $l_{11} = 4w_3 + 4s_3 + s_1$, $l_{21} = \max\{l_3 + s_3 + s_2 + 2w_3,\ w_1 + l_1\} + s_2$, whereas $w_0 = 1.7$ mm to ensure 50 Ohm input impedance (all parameters are in mm). The coupler footprint is defined as $A \times B$ rectangle where $A = 2(l_{21} + w_2 + w_1) + s_2$ and $B = 2(w_2 + s_3 + w_3 + l_{11} + w_1) + s_1$. The high-fidelity EM model of the structure is implemented in CST Microwave Studio (~860,000 mesh cells; LPW = 40). Its average simulation time is 18 min.

The design objectives are the same as for the first example: (i) equal power split between the transmission and coupling ports (i.e., $|S_{21}| = |S_{31}|$), (ii) minimization of the return loss $|S_{11}|$, and (iii) minimization of isolation $|S_{41}|$. The assumed operating frequencies are $f_1 = 0.9$ GHz and $f_2 = 2.0$ GHz. The following lower and upper bounds are assumed for the design variables: $\boldsymbol{l} = [4.0\ 4.0\ 4.0\ 0.2\ 0.2\ 0.2\ 0.5\ 0.5\ 0.5]^T$ and $\boldsymbol{u} = [9.0\ 9.0\ 9.0\ 0.6\ 0.6\ 0.6\ 2.0\ 2.0\ 2.0]^T$.

The low-fidelity model has been selected by considering various values of LPW in the range from 8 to 40. The r^2 plots are presented in Fig. 7.22. Based on this analysis, the model with LPW = 20 has been selected as the low-fidelity model for the variable-fidelity feature-based optimization as described in Sect. 7.2.3. Note that the condition $r^2 > 0.9$ is satisfied even for LPW = 16; however, a slightly higher value has been selected to be on a safe side.

The initial design $\boldsymbol{x}^{(0)} = [8.71\ 4.00\ 8.00\ 0.60\ 0.47\ 0.20\ 1.60\ 0.92\ 0.97]^T$ mm has been obtained by optimizing the equivalent circuit model of the circuit. Figure 7.23 shows the responses at $\boldsymbol{x}^{(0)}$. It should be emphasized that the response is severely detuned with respect to the assumed design specifications. Also, this indicates poor agreement between the equivalent circuit model and the full-wave EM simulation one. The final design $\boldsymbol{x}^* = [4.42\ 6.35\ 5.51\ 0.20\ 0.60\ 0.21\ 1.67\ 1.14\ 1.69]^T$ mm has

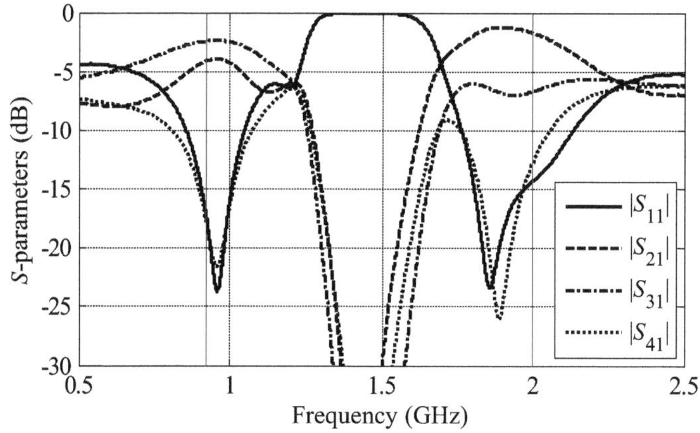

Fig. 7.23 Responses of the compact dual-band coupler (BLC II) at the initial design. Operating frequencies $f_1 = 0.9$ GHz and $f_2 = 2.0$ GHz are marked using vertical lines

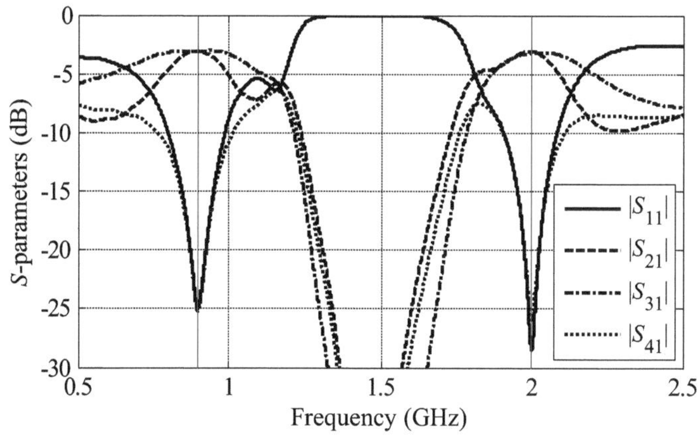

Fig. 7.24 Responses of the compact dual-band coupler (BLC II) at the design obtained by feature-based optimization using the low-fidelity model with LPW = 20. Operating frequencies $f_1 = 0.9$ GHz and $f_2 = 2.0$ GHz are marked using vertical lines

been obtained after seven iterations of the algorithm (7.3). The footprint of the coupler is $25.8 \times 26.9 = 694$ mm^2. The optimized circuit responses are shown in Fig. 7.24. They are perfectly centered at the operating frequencies. Furthermore, almost perfectly equal power split has been obtained ($|S_{21}| - |S_{31}| < 0.025$ dB for f_1, and $|S_{21}| - |S_{31}| < 0.12$ dB for f_2).

Direct optimization of the coupler structure at the level of S-parameters versus frequency has been conducted using a pattern search algorithm (Kolda et al., 2003) yields a slightly worse design at a much higher cost of 160 evaluations of the EM coupler model.

Table 7.6 Optimization of a dual-band coupler (BLC I). Design cost

Algorithm	Number of model evaluations		Optimization cost	
	Low-fidelity	High-fidelity	Absolute [hours]	Relative to high-fidelity model
Variable-fidelity feature-based (low-fidelity model with LPW = 20)	63	8	5.2	17.3
Single-fidelity feature-based (low-fidelity model with LPW = 40)	–	55	16.5	55.0
Variable-fidelity feature-based (low-fidelity model with LPW = 10)	N/A (algorithm is divergent)			
Pattern search	–	160	48.0	160.0

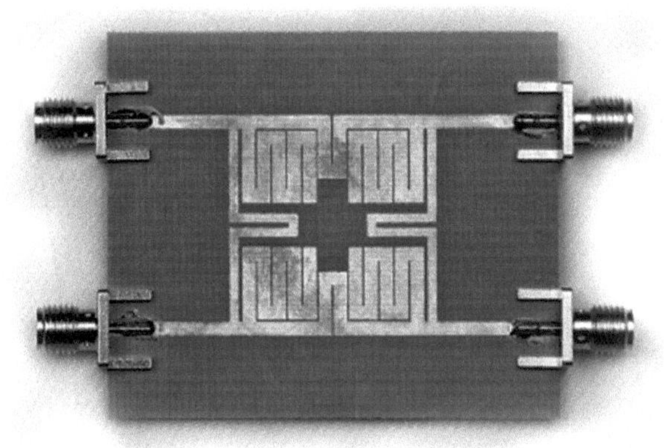

Fig. 7.25 Photograph of the fabricated prototype of the dual-band branch-line coupler of Fig. 7.21 (BLC II)

Similarly as for the previous example, feature-based optimization was also conducted for the low-fidelity model set to LPW = 40 (i.e., single-fidelity optimization), and for LPW = 10. In the first case, the final design was obtained at six iterations of the algorithm (the cost of 55 high-fidelity model evaluations). In the second case, the algorithm was divergent due to limited accuracy of the low-fidelity model. The cost breakdown for all considered optimization approaches has been gathered in Table 7.6.

The optimized BLC II has been fabricated and measured. A photograph of the manufactured circuit and comparison of the simulated and the measured responses are provided in Figs. 7.25 and 7.26, respectively. Detailed data concerning coupler is provided in Table 7.7. The obtained simulation and measurement results are in good agreement; however, a slight frequency shift for the upper operating frequency can be observed. Moreover, the power split level for the measured structure is slightly lower than for the EM model responses. Nonetheless, the results obtained by means

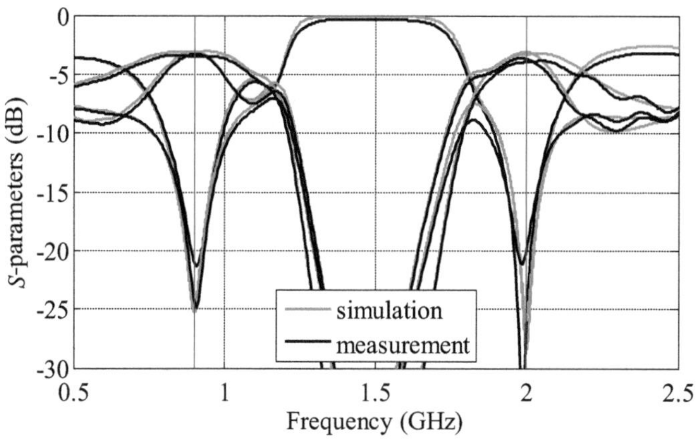

Fig. 7.26 The measured (black line) and simulated (gray dashed line) scattering parameters of the compact dual-band branch-line coupler (BLC II)

Table 7.7 Simulated versus measured BLC II responses

	Simulation		Measurement	
	f_1	f_2	f_1	f_2
S_{11} [dB]	−24.9	−28.4	−21.3	−27.4
S_{21} [dB]	−3.08	−3.08	−3.28	−3.67
S_{31} [dB]	−3.11	−3.27	−3.34	−3.76
S_{41} [dB]	−25.3	−25.9	−24.8	−21.1
BW_1 [GHz]	0.87 to 0.92	1.98 to 2.02	0.89 to 0.93	1.97 to 2.00
BW_2 [GHz]	0.86 to 0.93	1.87 to 2.04	0.86 to 0.95	2.00 to 2.04

of simulation exhibit similar power split imbalance. The discrepancies between simulations and measurements result from inaccuracy of the fabrication process, as well as from utilization of the EM model, which lacks connectors. The measurements indicate that the bandwidth BW_1 where $|S_{11}|$ and $|S_{41}|$ are both below −20 dB level is 4.5% and 1.5% for 0.9 GHz and 2 GHz, respectively. At the same time, the bandwidth BW_2 (defined as $||S_{21}| - |S_{31}|| \leq 0.2$ dB) is 10% for f_1 frequency and 2% for f_2 frequency.

The results of this section demonstrate the advantages of employing variable-fidelity feature-based optimization for rapid design closure of compact multi-band microstrip couplers. Design speedup has been achieved due to only slightly nonlinear dependence of the feature point coordinates on the geometry variables of the structures at hand. Further cost reduction has been obtained by using variable-fidelity EM models. An automated procedure for low-fidelity model setup has been utilized for a selection of the resolution of the low-fidelity model of the microwave circuits under design. Numerical results have been validated by physical measurements of the fabricated coupler prototypes.

7.3 Summary

This chapter considered accelerated feature-based optimization of high-frequency structures with the use of variable-fidelity EM simulation models. Alternative methods of expediting FBO techniques, employing adjoint sensitivities and sparse sensitivity updates, will be discussed in Chap. 8. In the examples of this chapter, dedicated procedures for an automated adjustment of the coarse model discretization have been used. In each presented case, considerable cost savings were obtained due to the incorporation of feature-based formulation of the optimization task, as well as variable-fidelity EM models for the evaluation of the device sensitivity required by the optimization algorithm.

References

Chen, J.-S. (2007). CPW-fed compact slot-ring coupled square patch antennas. In *IEEE Ant. Prop. Society Int. Symp.*, Honolulu, HI, USA, pp. 2297–2300.

Cheng, K.-L. M., & Wong, F.-L. (2004). A novel approach to the design and implementation of dual-band compact planar 90° branch-line coupler. *IEEE Transactions on Microwave Theory and Techniques, 52*(11), 2458–2463.

Conn, A. R., Gould, N. I. M., & Toint, P. L. (2000). *Trust region methods. MPS-SIAM series on optimization*. MPS-SIAM.

Draper, N. R., & Smith, H. (1998). *Applied regression analysis* (3rd ed.). Wiley Interscience.

Kolda, T. G., Lewis, R. M., & Torczon, V. (2003). Optimization by direct search: New perspectives on some classical and modern methods. *SIAM Review, 45*(3), 385–482.

Koziel, S. (2010). Computationally efficient multi-fidelity multi-grid design optimization of microwave structures. *Applied Computational Electromagnetics Society Journal, 25*(7), 578–586.

Koziel, S., & Bekasiewicz, A. (2015). Efficient design optimization of compact dual-band microstrip branch-line coupler using response features. In *IEEE MTT-S Int. Conf. Numerical Electromagn. Multiphysics Modeling Opt. (NEMO), Ottawa, ON, Canada, 2015.*

Koziel, S., & Bekasiewicz, A. (2018). Low-cost multiband compact branch-line coupler design using response features and automated EM model fidelity adjustment. *International Journal of RF and Microwave Computer-Aided Engineering, 28*(4), e21233.

Koziel, S., & Ogurtsov, S. (2014). *Antenna design by simulation-driven optimization*. Springer.

Pan, Y. M., Leung, K. W., & Lu, K. (2014). Compact quasi-isotropic dielectric resonator antenna with small ground plane. *IEEE Transactions on Antennas and Propagation, 62*(2), 577–585.

Park, M.-J., & Lee, B. (2005). Dual-band, cross coupled branch line coupler. *IEEE Microwave and Wireless Components Letters, 15*(10), 655–657.

Chapter 8
Expedited Feature-Based Local Optimization: Adjoint Sensitivities and Sparse Sensitivity Updates

The main contributor to the computational cost of solving local optimization tasks using gradient-based routines with numerical derivatives is the estimation of the response sensitivity through finite differentiation. Chapter 7 discussed procedures developed to reduce this cost by employing variable-fidelity EM models. This chapter focuses on alternative techniques that enable fast evaluation of the system response gradients. The considered methods may be incorporated into feature-based optimization routines. Two specific acceleration mechanisms are highlighted: adjoint sensitivities and sparse sensitivity updates. The first technique may be considered less versatile, as it is intrusive from the point of view of the EM simulation engine, and it is only accessible through a few commercial simulation packages. The second option is sparse sensitivity updates where costly finite-differentiation-based sensitivity updates throughout the algorithm run are suppressed to a certain degree. Several different mechanisms for restraining the updates have been developed. These include analysis of the design relocation between consecutive iterations (Koziel & Pietrenko-Dabrowska, 2019), monitoring of the magnitude of gradient changes during the optimization run (Pietrenko-Dabrowska & Koziel, 2020), and restricting updates to the directions of the most significant changes of the structure responses (Tomasson et al., 2020). Another approach involves the employment of updating formulas (Koziel & Pietrenko-Dabrowska, 2020a). In this chapter, FBO procedure is expedited by detecting patterns of response gradient variability throughout the algorithm run, and suppressing sensitivity updates for the variables that exhibit stable gradient behavior.

A. Pietrenko-Dabrowska, S. Koziel, *Response Feature Technology for High-Frequency Electronics. Optimization, Modeling, and Design Automation*,
https://doi.org/10.1007/978-3-031-43845-5_8

8.1 Feature-Based Optimization with Adjoint Sensitivities

This section delineates feature-based optimization with adjoint sensitivities (Koziel & Bekasiewicz, 2018a). As discussed in the previous chapters, FBO exploits a particular structure of the system response to "flatten" the functional landscape handled during the optimization process, which leads to reducing its computational complexity. This section delineates the incorporation of adjoint sensitivities into gradient-based optimization procedures, which allows for reducing the optimization cost becomes even further. The operation and performance of the algorithm is demonstrated using a waveguide filter and a miniaturized microstrip rat-race coupler (RRC). In the case of RRC, numerical results are supported by experimental validation.

8.1.1 Problem Formulation

To make this chapter self-contained, let us recall the conventional formulation of the design closure task (see Eq. 7.1)

$$\boldsymbol{x}^* = \arg\min_{\boldsymbol{x}} U\left(\boldsymbol{R}_f(\boldsymbol{x})\right) \tag{8.1}$$

Similarly, as in Chap. 7, in (8.1), the symbol $\boldsymbol{R}_f(\boldsymbol{x})$ stands for the response vector of a high-fidelity EM-simulated computational model of the structure of interest. In the case of filters, the responses would normally be return loss and transmission characteristics versus frequency; whereas in the case of couplers, the response could be, e.g., S_{k1} characteristics as well as a phase shift between the output ports. Again, the vector $\boldsymbol{x}$ represents designable parameters of the problem, typically geometry parameters of the structure. Finally, the scalar function U is a merit function that "measures" the quality of the design (problem dependent).

At a generic level, the optimization algorithm described further may utilize several EM models of various fidelities. The models will be denoted as $\{\boldsymbol{R}_j\}$, $j = 1, \ldots, K$, and are all evaluated using the same EM solver. The model discretization increases for increasing values of j and $\boldsymbol{R}_K = \boldsymbol{R}_f$. Utilization of a lower-fidelity model is motivated by potential advantages from the point of view of computational cost reduction: coarse-discretization simulation is faster so that larger steps in the design space can be made at a lower cost. Inaccuracy of the model is corrected by switching to finer discretization at later stages of the process.

Here, models with different levels of fidelity have been determined based on engineering experience. Given the high-fidelity model, discretization of the lower-fidelity one is reduced while ensuring that it properly captures effects of parametric changes on the structure response (e.g., the model reproduces all the resonances that are essential to define feature points). The user can decide to construct an even more relaxed model assuming that it still represents important features of the structure at hand. Quality of the lower-fidelity model responses with respect to high-fidelity one is assessed based on visual inspection.

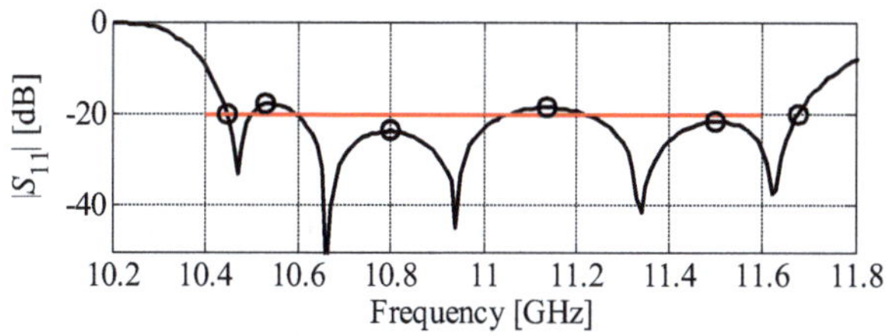

Fig. 8.1 Response features for microwave filters: filter response (—) and feature points (o), here, corresponding to the edges of the passband (at −20 dB level) as well as local maxima of the in-band ripples. All of these points (and only these) are sufficient to determine violation/satisfaction of design specifications imposed on return loss characteristic. Design specifications marked using a horizontal line

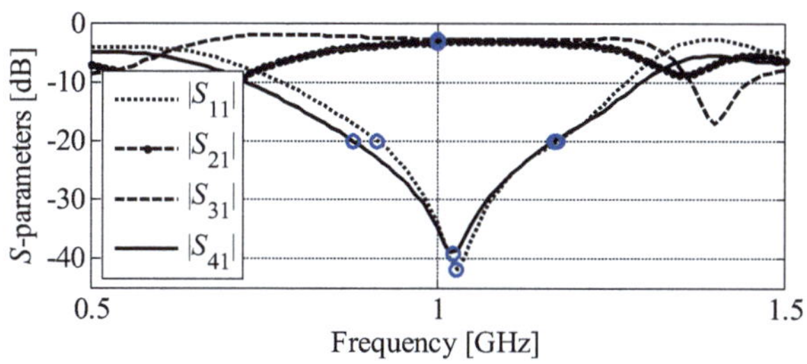

Fig. 8.2 Response features for coupler structures: minima and −20-dB-levels points of $|S_{11}|$ and $|S_{41}|$ as well as $|S_{21}|$ and $|S_{31}|$ values at the required operating frequency. These points are sufficient to determine satisfaction/violation of the design specs concerning the operating frequency, bandwidth, as well as power split

Responses of microwave components (typically, S-parameters vs. frequency) are highly nonlinear functions of frequency as well as geometry parameters of the structure at hand. This results in slow convergence of the optimization procedures directly controlling such characteristics. The main topic of this book is the employment of the response feature technology, which allows for achieving computational speedup due to less nonlinear dependencies of the features on circuit dimensions. Figure 8.1 shows an example of feature point selection for a microwave filter.

The points correspond to the edges of the passband (at −20 dB levels) as well as local in-band maxima of the return loss characteristic. Coordinates of these points are sufficient to determine whether the filter satisfies given design specifications, in the example of Fig. 8.1 defined as $|S_{11}| \leq -20$ dB for 10.4–11.6 GHz.

Figure 8.2 shows a selection of the feature points for a microwave coupler. Typically, one is interested in controlling the operating frequency, along with −20 dB bandwidth, as well as power split. Consequently, the feature points correspond to minimum of matching and isolation characteristics, $|S_{21}|$ and $|S_{31}|$ at the required operating frequency, as well as the frequencies corresponding to −20 dB levels of $|S_{11}|$ and $|S_{41}|$ (necessary for bandwidth calculation).

In more rigorous terms, the feature points of the response vector $\boldsymbol{R}_k(\boldsymbol{x})$ are denoted as $\boldsymbol{p}_k^j(\boldsymbol{x}) = \left[f_k^j(\boldsymbol{x}) l_k^j(\boldsymbol{x})\right]^T$, $j = 1, \ldots, p$. Here, f and l are the frequency and magnitude coordinates of the respective point. A practical way of extracting the characteristic points is a simple analysis of the EM-simulated response. Feature-based optimization benefits from the fact that dependence of the feature point coordinates on geometry parameters of the structure is less nonlinear than for the original characteristics. This leads to considerable simplification of the functional landscape to be optimized.

Observe that in this section, the joint vectors $\boldsymbol{p}_k^j(\boldsymbol{x})$ are used instead of employed in previous chapters separate vectors comprising frequency and level coordinates of the relevant feature points, i.e., $\boldsymbol{F}(\boldsymbol{x}) = [f_1(\boldsymbol{x})\, f_2(\boldsymbol{x}) \ldots f_p(\boldsymbol{x})]^T$ and $\boldsymbol{L}(\boldsymbol{x}) = [l_1(\boldsymbol{x})\ l_2(\boldsymbol{x}) \ldots l_p(\boldsymbol{x})]^T$. In practice, the decision whether to handle features jointly or separately should be made by the designer according to his or her preferences and is of secondary importance.

According to the FBO paradigm, the problem (8.1) can be reformulated in terms of the feature points as follows

$$\boldsymbol{x}^* = \arg\min_{\boldsymbol{x}} U_F\left(\boldsymbol{P}_f(\boldsymbol{x})\right) \tag{8.2}$$

In (8.2), the function U_F represents design objectives defined in the feature space, which is equivalent to U for the original response $\boldsymbol{R}_f(\boldsymbol{x})$ (see Eq. 8.1). Furthermore, $\boldsymbol{P}_f = \left[\left(\boldsymbol{p}_f^1\right)^T \ldots \left(\boldsymbol{p}_f^p\right)^T\right]^T$ represents the vector of aggregated feature points.

8.1.2 Optimization Algorithm with Adjoint Sensitivities

The exploitation of adjoint sensitivities (Koziel et al., 2012; Bakr & Nikolova, 2004; Bekasiewicz & Koziel, 2015) constitutes one of the available approaches to lowering the cost of EM-driven design. Adjoints have been extensively used in other engineering areas since 1980s (e.g., Jameson, 1988; Mader et al., 2008). However, commercial availability of this technology in EM simulation software packages is recent (e.g., HFSS (2021), or CST Microwave Studio (CST) (2022)). Yet, availability of cheap derivatives already revived interest in gradient-based optimization because it is a simple way of a substantial reduction of the optimization cost.

In general, the variable-fidelity feature-based optimization algorithm involving adjoint sensitivities is an iterative procedure that generates a sequence of approximate solutions to (1), $\boldsymbol{x}^{(j)}$, $j = 0, 1, \ldots, K$, as follows

$$\boldsymbol{x}^{(1)} = \arg\min_{\boldsymbol{x}} U(\boldsymbol{R}_1(\boldsymbol{x})) \tag{8.3}$$

and

$$\boldsymbol{x}^{(j)} = \arg\min_{\boldsymbol{x}} U_F\left(\boldsymbol{P}_j(\boldsymbol{x})\right) \tag{8.4}$$

for $j > 1$. More specifically, the design $\boldsymbol{x}^{(j)}$ is an optimum of the jth model $\boldsymbol{R}_j$. The final outcome of the optimization process is $\boldsymbol{x}^{(K)} = \boldsymbol{x}^*$. The sub-problem (8.3) is solved at the level of the original responses, which is due to the fact that when away from the optimum, the response structure can be different from that close to the final design and the feature points may not be well defined.

Both sub-problems (8.3) and (8.4) are solved in an iterative manner as follows

$$\boldsymbol{x}^{(1.i+1)} = \arg \min_{\boldsymbol{x}:\ \|\boldsymbol{x}-\boldsymbol{x}^{(1.i)}\| \le \delta^{(1.i)}} U\left(\boldsymbol{R}_{L.1}^{(i)}(\boldsymbol{x})\right) \tag{8.5}$$

and

$$\boldsymbol{x}^{(j.i+1)} = \arg \min_{\boldsymbol{x}:\ \|\boldsymbol{x}-\boldsymbol{x}^{(j.i)}\| \le \delta^{(j.i)}} U_F\left(\boldsymbol{P}_{L.j}^{(i)}(\boldsymbol{x})\right) \tag{8.6}$$

In the above equations, the vectors $\boldsymbol{x}^{(j.i)}$, $i = 0, 1, \ldots$, represent approximations to the solution of (8.3) (for $j = 1$) and (8.4) (for $j > 1$). The model $\boldsymbol{R}_{L.1}^{(i)}$ is the linear expansion of the EM model $\boldsymbol{R}_1$ at $\boldsymbol{x}^{(1.i)}$ defined as

$$\boldsymbol{R}_{L.j}^{(i)}(\boldsymbol{x}) = \boldsymbol{R}_j\left(\boldsymbol{x}^{(i)}\right) + \boldsymbol{J}_{\boldsymbol{R}_j}\left(\boldsymbol{x}^{(j.i)}\right) \cdot \left(\boldsymbol{x} - \boldsymbol{x}^{(j.i)}\right) \tag{8.7}$$

Similarly, the model $\boldsymbol{P}_{L.j}^{(i)}$ is the first-order expansion of the feature vector $\boldsymbol{P}_j$:

$$\boldsymbol{P}_{L.1}^{(i)}(\boldsymbol{x}) = \boldsymbol{P}_j\left(\boldsymbol{x}^{(i)}\right) + \nabla_{\boldsymbol{P}_j}\left(\boldsymbol{x}^{(j.i)}\right) \cdot \left(\boldsymbol{x} - \boldsymbol{x}^{(j.i)}\right) \tag{8.8}$$

In (8.7) and (8.8), the derivatives of the respective models (Jacobian $\boldsymbol{J}_{\boldsymbol{R}1}$ of $\boldsymbol{R}_1$ and feature point gradients $\nabla_{\boldsymbol{P}j}$ of $\boldsymbol{P}_j$) are obtained using adjoint sensitivities.

The results of the previous iteration are used as a starting point for the next optimization stage, i.e., we have $\boldsymbol{x}^{(j.0)} = \boldsymbol{x}^{(j-1)}$. The user-supplied initial design $\boldsymbol{x}^{(0)}$ is the starting point for optimizing the first model $\boldsymbol{R}_1$.

The optimization algorithm is embedded in the trust-region (TR) framework (Conn et al., 2000), where optimization of the linear expansion models is restricted to the vicinity of the current design of the size $\delta^{(j.i)}$. Updating of the TR radius follows the standard rules, and it is based on the gain ratio calculated as the ratio of the actual versus predicted improvement of the objective function value (Conn et al., 2000).

An important factor is the strategy for switching between the models. This is controlled by the termination condition $\|\boldsymbol{x}^{(j.i)} - \boldsymbol{x}^{(j.i-1)}\| < \varepsilon_j$, where $\varepsilon_j = M^{(K-j)}\,\varepsilon$, with ε being the overall termination threshold, here, $\varepsilon = 0.001$, and M being the scaling factor (here, we use $M = 10$). In other words, the termination condition for optimizing lower-fidelity models is more relaxed than for the higher-fidelity ones. The motivation is that due to the limited accuracy of coarse-discretization simulations, there is no need to aim at precise allocation of the optimum design for such models. Although the generic methodology described above utilized auxiliary coarse-discretization EM models, in principle, it is possible to utilize only one model (the high-fidelity one), i.e., $K = 1$, which is illustrated in Sect. 8.1.3.2.

8.1.3 Case Studies

In this section, the design optimization framework has been demonstrated using two cases: a coupler iris waveguide filter and a miniaturized rat-race coupler. For the first example, variable-fidelity models are utilized, whereas the second test case only exploits the high-fidelity model.

8.1.3.1 Four-Aperture Iris Coupled Bandpass Filter

The first considered design case is a four-aperture iris coupled bandpass filter shown in Fig. 8.3 (Hauth et al., 1993). The structure is implemented in WR-90 waveguide. The vector of design parameters is $\boldsymbol{x} = [z_1\ x_1\ x_2\ s_{x1}\ s_{x2}\ s_{y1}\ s_{y2}\ z_{t1}\ z_{t2}]^T$ (all in mm). The EM model of the filter is prepared and evaluated in HFSS. Optimization of the considered filter is performed using variable-fidelity EM model framework ($K = 3$; cf. Sect. 8.1.2). Feature points for the structure have been located at the edges of the passband (at −20 dB levels) as well as local in-band maximum of the reflection response (cf. Fig. 8.4).

We utilize two coarsely discretized EM surrogates, i.e., $\boldsymbol{R}_1$ that contains about 6000 tetrahedral cells and $\boldsymbol{R}_2$ with ~19,000 cells. Their average simulation times on a dual Xeon E5540 machine with 64 GB of RAM are about 3 min and 10 min, respectively. We also use the high-fidelity model $\boldsymbol{R}_f$ consisting of ~150,000 cells (simulation time 40 min). The design objective is to minimize the structure reflection below −20 dB in 14.9–15.1 GHz frequency range.

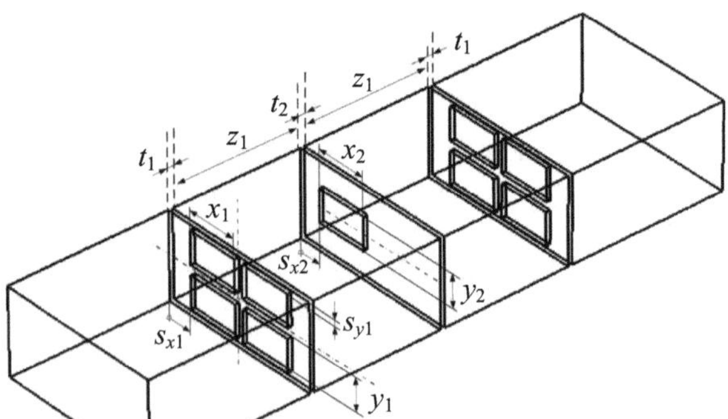

Fig. 8.3 Geometry of the four-aperture iris coupled bandpass filter (Hauth et al., 1993)

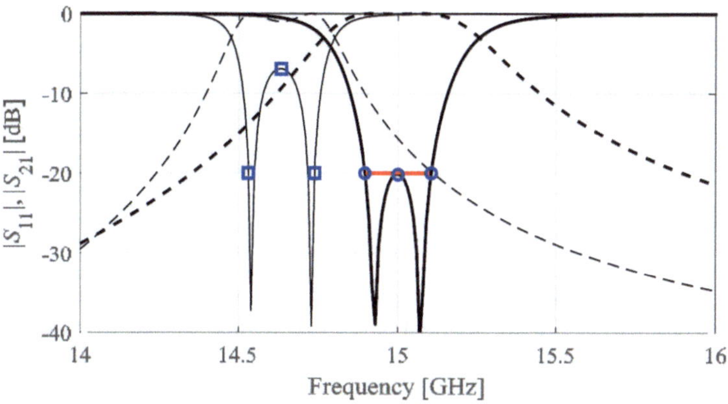

Fig. 8.4 High-fidelity model responses of the four-aperture iris coupled bandpass filter at the initial (thin lines) and optimized (thick lines) designs: $|S_{11}|$ (—) and $|S_{21}|$ (– –). Squares and circles represent feature points at the initial and final designs, respectively

Table 8.1 Four-aperture iris coupled bandpass filter: optimization results

Algorithm		Design cost: Number of evaluations	Design cost: Absolute (min)	Design cost: Relative to $\boldsymbol{R}_f$	Objective function value: max $\|S_{11}\|$
1	Variable-fidelity FBO	$11 \times \boldsymbol{R}_1$	33	0.8	−20.4 dB
		$5 \times \boldsymbol{R}_2$	50	1.2	
		$4 \times \boldsymbol{R}_f$	160	4.0	
		Total cost:	243	6.0	
2	Variable-fidelity trust-region[a]	$11 \times \boldsymbol{R}_1$	33	0.8	−23.5 dB
		$8 \times \boldsymbol{R}_2$	80	2.0	
		$6 \times \boldsymbol{R}_f$	240	6.0	
		Total cost:	353	8.8	
3	Trust-region[b]	$20 \times \boldsymbol{R}_f$	800	20.0	−23.0 dB

[a]Optimization with adjoint sensitivities (with $K = 3$)
[b]Optimization of $\boldsymbol{R}_f$ with adjoints (cf. Sect. 8.1.2 with $K = 1$)

The initial design is $\boldsymbol{x}^{(0)} = [12.5\ 5.0\ 5.0\ 2.5\ 2.5\ 0.5\ 0.0\ 0.25\ 0.25]^T$ mm. The filter has been optimized using method of Sect. 8.1.2 (Algorithm 1). The final dimensions of the structure $\boldsymbol{x}^* = [11.63\ 5.03\ 5.07\ 2.41\ 2.149\ 0.497\ {-0.068}\ 0.201\ 0.206]^T$ mm have been obtained at a cost corresponding to just six evaluations of the high-fidelity model (cf. Table 8.1). Comparison of the $\boldsymbol{R}_f$ model responses at the initial and optimized designs is shown in Fig. 8.4. It should be noted that, despite poor quality of the $\boldsymbol{x}^{(0)}$, "flattening" of the functional landscape by FBO allows for obtaining the final design at a low cost.

For the sake of comparison, the considered filter has been also optimized using trust-region frameworks with (i) variable-fidelity models (Algorithm 2), and (ii) only the high-fidelity model (Algorithm 3). Both benchmark approaches exploit

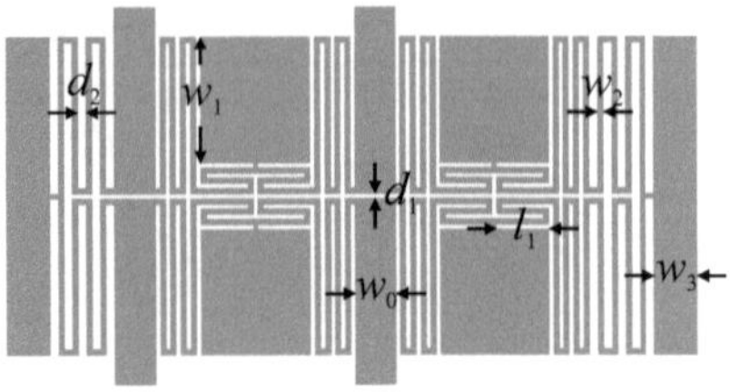

Fig. 8.5 Compact RRC composed of compact microstrip resonant cells (Koziel & Bekasiewicz, 2016)

sensitivity data. A summary of the optimization results for considered methods is given in Table 8.1. It should be noted that benchmark techniques yield comparable designs but at higher cost (by 47% and 233% for (i) and (ii), respectively).

8.1.3.2 Compact Microstrip Rat-Race Coupler

Our second example is a miniaturized microstrip 3 dB rat-race coupler (RRC) shown in Fig. 8.5 (Koziel & Bekasiewicz, 2016). The circuit is composed of two different compact microstrip resonant cells. RRC is designed on a 0.762 mm thick Taconic RF-35 dielectric substrate ($\varepsilon_r = 3.5$, $\tan\delta = 0.0018$). The vector of design parameters is $\boldsymbol{x} = [w_1\ w_2\ w_3\ d_1\ d_2\ l_1]^T$. Dimension $w_0 = 1.7$ remains fixed to ensure 50 ohm input impedance. The EM model of the structure is implemented in CST Microwave Studio and simulated using its transient solver. The model contains about 330,000 tetrahedral mesh cells and its simulation time is 6 min.

The following design requirements are assumed with respect to matching (S_{11}), transmission (S_{21}), coupling (S_{31}), and isolation (S_{41}):

- Maximization of bandwidth (BW) defined as a symmetric—with respect to the center frequency of 1 GHz—part of the response for which both $|S_{11}|$ and $|S_{41}|$ are below −20 dB.
- Reducing the power split error at the center frequency $d_S = |S_{21.f}(\boldsymbol{x}) - S_{31.f}(\boldsymbol{x})|$ to zero.
- Maintaining possibly small shift of the resonances $S_{11.f}(\boldsymbol{x})$ and $S_{41.f}(\boldsymbol{x})$ with respect to the operating frequency.

It should be noted that maximization of bandwidth is a primary objective, whereas the remaining requirements are handled as constraints. Here, they are incorporated into the objective using penalty functions.

The considered structure has been optimized using the technique of Sect. 8.1.2 assuming $K = 1$ (i.e., no auxiliary lower-fidelity models). The initial design is $\boldsymbol{x}^0 = [4.0\ 0.3\ 2.0\ 0.3\ 0.2\ 1.0]^T$. The optimized coupler design is $\boldsymbol{x}^* = [3.87\ 0.43\ 1.76\ 0.37\ 0.36\ 1.10]^T$. It should be noted that, due to "flattening" of the functional landscape, final RRC design has been obtained in just six iterations of the feature-based algorithm at a cost of only eight EM model evaluations. A comparison of the structure responses at the initial and final designs is shown in Fig. 8.6. The results indicate that, for the optimized design, the −20 dB bandwidth of 343 MHz has been

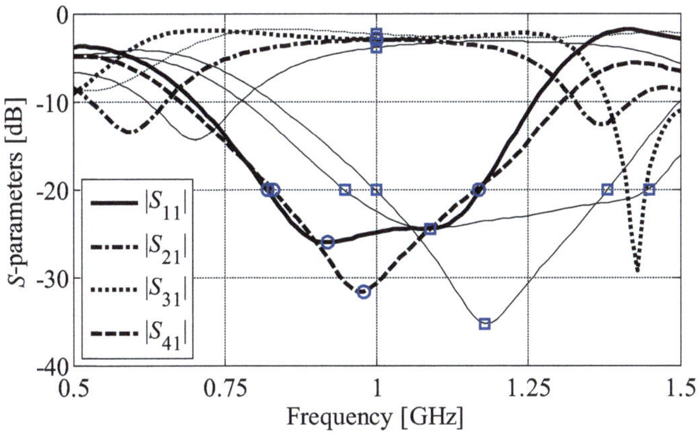

Fig. 8.6 Compact rat-race coupler: frequency responses at the initial (thin lines) and final designs (thick lines). Squares and circles represent feature points at the initial and final designs, respectively

obtained together with power split error of 0.15 dB and acceptable resonances shifts. The size of the optimized structure is 470 mm^2, which is over 90% smaller compared to conventional rectangular RRC.

The coupler has been also optimized at the level of frequency responses using adjoints-based gradient search embedded in trust-region framework (also with $K = 1$, cf. Sect. 8.1.2). Although the obtained design is similar (BW: 341 MHz, 3 dB error: 0.15 dB, and footprint: 472 mm^2), it has been obtained at a cost of 13 evaluations of the structure EM model (eight algorithm iterations). In other words, combining adjoints with FBO results in almost 40% computational cost savings.

Figure 8.7 shows the results of experimental validation of the optimized RRC. It can be observed that the agreement between simulation and measurements is good. Visible discrepancies between responses are mostly due to the utilization of simplified EM model that lacks SMA connectors and fabrication tolerances.

This section discussed feature-based optimization with adjoint sensitivities for rapid design closure of microwave components. Numerical results indicate that the aforementioned approach allows for a significant reduction of the computational cost of the optimization process. The next section will address the issue of accelerating FBO procedures with the use of sparse sensitivity updates on the example of multi-band antennas.

8.2 Feature-Based Optimization with Sparse Sensitivity Updates

This section discusses expedited FBO of multi-band antennas using sparse sensitivity updates (Koziel & Pietrenko-Dabrowska, 2020b). In the considered approach, costly finite-differentiation-based sensitivity updates are omitted for design variables

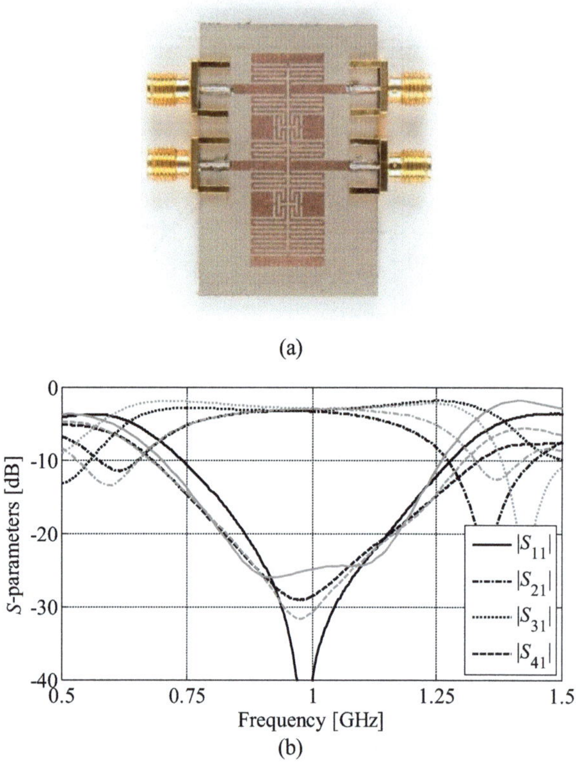

Fig. 8.7 Experimental validation of the RRC: (**a**) photograph of the fabricated coupler prototype, (**b**) simulated (gray) and measured (black) coupler characteristics

characterized by stable sensitivity patterns. In the examples of this section, the employment of the response feature technology enables a considerable improvement of the reliability of the optimization process. Despite poor initial designs, the discussed procedure allows for a proper allocation of the target operating frequencies of multi-band antennas, whereas the conventional approach fails to do so. Furthermore, conducting the optimization process at the level of the response features enhances its cost-efficiency. In consequence, the cost of the discussed optimization procedure is comparable to that of local routines while ensuring nearly global search capabilities.

8.2.1 Globalized Feature-Based Optimization with Sparse Sensitivity Updates

This section briefly recapitulates the response feature technology and its employment for solving design optimization tasks of multi-band antennas, as well as incorporating of response features into the accelerated local optimizer (here, trust-region gradient algorithm) with sparse sensitivity updates based on gradient change monitoring. The combination of the two techniques allows for a cost-efficient and quasi-global optimization of multi-band antennas as demonstrated by the numerical studies of Sect. 8.2.2.

8.2.1.1 Feature-Based Optimization of Multi-Band Antennas

In this section, we focus on multi-band antennas. Fulfilling stringent specifications imposed on electrical and field antenna characteristics requires careful tuning of its geometry parameters. This process is referred to as the design closure and its ultimate goal is to improve the performance of the device at hand according to the selected quality metric, typically being a scalar function of the parameters. A conventional formulation of the design closure problem has been given by (8.1).

Here, we will use $\boldsymbol{R}(\boldsymbol{x})$ to denote antenna characteristics (e.g., reflection coefficient, gain, axial ratio, or radiation pattern) that are of interest in a given design context. For multi-band antennas, a typical design problem is a frequency allocation of the antenna resonances and/or maximization of the fractional bandwidth at the desired operating frequencies. Here, the former case is considered; hence, the conventional scalar objective function U may be formulated as follows

$$U(\boldsymbol{R}(\boldsymbol{x})) = \max_{f \in F} \mid S_{11}(\boldsymbol{x},f) \mid \tag{8.9}$$

where $S_{11}(x,f)$ denotes the antenna reflection being a function of $\boldsymbol{x}$ and frequency f, whereas F stands for the frequency range of interest. Local optimization routines with the objective function U formulated as in (8.9), may or may not succeed in finding the optimal solution depending on the quality of the initial design. In particular, due to high nonlinearity of antenna responses typical for multi-band structures, the allocation of resonances at the initial design is critical for the success of the local search. Some typical scenarios have been illustrated in Fig. 8.8.

Employing global optimization routines to solve the antenna design closure task is associated with a high computational cost that can be conveniently reduced with the response feature approach. Let us consider, for the illustration purposes and without losing generality, the feature points corresponding to the antenna resonances, i.e., $\boldsymbol{P}_f(\boldsymbol{x}) = [f_1(\boldsymbol{x})\ f_2(\boldsymbol{x})\ \ldots\ f_p(\boldsymbol{x})\ l_1(\boldsymbol{x})\ l_2(\boldsymbol{x})\ \ldots\ l_p(\boldsymbol{x})]^T$, where f_k and l_k refer to the frequency and level coordinates of the respective p antenna resonances. (Here, we use aggregated feature vector similarly as in Sect. 8.1).

As the dependence of the feature point coordinates on the design variables is notably less nonlinear than for the responses themselves, thus, a local feature-based search most often brings satisfactory designs even starting from poor initial ones, where the routines solving the problem (8.1) or (8.8) in a conventional manner may fail. It should be emphasized that for the response-feature approach to work, it is sufficient that the initial design exhibits all the necessary features (i.e., clearly distinguished resonances in the case of a multi-band antenna), regardless of their specific frequency allocation or the levels.

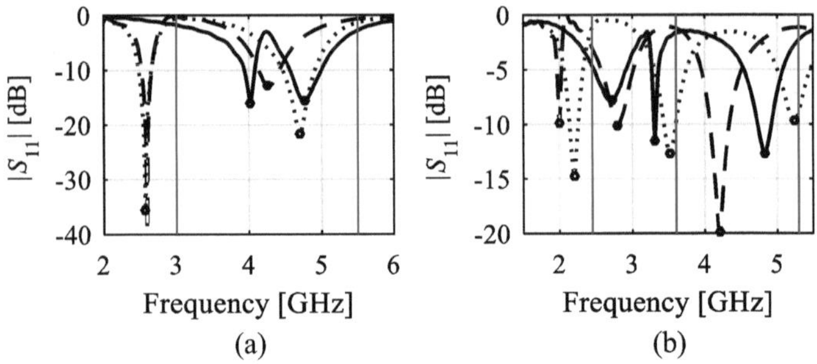

Fig. 8.8 Exemplary reflection characteristics of (**a**) dual-band antenna (Chen et al., 2006) and (**b**) triple-band antenna (Koziel & Bekasiewicz, 2018b) at three different designs; characteristic points corresponding to the coordinates of the antenna resonances marked with (o); intended operating frequencies marked with vertical lines. Local (gradient-based) optimization routines starting from the designs (- - -) and (—) fail in the case of minimax formulation of the objective function as in (8.1). Local optimization involving response features (8.8) finds satisfactory design starting from all the presented designs: (- - -), (—), and (. . .)

The antenna design problem may be reformulated for the objective function $U_F(\boldsymbol{P}_f(\boldsymbol{x}))$ defined in terms of the response features $\boldsymbol{P}_f(\boldsymbol{x})$ in the following manner

$$\boldsymbol{x}^* = \arg\min_{\boldsymbol{x}} U_F\left(\boldsymbol{P}_f(\boldsymbol{x})\right) \tag{8.10}$$

We intend to relocate the resonances of a multi-band antenna to the target frequencies of choice $f_{0.k}$, $k = 1, \ldots, p$. Observe that p is employed to denote the number of the antenna bands, as in this case, it coincides with the number of the features that are of interest. Thus, the objective function U_F is defined as

$$U_F\left(\boldsymbol{P}_f(\boldsymbol{x})\right) = \max\left\{l_1(\boldsymbol{x}), \ldots, l_p(\boldsymbol{x})\right\} + \beta \left\| \left[f_1(\boldsymbol{x}) \ldots f_p(\boldsymbol{x})\right] - \left[f_{0.1} \cdots f_{0.p}\right] \right\|^2 \tag{8.11}$$

with β being the penalty factor. Here, the antenna reflection is considered as the system output of interest and the level coordinates are $l_k(\boldsymbol{x}) = S_{11}(\boldsymbol{x}, f_k)$, $k = 1, \ldots, p$. In (8.11), minimization of the antenna reflection is the primary objective, whereas the second (penalty) term permits to control the resonant frequencies of the antenna. The value of the penalty factor β should ensure that the penalty term is commensurable to the principal objective in case of an unsatisfactory allocation of the operating frequencies (here, $\beta = 100$).

8.2.1.2 Trust-Region Search with Jacobian Changes Tracking

The optimization procedure of choice to solve the problem (8.10) is the trust-region gradient search algorithm (e.g., Conn et al., 2000) that iteratively yields

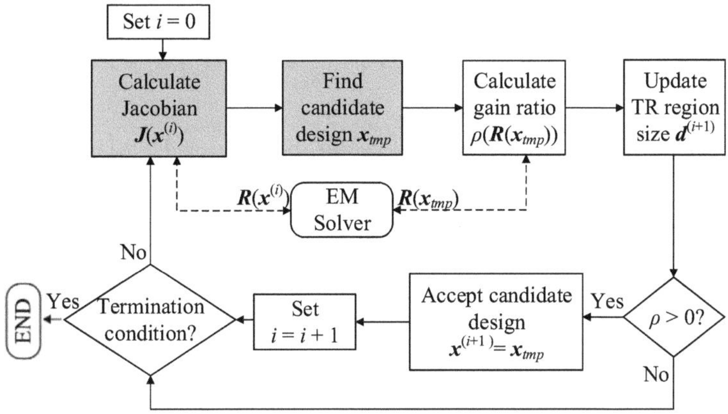

Fig. 8.9 Flow diagram of the trust-region (TR) algorithm. The gray boxes refer to the operations that are dissimilar for the conventional and feature-based versions of the algorithm: both with and without Jacobian change tracking

approximations $\boldsymbol{x}^{(i)}$, $i = 0, 1, \ldots$, to the optimum design $\boldsymbol{x}^*$. The flow diagram of the TR procedure is shown in Fig. 8.9. In each i-th iteration, a linear expansion model $\boldsymbol{P}_L^{(i)}$ of $\boldsymbol{P}_f(\boldsymbol{x})$ is defined $\boldsymbol{x}^{(i)}$ as

$$\boldsymbol{P}_L^{(i)}(\boldsymbol{x},f) = \boldsymbol{P}_f\left(\boldsymbol{x}^{(i)}\right) + \boldsymbol{J}_f\left(\boldsymbol{x}^{(i)}\right) \cdot \left(\boldsymbol{x} - \boldsymbol{x}^{(i)}\right) \tag{8.12}$$

Subsequently, the following sub-problem is solved

$$\boldsymbol{x}^{(i+1)} = \arg \min_{\boldsymbol{x};\ -\boldsymbol{\delta}^{(i)} \le \boldsymbol{x} - \boldsymbol{x}^{(i)} \le \boldsymbol{\delta}^{(i)}} U_F\left(\boldsymbol{P}_L^{(i)}(\boldsymbol{x})\right) \tag{8.13}$$

In (8.12), $\boldsymbol{J}_f(\boldsymbol{x}^{(i)}) = [\nabla f_1(\boldsymbol{x}^{(i)}) \ldots \nabla f_p(\boldsymbol{x}^{(i)})\ \nabla l_1(\boldsymbol{x}^{(i)}) \ldots \nabla l_p(\boldsymbol{x}^{(i)})]^T$ refers to the response feature vector Jacobian.

In addition, $\boldsymbol{\delta}^{(i)}$ stands for the search region size vector adjusted in conformance with the standard rules (Conn et al., 2000), based on the gain ratio $\rho = (U_F(\boldsymbol{P}_f(\boldsymbol{x}_{\text{tmp}})) - U_F(\boldsymbol{P}_f(\boldsymbol{x}^{(i)})))/(U_F(\boldsymbol{P}_L^{(i)}(\boldsymbol{x}_{\text{tmp}})) - U_F(\boldsymbol{P}_L^{(i)}(\boldsymbol{x}^{(i)})))$, where $\boldsymbol{x}_{\text{tmp}}$ is the candidate design obtained in the $(i + 1)$th iteration. The inequalities $-\boldsymbol{\delta}^{(i)} \le \boldsymbol{x} - \boldsymbol{x}^{(i)} \le \boldsymbol{\delta}^{(i)}$ in (8.13) are to be interpreted component-wise. This is to ensure similar handling of the variables with significantly different ranges, commonly occurring in the antenna design.

Unless adjoint sensitivities (see Sect. 8.1) are readily available, the Jacobian $\boldsymbol{J}_f$ has to be estimated through finite differentiation (FD) at the expense of n additional antenna EM model simulations. Here, in order to lessen this cost, an expedited procedure is utilized where some of the FD-based Jacobian updates are omitted for the variables that exhibit stable sensitivity pattern (Pietrenko-Dabrowska & Koziel, 2020). The variations of the Jacobian $\boldsymbol{J}_f$ columns $J_k^{(i)} = \left[g_{k.1}^{(i)} \ldots g_{k.n}^{(i)}\right]^T$, $k = 1, \ldots, p$, between iterations are assessed by the following metric:

$$\Delta_k^{(i+1)} = \frac{1}{2p}\sum_{j=1}^{2p}\left(2\cdot\frac{g_{j.k}^{(i)} - g_{j.k}^{(i-1)}}{g_{j.k}^{(i)} + g_{j.k}^{(i-1)}}\right) \tag{8.14}$$

In (8.14), averaging is performed over both the frequency and level coordinates of all relevant feature points. Let us define the following quantities:

- $\boldsymbol{\Delta}^{(i)} = \left[\Delta_1^{(i)} \ldots \Delta_n^{(i)}\right]^T$—a vector of Jacobian change factors at the i-th iteration
- $\Delta_{\min}^{(i)} = \min\left\{\Delta_1^{(i)}, \ldots, \Delta_n^{(i)}\right\}$ $\Delta_{\max}^{(i)} = \max\left\{\Delta_1^{(i)}, \ldots, \Delta_n^{(i)}\right\}$
- $\boldsymbol{N}^{(i)} = \left[N_1^{(i)} \ldots N_n^{(i)}\right]^T$—a vector of the numbers of upcoming iterations without FD, calculated in the i-th iteration with the use of the conversion function

$$N_k^{(i)} = [\![N_{\max} + a^{(i)}\left(\Delta_k^{(i)} - \Delta_{\min}^{(i)}\right)]\!] \tag{8.15}$$

with $N_{\min}$ and $N_{\max}$ being the algorithm control parameters that refer to the minimum and the maximum number of iterations without FD; $a^{(i)} = (N_{\max} - N_{\min})/\left(\Delta_{\min}^{(i)} - \Delta_{\max}^{(i)}\right)$, where the nearest integer function is denoted as [[.]]. The flow diagram of the gradient update procedure has been shown in Fig. 8.10.

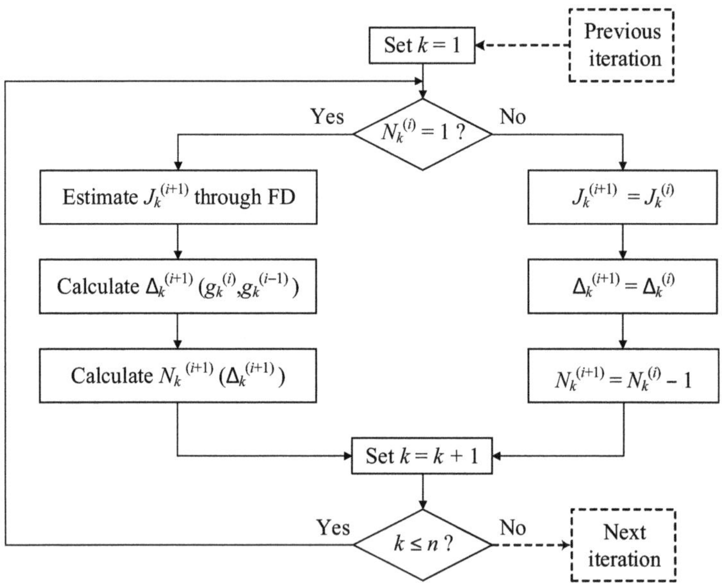

Fig. 8.10 Flow diagram of the gradient update procedure utilized for both the conventional and feature-based optimization

Table 8.2 Taxonomy of the considered optimization procedures

Algorithm	Problem formulation	Acceleration mechanism	Finite differentiation gradient update
1	Conventional	N/A	Performed in each iteration for each parameter
2	Conventional	Measure of changes of the response gradient between iterations	Omitted in some iterations for some parameters
3	Feature-based	N/A	Performed in each iteration for each parameter
4	Feature-based	Measure of changes of the response gradient between iterations	Omitted in some iterations for some parameters

In the considered algorithm, the following rules apply:

1. Factors $\Delta_k^{(i)}$ are preserved throughout all iterations without FD and employed to establish $\Delta_{\min}^{(i)}$ and $\Delta_{\max}^{(i)}$ (i.e., they affect $N_k^{(i)}$ for other parameters).
2. $N_k^{(i)}$ is proportional to the magnitude of the gradient variation as assessed by the factors $\Delta_k^{(i)}$. For the variables featuring the smallest Jacobian changes between subsequent iterations $\left(\Delta_k^{(i)} = \Delta_{\min}^{(i)}\right)$, the highest number of the suppressed updates is appointed, i.e., $N_k^{(i)} = N_{\max}$. Furthermore, following the definition of the conversion function (8.15), the FD-based updates are performed at least once per $N_{\max}$ iterations.
3. For the variables with the FD update, the components $N_k^{(i+1)}$ of the vector $\boldsymbol{N}^{(i+1)}$ are determined through the conversion function (8.15); otherwise, the previous number of iterations is decremented, i.e., $N_k^{(i+1)} = N_k^{(i)} - 1$.

In Table 8.2, a comparison of the main properties of the four algorithms considered in the article is provided. Algorithms 1 and 2 are the conventional (i.e., with full FD Jacobian update) and the expedited TR routine, respectively, both solving a conventionally formulated problem (8.1). Algorithm 3 (basic) and Algorithm 4 (expedited) solve the design task formulated in terms of the response features. Algorithms 1 through 3 are used for benchmarking purposes. The principal differences in operation of the fours algorithms include (see also Fig. 8.9):

1. Calculation of the Jacobian: either in terms of the entire response (Algorithms 1 and 2) or the response features (Algorithms 3 and 4).
2. Jacobian update procedure: either solely FD-based (Algorithms 1 and 3) or performed according to the gradient change tracking procedure (Algorithms 2 and 4).

The considered algorithm, involving the gradient change control (this pertains to both Algorithm 2 and 4), operates as the reference algorithm merely in the first two iterations, when the Jacobian is estimated through FD. Full Jacobian data is required to determine the initial values of the factors $\Delta_k^{(i)}$. Subsequently, FD is only applied to the components J_k of $\boldsymbol{J}_F$ selected according to the sensitivity pattern measure; for other parameters, the former values of the factors $\Delta_k^{(i)}$ are retained (cf. Fig. 8.10). The numerical results gathered in Sect. 8.2.2 confirm that the considered procedure secures considerable CPU cost savings. At the same time, a quasi-global search capability is maintained.

8.2.2 Verification Case Studies

This section provides a numerical verification of the discussed optimization framework and comparisons with conventionally formulated (not feature-based) procedures. The performance of the considered approach is verified by analyzing the results obtained for multiple independent algorithm runs from random initial designs. Our benchmark set comprises two antenna structures: dual-band (Chen et al., 2006) and triple-band (Koziel & Bekasiewicz, 2018b) uniplanar dipoles shown in Fig. 8.11. Both structures are implemented on the RO4350 substrate ($\varepsilon_r = 3.48$, $h = 0.762$ mm) and fed by a coplanar waveguide. The geometry of the dual-band antenna (Antenna I) is described by the parameters $\boldsymbol{x} = [l_1\ l_2\ l_3\ w_1\ w_2\ w_3]^T$, with the following variables fixed $l_0 = 30$, $w_0 = 3$, $s_0 = 0.18$ and $o = 5$ (all dimensions in mm). The antenna is supposed to operate at the center frequencies of 3 and 5.5 GHz. The triple-band antenna (Antenna II) is described by the design variable vector $\boldsymbol{x} = [l_1\ l_2\ l_3\ l_4\ l_5\ w_1\ w_2\ w_3\ w_4\ w_5]^T$; with $l_0 = 30$, $w_0 = 3$, $s_0 = 0.15$, and $o = 5$ being fixed (all dimensions in mm). The antenna is supposed to operate at the center frequencies of 2.45, 3.6 and 5.3 GHz. The EM antenna models are implemented in CST Microwave Studio and simulated using its time-domain solver.

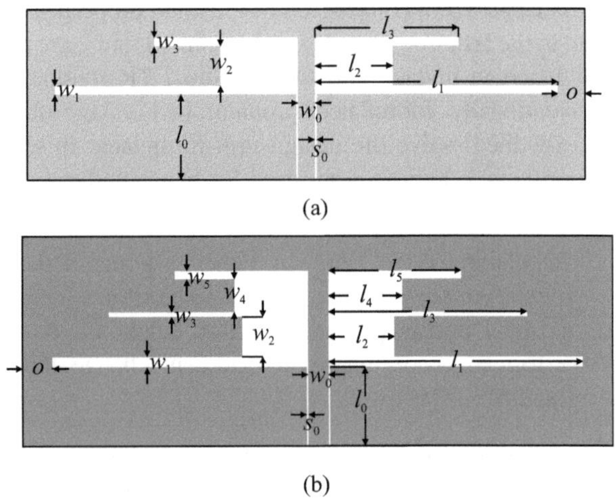

Fig. 8.11 Benchmark uniplanar dipole antenna structures used for verification of the considered algorithm: (**a**) dual-band (Chen et al., 2006) and (**b**) triple-band (Koziel & Bekasiewicz, 2018b) antenna

Table 8.3 Optimization results for Antenna I

Algorithm	Cost[a]	$f_{0.1}$[b] (GHz)	std ($f_{0.1}$)[c]	$f_{0.2}$[b] (GHz)	std ($f_{0.2}$)[c]	$\lvert S_{11}(f_{0.1})\rvert$[d] (dB)	$\lvert S_{11}(f_{0.2})\rvert$[d] (dB)
1	78.3	2.68	1.1	5.10	1.5	−8.4	−10.3
2	53.5	2.69	1.1	4.80	1.6	−9.6	−13.2
3	78.0	3.00	0.0	5.50	0.0	−38.8	−40.3
4	49.9	3.04	0.1	5.48	0.1	−27.1	−27.8

[a]Computational cost (average number of EM simulations; 20 algorithm runs)
[b]Resonance frequencies $f_{0.1}$ and $f_{0.2}$ averaged over the set of 20 algorithm runs
[c]Standard deviations of resonance frequencies across the set of 20 algorithm runs
[d]Antenna reflections at the resonance frequencies averaged over 20 algorithm runs

Both antennas have been optimized using the algorithms of Table 8.2: conventional and expedited procedures solving a traditionally formulated problem (8.1) (Algorithms 1 and 2, respectively) and conventional and expedited procedures solving problem (8.10) reformulated in terms of response features (Algorithms 3 and 4, respectively). The expedited Algorithms 2 and 4 were executed with the following values of the control parameters: $N_{\min} = 1$, $N_{\max} = 5$ (Pietrenko-Dabrowska & Koziel, 2020). These parameters refer to the minimum and the maximum number of iterations without FD, respectively. They decide upon the frequency of performing FD-based sensitivity updates: no more than once per $N_{\min}$ iterations, and at least once per $N_{\max}$ iterations. The adopted values allow us to achieve a substantial acceleration of the optimization process without compromising the solution quality in a significant manner.

In order to verify the robustness of the optimization process, each of the algorithms were executed 20 times from the same set of random initial designs. In Tables 8.3 and 8.4, the averaged performance measures for Antennas I and II across the set have been presented, and the antenna responses for the representative algorithm runs are shown in Figs. 8.12 and 8.13.

It should be emphasized that the values of the objective functions for the algorithms using conventional (Algorithms 1 and 2) and feature-based formulation (Algorithms 3 and 4) of the problem cannot be compared directly. Therefore, the comparison is carried out based on the feature point coordinates, which, in the case of Algorithms 1 and 2, are extracted from the optimum responses obtained with these procedures.

As a measure of the results quality, Tables 8.3 and 8.4 report the standard deviation of the resonant frequencies obtained for 20 algorithm runs. The computational cost in the form of total number of EM simulations required by the procedure to converge is also given in Tables 8.3 and 8.4.

The presented results (Tables 8.3 and 8.4, Figs. 8.12 and 8.13) allow us to draw certain conclusions concerning the algorithm performance. It can be observed that the feature-based algorithms (Algorithms 3 and 4) are superior over those solving a conventionally formulated design problem (Algorithms 1 and 2), both in terms of reliability and design quality. For Antenna I, the reference conventional TR

Table 8.4 Optimization results for Antenna II

Algorithm	Cost[a]	$f_{0.1}$[b] (GHz)	std($f_{0.1}$)[c]	$f_{0.2}$[b] (GHz)	std($f_{0.1}$)[c]	$f_{0.3}$[b] (GHz)	std($f_{0.1}$)[c]	$\|S_{11}(f_{0.1})\|$[d] (dB)	$\|S_{11}(f_{0.2})\|$[d] (dB)	$\|S_{11}(f_{0.3})\|$[d] (dB)
1	50.5	2.19	0.8	2.98	1.1	4.70	1.7	−4.7	−4.3	−7.5
2	25.7	2.23	0.8	3.06	1.2	4.58	1.6	−4.9	−3.9	−4.6
3	83.9	2.45	0.0	3.60	0.0	5.31	0.0	−30.5	−31.4	−29.0
4	46.5	2.45	0.0	3.60	0.0	5.30	0.0	−27.1	−28.4	−25.6

[a]Computational cost (average number of EM simulations; 20 algorithm runs)
[b]Resonance frequencies $f_{0.1}$, $f_{0.2}$, and $f_{0.3}$ averaged over the set of 20 algorithm runs
[c]Standard deviations of resonance frequencies across the set of 20 algorithm runs
[d]Antenna reflections at the resonance frequencies averaged over 20 algorithm runs

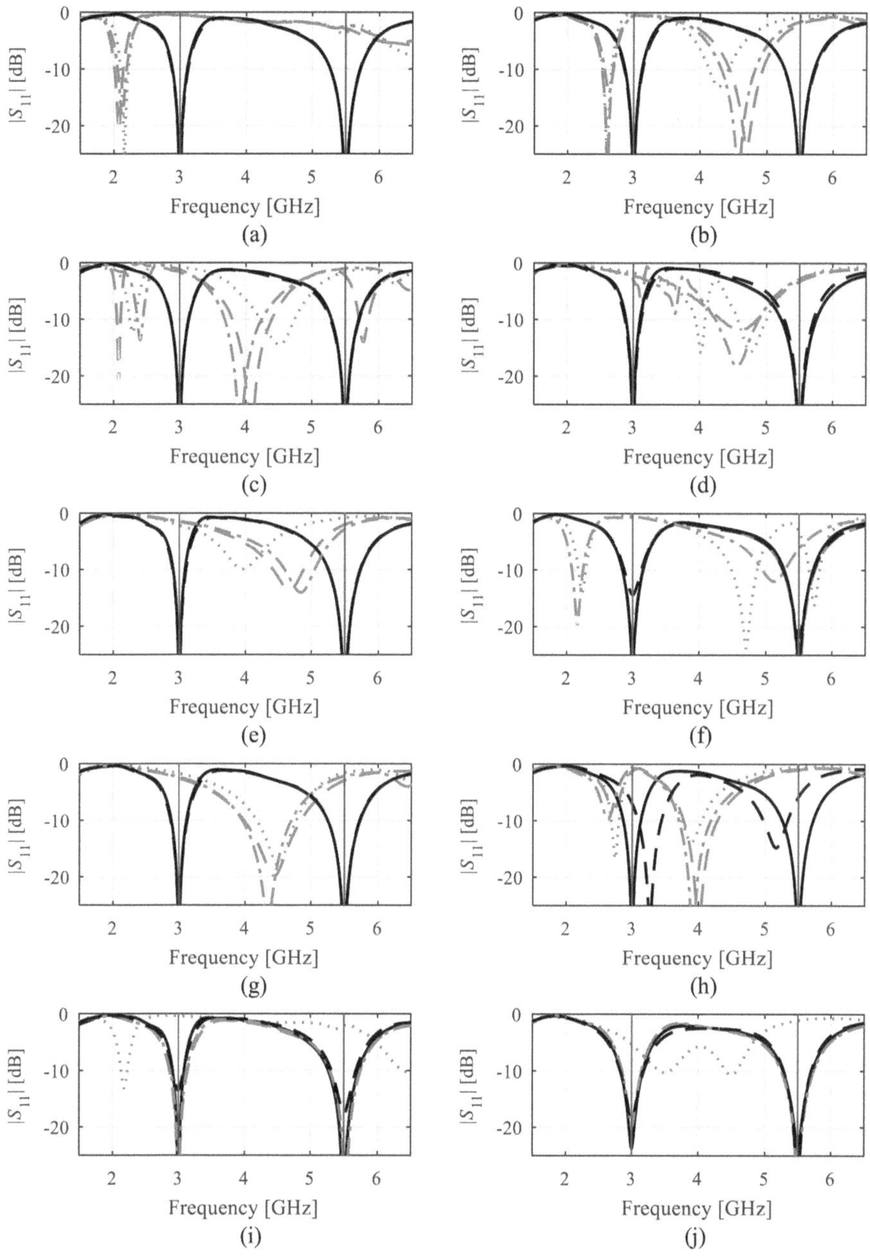

Fig. 8.12 Reflection characteristics of a dual-band antenna (Chen et al., 2006) for the representative algorithm runs ((**a**) through (**j**)) of Algorithms 1 through 4. The initial design (. . .) and the designs optimized in the conventional setup are marked gray: (- - -) Algorithm 1 (reference TR algorithm with full FD Jacobian update) and (-.-.) Algorithm 2 (expedited TR routine). The designs found within the response feature frameworks are marked black: (—) Algorithm 3 (conventional version) and (- - -) Algorithm 4 (expedited routine). Vertical lines mark the target operating frequencies. The conventional routines (Algorithms 1 and 2) fail to find satisfactory designs in most cases

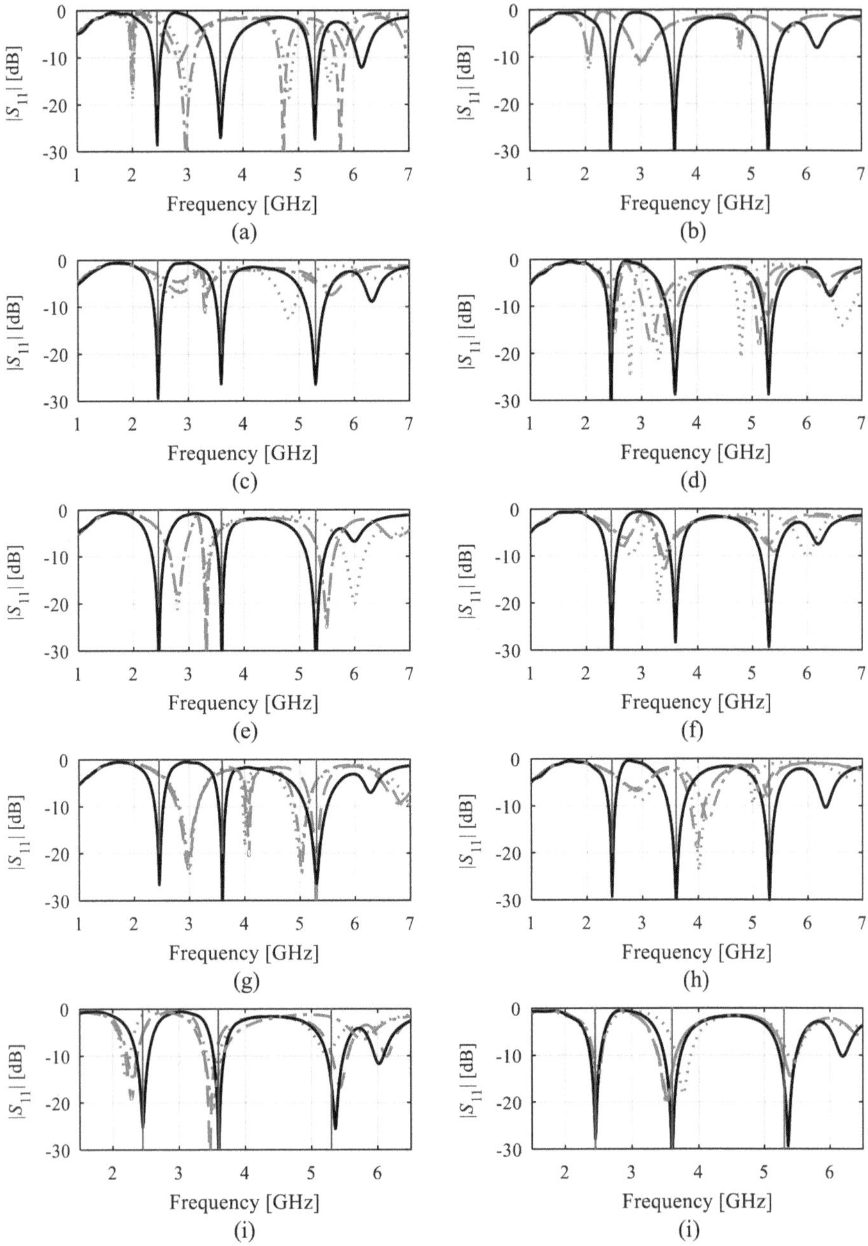

Fig. 8.13 Reflection characteristics of a triple-band antenna (Koziel & Bekasiewicz, 2018b) for the representative algorithm runs ((**a**) through (**j**)) of Algorithms 1 through 4. The initial design (. . .) and the designs optimized in the conventional setup are marked gray: (- - -) Algorithm 1 (reference TR algorithm with full FD Jacobian update) and (-.-.) Algorithm 2 (expedited TR routine). The designs found within the response feature frameworks are marked black: (—) Algorithm 3 (conventional version) and (- - -) Algorithm 4 (expedited routine). Vertical lines mark the target operating frequencies. The conventional routines (Algorithms 1 and 2) fail to find satisfactory designs in most cases

algorithm (Algorithm 1) fails to find satisfactory designs for a majority (14 out of 20) of the initial designs (cf. Fig. 8.12a–h). This is consistent with the results obtained for Antenna II, where Algorithm 1 fails to adequately allocate the antenna resonances for 18 out of 20 the considered starting points (see Fig. 8.13).

In the response features setup, however, Algorithm 3 accurately allocates the resonances for Antenna I and II in all cases (see Figs. 8.12 and 8.13). As far as the accelerated Algorithm 4 is concerned, for Antenna I, inadequate allocation of the resonant frequencies only occurs in four cases. For Antenna II, the designs satisfying the specifications are found in all cases.

Detailed scrutiny of the gathered results indicates that in the response feature framework, satisfactory designs can be found for a wide range of initial allocation of Antenna I resonances. This is also the case for Antenna II, where for all the initial designs the feature-based algorithm yields excellent solutions.

In fact, as illustrated in Fig. 8.13, for Antenna II, the resonant frequencies of the initial design have to be close to the target ones (cf. Fig. 8.13j), for the conventional procedure to find an acceptable solution. The aforementioned variety of initial resonant frequency configurations as well as a comprehensive validation using a large number of random starting points demonstrates the quasi-global capabilities of both feature-based algorithms. Moreover, the accelerated feature-based procedure (Algorithm 4) exhibits a remarkable computational efficiency.

The resonance depths in the conventional optimization setup for both antennas are poor. For Antenna I and Algorithm 1, the value -9 dB on average (for two resonances) is obtained; it is even worse for Antenna II (around -5.5 dB for three resonances). This is partially related to inadequate allocation of the resonant frequencies. It should be emphasized that the maximum acceptable level of antenna reflection at its operating frequencies is -10 dB. On the other hand, in the feature-based setup, the design quality is significantly better. For Antenna I, the average reflection level is around -40 dB for the non-accelerated routine (Algorithm 3), and -27.5 dB on average for the accelerated one (Algorithm 4). Whereas for Antenna II, the obtained average values are -30 dB and -27 dB for Algorithms 3 and 4, respectively.

The most important aspect of the optimization process for multi-band antennas is a proper allocation of the resonant frequencies. As indicated in Tables 8.3 and 8.4, the feature-based algorithms (Algorithms 3 and 4) are far superior over the conventional ones with this respect. In particular, the frequencies obtained in the considered approach are almost equal to the target frequencies (within the resolution important for practical purposes) for both considered antenna structures. This is not the case for the conventional frameworks.

In order to assess the computational speedup provided by the procedures involving smart Jacobian updates, one has to compare the conventional procedures with their accelerated counterparts, i.e., Algorithm 1 versus 2 (formulated in the entire response sense), as well as Algorithm 3 versus 4 (feature-based formulations). For Antenna I, the savings of around 32% and 37% are obtained in both cases, respectively. As for Antenna II, the savings are even more pronounced: 49% (Algorithm 2 w.r.t. Algorithm 1), and 45% (Algorithm 4 w.r.t. Algorithm 1). At this point, it should become clear that the computational cost of the considered approach is dramatically lower than the cost of population-based metaheuristics, routinely used for global search purposes, e.g., (BouDaher & Hoorfar, 2015; Kovaleva et al., 2017; Lalbakhsh et al., 2017). The latter, given any imaginable setup (population size and the number of iterations) would well exceed 1000 antenna evaluations, which is one of the reasons why these methods are most often applied to handle analytical representations (e.g., array factor models (Goudos & Sahalos, 2016; Banerjee & Mandal, 2018; Wang et al., 2018)). It is also well known that metaheuristics require careful tuning of control parameters and exhibit poor solution repeatability. Therefore, these sort of methods were not included in the benchmark pool.

This section discussed a computationally efficient algorithm for quasi-global optimization of input characteristics of multi-band antennas. The antenna design closure task has been handled by employing the response feature technique in conjunction with the Jacobian change monitoring procedure. The reliability of the framework has been comprehensively validated using uniplanar dual- and triple-band dipole antennas. It has been demonstrated to be significantly more reliable than the conventional gradient-based algorithms. In particular, the considered algorithm yields satisfactory designs for a wide range of initial designs, where the local routines fail. Another advantage of the approach is a high precision of allocating the resonant frequencies of the antenna. Apart from reliability, exploitation of the Jacobian variability tracking technique permits significant computational savings, which are as high as 40% (on average) as compared to the standard (non-accelerated) algorithm. The optimization cost is therefore comparable to that of local search routines. At the same time, nearly global search capabilities are secured.

8.3 Summary

This chapter considered two methods for expediting feature-based optimization of high-frequency structures: adjoint sensitivities and sparse sensitivity updates. The considered approaches, along with the employment of variable-fidelity EM models addressed in Chap. 7, permit for achieving a considerable computational speedup of local feature-based optimization procedures. Observe that adjoint sensitivities are not widespread within commercial EM solvers and constitute an intrusive technique. These limitations however do not pertain to sparse sensitivity updates. Furthermore, several mechanisms have been developed to suppress costly sensitivity updates. In this chapter, the approach based on detecting the stability patterns for the system response gradients has been discussed. The techniques have been verified with the use of several high-frequency structures: a waveguide filter and a microstrip rat-race coupler (FBO with adjoint sensitivities of Sect. 8.1), as well as a two- and a three-band antenna (FBO with sparse sensitivity updates discussed in Sect. 8.2). The presented results corroborate reliability of accelerated feature-based optimization: in all presented cases, the assumed design specifications have been achieved despite the fact that the initial designs have been severely misaligned with the targets.

References

Bakr, M. H., & Nikolova, N. K. (2004). An adjoint variable method for time domain TLM with wideband Johns matrix boundaries. *IEEE Transactions on Microwave Theory and Techniques, 52*(2), 678–685.

Banerjee, S., & Mandal, D. (2018). Array pattern optimization for steerable circular isotropic antenna array using cat swarm optimization algorithm. *Wireless Personal Communications, 99*(3), 1169–1194.

Bekasiewicz, A., & Koziel, S. (2015). Efficient multi-fidelity design optimization of microwave filters using adjoint sensitivity. *International Journal of RF and Microwave Computer-Aided Engineering, 25*(2), 178–183.

BouDaher, E., & Hoorfar, A. (2015). Electromagnetic optimization using mixed-parameter and multiobjective covariance matrix adaptation evolution strategy. *IEEE Transactions on Antennas and Propagation, 63*(4), 1712–1724.

Chen, Y. C., Chen, S. Y., & Hsu, P. (2006). Dual-band slot dipole antenna fed by a coplanar waveguide. In *IEEE antennas and propagation society international symposium*, Albuquerque, NM, USA (pp. 3589–3592). IEEE.

Conn, A. R., Gould, N. I. M., & Toint, P. L. (2000). *Trust region methods* (MPS-SIAM Series on Optimization). MPS-SIAM.

CST Microwave Studio. (2022). Dassault Systemes, France.

Goudos, S. K., & Sahalos, J. N. (2016). Design of large thinned arrays using different biogeography-based optimization migration models. *International Journal of Antennas and Propagation, 2016*, 5359298, 1–11.

Hauth, W., Keller, R., Papziner, U., Ihmels, R., Sieverding, T., & Arndt, F. (1993). Rigorous CAD of multipost coupled rectangular waveguide components. In *Proceedings – European microwave conference*, Madrid, pp. 611–614.

HFSS. (2021). *Release 19.0*. ANSYS, Canonsburg. http://www.ansoft.com/products/hf/hfss/

Jameson, A. (1988). Aerodynamic design via control theory. *Journal of Scientific Computing, 3*, 233–260.

Kovaleva, M., Bulger, D., Zeb, B. A., & Esselle, K. P. (2017). Cross-entropy method for electromagnetic optimization with constraints and mixed variables. *IEEE Transactions on Antennas and Propagation, 65*(10), 5532–5540.

Koziel, S., & Bekasiewicz, A. (2016). Novel structure and size-reduction-oriented design of microstrip compact rat-race coupler. In *IEEE/ACES international conference on wireless information technology and systems (ICWITS) and applied computational electromagnetics (ACES)*, Honolulu (pp. 1–2). IEEE.

Koziel, S., & Bekasiewicz, A. (2018a). Rapid design closure of microwave components by means of feature-based optimization and adjoint sensitivities. *International Journal of RF and Microwave Computer-Aided Engineering, 28*(2), e21182.

Koziel, S., & Bekasiewicz, A. (2018b). Fast re-design and geometry scaling of multi-band antennas using inverse surrogate modeling techniques. *International Journal of Numerical Modelling, 31*(3), e2287.

Koziel, S., & Pietrenko-Dabrowska, A. (2019). Reduced-cost electromagnetic-driven optimisation of antenna structures by means of trust-region gradient-search with sparse Jacobian updates. *IET Microwaves, Antennas & Propagation, 13*(10), 1646–1652.

Koziel, S., & Pietrenko-Dabrowska, A. (2020a). Expedited optimization of antenna input characteristics with adaptive Broyden updates. *Engineering Computations, 37*(3), 851–862.

Koziel, S., & Pietrenko-Dabrowska, A. (2020b). Expedited feature-based quasi-global optimization of multi-band antenna input characteristics with Jacobian variability tracking. *IEEE Access, 8*, 83907–83915.

Koziel, S., Mosler, F., Reitzinger, S., & Thoma, P. (2012). Robust microwave design optimization using adjoint sensitivity and trust regions. *International Journal of RF and Microwave Computer-Aided Engineering, 22*(1), 10–19.

Lalbakhsh, A., Afzal, M. U., & Esselle, K. P. (2017). Multiobjective particle swarm optimization to design a time-delay equalizer metasurface for an electromagnetic band-gap resonator antenna. *IEEE Antennas and Wireless Propagation Letters, 16*, 912–915.

Mader, C., Martins, J., Alonso, J. J., & van der Weide, E. (2008). ADjoint: An approach for the rapid development of discrete adjoint solvers. *AIAA Journal, 46*(4), 863–873.

Pietrenko-Dabrowska, A., & Koziel, S. (2020). Computationally-efficient design optimisation of antennas by accelerated gradient search with sensitivity and design change monitoring. *IET Microwaves, Antennas & Propagation, 14*(2), 165–170.

Tomasson, J. A., Koziel, S., & Pietrenko-Dabrowska, A. (2020). Expedited design closure of antenna input characteristics by trust region gradient search and principal component analysis. *IEEE Access, 8*, 8502–8511.

Wang, C., Wang, Y., Wang, Z., Wang, M., Yuan, S., & Wang, W. (2018). Structural–electrical coupling optimisation for radiating and scattering performances of active phased array antenna. *International Journal of Electronics, 105*(4), 586–597.

Chapter 9
Response Features for Reliability Improvement of Local Optimization Procedures

The focus of this chapter is the employment of the response feature technology for improving reliability of local optimization procedures. This chapter considers two different approaches: adaptively adjusted design specification strategy and frequency-based regularization. Both methodologies involve mechanisms aimed at enhancing the performance of the search process under challenging scenarios, primarily poor initial conditions or the necessity of re-designing the high-frequency structure under study for operating conditions distant from that of the available design. If a decent starting point is not readily available, local search procedures are usually prone to a failure, and, consequently, a designer may resort to the employment of global optimization algorithms, which allow to mitigate the aforementioned issues. However, their computational cost is exorbitant when carried out using full-wave electromagnetic (EM) simulations, otherwise necessary to ensure adequate accuracy. Typically, the computational cost of the global search is at least ten times higher than that of the local procedures. The techniques of this chapter allow to refrain from utilizing prohibitively costly global optimizers by enhancing the performance of the local search routines. Consequently, the computational cost of rendering the optimal design becomes acceptable. In this chapter, we also discuss the incorporation of various acceleration mechanisms into the considered procedures such as variable-fidelity simulations and sparse sensitivity updating schemes. The outline of the mentioned methods is accompanied by illustrative examples including microstrip antennas and passive microwave circuits.

A. Pietrenko-Dabrowska, S. Koziel, *Response Feature Technology for High-Frequency Electronics. Optimization, Modeling, and Design Automation*,
https://doi.org/10.1007/978-3-031-43845-5_9

9.1 Design Specification Adaptation for Reliability Improvement

This section outlines reliability-enhanced parameter tuning of high-frequency structures by means of design specification adaptation (Koziel & Pietrenko-Dabrowska, 2021; Koziel et al., 2022). We start by presenting the main concepts of the procedure that aims at improving the robustness of the local optimization process (Sect. 9.1.1). In order to make this chapter self-contained, Sect. 9.1.2 describes the gradient-based algorithm, which is employed as an underlying optimization engine. The overall optimization framework is summarized in Sect. 9.1.3. Sections 9.1.4 and 9.1.5 provide the results obtained using the considered approach for design optimization of a dual-band branch-line coupler and a dual-band power divider, respectively. Furthermore, Sect. 9.2 outlines an expedited version of the discussed design specification adaptation procedure by means of variable-fidelity simulations.

9.1.1 Design Specification Adaptation: Concept and Prerequisites

The major objective of the discussed methodology is to facilitate simulation-driven optimization processes of the high-frequency structures carried out under challenging scenarios. This, in particular, pertains to the lack of reasonable initial design, or the necessity of re-designing the structure for the operating frequencies/bandwidths that are severely misaligned with those at the available design. In these and similar situations, local optimizers are prone to a failure, whereas application of global methods (nature-inspired metaheuristics (Jeong et al., 2019; Luo et al., 2019), machine learning routines (Lalbakhsh et al., 2017; Xiao et al., 2018), and efficient global optimizers (Liu et al., 2019)) is usually unjustifiable due to the excessive computational expenses associated with such techniques.

9.1.1.1 Specification Adjustment Concept

In order to explain the automated design specification adjustment concept (Koziel et al., 2022), we start by formulating the EM-driven parameter tuning task. Let us assume that the structure of interest is supposed to operate at the target frequencies f_k, $k = 1, \ldots, N$, where N is the number of operating bands. Given the vector of design variables (typically, geometry parameters) $\boldsymbol{x}$, the EM-simulated system outputs $\boldsymbol{S}(\boldsymbol{x})$ (typically, S-parameters vs. frequency), and the target operating frequency vector $\boldsymbol{F} = [f_1 \ldots f_N]^T$, the design task is defined as a minimization problem

$$x^* = \arg\min_{x} U(S(x), F) \tag{9.1}$$

where U is an objective function that quantifies the design quality. For example, let us assume that the circuit at hand is a microstrip coupler supposed to operate at the frequency f_0. The circuit is to provide equal power split, while minimizing the matching and isolation, both at f_0. Thus, the relevant system outputs would be S_{11}, S_{21}, S_{31}, and S_{41}, and the objective function can be defined as

$$\begin{aligned} U(S(x), F) &= U([S_{11}(x,f), S_{21}(x,f), S_{41}(x,f), S_{41}(x,f)], [f_0]) \\ &= \max\{|S_{11}(x,f_0)|, |S_{41}(x,f_0)|\} + \beta[|S_{21}(x,f_0)| - |S_{31}(x,f_0)|]^2 \end{aligned} \tag{9.2}$$

In this example, minimization of matching and isolation is a primary objective, whereas the power split condition is treated as a design constraint, and handled by adding a penalty term with the proportionality coefficient β.

Figure 9.1 illustrates a common situation where local optimization is prone to a failure due to the poor quality of the initial design. Among the two designs shown in the picture, the operating frequency of the first one (black lines) is sufficiently close to the target, whereas the second design (gray lines) is too far away in terms of the frequency misalignment to make the target attainable when using, e.g., a gradient-based search procedure.

The primary goal is to develop algorithmic tools that permit reliable parameter tuning under challenging scenarios such as the one presented in Fig. 9.1 without defaulting to global optimization methods. Another objective is to avoid excessive

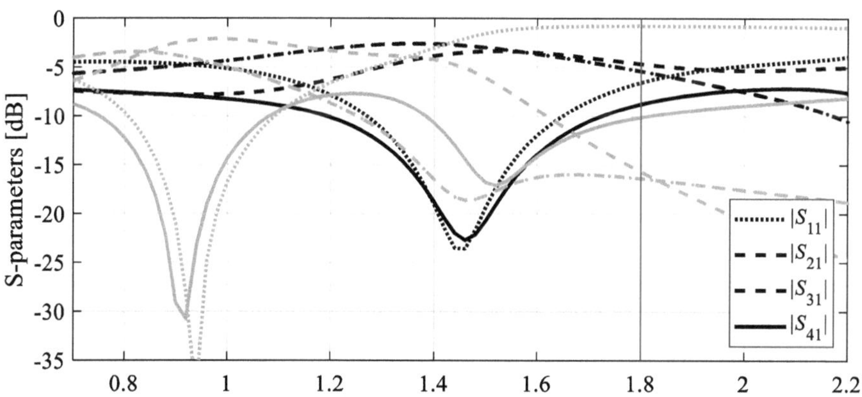

Fig. 9.1 Scattering parameters of an exemplary compact branch-line coupler. The target operating frequency is 1.8 GHz, marked using a vertical line. This target is attainable by means of local search when starting from the design marked with black lines; however, it is not attainable from the design marked using the gray lines. The latter is due to the fact that the operating bandwidth of the circuit is too far away from the target frequency

computational costs, which would be entailed by straightforward approaches such as multiple optimization runs initiated from random starting points. Here, we outline a design specification adjustment strategy, where the target operating frequencies (or bandwidths) are altered by taking into account the actual operating frequencies of the structure at hand at the currently available design. The prerequisite for such alterations is to ensure that the optimum design with respect to the current status of the specifications is attainable at each stage of the optimization process. Another prerequisite is that the specifications eventually converge to their original levels toward the end of the procedure.

The aforementioned concepts have been illustrated in Fig. 9.2 and will be rigorously formulated hereinafter. At the beginning of the optimization run, the target frequency is shifted to make the temporary goal attainable from a given initial design using a local algorithm. During the optimization process, the specifications are continuously adjusted based on the system outputs at the current design (cf. Fig. 9.2b, c). At the last stages, when the circuit operates sufficiently close to the original target frequency (or frequencies), the specifications are relocated accordingly so that the optimum design can be identified, again, using the local means. The adjustment of the specifications through automated decision-making process and the design update (e.g., within the descent type of algorithm) are interleaved so that the entire procedure is concluded within a single algorithm run. It should be noted that gradual alteration of the design specifications toward the ultimate targets throughout the optimization run is a common practice, when good initial designs are not available. This facilitates identification of the solutions meeting the assumed targets, which would otherwise be unattainable using the local search routines. Nevertheless, such an interactive process is heavily based on engineering insight, is time-consuming, and is prone to failure. The presented approach is fully automated and requires no interaction from the designer. The knowledge necessary to make decisions concerning the scale of the adjustments, and the time of launching them, is acquired from the convergence indicators of the optimization process, design quality (as compared to the current targets), and the initial assessment of the frequency spread of the system outputs.

9.1.1.2 Specification Adjustment Prerequisites

Section 9.1.1.1 discussed the motivation for, the purpose, and potential benefits of involving intelligent decision-making in design specification adjustment. When it comes to its rigorous formulation, one needs to consider the following issues:

- The necessary amount of specification adjustment has to be quantified based on the available (current) design rendered during the optimization process.
- The (local) attainability of the adjusted specifications has to be determined when using the current design as the starting point.
- The above endeavors should be accomplished at low computational cost, preferably without involving extra EM simulations, apart from those already entailed by the operation of the core optimization algorithm itself.

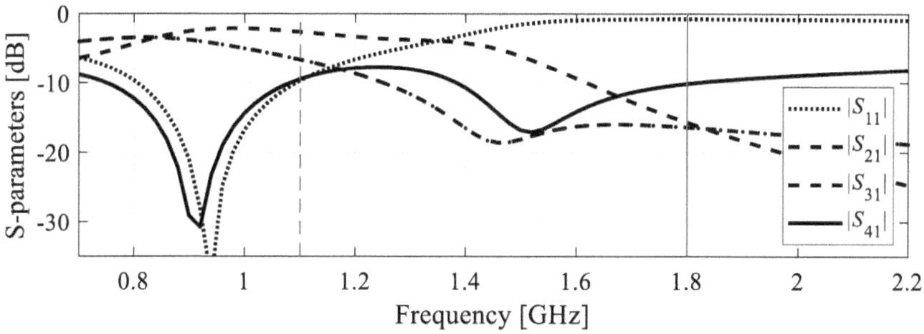

(a)

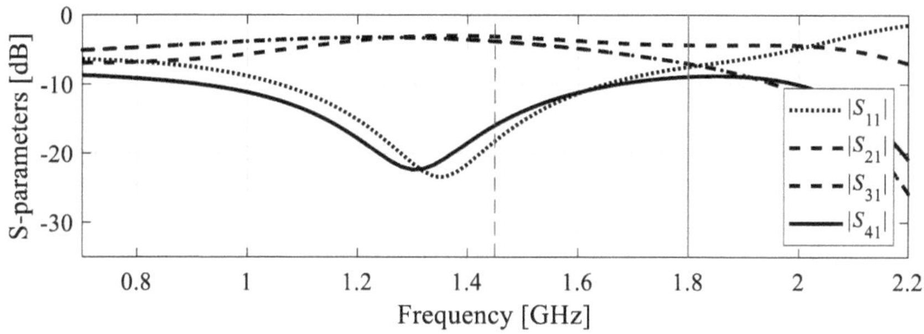

(b)

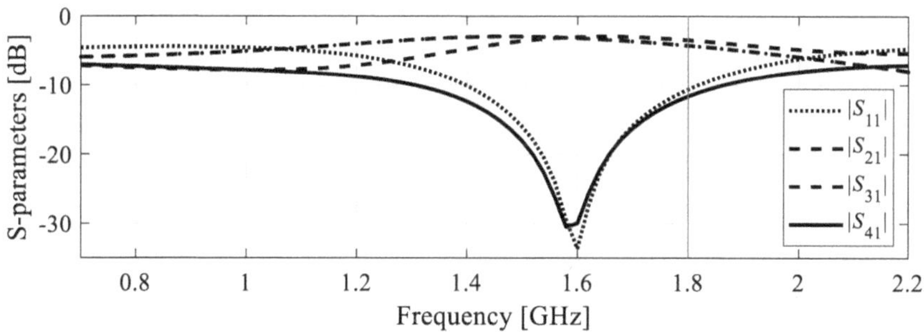

(c)

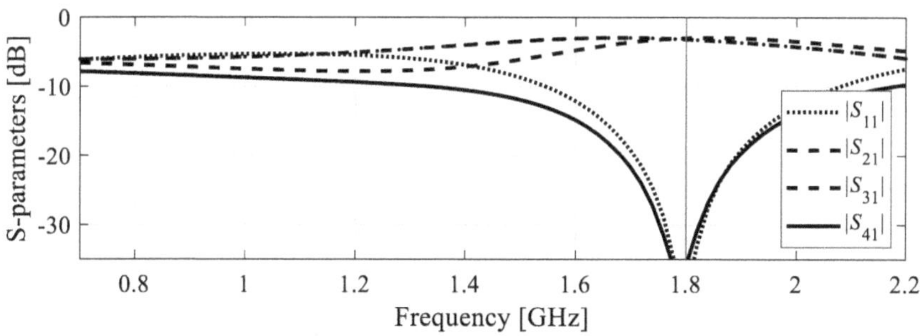

(d)

Fig. 9.2 Automated design specification adjustment concept explained using a branch-line coupler. The initial design and the target operating frequency are the same as in Fig. 9.1 (initial design marked gray): (**a**) target frequency relocated toward the operating frequency at the initial design in order to ensure that the current specifications (dashed line) are attainable from that design, (**b**) one of the iterations in the middle of the optimization run with the current design and current specifications, (**c**) final optimization stage: the target operating frequency converged to its original value, (**d**) final design optimized with respect to the original design requirement

The last factor is dictated by the practical utility of the procedure. Specific arrangements depend on the underlying optimization algorithm, which, in general, would be a type of a gradient-based routine. Here, we use trust-region gradient search; consequently, the fundamental tool utilized to address the aforementioned issues will be the first-order Taylor expansion of the system frequency characteristics constructed using the sensitivity matrix, which is available at no extra cost: Jacobian has to be updated before each iteration of the trust-region algorithm as a part of its operation.

Let $\boldsymbol{J}(\boldsymbol{x})$ denote the sensitivity matrix of all relevant system responses at the design $\boldsymbol{x}$. Without loss of generality, we can assume that the underlying optimization algorithm is an iterative procedure, which yields a series of approximations $\boldsymbol{x}^{(i)}$, $i = 0, 1, \ldots$, to the optimum design $\boldsymbol{x}^*$ of (9.1); here, $\boldsymbol{x}^{(0)}$ is the initial design. Consider the first-order Taylor model $L^{(i)}(\boldsymbol{x})$ of $\boldsymbol{S}(\boldsymbol{x})$ established at the current design $\boldsymbol{x}^{(i)}$. We have

$$L^{(i)}(\boldsymbol{x}) = \boldsymbol{S}\left(\boldsymbol{x}^{(i)}\right) + \boldsymbol{J}\left(\boldsymbol{x}^{(i)}\right) \cdot \left(\boldsymbol{x} - \boldsymbol{x}^{(i)}\right) \tag{9.3}$$

Further, let us consider a supplementary optimization problem

$$\boldsymbol{x}^{\text{tmp}} = \arg \min_{\|\boldsymbol{x} - \boldsymbol{x}^{(i)}\| \leq D} U\left(L^{(i)}(\boldsymbol{x}), \boldsymbol{F}\right) \tag{9.4}$$

In (9.4), D is the search radius in the vicinity of $\boldsymbol{x}^{(i)}$, typically set to $D = 1$.

The knowledge-based adjustment of the design specifications will be determined using the following factors:

- The improvement factor F_r defined as

$$F_r = \left| U\left(L^{(i)}(\boldsymbol{x}^{\text{tmp}}), \boldsymbol{F}\right) - U\left(L^{(i)}\left(\boldsymbol{x}^{(i)}\right), \boldsymbol{F}\right) \right| \tag{9.5}$$

- The distance D_c between the actual operating frequencies of the system at hand at the design $\boldsymbol{x}^{(i)}$, denoted as $\boldsymbol{F}_c = [f_{c.1} \ldots f_{c.N}]^T$, and the target frequencies $\boldsymbol{F} = [f_1 \ldots f_N]^T$, defined as

$$D_c = \|\boldsymbol{F}_c - \boldsymbol{F}\| \tag{9.6}$$

The first factor, F_r, allows us for assessing the potential for design improvement when starting from the current design $\boldsymbol{x}^{(i)}$. The second factor, D_c, is essentially a safeguard, introduced to ensure that the adjusted specifications are sufficiently close to the actual operating frequencies of the circuit under optimization at the current design. For both factors, the acceptance thresholds are defined: $F_{r.\min}$ and $D_{c.\max}$ (a practical procedure for setting them up will be discussed later). Using these, the design specification will be adjusted if:

- $F_r < F_{r.\min}$, i.e., the current design is unlikely to be improved to a sufficient extent, when starting from the current point $\boldsymbol{x}^{(i)}$.
- $D_c > D_{c.\max}$, i.e., the operating frequencies at the design $\boldsymbol{x}^{(i)}$ are too far away from the current targets.

If either of these conditions is satisfied, the current specifications are too strict to be attainable from $\boldsymbol{x}^{(i)}$ and should be relaxed. The acceptance threshold values are not critical; however, it is recommended that the system characteristics are taken into account, especially when deciding about the value of $D_{c.\max}$. In the following, we provide a simple yet practical procedure for establishing $F_{r.\min}$ and $D_{c.\max}$:

1. Set $D_{c.\max}$ to approximately half of the system bandwidth(s) at the initial design. This allows to allocate the target frequency (or frequencies) on the slopes of the frequency characteristics near the respective operating bandwidths if $D_c < D_{c.\max}$.
2. Relocate the target operating frequencies so that $D_c = D_{c.\max}$ (cf. Eq. 9.6).
3. Solve (9.4) with $D = 1$.
4. Set $F_{r.\min} = F_r$, with F_r calculated using the temporary design $\boldsymbol{x}_{\text{tmp}}$ obtained in Step 3.

Establishing the threshold $F_{r.\min}$ as described above accounts for a typical improvement of the objective function assuming that design requirements are adjusted to satisfy the condition $D_c < D_{c.\max}$. As mentioned before, the latter is set to ensure that the operating frequencies of the system can be relocated to the current target frequencies using local search. When using these thresholds for adjusting the target operating frequencies within the actual optimization procedure, the above attainability property is therefore secured at each iteration of the algorithm.

9.1.1.3 Specification Adjustment Procedure

We denote by $\boldsymbol{F}_{\text{current}}(a) = [f_{\text{current}.1}(a) \ \ldots \ f_{\text{current}.N}(a)]^T$ the updated specifications (target frequencies) for the subsequent, or $(i + 1)$th, iteration of the optimization procedure. The vector $\boldsymbol{F}_{\text{current}}$ is parameterized by a scalar a, $0 \leq a \leq 1$. We have

$$f_{\text{current}.k}(a) = (1 - a)f_{c.k} + af_k \qquad \text{for } k = 1, \ldots, N \tag{9.7}$$

Recall that $f_{c.k}$ are the entries of the vector $\boldsymbol{F}_c = [f_{c.1} \ldots f_{c.N}]^T$ of the actual operating frequencies at the design $\boldsymbol{x}^{(i)}$. The coefficient a is identified as the maximum number $a \leq 1$ for which the conditions $F_r \geq F_{r.\min}$, and $D_c \leq D_{c.\max}$, are satisfied at the design $\boldsymbol{x}^{\text{tmp}}$ found as (cf. Eq. 9.4)

$$\boldsymbol{x}^{\text{tmp}} = \arg \min_{\|\boldsymbol{x} - \boldsymbol{x}^{(i)}\| \leq 1} U\left(L^{(i)}(\boldsymbol{x}), \boldsymbol{F}_{\text{current}}(a)\right) \tag{9.8}$$

The meaning of the last statement is that the coefficient a is, as a matter of fact, obtained through an auxiliary optimization procedure, in which it is adjusted (lowered) as much as necessary to eventually ensure that $F_r \geq F_{r.\min}$ and $D_c \leq D_{c.\max}$

for the vector $\boldsymbol{x}^{\text{tmp}}$ generated by (9.8). At this point, the specifications have been relaxed to ensure that the target operating frequencies are sufficiently close to those at the design $\boldsymbol{x}^{(i)}$ (in the sense of (9.6)), and the local improvement of the design (when starting from $\boldsymbol{x}^{(i)}$) is at least equal to $F_{r.\min}$. As discussed before, satisfaction of these conditions make the adjusted specs attainable from $\boldsymbol{x}^{(i)}$, and, consequently, throughout the entire optimization process.

Formally speaking, the updated frequency $f_{\text{current}.k}$ is a convex combination of $f_{c.k}$ and the original frequency f_k. Reducing a leads to relaxing the specifications. When the current design approaches the optimum, the conditions will be satisfied for $a = 1$, i.e., the value to which the coefficient should eventually converge. The latter will take place if the original specifications are attainable. If this is not the case, the optimization algorithm will be terminated after approaching the target frequencies as closely as possible.

What was described so far in this section pertains to design specification adjustment in one algorithm iteration. The automated adjustment procedure is launched before each iteration, that is, at all points $\boldsymbol{x}^{(i)}$, $i = 0, 1, \ldots$. This results in a continuous modification of the target frequencies based on the current relationships between the system response at $\boldsymbol{x}^{(i)}$ and design targets. At the same time, it should be reiterated that the specification management is not associated with any extra computational overhead (here, entailed by additional EM simulations of the system under design): the only required information is the sensitivity matrix, which is available anyhow as its evaluation is a part of the algorithm operation (as mentioned before, gradient-based routines are assumed as the core optimization procedures).

9.1.2 Optimization Engine: Trust-Region Algorithm

The discussed design specification management strategy can be incorporated into various iterative search procedures. Here, for the sake of demonstration, it is combined with the trust-region (TR) gradient search algorithm (Conn et al., 2000). For the convenience of the reader, it is briefly outlined below. The problem at hand is the minimization task (9.1), and the algorithm produces a series of approximations $\boldsymbol{x}^{(i)}$, $i = 0, 1, \ldots$, to $\boldsymbol{x}^*$ (cf. Eq. 9.1), by solving

$$\boldsymbol{x}^{(i+1)} = \arg \min_{\|\boldsymbol{x} - \boldsymbol{x}^{(i)}\| \le d^{(i)}} U\left(\boldsymbol{L}^{(i)}(\boldsymbol{x}), \boldsymbol{F}_{\text{current}}\right) \tag{9.9}$$

where $\boldsymbol{L}^{(i)}$ is a linear model (9.3), $\boldsymbol{F}_{\text{current}}$ represents the current vector of target operating frequencies, whereas trust-region radius $d^{(i)}$ is updated after each iteration of the algorithm using the gain ratio $r = [U(\boldsymbol{S}(\boldsymbol{x}^{(i+1)}), \boldsymbol{F}_{\text{current}}) - U(\boldsymbol{S}(\boldsymbol{x}^{(i)}), \boldsymbol{F}_{\text{current}})]/[U(\boldsymbol{L}^{(i)}(\boldsymbol{x}^{(i+1)}), \boldsymbol{F}_{\text{current}}) - U(\boldsymbol{L}^{(i)}(\boldsymbol{x}^{(i)}), \boldsymbol{F}_{\text{current}})]$; $r > 0$ indicates the design improvement, in which case $\boldsymbol{x}^{(i+1)}$ is accepted.

Furthermore, if r is large (e.g., $r > 0.75$), $d^{(i+1)}$ is updated to $2d^{(i)}$; if r is too low (e.g., $r < 0.25$), $d^{(i+1)}$ is reduced to $d^{(i)}/3$; finally, if $r < 0$, the new design is rejected, and the iteration is repeated with a reduced TR. The example objective function has been discussed in Sect. 9.1.1.1 (cf. Eq. 9.2).

9.1.3 Optimization Algorithm

This section provides a description of the complete optimization procedure (Koziel et al., 2022) that combines local optimizer (here, the trust-region algorithm recalled in Sect. 9.1.2), and the design specification adjustment procedure of Sect. 9.1.1. The algorithm termination is based on the convergence in argument $||\boldsymbol{x}^{(i+1)} - \boldsymbol{x}^{(i)}|| < \varepsilon$, or reduction of the TR radius $d^{(i)} < \varepsilon$. The threshold ε is set to 10^{-3} in all numerical experiments presented in Sects. 9.1.4 and 9.1.5. It should be noted that in the case of the lack of improvement of the objective function (cf. Step 6), the candidate design is rejected, and the iteration is repeated upon reducing the trust-region radius. Interested readers can find more details concerning the trust-region algorithms in in the literature (Conn et al., 2000). The pseudocode of the considered optimization procedure and the flow diagram have been shown in Figs. 9.3 and 9.4, respectively.

9.1.4 Results: Dual-Band Branch-Line Coupler

The first verification case is a dual-band branch-line coupler (Xia et al., 2020) shown in Fig. 9.5, implemented on the RO4003 substrate ($\varepsilon_r = 3.5$, $h = 0.51$ mm, $\tan\delta = 0.0027$). The circuit geometry is described by nine independent parameters $\boldsymbol{x} = [L_s\ W_s\ l_{3r}\ w_1\ w_2\ w_3\ w_4\ w_5\ w_v]^T$ (all dimensions in mm, except l_{3r}, which is a relative parameter). Furthermore, we have the following relationships: $d_L = d_W = 10$ mm, $L = 2d_L + L_s$, $W = 2d_W + 2w_1 + (W_s - 2w_f)$, $l_1 = W_s/2$, $l_2 = l_3 2^{1/2}$, $l_3 = l_{3r}((L_s - w_3)/2 - w_4/2^{1/2})$, $l_{v1} = l_3/3$, and $l_{v3} = L_s/2 - w_3/2 - l_3 + l_{v1}$; $w_f = 1.15$ mm is fixed to ensure 50-ohm line impedance. The computational model is implemented in CST Microwave Studio and evaluated using its time-domain solver (~150,000 mesh cells, simulation time about 2 min).

For this example, the design objective is to optimize the geometry parameters so that the circuit operates at the frequencies $f_1 = 1.2$ GHz and $f_2 = 2.7$ GHz, with the input matching $|S_{11}|$, and isolation $|S_{41}|$ simultaneously minimized at both f_1 and f_2. Furthermore, the circuit has to ensure equal power split, i.e., $|S_{21}| = |S_{31}|$ at f_1 and f_2. The objective function for this problem is based on (9.2).

The initial design is shown in Fig. 9.6 using the gray lines. The operating frequencies at $\boldsymbol{x}^{(0)}$ (around 1.7 GHz and 3.5 GHz, respectively) are severely misaligned with the target ones, and conventional local optimization fails to yield satisfactory results. The considered approach renders the design $\boldsymbol{x}^* = [41.1\ 8.19\ 0.95\ 2.26\ 1.68\ 0.96\ 0.34\ 1.18\ 1.14]^T$ (responses marked black in Fig. 9.6) that is of high quality with respect to the assumed goals.

Input arguments:

1. $\boldsymbol{x}^{(0)}$ – initial design;
2. $\boldsymbol{F}$ – target operating frequency/band vector;
3. U_R – primary objective function encoding design goals;

Algorithm operating flow:

1. Set the iteration index $i = 0$;
2. Find the scalar a to determine current specification vector $\boldsymbol{F}_{currentr}(a)$ (cf. "Specification adjustment procedure");
3. Perform TR iteration (9.9) to find the new iteration point $\boldsymbol{x}^{(i+1)}$ according to $\boldsymbol{F}_{current}$;
4. Update the TR radius $d^{(i)}$;
5. If the termination condition is satisfied, go to 7;
6. **If** $U(\boldsymbol{R}(\boldsymbol{x}^{(i+1)}),\boldsymbol{F}_{current}) < U(\boldsymbol{R}(\boldsymbol{x}^{(i)}),\boldsymbol{F}_{current})$
   Set $i = i + 1$;
   Go to 2;
   else
   Go to 3;
   end
7. END

Fig. 9.3 Pseudocode of the trust-region algorithm with design specification adjustment scheme through intelligent decision-making delineated in Sect. 9.1

The evolution of the target operating frequency has been shown in Fig. 9.7. Observe that the specifications are considerably relocated and eventually converge to the original values after ten iterations of the algorithm. Figure 9.8 shows the selected intermediate designs along with the respective target operating frequencies.

9.1.5 Results: Dual-Band Power Divider

Our second example is a dual-band equal split power divider based on coupled lines (Lin & Chu, 2010), shown in Fig. 9.9. The circuit is implemented on AD250 dielectric substrate ($\varepsilon_r = 2.5$, $h = 0.81$ mm, $\tan\delta = 0.0018$). There are seven adjustable parameters $\boldsymbol{x} = [l_1\ l_2\ l_3\ l_4\ l_5\ s\ w_2]^T$ (all dimensions in mm); $w_1 = 2.2$ is fixed to ensure 50-ohm line impedance; $g = 1$ mm is also fixed. The computational model is implemented in CST Microwave Studio and evaluated using its time-domain solver (~200,000 mesh cells, simulation time approx. 2 min).

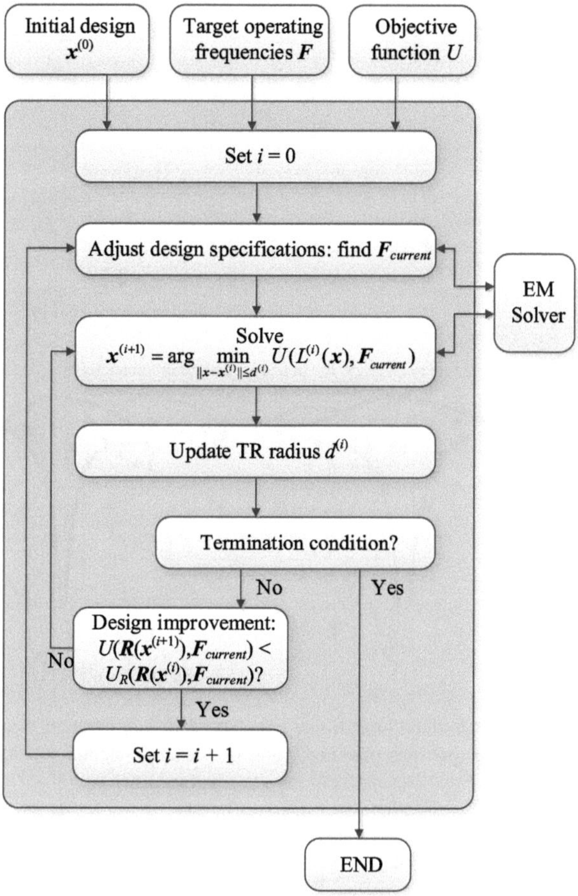

Fig. 9.4 Flow diagram of the considered optimization framework with design specification management

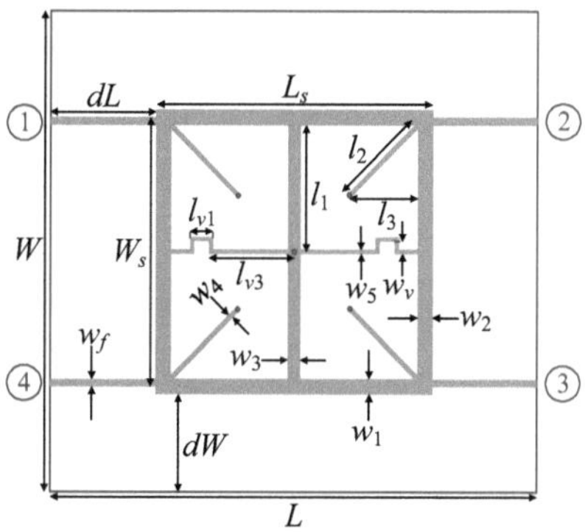

Fig. 9.5 Dual-band branch-line coupler (Xia et al., 2020): circuit topology; ports marked with numbers in circles

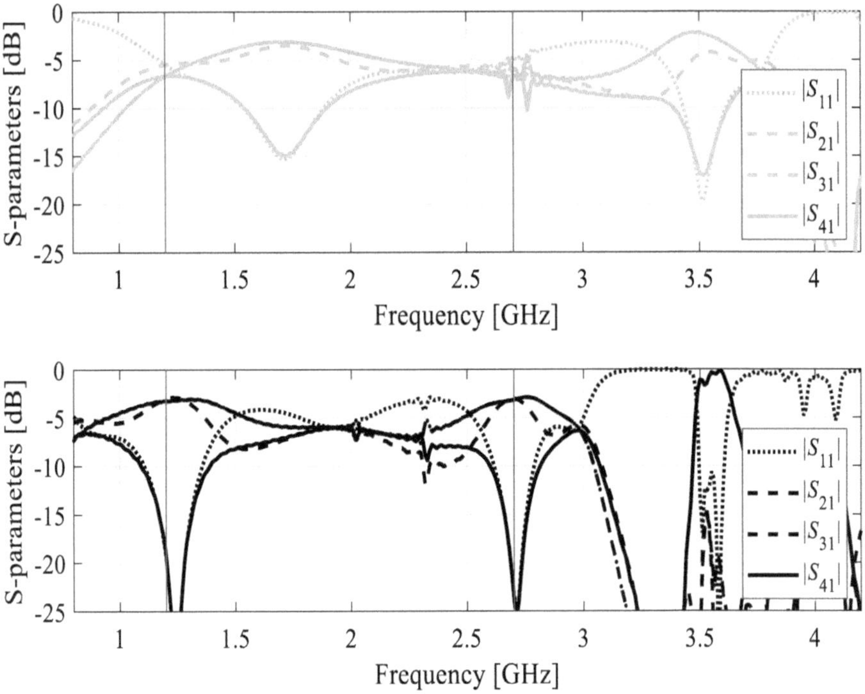

Fig. 9.6 Dual-band branch-line coupler: scattering parameters at the initial design (top), and the final design (bottom) obtained using the design specification management methodology. Target operating frequencies marked using the vertical lines

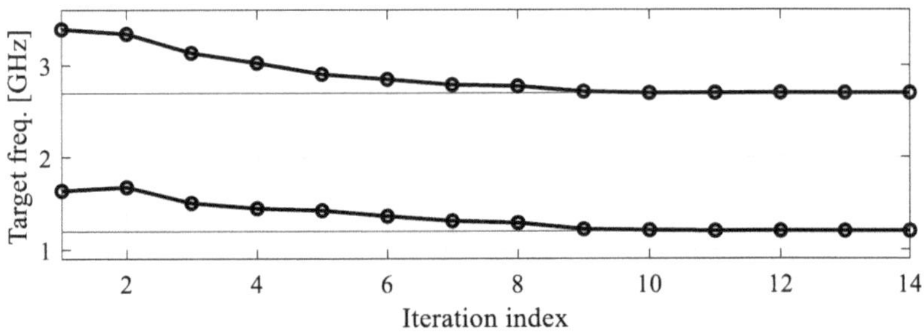

Fig. 9.7 Dual-band branch-line coupler: evolution of the target operating frequencies versus iteration index of the optimization algorithm. Original target frequencies marked using horizontal lines

The design objective is to optimize the geometry parameters so that the circuit operates at the frequencies $f_1 = 2.4$ GHz and $f_2 = 3.8$ GHz, with the input matching $|S_{11}|$, output matching $|S_{22}|$, $|S_{33}|$, as well as isolation $|S_{23}|$ simultaneously minimized at both f_1 and f_2. The equal power split condition is not directly handled in the optimization process because it is implied by the structure symmetry. The initial

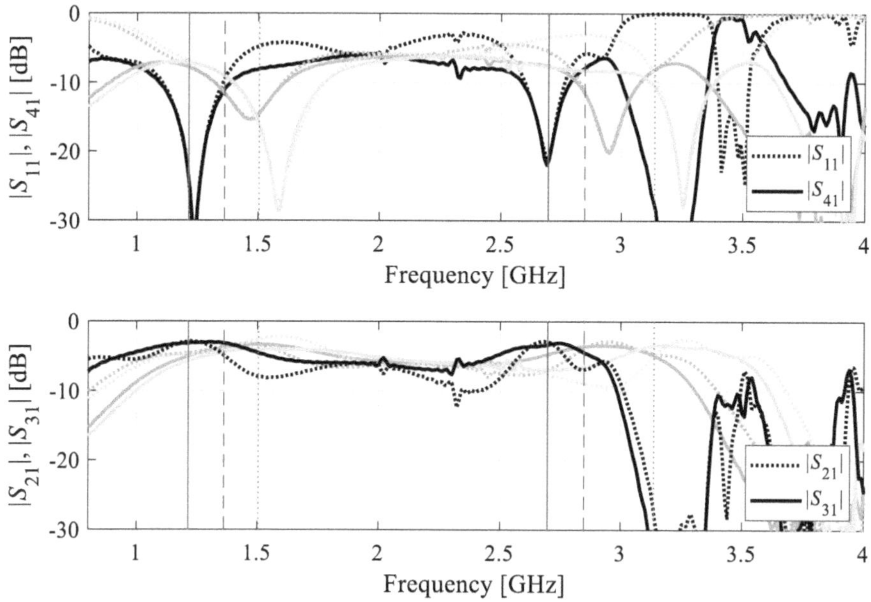

Fig. 9.8 Dual-band branch-line coupler: scattering parameters at the three intermediate designs, marked using the light-gray, dark-gray, and black colors (optimum), along with the corresponding target frequencies. For the sake of clarity, the matching/isolation characteristics are shown separate from the transmission characteristics

Fig. 9.9 Dual-band equal split power divider (Lin & Chu, 2010): circuit topology; ports marked using numbers in circles. Lumped resistor denoted as R

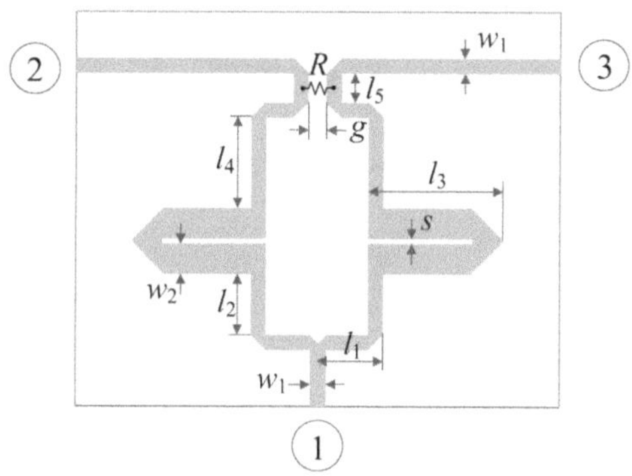

design is shown in Fig. 9.10 using the gray lines. The operating frequencies at $\boldsymbol{x}^{(0)}$ (around 1.4 GHz and 2.0 GHz, respectively) are away from the target ones. Similarly as for the previous example, conventional local optimization fails to produce satisfactory results. The discussed methodology yields the design $\boldsymbol{x}^* = [26.86\ 2.18\ 21.92\ 2.00\ 3.82\ 0.50\ 4.56]^T$ (marked black in Fig. 9.10), which meets the performance requirements imposed on the structure.

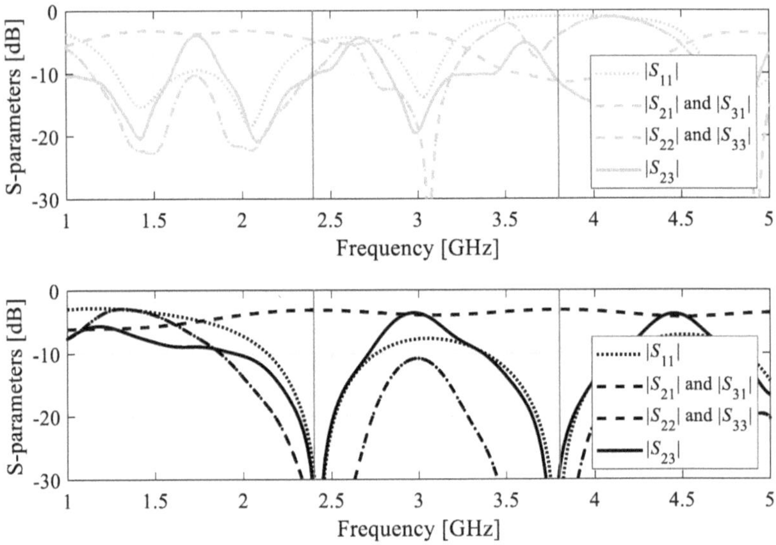

Fig. 9.10 Dual-band power divider: scattering parameters at the initial (top) and the final design (bottom) obtained using the design specification management methodology. Target operating frequencies marked using the vertical lines

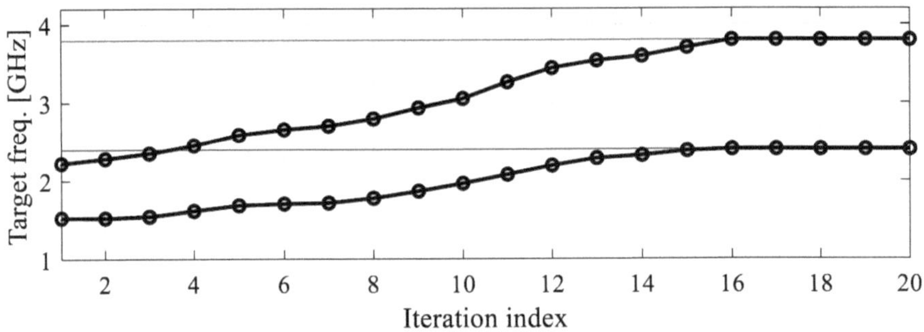

Fig. 9.11 Dual-band power divider: evolution of the target operating frequencies versus iteration index of the optimization algorithm. Original target frequencies marked using horizontal lines

The evolution of the target operating frequencies has been shown in Fig. 9.11. It is consistent with what was observed in the previous sections: following the initial (and significant) adjustment, the target frequencies converge to their original values toward the end of the optimization process. Figure 9.12 shows the selected intermediate designs along with the respective target operating frequencies.

This section discussed an approach employing design specification adjustment that aims at the improvement of the reliability of EM-driven optimization of high-frequency components. The procedure is intended to be used with local iterative search procedures (primarily gradient-based routines of a descent type), and to alleviate the difficulties related to the lack of reasonable initial designs. The

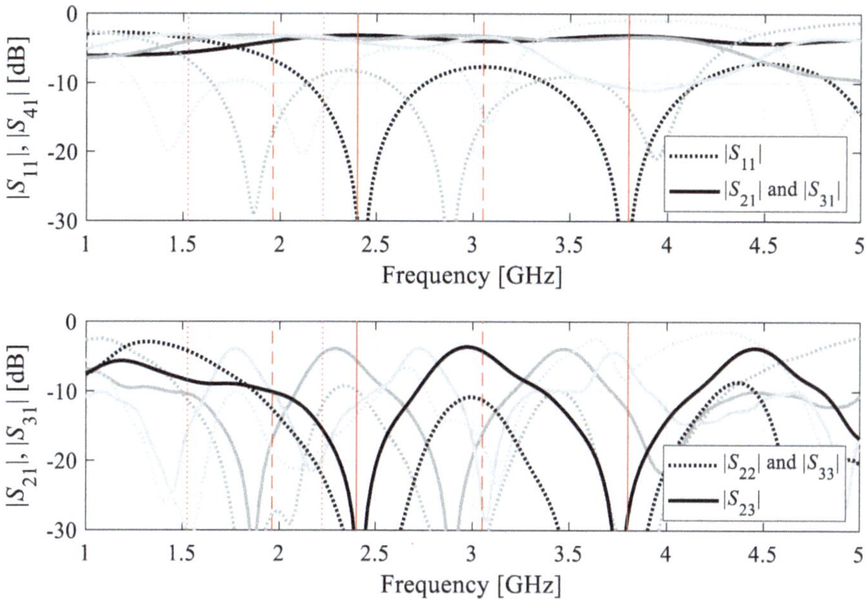

Fig. 9.12 Dual-band power divider: scattering parameters at the three intermediate designs, marked using the light-gray, dark-gray, and black colors (optimum), along with the corresponding target frequencies. For clarity, the input matching and transmission responses, as well as the output matching and isolation responses, are shown in separate panels

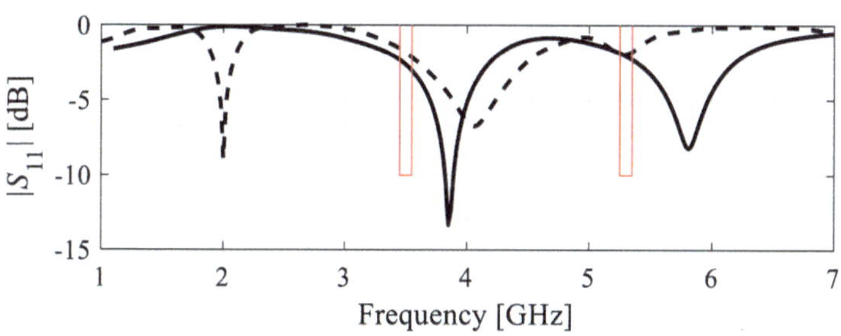

Fig. 9.13 Input characteristics of a dual-band antenna. The target bandwidths are 3.45–3.55 and 5.25–5.35 GHz, marked using vertical lines. The target is attainable using local optimization algorithm from the design featuring the response marked using the solid line, whereas it is not from the design represented by the dashed line

considered method allows for automated and adaptive modification of the design objectives in terms of the target operating frequencies or bandwidths, so that the relocated targets are attainable at any given stage of the optimization process from the currently available design. The discussed technique has been validated using two microstrip circuits: a dual-band branch-line coupler and a dual-band power divider. In all considered cases, the optimization process was demonstrably convergent to satisfactory designs despite of poor starting points, at which the operating frequencies of the respective structures have been severely misaligned with the target ones. Based on these results, it can be concluded that the considered methodology significantly improves reliability of the optimization procedure.

9.2 Design Specification Management with Variable-Fidelity Models

In pursuit of improving the efficacy of the design specification management scheme (Koziel et al., 2022), which was discussed in the previous section, a technique incorporating variable-resolution models into the procedure has been proposed (Koziel & Pietrenko-Dabrowska, 2023a). Therein, the computational model is selected from a continuous range of admissible resolutions (from the lowest usable fidelity to the high-fidelity representation), based on the misalignment between the current design specifications and the targets, as well as the convergence status of the search process. Because most of the operations are performed at the level of lower-resolution (therefore, faster) EM models, a significant reduction of the computational cost has been achieved.

9.2.1 *Problem Formulation: Multi-Band Antennas*

This section recalls the design specification adaptation framework introduced in (Koziel & Pietrenko-Dabrowska, 2021) as an algorithmic tool for enhancing the reliability of antenna parameter tuning using local search methods. The technique is briefly delineated below using a specific design task, which is antenna matching improvement over the target frequency ranges $[f_{j.1}\, f_{j.2}]$, $j = 1, \ldots, N$, where the goal is to solve

$$\boldsymbol{x}^* = \arg\min_{\boldsymbol{x}} U(\boldsymbol{x}, S_{11}, \boldsymbol{F}) \tag{9.10}$$

in which the entries of the vector $\boldsymbol{F} = [f_{1.1}\, f_{1.2} \ldots f_{N.1}\, f_{N.2}]^T$ are the aggregated target frequencies. The merit function U that quantifies the quality of design is defined as a minimax objective (see analogous definition (9.2) of the objective function U for a microwave coupler)

$$U(\boldsymbol{x}, S_{11}, \boldsymbol{F}) = \max_{j=1,\ldots,N} \{f_{j.1} \le f \le f_{j.2} : |S_{11}(\boldsymbol{x}, f)|\} \tag{9.11}$$

Let us consider an example similar to that described in Sect. 9.1.1. Figure 9.12 shows the input characteristics of a dual-band antenna, supposed to operate at 3.5 and 5.3 GHz. Local parameter tuning will be successful when starting from the design represented by the solid line, whereas it will fail when initiated from the design corresponding to the dashed line due to excessive misalignment between the actual and target operating frequencies. As discussed in Sect. 9.1, the adaptive design specification approach allows for mitigating this problem without defaulting to global search. A conceptual illustration of employing the specification adaptation concept for design optimization of the considered dual-band antenna can be found in Fig. 9.14.

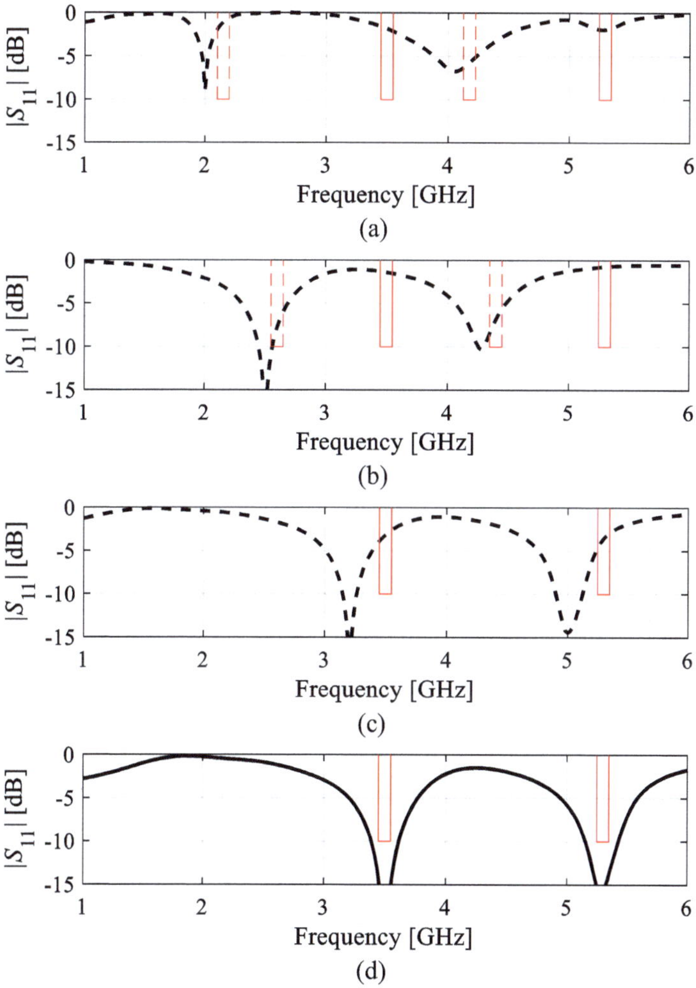

Fig. 9.14 Conceptual illustration of design specification management approach. Initial design and original design specifications (marked using solid lines in all pictures) are the same as in Fig. 9.13: (**a**) design specifications shifted toward the initial design resonances to allocate the design within the region of attraction of the optimum with respect to the current specs marked using dashed lines; (**b**) in the middle of the optimization run with current specs adjusted toward the original ones and the design following these; (**c**) before the last optimization stage: specifications set at their original values; (**d**) optimization accomplished with respect to the original specs

9.2.2 Improving Computational Efficiency: Variable-Resolution Models

This section delineates the incorporation of variable-resolution models into the considered optimization procedure. The two factors controlling the model fidelity employed at any given stage of the search process are the misalignment between the current and original design targets, and the convergence status of the algorithm.

9.2.2.1 Variable-Resolution Models

In the context of antenna design, variable-resolution models will be understood as computational (primarily, full-wave electromagnetic) models that differ in terms of discretization density of the structure at hand, as well as possibly other factors (e.g., simplifications obtained by neglecting losses, reducing computational domain, treating metallization as perfect electrical conductors, Koziel & Ogurtsov, 2019). The potential benefits of variable-fidelity models have been explored by a number of techniques such as space mapping (Koziel & Bandler, 2008), response correction frameworks (Koziel et al., 2016; Koziel & Unnsteinsson, 2018), or co-kriging (Pietrenko-Dabrowska & Koziel, 2020). In most cases, two resolution levels are utilized (low- and high-fidelity model, also referred to as coarse and fine ones (Koziel et al., 2008)). As indicated in the literature, appropriate model management is rather intricate (Koziel & Ogurtsov, 2012).

In practice, the adjustment of model resolution is realized using auxiliary parameters that control discretization density, such as lines-per-wavelength (LPW) in CST Microwave Studio. An example of relationships between LPW, the simulation time, and antenna responses (here, reflection characteristics) has been shown in Fig. 9.15, and a dual-band dipole antenna.

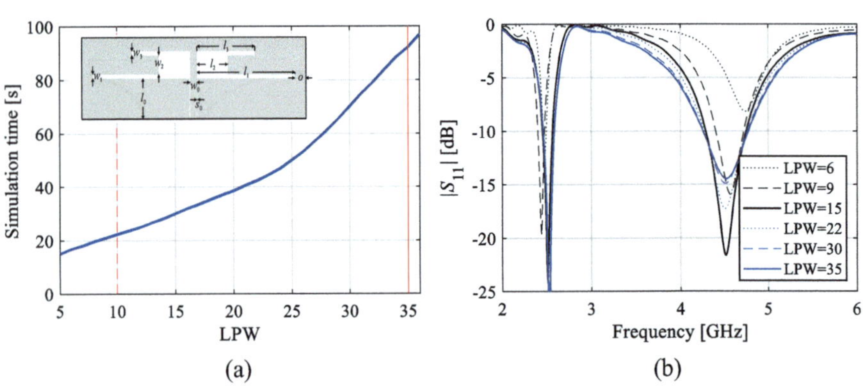

Fig. 9.15 Variable-resolution EM models for a dual-band dipole antenna: (**a**) average simulation time versus LPW, (**b**) reflection responses corresponding to various discretization densities. The vertical lines denote the values of LPW corresponding to the high-fidelity model (—), and the lowest usable low-fidelity model (- - -)

Clearly, the simulation time is monotonically dependent on LPW, and so is the accuracy, understood as the alignment between the antenna characteristics obtained for a given LPW and for the high-fidelity representation (large LPWs). The range of usable resolutions is established by selecting the lowest value $L_{\min}$, which corresponds to the situation where a degradation of antenna characteristics is still (but barely) acceptable as decided upon by the user, and the highest value $L_{\max}$, corresponding to the high-fidelity model (rendering responses with sufficient accuracy as per designer's requirements).

In the approach discussed in this section, two mechanisms to control the model resolution in the course of the optimization process are utilized. The first one is governed by the misalignment between the current and original design targets (Sect. 9.2.2.2). The second is controlled by the convergence status of the optimization process (Sect. 9.2.2.3). The resolution L is adjusted in the range $L_{\min} \le L \le L_{\max}$.

9.2.2.2 Model Resolution and Design Targets

In the considered approach, at the early stages of the optimization process, the model resolution is controlled by the distance $E_{\rm cr} = ||\boldsymbol{F}_{\rm cr} - \boldsymbol{F}_t||$ between the current and original design targets $\boldsymbol{F}_{\rm cr}$ and $\boldsymbol{F}_t$, respectively. The factor $E_{\rm cr}$ is similar to E_c, which was one of the decision factors discussed in Sect. 9.1.1.2, but computed with respect to the modified targets rather than the actual operating parameters of the antenna $\boldsymbol{F}_c$. Let $E_{\rm cr}^{(0)}$ be the value of $E_{\rm cr}$ after the first modification of the design specifications, upon which the model resolution is set to $L_{\min}$ (thus, the optimization process is initiated at the lowest admissible fidelity). Further, let us denote by $L^{(i)}$ the model resolution at the iteration i of the optimization process. It is set according to the formula

$$L^{(i)} = L_{\min} + \alpha(L_{\max} - L_{\min})\left[1 - \frac{E_{\rm cr}^{(i)}}{E_{\rm cr}^{(0)}}\right] \tag{9.12}$$

where $E_{\rm cr}^{(i)} = \left\|\boldsymbol{F}_{\rm cr}^{(i)} - \boldsymbol{F}_t\right\|$ is the distance between the design specifications $\boldsymbol{F}_{\rm cr}^{(i)}$ modified for iteration i and the original targets, whereas $0 \le \alpha \le 1$ is a control factor that decides upon the maximum model resolution to be used before the design specifications reach their target levels. In the numerical experiments of Sects. 9.2.4 and 9.2.5, $\alpha = 0.5$ is utilized.

9.2.2.3 Model Resolution and Convergence Status

Upon reducing $E_{\rm cr}^{(i)}$ to zero, i.e., after the design specifications converge to their original values (or after $\boldsymbol{F}_{\rm cr}^{(i)} = \boldsymbol{F}_t$), the second mechanism for model resolution adjustment is launched, which follows the procedure reported in (Pietrenko-Dabrowska & Koziel, 2021).

The algorithm termination is based on satisfying either of the following conditions:

- $||\boldsymbol{x}^{(i+1)} - \boldsymbol{x}^{(i)}|| < \varepsilon_x$ (convergence in argument)
- $|U(\boldsymbol{x}^{(i+1)}) - U(\boldsymbol{x}^{(i)})| < \varepsilon_U$ (convergence in the objective function value)

where ε_x and ε_U are the user-defined thresholds, dependent on the required resolution of the search process (here, the following values are used $\varepsilon_x = 10^{-3}$, and $\varepsilon_U = 10^{-2}$). The adjustment of model resolution is based on the convergence factor

$$Q^{(i)}(\varepsilon_x, \varepsilon_U) = \max\left\{\varepsilon_x/||\boldsymbol{x}^{(i+1)} - \boldsymbol{x}^{(i)}||, \varepsilon_U/\left|U_P\left(\boldsymbol{x}^{(i+1)}\right) - U_P\left(\boldsymbol{x}^{(i)}\right)\right|\right\} \tag{9.13}$$

Following (Pietrenko-Dabrowska & Koziel, 2021), we set the resolution level $L^{(i+1)}$ for at the iteration $i + 1$ as

$$L^{(i+1)} = \begin{cases} L_{\mathrm{cr}} & \text{if} \quad Q^{(i)}(\varepsilon_x, \varepsilon_U) \le M \\ \max\left\{L^{(i)}, L_{\mathrm{cr}} + (L_{\max} - L_{\mathrm{cr}})\left[1 - \dfrac{\log\left(Q^{(i)}(\varepsilon_x, \varepsilon_U)\right)}{\log M}\right]\right\} \end{cases} \tag{9.14}$$

where L_{cr} is the model resolution at the iteration when E_{cr} converged to zero. The (user-defined) parameter M establishes the convergence-related baseline for updating L (following [Pietrenko-Dabrowska & Koziel, 2021], we set $M = 10^2\varepsilon_x$).

Additionally, the model resolution is set to $L_{\max}$ upon convergence if the highest value of $L^{(i)}$ was lower than $L_{\max}$, in which case the search region size is increased by a factor M_d (here, $M_d = 10$), and the algorithm is restarted with $L^{(i+1)} = L_{\max}$ (Pietrenko-Dabrowska & Koziel, 2021). In order to further improve the computational efficiency, evaluation of response gradients at iteration i is realized as the resolution level L_{FD} (instead of $L^{(i)}$), which is $L_{\mathrm{FD}} = \max\{L_{\min}, \lambda L^{(i)}\}$. The control parameter λ is set to 2/3.

9.2.3 *Complete Optimization Framework*

This section explains the operating flow of the considered optimization algorithm. The core search routine of the discussed framework is the gradient-based procedure delineated in Sect. 9.1.2.

Section 9.2.1 provided a simple example of the objective function, associated with matching improvement of a multi-band antenna. Here, as another example, consider maximization of the average in-band gain $G(\boldsymbol{x},f)$. The task is constrained, so that $|S_{11}(\boldsymbol{x},f)| \le -10$ dB over all operating bands, $f_{k.1} \le f \le f_{k.2}$, $k = 1, \ldots, N$. The constraint can be translated into a condition that the function $U(\boldsymbol{x},S_{11},\boldsymbol{F})$ of (9.11) does not exceed -10 dB. The objective function for our exemplary task can be defined as

$$U_R(\boldsymbol{x},\boldsymbol{R},\boldsymbol{F}) = U_R(\boldsymbol{x},[G,S_{11}],\boldsymbol{F}) = -\frac{\sum_{k=1}^{N}\int_{f_{k.1}}^{f_{k.2}} G(\boldsymbol{x},f)df}{\sum_{k=1}^{N}(f_{k.2}-f_{k.1})} + \beta\left[\max\left\{\frac{U(\boldsymbol{x},S_{11},\boldsymbol{F})+10}{10},0\right\}\right]^2 \tag{9.15}$$

Note that function (9.15) contains two terms, the first representing the primary objective (average in-band gain), the second being a penalty function, which is to enforce satisfaction of the reflection constraint.

The optimization procedure incorporates the design specification management scheme of Sect. 9.1, and the model resolution management strategy of Sect. 9.2.2 into the TR algorithm of Sect. 9.1.2. The design specifications and the EM model resolution are simultaneously adjusted before each iteration of the optimization process. The former utilizes the auxiliary objective function U (9.11), whereas the latter involves the decision factor E_{cr} (Sect. 9.1.1.2), and convergence factor $Q^{(i)}$ (9.13). The design quality is evaluated using the primary objective function U_R (9.15). The pseudocode of the algorithm has been shown in Fig. 9.16. The data supplied by the user include the following: (i) the initial design $\boldsymbol{x}^{(0)}$; (ii) the target vector $\boldsymbol{F}$; (iii) the analytical formulation of the objective function U_R, which encodes the design goals; and (iv) the range of EM model resolutions L_{min} and L_{max}. The termination condition of Sect. 9.2.2.3 (convergence in argument and objective function value) is supplemented by a TR-specific condition $d^{(i)} < \varepsilon_x$ (sufficient reduction of the search radius). For additional clarification, Fig. 9.17 provides a flow diagram of the outlined procedure.

Sections 9.2.4 and 9.2.5 discuss the results of numerical verification of the considered optimization procedure with design specification management and variable-resolution models. It is applied to two microstrip antenna structures, a dual-band and triple-band dipoles optimized for best in-band impedance matching. In both cases, the optimization process starts from poor initial designs, where the operating parameters are away from the assumed targets. This is to validate the capability of the presented technique to succeed under challenging scenarios.

9.2.4 Results: Dual-Band Dipole

The first verification case is a dual-band uniplanar dipole antenna shown in Fig. 9.18a. The important parameters of the structure can be found in Fig. 9.18c. Table 9.1 gathers the optimization results obtained for the discussed algorithm, conventional gradient-based search (cf. Sect. 9.1.2), and single-fidelity adaptive design specification approach of (Koziel & Pietrenko-Dabrowska, 2021). Figure 9.19 shows the antenna responses as the initial design, as well as designs obtained using conventional gradient-based search, adaptive design specifications, as well as the discussed approach. The final design is $\boldsymbol{x}^* = [30.0\ 15.9\ 22.2\ 0.31\ 1.59\ 2.06]^T$ mm, and, as shown in Fig. 9.19, the antenna response is very close to that obtained using the method of the single-fidelity procedure. At the same time, the overall cost of the

1. Set the iteration index $i = 0$;
2. Set model resolution $L^{(i)} = L_{min}$;
3. Evaluate antenna responses $\boldsymbol{R}(\boldsymbol{x}^{(i)})$ and gradients $\boldsymbol{G}_R(\boldsymbol{x}^{(i)})$;
4. Find the scalar a to determine current specification vector $\boldsymbol{F}_{cr}^{(i)}(a)$; if the conditions $E_r \geq E_{r.min}$, $E_0 \leq E_{0.max}$, and $E_c \leq E_{c.max}$, cannot be satisfied even for $a = 0$, go to 7 (premature termination);
5. Perform TR iteration (9.9) to find the new iteration point $\boldsymbol{x}^{(i+1)}$ according to $\boldsymbol{F}_{cr}^{(i)}$;
6. Update the TR radius $d^{(i)}$;
7. **If** $E_{cr}^{(i)} = ||\boldsymbol{F}_{cr}^{(i)} - \boldsymbol{F}_t|| = 0$ (current design specifications coincide with original targets)

    Update model resolution $L^{(i)}$ using (9.12);

    else

    Update model resolution $L^{(i)}$ using (9.14);

    end
8. **If** termination condition is satisfied

    if $L^{(i)} = L_{max}$

    Go to 10;

    else

    Set $L^{(i)} = L_{max}$;

    Set $d^{(i)} = M_d d^{(i)}$;

    Go to 3;

    end

    end
9. **If** $U_R(\boldsymbol{x}^{(i+1)},\boldsymbol{R}(\boldsymbol{x}^{(i+1)}),\boldsymbol{F}_{cr}^{(i)}) < U_R(\boldsymbol{x}^{(i)},\boldsymbol{R}(\boldsymbol{x}^{(i)}),\boldsymbol{F}_{cr}^{(i)})$

    Set $i = i + 1$;

    Go to 3;

    else

    Go to 5;

    end
10. END

Fig. 9.16 Pseudocode of the discussed optimization procedure with design specification adaptation and variable-resolution EM models

optimization process corresponds to 63 high-fidelity EM simulations, which is equivalent to over 40% savings. Figure 9.20 shows the evolution of the design specifications, as well as the evolution of the model resolution. It can be observed that the high-fidelity model is only used as the last few iterations of the optimization process, which is the reason for the achieved speedup.

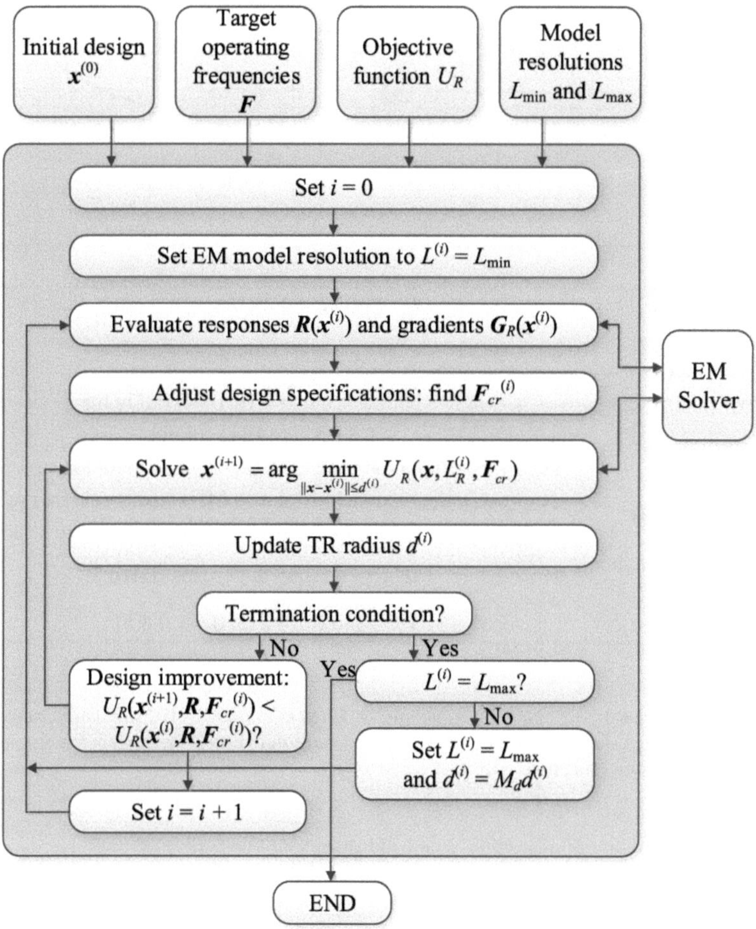

Fig. 9.17 Flow diagram of the optimization procedure with design specification adaptation and variable-resolution EM models

9.2.5 Results: Triple-Band Dipole

The second example is a triple-band dipole antenna shown in Fig. 9.21a. Figure 9.21c provides the relevant antenna parameters, whereas the optimization results can be found in Table 9.2. The antenna responses at the initial design, and the design obtained using the considered and the benchmark methods have been shown in Fig. 9.22. The final design produced by our algorithm is $x^* = [36.73\ 12.46\ 27.10\ 12.47\ 20.55\ 1.78\ 2.50\ 1.18\ 0.22\ 1.72]^T$ mm. It is close to that found by the adaptive design specification approach of (Pietrenko-Dabrowska & Koziel, 2021), which demonstrates that computational speedup achieved here is not detrimental to the design quality (Fig. 9.23). The relative savings are slightly lower than for the dual-

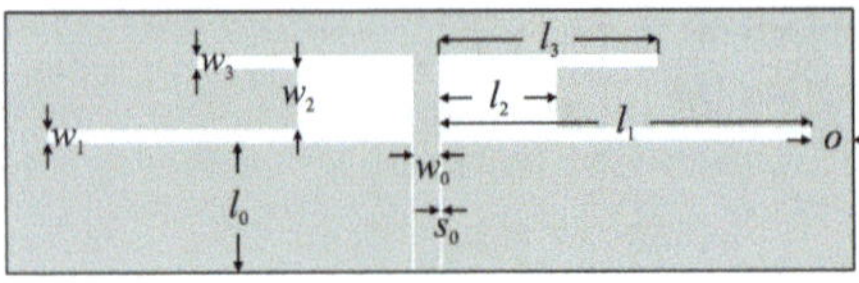

(a)

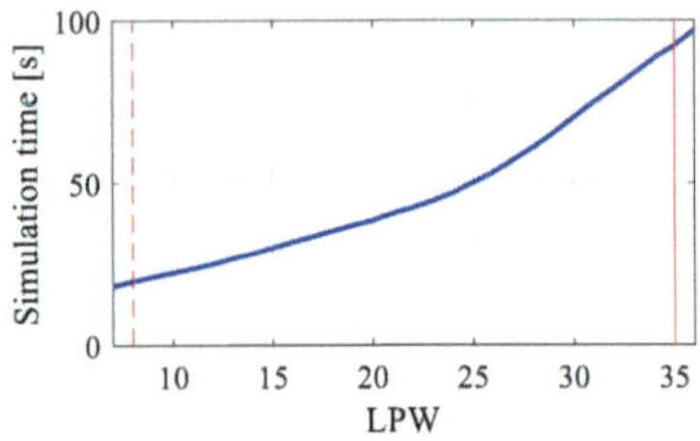

(b)

Parameter	Value
Substrate	RO4350 ($\varepsilon_r = 3.48$, $h = 0.762$ mm)
Designable parameters	$\boldsymbol{x} = [l_1\ l_2\ l_3\ w_1\ w_2\ w_3]^T$
Other parameters	$l_0 = 30$, $w_0 = 3$, $s_0 = 0.18$, $o = 5$ (all in mm)
EM models	CST Microwave Studio ($L_{\min} = 8$, simulation time 20s; $L_{\max} = 35$, simulation time 92s)
Design goals	Improve antenna matching at ±50 MHz bands centered at 3.5 GHz and 5.3 GHz
Operating frequency vector	$\boldsymbol{F} = [3.45\ 3.55\ 5.25\ 5.35]^T$ GHz
Initial design	$\boldsymbol{x}^{(0)} = [36.0\ 7.0\ 21.0\ 0.6\ 5.0\ 4.5]^T$

(c)

Fig. 9.18 Dual-band dipole antenna: (**a**) geometry (Chen et al., 2006); (**b**) simulation time versus model resolution, $L_{\min}$ and $L_{\max}$ marked using dashed and solid vertical lines, respectively; (**c**) main parameters and design objectives

Table 9.1 Optimization results for the dual-band antenna of Fig. 9.18

Algorithm	Objective function[a] (dB)	Optimization cost[b]
Conventional TR procedure (cf. Sect. 9.1.2)	−0.7[c]	33[c]
Single-fidelity adaptive design specifications	−20.5	106
Variable-resolution adaptive design specifications	−20.9	63.3

[a]Maximum value of in-band reflection $|S_{11}|$
[b]Cost expresses in equivalent number of high-fidelity EM simulations
[c]The algorithm failed to identify a satisfactory design

band antenna of Sect. 9.2.4, which can be attributed to a lower simulation time ratio for the high-fidelity and lowest-fidelity EM models (cf. Fig. 9.21b). Notwithstanding the optimization cost corresponds to 127 high-fidelity antenna simulations versus 178 for the single-fidelity method, which gives about 30% cost reduction. The evolution of the design specifications and model resolution can be found in Fig. 9.23.

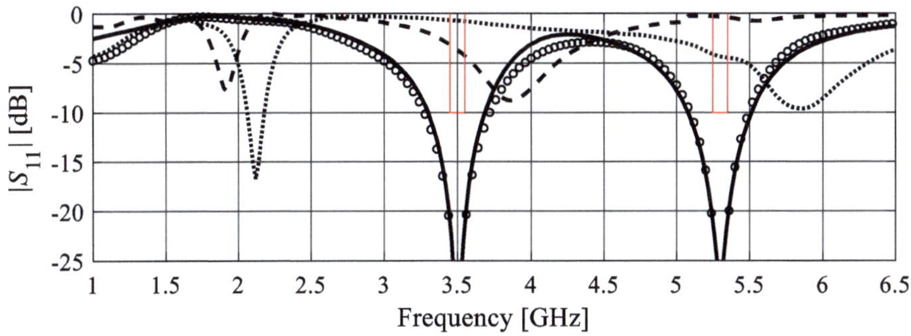

Fig. 9.19 Dual-band dipole antenna: reflection response at the initial design (- - -), design obtained using conventional gradient-based search (····), design produced by adaptive design specification approach (Koziel & Pietrenko-Dabrowska, 2021) (o), and the design obtained using the design specification adaptation with variable-resolution EM models, here, with $\alpha = 0.5$ (—). Design specifications marked using the thin solid lines

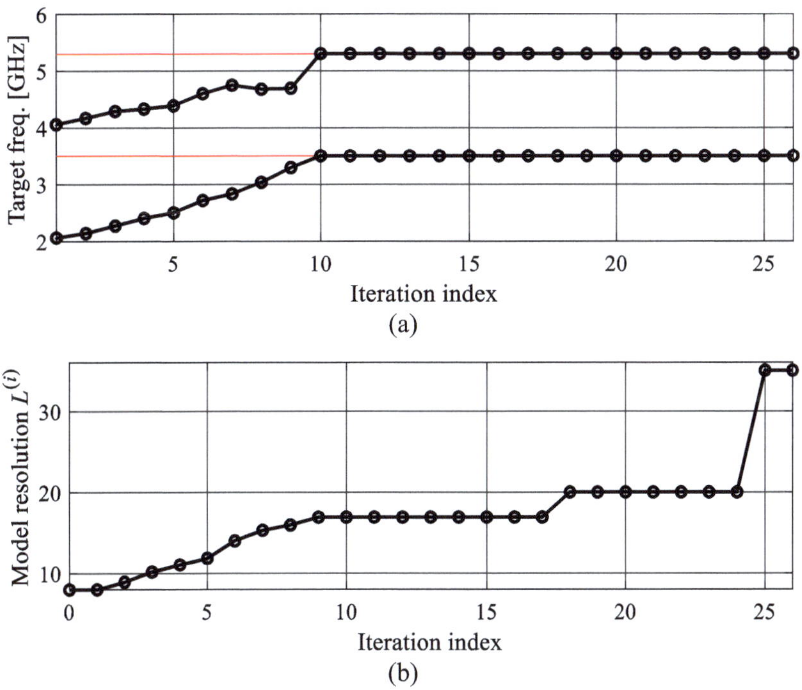

Fig. 9.20 Dual-band dipole antenna: (**a**) evolution of the target operating frequencies (original target frequencies marked using horizontal lines); (**b**) evolution of the model resolution

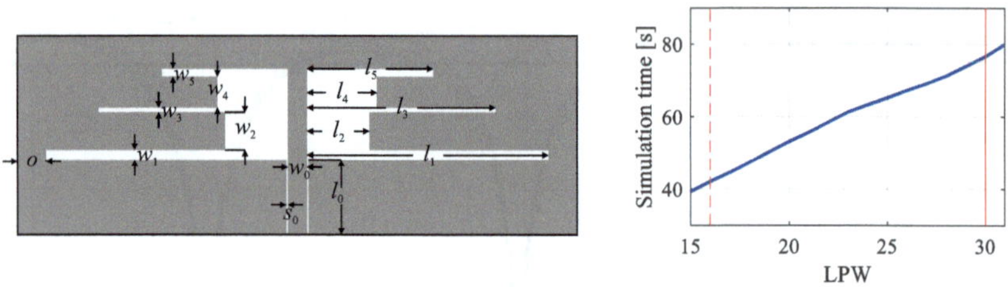

(a) (b)

Parameter	Value
Substrate	RO4350 ($\varepsilon_r = 3.48$, $h = 0.762$ mm)
Designable parameters	$\boldsymbol{x} = [l_1\ l_2\ l_3\ l_4\ l_5\ w_1\ w_2\ w_3\ w_4\ w_5]^T$
Other parameters	$l_0 = 30$, $w_0 = 3$, $s_0 = 0.18$, $o = 5$ (all in mm)
EM models	CST Microwave Studio (L_{min} = 16, simulation time 42s; L_{max} = 30, simulation time 78s)
Design goals	Improve antenna matching at ±50 MHz bands centered at 2.45 GHz, 3.6 GHz, and 5.3 GHz
Operating frequency vector	$\boldsymbol{F} = [2.40\ 2.50\ 3.55\ 3.65\ 5.25\ 5.35]^T$ GHz
Initial design	$\boldsymbol{x}^{(0)} = [39.9\ 9.42\ 30.0\ 9.69\ 20.9\ 2.2\ 4.2\ 0.62\ 0.38\ 2.00]^T$

(c)

Fig. 9.21 Triple-band dipole antenna: (**a**) geometry (Chen et al., 2006); (**b**) simulation time versus model resolution, L_{min} and L_{max} marked using dashed and solid vertical lines, respectively; (**c**) main parameters and design objectives

Table 9.2 Optimization results for the triple-band antenna of Fig. 9.21

Algorithm	Objective function[a] (dB)	Optimization cost[b]
Conventional TR procedure (cf. Sect. 9.1.2)	−2.5[c]	95[c]
Single-fidelity adaptive design specifications	−13.2	178
Variable-resolution adaptive design specifications	−11.0	127.1

[a]Maximum value of in-band reflection $|S_{11}|$
[b]Cost expresses in equivalent number of high-fidelity EM simulations
[c]The algorithm failed to identify a satisfactory design

This section discussed the technique for reduced-cost and reliable parameter tuning of antenna structures, which exploits the adaptively adjusted design specifications technique and variable-resolution EM models. In the considered technique, the model fidelity is controlled by the misalignment between the target operating frequencies set up for the current iteration of the optimization algorithm, and their original values, as well as by the convergence status of the process. The adjustment

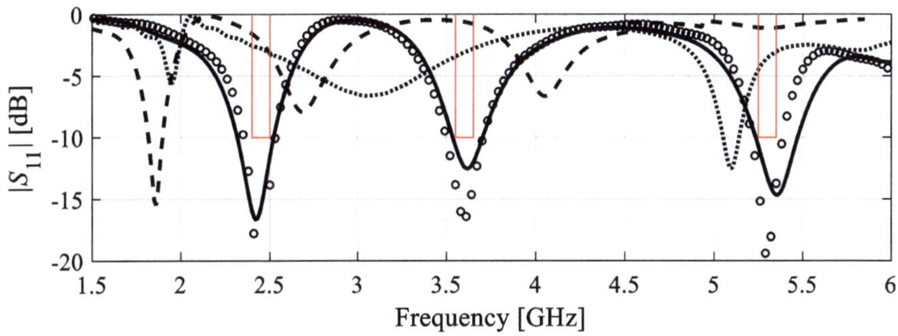

Fig. 9.22 Triple-band dipole antenna: reflection response at the initial design (- - -), design obtained using conventional gradient-based search (····), design produced by adaptive design specification approach (Koziel & Pietrenko-Dabrowska, 2021) (o), and the design obtained using the design specification adaptation with variable-resolution EM models, here, with $\alpha = 0.5$ (—). Design specifications marked using the thin solid lines

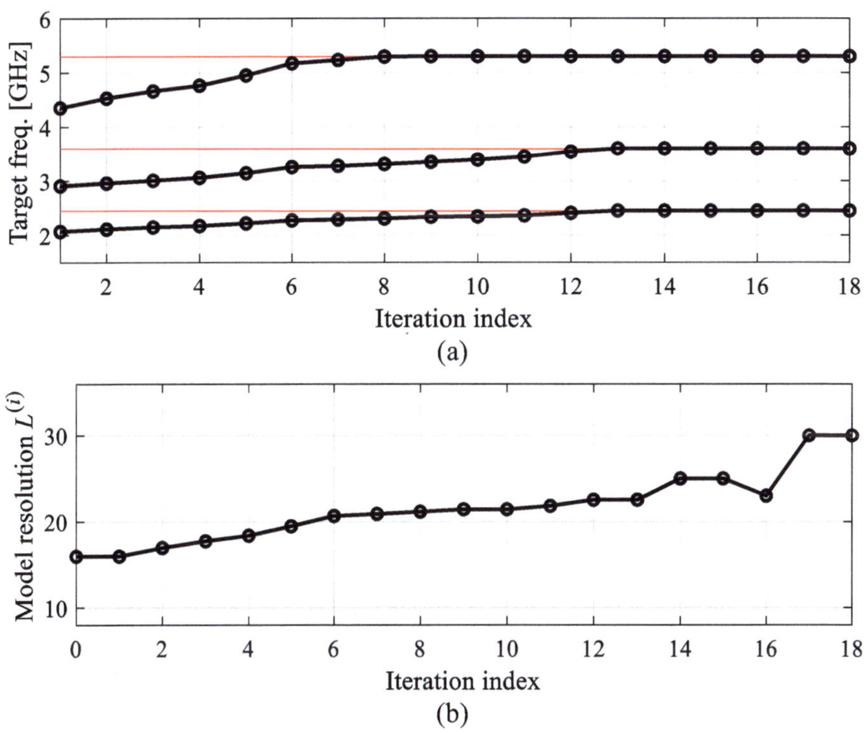

Fig. 9.23 Triple-band dipole antenna: (**a**) evolution of the target operating frequencies (original target frequencies marked using horizontal lines); (**b**) evolution of the model resolution

of design targets (modified to become reachable in each iteration) and the EM simulation fidelity is simultaneous and conducted throughout the entire run of the underlying gradient-based routine. As a consequence, satisfactory design can be identified even if the starting point is severely misaligned with the target, whereas the optimization cost is significantly lower than for the adaptively adjusted specification mechanism used alone. The performance of the discussed technique has been validated using two antennas: a dual- and triple-band dipole, and compared to conventional gradient-based procedure, as well as the adaptive design specifications approach. Numerical results demonstrate improved efficacy with the average speedup of over 40% over the benchmark (up to 60% for one of the test cases), without compromising the design quality.

9.3 Feature-Based Frequency-Based Regularization

The second approach to reliability improvement of local optimization procedures, i.e., frequency-based regularization (Koziel et al., 2020a, b), employs response features technology for smoothening the functional landscape of the assumed objective function encoding the design specifications. The regularization itself is implemented as a special penalty term complementing the primary objective and enforcing the alignment of the responses of the device under design with the target operating frequency. As a result, an improved reliability and reduced cost of the optimization process are achieved. Similarly as for the adjusted design specifications technique of Sect. 9.1, regularization makes the use of local routines sufficient even in situations where global search is otherwise imperative. The main concepts of the discussed approach are delineated in Sect. 9.3.1, whereas Sects. 9.3.2 and 9.3.3 present the results obtained for a quasi-Yagi and ultra wide-band antenna, respectively. In addition, Sect. 9.4 recalls an accelerated version of the considered procedure using variable-fidelity simulations and sparse sensitivity updates.

9.3.1 Optimization by Frequency-Based Regularization

One of the most important issues is an appropriate allocation of the operating bandwidth (or bandwidths) of the structure under design. From numerical optimization perspective, frequency-wise manipulation of the antenna characteristics is considerably more difficult than the level adjustment, especially for narrow- and multi-band antennas. This difficulty renders the application of local routines unreliable and computationally inefficient in many real-world cases.

An exemplary situation of this type is illustrated in Fig. 9.24, showing the reflection and realized gain characteristics of a quasi-Yagi antenna considered in Sect. 9.3.2. Re-designing the antenna for the target operating frequency of 5 GHz, and starting from the exemplary initial design shown in the picture, requires global

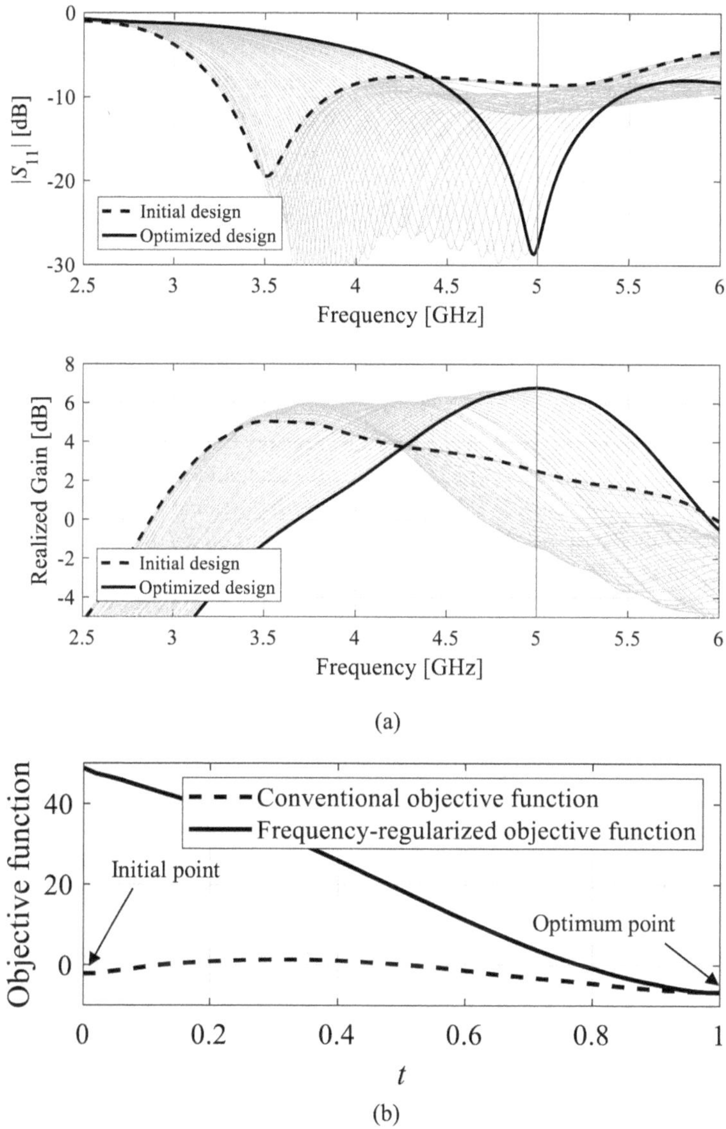

Fig. 9.24 Conceptual illustration of frequency-based regularization: (**a**) exemplary initial design for a quasi-Yagi antenna (thick dashed line), design optimized for the target frequency 5.0 GHz (thick solid line), and the family of antenna characteristics (reflection—top, realized gain—bottom) along the line segment connecting these two designs parameterized by $0 \leq t \leq 1$ (gray lines); (**b**) conventional objective function (3) (- - -) and frequency-regularized objective function (9.16) with the regularization function (9.17) (—) versus parameter t. It can be observed that the regularized objective function is monotonic, and the optimum design is attainable from the given initial point using local search routine, which is not the case for the conventional formulation

search. This is because the initial design and the optimum are separated by a local maximum. Although the picture is a simplified representation of the problem (the initial design and the optimum design are connected through a line segment), it gives a flavor of the practical challenges one may face when trying to apply numerical optimization.

To address this issue, a regularization function $f_r(\boldsymbol{x})$ is defined (Koziel et al., 2020a), which accounts for the discrepancies between the actual operating frequency (or frequencies) of the antenna and the target ones. The frequency-regularized objective function is then defined as

$$U_r(\boldsymbol{R}(\boldsymbol{x})) = U(\boldsymbol{R}(\boldsymbol{x})) + \beta_r \left[\max\left\{ \frac{f_r(\boldsymbol{x}) - f_{r.\max}}{f_{r.\max}}, 0 \right\} \right]^2 \tag{9.16}$$

where $f_{r.\max}$ is the maximum acceptable discrepancy between the actual and target operating frequency (or frequencies). The regularized objective function is defined to increase the original objective by the amount proportional (with the user-defined proportionality factor β_r) to the discrepancy quantified by f_r. However, the said contribution is zero if the discrepancy is sufficiently small (i.e., smaller than $f_{r.\max}$, also user-defined). This is to ensure that the original objective function is not distorted when the design approaches the optimum, and that the location of the optimum is the same with respect to both U and U_r, assuming of course that the design for which $f_r(\boldsymbol{x}) \le f_{r.\max}$ is attainable. Furthermore, the regularization factor is smooth (continuously differentiable) with respect to f_r, which makes it appropriate to work with gradient-based optimization routines.

The effect of regularization is shown in Fig. 9.24b: the functional landscape of the objective function is dramatically changed, and the reformulated problem becomes unimodal with the initial design being in the region of attraction of the optimum. Furthermore, the optimum design according to the original objective function coincides with that of the regularized objective.

A specific definition of a regularization function depends on a considered antenna structure and the type of response. Let us assume that the quasi-Yagi antenna of Fig. 9.24 is to operate within the frequency range defined by its center frequency f_0 and the fractional bandwidth B, so that $|S_{11}| \le -10$ dB and the realized gain is maximized within at f_0. Then, a possible formulation of the objective function (with implicit constraint handling) may be

$$U(\boldsymbol{R}(\boldsymbol{x})) = -G(\boldsymbol{x}, f_0) + \beta c(\boldsymbol{x})^2 \tag{9.17}$$

Here, $G(\boldsymbol{x},f)$ is the realized gain at the frequency f, whereas $\beta c(\boldsymbol{x})$ is a penalty term (with β being a penalty coefficient) introduced to enforce $|S_{11}| \le -10$ dB over the bandwidth B. The penalty function may be defined to quantify a relative violation of the matching condition, e.g., $c(\boldsymbol{x}) = \max\{\max\{f_0(1 - B/2) \le f \le f_0(1 + B/2):|S_{11}(\boldsymbol{x}, f)| + 10\},0\}$. In this case, the regularization function can be defined as

$$f_r(\boldsymbol{x}) = \left| f_0 - \frac{f_{L.1} + f_{L.0} + f_{L.2}}{3} \right| \tag{9.18}$$

where $f_{L.j}$, $j = 1, 2$, are the frequencies corresponding to -10 dB levels of $|S_{11}|$, and $f_{L.0}$ is the frequency of the antenna resonance. If, at a particular design $\boldsymbol{x}$, the resonance level is above -10 dB, we assign $f_{L.j} = f_{L.0}$. In other words, the function f_r measures the distance between the target operating frequency and the approximate allocation of the antenna bandwidth (the average of the three frequencies mentioned above).

If, however, we consider a multi-band antenna to be matched at the target frequencies $f_{0.k}$, $k = 1, \ldots, N$, then the minimax merit function may be defined as

$$U(\boldsymbol{R}(\boldsymbol{x})) = \max\{|S_{11}(\boldsymbol{x}, f_{0.1})|, \ldots, |S_{11}(\boldsymbol{x}, f_{0.N})|\} \tag{9.19}$$

where $S_{11}(\boldsymbol{x}, f_{0.k})$ is antenna reflection at the design $\boldsymbol{x}$ and frequency f. For this antenna, the regularization function can be defined as

$$f_r(\boldsymbol{x}) = \left\| [f_{0.1} \ \ \ldots \ \ f_{0.N}]^T - [f_1(\boldsymbol{x}) \ \ \ldots \ \ f_N(\boldsymbol{x})]^T \right\| \tag{9.20}$$

where f_k, $k = 1, \ldots, N$, are the actual resonant frequencies at the design $\boldsymbol{x}$, extracted from the EM simulation results.

9.3.2 Results: Quasi-Yagi Planar Antenna

This section demonstrates the operation of the frequency-based regularization using two examples of planar antennas exhibiting different characteristics. We compare the efficacy of the optimization process involving the conventional formulation of the objective function and the regularized one in terms of the ability to find the satisfactory design but also the computational cost of the algorithm.

The first example is a quasi-Yagi antenna with a parabolic reflector shown in Fig. 9.25 (Hua et al., 2015), implemented on FR4 substrate ($\varepsilon_r = 4.4$, $h = 1.5$ mm). The design variables are $\boldsymbol{x} = [W\, L\, L_m\, L_p\, S_d\, S_r\, W_2\, W_a\, W_d\, g]^T$ (all dimensions in mm). The computational model is implemented in CST Microwave Studio.

The design task is to optimize the antenna for a given center frequency f_0 while ensuring at least 8% impedance bandwidth (symmetric w.r.t. f_0). Furthermore, the realized gain is to be maximized at f_0. The conventional formulation of the optimization problem uses the objective function (9.1). The frequency-based regularization employs the regularization function (9.18) with $f_{r.\max} = 0.15$ GHz. The antenna was optimized for $f_0 = 2.5$ GHz, $f_0 = 4.5$ GHz, and $f_0 = 5.0$ GHz, in all cases starting from the initial design that corresponds to the antenna operating at the frequency around 3.5 GHz. Figure 9.26 shows the initial and optimized designs obtained using the standard and regularized formulations. It can be observed that the algorithm

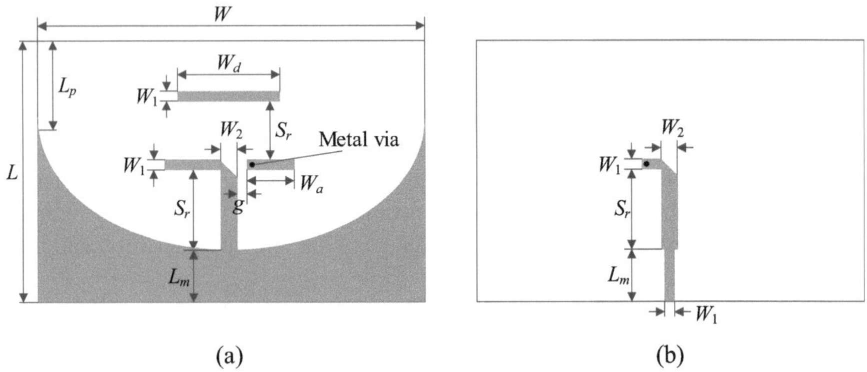

Fig. 9.25 Quasi-Yagi antenna with a parabolic reflector (Hua et al., 2015): (**a**) top layer, (**b**) bottom layer

using the standard formulation was unable to find high-quality designs in neither of the considered cases, whereas applying regularization ensures reaching satisfactory design for all scenarios. The performance comparison has been provided in Table 9.3. Thus, regularization allows for consistently yielding satisfactory design but also the average optimization cost is lower than for the standard formulation (assuming the same termination criteria, which was 10^{-2} for the convergence in argument and 10^{-2} for the trust-region size) (see Sect. 9.1.2).

9.3.3 Results: Ultra Wide-Band Antenna

The second example is the ultra wide-band monopole (Alsath & Kanagasabai, 2015) implemented on RF-35 substrate ($\varepsilon_r = 3.5$, $h = 0.762$ mm), shown in Fig. 9.27. The adjustable parameters are $\boldsymbol{x} = [L_0\ dR\ R\ r_{\text{rel}}\ dL\ dw\ L_g\ L_1\ R_1\ dr\ c_{\text{rel}}]^T$. The EM model is implemented in CST and incorporates the SMA connector. The operating frequency range of the antenna is 3.1–10.6 GHz. The design objective is to minimize the antenna footprint $A(\boldsymbol{x})$ while ensuring that $|S_{11}(\boldsymbol{x},f)| \leq -10$ dB within the operating band. The objective function is defined as

$$U(\boldsymbol{R}(\boldsymbol{x})) = A(\boldsymbol{x}) + \beta c(\boldsymbol{x})^2 \tag{9.21}$$

where the penalty function takes the form $c(\boldsymbol{x}) = \max\{\max\{3.1 \text{ GHz} \leq f \leq 10.6 \text{ GHz}: |S_{11}(\boldsymbol{x},f)| + 10\},0\}$. The regularization function is defined as

$$f_r(\boldsymbol{x}) = |f_L - f_{r.1}| \tag{9.22}$$

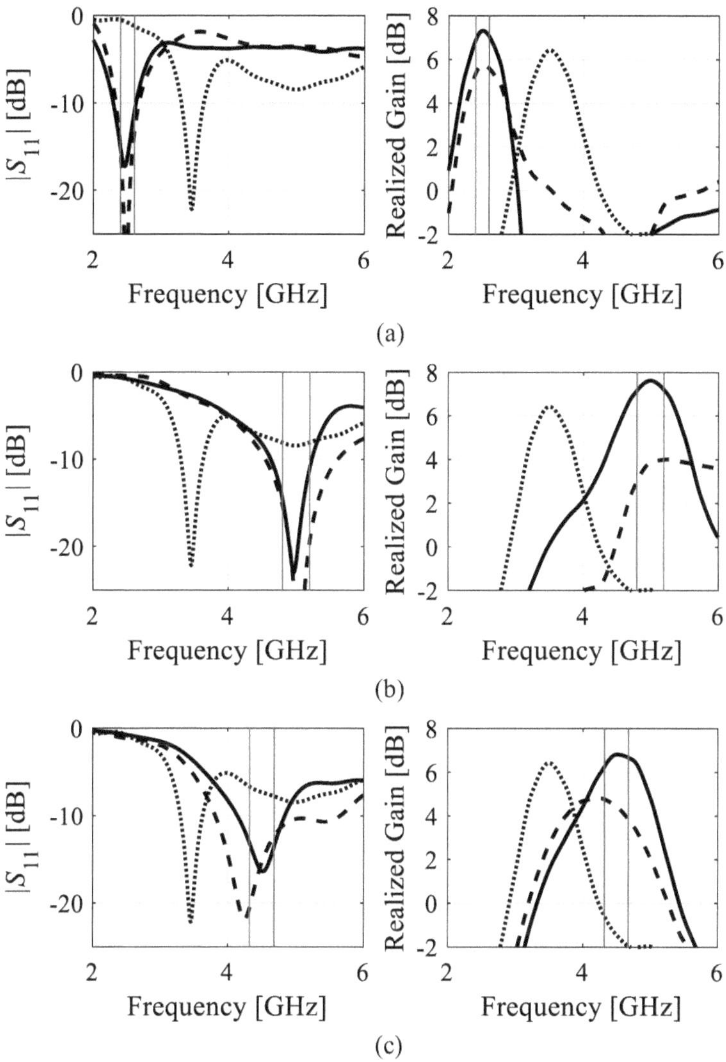

Fig. 9.26 Quasi-Yagi antenna optimization using standard formulation (cf. Eq. 9.1), and frequency-based regularization (cf. Eqs. 9.16 and 9.18). Initial design (····), optimum found using standard formulation (- - -), optimum found with regularization (—) are shown. Target frequencies: (**a**) 2.5 GHz, (**b**) 4.5 GHz, (**c**) 5.0 GHz. Vertical lines mark the intended operating bandwidth of the antenna

where $f_L = 3.1$ GHz and $f_{r.1}$ is the lowest frequency corresponding to -10 dB level of reflection (or the frequency of the first antenna resonance if the resonance level is above -10 dB). The reason for this setup is that the minimum-size design is normally such that the matching condition is barely met (or slightly violated when using implicit constraint handling as in (9.21)), with the first antenna resonance close

Table 9.3 Optimization results for the quasi-Yagi antenna

Problem formulation	Target frequency (GHz)	Results			
		Individual case		Average	
		Cost[a]	$U(\boldsymbol{R}(\boldsymbol{x}))$[b] (dB)	Cost[a]	$U(\boldsymbol{R}(\boldsymbol{x}))$[b] (dB)
Standard	2.5	165	−5.7	171	−4.6
	4.5	123	−3.7		
	5.0	224	−4.4		
Regularization	2.5	147	−7.2	131	−7.1
	4.5	135	−7.5		
	5.0	110	−6.7		

[a]Cost expressed in the number of EM simulations of the antenna
[b]Objective function value according to the function (9.17)

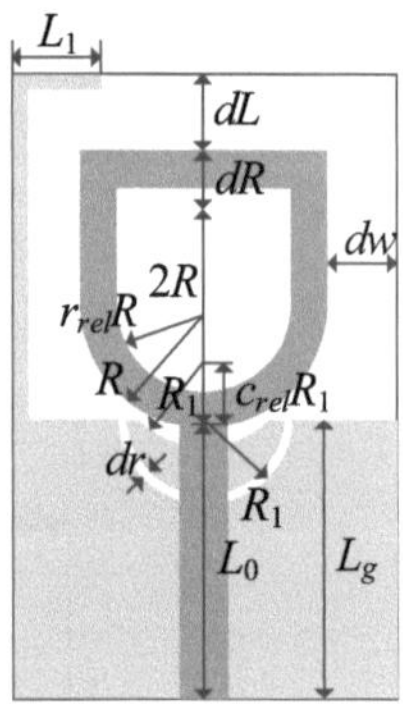

Fig. 9.27 Geometry of the ultra wide-band monopole antenna (Alsath & Kanagasabai, 2015). The ground plane marked using the light gray shade

to the lower edge of the operating band (here, 3.1 GHz). By enforcing this resonance to be close to that frequency, the problem of reducing the antenna size becomes numerically more tractable.

For the sake of demonstration, we optimize the antenna using three different starting points, as illustrated in Fig. 9.28. Regularizations (9.16) and (9.22) lead to considerably better results than the standard formulation (9.21), cf. Table 9.4. In all cases, the condition $|S_{11}(\boldsymbol{x},f)| \leq -10$ dB is satisfied in a satisfactory manner, but the obtained footprint area is smaller for the algorithm using regularization. It should be emphasized that optimization-based size reduction is a challenging problem from the numerical perspective because the reflection constraint is active at the optimum. Locating it requires traversing the boundary of the feasible region, and the results are typically dependent on the starting point, partially due to numerical noise issues. Notwithstanding, regularization noticeably improves repeatability of the results compared to the standard formulation because it allows for quick adjustment of the lower edge of the operating bandwidth, which yields a better position before further tuning. At the same time, the optimization cost with regularization is slightly higher as the algorithm using the standard formulation often converges prematurely.

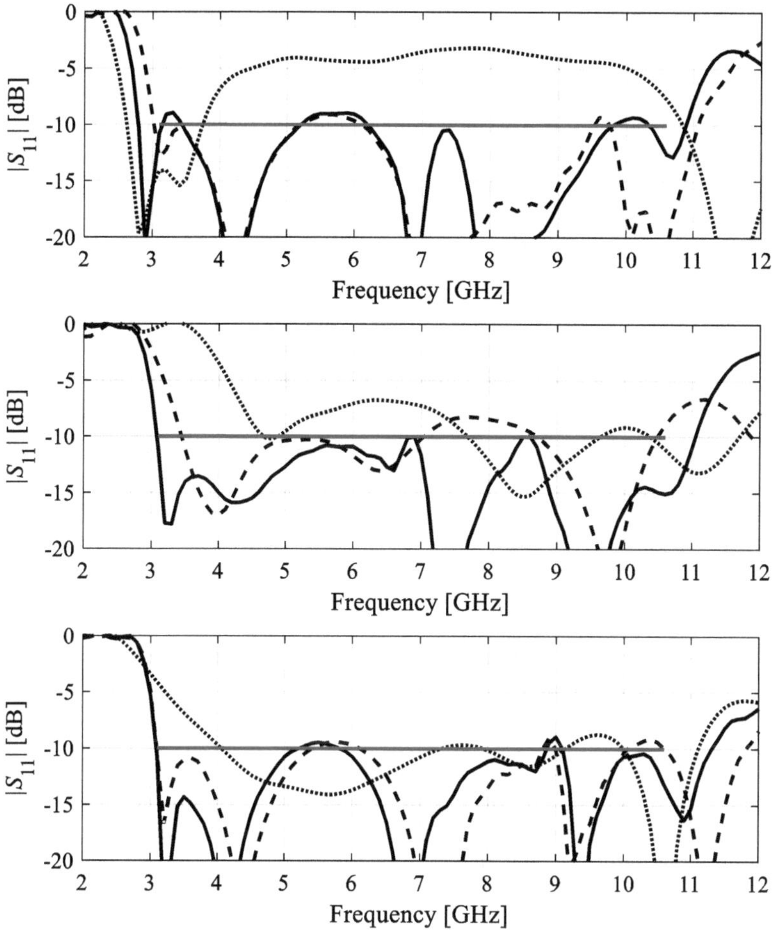

Fig. 9.28 Ultra wide-band monopole antenna optimization using standard formulation (cf. Eq. 9.1), and frequency-based regularization (cf. Eqs. 9.16 and 9.22). Initial design (····), optimum found using standard formulation (- - -), optimum found with regularization (—) for three different initial designs are shown. Target operating band 3.1–10.6 GHz is marked using the horizontal line

Table 9.4 Optimization results for the ultra wide-band antenna

		Results			
		Individual case		Average	
Problem formulation	Initial design	Cost[a]	Area[b]	Cost[a]	Area[b]
Standard	1	93	261 mm^2	113	293 mm^2
	2	101	401 mm^2		
	3	146	216 mm^2		
Regularization	1	125	238 mm^2	138	219 mm^2
	2	151	201 mm^2		
	3	138	219 mm^2		

[a]Cost expressed in the number of EM simulations of the antenna
[b]Footprint area of the optimized antenna

9.3.4 Summary

This section discussed a frequency-based regularization approach for reliable parameter tuning of antennas. It was incorporated into the objective function through a penalty term that quantifies the discrepancy between the target and the actual operating conditions of the antenna in terms of the relevant frequencies (e.g., allocation of the antenna resonances). The regularization function is formulated to be smooth (i.e., continuously differentiable) and only contributes to the primary objective if the aforementioned discrepancies are beyond the acceptance limit decided by the user. The results show that a significant improvement of the search process efficacy was achieved over the conventional design closure task formulations. This included reducing the sensitivity of the optimization outcome to the initial design, a possibility of re-designing antenna structures for the operating frequencies significantly different from those at the starting point, as well as improving repeatability of solutions. An additional benefit resulting from regularization-based problem reformulation is that local methods may be sufficient in the cases that otherwise require global search.

9.4 Regularization with Multi-Fidelity Models and Sparse Sensitivity Updates

In this section, an expedited version of the procedure considered in Sect. 9.3 is delineated. The discussed approach (Koziel & Pietrenko-Dabrowska, 2022) employs a frequency-based regularization to facilitate relocation of antenna operating parameters to their target values, which increases the chances of identifying a satisfactory design under challenging conditions. Computational efficiency of the tuning process is enhanced through the involvement of variable-resolution EM models, and restricting the finite-differentiation sensitivity updates to selected parameters only. The latter are decided upon based on the analysis of the variability of the response sensitivities between the subsequent algorithm iterations.

As the constituent parts of the discussed framework have been already outlined in this chapter, we do not reprise them in this section to avoid unnecessary repetitions. The main optimization procedure with frequency-based regularization has been outlined in Sect. 9.3. The employment of variable-fidelity EM simulations for expediting simulation-driven optimization procedures has been described in Sect. 9.2.2.1 where accelerated version of the framework with automatically adjusted design specifications has been considered. Finally, the trust-region gradient-based local optimization algorithm, which is shared as an optimization engine by all the discussed procedures of this chapter, has been depicted in Sect. 9.1.2.

Thus, this section starts by a description of the employment of restricted sensitivity updates for cost reduction in Sect. 9.4.1. Subsequently, overall optimization framework is outlined in Sect. 9.4.2, whereas Sect. 9.4.3 presents the results obtained for a quasi-Yagi and an ultra wide-band antenna.

9.4.1 Restricted Sensitivity Updates

Gradient-based local search routines can be accelerated in various ways. One of the available approaches are sparse sensitivity updating methods, where some finite-differentiation-based updates of the gradients of response of the component under design are not performed in order to reduce the overall numerical cost of the optimization process. Among various developed schemes, tracking of design relocation in the course of the optimization run (Koziel & Pietrenko-Dabrowska, 2019), the employment updating formulas (e.g., Broyden; Koziel & Pietrenko-Dabrowska, 2020a), detecting modification patterns of response gradients (Koziel & Pietrenko-Dabrowska, 2020b), but also performing updates along the selected directions within the parameter space determined either using principal component analysis (Tomasson et al., 2020) or a dedicated knowledge-based procedure (Koziel & Pietrenko-Dabrowska, 2023b). Sparse sensitivity update methods have been demonstrated to provide considerable computational speedup of up to 60% in comparison to reference routines (based on complete finite-differentiation (FD) updates), at the expense of a certain quality deterioration. In the approach discussed in this section, the response gradients are monitored, and FD is suppressed along the directions that exhibit stable patterns.

The explanation below is based on the example of antenna reflection characteristics S_{11}, but the description applies to all antenna (or other high-frequency components) characteristics that might be of interest. Let $\boldsymbol{G}_S = [G_1 \ \ldots \ G_n]^T$ be the gradient vector with $G_k = \partial S_{11}/\partial x_k$, $k = 1, \ldots, n$. Further, let $G_k^{(i)}(f)$ and $G_k^{(i-1)}(f)$ be the k-th entries of $\boldsymbol{G}_S$ in ith and (i–1)th iteration of the optimization algorithm, respectively, at frequency f.

The metric quantifying the relative difference between $G_k^{(i)}$ and $G_k^{(i-1)}$ used in (Koziel & Pietrenko-Dabrowska, 2020b) takes the form

$$\delta_k^{(i)} = \underset{f \in F}{\mathrm{mean}}\left(2 \cdot \left(\left|G_k^{(i)}(f)\right| - \left|G_k^{(i-1)}(f)\right|\right) / \left|G_k^{(i)}(f)\right| + \left|G_k^{(i-1)}(f)\right|\right) \tag{9.23}$$

for $k = 1, \ldots, n$. Having evaluated (9.23), we also find the minimum value of gradient change metric $\delta_{\min}^{(i)} = \min\left\{k = 1, \ldots, n : \delta_k^{(i)}\right\}$, as well as the maximum one $\delta_{\max}^{(i)} = \max\left\{k = 1, \ldots, n : \delta_k^{(i)}\right\}$. Based on these numbers, a number $N_k^{(i)}$ of future iterations without FD for the k-th parameter are assessed. FD will not be applied for parameters with the smallest values of $\delta_k^{(i)}$ for the maximum allowable number of iterations denoted as $N_{\max}$. Whereas for the variables featuring the highest $\delta_k^{(i)}$, the FD update is obligatory. These are the extreme cases, whereas the actual number $N_k^{(i+1)}$ of upcoming iterations without FD is computed for each parameter using the conversion function

$$N_k^{(i+1)} = \left[\!\left[N_{\max} - (N_{\max} - N_{\min}) \cdot \left(\delta_k^{(i)} - \delta_{\min}^{(i)}\right) \Big/ \left(\delta_{\max}^{(i)} - \delta_{\min}^{(i)}\right) \right]\!\right] \tag{9.24}$$

where [[.]] is the nearest integer function. Both $N_{\min}$ and $N_{\max}$ are the control parameters of the algorithm and refer to the minimum and the maximum number of iterations without FD. Their default values are $N_{\min} = 1$, and $N_{\max} = 5$.

Clearly, an increase in $\delta_k^{(i)}$ leads to a decrease in $N_k^{(i)}$, i.e., the higher the gradient variation, the less often FD may be skipped. The sensitivity update through FD is carried out at least once per $N_{\max}$ iterations, and not more often than every $N_{\min}$ iterations. The value of $N_k^{(i)}$ is based both on the "current" difference factors (calculated using (9.23) for the parameters with FD performed in the i-th iteration) and the "prior" ones (retained from the previous iterations for the parameters without FD in the i-th iteration).

9.4.2 *Optimization Framework*

Figure 9.29 shows the flow diagram of the entire optimization algorithm with multi-fidelity EM simulations and restricted sensitivity updates. The model resolution is set to its minimum value $L_{\min}$ at the beginning of the process. Evaluation of the objective function within the trust-region loop involves extraction of the operating frequencies of the antenna, followed by a computation of the regularization function f_r. Upon each iteration, the model resolution is updated, and the gradient variability factors are computed as outlined in Sect. 9.4.1, the updating process of the sensitivity matrix is only performed for selected variables, depending on the coefficients $N_k^{(i)}$ (cf. Eq. 9.24).

9.4.3 *Results*

This section provides verification results of the discussed accelerated optimization procedure with frequency regularization. It is based on two microstrip antennas, optimized under different scenarios (gain enhancement, impedance matching improvement, and size reduction). The obtained results are compared to benchmark techniques that include conventional gradient-based routine (Conn et al., 2000), optimization with frequency-based regularization (Koziel et al., 2020a), two frequency-regularized algorithms employing one of the two acceleration mechanisms (either variable-resolution models, or sparse sensitivity updates), and derivative-free pattern search algorithm (Kolda et al., 2003). We are interested in the reliability of the optimization process, as well as its computational efficiency. The antenna structures are described in Sect. 9.4.3.1. Section 9.4.3.2 outlines the experimental setup, whereas Sect. 9.4.3.3 discusses the results.

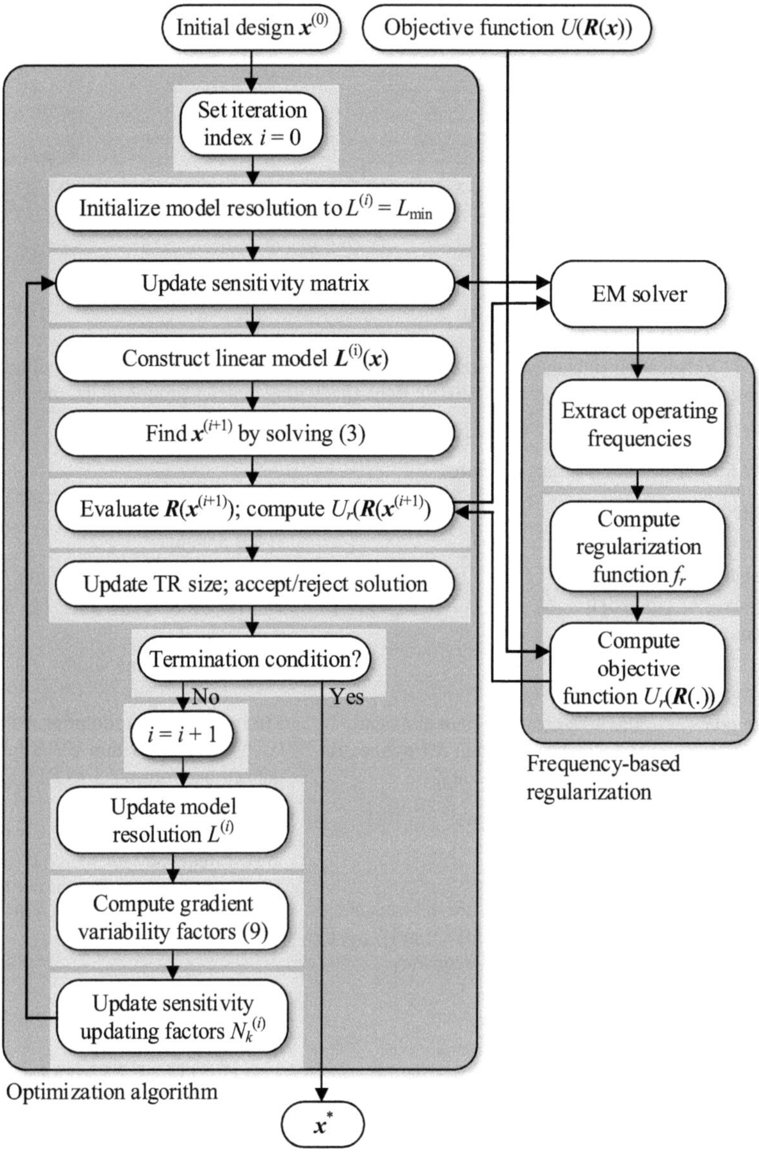

Fig. 9.29 Flow diagram of the optimization procedure with frequency regularization and acceleration mechanisms: multi-fidelity EM simulations and sparse sensitivity updates

9.4.3.1 Test Antenna Structures

The verification experiments involve the following antenna structures: a quasi-Yagi antenna with a parabolic reflector (Hua et al., 2015) shown in Fig. 9.30a; an ultra wide-band (UWB) monopole antenna (Alsath & Kanagasabai, 2015) shown in Fig. 9.30b.

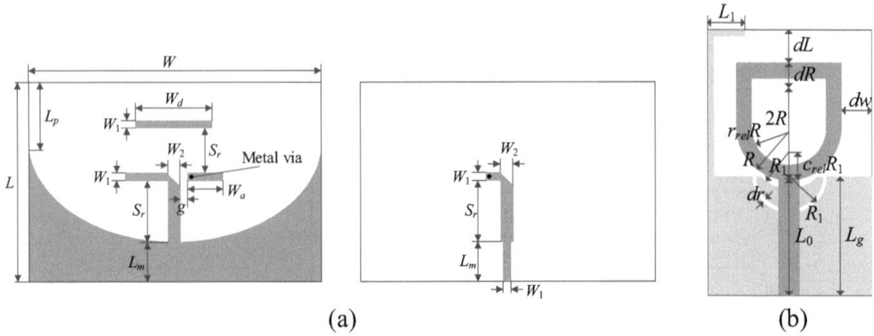

Fig. 9.30 Verification antenna structures: (**a**) quasi-Yagi antenna (Hua et al., 2015), (**b**) ultra wideband antenna (Lesath & Kanagasabai, 2015); light-shade gray denotes ground-plane metallization

Table 9.5 Verification antenna structures

Case study		
Antenna I	Quasi-Yagi antenna	UWB antenna
Substrate	FR4 ($\varepsilon_r = 4.4$, $h = 1.5$ mm)	RF-35 ($\varepsilon_r = 3.5$, $h = 0.76$ mm)
Design parameters	$\boldsymbol{x} = [W\ L\ L_m\ L_p\ S_d\ S_r\ W_2\ W_a\ W_d\ g]^T$	$\boldsymbol{x} = [L_0\ dR\ R\ r_{\text{rel}}\ dL\ dw\ L_g\ L_1\ R_1\ dr\ c_{\text{rel}}]^T$
Other parameters	$W_1 = 3.0$	$w_0 = 1.7$ (feed line width)
Design goals	Maximize realized gain at f_0, and $\|S_{11}\| \leq -10$ dB within 8% symmetric bandwidth centered at f_0	Minimize antenna footprint $A(\boldsymbol{x})$, and $\|S_{11}\| \leq -10$ dB within UWB frequency band from 3.1 to 10.6 GHz
Regularization function	$f_r(\boldsymbol{x}) = \left\| f_0 - \frac{\sum_{j=0}^{2} f_{L.j}(\boldsymbol{x})}{3} \right\|$ where $f_{L.j}$, $j = 1, 2$ are the frequencies of -10 dB levels of $\|S_{11}\|$, and $f_{L.0}$ is the antenna resonance frequency	$f_r(\boldsymbol{x}) = \|f_L - f_{r.1}(\boldsymbol{x})\|$ where $f_L = 3.1$ GHz and $f_{r.1}$ is the lowest frequency for which $\|S_{11}\| = -10$ dB (or the frequency of the first resonance if it is above -10 dB)
Design cases	Center frequencies: 1. $f_0 = 2.5$ GHz 2. $f_0 = 4.5$ GHz 3. $f_0 = 5.0$ GHz One initial design: $\boldsymbol{x}^{(0)} = [143.1\ 59.5\ 13.0\ 28.7\ 17.6\ 14.3\ 5.0\ 9.5\ 23.0\ 1.0]^T$	Three initial designs: 1. $\boldsymbol{x}^{(0)} = [14.5\ 0.7\ 3.2\ 0.2\ 3.9\ 4.9\ 8.5\ 4.3\ 3.3\ 0.22\ 0.25]^T$ 2. $\boldsymbol{x}^{(0)} = [7.2\ 0.7\ 3.9\ 0.5\ 1.3\ 1.3\ 7.2\ 2.6\ 2.6\ 0.16\ 0.52]^T$ 3. $\boldsymbol{x}^{(0)} = [9.5\ 2.3\ 3.7\ 0.7\ 4.3\ 6.3\ 11.7\ 5.8\ 4.0\ 0.8\ 0.22]^T$
Parameter space	$\boldsymbol{l} = [50\ 30\ 10\ 0\ 5\ 5\ 2\ 5\ 15\ 0.5]^T$ $\boldsymbol{u} = [150\ 90\ 35\ 30\ 30\ 30\ 5\ 40\ 50\ 1]^T$	$\boldsymbol{l} = [4\ 0\ 3\ 0.1\ 0\ 0\ 4\ 0\ 2\ 0.2\ 0.2]^T$ $\boldsymbol{u} = [15\ 6\ 8\ 0.9\ 5\ 8\ 15\ 6\ 5\ 1\ 0.9]^T$

The information concerning antennas used as verification examples can be found in Table 9.5. It includes data on the dielectric substrate antennas are implemented on, design parameters, design goals, as well as parameter spaces, defined by the lower bounds $\boldsymbol{l}$ and upper bounds $\boldsymbol{u}$ for geometry variables. The EM models are

Table 9.6 Variable-resolution EM models for antennas of Fig. 9.30

	Lowest-fidelity model		High-fidelity model	
Antenna	L_{min}	Simulation time (s)	L_{max}	Simulation time (s)
Quasi-Yagi antenna	7	66	25	261
UWB antenna	11	41	24	424

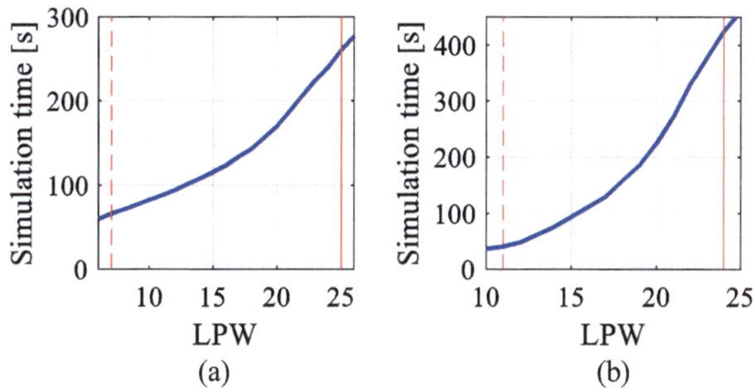

Fig. 9.31 Evaluation time versus EM model resolution for antennas of Fig. 9.30: (**a**) Quasi-Yagi antenna, (**b**) UWB antenna. Vertical lines denote the resolutions L_{min} (- - -) and L_{max} (high-fidelity) (—)

implemented in CST Microwave Studio and simulated using the time-domain solver. The model of UWB antenna incorporates the SMA connector. Table 9.6 provides data on the computational models, in particular, the range of model resolutions (L_{min} and L_{max}) as well as the simulation times. Graphical information on the relationships between the discretization density and the simulation time can be found in Fig. 9.31. The simulation time ratio between the high-fidelity model (L_{max}) and the lowest-fidelity one (L_{min}) is 4.0 and 10.3, for quasi-Yagi and UWB antenna, respectively. Higher ratio suggests potentially higher computational benefits due to the incorporation of variable-resolution models.

9.4.3.2 Experimental Setup and Results

Both antennas have been optimized using the considered algorithm. In each case, the optimization process was executed three times under different scenarios as listed in the second-but-last row of Table 9.5. In particular, quasi-Yagi antenna was optimized to improve the realized gain with a constraint imposed on reflection response over 8% bandwidth centered at the operating frequency f_0, with f_0 set to three different values, 2.5, 4.5, and 5.0 GHz. Whereas UWB antenna was optimized for minimum size with a reflection constraint $|S_{11}| \leq -10$ dB over UWB band (3.1–10.6 GHz), also using three initial designs. The reason for this setup is to verify the efficacy of the discussed technique for a variety of design tasks, and under challenging conditions, especially when the operating parameters of the antenna are away from the assumed targets at the initial design.

Table 9.7 presents the juxtaposition of the control parameters of the discussed algorithm. Moreover, the recommended values of all of the parameters are also given, as well as guidelines on how to adjust the procedure to a specific design task.

Table 9.7 Recommended values of algorithm control parameters

Parameter	Purpose	Guidelines	Utilized value
M	Launching of the LPW increase	$10 \cdot \varepsilon_x$	10^{-2}
M_d	Increase of the TR radius near convergence	$5 \div 10$	10
λ	Setting LPW for FD	$0.5 \div 0.75$	2/3

Table 9.8 Results for the quasi-Yagi antenna

Algorithm	Problem formulation	Target frequency (GHz)	Results: Individual case Cost[a]	Results: Individual case $U(\boldsymbol{R}(\boldsymbol{x}))$[b] (dB)	Results: Average Cost[a]	Results: Average $U(\boldsymbol{R}(\boldsymbol{x}))$[b] (dB)
1	Standard	2.5	165	−5.7	171	−4.6
		4.5	123	−3.7		
		5.0	224	−4.4		
2	Regularization	2.5	147	−7.2	131	−7.1
		4.5	135	−7.5		
		5.0	110	−6.7		
3	Regularization with variable-fidelity models	2.5	51	−6.7	57	−6.0
		4.5	67	−6.0		
		5.0	52	−5.3		
4	Regularization with sparse sensitivity updates	2.5	73	−5.1	69	−4.8
		4.5	75	−5.9		
		5.0	60	−3.4		
5	Pattern search algorithm	2.5	523	−6.7	510	−4.6
		4.5	362	−6.7		
		5.0	646	−6.7		
6	Multi-fidelity regularization with sparse updates	2.5	16	−5.1	25	−5.5
		4.5	29	−5.4		
		5.0	29	−6.1		

[a]Cost expressed in the number of EM simulations of the antenna
[b]Objective function value $U(\boldsymbol{R}(\boldsymbol{x})) = -G(\boldsymbol{x}, f_0) + \beta c(\boldsymbol{x})^2$

The considered procedure (Algorithm 6) has been compared to conventional gradient-based search with numerical derivatives (Algorithm 1; cf. Sect. 9.2.2.1), the algorithm with frequency-based regularization (Sect. 9.3) (Algorithm 2), two algorithms that employ one of the acceleration mechanisms: (i) variable-resolution models (Algorithm 3), and (ii) sparse sensitivity updates (Algorithm 4), as well as conventional derivative-free pattern search algorithm (Kolda et al., 2003) (Algorithm 5). This allows us to investigate the relevance and the benefits of incorporating both acceleration factors into the procedure.

The numerical results have been gathered in Tables 9.8 and 9.9 for quasi-Yagi and UWB antennas, respectively. The tables report the computational cost of the optimization process as well as the objective function values for all algorithms and all design cases. Figures 9.32 and 9.33 show the antenna responses at the starting

Table 9.9 Results for the UWB antenna

Algorithm	Problem formulation	Initial design	Results: Individual case Cost[a]	A[b]	Average Cost[a]	A[b]
1	Standard	1	93	261 mm^2	113	293 mm^2
		2	101	401 mm^2		
		3	146	216 mm^2		
2	Regularization	1	125	238 mm^2	138	219 mm^2
		2	151	201 mm^2		
		3	138	219 mm^2		
3	Regularization with variable-fidelity models	1	38	248 mm^2	36	238 mm^2
		2	21	193 mm^2		
		3	49	237 mm^2		
4	Regularization with sparse sensitivity updates	1	83	250 mm^2	83	251 mm^2
		2	75	269 mm^2		
		3	92	234 mm^2		
5	Pattern search algorithm	1	518	245 mm^2	660	284 mm^2
		2	641	395 mm^2		
		3	820	212 mm^2		
6	Multi-fidelity regularization with sparse updates	1	8	252 mm^2	16	229 mm^2
		2	16	208 mm^2		
		3	25	2229 mm^2		

[a]Cost expressed in the number of EM simulations of the antenna
[b]Footprint area of the optimized antenna

points, the designs obtained using the discussed variable-fidelity algorithm with frequency regularization and sparse sensitivity updates (Algorithm 6), as well as the two accelerated algorithms with variable-resolution models and sparse sensitivity updates (Algorithms 3 and 4).

9.4.3.3 Discussion

The results of Sect. 9.4.3.2 allow us to draw some conclusions concerning the performance of the considered algorithm in comparison to the reference routines, including the conventional trust-region procedure, as well as its versions exploiting frequency regularization: the basic one, and the two accelerated routines, with multi-fidelity simulations and sparse sensitivity updates, respectively.

The discussed algorithm (Algorithm 6) is demonstrably and consistently the fastest of all the considered procedures. The speedup over the reference TR routine (Algorithm 1) is remarkable: from 65 for Antenna 2, to as high as 88% for Antenna 3. The average speedup across the antenna benchmark set is 78%.

The obtained savings are correlated with the simulation time ratios between the high-fidelity model and the lowest-fidelity model (4.0 and 10.3 for quasi-Yagi and

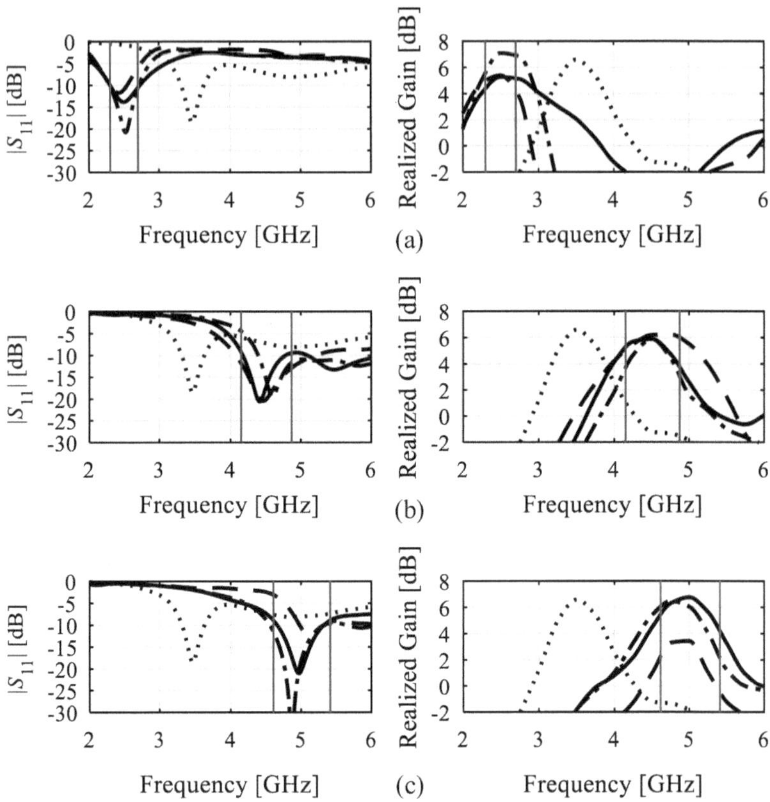

Fig. 9.32 Quasi-Yagi antenna: optimization using the accelerated frequency-based regularization for the following operating frequencies: (**a**) 2.5 GHz, (**b**) 4.5 GHz, (**c**) 5.0 GHz (cf. Table 9.5). The plots show the initial design (····), the optimal design found using Algorithm 3 (with variable-resolution models) (—), the optimized design found using Algorithm 4 (with sparse sensitivity updates) (- - -), and the optimum design obtained using Algorithm 5 (with variable-resolution models and sparse sensitivity updates) (- . - .). Vertical lines mark the target antenna operating bandwidth

UWB antenna, respectively). The computational cost of the sole derivative-free procedure from the benchmark pool, i.e., pattern search algorithm (Algorithm 5), is over 25 times higher than that of the discussed one.

As for the design quality, the considered algorithm yields the designs satisfying the assumed design specifications in all of the presented cases, which is not the case for the conventional procedures: TR routine (reference Algorithm 1) and pattern search algorithm (Algorithm 5). Although a minor degradation of the design quality in comparison to the procedure with frequency regularization (Algorithm 2) is being observed, its magnitude is insignificant and equals around 1 dB (quasi-Yagi antenna) and 10 mm^2 (UWB antenna). As compared to the basic TR routine, our approach is far less costly and much more reliable in terms of the ability to render the designs meeting the required targets in every case.

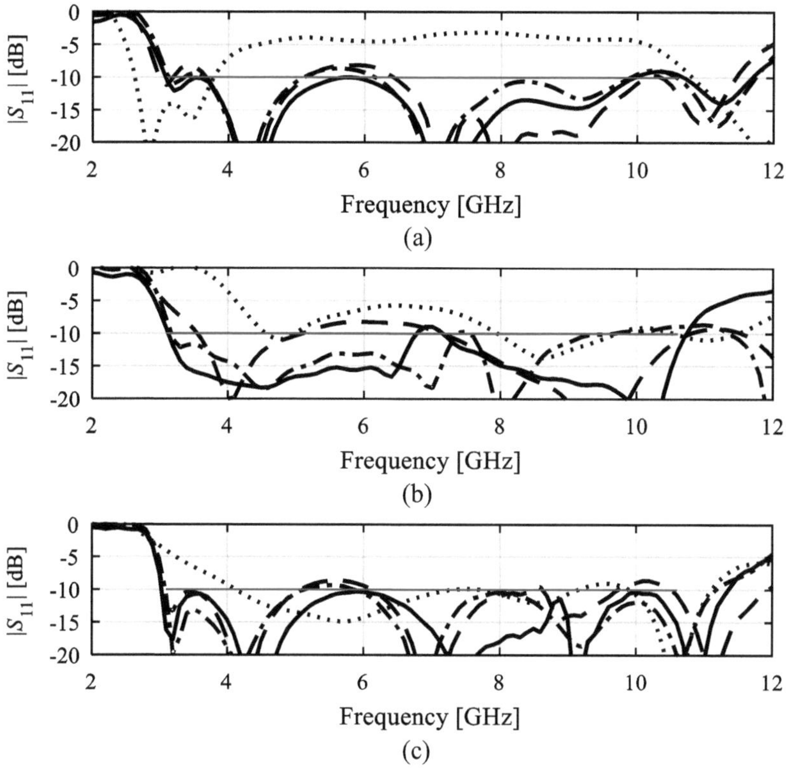

Fig. 9.33 UWB antenna: optimization using the accelerated frequency-based regularization from three different initial designs (cf. Table 9.5). The plots show the initial design (····), the optimized design found using Algorithm 1 (with variable-resolution models) (- . - .), the optimized design found using Algorithm 2 (with sparse sensitivity updates) (- - -), and the optimum design obtained using Algorithm 3 (with variable-resolution models and sparse sensitivity updates) (—). Horizontal line marks the target operating band (3.1–10.6 GHz)

The quality of the designs yielded by the considered algorithm is comparable to that of the Algorithms 3 and 4 (i.e., the procedures utilizing only one acceleration mechanism: either variable-fidelity simulations or sparse sensitivity updates). Overall, the discussed algorithm exploiting both acceleration mechanisms is significantly faster and more reliable than the conventional TR procedure, while ensuring satisfactory design quality, which makes it competitive from the point of view of both reducing time-to-market and ensuring reliability of design procedures.

It should be emphasized that utilization of frequency regularization as one of the underlying components of the considered procedure is instrumental in enabling quasi-global optimization capability. The acceleration mechanisms allow for maintaining this feature while providing unprecedented cost reduction. To the best knowledge of the authors, the literature has not yet offered an optimization engine of comparable efficacy in the category of direct optimization procedures.

9.5 Summary

This chapter discussed two methods for reliability enhancement of local optimization procedures: design specification adjustment and frequency-based regularization. In addition, accelerated versions of both considered frameworks were also outlined. In the case of the algorithm with automatically adjusted design specifications, the employed acceleration mechanism involved variable-fidelity EM simulations models. Whereas frequency regularization algorithm was expedited simultaneously using multi-fidelity EM variable-fidelity models and sparse sensitivity updates. The techniques of this chapter considerably improve the performance of the local search routines, thereby allowing for solving challenging optimization tasks without resorting to the use of costly global optimizers. This is because the discussed strategies feature quasi-global search capabilities, i.e., they increase the search process immunity to poor starting points and ensure reliable relocation of the operating parameters over broad ranges thereof. The algorithms of this chapter were verified using of several high-frequency structures: a dual-band coupler and a dual-band power divider (Sect. 9.1), multi-band antennas (Sect. 9.2), as well as narrow- and ultra wide-band antennas (Sects. 9.3 and 9.4). The presented examples corroborate the efficacy of the considered approaches in design optimization of high-frequency components.

References

Alsath, M. G. N., & Kanagasabai, M. (2015). Compact UWB monopole antenna for automotive communications. *IEEE Transactions on Antennas and Propagation, 63*(9), 4204–4208.

Chen, Y.-C., Chen, S.-Y., & Hsu, P. (2006). Dual-band slot dipole antenna fed by a coplanar waveguide. In *IEEE antennas and propagation society international symposium*, Albuquerque, NM (pp. 3589–3592). IEEE.

Conn, A. R., Gould, N. I. M., & Toint, P. L. (2000). *Trust region methods* (MPS-SIAM Series on Optimization). MPS-SIAM.

Hua, Z., Haichuan, G., Hongmei, L., Beijia, L., Guanjun, L., & Qun, W. (2015). A novel high-gain quasi-Yagi antenna with a parabolic reflector. In *International symposium on antennas and propagation (ISAP)*, Hobart, Australia. IEEE.

Jeong, S., Hester, J. G. D., Su, W., & Tentzeris, M. M. (2019). Read/interrogation enhancement of chipless RFIDs using machine learning techniques. *IEEE Antennas and Wireless Propagation Letters, 18*(11), 2272–2276.

Kolda, T. G., Lewis, R. M., & Torczon, V. (2003). Optimization by direct search: New perspectives on some classical and modern methods. *SIAM Review, 45*, 385–482.

Koziel, S., & Bandler, J. W. (2008). Space mapping with multiple coarse models for optimization of microwave components. *IEEE Microwave and Wireless Components Letters, 18*, 1–3.

Koziel, S., & Ogurtsov, S. (2012). Model management for cost-efficient surrogate-based optimization of antennas using variable-fidelity electromagnetic simulations. *IET Microwaves, Antennas & Propagation, 6*(15), 1643–1650.

Koziel, S., & Ogurtsov, S. (2019). *Simulation-based optimization of antenna arrays*. World Scientific.

Koziel, S., & Pietrenko-Dabrowska, A. (2019). Reduced-cost electromagnetic-driven optimization of antenna structures by means of trust-region gradient-search with sparse Jacobian updates. *IET Microwaves, Antennas & Propagation, 13*(10), 1646–1652.

Koziel, S., & Pietrenko-Dabrowska, A. (2020a). Expedited optimization of antenna input characteristics with adaptive Broyden updates. *Engineering Computations, 37*(3), 851–862.

Koziel, S., & Pietrenko-Dabrowska, A. (2020b). Expedited feature-based quasi-global optimization of multi-band antennas with Jacobian variability tracking. *IEEE Access, 8*, 83907–83915.

Koziel, S., & Pietrenko-Dabrowska, A. (2021). Robust parameter tuning of antenna structures by means of design specification adaptation. *IEEE Transactions on Antennas and Propagation, 69*(12), 8790–8798.

Koziel, S., & Pietrenko-Dabrowska, A. (2022). Rapid variable-resolution parameter tuning of antenna structures using frequency-based regularization and sparse sensitivity updates. *IEEE Transactions on Antennas and Propagation, 70*(12), 12177–12188.

Koziel, S., & Pietrenko-Dabrowska, A. (2023a). Improved-efficacy EM-driven optimization of antenna structures using adaptive design specifications and variable-resolution models. *IEEE Transactions on Antennas and Propagation, 71*(2), 1863–1874.

Koziel, S., & Pietrenko-Dabrowska, A. (2023b). Rapid antenna optimization with restricted sensitivity updates by automated dominant direction identification. *Knowledge-Based Systems, 268*, 110453.

Koziel, S., & Unnsteinsson, S. D. (2018). Expedited design closure of antennas by means of trust-region-based adaptive response scaling. *IEEE Antennas and Wireless Propagation Letters, 17*(6), 1099–1103.

Koziel, S., Cheng, Q. S., & Bandler, J. W. (2008). Space mapping. *IEEE Microwave Magazine, 9*(6), 105–122.

Koziel, S., Leifsson, L., & Yang, X. S. (Eds.). (2016). *Simulation-driven modeling and optimization*. Springer.

Koziel, S., Pietrenko-Dabrowska, A., & Al-Hasan, M. (2020a). Frequency-based regularization for improved reliability optimization of antenna structures. *IEEE Transactions on Antennas and Propagation, 69*(7), 4246–4251.

Koziel, S., Pietrenko-Dabrowska, A., & Al-Hasan, M. (2020b). Improved-efficacy optimization of compact microwave passives by means of frequency-related regularization. *IEEE Access, 8*, 195317–195326.

Koziel, S., Pietrenko-Dabrowska, A., & Plotka, P. (2022). Design specification management with automated decision-making for reliable optimization of miniaturized microwave components. *Scientific Reports, 12*, 829.

Lalbakhsh, A., Afzal, M. U., & Esselle, K. (2017). Multiobjective particle swarm optimization to design a time-delay equalizer metasurface for an electromagnetic band-gap resonator antenna. *IEEE Antennas and Wireless Propagation Letters, 16*, 912–915.

Lin, Z., & Chu, Q.-X. (2010). A novel approach to the design of dual-band power divider with variable power dividing ratio based on coupled-lines. *Progress In Electromagnetics Research, 103*, 271–284.

Liu, B., Akinsolu, M. O., Ali, N., & Abd-Alhameed, R. (2019). Efficient global optimisation of microwave antennas based on a parallel surrogate model-assisted evolutionary algorithm. *IET Microwaves, Antennas & Propagation, 13*(2), 149–155.

Luo, X., Yang, B., & Qian, H. J. (2019). Adaptive synthesis for resonator-coupled filters based on particle swarm optimization. *IEEE Transactions on Microwave Theory and Techniques, 67*(2), 712–725.

Pietrenko-Dabrowska, A., & Koziel, S. (2020). Surrogate modeling of impedance matching transformers by means of variable-fidelity EM simulations and nested co-kriging. *The International Journal of RF and Microwave Computer-Aided Engineering, 30*(8), e22268.

Pietrenko-Dabrowska, A., & Koziel, S. (2021). Accelerated gradient-based optimization of antenna structures using multi-fidelity simulation models. *IEEE Transactions on Antennas and Propagation, 69*(12), 8778–8789.

Tomasson, J. A., Koziel, S., & Pietrenko-Dabrowska, A. (2020). Quasi-global optimization of antenna structures using principal components and affine subspace-spanned surrogates. *IEEE Access, 8*, 50078–50084.

Xia, J., Li, B., Twumasi, A., Liu, P., & Gao, S. (2020). Planar dual-band branch-line coupler with large frequency ratio. *IEEE Access, 8*, 33188–33195.

Xiao, L., Shao, W., Ding, X., & Wang, B. (2018). Dynamic adjustment kernel extreme learning machine for microwave component design. *IEEE Transactions on Microwave Theory and Techniques, 66*(10), 4452–4461.

Chapter 10
Response Features for Global and Multi-objective Optimization

Previous chapters have discussed utilization of response feature technology for solving local optimization problems. Specifically, Chap. 6 addressed local feature-based optimization of high-frequency components for various types of structures and responses, whereas in Chap. 9, the employment of the response feature technology for reliability improvement of local optimization procedures has been considered. The examples described in Chaps. 6 and 9 corroborate that the incorporation of the response features technology into local search procedures enable quasi-global search capabilities. This is because the employment of feature-related reliability enhancements inherently regularizes the objective function landscape, thereby improving the immunity of the search process to poor initial conditions and permitting redesign of a given device for operational parameters significantly different than those at the starting point. Nevertheless, in some cases, the enhancement of local search procedures may not be sufficient, and a truly global search may be indispensable. Still, computational cost of simulation-driven global design optimization of high-frequency components is usually exorbitant. A workaround is offered by surrogate-based procedures; yet, the curse of dimensionality strongly limits their applicability to the devices described by few design parameters within the limited ranges thereof. This chapter outlines two feature-based surrogate-assisted frameworks for global optimization of antennas and microwave devices. The first one employs feature-based inverse surrogate models, whereas in the second, simplex-based regression surrogates have been exploited. The remaining part of this chapter outlines feature-based multi-objective design optimization. All the discussed methods have been extensively illustrated using examples of microwave and antenna structures.

A. Pietrenko-Dabrowska, S. Koziel, *Response Feature Technology for High-Frequency Electronics. Optimization, Modeling, and Design Automation*,
https://doi.org/10.1007/978-3-031-43845-5_10

10.1 Global Optimization Using Feature-Based Inverse Surrogates

This section discusses an algorithmic framework for globalized optimization of multiband antennas and microwave circuits. The considered methodology incorporates the inverse-modeling-based identification of the promising parameter space regions and the response feature technology for extracting the operating parameters of the high-frequency device of interest from the electromagnetic (EM) simulation data. Both mechanisms are utilized to render an initial design, which is further tuned by means of trust-region gradient search procedure. The inverse model is constructed based on preselected data obtained through stochastic search. During the selection process, relevant knowledge about the device operational parameters is extracted from its simulated responses. This knowledge is subsequently utilized to facilitate a decision-making process of a cost-efficient identification of the parameter space regions containing high-quality designs. The performance of the considered framework has been showcased using two verification structures: a triple band U-slotted patch antenna with L-slot DGS (defected ground structure) optimized to allocate its operating frequencies at specific target values and improve matching therein, and a rat-race coupler with folded transmission lines. The results indicate that the considered algorithm consistently yields satisfactory designs, whereas its computational complexity is comparable to that of local optimization routines.

10.1.1 Design Problem Formulation

Let us start by formulating a general simulation-driven design problem as a nonlinear minimization task of the form

$$\boldsymbol{x}^* = \arg \min_{\boldsymbol{x}} U(\boldsymbol{x}, \boldsymbol{f}_t) \tag{10.1}$$

where U is a scalar merit function to be minimized, whereas $\boldsymbol{f}_t = [f_{t.1} \ldots f_{t.K}]^T$ is a target vector of operating parameters, which may include operating frequencies of an antenna or a microwave circuit at hand, but also other parameters, such as power split ratio.

Let us first consider global optimization of input characteristics of multiband antennas, where the design task is to allocate the antenna resonances at the target frequencies $f_{0.j}$, $j = 1, \ldots, K$, with K being the number operating bands. The merit function is defined in a minimax sense

$$U(\boldsymbol{x}, \boldsymbol{f}_t) = \max_{\boldsymbol{x}} \{|S_{11}(\boldsymbol{x}, f_{t.1})|, \ldots, |S_{11}(\boldsymbol{x}, f_{t.K})|\} \tag{10.2}$$

where $S_{11}(\boldsymbol{x}, f)$ denotes the EM-simulated antenna reflection response at the frequency f and $\boldsymbol{x}$ refers to a vector of designable variables.

In the case of the antennas, the formulation (10.1) and (10.2) can be generalized to consider, e.g., matching improvement over bandwidths, as well as to include other antenna characteristics. Nevertheless, the objective function (10.2) constitutes a representative case study, which will be further used to explain the considered globalized search procedure.

Let us consider as an example a microwave coupler, which is to operate at the frequency f_0 so that its matching and isolation characteristics, $|S_{11}|$ and $|S_{41}|$, are minimized at f_0, and the power split ratio $d_S(\boldsymbol{x},f_0) = |S_{21}(\boldsymbol{x},f_0)| - |S_{31}(\boldsymbol{x},f_0)|$ reaches a target value K_P (e.g., 0 dB for equal power split). In this case, the operating parameter vector would be $\boldsymbol{F}_t = [f_0\ K_P]^T$, whereas the objective function may be defined as

$$U(\boldsymbol{x},\boldsymbol{f}_t) = \max\{|S_{11}(\boldsymbol{x},f_0)|, |S_{41}(\boldsymbol{x},f_0)|\} + \beta[d_S(\boldsymbol{x},f_0) - K_P]^2 \tag{10.3}$$

where the second term is a penalty function enforcing the required power split ratio, with β being the penalty factor controlling the contribution of the penalty term to the overall objective function. In this case, we have the following operating parameter vector $\boldsymbol{f}_t = [f_0\ K_P]^T$.

Another example is design of a dual-band coupler for a substrate characterized by a specific relative permittivity ε_r, so that the circuit minimizes both $|S_{11}|$ and $|S_{41}|$ responses at the two operating frequencies $f_{0.1}$ and $f_{0.2}$, while providing equal power split at these frequencies. Here, the operating parameter vector is $\boldsymbol{f}_t = [f_{0.1}\ f_{0.2}\ \varepsilon_r]^T$, and the objective function may be defined as

$$\begin{aligned} U(\boldsymbol{x},\boldsymbol{f}_t) = \ & \max\{|S_{11}(\boldsymbol{x},f_{0.1})|, |S_{41}(\boldsymbol{x},f_{0.1})|, |S_{11}(\boldsymbol{x},f_{0.2})|, |S_{41}(\boldsymbol{x},f_{0.2})|\} \\ & +\beta\left[d_S(\boldsymbol{x},f_{0.1})^2 + d_S(\boldsymbol{x},f_{0.2})^2\right] \end{aligned} \tag{10.4}$$

Other design scenarios can be treated in a similar manner. This includes cases where the operating bandwidth of the device is handled explicitly (e.g., bandwidth enhancement).

10.1.2 Response Features for Initial Design Quality Assessment

Appropriate tuning of geometry parameters is instrumental in achieving satisfactory performance of high-frequency components. For reliability reasons, it is normally conducted at the level of EM simulation models. In many cases, it has to be carried out in a global sense, for example, if the problem at hand is multimodal (i.e., multiple local optimal are present), or a sufficiently good initial design is not available. In practice, both issues may occur at the same time. Common examples include design of compact structures, where the introduction of various topological modifications leads to parameter redundancy, or redesign of a device at hand for operating frequencies or substrate considerably different from those at the current design. The primary challenge of globalized search is to conduct exploration of the entire parameter space, which is hindered by considerable nonlinearity and variability of system responses, both as a function of geometry parameters and frequency. Whereas dimensionality-related issues make a construction of globally accurate surrogates virtually impossible.

In the case of multiband antennas, the primary objective is to allocate the antenna resonances at the target operating frequencies. If the parameter space is large in terms of the considered parameter ranges, the resonances will be distributed outside the frequency range of simulation or allocated too far away from the target values for the vast majority of randomly picked variable vectors. In either case, local optimization is unable to yield a satisfactory design. This is illustrated in Fig. 10.1 for an exemplary dual-band dipole antenna.

This book is devoted to the employment of response features technology for expediting various design and modeling tasks. In the technique discussed in this section, the extracted information about the allocation of antenna resonances (both in

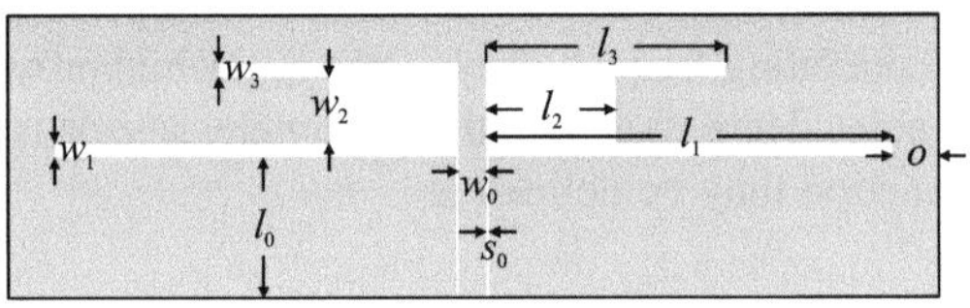

(a)

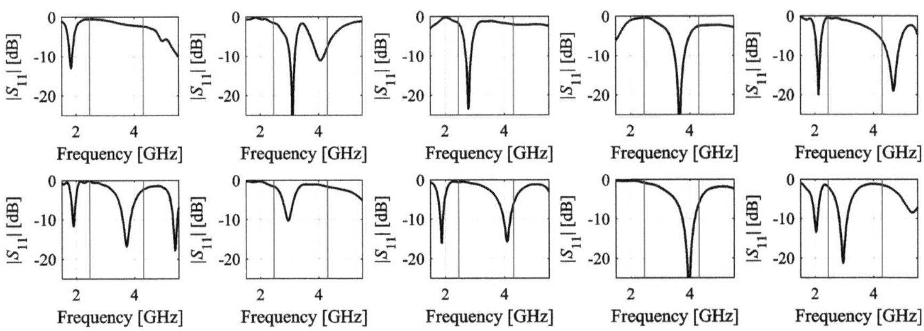

(b)

Fig. 10.1 Exemplary dual-band antenna and its reflection characteristics: (**a**) antenna geometry, (**b**) reflection responses at selected designs randomly assigned within the assumed parameter space. The vertical lines mark the target operating frequencies (here, 2.45 and 4.3 GHz). Initiating the local search, e.g., in the mini-max sense of reducing the reflection level at and around the target frequencies, from the majority of the shown designs would lead to a failure due to poor initial allocation of the antenna resonances terms of their frequency and level coordinates) is used to facilitate the global search procedures. As any other feature-based framework, the considered approach capitalizes on slightly nonlinear dependence between the antenna geometry parameters and the feature point coordinates (Koziel & Bandler, 2015a). An example of such a relationship has been shown in Fig. 10.2 for the dual-band antenna of Fig. 10.1.

For the considered dual-band antenna example, if the goal is to allocate the antenna resonances at the given target frequencies and to minimize the reflection level therein, the suitable response features are just the frequency and level coordinates of the resonances. If the objective is to maximize the fractional bandwidth around the target frequencies, a more convenient characteristics points would be the frequencies corresponding to -10 dB reflection level. Here, the feature points corresponding to antenna resonances will be used for design quality assessment in the first stage of the considered global search algorithm.

The response features may be also employed for assessing the design of microwave components such as coupler structure. In this case, a possible option is to select the response features corresponding to minimum of matching and isolation characteristics, as well as the power split ratio.

10.1.3 Global Search Using Inverse Surrogates

As mentioned in Sect. 10.1.2, extracting and processing characteristic points of the responses of antennas or microwave components is a powerful tool for evaluating the design quality, especially in terms of the misalignment between the actual and target operating frequencies. In the discussed technique, the weakly nonlinear relationships between the operating frequencies and geometry parameters (as illustrated in Fig. 10.2 for the dual-band antenna example) are exploited.

A similar concept was a foundation of performance-driven modeling frameworks (Koziel & Pietrenko-Dabrowska, 2019a, 2020a). Therein, the domain of the surrogate was established by approximating the optimum design manifold using an auxiliary inverse model constructed using a set of reference designs optimized for selected vectors of target operating conditions (e.g., operating frequencies in the case of multiband antennas). This allowed for restricting the volume of the domain and a rendition of accurate surrogates using very limited number of training data samples (Koziel & Pietrenko-Dabrowska, 2019a). In Koziel and Pietrenko-Dabrowska (2019a), the inverse model was constructed using kriging interpolation (Queipo et al., 2005). The employment of interpolative models was justified by the fact that the reference points were allocated on the aforementioned optimum design manifold. This is clearly convenient but not directly applicable here, because pre-optimized designs are not available in a global optimization context.

The approach utilized in the considered methodology is loosely related to performance-driven modeling concepts, but it does not assume availability of any pre-optimized points. Instead, it only uses a set of random observables generated over the antenna parameter space, which may or may not be of good quality. The

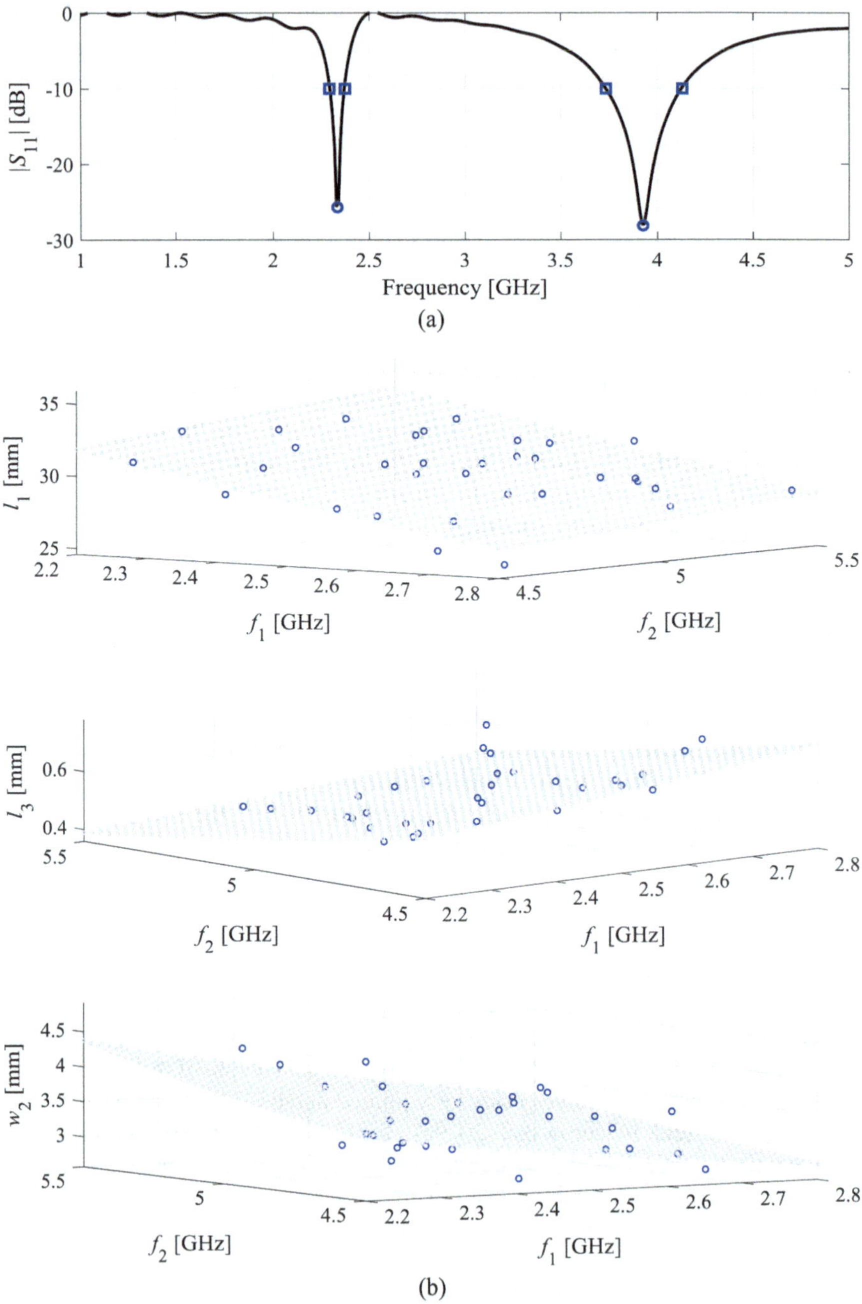

Fig. 10.2 Dual-band dipole antenna: (**a**) response features corresponding to antenna resonances (○) and −10 dB reflection levels (□); note that some of the feature points may not exist depending on the design (e.g., one of the resonances being outside the simulation frequency range); (**b**) relationship between the operating frequencies and the three selected geometry parameters. The circles represent specific antenna designs, whereas the gray points denote the regression model of

quality assessment is realized using the problem-specific knowledge in the form of characteristic points of antenna reflection characteristics outlined in Sect. 10.1.2. Selected observables are used for a construction of a regression-based inverse surrogate, which is subsequently employed to generate the infill points in the most promising regions of the parameter space. The decision-making procedure is iterated until a sufficiently good initial design is found for further local tuning. In each iteration, the current set of observables is refined by replacing the worst designs by those that are of higher quality with respect to the assumed design objectives. The remaining part of this subsection provides a general outline of the method, followed by a detailed pseudocode.

The following notation will be used throughout:

- $\boldsymbol{F}(\boldsymbol{x}) = [f_1(\boldsymbol{x}) \ldots f_K(\boldsymbol{x})]^T$—a vector of the operating parameters of a high-frequency component of interest at the design $\boldsymbol{x}$ (in the example of this section, we will assume that vector $\boldsymbol{F}(\boldsymbol{x})$ comprises frequency coordinates of the feature points; yet, other entities are also possible, e.g., center frequency, bandwidth, power split ratio); if some of the parameters f_ks cannot be extracted (e.g., some of the relevant feature points cannot be distinguished or are allocated outside the frequency range of simulation), a zero vector $\boldsymbol{F}(\boldsymbol{x}) = [0 \ldots 0]^T$ is assigned;
- $\boldsymbol{L}(\boldsymbol{x}) = [l_1(\boldsymbol{x}) \ldots l_K(\boldsymbol{x})]^T$—a vector of auxiliary coefficients reflecting the design quality and corresponding to the entries of the vector $\boldsymbol{F}(\boldsymbol{x})$. In the case of the antennas, the entries of the vector $\boldsymbol{L}(\boldsymbol{x})$ may include antenna reflection levels at the operating frequencies at the design $\boldsymbol{x}$ (i.e., level coordinates of the feature points). If a microwave coupler is of interest, with the objective to reduce the level of $|S_{11}|$ and $|S_{41}|$ at the operating frequency, the corresponding l_k can be the average of $|S_{11}|$ and $|S_{41}|$ at their respective minima: the lower value indicates that the design $\boldsymbol{x}$ is of higher quality. Similarly, if the operating parameter is a power split ratio, the corresponding l_k might be a deviation from the estimated power split and its target value. If, for any reason some of l_ks cannot be extracted, a zero vector $\boldsymbol{L}(\boldsymbol{x}) = [0 \ldots 0]^T$ is assigned;
- $D(\boldsymbol{F}, \boldsymbol{f}_t)$—a distance function determining the misalignment between the target vector of operating parameters $\boldsymbol{f}_t$ (cf. Sect. 10.1.2) and a target operating parameter vector $\boldsymbol{F}$; a simple example is an Euclidean norm $D(\boldsymbol{F},\boldsymbol{f}_t) = ||\boldsymbol{F} - \boldsymbol{f}_t||$, but here we use the infinity norm, i.e., $D(\boldsymbol{F}, \boldsymbol{f}_t) = ||\boldsymbol{F} - \boldsymbol{f}_t||_\infty$ to account for the worst-case scenario;
- D_{accept}—user-defined threshold distance to determine whether the best design obtained so far with the corresponding operating parameter vector $\boldsymbol{F}$ is sufficiently close to the target, i.e., if $D(\boldsymbol{F}, \boldsymbol{f}_t) \le D_{\text{accept}}$.

The global search procedure employs the following steps (a rigorous description is provided later on):

Fig. 10.2 (continued) the form $a_0 + a_1 \exp(a_2 f_1 + a_3 f_2)$, representing the trends between the operating frequencies and antenna dimensions. It should be noted that the trends are smooth and weakly nonlinear

1. Generate a set of random vectors $\boldsymbol{x}^{(j)}, j = 1, \ldots, N$, over the assumed parameter space X (typically, determined by the lower and upper bounds for design variables), such that $||\boldsymbol{F}(\boldsymbol{x}^{(j)})|| > 0$ for all j. The set is found by sequentially generating random observables until N vectors satisfying the above condition have been found.
2. Construct an inverse regression model $r_I(\boldsymbol{F})$ with the values in X, which represents the relationship between the operating and geometry parameters; the model is established using the feature vectors $\boldsymbol{F}(\boldsymbol{x}^{(j)})$ and $\boldsymbol{L}(\boldsymbol{x}^{(j)})$, as well as the corresponding parameter vectors $\boldsymbol{x}^{(j)}$.
3. Use r_I as a predictor to find a candidate design $\boldsymbol{x}_{\text{tmp}} = r_I(\boldsymbol{f}_t)$, where $\boldsymbol{f}_t$ is the vector of target operating parameters. If $||\boldsymbol{F}(\boldsymbol{x}_{\text{tmp}})|| > 0$ and $D(\boldsymbol{F}(\boldsymbol{x}_{\text{tmp}}), \boldsymbol{f}_t) < \max\{j = 1, \ldots, N : D(\boldsymbol{F}(\boldsymbol{x}^{(j)}), \boldsymbol{f}_t)\}$, replace the vector ensuring the above maximum by $\boldsymbol{x}_{\text{tmp}}$.

Steps 2 and 3 are iterated until identifying the candidate satisfying $D(\boldsymbol{F}(\boldsymbol{x}_{\text{tmp}}), \boldsymbol{f}_t) < D_{\max}$ (a user-defined acceptance threshold), upon which the local refinement is executed. In plain words, the procedure generates random observables until a sufficient number of designs have been found for which clearly defined feature points (e.g., antenna resonances) can be extracted. Subsequently, the inverse model (similar to what has been visualized in Fig. 10.2b) is constructed and used to predict a location of the design for which the operating parameters are possibly close to the target $\boldsymbol{f}_t$. This design replaces one of the base designs $\boldsymbol{x}^{(j)}$ assuming it is of sufficient quality in terms of the metric $D(\boldsymbol{F}, \boldsymbol{f}_t)$. Over the iterations, the inverse model will be gradually more and more focused in the vicinity of the parameter space region that contains the designs featuring low values of $D(\boldsymbol{F}, \boldsymbol{f}_t)$. Concentration of these base designs would also lead to the improved local accuracy of the inverse model. Figure 10.3 shows generation of random observables, whereas the entire optimization procedure is presented in Fig. 10.4.

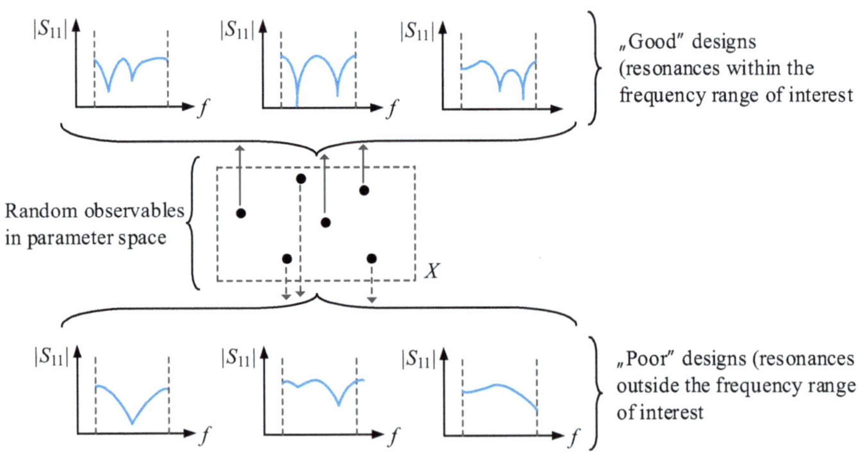

Fig. 10.3 Graphical illustration of the generation of random observables: the decision-making process allows for selecting only the parameter vectors for which the corresponding antenna responses have resonances within the simulation frequency range, which are further utilized for inverse model construction

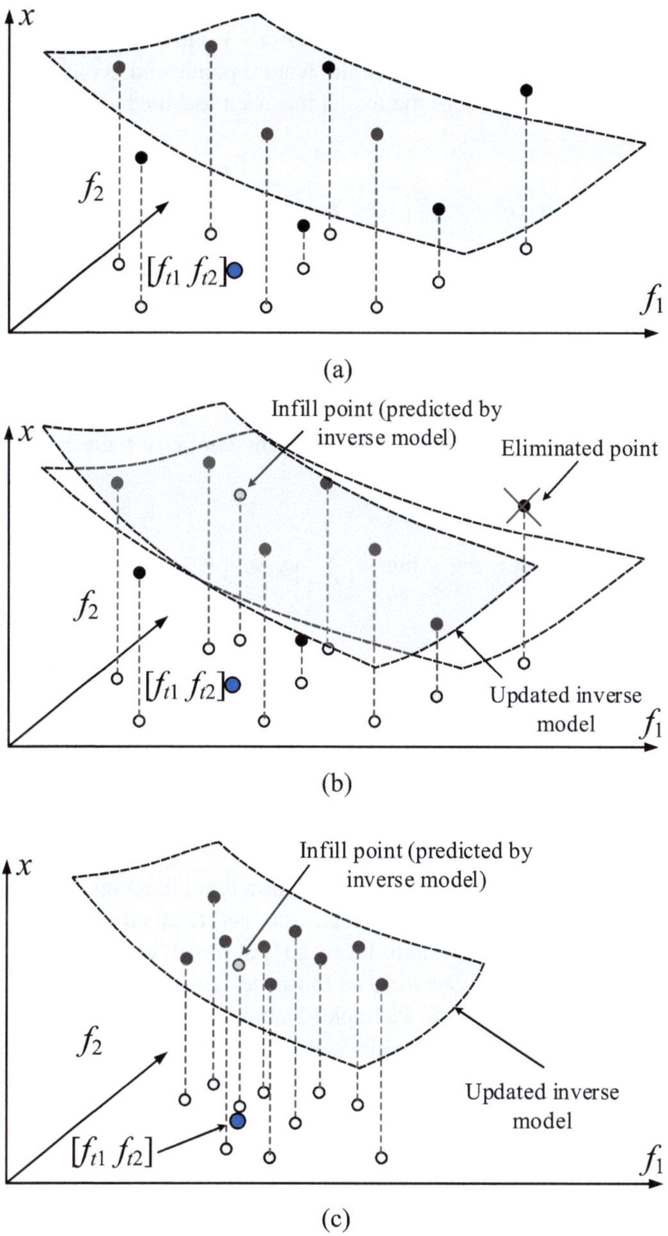

Fig. 10.4 Graphical illustration of the basic components of the discussed knowledge-based global optimization procedure: (**a**) observables (•) in the two-dimensional operating frequency space f_1, f_2 for a selected geometry parameter x along with their projections onto the f_1–f_2 plane, and the initial inverse model (gray surface); target operating frequencies marked using blue circle; (**b**) the first iteration of the global search process: the infill point predicted by the inverse model (marked using gray circle) replaces the worst observable and the inverse model is updated; the procedure is then repeated; (**d**) last iteration: the observables are concentrated near the target frequencies and the updated inverse model yields the design which is sufficiently close to the target; the procedure is

As mentioned earlier, a critical component of the procedure is the inverse model $r_I(\boldsymbol{F})$ established using the feature vectors $\boldsymbol{F}(\boldsymbol{x}^{(j)})$ and $\boldsymbol{L}(\boldsymbol{x}^{(j)})$. Due to typically weakly nonlinear dependence between the feature points and geometry parameters, the following analytical form of the model has been assumed:

$$\boldsymbol{r}_I(\boldsymbol{F}) = \boldsymbol{r}_I\left([f_1 \ \ldots \ f_K]^T\right) = \begin{bmatrix} r_{I.1}(\boldsymbol{F}) \\ \cdots \\ r_{I.n}(\boldsymbol{F}) \end{bmatrix}$$
$$= \begin{bmatrix} p_{1.0} + p_{1.1}\exp\left(\sum_{k=1}^{K} p_{1.k+1} f_k\right) \\ \cdots \\ p_{n.0} + p_{n.1}\exp\left(\sum_{k=1}^{K} p_{n.k+1} f_k\right) \end{bmatrix} \tag{10.5}$$

The model coefficients are obtained by solving the nonlinear regression problems of the form

$$[p_{j.0}\ p_{j.1}\ \cdots\ p_{j.K+1}] = \arg \min_{[b_0 \ \ldots \ b_{K+1}]} \sum_{k=1}^{N} w_k \left[r_{I.j}\left(\boldsymbol{F}\left(\boldsymbol{x}^{(k)}\right)\right) - x_j^{(k)} \right]^2, j = 1, \ldots, n \tag{10.6}$$

where $x_j^{(k)}$ is the jth component of the parameter vector $\boldsymbol{x}^{(j)}$, whereas the weighting factors w_k are related to the level coordinates of the feature points as

$$w_k = \left[1 - \max\left\{l_1\left(\boldsymbol{x}^{(j)}\right), \ldots, l_k\left(\boldsymbol{x}^{(j)}\right)\right\}\right]^2, k = 1, \ldots, N \tag{10.7}$$

In other words, the inverse model is set up as a trend function approximating the observables $\boldsymbol{x}^{(j)}$ in the sense of (10.2), i.e., weighted L-square.

The weighting factors are introduced to put more emphasis on the points corresponding to deeper resonances, as those designs are closer to the optimum design manifold (cf. Koziel & Pietrenko-Dabrowska, 2019a). Examples of the surfaces representing the inverse model can be found in Fig. 10.2b.

Now we are in a position to outline a rigorous formulation of the global search procedure outline above. It works as follows:

1. Set $j = 1$;
2. Generate a random vector $\boldsymbol{x}^{(j)} \in X$;
3. **if** $\|\boldsymbol{F}(\boldsymbol{x}^{(j)})\| > 0$

   Accept $\boldsymbol{x}(j)$; set $j = j + 1$;

Fig. 10.4 (continued) terminated and followed by local refinement (Sect. 10.1.4). Note that the presented pictures refer to a single geometry parameter x, in practice the same correction–prediction scheme is applied to all relevant geometry parameters simultaneously

end

4. **if** j ≤ N AND computational budget has not been exceeded

   Go to 2;

   else

   Go to 5;

   end
5. Construct the inverse regression model $r_I(\boldsymbol{F})$ as in (10.5)–(10.7);
6. Generate a candidate design $\boldsymbol{x}_{\text{tmp}} = r_I(\boldsymbol{f}_t)$, where $\boldsymbol{f}_t$ is the vector of target parameters frequencies (cf. Sect. 10.1.1);
7. **if** $\|\boldsymbol{F}(\boldsymbol{x}_{\text{tmp}})\| > 0$ AND $D(\boldsymbol{F}(\boldsymbol{x}_{\text{tmp}}),\boldsymbol{f}_t) < D_{\max} = \max\{j = 1, \ldots, N : D(\boldsymbol{F}(\boldsymbol{x}^{(j)}),\boldsymbol{f}_t)\}$

   Replace the vector realizing $D_{\max}$ in $\{\boldsymbol{x}^{(j)}\}_{j=1,\ldots,N}$ by $\boldsymbol{x}_{\text{tmp}}$;

   else

   Generate random observables $\boldsymbol{x}_{\text{tmp}}$ until satisfying $D(\boldsymbol{F}(\boldsymbol{x}_{\text{tmp}}), \boldsymbol{f}_t) < D_{\max}$, in which case the vector realizing $D_{\max}$ in $\{\boldsymbol{x}^{(j)}\}_{j=1,\ldots,N}$ is replaced by $\boldsymbol{x}_{\text{tmp}}$; in case of exceeding the computational budget go to 9;

   end
8. Update the inverse regression model $r_I(\boldsymbol{F})$ using current set $\{\boldsymbol{x}^{(j)}\}_{j=1,\ldots,N}$;
9. Find $\boldsymbol{x}^{(0)} = \boldsymbol{x}^{(j\text{min})}$, where $j_{\min} = \text{argmin}\{j = 1, \ldots, N : D(\boldsymbol{F}(\boldsymbol{x}^{(j)}), \boldsymbol{f}_t)\}$;
10. **if** $D(\boldsymbol{F}(x^{(0)}), \boldsymbol{f}_t) \le D_{\text{accept}}$ OR computational budget has been exceeded

    Go to 11;

    else

    Go to 6;

    end
11. Return $\mathbf{x}^{(0)}$; END.

The first four steps of the procedure describe the initial sampling that leads to identification of N designs featuring clearly defined characteristics points. These vectors are used to construct the first inverse surrogate r_I (Step 5). In Steps 6 through 10, the inverse model is employed as a predictor tool to yield the candidate design $\boldsymbol{x}_{\text{tmp}}$, at which the antenna is intended to have its resonances allocated as close to the target values $\boldsymbol{f}_t$ as possible. The candidate design—provided it is of sufficient quality—replaces the worst observable vector, which is followed by rebuilding the inverse model. The process continues until a sufficiently good design $\boldsymbol{x}^{(0)}$ has been found (i.e., such that $D(\boldsymbol{F}(\boldsymbol{x}^{(0)}),\boldsymbol{f}_t) \le D_{\text{accept}}$) or the computational budget is exceeded (in which case the best design found so far is returned as $\boldsymbol{x}^{(0)}$). The design $\boldsymbol{x}^{(0)}$ will undergo local tuning, as described in Sect. 10.1.4.

10.1.4 Local Refinement by Trust-Region Gradient Search

The acceptance threshold D_{accept} in Sect. 10.1.3 is set to ensure that the operating parameters of the design satisfying the condition $D(\boldsymbol{F}(\boldsymbol{x}^{(0)}), \boldsymbol{f}_t) \leq D_{\text{accept}}$ are sufficiently close to $\boldsymbol{f}_t$ so that the target values are attainable from $\boldsymbol{x}^{(0)}$ through local search. In the discussed procedure, trust-region (TR) gradient-based optimizer with numerical derivatives is employed as the optimization engine (Conn et al., 2000). The TR procedure yields a series of approximations to $\boldsymbol{x}^*$, denoted as $\boldsymbol{x}^{(i)}$, $i = 0, 1, \ldots$, by solving

$$\boldsymbol{x}^{(i+1)} = \arg \min_{\boldsymbol{x};\ -\boldsymbol{d}^{(i)} \leq \boldsymbol{x} - \boldsymbol{x}^{(i)} \leq \boldsymbol{d}^{(i)}} U_L(\boldsymbol{x}, \boldsymbol{f}_t) \tag{10.8}$$

where U_L is the objective function of the same form as the original function U (cf. 10.1), but obtained using the first-order Taylor expansion model $\boldsymbol{G}^{(i)}(\boldsymbol{x}, f)$ of antenna responses (rather than directly from EM-simulated responses $S_{11}(\boldsymbol{x}, f)$). The linear model is defined as

$$\boldsymbol{G}^{(i)}(\boldsymbol{x}, f) = S_{11}\left(\boldsymbol{x}^{(i)}, f\right) + \nabla_S\left(\boldsymbol{x}^{(i)}, f\right) \cdot \left(\boldsymbol{x} - \boldsymbol{x}^{(i)}\right) \tag{10.9}$$

with the gradients estimated using finite differentiation. Note that the sub-problem (10.8) is solved within the interval $\boldsymbol{x}^{(i)} - \boldsymbol{d}^{(i)} \leq \boldsymbol{x} \leq \boldsymbol{x}^{(i)} + \boldsymbol{d}^{(i)}$, referred to as the trust region (here, the inequalities are understood component-wise). The size vector $\boldsymbol{d}^{(i)}$ is adjusted using the standard rules (Conn et al., 2000). If the iteration is successful, i.e., $U(\boldsymbol{x}^{(i+1)}, \boldsymbol{f}_t) < U(\boldsymbol{x}^{(i)}, \boldsymbol{f}_t)$, the design yielded by (10.8) is accepted and the next iteration begins. The computational cost of updating the linear model is $n + 1$ EM antenna simulation.

The termination condition is convergence in argument $||\boldsymbol{x}^{(i+1)} - \boldsymbol{x}^{(i)}|| < \varepsilon$, or shrinking the trust region $||\boldsymbol{d}^{(i)}|| < \varepsilon$ (whichever occurs first). Here, we assume $\varepsilon = 10^{-3}$. The local tuning can be sped up by means of sparse sensitivity updates (e.g., Koziel & Pietrenko-Dabrowska, 2020b). In the considered approach, a very simple mechanism is employed, where the finite differentiation is replaced by the Broyden updating formula (Broyden, 1965) when the optimization process is sufficiently close to convergence, or $||\boldsymbol{x}^{(i+1)} - \boldsymbol{x}^{(i)}|| < 10\varepsilon$.

10.1.5 Optimization Framework

This section briefly summarizes the entire optimization framework considered here. The two major stages thereof are the global search process of Sect. 10.1.3, and local refinement of Sect. 10.1.4. The control parameters of the algorithm are juxtaposed again below for the convenience of the reader:

- N—the number of observables for inverse model construction;
- $N_{max.1}$—computational budget: maximum number of EM antenna evaluations for initial sampling;
- $N_{max.2}$—computational budget: maximum number of EM antenna evaluations for global search stage;
- $N_{max.3}$—computational budget: maximum number of EM antenna evaluations for local tuning stage;
- D_{accept}—the threshold for accepting designs produced by the global search stage (cf. Sect. 10.1.3);
- ε—termination threshold (for convergence in argument and trust-region size, cf. Sect. 10.1.4).

It should be noted that the algorithm only has two important control parameters, which are N and D_{accept}. The remaining parameters are standard. In particular, the termination threshold is set to determine the required resolution of the optimization process, whereas the computational budget numbers are typically set up with some margin (in practice, the termination of both global and local search stages is due to convergence rather than exceeding the budget). Typically, the number N of samples for inverse model construction is normally kept small, say 10 or 20, which is because the inverse model is set up over low-dimensional operating frequency space, and the number of model coefficients is also low (specifically, $K + 2$). Selection of D_{accept} is somehow problem dependent: in order to ensure that the target parameters are attainable from $\boldsymbol{x}^{(0)}$ using local search, D_{accept} should be set so that the distances between the targets and the actual operating parameters at the initial design are not larger than the corresponding impedance bandwidths.

The operating flow of the entire algorithm can be described as follows:

1. Input arguments:
   - Target operating frequencies $\boldsymbol{f}_t$,
   - Objective function U,
   - Parameter space X;
2. Produce initial design $\boldsymbol{x}^{(0)}$ by performing global search (Sect. 10.1.3);
3. Obtain final design $\boldsymbol{x}^*$ through local tuning (Sect. 10.1.4), using $\boldsymbol{x}^{(0)}$ as the starting point.

Figure 10.5 shows the flow diagram of the process. Therein, the global search part has been broken down into several steps, according to the description provided in Sect. 10.1.3. The local tuning is represented as a single block.

At this point, it should also be emphasized that the presented approach may be utilized for optimization of multiband antennas in the sense of the objective function (10.2), i.e., matching improvement at the center frequencies. It can also be employed to improve matching over the bandwidths centered at the target frequencies, or even bandwidth enhancement. Other possible design tasks include minimization of coupler matching and isolation characteristics at the required frequency, and maintaining the assumed power split ratio (as in 10.3). Yet another possibility is to

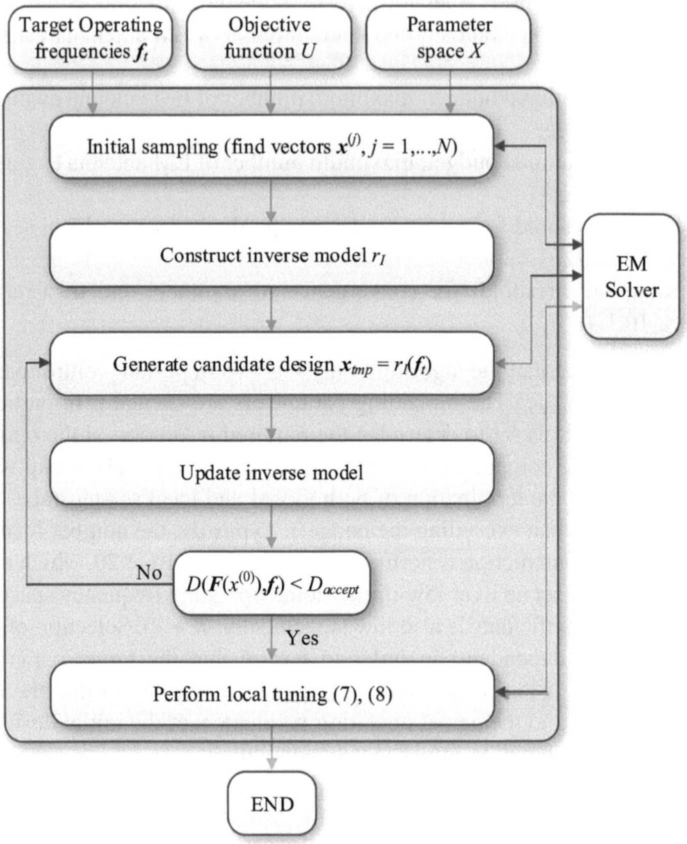

Fig. 10.5 Flow diagram of the considered framework for globalized optimization of high-frequency components

design a coupler for a given substrate, so that matching and isolation are minimized at its operating frequencies, while providing target power split at these frequencies (see 10.4). All these design scenarios would require a different definition of the objective function at the local tuning stage; however, the global search part would remain essentially intact.

10.1.6 Demonstration Case Study I: Triple-Band Antenna

This section provides numerical verification of the knowledge-based optimization framework of Sects. 10.1, 10.2, 10.3 and 10.4 with the use of a triple-band antenna (Koziel & Pietrenko-Dabrowska, 2021a). The results are benchmarked against multiple-start local optimization as well as a representative nature-inspired algorithm

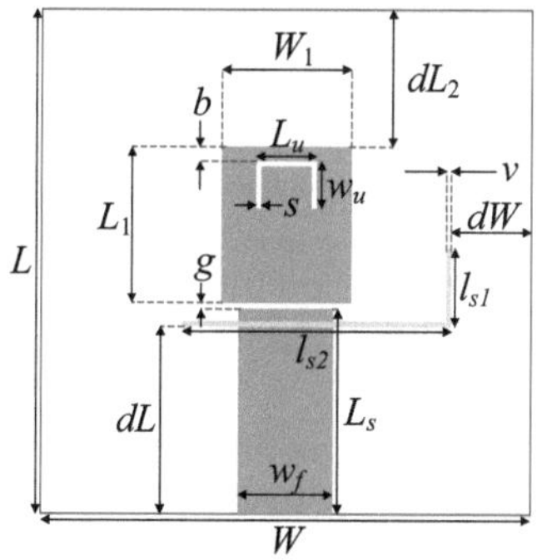

Fig. 10.6 Verification antenna structure: triple-band antenna with L-slot DGS (Consul, 2015); the light-shade gray denotes a ground-plane slot

(here, the particle swarm optimizer). The major factors considered in the analysis include reliability of the optimization process, quality of the final design, as well as the computational cost.

Figure 10.6 shows the triple band U-slotted patch with L-slot DGS (Consul, 2015), implemented on 3.064-mm-thick substrate of relative permittivity equal to 3.2. The independent geometry parameters are $\boldsymbol{x} = [L_1\ L_s\ L_{ur}\ W\ W_1\ dL_r\ dW_r\ g\ l_{s1r}\ l_{s2r}\ w_{ur}]^T$; the following dimensions are fixed: $b = 1$, $w_f = 7.4$, $s = 0.5$, $w = 0.5$, $dL_2 = L_1$; we also have additional relationships: $L = L_s + g + L_1 + dL_2$, $L_u = L_{ur}W_1$, $dL = dL_rL$, $dW = dW_rW$, $l_{s1} = l_{s1r}(L - dL)$, $l_{s2} = l_{s2r}(W - dW)$, $w_u = w_{ur}(L_1 - b - s)$. The computational model is implemented in CST Microwave Studio, and simulated using its time-domain solver.

The design tasks are formulated as in Sect. 10.1.1 with the objective function in the form of (10.2). The target operating frequencies for the antenna under design are: $f_t = [3.5\ 5.8\ 7.5]^T$ (all in GHz). The parameter space X is delimited by the following lower and upper bound on design variables: $\boldsymbol{l} = [10\ 17\ 0.2\ 45\ 5\ 0.4\ 0.15\ 0.2\ 0.1\ 0.5\ 0.1]^T$ and $\boldsymbol{u} = [16\ 25\ 0.6\ 55\ 15\ 0.5\ 0.3\ 0.8\ 0.4\ 0.65\ 0.5]^T$, respectively. It should be noticed that the parameter space dimensionality is relatively large (11 geometry parameters).

The optimization process is executed 10 times using the following algorithms:

- The considered inverse-surrogate-based framework of Sect. 10.1.2. The control parameters are set as follows (the same for all cases): $N = 10$, $N_{\max.1} = 100$, $N_{\max.2} = 100$, $N_{\max.3} = 500$, $\varepsilon = 10^{-3}$, $D_{\text{accept}} = 0.2$;
- Particle swarm optimizer (PSO) (Kennedy & Eberhart, 2001) using the swarm size of 10, with maximum number of iterations 50 (Version I) and 100 (Version II), and the standard control parameter setup ($\chi = 0.73$, $c_1 = c_2 = 2.05$);
- Trust-region gradient-based optimization (Conn et al., 2000) using the algorithm of Sect. 10.1.4; each run is executed using a random starting point.

PSO is employed as a representative nature-inspired global optimization routine, whereas local optimization is included in order to demonstrate the need for global optimization for the considered test problems. The optimization results have been shown in Table 10.1, and Fig. 10.7 shows—for the selected algorithm runs—the antenna responses at the final designs obtained using the discussed framework.

Table 10.1 Optimization results for triple-band antenna

Optimization method	Algorithm with inverse surrogates	PSO 50 iterations	PSO 100 iterations	Trust-region gradient-based algorithm
Average objective function value (dB)	−20.2	−12.3	−14.2	−12.1
Computational cost[a]	142.6	500	1000	125.4
Success rate[b]	10/10	6/10	8/10	4/10

[a]The cost expressed in terms of the number of EM simulations of the antenna structure under design
[b]Number of algorithms runs at which the operating frequencies were allocated with sufficient accuracy, i.e., to satisfy the condition $D(\boldsymbol{F}(\boldsymbol{x}^*), f_t) \le D_{\text{accept}}$

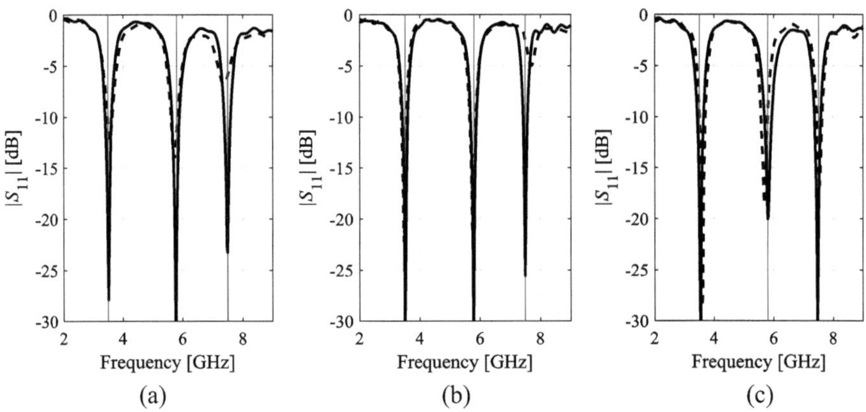

Fig. 10.7 Antenna responses at the designs obtained using the considered global optimization framework for selected algorithm runs: (**a**) run 1, (**b**) run 2, (**c**) run 3. Dashed line shows the initial design $\boldsymbol{x}(0)$ obtained after the global search stage, and solid line represents the antenna response at the final design. Vertical lines mark the target operating frequencies, here 3.5, 5.8, and 7.5 GHz

The results provided in Table 10.1 demonstrate truly global optimization capabilities of the considered inverse-modeling-based procedure. In the analysis below we focus on the three aspects of the optimization process:

- *Reliability*: The discussed algorithm consistently finds satisfactory design (i.e., appropriately allocates the antenna operating frequencies) at all of its runs. A comparison with local search indicates that the design task is indeed multimodal and inadequate choice of the initial design results in poor performance (satisfactory designs found in less than 50% of the cases). Reliability of PSO is noticeably better, although it is not as good as for the considered approach. Also, restricting the computational budget to 500 antenna simulations has a detrimental effect on the reliability, which suggests that 1000 objective function calls is about a minimum number required to maintain practical result repeatability.

- *Design quality*: The discussed algorithm clearly ensures the best design quality as measured by the average objective function value. It is considerably better than both the local routine and PSO (in both 50 and 100 iteration versions).
- *Computational cost*: The computational efficiency of the considered method is considerably better than PSO. What is more important, the average cost of our procedure is comparable to that of the gradient search. The average expenses are only around 15% higher, which is negligible having in mind that this small extra cost enables global search capabilities. It should also be mentioned that the average cost of the global search stage equals 36 antenna simulations, which is about 25% of the overall costs, whereas the number of iterations of the local search is equal to 9. This indicates that a combination of response features and inverse surrogates constitutes a quite powerful tool for exploiting the dependence between the antenna operating frequencies and its geometry parameters at minimum computational budget.

10.1.7 Demonstration Case Study II: Rat-Race Coupler

Numerical verification of the discussed global optimization procedure has been also carried out using a compact microstrip rat-race coupler (RRC) (Pietrenko-Dabrowska & Koziel, 2021) shown in Fig. 10.8, implemented on RO4003 substrate ($\varepsilon_r = 3.38$, $h = 0.762$ mm). The adjustable geometry parameters are $\boldsymbol{x} = [l_1\ l_2\ l_3\ d\ w\ w_1]^T$; the remaining parameters are $d_1 = d + |w - w_1|$, $d = 1.0$, $w_0 = 1.7$, and $l_0 = 15$ fixed (dimensions in mm). For this circuit, the goal is to minimize the matching and isolation characteristics, $|S_{11}|$ and $|S_{41}|$, at the intended operating frequency f_0, while ensuring a required power split ratio K_P. The objective function is defined as in (10.3). The computational model is implemented in CST Microwave Studio, and simulated using the time-domain solver. Table 10.2 provides information about the specific design tasks considered (i.e., the target values of the operating parameters), as well as the lower and upper bounds for parameters. It should be noted that parameter ranges are very wide with the average ratio of the upper-to-lower bound equal to 11.7. The reason for selecting such broad ranges of geometry parameters was to ensure that the considered design tasks are sufficiently challenging, as well as to emulate the scenario under which the designer does not have a clear indication of what a good initial design should be, thereby to shift most of the design decisions concerning the starting point allocation to the optimization algorithm itself, rather than to engage the expert knowledge.

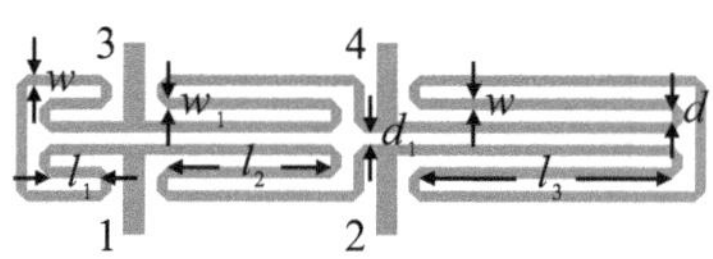

Fig. 10.8 Rat-race coupler (RRC) with folded transmission lines. (Koziel & Pietrenko-Dabrowska, 2019b)

Table 10.2 Target operating parameters and parameter spaces for RRC

Target operating parameters		
Symbols	Specific values	Parameter space X
$F_t = [f_0\ K_P]^T$	Case 1: $f_0 = 1.8$ GHz, $K_P = -3$ dB	$l = [0.5\ 5.0\ 5.0\ 0.2\ 0.2\ 0.2]^T$
	Case 2: $f_0 = 1.2$ GHz, $K_P = 0$ dB	$u = [15.0\ 30.0\ 50.0\ 2.0\ 2.0\ 2.0]^T$

[a]The circuit is to be optimized for a given substrate of permittivity ε_r

Table 10.3 Optimization results for rat-race coupler

Case	Optimization method	Algorithm with inverse surrogates	PSO 50 iterations	PSO 100 iterations	Trust-region gradient-based algorithm
1	Average objective function value (dB)	−36.7	−24.8	−34.0	−18.7
	Computational cost[a]	125.8	500	1000	102.8
	Success rate[b]	10/10	9/10	10/10	6/10
2	Average objective function value (dB)	−39.9	−23.7	−36.2	48.3
	Computational cost[a]	130.5	500	1000	68.7
	Success rate[b]	10/10	9/10	10/10	5/10

[a]The cost expressed in terms of the number of EM simulations of the coupler
[b]Number of algorithms runs at which the operating parameters were allocated with sufficient accuracy, i.e., to satisfy the condition $D(\boldsymbol{F}(\boldsymbol{x}^*), \boldsymbol{f}_t) \le D_{\text{accept}}$

The experimental setup for the discussed optimization algorithm and the benchmark methods is the same as in Sect. 10.1.6. The numerical results are provided in Table 10.3. The scattering parameters at the designs obtained using the considered framework for selected algorithm runs can be found in Fig. 10.9.

The breakdown of the computational cost of the global optimization algorithm is the following (averaging over 10 algorithm runs): average cost of global search stage was 40 EM simulations; average number of TR iterations was 10, with the average cost of each iteration of about 8.8 EM simulations (each iteration involved six extra analyses for Jacobian estimation). For the TR gradient-based algorithm (last column of Table 10.3), the breakdown of the cost is the following: Circuit I: average number of TR iterations was 11, with the average cost of each iteration of about eight EM simulations.

The obtained results allow us to draw several conclusions concerning the presented optimization strategy, not only in terms of its efficacy and the computational complexity but also how its performance compares to the benchmark methods. These are the main points:

- Similarly, as in Sect. 10.1.6, the considered approach does exhibit a global search capability, which is corroborated by the fact that satisfactory designs have been found in all runs of the algorithm (10 per case). Also, as illustrated in Fig. 10.8, the parameter vectors $\boldsymbol{x}(0)$ found by the global search stage are of good quality (in particular, with the operating frequency being close to the target), which

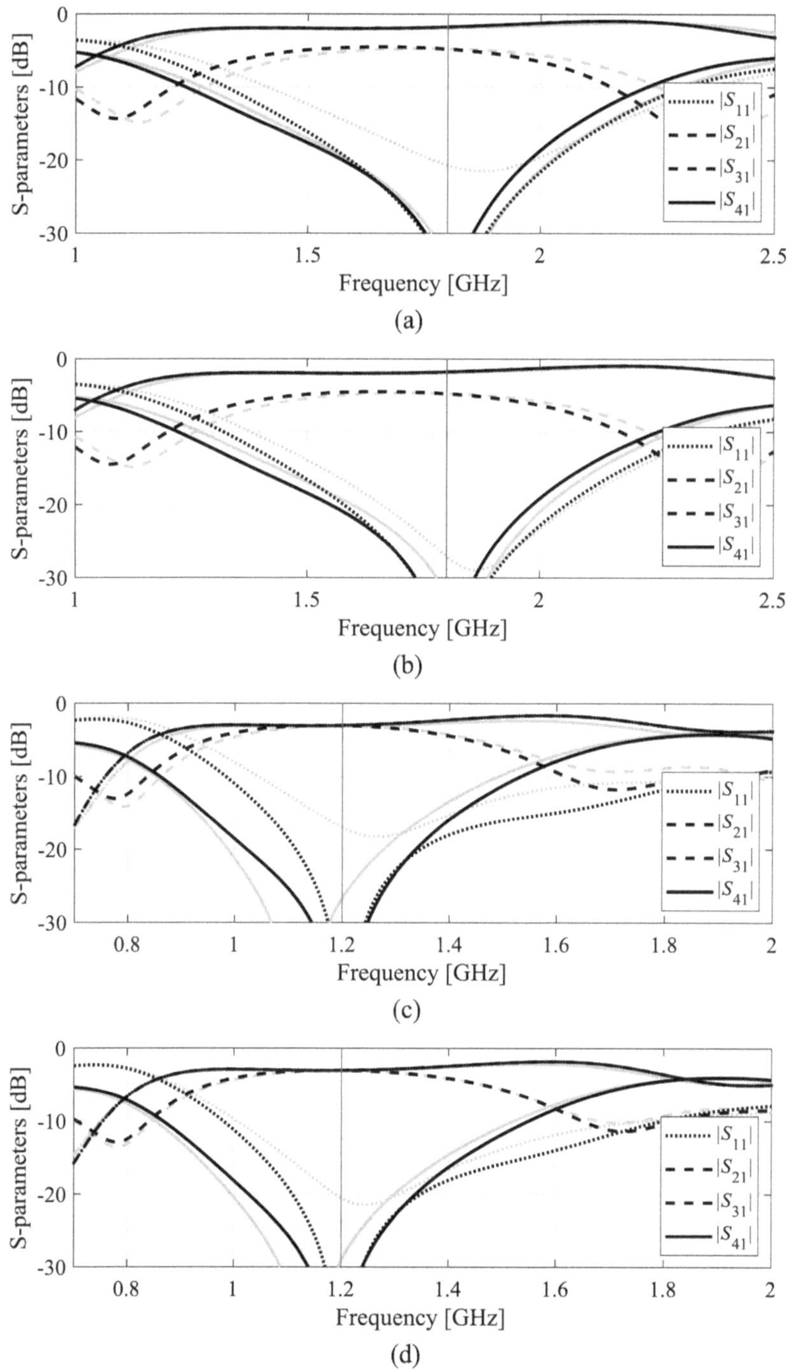

Fig. 10.9 Rat-race coupler: S-parameter characteristics at the optimized designs found by the discussed global optimization framework for two selected algorithm runs, Case 1: (**a**), (**b**) Designs 1 and 2, respectively, Case 2: (**c**), (**d**) Designs 1 and 2, respectively. Gray lines correspond to the initial design $\boldsymbol{x}^{(0)}$ obtained using the global search stage, black lines represent the responses at the final design. Vertical lines mark the target operating frequencies

makes local optimization sufficient. On the contrary, the benchmark local search from random initial designs often fails (at about 50% of the cases), and its performance is highly dependent on the initial design quality. Consequently, the average value of the optimized objective function is considerably worse than for the discussed method. Population-based metaheuristics (here, PSO) performs much better, but its computational complexity is high. It can also be observed that there is a noticeable difference in the design quality produced after 50 and 100 PSO iterations. This indicates that computational budget set at 500 EM simulations is insufficient for this method.

- The quality of the designs obtained using the considered technique surpasses that obtained using both local search and PSO. Although local optimization may produce a design of similar quality, it only happens when the initial points was sufficiently close to the target.
- In terms of computational efficiency, the discussed approach compares favorably with the population-based algorithm as the average running cost is only about 128 EM analyses of the circuit.
- The cost of the first (global) search stage is low, only 40 EM simulations on an average. This level of efficiency is a result of combining the response feature technology with inverse modeling, in particular, establishing the model over low-dimensional operating condition space.

As demonstrated in Sects. 10.1.6 and 10.1.7, the discussed technique offers both global search capability and computational efficiency. The aforementioned features of the introduced optimization procedure make it a potentially useful tool for handling global optimization tasks for antennas and microwave circuits. The repeatability of solutions produced by the procedure is remarkable while maintaining the computational efficiency comparable to that of gradient-based local optimizers. The considered procedure can be employed for solving optimization tasks, especially in the cases when the objective function is expensive to calculate. The sole limitation is that the high-frequency device under design featured suitably shaped responses with easily distinguishable characteristic points.

10.2 Global Optimization Using Simplex-Based Regression Surrogates

The technique of this section, similarly as the technique exploiting inverse surrogates delineated in Sect. 10.1, is also aimed at global parameter tuning of high-frequency structures. Yet, in the approach of this section, response features are utilized to facilitate the search process in a different manner. Specifically, simplex-based search is carried out at the level of operating and performance figures of the system at hand. This allows for construction of a fast predictor that generates subsequent candidate designs using an automated decision-making procedure that capitalizes on weakly-nonlinear dependence between the operating figures and antenna geometry parameters, as in any other response feature frameworks. An automated rendition of new design points is computationally efficient, and requires a single EM analysis per iteration. At the same time, the algorithm incorporates a mechanism for reducing the simplex size upon detecting the lack of objective function improvement, which guarantees a formal convergence of the optimization procedure. The final design is identified by gradient-based parameter tuning that follows the global search stage.

The section is organized as follows. Section 10.2.1 formulates the design problem as a constrained minimization task. Section 10.2.2 introduces simplex-based models, whereas Sect. 10.2.3 discusses their incorporation into a global optimization stage. The procedure for final parameter tuning is outlined in Sect. 10.2.4. Section 10.2.5 sums up the entire optimization flow. The results obtained using the considered framework for a microstrip antenna are presented in Sect. 10.2.6, whereas Sect. 10.2.7 provides the validation of the algorithm with the use of a power-divider.

10.2.1 Problem Formulation

Various formulations of optimization task of high-frequency components are conceivable, depending on the design goals, constraints, the number and type of characteristics to undergo adjustments, and so on. Here, we assume a generic scalar formulation, in which the optimal solution $\boldsymbol{x}^*$ is found as

$$\boldsymbol{x}^* = \arg\ \min_{\boldsymbol{x}} U(\boldsymbol{x}, \boldsymbol{f}_t) \tag{10.10}$$

where U is a merit function, and $\boldsymbol{f}_t = [f_{t.1}\ \ldots\ f_{t.K}]^T$ is a vector of the intended operating parameters. In the case of a K-band antenna, $\boldsymbol{f}_t$ will be a vector of the intended operating frequencies. For microwave circuits, the vector $\boldsymbol{f}_t$ may also include operating frequency (or frequencies), bandwidth, power split ratio (for coupling structures), but also material parameters, e.g., dielectric permittivity of the substrate the device is to be implemented on, whereas the vector $\boldsymbol{x} = [x_1\ \ldots\ x_n]^T$ denotes adjustable (usually geometry) parameters of the structure at hand. The problem (10.10) may be subject to inequality constraints $g_k(\boldsymbol{x}) \le 0$, $k = 1, \ldots, n_g$ and equality constraints $h_k(\boldsymbol{x}) = 0$, $k = 1, \ldots, n_h$.

Table 10.4 provides a few examples of typical antenna optimization tasks along with the corresponding objective functions, constraints, and target frequency vectors. The reason for distinguishing the operating frequencies is that appropriate allocation thereof is the major challenge in global optimization of antenna structures. The algorithm discussed in this section when applied to global optimization of antennas is largely based on enforcing required allocation of those frequencies, before proceeding to the local tuning phase.

Several examples of design optimization tasks for various microwave components are shown in Table 10.5. It can be noted that in most cases we have several objectives, some of which are handled directly, with others considered as design

Table 10.4 Examples of antenna design optimization scenarios

Design objectives	Target frequency vector	Primary objective function $U(\boldsymbol{x})$	Constraints
In-band matching improvement in the frequency range $F = [f_t - B/2, f_t + B/2]$ centered at the target frequency f_t	$f_t = f_t$	$U(\boldsymbol{x}) = S(\boldsymbol{x}) = \max\{f \in F: \lvert S_{11}(\boldsymbol{x}, f)\rvert\}$	–
Matching improvement at K target operating frequencies $f_{t.j}$, $j = 1, \ldots, K$	$f_t = [f_{t.1} \ldots f_{t.K}]^T$	$U(\boldsymbol{x}) = \max\{\boldsymbol{x} : \max\{\lvert S_{11}(\boldsymbol{x}, f_{t.1})\rvert, \ldots, \lvert S_{11}(\boldsymbol{x}, f_{t.K})\rvert\}\}$	$S(\boldsymbol{x}) \leq -10$ dB
Maximizing average in-band gain while ensuring reflection of −10 dB or better in the frequency range $F = [f_t - B/2, f_t + B/2]$, centered at the target frequency f_t	$f_t = f_t$	$U(\boldsymbol{x}) = -F^{-1}\int_F G(\boldsymbol{x}, f)\, df$	$S(\boldsymbol{x}) \leq -10$ dB

Explanation of terms:
$\lvert S_{11}(\boldsymbol{x}, f)\rvert$—antenna reflection at design $\boldsymbol{x}$ and frequency f
$G(\boldsymbol{x}, f)$—realized gain at design $\boldsymbol{x}$ and frequency f

Table 10.5 Examples of microwave design optimization scenarios

Design task	Target vector	Objective function[a]
Minimize maximum in-band reflection of an impedance matching transformer within operating band $F = [f_1\ f_2]$	$\boldsymbol{F}_t = [f_1\ f_2]^T$	$U(\boldsymbol{x}, \boldsymbol{F}_t) = U(\boldsymbol{x}, [f_1\ f_2]^T) = \max\{f \in [f_1\ f_2] : \lvert S_{11}(\boldsymbol{x}, f)\rvert\}$
Minimize matching and isolation of a microstrip coupler at the target operating frequency f_0 and ensure the required power split K_P at f_0	$\boldsymbol{F}_t = [f_0\ K_P]^T$	$U(\boldsymbol{x}, \boldsymbol{F}_t) = U(\boldsymbol{x}, [f_0\ K_P]^T) = \max\{\lvert S_{11}(\boldsymbol{x}, f_0)\rvert, \lvert S_{41}(\boldsymbol{x}, f_0)\rvert\} + \beta[\lvert S_{21}(\boldsymbol{x}, f_0)\rvert - \lvert S_{31}(\boldsymbol{x}, f_0)\rvert - K_P]^2$
Ensure equal power split of a triple-band power divider at the operating frequencies f_1 through f_3 and minimize input matching $\lvert S_{11}\rvert$, output matching $\lvert S_{22}\rvert$ and $\lvert S_{33}\rvert$, and isolation $\lvert S_{23}\rvert$ at f_1, f_2, and f_3	$\boldsymbol{F}_t = [f_1\ f_2\ f_3]^T$	$U(\boldsymbol{x}, \boldsymbol{F}_t) = U(\boldsymbol{x}, [f_1\ f_2\ f_3]^T) = \max\left\{\max_{k, l \in \{1,2,3\}} \lvert S_{kk}(\boldsymbol{x}, f_l)\rvert, \max_{l \in \{1,2,3\}} \lvert S_{23}(\boldsymbol{x}, f_l)\rvert\right\} + \beta \sum_{l=1}^{3} (\lvert S_{21}(\boldsymbol{x}, f_l)\rvert - \lvert S_{31}(\boldsymbol{x}, f_l)\rvert)^2$

[a]The terms containing β are penalty factors introduced to enforce the design constraints (e.g., required power split or matching/isolation bandwidth)

constraints enforced using a penalty function approach (Nocedal & Wright, 2000). This is not the only way of formulating the parameter adjustment problem, yet it is convenient due to rendering a scalar merit function. Other possibilities, not considered here, include objective aggregation (Marler & Arora, 2010) or genuine multi-objective design (Mirjalili & Dong, 2019; Mandal et al., 2018; Koziel & Pietrenko-Dabrowska, 2021b).

The treatment of constraints requires a separate note as it presents a challenge on its own. In particular, majority of constraints are expensive, i.e., their evaluation is based on EM analysis results. A convenient handling thereof is an implicit one, by means of a penalty function approach (Nocedal & Wright, 2000), which is assumed in the considered approach. More specifically, the original problem (10.10) is reformulated as (Koziel & Pietrenko-Dabrowska, 2022)

$$\boldsymbol{x}^* = \arg\min_{\boldsymbol{x}} U_P(\boldsymbol{x}) \tag{10.11}$$

with

$$U_P(\boldsymbol{x}) = U(\boldsymbol{x}) + \sum\nolimits_{k=1}^{n_g+n_h} \beta_k c_k(\boldsymbol{x}) \tag{10.12}$$

The penalty functions $c_k(\boldsymbol{x})$ quantify constraints' violations, whereas $\beta_k > 0$ are the penalty factors. It is convenient to define the functions $c_k(\boldsymbol{x})$ to measure relative constraint violations with respect to the assumed threshold, e.g., $c_k(\boldsymbol{x}) = [(S(\boldsymbol{x}) + 10)/10]^2$ for impedance matching (cf. Table 10.3 for the explanation of $S(\boldsymbol{x})$). The use of the second power $[\cdot]^2$ serves two purposes: (1) ensuring that U_P at the feasible region boundary is smooth, and (2) providing a leeway for small violations.

10.2.2 Simplex-Based Predictors

Attaining the best achievable performance of antenna or microwave structures requires meticulous and simultaneous tuning of all relevant geometry parameters, which is a challenging endeavor due to high computational expenses entailed by repetitive EM analyses involved in the process. As explained in Sect. 10.2.1, global search is required in many cases, either due to inherent multimodality of the problem or the lack of sufficiently good initial design. Global optimization requires exploring the parameter space in its entirety, which is impeded by nonlinearity of component responses and significant relocations of operating frequencies across the space.

To illustrate the issue, Fig. 10.10 presents responses of an exemplary dual-band antenna evaluated at a number of random parameter vectors. It is clear that local search initiated from the majority of these points would fail to place the antenna resonances at the intended values (marked red in the plots of Fig. 10.10). At the same time, rendering accurate surrogate for this sort of characteristics requires a large

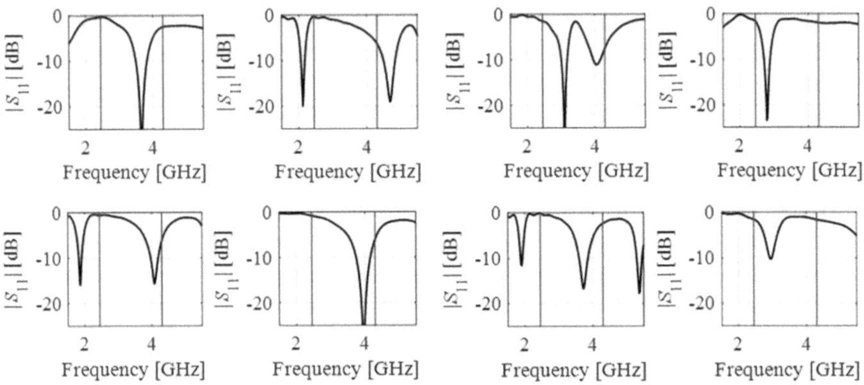

Fig. 10.10 Exemplary reflection characteristics of a dual-band antenna at random designs belonging to the assumed design space with the intended operating frequencies 2.45 and 4.3 GHz indicated using vertical lines. Local search initiated from over a half of the presented designs would fail as a result of a considerable misalignment of the target and existing antenna resonances

number of training data points, acquisition of which may be prohibitive in computational terms. A similar situation occurs for microwave devices.

The situation changes dramatically when the problem is considered from the perspective of operating parameters of the structure. Figure 10.11 shows the relationships between the resonant frequencies f_1 and f_2 of the antenna of Fig. 10.10a and its selected geometry parameters. The plots were obtained from a number of random trial points. Despite the fact that the parameter vectors are not optimized, clear patterns can be observed, along with well-pronounced correlations between the spaces of operating and geometry parameters. As demonstrated in the previous chapters, this sort of relationship is rather universal.

With Fig. 10.11 in mind, the problem of global optimization of high-frequency devices is approached from the standpoint of the relationships illustrated therein. Toward this end, problem-specific knowledge embedded in component responses has to be employed. Due to geometrical simplicity of the said relations, it is sufficient—for the purpose of globalized optimization—to construct relatively simple surrogate models (predictors), defined at the level of operating and performance figures rather than directly geometry parameters and device characteristics in their entirety. In order to account for all n directions within the parameter space, the model has to involve at least $n + 1$ distinct vectors $\boldsymbol{x}^{(j)}$. The simplest structure that enables this, assuming almost arbitrary point allocation, is a simplex.

Before proceeding further, let us introduce the notion of operating figure vectors $\boldsymbol{f} = [f_1 \ldots f_N]^T$ and performance figure vectors $\boldsymbol{l} = [l_1 \ldots l_M]^T$. The first will be used to denote the target quantities (e.g., center frequency for a narrow-band antenna, bandwidth, power division ratio for a coupling structure, or even material parameters, e.g., substrate permittivity the device is fabricated on). For a multiband antenna with target operating frequency vector $\boldsymbol{f}_t$ discussed earlier, the vector $\boldsymbol{f}$ would coincide with $\boldsymbol{f}_t$. The performance figure vector $\boldsymbol{l}$ contains quantities used to

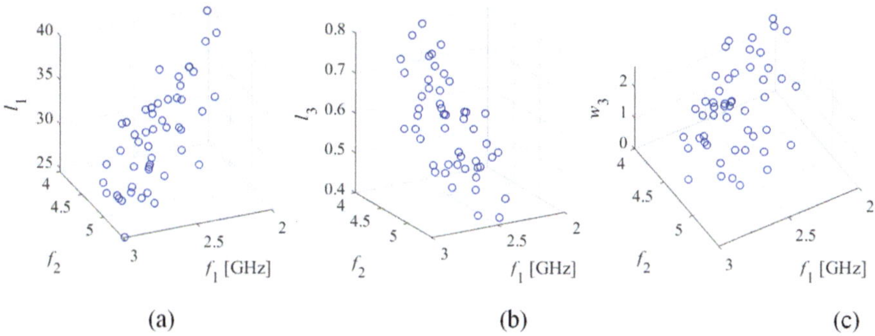

Fig. 10.11 Relationship between operating parameters (here, resonant frequencies f_1 and f_2) and selected geometry parameters of the antenna of Fig. 10.10a obtained for a number of random trial points. To create the plots, only those points were chosen, for which the corresponding characteristics show clearly visible resonances allocated within the ranges of interest, as shown in axis description. Clear patterns are visible even though the trial points were not optimized whatsoever

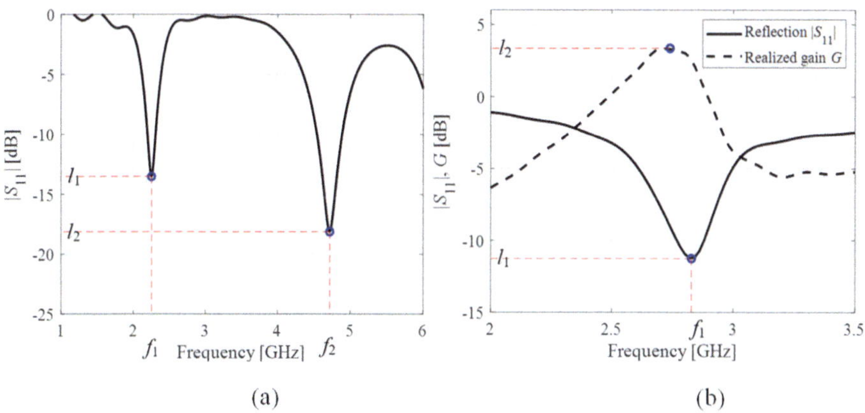

Fig. 10.12 Examples of the operating and performance parameters: (**a**) exemplary dual-band antenna with the operating figure vector consisting of antenna resonant frequencies $\boldsymbol{f} = [f_1\ f_2]^T$, and the performance figure vector consisting of reflection levels at the center frequencies $\boldsymbol{l} = [l_1\ l_2]^T$; (**b**) exemplary quasi-Yagi antenna with the operating figure vector containing the center frequency $\boldsymbol{f} = f_1$, and the performance figure vector containing reflection level at f_1 and maximum realized gain value l_2, i.e., $\boldsymbol{l} = [l_1\ l_2]^T$

determine design quality (other than those already allocated in the vector $\boldsymbol{f}$). These may include reflection levels at the antenna resonances (or over specific frequency ranges), the value of gain or axial ratio at the center frequency, side lobe levels, and so on. Figure 10.12 shows examples of operating and performance figure vectors for an exemplary dual-band and quasi-Yagi antennas, whereas Fig. 10.13 presents an example of the operating and performance figure vector for a microwave coupler.

Let X be the space of design parameters delimited by the lower and upper bounds for antenna parameters. Furthermore, let $\boldsymbol{f}_L = [f_{L.1} \ldots f_{L.N}]^T$ and $\boldsymbol{f}_U = [f_{U.1} \ldots f_{U.N}]^T$

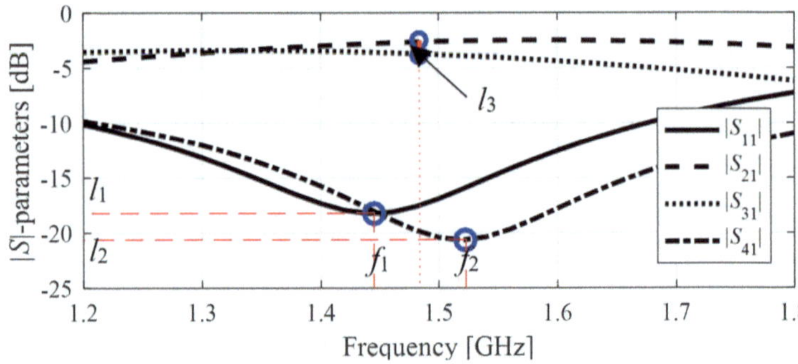

Fig. 10.13 An example of the operating and performance parameters for a microwave coupler: the operating figure vector consists of the frequencies $\boldsymbol{f} = [f_1\ f_2]^T$ corresponding to the minima of $|S_{11}|$ and $|S_{41}|$; the performance figure vector consists of $|S_{11}|$ and $|S_{41}|$ levels l_1 and l_2 at f_1 and f_2, respectively, as well as the power split at the frequency $(f_1 + f_2)/2$, i.e., $\boldsymbol{l} = [l_1\ l_2\ l_3]^T$

be the acceptance bounds for the operating figure vector, meaning that we are only interested in vectors that satisfy $f_{L.j} \leq f_j \leq f_{U.j}$, $j = 1, \ldots, N$. Similarly, let $\boldsymbol{l}_L = [l_{L.1} \ldots l_{L.M}]^T$ and $\boldsymbol{l}_U = [l_{U.1} \ldots l_{U.M}]^T$ be the acceptance bounds for the performance figure vector; we want to maintain $l_{L.j} \leq l_j \leq l_{U.j}$, $j = 1, \ldots, M$. Some of the $l_{L.j}$ may be equal to $-\infty$, meaning that there is no lower bound for l_j. We may also have $l_{U.j} = \infty$, meaning that there is no upper bound for l_j.

Let $\boldsymbol{x}^{(j)} = [x_1^{(j)} \ldots x_N^{(j)}]^T$, $j = 0, \ldots, n$, be $n + 1$ affinely independent points in X, $\boldsymbol{f}^{(j)} = \boldsymbol{f}(\boldsymbol{x}^{(j)}) = [f_1^{(j)} \ldots f_N^{(j)}]^T$ be the corresponding operating figure vectors, and $\boldsymbol{l}^{(j)} = \boldsymbol{l}(\boldsymbol{x}^{(j)}) = [l_1^{(j)} \ldots l_K^{(j)}]^T$ be the performance figure vectors. The vectors $\boldsymbol{x}^{(j)}$ are obtained to ensure that $\boldsymbol{f}_L \leq \boldsymbol{f}^{(j)} \leq \boldsymbol{f}_U$ and $\boldsymbol{l}_L \leq \boldsymbol{l}^{(j)} \leq \boldsymbol{l}_U$. In practice, an automated decision-making procedure is employed, in which random vectors are generated sequentially, and only those that satisfy the above conditions (the operating and performance vectors are extracted from EM simulation data) are accepted. The procedure is terminated after $n + 1$ points have been accepted, which are additionally affinely independent. A conceptual illustration of the random sampling procedure has been shown in Fig. 10.14.

We are now in a position to define the simplex-based predictors. Let $\boldsymbol{x} \in X$. Linear independence of vectors $\boldsymbol{x}^{(j)} - \boldsymbol{x}^{(0)}$ implies that we have a unique expansion

$$\boldsymbol{x} = \boldsymbol{x}^{(0)} + \sum_{j=1}^{n} a_j \left(\boldsymbol{x}^{(j)} - \boldsymbol{x}^{(0)} \right) \tag{10.13}$$

The expansion coefficient vector $\boldsymbol{a} = [a_1 \ldots a_n]^T$ can be found as

$$\boldsymbol{a}(\boldsymbol{x}) = \boldsymbol{X}^{-1} \left(\boldsymbol{x} - \boldsymbol{x}^{(0)} \right) \tag{10.14}$$

where $\boldsymbol{X}$ is a non-singular $n \times n$ matrix defined as

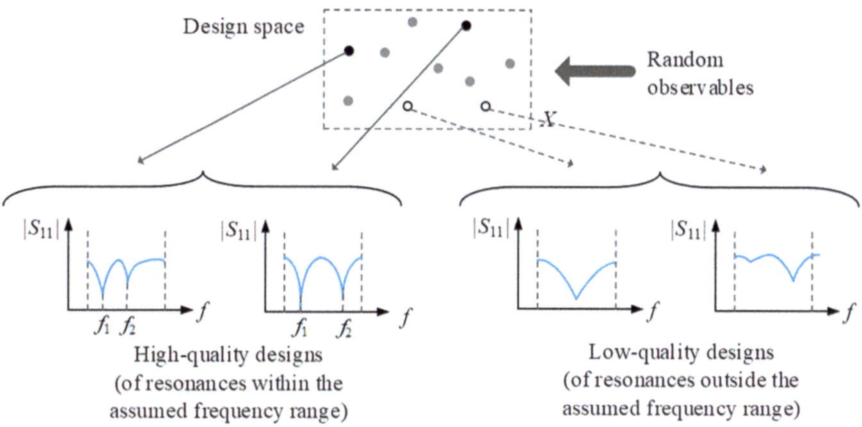

Fig. 10.14 Generating random trial points for the purpose of simplex-based predictor construction. The illustration is based on the antenna of Fig. 10.10a. The operating parameters are resonant frequencies f_1 and f_2. Only parameter vectors with the corresponding antenna responses featuring clearly visible resonances allocated within the prescribed ranges are accepted to become the basis to build the predictor

$$\boldsymbol{X} = \left[\boldsymbol{x}^{(1)} - \boldsymbol{x}^{(0)} \quad \cdots \quad \boldsymbol{x}^{(n)} - \boldsymbol{x}^{(0)}\right] \tag{10.15}$$

The simplex-based models of the operating parameters $\boldsymbol{F}(\boldsymbol{x}) : X \rightarrow F$ and performance parameters $\boldsymbol{L}(\boldsymbol{x}) : X \rightarrow R^M$ to be used as predictors of the operating and performance vectors over X and R^M, respectively, are set up with the use of the problem-specific knowledge extracted from the component responses. The predictors are defined as

$$\begin{aligned} \boldsymbol{F}(\boldsymbol{x}) = & \quad \boldsymbol{f}^{(0)} + \sum_{j=1}^{n} a_j\left(\boldsymbol{f}^{(j)} - \boldsymbol{f}^{(0)}\right) \\ = & \quad \boldsymbol{f}^{(0)} + \boldsymbol{X}_f\boldsymbol{a}(\boldsymbol{x}) = \boldsymbol{f}^{(0)} + \boldsymbol{X}_f\boldsymbol{X}^{-1}\left(\boldsymbol{x} - \boldsymbol{x}^{(0)}\right) \end{aligned} \tag{10.16}$$

$$\begin{aligned} \boldsymbol{L}(\boldsymbol{x}) = \boldsymbol{l}^{(0)} + \sum_{j=1}^{n} a_j\left(\boldsymbol{l}^{(j)} - \boldsymbol{l}^{(0)}\right) = \boldsymbol{l}^{(0)} + \boldsymbol{X}_l\boldsymbol{a}(\boldsymbol{x}) = \boldsymbol{l}^{(0)} \\ + \boldsymbol{X}_l\boldsymbol{X}^{-1}\left(\boldsymbol{x} - \boldsymbol{x}^{(0)}\right) \end{aligned} \tag{10.17}$$

with $\boldsymbol{a}$ computed as (10.14) and

$$\boldsymbol{X}_f = \left[\boldsymbol{f}^{(1)} - \boldsymbol{f}^{(0)} \quad \cdots \quad \boldsymbol{f}^{(n)} - \boldsymbol{f}^{(0)}\right] \tag{10.18}$$

$$X_l = \begin{bmatrix} l^{(1)} - l^{(0)} & \cdots & l^{(n)} - l^{(0)} \end{bmatrix} \tag{10.19}$$

The models $\boldsymbol{F}(\boldsymbol{x})$ and $\boldsymbol{L}(\boldsymbol{x})$ will be employed to yield predictions about the operating and performance vectors over the parameter space X as explained in Sect. 10.2.3.

10.2.3 Global Search by Means of Simplex-Based Predictors

The models $\boldsymbol{F}(\boldsymbol{x})$ and $\boldsymbol{L}(\boldsymbol{x})$ defined in Sect. 10.2.2 are the basis for performing the global search step of the considered optimization algorithm. As explained earlier (cf. Sect. 10.2.1), the dependence between the operating and performance vectors $\boldsymbol{f}$ and $\boldsymbol{l}$ and the component geometry parameters is weakly nonlinear (especially for $\boldsymbol{f}$). This means that the models (10.16) and (10.17) are likely to act as reliable predictors of performance of the device of interest over X, particularly within the simplex.

10.2.3.1 Design Quality Assessment

Evaluation of design quality is an important consideration in any optimization process. Let U_F be the objective function defined to compute the design quality by taking into account the vectors $\boldsymbol{f}(\boldsymbol{x})$ and $\boldsymbol{l}(\boldsymbol{x})$. Therein, the priority is given to reducing the distance between the existing operating vector and the target $\boldsymbol{f}_t$. Note that we use the same symbol $\boldsymbol{f}_t$ to denote the target operating figure vector and the target operating frequency vector of Sect. 10.2.1, which is for notational simplicity. The function U_F is defined as

$$U_F(\boldsymbol{x}) = U(\boldsymbol{f}(\boldsymbol{x}), \boldsymbol{l}(\boldsymbol{x})) = U_L(\boldsymbol{l}(\boldsymbol{x})) + \beta_F \|\boldsymbol{f}(\boldsymbol{x}) - \boldsymbol{f}_t\|^2 \tag{10.20}$$

Here, U_L is the merit function defined similarly as the function U of Sect. 10.2.1, however, computed based on $\boldsymbol{l}(\boldsymbol{x})$ rather than the entire component characteristics. For example, if $\boldsymbol{l}(\boldsymbol{x}) = [l_1 \ldots, l_M]^T$ represents reflection levels of a M-band antenna and the aim is to improve the matching at all operational frequencies f_1 through f_M, then we have $U_L(\boldsymbol{l}(\boldsymbol{x})) = \max\{l_1, \ldots, l_M\}$. If we intend to increase the antenna gain at the operating frequency f_1, with the performance vector defined as for the quasi-Yagi antenna of Fig. 10.12b, i.e., $\boldsymbol{l} = [l_1\ l_2]^T$, with l_2 being the maximum gain, we may define $U_L(\boldsymbol{l}(\boldsymbol{x})) = -l_2$. These functions may not be exact equivalents of functions $U(\boldsymbol{x})$, especially if certain response levels are to be minimized/maximized over bandwidths, but at this stage we are mainly focused on enforcing $\boldsymbol{f}(\boldsymbol{x}) \rightarrow \boldsymbol{f}_t$. The latter is achieved by the second (penalty) term in (10.20), where β_F is a penalty factor.

10.2.3.2 Global Search: Automated Simplex Update

The global search stage is an iterative process, in which the fundamental step is minimization of the function U_F using the simplex-based predictors $\boldsymbol{F}(\boldsymbol{x})$ and $\boldsymbol{L}(\boldsymbol{x})$. The candidate design is produced as

$$\boldsymbol{x}_{\text{tmp}} = \arg\min_{\boldsymbol{x}\in X} U_F(\boldsymbol{F}(\boldsymbol{x}), \boldsymbol{L}(\boldsymbol{x})) \tag{10.21}$$

The problem is constrained to ensure that the search process is carried out solely inside of the simplex defined by $\{\boldsymbol{x}^{(j)}\}_{j\,=\,0,\,\ldots,\,n}$ and its small vicinity. The constraints are imposed on the expansion coefficients $\boldsymbol{a}(\boldsymbol{x})$ of (10.14). We have

$$\sum_{j=1}^{n} a_j = 1 \tag{10.22}$$

$$-\alpha \le a_j \le 1 + \alpha, j = 1, \ldots, n \tag{10.23}$$

where $\alpha > 0$ is a small real number (e.g., $\alpha = 0.2$). The simplex vertices are ordered with respect to increasing norm $||\boldsymbol{f}^{(j)} - \boldsymbol{f}_t||$, i.e., the best vertex $\boldsymbol{x}^{(0)}$ is the one that features the smallest value of the mentioned norm. The starting point for (10.21) is $\boldsymbol{x}^{(0)}$ because this is the design that is closest to the target in the sense explained earlier. Note that its corresponding expansion vector $\boldsymbol{a}(\boldsymbol{x}^{(0)}) = [0 \ldots 0]^T$. Figure 10.15 shows graphically the concepts related to the simplex-based predictor and generation of the candidate design $\boldsymbol{x}_{\text{tmp}}$.

The candidate design is accepted assuming that it leads to the overall improvement of the simplex quality, i.e., if

$$\left\|\boldsymbol{f}_{\text{tmp}} - \boldsymbol{f}_t\right\| < \max\left\{j \in \{0, 1, \ldots, n\} : \left\|\boldsymbol{f}^{(j)} - \boldsymbol{f}_t\right\|\right\} \tag{10.24}$$

where $\boldsymbol{f}_{\text{tmp}} = \boldsymbol{f}(\boldsymbol{x}_{\text{tmp}})$. If this is the case, $\boldsymbol{x}_{\text{tmp}}$ replaces the worst vertex $\boldsymbol{x}^{(j\text{worst})}$, where

$$j_{\text{worst}} = \arg\max\left\{ j \in \{0, 1, \ldots, n\} : \left\|\boldsymbol{f}^{(j)} - \boldsymbol{f}_t\right\|\right\} \tag{10.25}$$

Otherwise, the vector is rejected, and the simplex is reduced toward the best vertex $\boldsymbol{x}^{(0)}$ using the following transformation:

$$\boldsymbol{x}^{(j)} \leftarrow \gamma\boldsymbol{x}^{(j)} + (1-\gamma)\boldsymbol{x}^{(0)} \ \text{ for } j = 1, \ldots, n \tag{10.26}$$

Here, γ is the reduction factor, normally set up to $\gamma = 0.5$, meaning that the simplex is reduced by half with $\boldsymbol{x}^{(0)}$ remaining intact. Note that this operation will generally improve the quality of all vertices, i.e., reduce the factors $||\boldsymbol{f}^{(j)} - \boldsymbol{f}_t||$. In Sect. 10.2.3.3, we prove that under mild assumptions, reduction of the simplex size

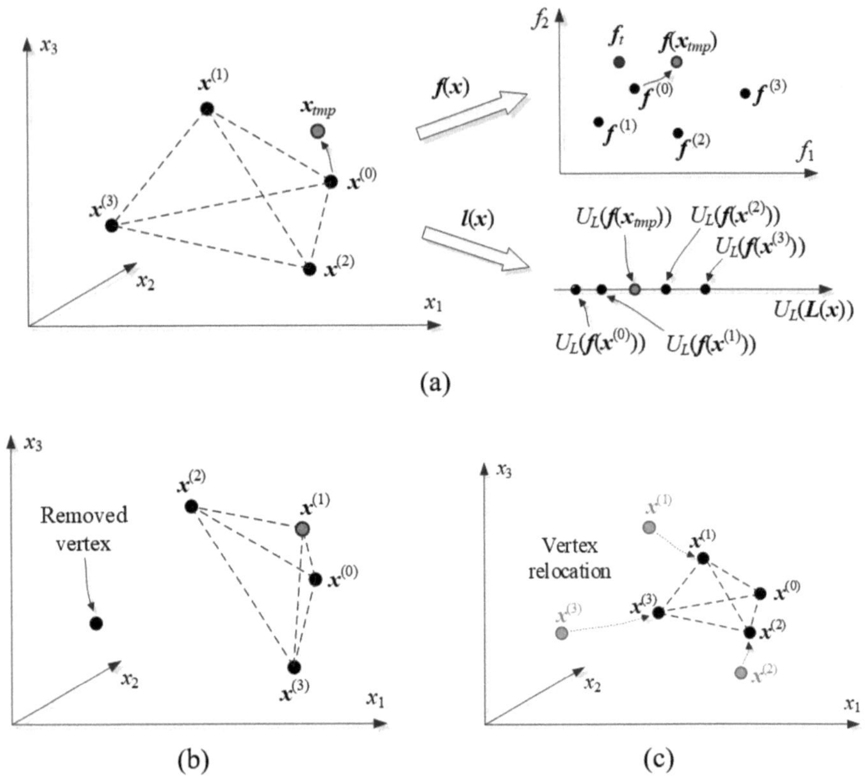

Fig. 10.15 Global search stage by simplex updates: (**a**) exemplary simplex in a three-dimensional design space. The predictions made using the simplex-based models $\boldsymbol{F}(\boldsymbol{x})$ and $\boldsymbol{L}(\boldsymbol{x})$ are validated against actual functions $\boldsymbol{f}(\boldsymbol{x})$ and $\boldsymbol{l}(\boldsymbol{x})$ at the candidate design $\boldsymbol{x}_{\text{tmp}}$ produced by (10.21). In the situation shown in the picture, the candidate design will be accepted due to improving both the antenna performance function U_L and the factor $||\boldsymbol{f}(\boldsymbol{x}) - \boldsymbol{f}_t||$; (**b**) simplex updating for accepted candidate design. In the situation shown, $\boldsymbol{x}_{\text{tmp}}$ was better than $\boldsymbol{x}^{(1)}$ but not as good as $\boldsymbol{x}^{(0)}$; (**c**) simplex reduction upon rejecting candidate design: all vertices $\boldsymbol{x}^{(j)}$, $j = 1, \ldots, n$, are moved toward $\boldsymbol{x}^{(0)}$

will eventually lead to generating candidate designs that will be accepted due to reducing $||\boldsymbol{f}(\boldsymbol{x}_{\text{tmp}}) - \boldsymbol{f}_t||$.

The above procedure is continued until we find a design that satisfies the condition

$$\left\|\boldsymbol{f}\left(\boldsymbol{x}_{\text{tmp}}\right) - \boldsymbol{f}_t\right\| \le F_{\max} \tag{10.27}$$

where $F_{\max}$ is a user-defined threshold. It is set up to ascertain that the optimum design is within the reach of the subsequent local tuning. This means that $F_{\max}$ has to be equal to a fraction (e.g., no more than half) of the expected component operating bandwidths. The value of this parameter is not critical as long as it is sufficiently small, e.g., 0.1 or 0.2 GHz for antennas working in the ranges of up to a few GHz or so. Other termination conditions are based on exceeding the assumed computational budget (N_{global} EM analyzes) or reducing the simplex size $D = \max\{j \in \{1,2, \ldots, n\} : ||\boldsymbol{x}^{(j)} - \boldsymbol{x}^{(0)}||\}$ beyond the user-defined threshold D_{min}. Both are used to ensure convergence of the optimization process even if the condition (10.27) is unattainable.

10.2.3.3 Objective Function Improvement

In this section, we provide a formal proof that sufficient reduction of the simplex size, according to (10.27), will ensure improvement of the objective function U_F as well as the factor $||\boldsymbol{f}(\boldsymbol{x}) - \boldsymbol{f}_t||$ (both over their values at $\boldsymbol{x}^{(0)}$). The sole assumption is made of the smoothness of the involved functions.

Let functions $\boldsymbol{f}(\boldsymbol{x})$ and $\boldsymbol{l}(\boldsymbol{x})$ be continuously differentiable in X. Then, we have

$$\boldsymbol{f}(\boldsymbol{x}) \approx \boldsymbol{f}\left(\boldsymbol{x}^{(0)}\right) + \boldsymbol{J}_f\left(\boldsymbol{x}^{(0)}\right) \cdot \left(\boldsymbol{x} - \boldsymbol{x}^{(0)}\right) \tag{10.28}$$

$$\boldsymbol{l}(\boldsymbol{x}) \approx \boldsymbol{l}\left(\boldsymbol{x}^{(0)}\right) + \boldsymbol{J}_l\left(\boldsymbol{x}^{(0)}\right) \cdot \left(\boldsymbol{x} - \boldsymbol{x}^{(0)}\right) \tag{10.29}$$

in a sufficiently small vicinity of the vector $\boldsymbol{x}^{(0)}$, where

$$\boldsymbol{J}_f(\boldsymbol{x}) = \begin{bmatrix} \frac{\partial f_1(\boldsymbol{x})}{\partial x_1} & \cdots & \frac{\partial f_1(\boldsymbol{x})}{\partial x_n} \\ \vdots & \ddots & \vdots \\ \frac{\partial f_N(\boldsymbol{x})}{\partial x_1} & \cdots & \frac{\partial f_N(\boldsymbol{x})}{\partial x_n} \end{bmatrix} \text{ and } \boldsymbol{J}_l(\boldsymbol{x}) = \begin{bmatrix} \frac{\partial l_1(\boldsymbol{x})}{\partial x_1} & \cdots & \frac{\partial l_1(\boldsymbol{x})}{\partial x_n} \\ \vdots & \ddots & \vdots \\ \frac{\partial l_M(\boldsymbol{x})}{\partial x_1} & \cdots & \frac{\partial l_M(\boldsymbol{x})}{\partial x_n} \end{bmatrix} \tag{10.30}$$

By applying (10.28) and (10.29) to all simplex vertices, we get

$$\boldsymbol{f}^{(j)} = \boldsymbol{f}\left(\boldsymbol{x}^{(j)}\right) \approx \boldsymbol{f}\left(\boldsymbol{x}^{(0)}\right) + \boldsymbol{J}_f\left(\boldsymbol{x}^{(0)}\right) \cdot \left(\boldsymbol{x}^{(j)} - \boldsymbol{x}^{(0)}\right) \tag{10.31}$$

$$\boldsymbol{l}^{(j)} = \boldsymbol{l}\left(\boldsymbol{x}^{(j)}\right) \approx \boldsymbol{l}\left(\boldsymbol{x}^{(0)}\right) + \boldsymbol{J}_l\left(\boldsymbol{x}^{(0)}\right) \cdot \left(\boldsymbol{x}^{(j)} - \boldsymbol{x}^{(0)}\right) \tag{10.32}$$

for $j = 1, \ldots, n$. By applying (10.31)–(10.18) and (10.32)–(10.19), we obtain

$$\begin{aligned} X_f &\approx \left[J_f\left(x^{(0)}\right) \cdot \left(x^{(1)} - x^{(0)}\right) \quad \cdots \quad J_f\left(x^{(0)}\right) \cdot \left(x^{(n)} - x^{(0)}\right) \right] \\ &= J_f\left(x^{(0)}\right) X \end{aligned} \tag{10.33}$$

$$\begin{aligned} X_l &\approx \left[J_l\left(x^{(0)}\right) \cdot \left(x^{(1)} - x^{(0)}\right) \quad \cdots \quad J_l\left(x^{(0)}\right) \cdot \left(x^{(n)} - x^{(0)}\right) \right] \\ &= J_l\left(x^{(0)}\right) X \end{aligned} \tag{10.34}$$

This results in (cf. 10.16 and 10.17)

$$\begin{aligned} F(x) &\approx f^{(0)} + J_f\left(x^{(0)}\right) X X^{-1} \left(x - x^{(0)}\right) = f^{(0)} + J_f\left(x^{(0)}\right)\left(x - x^{(0)}\right) \\ &\approx f(x) \end{aligned} \tag{10.35}$$

and

$$\begin{aligned} L(x) &\approx l^{(0)} + J_l\left(x^{(0)}\right) X X^{-1} \left(x - x^{(0)}\right) = l^{(0)} + J_l\left(x^{(0)}\right)\left(x - x^{(0)}\right) \\ &\approx l(x) \end{aligned} \tag{10.36}$$

in a sufficiently small neighborhood of $x^{(0)}$, i.e., when the simplex size $D = \max\{j \in \{1,2, \ldots, n\} : \|x^{(j)} - x^{(0)}\|\}$ is close to zero. In particular, the Jacobian matrices J_F and J_L of F and L at $x^{(0)}$ coincide with the respective matrices J_f and J_l of f and l. Consequently, when $D \to 0$, the predictions of the simplex-based models (10.16) and (10.17) coincide with the predictions of the Taylor models (10.28) and (10.29), i.e., the outcome of the simplex updating iteration x_{tmp} (cf. 10.21) will result in the improvement of the objective function U_F of (10.20). The latter is implied by the following observation. The first-order expansion model of U_F is given as

$$\begin{aligned} U_F(x) = & \quad U_F(f(x), l(x)) \\ \approx & \quad U_F\left(x^{(0)}\right) + \left[\nabla_f^T J_f\left(x^{(0)}\right) + \nabla_l^T J_l\left(x^{(0)}\right)\right]\left(x - x^{(0)}\right) \end{aligned} \tag{10.37}$$

where ∇_f and ∇_l are the gradients of U_F at $x^{(0)}$. Then, any descent direction h, i.e., such that $\left[\nabla_f^T J_f\left(x^{(0)}\right) + \nabla_l^T J_l\left(x^{(0)}\right)\right] h < 0$ is also a descent direction according to the simplex-based models because $\left[\nabla_f^T J_f\left(x^{(0)}\right) + \nabla_l^T J_l\left(x^{(0)}\right)\right] \approx \left[\nabla_f^T J_F\left(x^{(0)}\right) + \nabla_l^T J_L\left(x^{(0)}\right)\right]$. In a similar manner, one can show a reduction of $\|f(x) - f_t\|$.

10.2.4 Local Tuning Using Gradient-Based Trust-Region Search

Upon finding the design $\boldsymbol{x}^{(0)}$ that is close enough to the optimum (cf. 10.27), a local tuning procedure is launched to refine the antenna parameters. Here, it carried out using the trust-region (TR) gradient-based routine with numerical derivatives (Conn et al., 2000). The algorithm is an iterative one, and it produces a sequence $\boldsymbol{x}^{(i)}$, $i = 0$, 1, . . ., of approximations of the optimal solution $\boldsymbol{x}^*$. These designs are rendered by solving

$$\boldsymbol{x}^{(i+1)} = \arg \min_{\left\| \boldsymbol{x} - \boldsymbol{x}^{(i)} \right\| \leq d^{(i)}} U_L(\boldsymbol{x}, \boldsymbol{f}_t) \tag{10.38}$$

The objective function U_L is identical to the function U_P of (10.12), except that it is evaluated based on the first-order linear expansion model $\boldsymbol{L}^{(i)}(\boldsymbol{x}, f)$ of antenna characteristics, instead of original, EM-simulated responses. If, for example, we consider the reflection response $S_{11}(\boldsymbol{x}, f)$, the linear model becomes

$$\boldsymbol{L}^{(i)}(\boldsymbol{x}, f) = S_{11}\left(\boldsymbol{x}^{(i)}, f\right) + \nabla_S\left(\boldsymbol{x}^{(i)}, f\right) \cdot \left(\boldsymbol{x} - \boldsymbol{x}^{(i)}\right) \tag{10.39}$$

The gradient in (10.39) is evaluated through finite differentiation. The solution of the task (10.38) is sought for in a small vicinity of the actual design with its size $d^{(i)}$ set using the conventional TR rules (Conn et al., 2000). The new point $\boldsymbol{x}^{(i+1)}$ is accepted if it improves the original (EM-evaluated) merit function, i.e., $U_P(\boldsymbol{x}^{(i+1)}, \boldsymbol{f}_t) < U_P(\boldsymbol{x}^{(i)}, \boldsymbol{f}_t)$. Otherwise, it is dismissed and the iteration is relaunched within a smaller $d^{(i)}$. The CPU cost of constructing the model (10.39) is equivalent to $n + 1$ EM antenna simulations. In order to diminish this cost, in our implementation, the rank-one Broyden formula (Broyden, 1965), is employed instead of finite differentiation when the search approaches convergence, i.e., $||\boldsymbol{x}^{(i+1)} - \boldsymbol{x}^{(i)}|| < M_c\varepsilon$, where ε is the termination threshold, whereas M_c is the multiplication factor, e.g., $M_c = 10$. The algorithm is terminated either due to the convergence in argument $||\boldsymbol{x}^{(i+1)} - \boldsymbol{x}^{(i)}|| < \varepsilon$, or a sufficient diminution of the trust region $d^{(i)} < \varepsilon$ (whichever occurs first). In our experiments, we assume $\varepsilon = 10^{-3}$.

10.2.5 Optimization Procedure

Here, we put together all constituent parts of the global optimization procedure in a form of a pseudocode. The input variables are the following:

- Parameter space X;
- Definitions of operating and performance vectors $\boldsymbol{f}$ and $\boldsymbol{l}$;

- Target operating parameter vector $\boldsymbol{f}_t$;
- Merit function U_L (cf. 10.20).

The first two inputs depend on the high-frequency structure to be optimized, whereas the remaining two are governed by the design problem at hand.

The control parameters of our framework are the following:

- $F_{\max}$—a user-defined threshold for termination of the global search stage taking into account the distance between the target and actual operating parameter vector (cf. 10.27);
- α—search region extension parameter (cf. 10.23), normally set to small positive number, e.g., $\alpha = 0.2$;
- γ—simplex reduction ratio (cf. 10.26), normally set to $\gamma = 0.5$;
- $D_{\min}$—termination threshold (minimum simplex size leading to termination of global search stage), normally set to a small fraction (e.g., 1%) of the parameter space size;
- ε—termination threshold for local tuning stage, normally set to an assumed resolution of the optimization process, e.g., 10^{-3} (cf. Sect. 10.2.4).

Observe that none of the above parameters is critical for the performance of the algorithm, except $F_{\max}$, which needs to be established as suggested in Sect. 10.2.3.2, i.e., to a fraction of the expected component operating bandwidths. Figure 10.16 presents a flow diagram of the discussed framework.

10.2.6 Verification Case I: Quasi-Yagi Antenna

This section discusses validation of the global optimization using a quasi-Yagi antenna. The numerical experiments aim to verify the global search capability of the considered framework, as well as to compare it with benchmark methods, specifically, multiple-start local optimization and nature-inspired routines. The latter is represented by a particle swarm optimizer (PSO) (Kennedy & Eberhart, 2001), which is arguably the most widely utilized population-based algorithm nowadays. The primary factors that are of interest are the optimization process reliability, quality of the design produced in the course of optimization, and also computational cost.

The quasi-Yagi antenna employed for verification purposes has been presented in Fig. 10.17, where all important details on the antenna structure have been shown as well. The performance vectors include antenna reflection coefficients at the resonant frequencies, and maximum gain (cf. Fig. 10.12). The antenna EM model is simulated with the use of the time-domain solver of CST Microwave Studio. Observe that the parameter space is extensive both in terms of dimensionality and, most importantly, with respect to the parameter ranges. The average ratio of the upper to lower bounds equals 3.5.

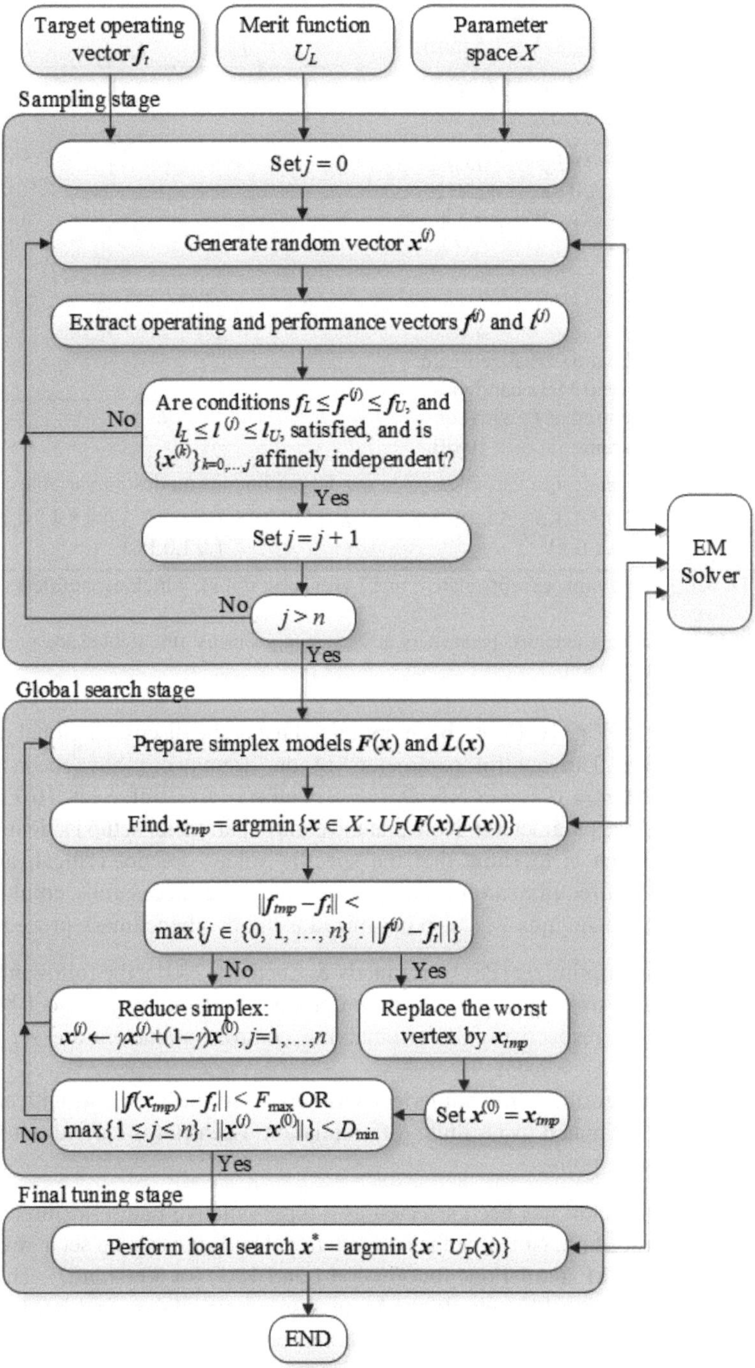

Fig. 10.16 Flowchart of the considered globalized optimization procedure using simplex-based predictors

Substrate	RO4003 ($\varepsilon_r = 3.38$, $h = 1.5$ mm)
Design parameters[$]	$\boldsymbol{x} = [L_a\ L_b\ L_c\ L_d\ W\ w_a\ D_a\ D_b\ D_c\ D_{lr}\ D_{rr}\ S_r\ w_{br}\ w_{cr}]^T$
Other parameters[$]	$D_l = D_{lr}L_a$, $D_r = D_{rr}L_a$, $S = S_rW$, $w_b = w_{br}W/2$, $w_c = w_{cr}W$, $w_0 = 3.4$
Target operating frequency [GHz]	$f_t = 2.5$
Design goals	Maximize realized gain in ±50 MHz bandwidth centred at f_t; with constraint: $\|S_{11}\| \le -10$ dB

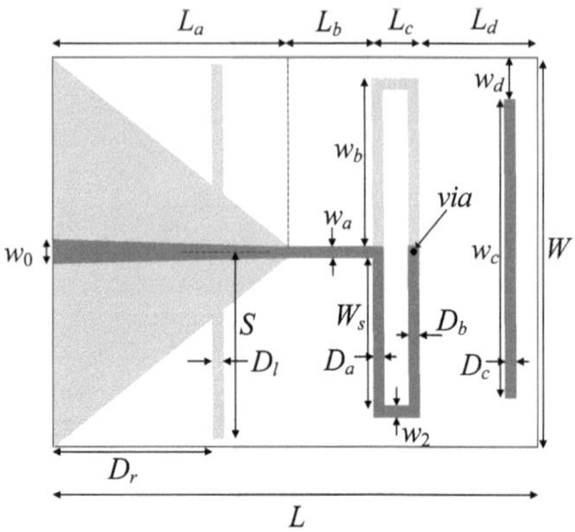

Lower bounds on design variables:	Upper bounds on design variables:
$\boldsymbol{l} = [15\ 5\ 1\ 15\ 25\ 0.5\ 1\ 1.5\ 1.5\ 0.05\ 0.4\ 0.5\ 0.5\ 0.5]^T$	$\boldsymbol{u} = [35\ 25\ 8\ 40\ 60\ 2.5\ 3.0\ 4.5\ 4.5\ 0.25\ 0.9\ 1.0\ 1.0\ 1.0]^T$

[$] Dimensions in mm, except relative one (with subscript r), which are unitless.

Fig. 10.17 Quasi-Yagi antenna: parameters and antenna geometry (the light-shade gray marks ground-plane metallization)

The antenna of Fig. 10.17 has been optimized using the discussed global optimization algorithm. The control parameters of our framework have been set to $F_{\max} = 0.2$ GHz, $\alpha = 0.2$, $\gamma = 0.5$, $D_{\min} = 1$, and $\varepsilon = 10^{-3}$ (cf. Sect. 10.2.5 for a description of these parameters). Note that the control parameter setup is identical for all antennas, which is to show that the specific values are not critical, and the algorithm does not require careful tuning before it can be successfully employed.

The verification antenna was also optimized using two benchmark procedures:

- Particle swarm optimizer (PSO) (Kennedy & Eberhart, 2001); the following setup is used: swarm size equals 10, maximum number of iterations 50 and 100 (Versions I and II, respectively), the remaining control parameters are set up to: $\chi = 0.73$, $c_1 = c_2 = 2.05$;
- Trust-region algorithm with numerical derivatives (see Sect. 10.2.4) with antenna sensitivities estimated using finite differentiation. Each run is carried out from a random starting point.

The PSO algorithm has been selected as a representative nature-inspired global optimization procedure. Note that the computational budget has been set at relatively low values (500 EM simulations for Version I and 1000 for Version II). This is to avoid excessive costs, although 1000 EM analyses is already beyond borderline in terms of practical utility. Local optimization has been included into the benchmark set to show the importance for global optimization in the case of the considered test problems.

Table 10.6 Optimization results for quasi-Yagi antenna

Case	Optimization method	Algorithm with inverse surrogates	PSO		Trust-region gradient-based algorithm
			50 iterations	100 iterations	
1	Average objective function value (dB)[a]	7.4	6.1	6.8	−1.1
	Computational cost[b]	144.3	500	1000	138.4
	Success rate[c]	10/10	9/10	10/10	1/10

[a]The values reported in the table refer to the realized gain at $f_0 = 2.5$ GHz
[b]The cost expressed in terms of the number of EM simulations of the antenna structure under design
[c]Number of algorithms runs at which the operating frequencies were allocated with sufficient accuracy, i.e., $||\boldsymbol{f}(\boldsymbol{x}^*) - \boldsymbol{f}_t|| < F_{\max}$

Table 10.6 presents the results rendered by the discussed and the benchmark methods, whereas Fig. 10.18 shows the antenna responses for the selected runs of the algorithm of Sect. 10.2. In the course of the experiments, each algorithm was executed 10 times, which was done in order to reduce a possible bias due to stochastic components within the considered procedures. The values reported in Table 10.6 refer to the average performance, including the objective function value, CPU cost, and success rate. The latter is a number of algorithm executions with the antenna operating parameters allocated sufficiently close to their target values, i.e., the condition $||\boldsymbol{f}(\boldsymbol{x}^*) - \boldsymbol{f}_t|| < F_{\max}$ was satisfied. Below, the analysis of the results from the perspective of computational efficacy of the respective optimization algorithms, their reliability, as well as design quality is provided.

It can be observed that the considered procedure ensures perfect success rate, i.e., satisfactory design was found in all of its runs. This can be considered evidence for the global search capabilities of the algorithm. Moreover, the success rate of multiple-start local optimization is considerably worse (only 1 per 10 algorithm runs), which indicates that the design problems are indeed multimodal.

Finally, the nature-inspired algorithm (here, PSO) performs considerably better, with the success rate better than 9/10 for the computational budget of 500 EM simulations and equal to 10/10 for the budget of 1000 simulations. This indicates that PSO is capable of identifying satisfactory designs, and it is likely that the success rate would become perfect if the budget is increased further. Yet, this is not a practical option due to excessive computational expenses. Notwithstanding, the average quality of the designs produced by both PSO and gradient search is inferior in comparison with those produced by the considered technique.

Computational efficacy of the discussed algorithm can be assessed as excellent. Given its global search capability, the average number of EM simulations necessary to conclude the optimization process is only around 144. This cost is only somewhat higher than that of the gradient search (138 EM analyzes) and dramatically lower than the expenses incurred by PSO. From the point of view of practical utility, this is one of the most important advantages of the discussed procedure. As mentioned before, this has been made possible by exploiting problem-specific knowledge

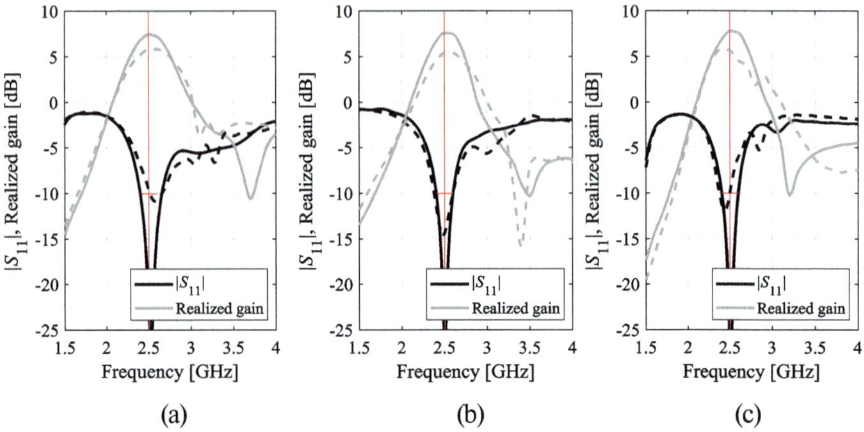

Fig. 10.18 Selected responses of the quasi-Yagi antenna for the designs rendered by the developed globalized simplex-based optimization algorithm: (**a**) design 1, (**b**) design 2, (**c**) design 3. Antenna response for the initial design $\boldsymbol{x}^{(0)}$ (produced by the global search stage) is shown using dashed line. Antenna response for the final design is shown by solid line. The intended center frequency 2.5 GHz is shown using vertical line, and the target impedance bandwidth (at the level of −10 dB) is marked using the horizontal line

embedded in antenna responses and incorporating it to construct the simplex-based inverse models defined over the operating and performance figures of the structure at hand. Another advantage of the considered approach is that all the basic steps are fully automated, i.e., no designer's interaction is required apart from providing the initial data.

The properties of the considered optimization algorithm make it a potentially attractive solution for quasi-global antenna optimization, which might be preferred over more conventional approaches, including surrogate-assisted procedures of the EGO type (efficient global optimization; Jones, 1998) and similar, let alone nature-inspired routines. It should be mentioned that a potential limitation of the discussed method is related to the size of the design space, especially with respect to the parameter ranges. More specifically, if the space is excessively large, the computational cost of identifying a useful set of trial points (i.e., those that are accepted in the sense discussed in Sect. 10.2.2) may be high because most of random observables will likely be rejected. On the other hand, if the designer is able to establish reasonable parameter bounds based on the engineering insight, this worst-case scenario would never occur. As a matter of fact, the design space of the antenna of interest considered is already large: the upper-to-lower parameter bound are 3.4 (on average).

10.2.7 Verification Case II: Power Divider

Figure 10.19 shows the dual-band equal-split power divider (PD) (Lin & Chu, 2010) employed for verification of the discussed optimization procedure. The structure is implemented on AD250 substrate with $\varepsilon_r = 2.5$ and $h = 0.81$ mm. The vector of design variables is $\boldsymbol{x} = [l_1\ l_2\ l_3\ l_4\ l_5\ s\ w_2]^T$, the remaining parameters are fixed: $w_1 = 2.2$, $g = 1$. The lower and upper bound on variables are $\boldsymbol{l} = [10\ 1\ 10\ 0.5\ 1\ 0.1\ 1.5]^T$ and $\boldsymbol{u} = [40\ 20\ 40\ 15\ 6\ 1.5\ 8]^T$, respectively. The design goal is to minimize input matching $|S_{11}|$, output matching $|S_{22}|$, $|S_{33}|$, and isolation $|S_{23}|$ at operating frequencies f_1 and f_2, as well as to ensure equal power split ratio. Two design cases are considered: $f_1 = 3.0$ GHz, $f_2 = 4.8$ GHz (Case 1) and $f_1 = 2.0$ GHz, $f_2 = 3.3$ GHz (Case 2). It can be observed that, for the considered structure, the parameter ranges are broad: the average upper-to-lower bound ratio equals 12.

As in Sect. 10.2.6, the circuit of Fig. 10.19 has been optimized using the discussed procedure, as well as two benchmark algorithms (PSO and trust-region gradient-based algorithm). Each algorithm was run 10 times for each circuit in order to investigate the global search capabilities and evaluate repeatability of solutions. The setup of the algorithms is the same as in Sect. 10.2.6 (i.e., $F_{\max} = 0.2$ GHz, $\alpha = 0.2$, $\gamma = 0.5$, $D_{\min} = 1$, and $\varepsilon = 10^{-3}$).

The numerical results have been shown in Table 10.7, whereas Fig. 10.20 shows the responses of the considered circuit at the designs obtained for the selected runs of the considered algorithm. As in the previous section, the results of Table 10.7 include the average values of the objective function and the computational cost of the optimization process, obtained for 10 independent runs of each algorithm.

The first observation is that the considered algorithm exhibits the perfect success rate and both considered design cases. It can also be noted that the success rate of the multiple-start local search is lower, and equals on average 4.5/10, which indicates multimodality of the underlying design problems. The reliability of the nature-inspired procedure is considerably better, although not perfect. We have 8/10 for the computational budget of 500, and 9/10 for 1000 EM simulations. This corroborates global search capability of PSO, but also the fact that the computational budget ensuring perfect success rate is considerably higher, perhaps 2000 objective

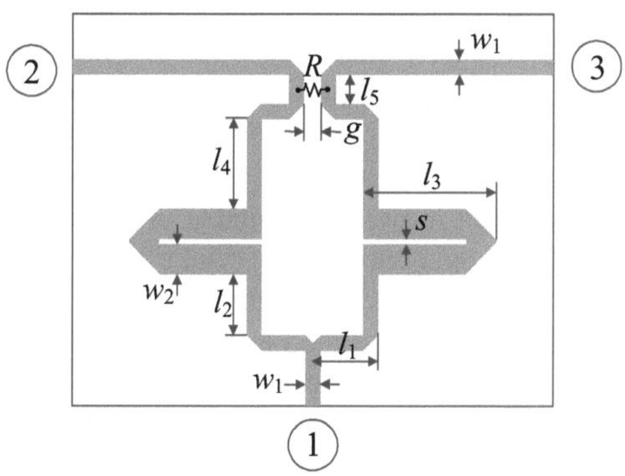

Fig. 10.19 Verification circuit: dual-band power divider (PD), lumped resistor denoted as R. (Lin & Chu, 2010)

Table 10.7 Optimization results for power divider

Case	Optimization method	Algorithm with inverse surrogates	PSO 50 iterations	PSO 100 iterations	Trust-region gradient-based algorithm
1	Average objective function value (dB)	−33.8	−19.6	−18.8	−12.3
	Computational cost[a]	82.0	500	1000	95.1
	Success rate[b]	10/10	8/10	9/10	2/10
2	Average objective function value (dB)	−25.9	−18.8	−19.7	−20.6
	Computational cost[a]	84.6	500	1000	93.8
	Success rate[b]	10/10	8/10	9/10	7/10

[a]The cost expressed in terms of the number of EM simulations of the circuit
[b]Number of algorithms runs at which the operating parameters were allocated with sufficient accuracy, i.e., to satisfy the condition $D(\boldsymbol{F}(\boldsymbol{x}^*), \boldsymbol{f}_t) \le D_{\text{accept}}$

function evaluation or so. Needless to say, extending the budget beyond a few hundred EM simulations is not practical, unless there are no alternatives.

In terms of design quality, measured by the average values of the objective function, the simplex-based algorithm produces the results that are comparable to PSO, and to conventional local optimization (for the runs at which the latter was successful). Notwithstanding, as mentioned earlier, the fundamental advantage of the discussed methodology is that it makes the search process considerably more robust. Because the procedure is global, the problem of potentially poor initial designs is eliminated: there are simply no initial designs required whatsoever.

Regarding computational efficiency, the performance of the discussed approach is excellent. The average optimization cost is slightly over 80 EM analyses of the circuit at hand, which is a small fraction of what would normally be incurred by global optimization routines. PSO is a representative example here: reaching perfect success rate requires a computational budged of well over 1000 EM simulations, which is a huge difference from the practical perspective.

Again, the major acceleration factor is the incorporation of the simplex-based surrogates defined over the operating figures of the circuits under optimization. These two factors are critical. On the one hand, relocation of the simplex is computationally cheap by itself. On the other hand, the relationship between the operating figures and geometry parameters is less nonlinear than a similar dependence for the complete frequency characteristics.

The overall features of the optimization framework discussed in this section indicate that it might be suitable as a replacement of the state-of-the-art techniques for globalized optimization, primarily nature-inspired algorithms, but also surrogate-assisted methods. Still, some limitations of the technique should be mentioned, which are mainly related to the size of the parameter space. For excessively broad ranges of geometry/material parameters, it might be difficult to find trial points of sufficient quality; in particular, the majority of randomly generated samples would

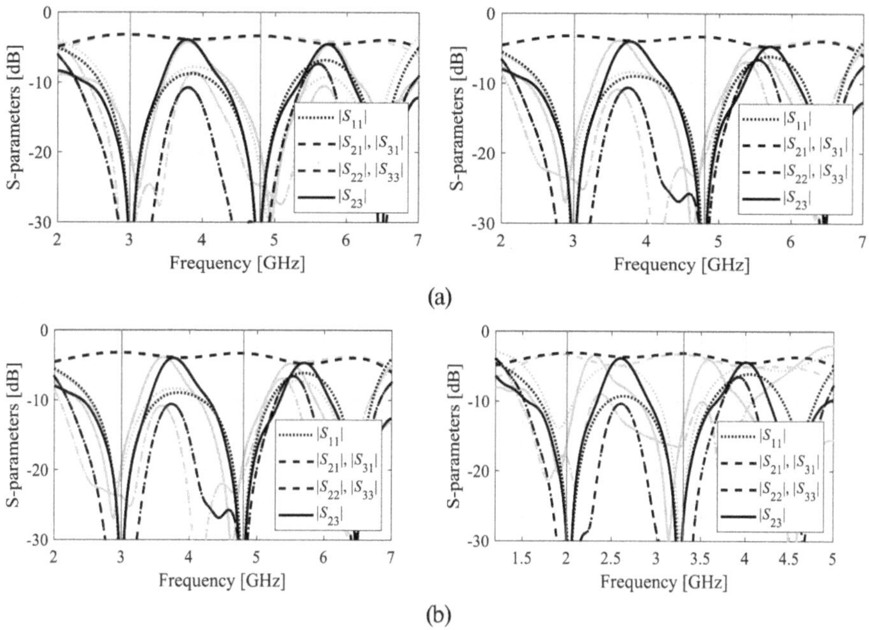

Fig. 10.20 Power divider: S-parameter characteristics at the optimized designs found by the considered global simplex-based optimization framework for two selected algorithm runs: (**a**) Case 1, (**b**) Case 2. The gray lines correspond to the initial design $\boldsymbol{x}^{(0)}$ obtained using the simplex-based search stage and black lines represent the responses at the final design. Vertical lines mark the target operating frequencies

not be accepted, which, in turn, would lead to generating excessively high computational expenses. However, in most practical scenarios, the parameter space is established based on the engineering insight, so that the aforementioned scenario is less likely to occur.

10.3 Multi-objective Optimization with Features

This section describes a technique for multi-objective design optimization of high-frequency components, based on the concept of Pareto front exploration, where subsequent Pareto-optimal designs are found using constrained feature-based local search starting from the previously found solution (Koziel & Bekasiewicz, 2018). The discussed approach is illustrated using an ultra-wideband antenna.

The section is organized as follows. The formulation of a multi-objective optimization problem is provided in Sect. 10.3.1. Section 10.3.2 describes a deterministic procedure for Pareto set identification. The feature-based procedure for finding subsequent Pareto-optimal designs is outlined in Sect. 10.3.3. The illustration example is presented in Sect. 10.3.4.

10.3.1 Multi-objective Design Problem Formulation

We will denote by $\boldsymbol{R}_f(\boldsymbol{x})$ a response of a high-fidelity computational model of the structure under design and by vector $\boldsymbol{x}$ a set of adjustable parameters (typically, geometrical dimensions and/or material parameters). Here, the model is normally evaluated using full-wave electromagnetic (EM) simulations. Moreover, $F_k(\boldsymbol{R}_f(\boldsymbol{x}))$, $k = 1, \ldots, N_{\text{obj}}$, will be a kth design objective (e.g., reduction of the area, improvement of performance characteristics of the structure, etc.).

If the number of objectives is larger than one, i.e., $N_{\text{obj}} > 1$, then any two designs $\boldsymbol{x}^{(1)}$ and $\boldsymbol{x}^{(2)}$ for which $F_k(\boldsymbol{R}_f(\boldsymbol{x}^{(1)})) < F_k(\boldsymbol{R}_f(\boldsymbol{x}^{(2)}))$ and $F_l(\boldsymbol{R}_f(\boldsymbol{x}^{(2)})) < F_l(\boldsymbol{R}_f(\boldsymbol{x}^{(1)}))$ for at least one pair $k \neq l$ are not commensurable (i.e., none is better than the other in multi-objective sense). In order to assess the designs, we use a Pareto dominance relation $\prec$ (Deb, 2001) defined as follows: for the two designs $\boldsymbol{x}$ and $\boldsymbol{y}$, we have $\boldsymbol{x} \prec \boldsymbol{y}$ ($\boldsymbol{x}$ dominates $\boldsymbol{y}$) if $F_k(\boldsymbol{R}_f(\boldsymbol{x})) \leq F_k(\boldsymbol{R}_f(\boldsymbol{y}))$ for all $k = 1, \ldots, N_{\text{obj}}$ and $F_k(\boldsymbol{R}_f(\boldsymbol{x})) < F_k(\boldsymbol{R}_f(\boldsymbol{y}))$ for at least one k.

Our goal is to obtain the Pareto-optimal set (a representation of the Pareto front) X_P in the search space X, such that for any $\boldsymbol{x} \in X_P$, there is no $\boldsymbol{y} \in X$ for which $\boldsymbol{y} \prec \boldsymbol{x}$ (Deb, 2001). Such a representation is considered to be the final outcome of the optimization process. Given available trade-offs as well as performance requirements, the designer can select a specific design for further processing, e.g., prototyping.

10.3.2 Optimization Algorithm

The multi-objective optimization algorithm considered in this section is formulated for two design objectives, which is the most practical case for typical problems in microwave and antenna engineering (e.g., structure size versus its electrical performance). It is possible to generalize it for a larger number of criteria, although this version is not discussed here. The algorithm generates a sequence of designs $\boldsymbol{x}^{(k)}$, $k = 1, 2, \ldots$, where $\boldsymbol{x}^{(1)}$ is a solution to the single-objective optimization problem of the form

$$\boldsymbol{x}^{(1)} = \arg\min_{\boldsymbol{x}} F_1\left(\boldsymbol{R}_f(\boldsymbol{x})\right) \tag{10.40}$$

which is the optimum design with respect to the first objective. For the sake of computational efficiency, the solution to (10.40) may be obtained, e.g., using surrogate-assisted techniques using an underlying low-fidelity model $\boldsymbol{R}_c$ (e.g., coarse-discretization version of the high-fidelity model $\boldsymbol{R}_f$) and suitable correction method such as space mapping (Koziel & Yang, 2011; Bandler et al., 2004). By $F_2{}^{(1)} = F_2(\boldsymbol{R}_f(\boldsymbol{x}^{(1)}))$, we will denote the corresponding value of the second objective.

The optimization algorithm generates subsequent Pareto-optimal designs using a series of constrained single-objective optimizations. In each iteration, the design obtained in the previous one is used as a starting point. More specifically, the kth element of the Pareto set $\boldsymbol{x}^{(k)}$ is obtained as

$$\boldsymbol{x}^{(k)} = \arg \min_{\boldsymbol{x},\ F_2(\boldsymbol{R}_f(\mathbf{x})) \le F_2^{(k)}} F_1(\boldsymbol{R}_f(\boldsymbol{x})) \tag{10.41}$$

where $F_2^{(k)}$ is a user-defined threshold value for the second objective. The process is terminated if $F_1(\boldsymbol{R}_f(\boldsymbol{x}^{(k)}))$ is no longer satisfactory from the point of view of given design specifications. This way, only a feasible part of the Pareto front (from the specification standpoint) is covered. There are various ways of establishing the sequence of thresholds $F_2^{(k)}$ (Koziel & Bekasiewicz, 2016), e.g., $F_2^{(k)} = \alpha \cdot F_2^{(k-1)}$ with $\alpha < 1$ or $F_2^{(k)} = F_2^{(k-1)}$ with $\beta > 0$. Figure 10.21 shows a conceptual illustration of the algorithm.

It should be noted that the algorithm (10.40)–(10.41) is deterministic and it does not rely on any form of population-based metaheuristic. Therefore, it may be computationally efficient, particularly if surrogate-based methods are used for solving (10.41). This can be realized using local methods because of the fact that subsequent Pareto-optimal designs are typically close to each other.

10.3.3 Pareto Front Exploration Using Response Features

For many classes of microwave and antenna circuits, it is possible to reduce the cost of the design process by exploiting feature-based optimization. As in any other feature-based framework discussed in this book, the technique of this section takes advantage of the fact that reformulating the design problem in terms of suitably defined response features leads to less nonlinear functional landscapes and, therefore, to reduction of the computational cost of the optimization process. Figure 10.22 shows the family of responses of a microwave filter (Koziel & Bandler, 2015b) evaluated along a selected line segment in the design space. It can be observed that the responses are highly nonlinear and change considerably when moving from one design to another. On the other hand, from the point of view of, e.g., minimax optimization as illustrated in Fig. 10.22b, only specific characteristic points of the filter responses are important to verify whether the design satisfies or violates given performance specifications. The dependence of these points (both frequency and level coordinates) is much less nonlinear as illustrated in Fig. 10.22c.

The feature-based reformulation of the original problem (10.41) is as follows:

$$\boldsymbol{x}^{(k)} = \arg \min_{\boldsymbol{x},\ F_{1.F}(\boldsymbol{R}_{F.f}(\boldsymbol{x})) \le F_{1.F}^{(k)}} F_{2.F}(\boldsymbol{R}_{F.f}(\boldsymbol{x})) \tag{10.42}$$

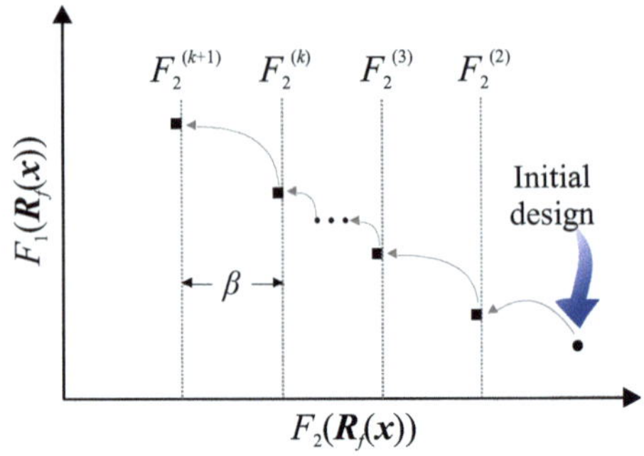

Fig. 10.21 Conceptual illustration of point-by-point exploration of the Pareto front using the algorithm (10.40)–(10.41). The initial design (black circle) is determined using (10.40). Subsequently, consecutive Pareto-optimal designs (black squares) are obtained sequentially using (10.41). Vertical dotted lines represent thresholds along objective F_2 (here, $F_2^{(k)} = F_2^{(k-1)} - \beta$)

Here, $\boldsymbol{R}_{F.f}(\boldsymbol{x})$ denotes the vector of the response features extracted from $\boldsymbol{R}_f(\boldsymbol{x})$, whereas $F_{k.F}$ denotes design objectives redefined in terms of the feature points. Feature-based optimization produces a sequence $\boldsymbol{x}^{(j.k)}$ of approximations to $\boldsymbol{x}^{(j)}$:

$$\boldsymbol{x}^{(j.k+1)} = \arg \min_{\substack{\boldsymbol{x},\ F_{1.F}\left(\boldsymbol{R}_{F.f}(\boldsymbol{x})\right) \le F_{1.F}^{(k)} \\ \left\|\boldsymbol{x} - \boldsymbol{x}^{(j.k)}\right\| \le \delta^{(k)}}} F_{2.F}\left(\boldsymbol{R}_l^{(k)}(\boldsymbol{x})\right) \tag{10.43}$$

where $\boldsymbol{R}_l^{(k)}$ is a linear approximation model of $\boldsymbol{R}_{F.f}(\boldsymbol{x})$ at $\boldsymbol{x}^{(j.k)}$ obtained using finite differentiation (Conn et al., 2000). As a convergence safeguard, $\boldsymbol{R}_l^{(k)}$ is optimized within the trust region of the radius $\delta^{(k)}$. The new design $\boldsymbol{x}^{(j.k+1)}$ is only accepted if $\boldsymbol{R}_{F.f}(\boldsymbol{x}^{(j.k+1)}) < \boldsymbol{R}_{F.f}(\boldsymbol{x}^{(j.k)})$. The trust region radius is updated using the standard rules (Conn et al., 2000). To achieve additional speedup, the gradient of $\boldsymbol{R}_{F.f}$ is estimated using the low-fidelity model so that we have

$$\boldsymbol{R}_l^{(k)}(\boldsymbol{x}) = \boldsymbol{R}_f\left(\boldsymbol{x}^{(k)}\right) + \begin{bmatrix} \left[\boldsymbol{R}_c\left(\boldsymbol{x}^{(k)} + \boldsymbol{h}_1\right) - \boldsymbol{R}_c\left(\boldsymbol{x}^{(k)}\right)\right]^T / d_1 \\ \cdots \\ \left[\boldsymbol{R}_c\left(\boldsymbol{x}^{(k)} + \boldsymbol{h}_n\right) - \boldsymbol{R}_c\left(\boldsymbol{x}^{(k)}\right)\right]^T / d_n \end{bmatrix}^T \cdot \left(\boldsymbol{x} - \boldsymbol{x}^{(k)}\right) \tag{10.44}$$

where $\boldsymbol{h}_k = [0 \ldots d_k \ldots 0]^T$ (d_k on kth position). This is justified because even if the low- and high-fidelity models are misaligned in absolute terms, they are normally well correlated so that gradient estimation obtained at the level of the low-fidelity model is a reliable representation of the high-fidelity one (Glubokov & Koziel, 2014).

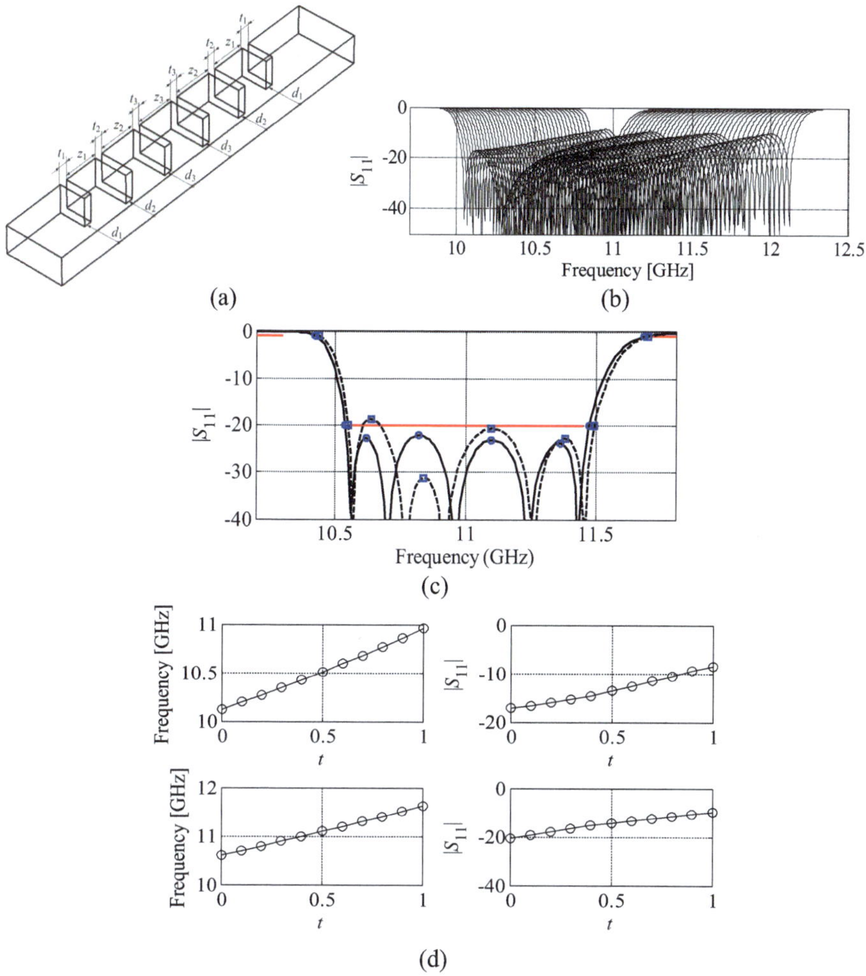

Fig. 10.22 Feature-based optimization concept: (**a**) example of a microwave waveguide filter, (**b**) the family of filter responses evaluated along the line segment parameterized by $0 \le t \le 1$, (**c**) minimax design specifications (response should be above the left and right horizontal lines and below the middle one), (**d**) variability of selected response features as function of parameter t (cf. Fig. 10.22b)

10.3.4 Verification Case: Ultra-wideband Monopole Antenna

This section illustrates multi-objective optimization with response features (Koziel & Bekasiewicz, 2018) using a compact ultra-wideband (UWB) monopole antenna implemented on a 0.762-mm-thick Taconic RF-35 substrate ($\varepsilon_r = 3.5$, $h = 0.762$ mm). The structure (Koziel & Bekasiewicz, 2016) is shown in Fig. 10.23. The set of adjustable parameters is $\boldsymbol{x} = [w_0\ g\ a_1\ a_2\ l_1\ l_2\ w_1\ o]^T$, parameter $w_0 = 2o + a_2$, whereas $w_i = 1.7$ is fixed (all dimensions in mm). The EM model is

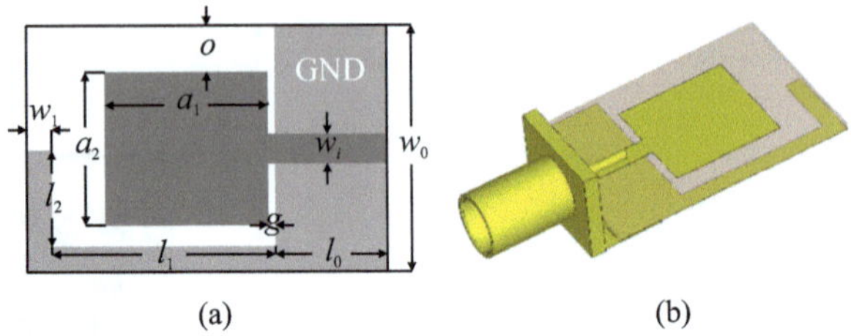

Fig. 10.23 UWB monopole antenna (Koziel & Bekasiewicz, 2016): (**a**) top view with highlighted design parameters, and (**b**) 3D view

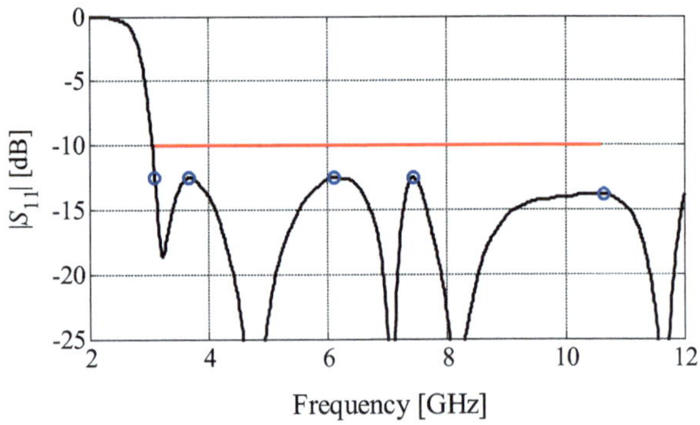

Fig. 10.24 UWB monopole antenna: feature points (o) utilized in the multi-objective optimization process

implemented in CST Microwave Studio ($\boldsymbol{R}_f$: ~4,600,000 mesh cells, simulation: 40 min). The model includes the SubMiniature version A (SMA) connector (see Fig. 10.23b).

Two objectives are assumed: F_1—minimization of the reflection in 3.1–10.6 GHz band and F_2—footprint reduction $S(\boldsymbol{x})$ defined as $A \times B$ rectangle, where $A = 0.5a_2 + o_1 + o_3$ and $B = l_g + l_1 + w_1$.

The problem (10.41) is solved using the feature-based optimization approach of the previous section. The feature points utilized in the optimization process include the local maxima of the reflection response as well as the point corresponding to the required reflection level (the antenna is optimized to) in the vicinity of the lower edge of the operational bandwidth (cf. Fig. 10.24).

The initial Pareto-optimal design has been obtained as in (10.40) using a combination of pattern search (Kolda et al., 2003) and feature-based optimization. The maximum in-band reflection at this design is -13.5 dB. The remaining Pareto-optimal designs have been obtained by applying the feature-based approach (cf. Sect. 10.3.3) for solving (10.41). The variability of the Pareto front along objectives F_1 and F_2 is 46 mm^2 and -3.5 dB, respectively. Detailed dimensions of selected Pareto designs are shown in Table 10.8.

Figure 10.25 shows the Pareto set representation obtained using the considered approach as well as the antenna reflection responses at the selected Pareto-optimal designs from Table 10.8. The total optimization cost corresponds to only about 28 evaluations of $\boldsymbol{R}_f$ (~25 h of CPU-time), whereas the cost of the surrogate-based optimization (Koziel & Bekasiewicz, 2015) is about 67 evaluations of the high-fidelity model.

Table 10.8 Selected Pareto-optimal antenna designs

Design	Size (mm^2)	max $\lvert S_{11}\rvert$ (dB)	Antenna dimensions in mm							
			w_0	g	a_1	a_2	l_1	l_2	w_1	o
$\boldsymbol{x}^{(1)}$	343	−13.5	6.00	0.77	8.93	10.29	12.75	5.20	2.45	2.94
$\boldsymbol{x}^{(3)}$	304	−12.5	5.69	0.80	8.99	9.96	12.63	5.78	1.89	2.52
$\boldsymbol{x}^{(6)}$	269	−11.0	5.13	0.80	9.14	9.04	12.40	6.35	1.75	2.46
$\boldsymbol{x}^{(8)}$	258	−10.0	5.06	0.78	9.12	8.94	12.36	6.38	1.64	2.30

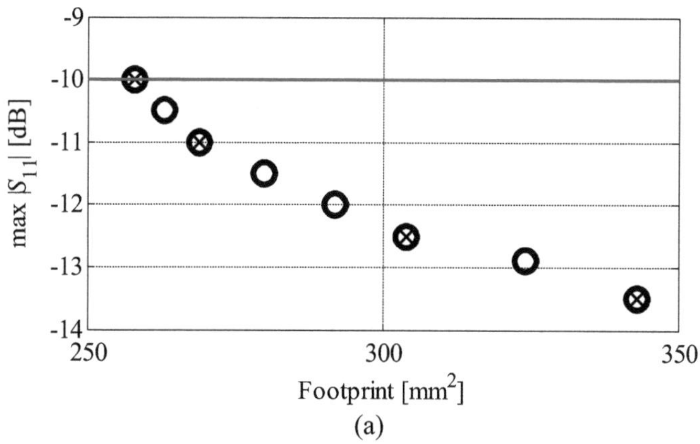

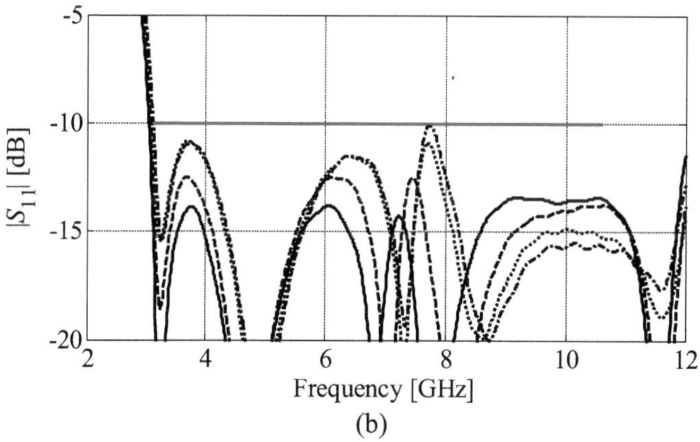

Fig. 10.25 Optimization results of the UWB monopole antenna of Fig. 10.23: (**a**) Pareto set representation obtained using the feature-based method; and (**b**) frequency characteristics of the antenna at the selected Pareto-optimal designs $\boldsymbol{x}^{(1)}$ (—), $\boldsymbol{x}^{(3)}$ (- - -), $\boldsymbol{x}^{(6)}$ (····), and $\boldsymbol{x}^{(8)}$ (-·-·-)

This section outlined a deterministic approach for rapid feature-based multi-objective optimization of antenna structures. The methodology exploits point-by-point exploration of the Pareto front using local search. The technique allows for quick identification of the set of Pareto-optimal designs. The cost of generating a Pareto set equals a few dozen of high-fidelity simulations, which is very low compared to techniques based on metaheuristics. Finally, it should be mentioned that although the presented technique is not generic (e.g., it might not work well if the Pareto front is not a connected set in the design space), it is well suited for problems in computational electromagnetics.

10.4 Summary

This chapter outlined several feature-based techniques for global and multi-objective optimization of high-frequency components. Specifically, two surrogate-based global optimization frameworks were delineated: the algorithm incorporating inverse surrogate models as well as the algorithm using simplex-based regression predictors. Furthermore, a procedure for expedited feature-based multi-objective antenna design optimization was described. The procedures considered in this chapter employ response feature technology to accelerate the process of solving the design tasks that are normally extremely expensive in terms of the required number of EM simulations. The cost of the metaheuristic-based algorithms, commonly utilized for global or multi-objective design, is usually prohibitive when carried out at the level of EM analysis. The use of the techniques of this section allows for circumventing this issue.

References

Bandler, J. W., Cheng, Q. S., Dakroury, S. A., Mohamed, A. S., Bakr, M. H., Madsen, K., & Søndergaard, J. (2004). Space mapping: The state of the art. *IEEE Transactions on Microwave Theory and Techniques, 52*(1), 337–361.

Broyden, C. G. (1965). A class of methods for solving nonlinear simultaneous equations. *Mathematics of Computation, 19*, 577–593.

Conn, A. R., Gould, N. I. M., & Toint, P. L. (2000). *Trust region methods* (MPS-SIAM series on optimization). MPS-SIAM.

Consul, P. (2015). Triple band gap coupled microstrip U-slotted patch antenna using L-slot DGS for wireless applications. In *Communication, Control and Intelligent Systems (CCIS)*, Mathura, India, pp. 31–34.

Deb, K. (2001). *Multi-objective optimization using evolutionary algorithms*. Wiley.

Glubokov, O., & Koziel, S. (2014). EM-driven tuning of substrate integrated waveguide filters exploiting feature-space surrogates. In *International Microwave Symposium*, Tampa Bay, FL, USA, pp. 1–3.

Jones, D. R. (1998). Efficient global optimization of expensive black-box functions. *The Journal of Global Optimization, 13*, 455–492.

Kennedy, J., & Eberhart, R. C. (2001). *Swarm intelligence*. Morgan Kaufmann.

Kolda, T. G., Lewis, R. M., & Torczon, V. (2003). Optimization by direct search: New perspectives on some classical and modern methods. *SIAM Review, 45*(3), 385–482.

Koziel, S., & Bandler, J. W. (2015a). Reliable microwave modeling by means of variable-fidelity response features. *IEEE Transactions on Microwave Theory and Techniques, 63*(12), 4247–4254.

Koziel, S., & Bandler, J. W. (2015b). Rapid yield estimation and optimization of microwave structures exploiting feature-based statistical analysis. *IEEE Transactions on Microwave Theory and Techniques, 63*(1), 107–114.

Koziel, S., & Bekasiewicz, A. (2015). Recent developments in simulation-driven multi-objective design of antennas. *Bulletin of the Polish Academy of Sciences Technical Sciences, 63*(3), 781–789.

Koziel, S., & Bekasiewicz, A. (2016). Low-cost multi-objective optimization of antennas using Pareto front exploration and response features. In *International Symposium on Antennas and Propagation (APSURSI)*, Fajardo, PR, USA, 2016, pp. 571–572.

Koziel, S., & Bekasiewicz, A. (2018). Recent advances in rapid multiobjective optimization of expensive simulation models in microwave and antenna engineering by Pareto front exploration. *International Journal of RF and Microwave Computer-Aided Engineering, 28*, art. no. e21190. [p36].

Koziel, S., & Pietrenko-Dabrowska, A. (2019a). Performance-based nested surrogate modeling of antenna input characteristics. *IEEE Transactions on Antennas and Propagation, 67*(5), 2904–2912.

Koziel, S., & Pietrenko-Dabrowska, A. (2019b). Reduced-cost surrogate modeling of compact microwave components by two-level kriging interpolation. *Engineering Optimization, 52*, 960–972.

Koziel, S., & Pietrenko-Dabrowska, A. (2020a). *Performance-driven surrogate modeling of high-frequency structures*. Springer.

Koziel, S., & Pietrenko-Dabrowska, A. (2020b). Expedited feature-based quasi-global optimization of multi-band antennas with Jacobian variability tracking. *IEEE Access, 8*, 83907–83915.

Koziel, S., & Pietrenko-Dabrowska, A. (2021a). Global EM-driven optimization of multi-band antennas using knowledge-based inverse response-feature surrogates. *Knowledge-Based Systems, 227*(12), 107189.

Koziel, S., & Pietrenko-Dabrowska, A. (2021b). Recent advances in accelerated multi-objective design of high-frequency structures using knowledge-based constrained modeling approach. *Knowledge-Based Systems, 214*, 106726.

Koziel, S., & Pietrenko-Dabrowska, A. (2022). Reliable EM-driven size reduction of antenna structures by means of adaptive penalty factors. *IEEE Transactions on Antennas and Propagation, 70*(2), 1389–1401.

Koziel, S., & Yang, X. S. (Eds.). (2011). *Computational optimization, methods and algorithms* (Series: Studies in computational intelligence) (Vol. 356). Springer.

Lin, Z., Chu, Q.-X. (2010). A novel approach to the design of dual-band power divider with variable power dividing ratio based on coupled-lines. *Prog. Electromagn. Res, 103*, 271–284.

Mandal, J. K., Mukhopadhyay, S., & Dutta, P. (Eds.). (2018). *Multi-objective optimization: Evolutionary to hybrid framework*. Springer.

Marler, R. T., & Arora, J. S. (2010). The weighted sum method for multi-objective optimization: New insights. *Structural and Multidisciplinary Optimization, 41*, 853–862.

Mirjalili, S., & Dong, J. S. (2019). *Multi-objective optimization using artificial intelligence techniques* (Springer briefs in applied sciences and technology). Springer.

Nocedal, J., & Wright, S. J. (2000). *Numerical optimization* (Springer series in operations research). Springer.

Pietrenko-Dabrowska, A., & Koziel, S. (2021). Globalized parametric optimization of microwave components by means of response features and inverse metamodels. *Scientific Reports, 11*, art. no. 23718. [p06].

Queipo, N. V., Haftka, R. T., Shyy, W., Goel, T., Vaidynathan, R., & Tucker, P. K. (2005). Surrogate-based analysis and optimization. *Progress in Aerospace Sciences, 41*(1), 1–28.

Chapter 11
Response Features for Warm-Start Optimization and Rapid Database Acquisition

Tuning of geometry parameters is one of the essential stages of contemporary high-frequency design. It is necessary because available design methods, whether based on theoretical considerations or on engineering experience, are only capable of yielding initial designs that need further adjustment in order to boost the performance parameters as much as possible. Numerical optimization is also imperative for system redesign with respect to different operating conditions (e.g., center frequency, bandwidth) or material parameters (e.g., substrate permittivity/thickness). As the tuning process normally involves full-wave electromagnetic (EM) analysis, it may incur considerable computational expenses. Reducing these costs is highly desirable from the perspective of both academic research and industry. This chapter discusses several surrogate-assisted frameworks for accelerated parameter tuning of high-frequency structures that allows for reusing the previously acquired design data. The major role is played by response features, which are used to identify the operating parameters of the systems under design and to construct forward and inverse surrogate models, later used as reliable predictors for generating reasonable initial designs, as well as further parameter tuning. The involvement of the response feature technology also leads to a more accurate rendition of the system gradients, which speeds up the design refinement as compared to models constructed at the level of original frequency characteristics. The presented methodologies are demonstrated using a number of microwave and antenna components, as well as compared to state-of-the-art parameter tuning procedures.

A. Pietrenko-Dabrowska, S. Koziel, *Response Feature Technology for High-Frequency Electronics. Optimization, Modeling, and Design Automation*,
https://doi.org/10.1007/978-3-031-43845-5_11

11.1 Expedited Parameter Tuning Using Inverse/Forward Surrogates and Response Features

This section discusses the parameter tuning procedure reported in Pietrenko-Dabrowska et al. (2021), which is an advancement over the technique proposed in Koziel and Pietrenko-Dabrowska (2020a). Therein, the initial design is generated using an inverse surrogate constructed from a set of database designs, available from the previous design work on the same structure or pre-optimized to set up the design framework. The tuning process itself has been conducted using (local) gradient-based search, which is jump-started by means of the forward model of the antenna sensitivities. The forward surrogate is the weakest point of the method of Koziel and Pietrenko-Dabrowska (2020a), because modeling of complex characteristics (here, Jacobian matrices) is very challenging, especially in higher-dimensional parameter spaces. The major advancement of the methodology presented here is to incorporate the response feature technology (Koziel, 2015) in two ways: (1) to model the system sensitivities at the level of properly selected feature points instead of the entire characteristics, and (2) reformulating the design problem in terms of response features. Both factors greatly improve the reliability of the optimization process and contribute to a reduction of its computational cost as demonstrated later in this section.

The remaining part of this section is organized as follows. Section 11.1.1 formulates the design problem. Section 11.1.2 briefly recalls the concept of response features. The inverse and feature-based forward surrogates are defined in Sect. 11.1.3, whereas the details concerning the design refinement (tuning) are provided in Sect. 11.1.4. Section 11.1.5 summarizes the entire procedure and explains it using a flow diagram. Demonstration case studies are discussed in Sects. 11.1.6, 11.1.7 and 11.1.8.

11.1.1 Problem Formulation

The primary computational model of the system under design is the full-wave electromagnetic (EM) simulation one, denoted as $\boldsymbol{R}(\boldsymbol{x})$, where $\boldsymbol{x} \in X$ is the vector of independent adjustable parameters. The model outputs are the system characteristics that are relevant to the design process (e.g., reflection response, gain, axial ratio, and power split ratio). In order to formulate the design framework, we will use the notion of the objective space F (Koziel & Pietrenko-Dabrowska, 2020a), which is a vector space of the figures of interest. The objective vectors have the form of $\boldsymbol{F} = [F_1 \ldots F_N]^T \in F$, where F_k, $k = 1, \ldots, N$, are individual figures, e.g., target operating frequency of the antenna (or frequencies in the case of multiband structures), parameters of the substrate (permittivity, thickness), the antenna is to be implemented on, and so on. Here, F is assumed to be an interval defined by the lower bounds $F_{k.\min}$ and upper bounds $F_{k.\max}$, i.e., we have $F_{k.\min} \leq F_k \leq F_{k.\max}$, $k = 1, \ldots, N$.

The design quality is quantified using the merit function $U(\boldsymbol{R}(\boldsymbol{x}), \boldsymbol{F})$. Its analytical form is problem dependent; however, it is defined so that better designs $\boldsymbol{x}$ correspond to lower values of $U(\boldsymbol{R}(\boldsymbol{x}), \boldsymbol{F})$. Using U, the parameter tuning problem can be formulated as a nonlinear minimization task of the form

$$\boldsymbol{x}^* = U^*(\boldsymbol{R}(\boldsymbol{x}), \boldsymbol{F}) = \arg \min_{\boldsymbol{x}} U(\boldsymbol{R}(\boldsymbol{x}), \boldsymbol{F}) \tag{11.1}$$

where $\boldsymbol{x}^*$ is the optimum design to be found. The problem (11.1) may be subject to additional constraints, which can be handled explicitly or implicitly (e.g., using the penalty function approach) (Pietrenko-Dabrowska & Koziel, 2019).

In order to clarify these concepts, let us consider an example of a multiband antenna that is to be matched as well as possible at the intended operating frequencies $f_{0.k}$, $k = 1, \ldots, N$. In this case, the objective space will consists of vectors $\boldsymbol{F} = [F_1 \ldots F_N]^T = [f_{0.1} \ldots f_{0.N}]^T$, i.e., $F_k = f_{0.k}$, $k = 1, \ldots, N$. The response of the computational model is the antenna reflection $|S_{11}(\boldsymbol{x}, f)|$, where f stands for the frequency. The minimax formulation of the merit function could be

$$U(\boldsymbol{R}(\boldsymbol{x}), \boldsymbol{F}) = \max\{|S_{11}(\boldsymbol{x}, F_1)|, \ldots, |S_{11}(\boldsymbol{x}, F_N)|\} \tag{11.2}$$

As another example, consider an antenna that is supposed to be matched at and around the operating frequency f_0, within the fractional bandwidth B (symmetric w.r.t. f_0). Furthermore, the realized gain $G(\boldsymbol{x}, f)$ should be maximized at f_0. Assuming that both f_0 and B are independent operating conditions, the objective vector would take the form of $\boldsymbol{F} = [F_1\ F_2]^T = [f_0\ B]^T$ and the merit function can be defined as

$$U(\boldsymbol{R}(\boldsymbol{x}), \boldsymbol{F}) = -G(\boldsymbol{x}, F_1) + \beta c(\boldsymbol{x}, \boldsymbol{F})^2 \tag{11.3}$$

where

$$c(\boldsymbol{x}, \boldsymbol{F}) = \max\left\{\frac{\max\{F_1(1 - F_2/2) \leq f \leq F_1(1 + F_2/2)\} + 10}{10}, 0\right\} \tag{11.4}$$

is the penalty function introduced to enforce the condition $|S_{11}| \leq -10$ dB over the bandwidth B (here, β is a penalty factor controlling the contribution of the penalty term).

11.1.2 Response Features

Recall that the response feature technology has been introduced to improve the computational efficiency of EM-driven optimization by exploring a specific structure of the system responses (Koziel, 2015). As the dependence between the coordinates of appropriately chosen characteristic (feature) points and the antenna geometry parameters is less nonlinear than a similar dependence for the original responses (e.g., frequency characteristics), reformulating the design problem in terms of response features normally leads to smoothening the functional landscape to be handled. The result is improved reliability and reduced cost of the optimization process (Koziel, 2015). This has been demonstrated, among others, for multiband antennas, microwave structures, or antenna arrays (Koziel & Pietrenko-Dabrowska, 2020b; Koziel & Ogurtsov, 2018; Koziel & Bandler, 2015). Here, we briefly recall the concept and discuss reformulation of the design problem (11.1) in terms of features.

For further use, the vector of the feature points of the antenna structure will be denoted as $\boldsymbol{p}(\boldsymbol{x}) = [p_1(\boldsymbol{x}) \ldots p_L(\boldsymbol{x})]^T$. As an example, let us consider a dual-band antenna shown in Fig. 11.1a. The feature points of the input characteristic are normally related to the location of the antenna resonances and the frequencies corresponding to -10 dB levels of $|S_{11}|$ (e.g., for bandwidth enhancement) (Fig. 11.1b). The locations of the resonances are sufficient to assess the design quality in terms of (11.2). Thus, we consider the resonant frequencies $f_{r.k}$ and the levels $l_{r.k}$, $k = 1, 2$. Then, the feature vector becomes $\boldsymbol{p}(\boldsymbol{x}) = [f_{r.1}(\boldsymbol{x})\ f_{r.2}(\boldsymbol{x})\ l_{r.1}(\boldsymbol{x})\ l_{r.2}(\boldsymbol{x})]^T$.

The first benefit of the feature-based approach has been illustrated in Fig. 11.1c, d: the dependence of the feature point coordinates on the antenna dimensions is quite regular, which indicates that the problem (11.2) reformulated in terms of the response features may be easier to solve. The feature-based objective function can be defined as

$$U_F(\boldsymbol{R}(\boldsymbol{x}), \boldsymbol{F}) = \max\{l_{r.1}(\boldsymbol{x}), l_{r.2}(\boldsymbol{x})\} + \beta \left\| \begin{bmatrix} F_1 \\ F_N \end{bmatrix} - \begin{bmatrix} f_{r.1}(\boldsymbol{x}) \\ f_{r.2}(\boldsymbol{x}) \end{bmatrix} \right\|^2 \tag{11.5}$$

As shown in Fig. 11.1e, the optimum design according to (11.2) is the same as that according to (11.5). Yet, the landscape of $U_F(\cdot)$ is monotonic so that the optimum can be reached using the local search routine. This may not be the case for $U(\cdot)$, which is more problematic. A consequence is that solving (11.5) generally requires less computational effort than solving (11.2). This is another benefit of the feature-based approach.

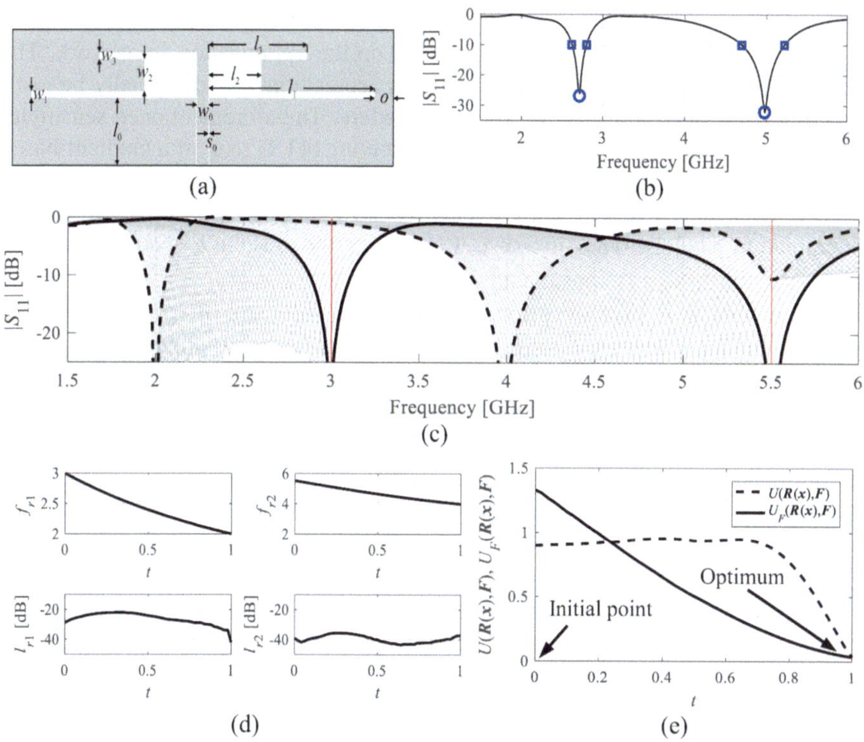

Fig. 11.1 The concept and the benefits of employing response features: (**a**) exemplary dual-band antenna geometry; (**b**) exemplary reflection characteristic (—) and the feature points corresponding to the antenna resonances (○), feature points corresponding to -10 dB levels (□); (**c**) exemplary initial design (thick dashed line) and the design optimized for target frequencies 3.0 and 5.5 GHz (thick solid line), and the family of antenna characteristics along the line segment connecting these two designs parameterized by $0 \leq t \leq 1$ (thin solid lines); (**d**) dependence between the parameter t and the feature point coordinates; (**e**) minimax objective function (11.2) (- - -) and feature-based objective function (11.5) (—) versus parameter t. The dependence of the feature points on the antenna dimensions is more regular than the dependence of the original responses; also, the feature-based objective function is more regular, and the target design can be reached from the given initial point through local search routine, which is not the case for the minimax formulation

11.1.3 Inverse Model: Feature-Based Forward Surrogate—Initial Design

The warm-start optimization procedure (Koziel & Pietrenko-Dabrowska, 2020a) assumes the existence of the database designs $\boldsymbol{x}_b^{(j)}$, $j = 1, \ldots, p$, obtained by solving the design problem (11.1) for several objective vectors $\boldsymbol{F}^{(j)} = [F_1^{(j)} \ldots F_N^{(j)}]^T$. Uniform allocation of $\boldsymbol{F}^{(j)}$ in F is preferred to maximize the amount of information carried by the surrogate models to be constructed. We will also use $\boldsymbol{J}_b^{(j)} = \boldsymbol{J}(\boldsymbol{x}_b^{(j)})$, $j = 1, \ldots, p$, to denote the Jacobian matrices of the antenna responses at $\boldsymbol{x}_b^{(j)}$. The designs $\boldsymbol{x}_b^{(j)}$ may be available from the previous work with the same structure but they may be also acquired specifically to set up the optimization framework. The initial computational cost associated with this acquisition would normally be justified by the intended multiple use of the procedure. The aforementioned sensitivity data is typically available as a by-product of solving (11.1) (e.g., if a gradient-based algorithm is used for this purpose).

The procedure of Koziel and Pietrenko-Dabrowska (2020a) utilizes two kriging interpolation models. The first one, $s_x(\cdot)$, maps the objective space F into the parameter space X. It is constructed using the set $\{\boldsymbol{F}^{(j)}, \boldsymbol{x}_b^{(j)}\}_{j = 1,...,p}$. It should be noted that this is an inverse surrogate because the argument of s_x is the objective vector $\boldsymbol{F}$ and the model returns the parameter vector $\boldsymbol{x}$ (corresponding to $\boldsymbol{F}$). Because s_x is constructed using the optimum designs $\boldsymbol{x}_b^{(j)}$, in effect, it provides an approximation of the set $U^*(\boldsymbol{R}(\boldsymbol{x}), F)$ of all designs that are optimum with respect to $\boldsymbol{F} \in F$ in the sense of (11.1). Furthermore, $s_x(\boldsymbol{F}^{(j)}) = U^*(\boldsymbol{R}(\boldsymbol{x}), \boldsymbol{F}^{(j)})$ for $j = 1, \ldots, p$, because the surrogate is interpolative. This suggests that the best possible approximation of the design $\boldsymbol{x}^* = U^*(\boldsymbol{R}(\boldsymbol{x}), \boldsymbol{F}_t)$ for a target objective vector $\boldsymbol{F}_t \in F$ would be

$$\boldsymbol{x}^{(0)} = s_x(\boldsymbol{F}_t) \tag{11.6}$$

The second surrogate is a forward model $s_J(\cdot)$, which maps the objective space F into the space of the antenna response sensitivities. The training data for this model is $\{\boldsymbol{F}^{(j)}, \boldsymbol{J}_b^{(j)}\}_{j = 1,...,p}$. As mentioned before, the Jacobians $\boldsymbol{J}_b^{(j)}$ are normally available from the optimization of the designs $\boldsymbol{x}_b^{(j)}$. The model s_J is used to provide an approximation of the system response sensitivities at the starting point $\boldsymbol{x}^{(0)}$ as

$$\boldsymbol{J}^{(0)} = s_J(\boldsymbol{F}_t) \tag{11.7}$$

The Jacobian matrix $\boldsymbol{J}^{(0)}$ is subsequently used to jump-start the design refinement process (cf. Sect. 11.1.4).

The most serious issue related to s_J is that forward modeling of sensitivities is very challenging due to their nonlinearity. Using the limited amount of training data $\{\boldsymbol{F}^{(j)}, \boldsymbol{J}_b^{(j)}\}_{j = 1,...,p}$ (in practice, the number p of the reference points is a few, usually less than 10) does not allow for rendering an accurate model. To improve the reliability, the forward model is constructed at the level of response features, i.e., it is identified using $\{\boldsymbol{F}^{(j)}, \boldsymbol{J}_P^{(j)}\}_{j = 1,...,p}$, where $\boldsymbol{J}_P^{(j)} = \boldsymbol{J}_P(\boldsymbol{x}_b^{(j)})$ is the Jacobian of the response features at $\boldsymbol{x}_b^{(j)}$, defined as

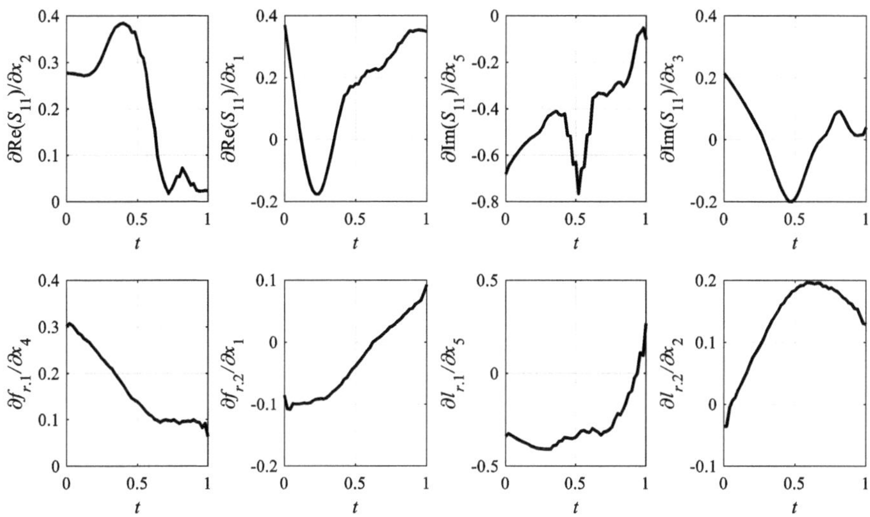

Fig. 11.2 Typical sensitivity values for the dual-band antenna of Fig. 11.1a, evaluated along the line segment connecting the designs shown in Fig. 11.1c and parameterized by $0 \le t \le 1$. The top panel shows the selected sensitivities of the antenna reflection at the selected frequencies. The bottom panel shows the sensitivities of the selected response feature coordinates. It can be observed that the feature-point sensitivities are more regular, therefore easier to model than the sensitivities of the antenna frequency characteristics

$$\boldsymbol{J}_P(\boldsymbol{x}) = \begin{bmatrix} \dfrac{\partial p_1(\boldsymbol{x})}{\partial x_1} & \cdots & \dfrac{\partial p_1(\boldsymbol{x})}{\partial x_n} \\ \vdots & \ddots & \vdots \\ \dfrac{\partial p_L(\boldsymbol{x})}{\partial x_1} & \cdots & \dfrac{\partial p_L(\boldsymbol{x})}{\partial x_n} \end{bmatrix} \tag{11.8}$$

The entries of $\boldsymbol{J}_P$ are estimated using finite differentiation as $\partial p_j(\boldsymbol{x})/\partial x_k \approx h^{-1}[p_j(\boldsymbol{x} + [0 \ldots 0 \ h \ 0 \ldots 0]^T) - p_j(\boldsymbol{x})]$ (h at the kth position in the perturbation vector). The design refinement described in the next section will be realized at the level of response features for which $\boldsymbol{J}_P$ is more relevant than $\boldsymbol{J}$. Figure 11.2 shows the exemplary components of the Jacobian matrices $\boldsymbol{J}_P$ and $\boldsymbol{J}$ of the dual-band antenna of Fig. 11.1a, evaluated along the same line segment as considered in Fig. 11.1c. It can be observed that the sensitivities of the response features are more regular as a function of the parameter t, therefore easier to model than the sensitivities of the frequency characteristics. Moreover, the number of Jacobian components is dramatically smaller for $\boldsymbol{J}_P$ than for $\boldsymbol{J}$, specifically $n \cdot L$ versus $n \cdot m$, where m is the number of frequencies considered in the simulation sweep (typically, a few dozen to a few hundred, whereas in most cases $L < 10$).

11.1.4 Design Refinement by Fast Gradient Search

The design refinement follows a generation of the initial point according to (11.5). It is required due to a limited accuracy of the inverse surrogate. The refinement is realized here using the trust-region gradient search (Conn et al., 2000), which is convenient because the expected design relocation is not significant as the initial design is typically of sufficient quality to apply local optimization. We use the standard form of the algorithm; however, the optimization process is carried out at the level of response features. The series of approximations $\boldsymbol{x}^{(i)}$, $i = 0, 1, \ldots$, is produced using the first-order Taylor model of the response features

$$\boldsymbol{L}_p^{(i)}(\boldsymbol{x}) = \boldsymbol{p}\left(\boldsymbol{x}^{(i)}\right) + \boldsymbol{J}_P\left(\boldsymbol{x}^{(i)}\right) \cdot \left(\boldsymbol{x} - \boldsymbol{x}^{(i)}\right) \tag{11.9}$$

by solving

$$\boldsymbol{x}^{(i+1)} = \arg \min_{\boldsymbol{x};\ -\boldsymbol{d}^{(i)} \le \boldsymbol{x} - \boldsymbol{x}^{(i)} \le \boldsymbol{d}^{(i)}} U\left(\boldsymbol{L}_P^{(i)}(\boldsymbol{x}), \boldsymbol{F}_t\right) \tag{11.10}$$

Normally, the Jacobian matrix is estimated using finite differentiation, which incurs the cost of additional n EM analyses of the antenna. Here, following Koziel and Pietrenko-Dabrowska (2020a), this was replaced by acquiring the initial estimation of $\boldsymbol{J}$ from the surrogate model s_J, and updating it using the Broyden formula (Broyden, 1965) in subsequent iterations. In this section, the forward surrogate represents the sensitivities of the response features, which—as argued before (cf. Fig. 11.2)—are less challenging to model due to their reduced nonlinearity (as compared to the gradients of the original antenna characteristics). As demonstrated in Sects. 11.1.6, 11.1.7 and 11.1.8, this has a considerable effect on the optimization procedure performance.

11.1.5 Complete Procedure

The flowchart of the discussed design optimization framework has been shown in Fig. 11.3. It consists of three major blocks: (1) construction of the inverse model and the forward feature-based surrogate (including the extraction of the response features; Sect. 11.1.3), (2) yielding the initial design and jump-starting the feature-based Jacobian (Sect. 11.1.3), and (3) design refinement (Sect. 11.1.4).

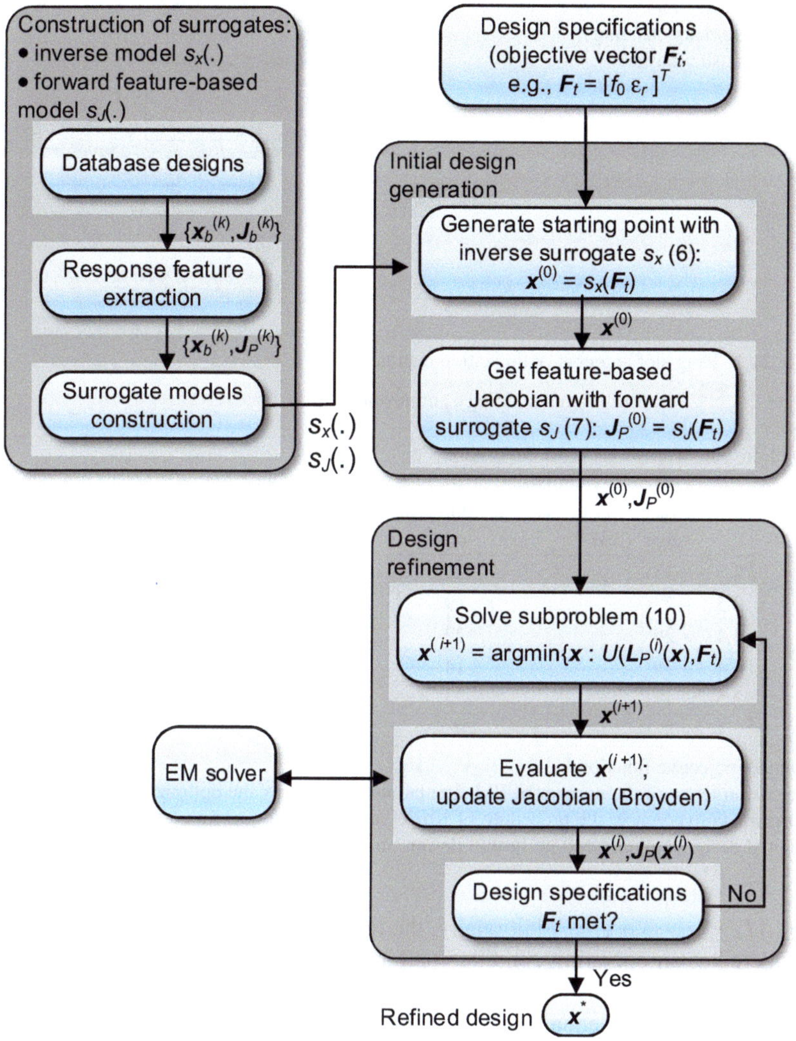

Fig. 11.3 Flowchart of the complete design optimization procedure and its three major blocks: construction of the inverse model and forward feature-based surrogate (including the extraction of the response features; Sect. 11.1.3—top left block), yielding the initial design and jump-starting the feature-based Jacobian (Sect. 11.1.3—top right block), and design refinement (Sect. 11.1.4—bottom right block)

11.1.6 Demonstration Examples—Case I: Ring-Slot Antenna

The first demonstration example is a slot antenna (Pietrenko-Dabrowska & Koziel, 2020a) implemented on the substrate of the height $h = 0.76$ mm. The antenna geometry has been shown in Fig. 11.4. The adjustable geometry parameters are

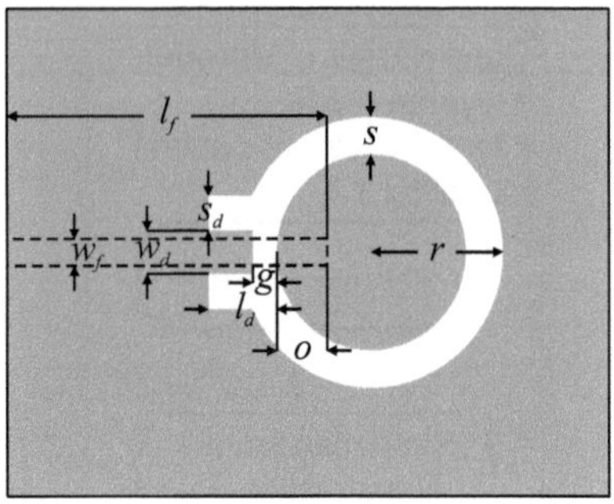

Fig. 11.4 Ring-slot antenna: geometry (feeding line shown using dashed lines)

Table 11.1 Ring slot antenna: optimization results

Target operating conditions $\boldsymbol{F}_t$		Optimization methodology and results			
		Warm-start procedure (Koziel & Pietrenko-Dabrowska, 2020a)		Feature-based warm-start procedure	
f_0 (GHz)	ε_r	Objective function value[a] (dB)	Optimization cost[b]	Objective function value[a] (dB)	Optimization cost[b]
3.3	2.5	−33	26	−54	17
5.3	2.5	−23	19	−55	15
3.0	4.4	−29	22	−33	17
5.8	4.4	−39	14	−56	11
5.1	2.5	−17	23	−63	20
5.8	2.1	−8	9[c]	−45	7

[a]Minimax objective function (11.2)
[b]The cost in terms of the number of EM simulations required by the optimization process
[c]Premature convergence: the procedure of Koziel and Pietrenko-Dabrowska (2020a) was stuck at the initial design and rendered no further design improvement

$\boldsymbol{x} = [l_f\ l_d\ w_d\ r\ s\ s_d\ o\ g]^T$. Additionally, the relative permittivity ε_r of the substrate is considered as an operating condition and a part of the objective space F. Given ε_r, the feed line width w_f is computed to ensure 50 Ω input impedance. The computational model of the antenna is implemented and simulated in CST Microwave Studio (~300,000 cells, simulation 90 s).

The design task is to allocate the antenna resonance at a target operating frequency f, assuming that the structure is implemented on the substrate of the permittivity ε_r. The objective space is defined by the ranges 2.5 GHz $\leq f \leq$ 6.5 GHz and $2.0 \leq \varepsilon_r \leq 5.0$. It should be noted that the ranges for both the f and ε_r are broad, which makes the optimization problem challenging. The inverse surrogate s_x and the forward model s_J are constructed using only six reference designs $\boldsymbol{x}_b^{(j)}$. These correspond to the pairs $\{f, \varepsilon_r\}$ = {2.5, 2.0}, {2.5, 5.0}, {6.5, 2.0}, {6.5, 5.0}, {4.0, 3.5}, and {5.0, 3.5} (frequency in GHz).

The considered methodology has been validated by optimizing the antenna of Fig. 11.4 for a number of target objective vectors $\boldsymbol{F}_t$. The detailed information has been gathered in Table 11.1. Optimized antenna dimensions can be found in

Table 11.2 Ring slot antenna: geometry parameter values

Target vector $\boldsymbol{F}_t$		Geometry parameter values (mm)							
f_0 (GHz)	ε_r	l_f	l_d	w_d	r	s	s_d	o	g
3.3	2.5	25.79	5.84	1.22	12.3	3.23	1.64	3.99	0.49
5.3	2.5	23.23	3.31	0.20	8.10	6.00	2.99	3.90	0.85
3.0	4.4	25.57	5.31	1.78	12.21	4.46	2.07	4.14	0.73
5.8	4.4	23.01	3.97	0.54	7.37	3.37	3.97	4.31	1.43
5.1	2.5	23.29	3.82	0.20	8.40	4.56	2.40	3.41	0.58
5.8	2.1	23.85	3.41	0.20	7.62	4.85	2.71	3.67	0.77

Table 11.2. Recall that the initial design is generated from the surrogate model s_x (cf. 11.5), whereas the refinement is carried out as described in Sect. 11.1.4 at the level of response features, with the antenna response Jacobian $\boldsymbol{J}_P$ (also at the level of features) jump-started using the surrogate s_J.

Figure 11.5 shows the antenna responses at the initial and the optimized designs. In order to demonstrate the importance of the feature-based approach, the antenna was also optimized using the methodology of Koziel and Pietrenko-Dabrowska (2020a), where the refinement (and the forward surrogate) involve sensitivities of the frequency characteristics of the antenna. It can be observed that incorporating response features into the warm-start framework considerably improves reliability of the optimization process and makes is faster. The average objective function value is −25 dB for the reference method and −51 dB for the considered approach. At the same time the average computational costs are 19 and 14 EM analyses for the reference and the considered procedures.

11.1.7 Demonstration Examples—Case II: Quasi-Yagi Antenna

The second example is a quasi-Yagi antenna with integrated balun (Farran et al., 2016) shown in Fig. 11.6. The antenna is implemented on the 1.55-mm-thick substrate. Its computational model is implemented and simulated in CST Microwave Studio (~430,000 mesh cells, simulation time 2 min). The independent design parameters are $\boldsymbol{x} = [L_a\ L_b\ L_c\ L_d\ W\ w_a\ D_a\ D_b\ D_c\ D_{lr}\ D_{rr}\ S_r\ w_{br}\ w_{cr}]^T$. The parameters with subscript r are relative, and the following relations hold: $D_l = D_{lr}\ L_a$, $D_r = D_{rr}\ L_a$, $S = S_r\ W$, $w_b = w_{br}\ W/2$, and $w_c = w_{cr}\ W$. The unit for the absolute dimensions is mm. Given the substrate permittivity ε_r, the input line width w_0 is adjusted to ensure 50 Ω input impedance.

The design objective is to ensure 8% impedance bandwidth centered at the target operating frequency f_0 and to maximize the antenna gain at f_0. All of this is to be achieved for a given substrate permittivity ε_r. A suitable objective function for this case is (11.3) and (11.4) (cf. Sect. 11.1.1). The optimization procedure should work for the following ranges of the operating frequency and permittivity:

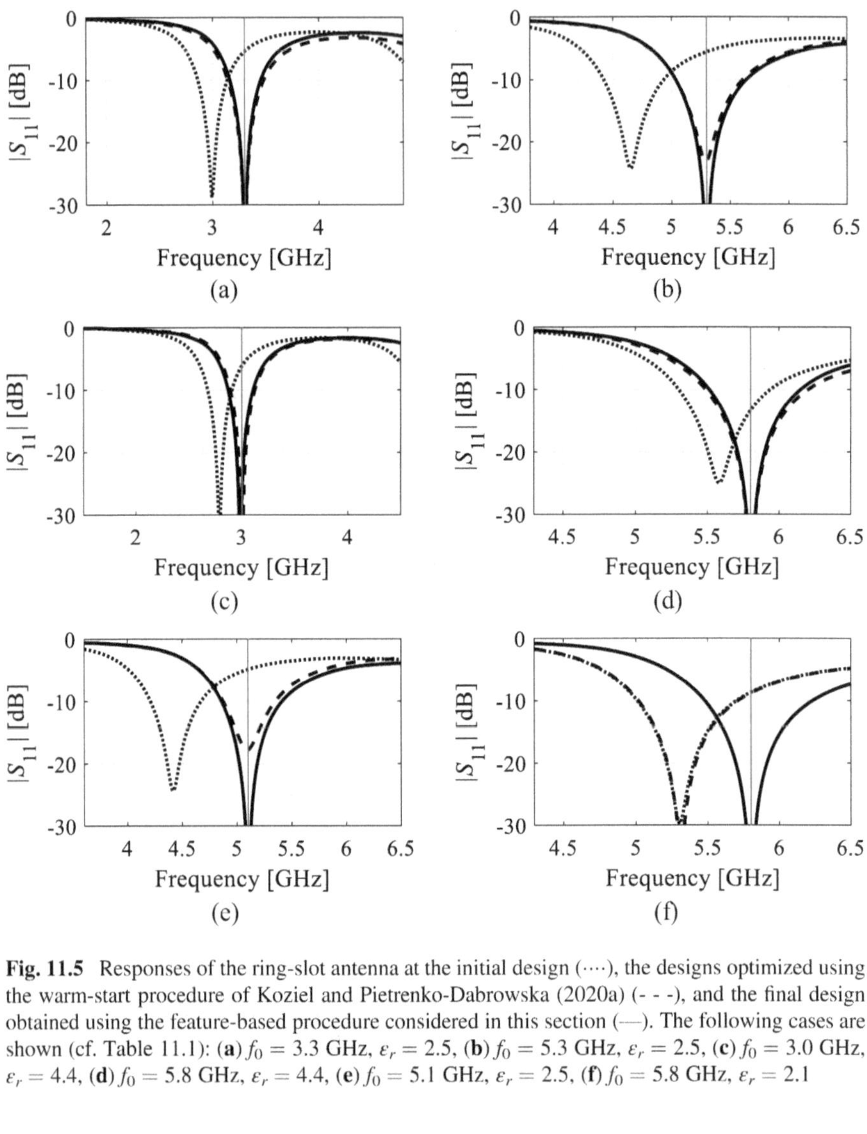

Fig. 11.5 Responses of the ring-slot antenna at the initial design (····), the designs optimized using the warm-start procedure of Koziel and Pietrenko-Dabrowska (2020a) (- - -), and the final design obtained using the feature-based procedure considered in this section (—). The following cases are shown (cf. Table 11.1): (**a**) $f_0 = 3.3$ GHz, $\varepsilon_r = 2.5$, (**b**) $f_0 = 5.3$ GHz, $\varepsilon_r = 2.5$, (**c**) $f_0 = 3.0$ GHz, $\varepsilon_r = 4.4$, (**d**) $f_0 = 5.8$ GHz, $\varepsilon_r = 4.4$, (**e**) $f_0 = 5.1$ GHz, $\varepsilon_r = 2.5$, (**f**) $f_0 = 5.8$ GHz, $\varepsilon_r = 2.1$

2.5 GHz $\leq f_0 \leq$ 4.5 GHz and $2.0 \leq \varepsilon_r \leq 5.0$. There are nine reference designs $\boldsymbol{x}_b^{(j)}$ that correspond to the pairs $\{f_0, \varepsilon_r\}$: {2.5, 3.5}, {3.5, 3.5}, {4.5, 3.5}, {3.5, 2.0}, {3.5, 5.0}, {4.5, 2.0}, {4.5, 5.0}, {2.5 2.0}, and {2.0, 5.0} (frequencies in GHz). One can observe that the problem is challenging primarily due to a large number of antenna parameters (14) but also broad ranges of both f_0 and ε_r.

The optimization methodology considered in this section has been validated by optimizing the antenna of Fig. 11.6 for several target objective vectors $\boldsymbol{F}_t$, as indicated in Table 11.3. Table 11.4 shows the optimized antenna dimensions. The antenna responses (reflection and realized gain) at the initial and the optimized

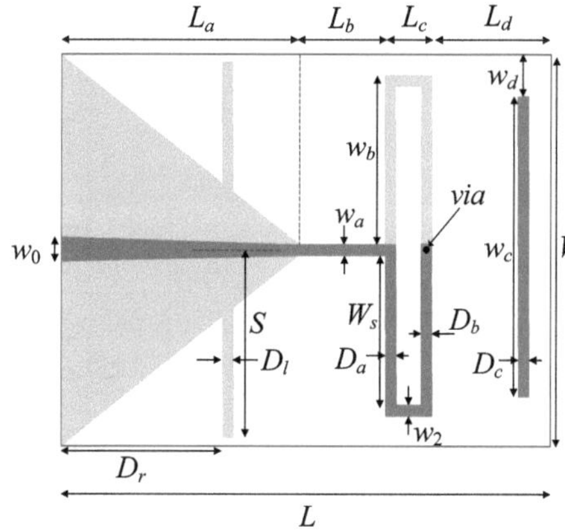

Fig. 11.6 Geometry of quasi-Yagi antenna with integrated balun (Farran et al., 2016). Ground-plane metallization shown using light-gray shade

Table 11.3 Quasi-Yagi antenna: optimization results

Target operating conditions F_t		Optimization methodology and results			
		Warm-start procedure (Koziel & Pietrenko-Dabrowska, 2020a)		Feature-based warm-start procedure	
f_0 (GHz)	ε_r	Realized gain value[a]	Optimization cost[b]	Realized gain value[a]	Optimization cost[b]
2.8	2.5	7.7	6	7.9	6
3.3	2.5	4.5	9[c]	7.5	10
3.0	3.0	6.4	8[c]	7.5	6
4.5	3.0	6.8	23	7.7	10
4.5	4.4	7.5	26	7.5	15
4.1	3.5	5.5	6[c]	7.2	6

[a]Realized gain at the target operating frequency f_0
[b]The cost in terms of the number of EM simulations required by the optimization process
[c]Premature convergence: the procedure of Koziel and Pietrenko-Dabrowska (2020a) was stuck at the initial design and rendered no further design improvement

designs are illustrated in Fig. 11.7. Here, also a non-feature methodology of Koziel and Pietrenko-Dabrowska (2020a) was used as a benchmark technique. The results of Table 11.4 and Fig. 11.7 demonstrate superiority of the approach discussed here. Similarly, as for the previous example, the response feature approach incorporated into the warm-start framework improves the reliability of the design process in a significant manner. For three out of six considered verification cases, the original procedure of Koziel and Pietrenko-Dabrowska (2020a) was stuck at the initial design produced by the inverse surrogate with no further improvement achieved, which is due to poor representation of the antenna sensitivities by the forward surrogate. On the other hand, the approach considered here was able to improve the design in all cases.

The average realized gain value at the center frequency is 6.4 dB for the reference method and 7.6 dB for the considered approach. The average computational costs are

Table 11.4 Quasi-Yagi antenna: geometry parameter values

Target vector $\boldsymbol{F}_t$		Geometry parameter values (absolute in mm, relative unitless)													
f_0 (GHz)	ε_r	L_a	L_b	L_c	L_d	W	w_a	D_a	D_b	D_c	D_{lr}	D_{rr}	S_r	w_{br}	D_a
2.8	2.5	19.87	11.00	15.44	27.54	51.47	1.52	1.27	4.26	40.34	0.37	0.41	0.99	0.64	0.72
3.3	2.5	18.63	9.33	13.10	25.42	48.81	1.77	1.64	3.85	4.16	0.42	0.49	0.90	0.60	0.63
3.0	3.0	19.13	10.20	14.12	26.34	49.57	1.35	1.49	3.97	4.38	0.38	0.45	0.99	0.62	0.66
4.5	3.0	14.59	7.46	7.33	19.34	40.60	1.08	1.93	2.44	4.35	0.05	0.40	0.66	0.53	0.50
4.5	4.4	12.52	5.45	6.81	19.00	36.16	1.22	1.17	2.25	4.32	0.05	0.40	0.50	0.51	0.50
4.1	3.5	15.87	6.33	8.16	21.17	41.88	1.17	1.25	3.11	3.65	0.39	0.50	0.83	0.52	0.53

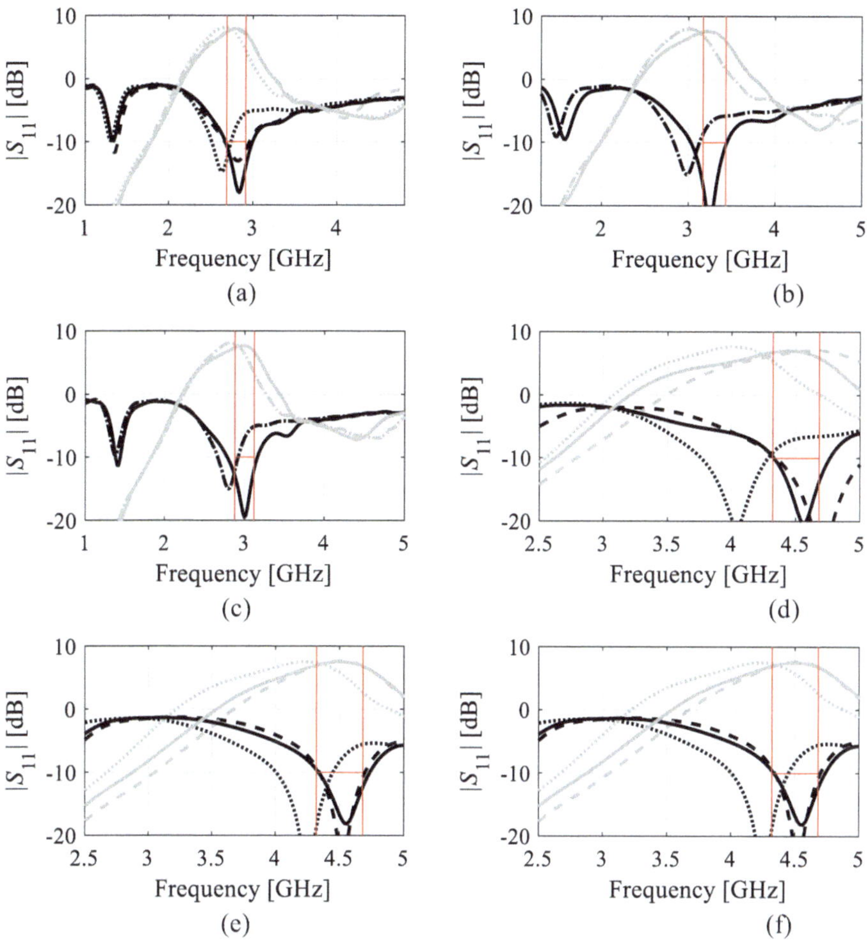

Fig. 11.7 Responses of the quasi-Yagi antenna of Fig. 11.6 at the initial design (····), the designs optimized using the warm-start procedure of Koziel and Pietrenko-Dabrowska (2020a) (- - -), and the final design obtained using the feature-based procedure considered in this section (—). The following cases are shown (cf. Table 11.3): (**a**) $f_0 = 2.8$ GHz, $\varepsilon_r = 2.5$, (**b**) $f_0 = 3.3$ GHz, $\varepsilon_r = 2.5$, (**c**) $f_0 = 3.0$ GHz, $\varepsilon_r = 3.0$, (**d**) $f_0 = 4.5$ GHz, $\varepsilon_r = 3.0$, (**e**) $f_0 = 4.5$ GHz, $\varepsilon_r = 4.4$, (**f**) $f_0 = 4.1$ GHz, $\varepsilon_r = 3.5$. The antenna reflection and realized gain characteristics are shown using black and gray, respectively

13 and 9 EM analyses for the reference and the feature-based procedures. However, one needs to have in mind that for three out of six cases, the reference method faces premature convergence, whereas in the remaining (successful) runs, the average cost was as high as 18 EM simulations.

11.1.8 Demonstration Examples—Case III: Rat-Race Coupler

The last demonstration example is a compact rat-race coupler (RRC) (Phani Kumar & Karthikeyan, 2013) shown in Fig. 11.8. The circuit employs a defected microstrip structure (meander spurline) within a folded transmission line. The slow-wave effect of the meander spurline allows for reducing the size of the coupler. Further size reduction is obtained by folding the part of the structure to the inside. The RRC is described by the following independent parameters $\boldsymbol{x} = [L_1\ b_r\ g\ h_{fr}\ s\ l_{fr}]^T$. All dimensions are expressed in mm except for the relative quantities with r-subscript, which are unitless. The following relationships hold: $L_2 = L_1 - g - w_0$, $a = (l_f - 17\,s)/16$, $b = (h_f - s)b_r$, $l_f = L_2\,l_{fr}$, $l_v = L_1 - 2g - 2w_0$, and $h_f = s + (w_0 - s)h_{fr}$. The fixed parameters are $dW = dL = 10$ mm. Whereas the input line width w_0 is computed for a given substrate permittivity ε_r so as to ensure 50 Ω input impedance. The computational model is implemented in CST Microwave Studio.

The design objectives include: (1) ensuring the coupler matching and isolation below −20 dB at the operating frequency f_0, and (2) achieving equal power split. The same ranges of the design objectives are considered, i.e., 1.0 GHz $\le f_0 \le$ 2.0 GHz and $2.0 \le \varepsilon_r \le 5.0$. Also, nine database designs have been used to set up the entire framework.

Table 11.5 gathers the selected target objective vectors $\boldsymbol{F}_t$, as well as the optimization results and computational cost. The geometry parameters of the optimized designs are provided in Table 11.6. Figure 11.9 shows the coupler S-parameters for the initial and optimized designs. The average computational cost of design optimization for this verification case is as low as 6 full-wave EM analyses. The quality of the optimal designs is quantified by the higher of the values of the matching and isolation characteristics at the target operating frequency f_0, the average value of which for the presented example and for all the target objective vectors is slightly below −20 dB. In this case, the average value of the power split at f_0 is around 0.1 dB. As for the previous examples, the framework of Koziel and Pietrenko-Dabrowska (2020a), in which the forward sensitivity model has been based on the

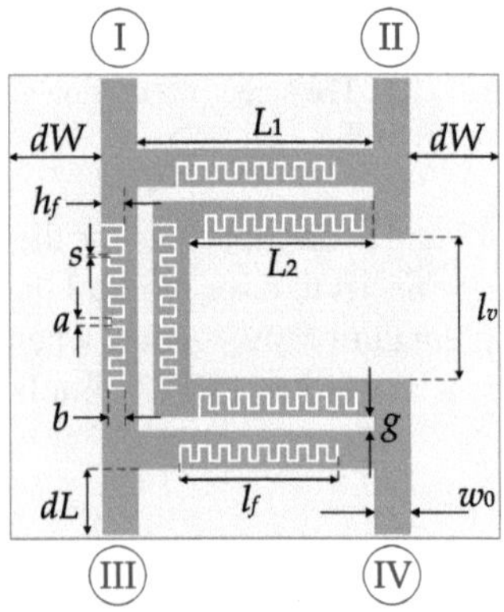

Fig. 11.8 Rat-race coupler (RRC) (Phani Kumar & Karthikeyan, 2013) with meander spurline and folded transmission line. The circuit ports are indicated by the numbered circles

Table 11.5 Optimization cost breakdown and the objective function values for rat-race coupler of Fig. 11.8

Target operating conditions $\boldsymbol{F}_t$				
f_0 (GHz)	ε_r	$\max(S_{11}(f_0), S_{41}(f_0))$ (dB)	dS (dB)	Optimization cost[a]
1.2	4.4	−22.1	0.03	5
1.5	2.7	−20.1	0.01	7
1.5	4.4	−19.9	0.30	5
1.6	2.5	−20.9	0.03	6
1.7	3.5	−18.9	0.25	5
1.8	2.5	−19.6	0.01	8

[a]The cost in terms of the number of EM simulations required by the optimization process

Table 11.6 Rat-race coupler: designs optimized for the target operating conditions of Table 11.5

Target vector $\boldsymbol{F}_t$		Geometry parameter values (mm)					
f_0 (GHz)	ε_r	L_1	b_r	g	h_{fr}	s	f_{fr}
1.2	4.4	32.46	0.55	2.63	0.69	0.33	0.53
1.5	2.7	31.50	0.51	2.06	0.55	0.26	0.63
1.5	4.4	28.68	0.52	2.95	0.54	0.23	0.47
1.6	2.5	30.82	0.57	2.46	0.52	0.24	0.62
1.7	3.5	27.46	0.50	2.52	0.52	0.20	0.48
1.8	2.5	28.55	0.59	1.63	0.44	0.31	0.58

entire frequency response of the coupler, failed to yield satisfactory results. Therefore, only an indirect comparison may be made with the results of optimizing the same structure with the use of the classical approach (Koziel et al., 2020). Using Koziel et al. (2020), the computational cost of yielding optimal RRC geometries exceeded 50 EM analyses.

11.2 Computationally-Efficient Reference Design Acquisition for Reduced-Cost Constrained Modeling and Redesign of Compact Microwave Passives

This section presents an automated procedure for accelerated acquisition of reference designs for user-defined sets of target values of performance parameters (operating frequencies, power split ratios) as well as the material parameters (substrate permittivity and/or thickness). The key component of the discussed methodology is the inverse model constructed at the level of response features (extracted from EM simulation results of the microwave component of interest). The inverse surrogate is set up using the estimated sensitivities of the system geometry parameters with respect to the performance figures. It allows for rendering a good initial design for further tuning, which is accomplished using local optimization, specifically, gradient search with sparse Jacobian updates. In order to maintain optimization reliability, a mechanism for adapting the design requirements is implemented and utilized if the

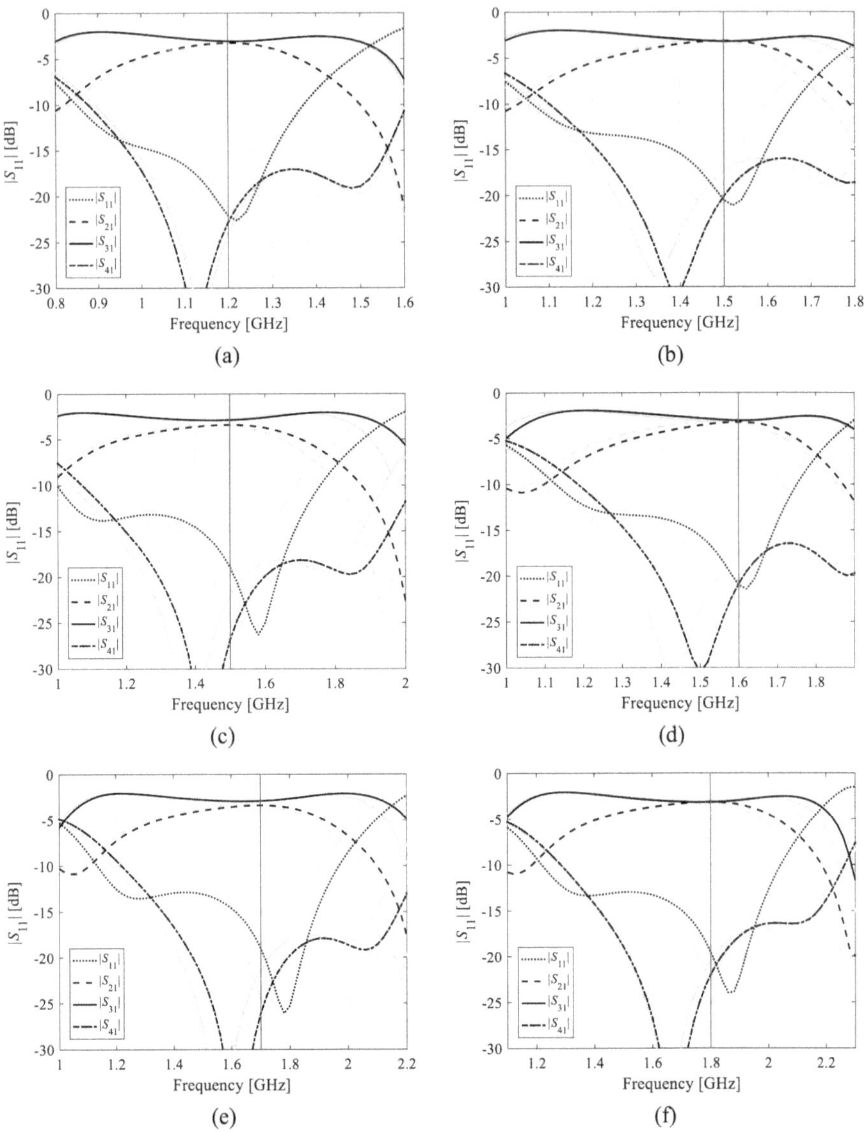

Fig. 11.9 The *S*-parameters of the rat-race coupler of Fig. 11.8: initial designs are marked gray, the designs optimized according to the feature-based procedure for the target operating conditions of Table 11.5: (**a**) $f_0 = 1.2$ GHz, $\varepsilon_r = 4.4$, (**b**) $f_0 = 1.5$ GHz, $\varepsilon_r = 2.7$, (**c**) $f_0 = 1.5$ GHz, $\varepsilon_r = 4.4$, (**d**) $f_0 = 1.6$ GHz, $\varepsilon_r = 2.5$, (**e**) $f_0 = 1.7$ GHz, $\varepsilon_r = 3.5$, (**f**) $f_0 = 1.8$ GHz, $\varepsilon_r = 2.5$ are marked black

initial point produced by the inverse surrogate is of insufficient quality. The approach is demonstrated using a miniaturized microstrip coupler. Application case studies concerning the employment of the reference sets for constructing performance-driven surrogates and warm-start design optimization are discussed as well.

11.2.1 Design Optimality: Database Design Acquisition Problem

The database design acquisition task is formalized using the following two spaces. The first one is the parameter space X of adjustable variables of the microwave circuit at hand. The variable vector is denoted as $\boldsymbol{x} = [x_1 \ldots x_n]^T$ and its entries normally represent the circuit dimensions. The space X is determined using the lower bounds $\boldsymbol{l} = [l_1 \ldots l_n]^T$ and the upper bound $\boldsymbol{u} = [u_1 \ldots u_n]^T$ for the parameters, i.e., $l_k \le x_k \le u_k$, $k = 1, \ldots, n$.

The second space is the objective space F that represents the figures of interest F_k, $k = 1, \ldots, N$, relevant to the design problem considered for the microwave component. The examples include the target operating frequency, the intended power split ratio for the couplers, but also material parameters such as permittivity of the substrate assigned to realize the circuit. The figures F_k form the objective vectors $\boldsymbol{F} = [F_1 \ldots F_N]^T$. Similarly as for the parameter space, it is assumed that the space F is an interval, i.e., it is determined by the lower bounds $\boldsymbol{l}_F = [l_{F.1} \ldots l_{F.N}]^T$ and the upper bounds $\boldsymbol{u}_F = [u_{F.1} \ldots u_{F.N}]^T$. For each $\boldsymbol{F} \in F$, we have $\boldsymbol{l}_F \le \boldsymbol{F} \le \boldsymbol{u}_F$ (component-wise).

Another important consideration is design optimality. It is understood here as the solution to the minimization problem

$$\boldsymbol{x}^* = U^*(\boldsymbol{F}) = \arg\min_{\boldsymbol{x}} U(\boldsymbol{x}, \boldsymbol{F}) \tag{11.11}$$

where U is a scalar merit function to be minimized. The function U quantifies the design quality with respect to the objective vector $\boldsymbol{F}$. A particular analytical formulation of U is problem dependent. For the sake of clarification, we discuss a specific example. Consider a microstrip coupler to be operating at the frequency f_0. The objective is to minimize $|S_{11}|$ and $|S_{41}|$ at f_0, as well as to ensure that the power split error $d_S(\boldsymbol{x},f) = |S_{21}(\boldsymbol{x},f)| - |S_{31}(\boldsymbol{x},f)| = K$ dB at f_0. The circuit is to be implemented on a dielectric substrate of permittivity ε_r. The figures of interest are $F_1 = f_0$, $F_2 = K$, and $F_3 = \varepsilon_r$. The merit function is defined as

$$U(\boldsymbol{x}, \boldsymbol{F}) = \max\{|S_{11}(\boldsymbol{x}, F_1)|, |S_{41}(\boldsymbol{x}, F_1)|\} + \beta_S c_S(\boldsymbol{x})^2 \tag{11.12}$$

where the penalty function $c_S(\boldsymbol{x})$ is

$$c(\boldsymbol{x}) = d_S(\boldsymbol{x}, F_1) - F_2 \tag{11.13}$$

Note that the penalty term is always positive but it becomes negligibly small whenever the power split error becomes sufficiently small.

11.2.2 Assessing Design Quality Using Response Features

For the sake of clarification, we consider an example branch-line coupler shown in Fig. 11.10a along with its *S*-parameter responses and a set of response features. As indicated in Fig. 11.10b, the feature points correspond to the minima of matching and isolation characteristics $|S_{11}|$ and $|S_{41}|$, respectively, the points allocated at –20 dB levels for $|S_{11}|$ and $|S_{41}|$, as well as the local maxima of the transmission responses $|S_{21}|$ and $|S_{31}|$. A particular selection of the feature points depends on the design context. For example, the transmission response maxima are useful to control the power split of the coupler; −20 dB (or other level) points can be used to control the operating bandwidth of the circuit.

For the purpose of further considerations, the feature points will be denoted as $\boldsymbol{p}_f(\boldsymbol{x}) = [f_{p1}(\boldsymbol{x}) \ldots f_{pK}(\boldsymbol{x})]^T$ (frequency coordinates) and $\boldsymbol{p}_l(\boldsymbol{x}) = [l_{p1}(\boldsymbol{x}) \ldots l_{pK}(\boldsymbol{x})]^T$ (level coordinates). As an illustration, let us consider again the problem discussed in Sect. 11.2.1. The objective function (11.12) pertinent to this task can be reformulated in terms of response features as

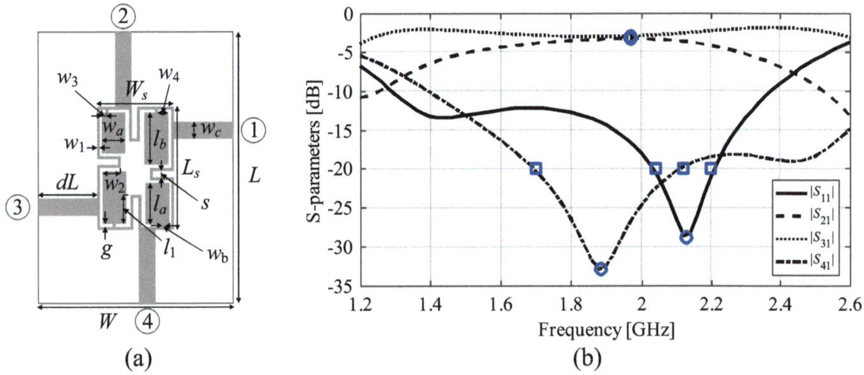

Fig. 11.10 Exemplary microwave coupler and its response features: (**a**) circuit geometry; (**b**) response features: exemplary *S*-parameter characteristic (—), feature points corresponding to $|S_{11}|$ and $|S_{41}|$ minima (○), feature points corresponding to −20 dB levels of $|S_{11}|$ and $|S_{41}|$ (□), feature points corresponding to local maxima of $|S_{21}|$ and $|S_{31}|$ (*)

$$U_F(\boldsymbol{x},\boldsymbol{F}) = \max\{l_{p1}(\boldsymbol{x}), l_{p2}(\boldsymbol{x})\} + \\ +\beta_1\left(l_{p3}(\boldsymbol{x}) - l_{p4}(\boldsymbol{x}) - F_2\right)^2 + \beta_2\left\|\begin{bmatrix} F_1 \\ F_1 \end{bmatrix} - \begin{bmatrix} f_{p1}(\boldsymbol{x}) \\ f_{p2}(\boldsymbol{x}) \end{bmatrix}\right\|^2 \tag{11.14}$$

where the feature points correspond to the minimum of $|S_{11}|$ and $|S_{41}|$ (points 1 and 2) as well as $|S_{21}|$ and $|S_{31}|$ at the target center frequency f_0 (points 3 and 4). Here, the primary objective is minimization of matching and isolation, whereas the penalty terms are to ensure the power split of K dB, as well as appropriate frequency alignment of the coupler responses with the target operating frequency f_0. It should be noted that the minimum of (11.12) and (11.14) coincide (assuming that perfect frequency alignment is attainable). Notwithstanding, the landscape of $U_F(\cdot)$ is more regular than that of $U(\cdot)$. Consequently, optimizing the coupler using (11.14) is, in generally, computationally cheaper than when using (11.12). As a matter of fact, the second penalty term in (11.14) can enforce monotonicity of the objective function profile so that the optimum solution is attainable even when starting from a relatively remote initial design (cf. Fig. 11.11).

The utilization of response features can also be used to exercise a better control over the circuit characteristics. If the third and the fourth feature points are the maxima of $|S_{21}|$ and $|S_{31}|$, respectively, the objective function is defined as

$$U_F(\boldsymbol{x},\boldsymbol{F}) = \max\{l_{p1}(\boldsymbol{x}), l_{p2}(\boldsymbol{x})\} + \\ +\beta_1\left(l_{p3}(\boldsymbol{x}) - l_{p4}(\boldsymbol{x}) - F_2\right)^2 + \beta_2\left\|\begin{bmatrix} F_1 \\ F_1 \\ F_1 \\ F_1 \end{bmatrix} - \begin{bmatrix} f_{p1}(\boldsymbol{x}) \\ f_{p2}(\boldsymbol{x}) \\ f_{p3}(\boldsymbol{x}) \\ f_{p4}(\boldsymbol{x}) \end{bmatrix}\right\|^2 \tag{11.15}$$

It is also possible (assuming that such a design exists) to enforce alignment of the transmission responses maxima with f_0, which generally leads to an improved power split bandwidth.

One of the important prerequisites of database design acquisition is to scale the circuit parameters for different operating frequencies. Although other performance criteria are often imposed as well (e.g., enforcing particular power split or ensuring a certain minimum bandwidth), the primary and the most challenging objective is to relocate the system characteristic to required frequency of operation. Once this is accomplished, the adjustment of other figures of interest is typically more straightforward. In order to efficiently handle frequency relocation, a frequency assessment function $f_a(\boldsymbol{x}, \boldsymbol{F}) = f_a(\boldsymbol{p}_f(\boldsymbol{x}), \boldsymbol{p}_l(\boldsymbol{x}), \boldsymbol{F})$ is utilized. It is defined at the level of response features and quantifies the distance between the actual and target operating frequency of the circuit at hand. The analytical form of f_a is problem dependent. Below, a specific example is provided to clarify the matter.

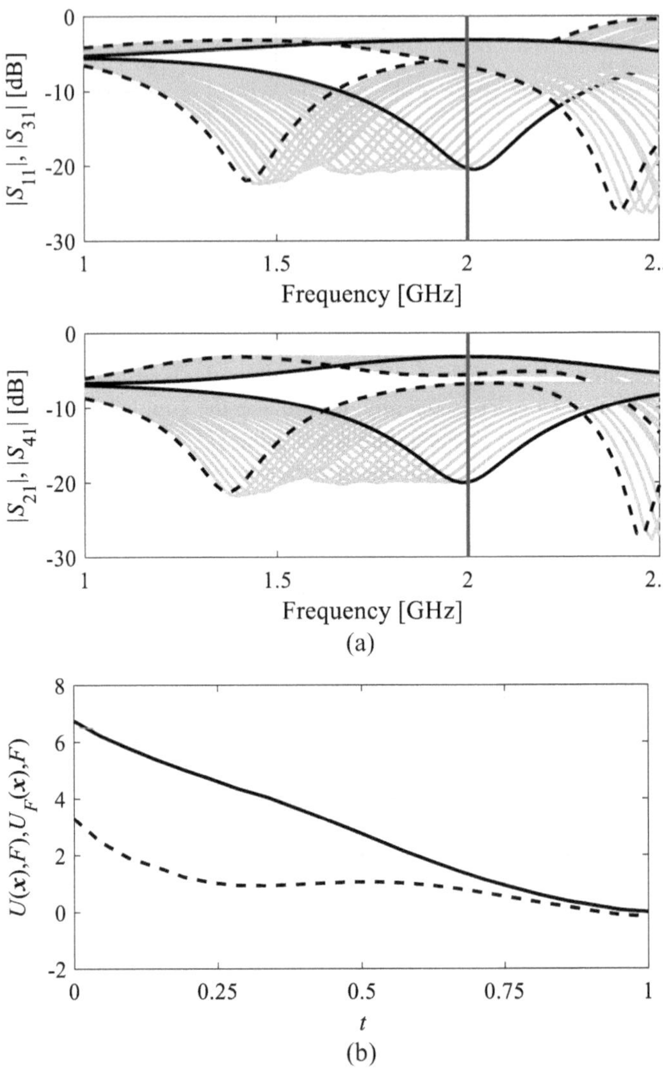

Fig. 11.11 Advantages of feature-based optimization: (**a**) responses of the coupler circuit of Fig. 11.10a at a certain initial design (- - -), the design optimized for the operating frequency of 2.0 GHz (—), as well as responses evaluated along the line segment connecting these two designs and parameterized using $0 \le t \le 1$ (gray lines); (**b**) conventional objective function (11.12) (- - -) and feature-based objective function (11.14) (—) as a function of t. It can be observed that the feature-based cost function is monotonic so that the optimum design is attainable through local search. This is not the case for the formulation (11.12)

Let us consider again the example of Sect. 11.2.1 and the feature-based objective function (11.14) determined using the same set of feature points: the minimum of $|S_{11}|$ and $|S_{41}|$ (points 1 and 2) as well as $|S_{21}|$ and $|S_{31}|$ at the target center frequency f_0

(points 3 and 4). We denote as $\boldsymbol{F}_{\text{extr}}(\boldsymbol{x})$ the actual objective vector extracted from EM-simulated circuit characteristics. In this case, the original figures of interest were (cf. Sect. 11.2.1) $F_1 = f_0$, $F_2 = K$, and $F_3 = \varepsilon_r$ (permittivity of the substrate the circuit is to be implemented on). The approximated operating frequency at the design $\boldsymbol{x}$ can be obtained, for example, as the average of the matching and isolation response minima, so that we have $\boldsymbol{F}_{\text{extr}}(\boldsymbol{x}) = [(f_{p1} + f_{p2})/2]^T$. Using this, the assessment function can be defined as

$$f_a(\boldsymbol{x}, \boldsymbol{F}) = \|\boldsymbol{F} - \boldsymbol{F}_{\text{extr}}(\boldsymbol{x})\|^2 = \left| F_1 - \frac{f_{p1} + f_{p2}}{2} \right| \tag{11.16}$$

11.2.3 Inverse Sensitivity

Another important part of the presented framework for database design acquisition is the so-called inverse sensitivity. It is denoted as

$$\boldsymbol{J}^x(\boldsymbol{F}) = \left[J_{jk}^x \right]_{\substack{j=1,\ldots,n \\ k=1,\ldots,N}} = \frac{\partial \boldsymbol{x}}{\partial \boldsymbol{F}} = \frac{\partial U^*(\boldsymbol{F})}{\partial \boldsymbol{F}} \tag{11.17}$$

The entries $J_{jk}^x = \partial x_j / \partial F_k$ of the sensitivity matrix $\boldsymbol{J}^x(\boldsymbol{F})$ are the partial derivatives of the system parameters x_j optimized for the objective vector $\boldsymbol{F}$ with respect to the performance figures F_k.

The knowledge of $\boldsymbol{J}^x$ allows for making reliable predictions of the designs optimized for the objective vectors in the vicinity of $\boldsymbol{F}$; however, the matrix $\boldsymbol{J}^x$ cannot be directly approximated using, e.g., finite differentiation. On the other hand, its estimation can be rendered using a mixture of analytical considerations and rapid numerical optimization as explained below. To this end, we start from the Jacobian $\boldsymbol{J}(\boldsymbol{x})$ of the EM-simulation response $\boldsymbol{R}$ of the circuit under design, evaluated at the design $\boldsymbol{x}$. Given the performance figure perturbations $\boldsymbol{d} = [d_1 \ldots d_N]^T$, the designs $\boldsymbol{x}^{(k)}$ corresponding to vectors $[F_1 \ldots F_k + d_k \ldots F_N]^T$ are found as

$$\boldsymbol{x}^{(k)} = \arg \min_{\boldsymbol{x}} U\left(\boldsymbol{x}, [F_1 \ \ldots \ F_k + d_k \ F_N]^T\right) \tag{11.18}$$

This is a two-step process with the initial approximation $\boldsymbol{x}^{(k.0)}$ of $\boldsymbol{x}^{(k)}$ obtained using the linear expansion $\boldsymbol{R}_L$ of the EM-simulation model $\boldsymbol{R}$

$$\boldsymbol{R}_L(\boldsymbol{y}) = \boldsymbol{R}(\boldsymbol{x}) + \boldsymbol{J}(\boldsymbol{x}) \cdot (\boldsymbol{y} - \boldsymbol{x}) \tag{11.19}$$

More specifically, $\boldsymbol{x}^{(k.0)}$ is the minimum of $U(\boldsymbol{x}, [F_1 \ \ldots \ F_k + d_k \ F_N]^T)$ computed using $\boldsymbol{R}_L$ in the neighborhood of $\boldsymbol{x}$. In the next step, the design is refined by means of a trust region (TR) optimization procedure

$$\boldsymbol{x}^{(k.i)} = \arg \min_{\boldsymbol{x}, \ \left\|\boldsymbol{x} - \boldsymbol{x}^{(k.i-1)}\right\| \leq \delta^{(i)}} U_L^{(i)}\left(\boldsymbol{x}, [F_1 \ \ldots \ F_k + d_k \ F_N]^T\right) \tag{11.20}$$

Here, the series $\boldsymbol{x}^{(k.i)}$, $i = 1, 2, \ldots$, approximates $\boldsymbol{x}^{(k)}$, whereas $\delta^{(i)}$ is the TR radius. The latter is adaptively adjusted using the standard TR rules (Conn et al., 2000). In (11.20), the cost function $U_L{}^{(i)}$ is defined using $\boldsymbol{R}_L$ with the Jacobian $\boldsymbol{J}$ updated using the Broyden formula (Broyden, 1965). Owing to this setup, the CPU cost of finding $\boldsymbol{x}^{(k)}$ is at the level of only $2n$ EM simulations (n being the parameter space dimensionality).

In practice, the actual performance figures $[F_1^{(k)} \ \ldots \ F_N^{(k)}]^T$ at the designs $\boldsymbol{x}^{(k)}$ obtained using the above procedure are only approximations of the target values $[F_1 \ldots F_k + d_k \ldots F_N]^T$, for $k = 1, \ldots, N$. Yet, because the perturbations d_k are small, the following relation holds:

$$x_l^{(k)} \approx x_l + \sum_{r=1}^{N} J_{lr}^x(\boldsymbol{x})\left[F_r^{(k)} - F_r\right] \tag{11.21}$$

In the matrix form, it can be rewritten as

$$\boldsymbol{X} = \boldsymbol{J}^x \boldsymbol{A}_F \tag{11.22}$$

where

$$\boldsymbol{X} = \left[\boldsymbol{x}^{(1)} - \boldsymbol{x} \quad \cdots \quad \boldsymbol{x}^{(N)} - \boldsymbol{x}\right] \tag{11.23}$$

and

$$\boldsymbol{A}_F = \begin{bmatrix} F_1^{(1)} - F_1 & \cdots & F_1^{(N)} - F_1 \\ \vdots & \ddots & \vdots \\ F_N^{(1)} - F_N & \cdots & F_N^{(N)} - F_N \end{bmatrix} \tag{11.24}$$

The unknown matrix $\boldsymbol{J}^x$ can be calculated from (11.22) as

$$\boldsymbol{J}^x = \boldsymbol{X}\boldsymbol{A}_F^{-1} \tag{11.25}$$

It should be noted that the matrix $\boldsymbol{A}_F$ is nonsingular because it is diagonally dominant (a consequence of the perturbed design construction).

The inverse sensitivity matrix $\boldsymbol{J}^x$ is the essential component for generating the initial estimates of the database designs as elaborated in Sect. 11.2.4.

11.2.4 Rapid Database Design Acquisition by Inverse Sensitivity and Response Features

The purpose of this section is to outline the algorithm for database design optimization. The procedure employs response features, assessment functions, and inverse sensitivity, as described in Sects. 11.2.1, 11.2.2 and 11.2.3, respectively. We start by discussing the process of generating the initial designs (Sect. 11.2.4.1), followed by the refinement procedure (Sect. 11.2.4.2). The entire algorithm is presented in Sect. 11.2.4.3 and illustrated using the flow diagram. Numerical verification using a microwave coupler will be provided in Sect. 11.2.5.

11.2.4.1 Initial Design Rendition Using Inverse Metamodels

The goal is to find a set of database designs corresponding to the predefined objective vectors $\boldsymbol{F}^{(k)} \in F$, $k = 1, \ldots, p$. These designs will be denoted as $\boldsymbol{x}_F^{(k)} = U^*(\boldsymbol{F}^{(k)})$ (cf. 11.11). It is assumed that the objective vectors are arranged in such a way that $\boldsymbol{F}^{(1)}$ represents the vector being the closest to $\boldsymbol{F}_c = p^{-1}\sum_k \boldsymbol{F}^{(k)}$ (i.e., the geometric center of $\{\boldsymbol{F}^{(k)}\}_{k=1,\ldots,p}$). The design $\boldsymbol{x}_F^{(1)} = U^*(\boldsymbol{F}^{(1)})$ is found in a conventional way, i.e., through direct optimization; however, to reduce the cost, feature-based optimization (FBO) is used whenever possible (Koziel, 2015). Having $\boldsymbol{x}_F^{(1)}$, the inverse sensitivity matrix $\boldsymbol{J}^x(\boldsymbol{F}^{(1)})$ is estimated using the procedure outlined in Sect. 11.2.3. $\boldsymbol{J}^x(\boldsymbol{F}^{(1)})$ allows us to establish an inverse linear model

$$\boldsymbol{s}(\boldsymbol{F}) = \boldsymbol{x}_F{}^{(1)} + \boldsymbol{J}^x\left(\boldsymbol{F}^{(1)}\right) \cdot \left(\boldsymbol{F} - \boldsymbol{F}^{(1)}\right) \tag{11.26}$$

The model (11.26) is subsequently used to generate the vectors $\boldsymbol{x}_F^{(k.0)}$, being the initial approximations of the remaining database designs $\boldsymbol{x}_F^{(k)}$. We have

$$\boldsymbol{x}_F^{(k.0)} = \boldsymbol{s}\left(\boldsymbol{F}^{(k)}\right) \tag{11.27}$$

It should be noted that the inverse model directly generates predictions of the optimum parameter sets corresponding to the objective vector being its input argument (which follows from the very definition of the inverse sensitivity). Reliability of these predictions has the same origin as what was observed in the case of response feature technology, i.e., less nonlinear relationship between the figures of interest (e.g., the circuit operating frequency) and the system geometry parameters

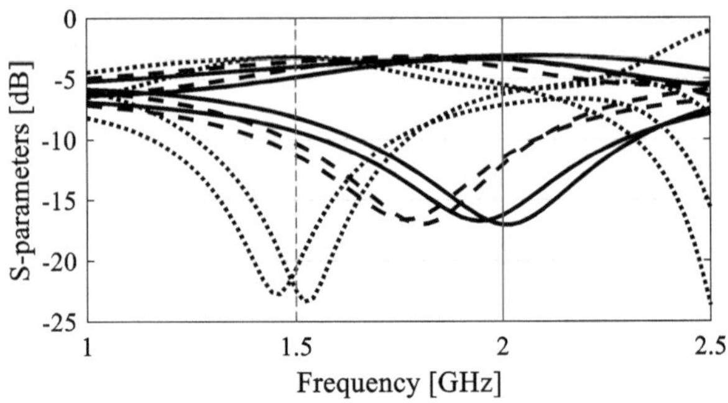

Fig. 11.12 Using inverse surrogate (11.26) for initial design prediction. The data corresponds to the coupler of Fig. 11.10a. Shown are the *S*-parameter characteristic for the center design $x_F^{(1)} = U^*(F^{(1)})$ (····), here, corresponding to $F^{(1)} = [f_0\ \varepsilon_r]^T = [1.5\ 3.5]^T$ (frequency in GHz), the initial design obtained using the inverse surrogate (22) for $F^{(2)} = [f_0\ \varepsilon_r]^T = [2.0\ 5.0]^T$ (- - -), and the corrected design generated by (11.29) (—). Note that the initial design retains adequate shape of the electrical characteristics, whereas the correction aligns the responses almost perfectly with the target frequency of 2.0 GHz

(as compared to the forward relationships between geometry parameters and the frequency characteristics). An illustration is provided in Fig. 11.12, showing the *S*-parameters of the coupler of Fig. 11.10a at $\boldsymbol{x}_F^{(1)}$ optimized to operate at the frequency $f_0 = 1.5$ GHz and realized on the substrate of relative permittivity $\varepsilon_r = 3.5$ (thus, $\boldsymbol{F}^{(1)} = [f_0\ \varepsilon_r]^T = [1.5\ 3.5]^T$). The inverse model is used to obtain the initial design $\boldsymbol{x}_F^{(2.0)}$ for $\boldsymbol{F}^{(2)} = [f_0\ \varepsilon_r]^T$ with $f_0 = 2.0$ GHz and $\varepsilon_r = 5.0$. It can be observed that this design, marked with the dashed line, is of good quality, yet with the operating frequency of around 1.8 GHz instead of the required 2.0 GHz. At this point, another advantage of the inverse surrogate becomes handy, which is a possibility of implementing a rapid design correction at the cost of only one EM simulation of the circuit under consideration. Assuming that $\boldsymbol{F}_{\text{extr}}^{(k)}$ denotes the actual objective vector extracted from EM-simulated circuit characteristics at the design $\boldsymbol{x}_F^{(k.0)}$ (cf. 11.27), the objective error $\Delta\boldsymbol{F}$ of the initial design can be calculated as

$$\Delta\boldsymbol{F} = \boldsymbol{F}^{(k)} - \boldsymbol{F}_{\text{extr}}^{(k)} \tag{11.28}$$

Then, the corrected design is obtained as

$$\boldsymbol{x}_{F.\text{corr}}^{(k.0)} = \boldsymbol{s}(\boldsymbol{F}^{(k)} + \Delta\boldsymbol{F}) \tag{11.29}$$

Equation (11.29) incorporates $\Delta\boldsymbol{F}$ and makes an improved prediction by reevaluating the inverse model at $\boldsymbol{F}^{(k)} + \Delta\boldsymbol{F}$. The source of the error $\Delta\boldsymbol{F}$ is a nonlinear relationship between the input argument $\boldsymbol{F}$ and the optimum parameter vector that corresponds to $\boldsymbol{F}$. This relationship is only approximated when using the linear inverse surrogate. Figure 11.12 shows the effect of correction (11.29) for the example discussed in the previous paragraph. It can be observed that the improved prediction (solid line) is considerably better and well aligned with the target operating frequency of 2.0 GHz.

11.2.4.2 Design Refinement Using Response Features

The refinement of the initial design $\boldsymbol{x}_F^{(k.0)}$ is executed using the trust-region (TR) gradient-based algorithm. If possible, the standard form of the cost function is replaced by the feature-based formulation as explained in Sect. 11.2.2. More specifically, the series $\boldsymbol{x}_F^{(k.i)}$, $i = 1, \ldots,$ approximating the design $\boldsymbol{x}_F^{(k)}$ is found as

$$\boldsymbol{x}_F^{(k.i+1)} = \arg \min_{\boldsymbol{x};\quad -\boldsymbol{d}^{(i)} \le \boldsymbol{x} - \boldsymbol{x}_F^{(k.i)} \le \boldsymbol{d}^{(i)}} U_F^{(i)}\left(\boldsymbol{x}, \boldsymbol{F}^{(k)}\right) \tag{11.30}$$

In (11.30), $U_F^{(i)}$ is the feature-based cost function evaluated at the level of the first-order expansion model $\boldsymbol{L}_F^{(k.i)}$ of $\boldsymbol{R}$

$$\boldsymbol{L}_F^{(k.i+1)} = \boldsymbol{R}\left(\boldsymbol{x}_F^{(k.i)}\right) + \boldsymbol{J}\left(\boldsymbol{x}_F^{(k.i)}\right) \cdot \left(\boldsymbol{x} - \boldsymbol{x}_F^{(k.i)}\right) \tag{11.31}$$

In the first iteration, the sensitivity matrix $\boldsymbol{J}$ is computed through finite differentiation. Subsequently, it is updated by means of the Broyden formula (Broyden, 1965). The optimization process in (11.30) is constrained to the region $\boldsymbol{x} \in [\boldsymbol{x}^{(i)} - \boldsymbol{d}^{(i)}, \boldsymbol{x}^{(i)} + \boldsymbol{d}^{(i)}]$ with the size vector $\boldsymbol{d}^{(i)}$ adjusted according to the usual TR rules (Conn et al., 2000).

11.2.4.3 Optimization Algorithm

The optimization algorithm generating the database designs involves—as its basic functional units—the components introduced in Sects. 11.2.1, 11.2.2 and 11.2.3 as well as Sects. 11.2.4.1 and 11.2.4.2. As before, we use the symbol $\boldsymbol{R}$ to denote the response of the EM simulation model of the structure under design; $\boldsymbol{p}_f$ and $\boldsymbol{p}_l$ will stand for the frequency and level coordinates of the feature points. Furthermore, we have p objective vectors $\boldsymbol{F}^{(k)} \in F$, $k = 1, \ldots, p$, that determine the targets for database design acquisition. The assessment function $f_a(\boldsymbol{x}, \boldsymbol{F}) = f_a(\boldsymbol{p}_f(\boldsymbol{x}), \boldsymbol{p}_l(\boldsymbol{x}), \boldsymbol{F})$ and the acceptance threshold $f_{a.\max}$ for f_a are defined by the user to determine whether the initial design obtained from the inverse surrogate can be accepted or rejected. Two other control parameters are used as well: $j_{\max}$—the maximum number of objective vector relaxations and $U_{F.\max}$—the maximum acceptable value of the objective function (see the explanation below).

1. Assign $\boldsymbol{F}^{(1)}$ (objective vector closest to $\boldsymbol{F}_c = p^{-1}\Sigma_k \boldsymbol{F}^{(k)}$);
2. Find the database design $\boldsymbol{x}_F^{(1)} = U^*(\boldsymbol{F}^{(1)})$ by direct search (typically, using FBO);
3. Estimate the inverse sensitivity matrix $\boldsymbol{J}^x(\boldsymbol{F}^{(1)})$ (cf. Section 11.2.3);
4. Using $\boldsymbol{J}^x(\boldsymbol{F}^{(1)})$, set up the inverse surrogate $\boldsymbol{s}(\boldsymbol{F})$ (11.26);
5. Set the design counter: $k = 2$;
6. Using $\boldsymbol{s}(\boldsymbol{F})$, find the initial design $\boldsymbol{x}_F^{(k.0)} = \boldsymbol{s}(\boldsymbol{F}^{(k)})$;;
7. **If** $f_a(\boldsymbol{x}_F^{(k.0)}) \le f_{a.\max}$,
        Go to Step 15;
    else
        Calculate the error $\Delta\boldsymbol{F} = \boldsymbol{F}^{(k)} - \boldsymbol{F}_{extr}^{(k)}$ (11.28);
        Find corrected design $\boldsymbol{x}_{corr}^{(k.0)} = \boldsymbol{s}(\boldsymbol{F}^{(k)} + \Delta\boldsymbol{F})$ (11.29);
        If $f_a(\boldsymbol{x}_{F.corr}^{(k.0)}) < f_a(\boldsymbol{x}_F^{(k.0)})$, set $\boldsymbol{x}_F^{(k.0)} = \boldsymbol{x}_{F.corr}^{(k.0)}$;
    end
8. **If** $f_a(\boldsymbol{x}_F^{(k.0)}) \le f_{a.\max}$,
        Go to Step 15;
    end
9. Set $j = 1$ (local counter);
10. Extract the actual objective values $\boldsymbol{F}_{extr}$ at $\boldsymbol{x}_F^{(k.0)}$;
    Calculate the updating factor $t = \min\{1, f_{a.\max}/||\boldsymbol{F}^{(k)} - \boldsymbol{F}_{extr}||\}$;
11. Update the objective vector as $\boldsymbol{F}_{tmp}^{(k)} = \boldsymbol{F}_{extr} + t(\boldsymbol{F}^{(k)} - \boldsymbol{F}_{extr})$;
12. Update the initial design as $\boldsymbol{x}_F^{(k.0)} = \text{argmin}\{\boldsymbol{x} : U_F(\boldsymbol{x}, \boldsymbol{F}_{tmp}^{(k)})$; the problem is solved using TR gradient search;
13. Set $j = j + 1$;
14. **If** ($f_a(\boldsymbol{x}_F^{(k.0)}) \le f_{a.\max}$ AND $t = 1$) OR $j > j_{\max}$ (use $\boldsymbol{F}_{tmp}^{(k)}$ in place of $\boldsymbol{F}^{(k)}$ to evaluate f_a),
        Set $\boldsymbol{x}_F^{(k)} = \boldsymbol{x}_F^{(k.0)}$ and go to Step 16;
    else
        Go to Step 10;
    end
15. Find $\boldsymbol{x}_F^{(k)} = \text{argmin}\{\boldsymbol{x} : U_F(\boldsymbol{x}, \boldsymbol{F}^{(k)})$ using TR gradient search;
16. Store $\boldsymbol{x}_F^{(k)}$ along with its objective vector $\boldsymbol{F}_{extr}^{(k)}$;
    Set $k = k + 1$;
17. If $k \le p$ go to Step 6;
18. END.

Fig. 11.13 Pseudocode of the database design acquisition algorithm

The assessment function and the assessment threshold are employed to decide about the sufficiency of the (intermediate) design obtained at a particular step of the optimization run. If the design is rejected, a correction is typically made or the performance specifications are temporarily relaxed. The latter can be done no more than $j_{\max}$ times; such a procedure is incorporated to allow approaching the target objective vector in smaller steps. The design produced at the refinement step is considered acceptable if its corresponding objective function value is not larger than $U_{F.\max}$.

The overall optimization algorithm has been presented in Fig. 11.13 in the form of a pseudocode. The purpose of Steps 1–4 is to construct the inverse surrogate model

(11.26) as elaborated in Sect. 11.2.4.1. The remaining part of the algorithm is a loop arranged to obtain all designs $\boldsymbol{x}_F^{(k)}$ corresponding to the target objective vectors $\boldsymbol{F}^{(k)}$. The initial approximation to $\boldsymbol{x}_F^{(k)}$ is produced in Step 6. In case it is accepted (i.e., if the corresponding assessment function value is below the acceptance threshold $f_{a.\max}$), it is further refined in Step 15. Otherwise, it is corrected (cf. 11.28 and 11.29). If the correction fails, an additional parameter tuning is executed upon relaxing the performance requirements (Steps 10 and 11). The tuning is realized using the TR algorithm (Step 12), initially using the Broyden updates (the cost of which is only one EM circuit simulation per iteration) or using full finite differentiation updates if the former procedure fails (i.e., $U_{\mathrm{F}}(\boldsymbol{x}_F^{(k.0)}) > U_{F.\max}$). The auxiliary tuning runs are terminated if the relaxed specifications are sufficiently close to the original target or their maximum number is exceeded.

Figure 11.14 provides a graphical explanation of the algorithm operation in the form of a flow diagram. At this point, it should be mentioned that the underlying assumption of the discussed optimization procedure is that the target objective vectors are attainable; in other words, the considered microwave structure can be optimized for all vectors $\boldsymbol{F}^{(k)}$. The algorithm will still operate even if this is not the case; however, some of the acquired designs may not satisfy the imposed performance requirements.

11.2.5 Demonstration Example: Miniaturized Branch-Line Coupler

For the sake of demonstration, consider a miniaturized branch-line coupler (BLC) (Tseng & Chang, 2012). The circuit, shown in Fig. 11.15, is implemented on a 0.76-mm-thick substrate of the permittivity ε_r, which is one of the components of the objective space. The design variables are $\boldsymbol{x} = [g\ l_{1r}\ l_a\ l_b\ w_1\ w_{2r}\ w_{3r}\ w_{4r}\ w_a\ w_b]^T$. Other parameters are described by the following relations: $L = 2dL + L_s$, $L_s = 4w_1 + 4g + s + l_a + l_b$, $W = 2dL + W_s$, $W_s = 4w_1 + 4g + s + 2w_a$, $l_1 = l_b l_{1r}$, $w_2 = w_a w_{2r}$, $w_3 = w_{3r} w_a$, and $w_4 = w_{4r} w_a$. The computational model of the structure is implemented in CST Microwave Studio.

The design objectives are as follows. The coupler is to be implemented on the substrate of permittivity ε_r within the range $2.0 \le \varepsilon_r \le 5.0$. The power split $d_S(\boldsymbol{x}, f) = |\,|S_{21}(\boldsymbol{x}, f)| - |S_{31}(\boldsymbol{x}, f)|\,|$ is supposed to be 0 dB at the target operating frequency f_0, with the matching and isolation characteristics minimized at f_0 (preferably both being below −20 dB). The operating frequency range of interest is 1.0 GHz $\le f_0 \le$ 2.0 GHz. The database designs are selected to correspond to the following pairs of f_0 and ε_r (frequency in GHz): $[f, \varepsilon_r]$ = [1.0 2.0], [1.5 2.0], [2.0 2.0], [1.0 3.5], [1.5 3.5], [2.0 3.5], [1.0 5.0], [1.5 5.0], and [2.0 5.0]. These points will be used to set up a warm-start optimization procedure (Pietrenko-Dabrowska & Koziel, 2020b) that permits rapid parameter tuning of the coupler geometry parameters, so that the structure can be optimized for f_0 and ε_r within the aforementioned objective

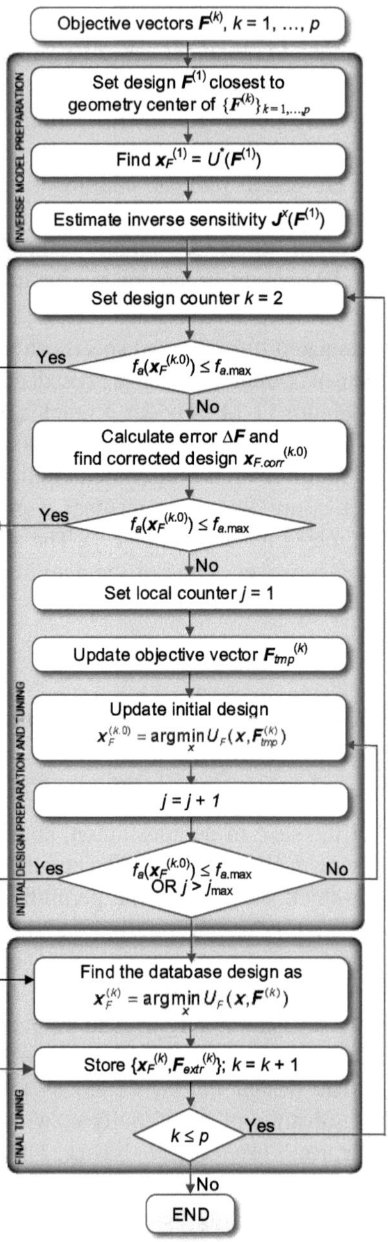

Fig. 11.14 Fast database design acquisition using inverse surrogates and response features: flow diagram

space ranges. More specifically, the database designs are utilized to construct two kriging metamodels: (1) an inverse model employed to generate the initial design for further tuning, and (2) a forward model of the coupler response sensitivities, here, evaluated at the level of response features. The latter allows us to jump-start the

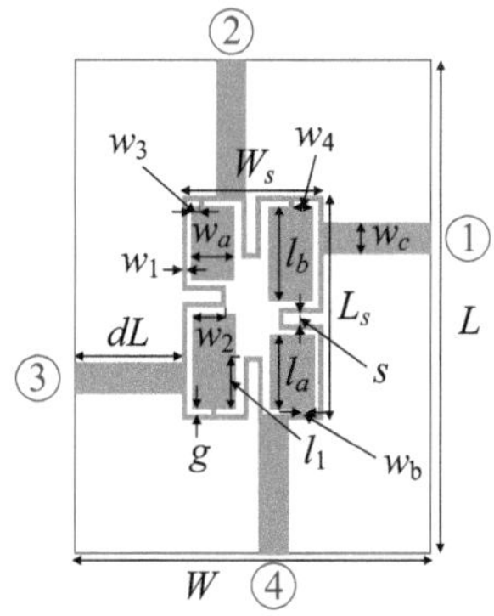

Fig. 11.15 Miniaturized branch-line coupler (BLC) (Tseng & Chang, 2012). The circuit ports marked using numbered circles

Table 11.7 Miniaturized BLC: optimization cost of database design generation

Optimization technique	Computational cost[a]	
	Total[d]	Per design
Conventional (minimax)[b]	1014	113
Conventional (feature-based)[c]	858	95
Algorithm of Sect. 11.2	389	43

[a]Computational cost (number of EM analyses of the BLC)
[b]Gradient-based optimization using minimax formulation; starting points adjusted by parameter sweeping whenever necessary
[c]Feature-based optimization (cf. 11.14 with $K = 0$ dB); starting points adjusted by parameter sweeping whenever necessary
[d]The database set consists of nine designs

gradient-based tuning process, executed with the trust-region framework and the rank-one Jacobian updating formula (Broyden, 1965).

For this example, the objective pair [1.5 3.5] was selected to be $\boldsymbol{F}^{(1)}$ because it coincides with $\boldsymbol{F}_c = p^{-1}\sum_k \boldsymbol{F}^{(k)}$. The design $\boldsymbol{x}_F^{(1)} = U^*(\boldsymbol{F}^{(1)})$ has been found using FBO (Koziel, 2015) (cost: 69 EM simulations of BLC). The cost of estimating the inverse sensitivity matrix $\boldsymbol{J}^x$ at $\boldsymbol{x}_F^{(1)}$ was 41 simulations of the coupler. Similarly, as for the first example, the assessment function (11.16) was utilized with the acceptance threshold $f_{a.\max} = 0.2$. In this case, the quality of the initial design was sufficient (i.e., $f_a(\boldsymbol{x}_F^{(k.0)}) \le f_{a.\max}$) for most of the designs, but in some cases, the correction step and objective relaxation was necessary. The design refinement was realized using FBO. Table 11.7 shows the overall and the average cost of generating the database points: 389 (total) and 43 (per design) EM analyses of the BLC, respectively. As before, the overall cost includes optimization of $\boldsymbol{x}_F^{(1)}$, identification of $\boldsymbol{J}^x$, and optimization of $\boldsymbol{x}_F^{(k)}$, $k = 2, \ldots, 9$. The cost of conventional optimization is much higher: 858 and 1014 EM simulations for FBO and minimax formulation, respectively. Figure 11.16 shows the simulated BLC characteristics at the selected database designs.

As discussed before, the application example for the BLC is the implementation of the warm-start optimization procedure (Pietrenko-Dabrowska & Koziel, 2020b), briefly outlined above. The procedure was used to optimize the coupler for several

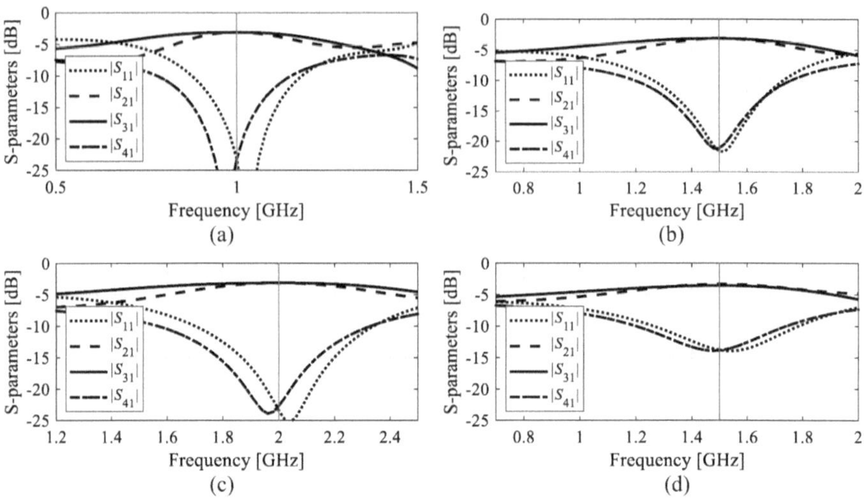

Fig. 11.16 Miniaturized BLC: EM-simulated *S*-parameters at the selected database designs found using the considered optimization procedure: (**a**) $[f_0\ \varepsilon_r] = [1.0\ 5.0]$, (**b**) $[f_0\ \varepsilon_r] = [1.5\ 5.0]$, (**c**) $[f_0\ \varepsilon_r] = [2.0\ 3.5]$, (**d**) $[f_0\ \varepsilon_r] = [1.5\ 2.0]$ (frequencies in GHz). Target operating frequencies marked using vertical lines

target objective vectors consisting of the pairs of the intended operating frequency of the circuit and the permittivity of the dielectric substrate the coupler is to be realized on. Figure 11.17 shows the obtained results, including the initial designs generated by the inverse surrogate and the final designs rendered through the refinement process. In all cases, the designs are of high quality with almost equal power split and well centered at the target frequency f_0. Owing to the problem-specific knowledge embedded into the framework (in the form of the database designs), the optimization cost is only a few EM coupler simulations on the average.

11.3 Expedited Microwave Design Closure Using Kriging Surrogates and Iterative Correction

This section discusses an approach to expedited design closure of microwave circuits. The procedure incorporates available designs (e.g., existing from the previous design work on the same structure) in the form of the kriging interpolation models, utilized to yield a reasonable initial design and to accelerate its further refinement. An important component of the framework is an iterative correction procedure that feeds the accumulated discrepancies between the target and the actual design objective values back to the kriging surrogate to produce improved predictions. The efficacy of the considered methodology is demonstrated using a miniaturized impedance-matching transformer with the optimized designs obtained at the cost of a few EM simulations of the respective circuits. The relevance of the

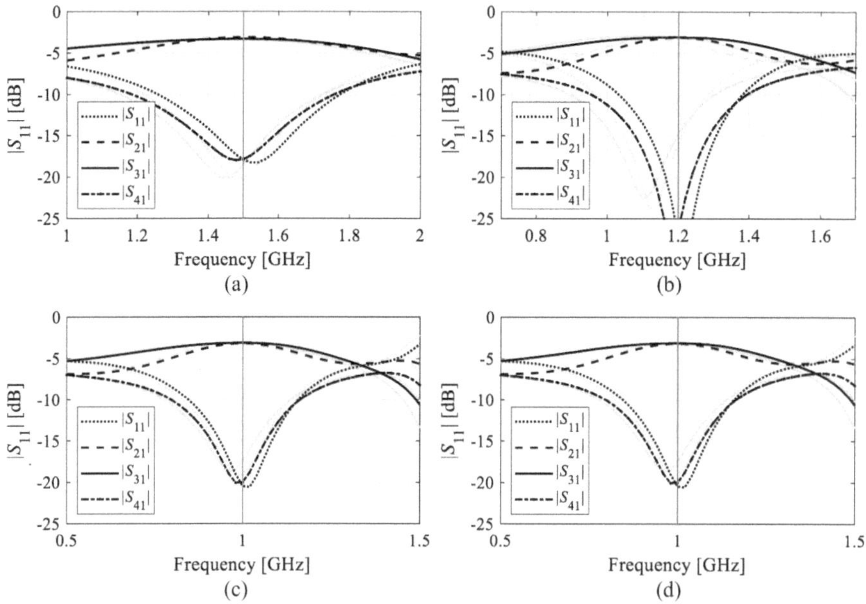

Fig. 11.17 Miniaturized BLC: EM-simulated S-parameters at the initial (gray) and the optimized (black) designs corresponding to the target objective vectors: (**a**) $f_0 = 1.5$ GHz, $\varepsilon_r = 2.7$, (**b**) $f_0 = 1.2$ GHz, $\varepsilon_r = 4.4$, (**c**) $f_0 = 1.0$ GHz, $\varepsilon_r = 3.0$, (**d**) $f_0 = 2.0$ GHz, $\varepsilon_r = 4.4$. The initial designs are generated by the inverse model constructed using the database designs acquired with the discussed optimization procedure. Target operating frequencies marked using vertical lines

iterative correction is corroborated through the comparative studies showing its superiority over rudimentary gradient-based refinement.

We use the same notation as in Sect. 11.1 concerning the vector of design parameters $\boldsymbol{x}$, objective vector $\boldsymbol{F}$, parameter and objective spaces X and F, respectively, as well as computational model $\boldsymbol{R}(\boldsymbol{x})$ (see Sect. 11.1.1). The design closure task is formulated as in (11.1).

11.3.1 Preexisting Designs and Kriging Surrogates

The prerequisite for the considered design closure framework is the existence of designs $\boldsymbol{x}_b^{(j)}$, $j = 1, \ldots, p$, pre-optimized (according to (11.1)) with respect to the selected objective vectors $\boldsymbol{F}^{(j)} = [F_1^{(j)} \ \ldots \ F_N^{(j)}]^T$. These vectors may be known beforehand, e.g., from the previous design work on a given component or obtained specifically to initialize the design procedure.

Furthermore, we denote by $\boldsymbol{J}_b^j = \boldsymbol{J}(\boldsymbol{x}_b^{(j)})$, $j = 1, \ldots, p$, the Jacobians of $\boldsymbol{R}(\boldsymbol{x}_b^{(j)})$. These are normally obtained as a by-product of solving (11.1) when $\boldsymbol{x}_b^{(j)}$ were found using gradient search; otherwise, sensitivity can be estimated through finite differentiation.

The reference designs are used to construct two kriging interpolation surrogates (Queipo et al., 2005):

- $s_x(\boldsymbol{F})$—the model defined over the objective space F with the outputs in the parameter space X; the training data set for $s_x(\cdot)$ is $\{\boldsymbol{F}^{(j)}, \boldsymbol{x}_b^{(j)}\}_{j=1,\ldots,p}$;
- $s_J(\boldsymbol{F})$—the model defined over the objective space F with the values in the space of the component response sensitivities; here, the training data is $\{\boldsymbol{F}^{(j)}, \boldsymbol{J}_b^j\}_{j=1,\ldots,p}$.

A conceptual illustration of the objective space, the meaning of the reference designs, as well as the kriging surrogates have been shown in Fig. 11.18 for an exemplary miniaturized impedance-matching transformer.

11.3.2 Initial Design

The set $s_x(F)$ of all points of the form $s_x(\boldsymbol{F})$, $\boldsymbol{F} \in F$ (also referred to as the image of the objective space through s_x) approximates the set $O(F)$ consisting of the designs that are optimum in the sense of (11.1) for all $\boldsymbol{F} \in F$. Consequently, the point

$$\boldsymbol{x}^{(0)} = s_x(\boldsymbol{F}_t) \tag{11.32}$$

is the best initial design corresponding to a target vector $\boldsymbol{F}_t$, which can be obtained from the information contained in the designs $\boldsymbol{x}_b^{(j)}$.

The approximated sensitivity (Jacobian matrix) of the component outputs at $\boldsymbol{x}^{(0)}$ can be found in a similar manner as

$$\boldsymbol{J}^{(0)} = s_J(\boldsymbol{F}_t) \tag{11.33}$$

Figure 11.19 illustrates the introduced concepts for an exemplary two-dimensional objective space and three-dimensional parameter space. It should be noted that the initial point $\boldsymbol{x}^{(0)}$ obtained from (11.32) does not coincide with $\boldsymbol{x}^*$ (the design optimum in the sense of (11.1)) for $\boldsymbol{F} = \boldsymbol{F}_t$. This is because $s_x(F)$ is only an approximation of $O(F)$. Consequently, a design refinement is necessary.

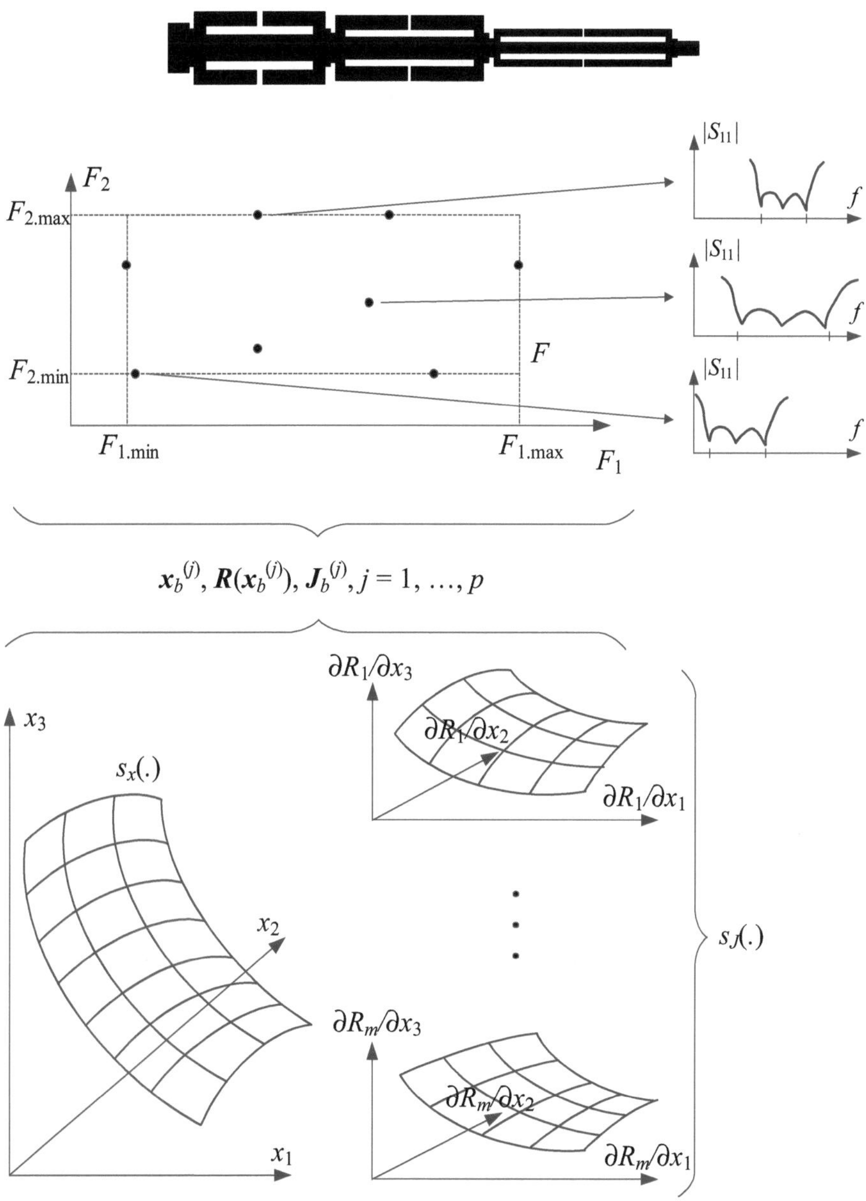

Fig. 11.18 Conceptual illustration of the objective space (here, with the operating conditions being the lower and upper ends of the operating frequency band) for a compact three-section impedance-matching transformer shown at the top. The frequency characteristics exemplify the transformer reflection optimized for various operating bandwidths according to the objective vectors allocated within the objective space F. The reference designs, corresponding transformer outputs, and their sensitivities are subsequently used to construct the surrogates s_x and s_J, shown as the response surfaces at the bottom part of the picture

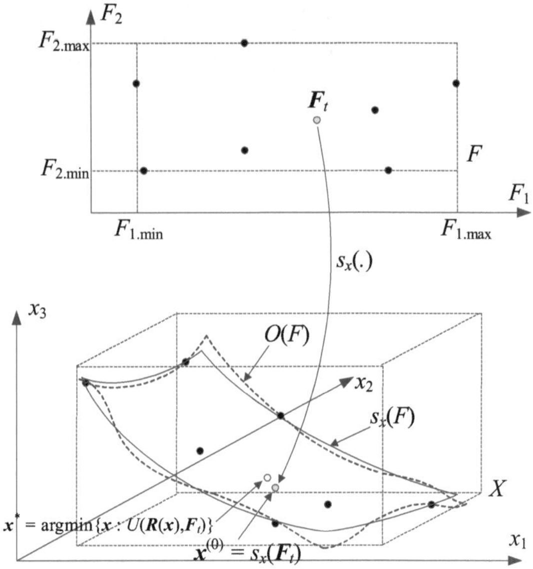

Fig. 11.19 Basic concepts of the expedited design optimization procedure. Top: the objective space F and the exemplary target objective vector $\boldsymbol{F}_t$. Bottom: $s_x(F)$ (solid lines), the set of optimum designs $O(F)$ (dashed lines), and the exemplary initial design $s_x(\boldsymbol{F}_t)$ (gray-shaded circle) correspond to the target objective vector $\boldsymbol{F}_t$. Designs $\boldsymbol{x}_b^{(j)}$ and $s_x(\boldsymbol{x}_b^{(j)})$ are marked using black circles. As $s_x(F)$ is an approximation of $x^* = \arg\min\{x: U(\boldsymbol{R}(x), \boldsymbol{F}_t)\}$, x^* (white circle) does not coincide with $s_x(\boldsymbol{F}_t)$

11.3.3 Basic Design Refinement: Fast Gradient Search

In this section, a simple design refinement procedure using trust-region (TR)-embedded gradient search is described (Conn et al., 2000). This procedure will be used as a reference for the iterative correction scheme of Sect. 11.3.4.

The TR algorithm employs an auxiliary linear model

$$\boldsymbol{L}^{(i)}(\boldsymbol{x}) = \boldsymbol{R}\left(\boldsymbol{x}^{(i)}\right) + \boldsymbol{J}_{\boldsymbol{R}}\left(\boldsymbol{x}^{(i)}\right) \cdot \left(\boldsymbol{x} - \boldsymbol{x}^{(i)}\right) \tag{11.34}$$

to generate a series $\boldsymbol{x}^{(i)}$, $i = 0, 1, \ldots$, of approximations to the optimum point $\boldsymbol{x}^*$ as

$$\boldsymbol{x}^{(i+1)} = \arg \min_{\boldsymbol{x};\ -\boldsymbol{d}^{(i)} \le \boldsymbol{x} - \boldsymbol{x}^{(i)} \le \boldsymbol{d}^{(i)}} U\left(\boldsymbol{L}^{(i)}(\boldsymbol{x}), \boldsymbol{F}_t\right) \tag{11.35}$$

The construction of model (11.34) requires the Jacobian matrix $\boldsymbol{J}_{\boldsymbol{R}}$, which is normally estimated through finite differentiation (FD), unless other means, e.g., adjoints sensitivities (Malhi & Bakr, 2015) are available. Here, two mechanisms are employed to reduce the computational cost of the process:

- The sensitivity matrix is initialized using the kriging surrogate (cf. 11.33) instead of being estimated by FD.
- For $i > 0$, $\boldsymbol{J}_{\boldsymbol{R}}(\boldsymbol{x}^{(i)})$ is updated from $\boldsymbol{J}_{\boldsymbol{R}}(\boldsymbol{x}^{(i-1)})$ using a rank-one Broyden formula (Broyden, 1965).

Owing to these, the cost of individual iterations of (11.35) can be as low as one EM analysis, whereas the entire refinement process can be accomplished using a small number of EM simulations. On the other hand, reliability of (11.35) depends on the accuracy of $\boldsymbol{J}^{(0)}$ and the distance between $\boldsymbol{x}^{(0)}$ and $\boldsymbol{x}^*$.

11.3.4 Iterative Design Refinement

This section formulates an alternative design refinement process involving multiple evaluations of the kriging surrogate model s_x.

Let $\boldsymbol{F}_{\text{tmp.0}}$ be the objective vector found by extracting the merit function values from the EM-simulated response of the microwave component of interest at $\boldsymbol{x}^{(0)}$ (cf. 11.33). The design inaccuracy, i.e., the difference between the performance figure or operating condition values at $\boldsymbol{x}^{(0)}$ and the target ones can be quantified as $\Delta\boldsymbol{F}^{(0)} = \boldsymbol{F}_{\text{tmp.0}} - \boldsymbol{F}_t$.

Using $\Delta\boldsymbol{F}^{(0)}$, the corrected design can be obtained from the surrogate model s_x as

$$\boldsymbol{x}^{(1)} = s_x\left(\boldsymbol{F}_t - \Delta\boldsymbol{F}^{(0)}\right) \tag{11.36}$$

Unfortunately, the optimum design $\boldsymbol{x}^*$ is generally a nonlinear function of the target vector $\boldsymbol{F}_t$; therefore, the correction (11.36) needs to be repeated. The iterative procedure employs the accumulated errors $\Delta\boldsymbol{F}^{(k)} = \boldsymbol{F}_{\text{tmp}.k} - \boldsymbol{F}_t$, where $\boldsymbol{F}_{\text{tmp}.k}$ is the vector of objectives extracted from the EM-simulated response of the component under design at the current point $\boldsymbol{x}^{(k)}$. We have

$$\boldsymbol{x}^{(i+1)} = s_x\left(\boldsymbol{F}_t - \sum\nolimits_{k=0}^{i} \Delta\boldsymbol{F}^{(k)}\right) \tag{11.37}$$

Assuming that the initial design $\boldsymbol{x}^{(0)}$ is of sufficient quality, the process (11.37) converges quickly. In practice, only a few (two to four) iterations are required. The gradient-based refinement (11.35) may be subsequently applied as the final stage of the optimization procedure. This is to correct the figures of interest that are not directly controlled by (11.37). For example, if the figures F_k correspond to the lower/upper ends of the operating bandwidth, the correction (11.37) would primarily affect the frequency allocation of the bandwidth but not the level of the response (e.g., return loss values). In other words, further design improvement w.r.t. the merit function U requires explicit minimization of $U(\boldsymbol{R}(\boldsymbol{x}), \boldsymbol{F}_t)$ using, e.g., (11.35).

Notwithstanding, the iterative correction (11.37) enables relatively large design relocations, which is particularly suitable when correcting the frequency shifts (with respect to, e.g., the target operating bandwidths) of the system outputs at the initial design. This is not straightforward for the gradient-based refinement (11.35) because the latter only employs roughly estimated sensitivity data. As mentioned earlier, the important advantage of (11.37) is its low computational cost of only one EM

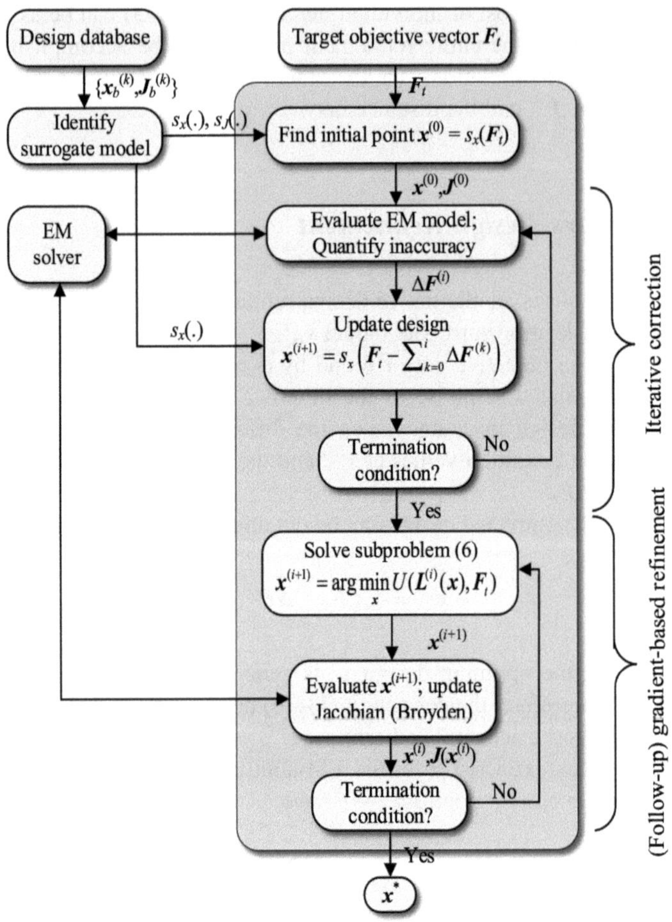

Fig. 11.20 Flow diagram of the design closure procedure with iterative refinement scheme

simulation per iteration. Figure 11.20 shows the flow diagram of the entire design closure procedure including the iterative refinement scheme and the follow-up gradient-based optimization.

11.3.5 Demonstration Case Study: Three-Section Impedance-Matching Transformer

To demonstrate the procedure discussed earlier in this section, consider a compact three-section 50–100 Ω impedance-matching transformer shown in Fig. 11.21a. A reduction of the physical length of the structure is obtained by replacing

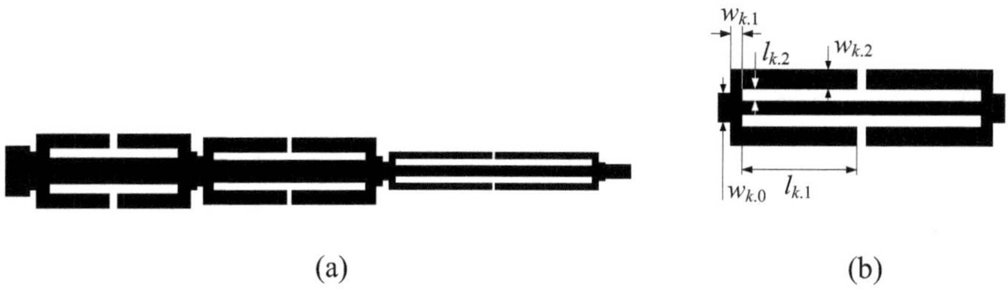

Fig. 11.21 CMRC-based compact three-section impedance-matching transformer: (**a**) topology of a three-section structure, (**b**) compact cell (CMRC)

conventional transmission lines with the compact microstrip resonant cells (CMRCs) shown in Fig. 11.21b. The circuit is implemented on RF-35 substrate ($\varepsilon_r = 3.5$, $h = 0.762$ mm) and described by 15 parameters $\boldsymbol{x} = [l_{1.1}\ l_{1.2}\ w_{1.1}\ w_{1.2}\ w_{1.0}\ l_{2.1}\ l_{2.2}\ w_{2.1}\ w_{2.2}\ w_{2.0}\ l_{3.1}\ l_{3.2}\ w_{3.1}\ w_{3.2}\ w_{3.0}]^T$.

The objective space is determined by the intended operating bands $[f_1\ f_2]$ within which the reflection $|S_{11}|$ is to be minimized. The ranges of interest for f_1 and f_2 are as follows: 1.5 GHz $\leq f_1 \leq$ 3.5 GHz and 4.5 GHz $\leq f_2 \leq$ 6.5 GHz. The kriging surrogates are constructed using five reference designs $\boldsymbol{x}_b^{(j)}$ corresponding to $\{f_1, f_2\} = \{1.5, 4.5\}, \{1.5, 6.5\}, \{2.5, 5.5\}, \{3.5, 4.5\}$, and $\{3.5, 6.5\}$ (frequencies in GHz). Note that the number of reference designs is very small. This, in conjunction with high dimensionality of the parameter space, contributes to the major challenges of the design task.

The presented design optimization procedure has been applied to optimize the transformer circuit for six target objective vectors as listed in Table 11.8. Figure 11.22 shows the transformer characteristics at the initial designs obtained using (11.32), as well as upon iterative correction (11.37) and the follow-up gradient search (11.35). The data in Table 11.8 indicates that the iterative correction scheme ensures a better design quality than the gradient search alone. The average improvement in terms of the maximum in-band reflection is almost 2 dB. At the same time, the average cost of (11.37) is four EM simulations of the transformer circuit, whereas the average cost of the follow-up gradient refinement is 4.5 EM analyses. Thus, the total cost of the optimization process is less than nine transformer simulations on average.

11.4 Automated Inverse Design of Bandpass Filters Through Linear Approximation of Physical Dimensions and Response Features

This section elaborates on an efficient methodology for finding initial designs of microwave bandpass filters directly from specification (center frequency f and bandwidth BW) (Glubokov & Koziel, 2014). The technique exploits an inverse linear

Table 11.8 Optimization results for impedance transformer of Fig. 11.21

Test case		1	2	3	4	5	6
Target operating band	f_1 (GHz)	2.0	1.8	2.7	2.7	3.5	2.8
	f_2 (GHz)	5.0	6.0	5.2	5.8	6.0	4.9
max$\|S_{11}\|$ (dB) (gradient-based refinement (11.35))[a]		−23.3	−19.3	−30.1	−28.9	−36.0	−31.4
max$\|S_{11}\|$ (dB) (iterative correction (11.37))[b]		−24.8	−21.3	−32.7	−28.9	−39.1	−32.3
Geometry parameters (mm)	$l_{1.1}$	2.95	2.91	2.76	2.42	2.12	2.33
	$l_{1.2}$	0.30	0.18	0.21	0.27	0.34	0.34
	$w_{1.1}$	0.75	0.80	0.80	0.80	0.80	0.79
	$w_{1.2}$	0.50	0.44	0.55	0.53	0.55	0.49
	$w_{1.0}$	1.14	0.53	0.72	0.61	0.87	1.00
	$l_{2.1}$	3.60	3.48	3.62	3.31	3.12	3.19
	$l_{2.2}$	0.33	0.18	0.17	0.17	0.20	0.21
	$w_{2.1}$	0.55	0.64	0.57	0.62	0.66	0.55
	$w_{2.2}$	0.23	0.15	0.25	0.24	0.18	0.30
	$w_{2.0}$	0.73	0.31	0.81	0.51	0.75	0.98
	$l_{3.1}$	4.03	3.67	3.77	3.53	3.21	3.60
	$l_{3.2}$	0.19	0.16	0.15	0.16	0.15	0.16
	$w_{3.1}$	0.36	0.42	0.34	0.36	0.35	0.32
	$w_{3.2}$	0.20	0.15	0.15	0.15	0.15	0.18
	$w_{3.0}$	0.70	0.81	1.00	0.77	0.30	0.56

[a]Maximum in-band reflection obtained when using gradient-based refinement (11.35)
[b]Maximum in-band reflection obtained using iterative correction (11.37) followed-up by gradient search (11.35)

surrogate model of the designable parameters with respect to the filter specifications, created using an available accurate design. Fast calculation of the model gradients is possible through efficient optimization in the feature space using a coarse-discretization model. As a result, the discussed methodology allows obtaining an accurate approximation of the initial filter dimensions at the cost of a few fine model simulations. The approach is illustrated with a 4th-order substrate-integrated wave-guide (SIW) filter design example. A study of the relative model error within the design space is included as well.

11.4.1 Filter Optimization Using Response Features

Filter design intends to achieve prearranged performance specifications, usually given as minimum (or maximum) levels of insertion and return losses for specific frequency ranges. Consider a return loss response of a typical 3rd-order detuned

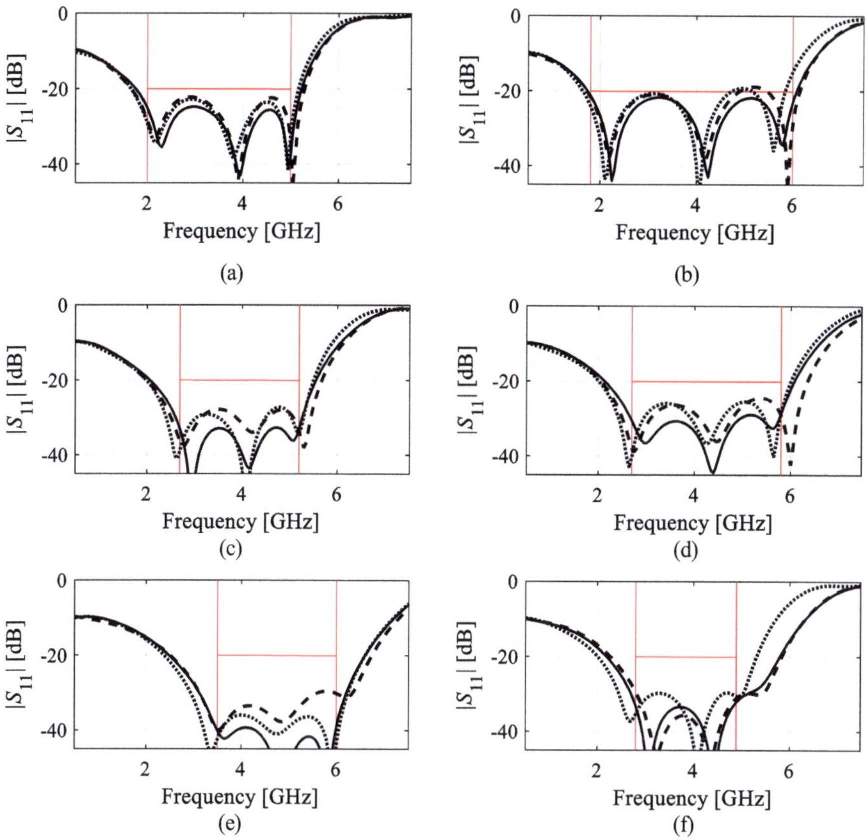

Fig. 11.22 Impedance-matching transformer responses at the initial design (····), the designs optimized using the iterative correction scheme (- - -), and the final design obtained using the gradient search of Sect. 11.3.3 (as a follow up) (—). Shown are designs corresponding to the target vectors of Table 11.8: **(a)** $f_1 = 2.0$ GHz, $f_2 = 5.0$ GHz, **(b)** $f_1 = 1.8$ GHz, $f_2 = 6.0$ GHz, **(c)** $f_1 = 2.7$ GHz, $f_2 = 5.2$ GHz, **(d)** $f_1 = 2.7$ GHz, $f_2 = 5.8$ GHz, **(e)** $f_1 = 3.5$ GHz, $f_2 = 6.0$ GHz, **(f)** $f_1 = 2.8$ GHz, $f_2 = 4.9$ GHz

filter with the specifications given as $|S_{11}| \leq -20$ dB for $f = 9.9$–10.1 GHz, as shown in Fig. 11.23. In order to satisfy the specifications, it is sufficient to ensure that the slopes of the return loss reach -20 dB at 9.9 and 10.1 GHz, as well as to bring the local maxima of the response to the same level.

Instead of handling the entire responses, it is more efficient to identify locations of a set of points on a filter response (referred to as characteristic or feature points) that determine satisfaction/violation of the specifications. These include the frequencies (f_{min} and f_{max}) where the return loss crosses specific level (RL) as well as the levels L of the local maxima of the return loss between the crossing points. Therefore, for a bandpass filter of order N, a set of useful $N + 1$ response features can be written as a response features vector $\boldsymbol{F} = [f_{min}\, f_{max}\, L_1 \ldots L_{N\text{-}1}]^T$.

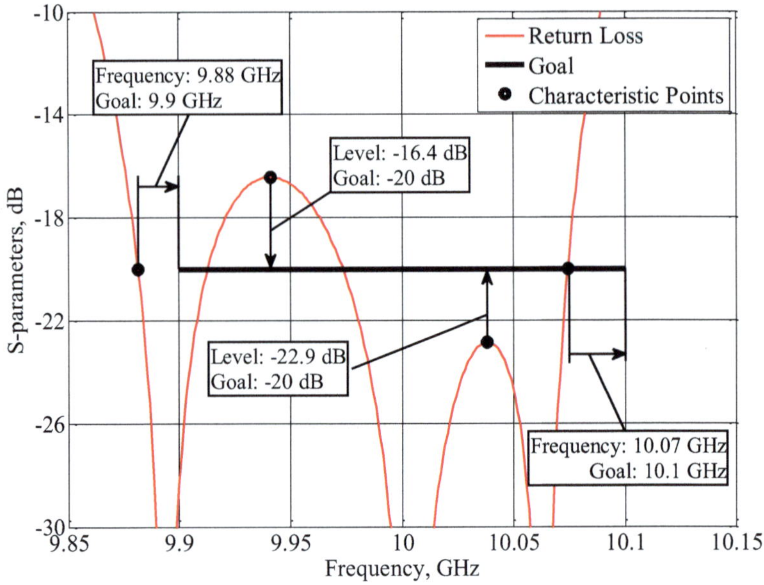

Fig. 11.23 Return loss of a detuned 3rd-order microwave filter with design specifications, characteristic points, and response features

The microwave filter optimization problem is formulated as

$$\boldsymbol{x}^* = \arg \min_{\boldsymbol{x}} U(\boldsymbol{R}_f(\boldsymbol{x})) \tag{11.38}$$

where $\boldsymbol{x} = [x_1, x_2, \ldots, x_P]^T$ is a vector of design parameters, $\boldsymbol{R}_f(\boldsymbol{x})$ is a response vector of the high-fidelity filter model, and U is the objective function encoding given design specifications.

The feature vector $\boldsymbol{F}_f(\boldsymbol{x})$ of the response $\boldsymbol{R}_f(\boldsymbol{x})$ is extracted by a function denoted as G: $\boldsymbol{F}_f(\boldsymbol{x}) = G(\boldsymbol{R}_f(\boldsymbol{x}))$. Now the optimization problem is reformulated as follows:

$$\boldsymbol{x}^* = \arg \min_{\boldsymbol{x}} \left\| \boldsymbol{F}_f(\boldsymbol{x}) - \boldsymbol{F}_{obj} \right\| \tag{11.39}$$

Here, $\boldsymbol{F}_{obj}$ is the objective vector extracted from specifications.

The optimization problem (11.39) is solved by a standard gradient-based approach (Conn et al., 2000), exploiting a linear model of the feature vector at a current design for ith iteration $\boldsymbol{x}^{(i)}$:

$$\boldsymbol{F}_{Lf}(\boldsymbol{x}) = \boldsymbol{F}_f\left(\boldsymbol{x}^{(i)}\right) + \boldsymbol{J}_f\left(\boldsymbol{x}^{(i)}\right) \cdot \left(\boldsymbol{x} - \boldsymbol{x}^{(i)}\right) \tag{11.40}$$

where the Jacobian of the feature vector is estimated using finite differentiation. Solving of (11.39) is then replaced by an iterative process

$$\boldsymbol{x}^{(i+1)} = \arg \min_{\left\|\boldsymbol{x}-\boldsymbol{x}^{(i)}\right\| \le \delta^{(i)}} \left\|\boldsymbol{F}_{L.f}(\boldsymbol{x}) - \boldsymbol{F}_{obj}\right\| \tag{11.41}$$

where $\boldsymbol{x}^{(i)}$, $i = 1, 2, \ldots$, is a sequence of designs approximating $\boldsymbol{x}^*$, whereas $\delta^{(i)}$ is the trust-region radius at iteration i, which is updated using conventional rules (Conn et al., 2000). Upon solving (11.41), the new design $\boldsymbol{x}^{(i+1)}$ is accepted if $\|\boldsymbol{F}_{L.f}(\boldsymbol{x}^{(i+1)}) - \boldsymbol{F}_{obj}\| < \|\boldsymbol{F}_{L.f}(\boldsymbol{x}^{(i)}) - \boldsymbol{F}_{obj}\|$.

Estimation of the Jacobian in (11.40) through finite differentiation requires several evaluations of a CPU-intensive high-fidelity filter model $\boldsymbol{R}_f$. Therefore, it is reasonable to use a less accurate but faster low-fidelity filter model $\boldsymbol{R}_c$ and refer to the expensive $\boldsymbol{R}_f$ only occasionally. The low-fidelity features vector is again extracted by the same function G: $\boldsymbol{F}_c(\boldsymbol{x}) = G(\boldsymbol{R}_c(\boldsymbol{x}))$.

Efficiency of the optimization procedure can be significantly increased by replacing the high-fidelity model Jacobian $\boldsymbol{J}_f(\boldsymbol{x})$ with the low-fidelity one $\boldsymbol{J}_c(\boldsymbol{x})$ in (11.40). In order for $\boldsymbol{J}_c(\boldsymbol{x})$ being reliable representation of $\boldsymbol{J}_f(\boldsymbol{x})$, both low- and high-fidelity models should display similar dependence of the response features on the design variables. This is the case when both models are physically related, e.g., when the models are evaluated using the same EM solver at different levels of discretization. In the considered procedure, we build low-fidelity models of SIW structures using coarse-discretization EM simulations. The models can be easily obtained from a fine-discretization model by changing standard mesh properties, such as number of lines per wavelength (*LPW*).

11.4.2 Coupling Coefficients as Filter Features

In the cases when a filter has a severely detuned response, the response features cannot be frequently extracted using a function G (see Sect. 11.4.1). For example, this may happen if $|S_{11}|$ response does not cross the specified return loss level RL at any of the points. Therefore, we use an alternative representation as proposed in Amari et al. (2010), which enables a continuous errorless performance of the filter optimization algorithm.

In Amari et al. (2010), a coupled resonator filter structure is considered as a "black box" with several global eigenresonances; based on this notion, a transversal circuit model is constructed, which generally consists of $N + 1$ parallel paths: each of N contains a resonator k with a resonant angular frequency $\omega_{\mathrm{res},k}$ connected to the source and the load through admittance inverters J_{Sk} and J_{kL}, and the last path represents the direct source-to-load coupling and consists of an inverter J_{SL} between the two nodes. Here, N is the number of modes whose contribution into the power transfer between the source and the load cannot be neglected.

The parameters of the transversal circuit model constitute a vector of coupling coefficients $\boldsymbol{m} = [J_{S1} \ldots J_{\mathrm{Sk}}\ J_{1L} \ldots J_{kL}\ J_{SL}\ \omega_{\mathrm{res},1} \ldots \omega_{\mathrm{res},k}]^T$. The vector $\boldsymbol{m}$ is consequently converted into a coupling matrix; the processed matrix reproduces the frequency response of the filter.

Transversal circuit parameters are extracted by fitting the EM-simulated and schematic circuit model responses by the least-square criterion. The cost function to be minimized is:

$$U(\boldsymbol{m}) = \sum_{i=1}^{K} \|\boldsymbol{R}(\omega_i, \boldsymbol{x}) - \boldsymbol{R}(\omega_i, \boldsymbol{m})\| \tag{11.42}$$

where $\boldsymbol{R}(\omega_i, \boldsymbol{x})$ and $\boldsymbol{R}(\omega_i, \boldsymbol{m})$ are the high-fidelity EM-simulated model response and a transversal circuit model response, respectively; ω_i are the frequency samples.

Optimization of the filter through the vector of coupling coefficients is carried out according to the algorithm described in Sect. 11.4.1. A filter may be optimized using both types of the features during the procedure. However, it should be noted that only one type of the features may be used within a single iteration and the decision regarding acceptance of the new design must be made by comparison of the metrics based on the same filter features type.

11.4.3 Filter Implementation Using Inverse Model of Design Parameters

Let us consider filters of order N with a return loss level RL within the passband, designed to satisfy the specified center frequency f, bandwidth BW, and having a response features vector $\boldsymbol{F}$ such that $\|\boldsymbol{F} - \boldsymbol{F}_{obj}\| \le \varepsilon$, where $\boldsymbol{F}_{obj} = [f \cdot (1 - 50 \cdot \mathrm{BW})\, f \cdot (1 + 50 \cdot \mathrm{BW})\ \mathrm{RL}\ \mathrm{RL} \ldots \mathrm{RL}]^T$ and ε is a predefined small constant, as prototypical. A vector of design parameters, corresponding to this prototypical filter, is a function of specification parameters f and BW only: $\boldsymbol{x}_p = \boldsymbol{x}_p(\boldsymbol{s})$, where $\boldsymbol{s} = [f, \mathrm{BW}]^T$. A set of dimensions creates a subspace of all the design parameter vectors $\boldsymbol{x}$. The vector of the response features of a prototypical design is defined uniquely through the specification vector: $\boldsymbol{F}_{obj} = \boldsymbol{F}(\boldsymbol{s}^*)$.

Consider an available prototypical filter with design parameters $\boldsymbol{x}_{p,0} = \boldsymbol{x}_p(\boldsymbol{s}_0)$. The prototypical filter designs corresponding to small perturbations of the center frequency Δf and bandwidth $\Delta\mathrm{BW}$ are $\boldsymbol{x}_p(\boldsymbol{s}_0 + \Delta\boldsymbol{s}_f)$ and $\boldsymbol{x}_p(\boldsymbol{s}_0 + \Delta\boldsymbol{s}_{\mathrm{BW}})$, respectively, where $\Delta\boldsymbol{s}_f = [\Delta f\ 0]^T$ and $\Delta\boldsymbol{s}_{\mathrm{BW}} = [0\ \Delta\mathrm{BW}]^T$. Taking this into account, a numerical representation of the design parameter's gradient calculated through the finite differences at $\boldsymbol{s}_0$ is expressed as follows:

$$\nabla\boldsymbol{x}_p(\boldsymbol{s}_0) = \left[\frac{\partial\boldsymbol{x}_p(\boldsymbol{s}_0)}{\partial f} \quad \frac{\partial\boldsymbol{x}_p(\boldsymbol{s}_0)}{\partial\mathrm{BW}}\right]^T \tag{11.43}$$

$$\frac{\partial \boldsymbol{x}_p(\boldsymbol{s}_0)}{\partial f} \approx \frac{\boldsymbol{x}_p(\boldsymbol{s}_0 + \Delta \boldsymbol{s}_f) - \boldsymbol{x}_p(\boldsymbol{s}_0)}{\Delta f} \tag{11.44}$$

$$\frac{\partial \boldsymbol{x}_p(\boldsymbol{s}_0)}{\partial \mathrm{BW}} \approx \frac{\boldsymbol{x}_p(\boldsymbol{s}_0 + \Delta \boldsymbol{s}_{\mathrm{BW}}) - \boldsymbol{x}_p(\boldsymbol{s}_0)}{\Delta \mathrm{BW}} \tag{11.45}$$

The design parameter vectors $\boldsymbol{x}_p(\boldsymbol{s}_0 + \Delta \boldsymbol{s}_f)$ and $\boldsymbol{x}_p(\boldsymbol{s}_0 + \Delta \boldsymbol{s}_{\mathrm{BW}})$ can be easily found by the feature space optimization described in Sect. 11.4.1 for one or, rarely, two evaluations of the high-fidelity model of the filter.

Consequently, we estimate the initial design $\boldsymbol{x}_{\mathrm{in}}(\boldsymbol{s}^*)$ for a prototypical filter $\boldsymbol{x}_p(\boldsymbol{s}^*)$, where $\boldsymbol{s}^* = [f^* \ \mathrm{BW}^*]^T$, as:

$$\boldsymbol{x}_{\mathrm{in}}(\boldsymbol{s}^*) = \boldsymbol{x}_p(\boldsymbol{s}_0) + \nabla \boldsymbol{x}_p(\boldsymbol{s}_0) \cdot (\boldsymbol{s}^* - \boldsymbol{s}_0) \tag{11.46}$$

Note that the estimated initial design $\boldsymbol{x}^*$ generally does not lead to a prototypical design, i.e., $\boldsymbol{R}_f(\boldsymbol{x}^*) \neq \boldsymbol{R}_f(\boldsymbol{x}_p(\boldsymbol{s}^*))$.

The design procedure to find the dimensions $\boldsymbol{x}_p(\boldsymbol{s}^*)$ of a prototypical filter is described as follows:

1. Find dimensions $\boldsymbol{x}_p(\boldsymbol{s}_0)$, where $\boldsymbol{s}_0 = [f_0 \ \mathrm{BW}_0]^T$, of a prototypical filter with any available technique or get them from a library;
2. Set perturbation values Δf and $\Delta\mathrm{BW}$ and find the dimensions $\boldsymbol{x}_p(\boldsymbol{s}_0 + \Delta \boldsymbol{s}_f)$ and $\boldsymbol{x}_p(\boldsymbol{s}_0 + \Delta \boldsymbol{s}_{\mathrm{BW}})$ of the prototypic filter with $\Delta \boldsymbol{s}_f = [\Delta f \ 0]^T$ and $\Delta \boldsymbol{s}_{\mathrm{BW}} = [0 \ \Delta\mathrm{BW}]^T$ using the optimization in the response features space (cf. Sect. 11.4.1);
3. Find the gradient of the dimensions $\nabla \boldsymbol{x}_p(\boldsymbol{s}_0)$ using (11.43)–(11.45);
4. Calculate the vector of initial design parameters $\boldsymbol{x}_{\mathrm{in}}(\boldsymbol{s}^*)$ for a prototypical design $\boldsymbol{x}_p(\boldsymbol{s}^*)$ using a linear model (11.46);
5. Obtain the prototypical design of interest by optimizing the high-fidelity model of the filter using response features starting from the initial design $\boldsymbol{x}_{\mathrm{in}}(\boldsymbol{s}^*)$: $\boldsymbol{x}_p(\boldsymbol{s}^*) = \arg\min\{\boldsymbol{x}, \|\boldsymbol{x} - \boldsymbol{x}^{(i)}\| \leq \delta^{(i)}: \|\boldsymbol{F}_f(\boldsymbol{x}) - \boldsymbol{F}(\boldsymbol{s}^*)\|\}$.

11.4.4 Illustration and Application Example

The discussed technique is illustrated by designing a 4th-order SIW bandpass filter whose configuration is shown in Fig. 11.24. The filter consists of four direct-coupled SIW cavities with side walls and coupling irises composed of metallic posts. To reduce the complexity of the problem, the filter structure is chosen to be symmetric with respect to all three symmetry planes. The vector of design parameters is $\boldsymbol{x} = [L_{\mathrm{res1}} \ L_{\mathrm{res2}} \ W_{12} \ W_{213} \ W_{\mathrm{in}}]^T$. The width of the waveguide W_{wg} is set to 15 mm throughout the entire process. The height of the substrate is $H_{\mathrm{sub}} = 1$ mm and its relative permittivity is $\varepsilon_r = 2.2$. The return loss level of interest in passband is set to $\mathrm{RL} = -20$ for all the filters.

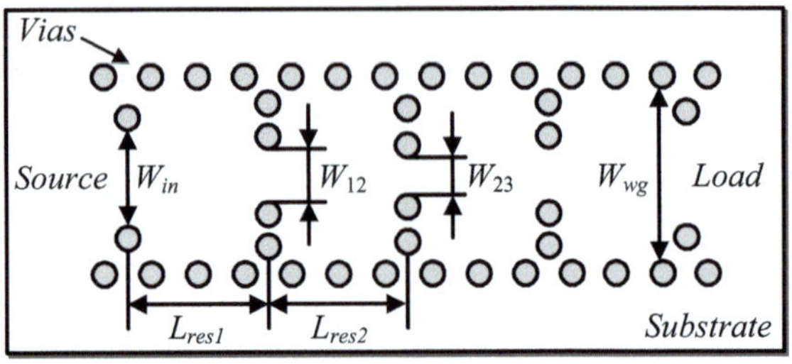

Fig. 11.24 Configuration of the 4th-order SIW bandpass filter

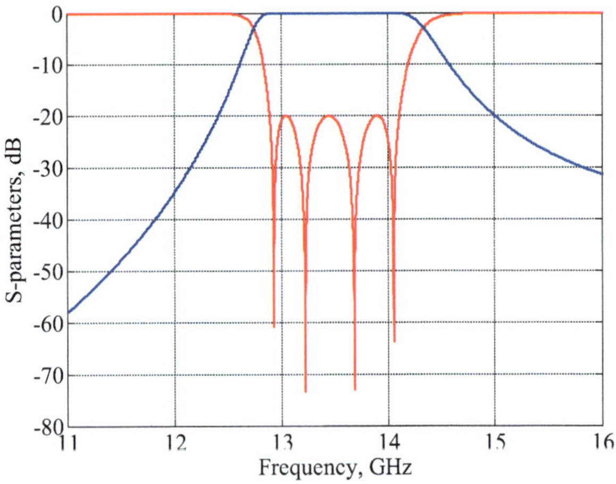

Fig. 11.25 Frequency response of the initial prototypical filter $\boldsymbol{R}_f(\boldsymbol{x}_p(\boldsymbol{s}_0))$ for $\boldsymbol{s}_0 = [13.5 \text{ GHz } 9\%]^T$

The initial prototypical filter is designed for $\boldsymbol{s}_0 = [13.5 \text{ GHz } 9\%]^T$ using manual tuning and further optimization in filter features space. The dimensions vector corresponding to the design is $\boldsymbol{x}_p(\boldsymbol{s}_0) = [7.406 \ 8.317 \ 5.694 \ 5.346 \ 7.614]^T$ mm. The frequency response of the filter $\boldsymbol{R}_f(\boldsymbol{x}_p(\boldsymbol{s}_0))$ is shown in Fig. 11.25. To illustrate the approach, we estimated the initial designs $\boldsymbol{x}_{\text{in}}(\boldsymbol{s})$ for $N_s = 400$ training samples randomly distributed within a space of specifications constrained by $\boldsymbol{s}_{\min} = [12 \text{ GHz } 4\%]^T$ and $\boldsymbol{s}_{\max} = [15 \text{ GHz } 14\%]^T$. The quality of the initial designs $\boldsymbol{x}_{\text{in}}(\boldsymbol{s})$ is evaluated through a relative error $||\boldsymbol{s}^* - \boldsymbol{s}_{\text{act}}||/||\boldsymbol{s}^*||$, where $\boldsymbol{s}^*$ corresponds to the specifications of interest, while $\boldsymbol{s}_{\text{act}}$ is the result actually obtained by means of the model (11.46). The distribution of the relative error values over the modeling space is illustrated in Fig. 11.26. It is evident that the error does not exceed 10% for the majority of the specification space.

The procedure of obtaining the prototypical response is demonstrated by an example with $\boldsymbol{s}^* = [14.447 \text{ GHz } 12.048\%]$. The initial design obtained for this specification is $\boldsymbol{x}_{\text{in}}(\boldsymbol{s}^*) = [6.418 \ 7.222 \ 6.091 \ 5.75 \ 7.871]$ mm; its corresponding relative error is calculated to be 8.39%. The filter has been optimized using the response features within a trust region framework with the total cost corresponding to about 2.5 evaluations of the high-fidelity model. The optimized design parameters vector is $\boldsymbol{x}_p(\boldsymbol{s}^*) = [6.534 \ 7.398 \ 5.898 \ 5.551 \ 7.695]$ mm.

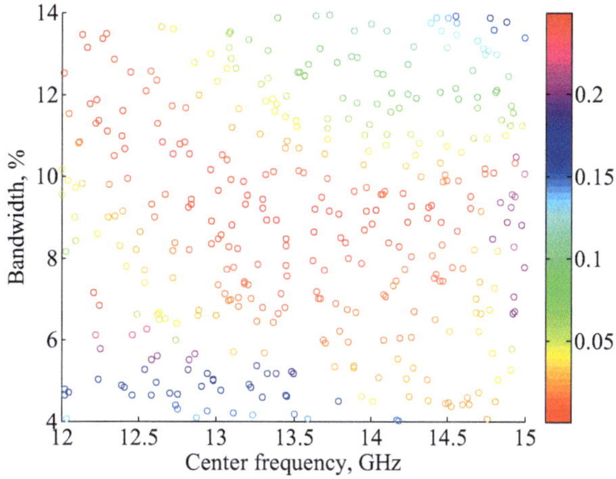

Fig. 11.26 Distribution of the relative error for the initial design estimations within the constrained specification space for the 4th-order SIW bandpass filter

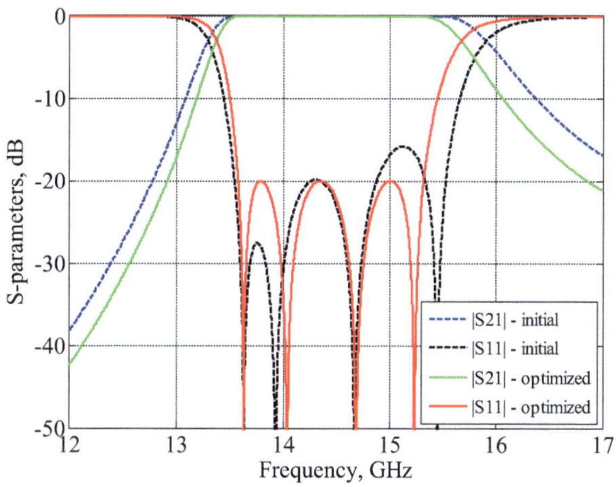

Fig. 11.27 Filter optimization (high-fidelity model responses): $|S_{11}|$ and $|S_{21}|$ at the initial design $\boldsymbol{x}_{\text{in}}(\boldsymbol{s}^*)$ obtained with the inverse linear model and at the design $\boldsymbol{x}_p(\boldsymbol{s}^*)$ obtained by optimizing the structure using the response features

Comparison of the initial response $\boldsymbol{R}_f(\boldsymbol{x}_{\text{in}}(\boldsymbol{s}^*))$ obtained through the discussed implementation technique and the optimized prototypical response $\boldsymbol{R}_f(\boldsymbol{x}_p(\boldsymbol{s}^*))$ is shown in Fig. 11.27. It is observed that the considered implementation technique produces a good initial design, which can be quickly converted into an optimized solution. Generally, the studies of the selected initial designs optimization from Fig. 11.26 indicate that a desirable prototypical design can be obtained for up to three evaluations of $\boldsymbol{R}_f$ for initial designs with the relative errors below 11%.

References

Amari, S., Seyfert, F., & Bekheit, M. (2010). Theory of coupled resonator microwave bandpass filters of arbitrary bandwidth. *IEEE Transactions on Microwave Theory and Techniques, 58*(8), 2188–2203.

Broyden, C. G. (1965). A class of methods for solving nonlinear simultaneous equations. *Mathematics of Computation, 19*, 577–593.

Conn, A. R., Gould, N. I. M., & Toint, P. L. (2000). *Trust region methods* (MPS-SIAM series on optimization).

Farran, M., Boscolo, S., Locatelli, A., Capobianco, A. D., Mirio, M., Ferrari, V., & Modotto, D. (2016). Compact quasi-Yagi antenna with folded dipole fed by tapered integrated balun. *Electronics Letters, 52*(10), 789–790.

Glubokov, O., & Koziel, S. (2014). Automated inverse design of bandpass filters with invariable layout through linear approximation of physical dimensions. In *European microwave conference*.

Koziel, S. (2015). Fast simulation-driven antenna design using response-feature surrogates. *International Journal of RF and Microwave Computer-Aided Engineering, 25*(5), 394–402.

Koziel, S., & Bandler, J. W. (2015). Reliable microwave modeling by means of variable-fidelity response features. *IEEE Transactions on Microwave Theory and Techniques, 63*(12), 4247–4254.

Koziel, S., & Ogurtsov, S. (2018). Rapid design closure of linear microstrip antenna array apertures using response features. *IEEE Antennas and Wireless Propagation Letters, 17*(4), 645–648.

Koziel, S., & Pietrenko-Dabrowska, A. (2020a). Kriging metamodels and design re-utilization for fast parameter tuning of antenna structures. *The International Journal of Numerical Modelling, 34*(12), art. no. e02811.

Koziel, S., & Pietrenko-Dabrowska, A. (2020b). Expedited feature-based quasi-global optimization of multi-band antennas with Jacobian variability tracking. *IEEE Access, 8*, 83907–83915.

Koziel, S., Pietrenko-Dabrowska, A., & Al-Hasan, M. (2020). Improved-efficacy optimization of compact microwave passives by means of frequency-related regularization. *IEEE Access, 8*, 195317–195326.

Malhi, H., & Bakr, M. H. (2015). Geometry evolution of microwave filters exploiting self-adjoint sensitivity analysis. In *International conference on numerical electromagnetic and multiphysics modeling and optimization (NEMO)*, Ottawa, Canada, August 11–14.

Phani Kumar, K. V., & Karthikeyan, S. S. (2013). A novel design of rat race coupler using defected microstrip structure and folding technique. In *2013 IEEE applied electromagnetics conference (AEMC)*, Bhubaneswar, India, December 18–20, pp. 1–2.

Pietrenko-Dabrowska, A., & Koziel, S. (2019). Expedited antenna optimization with numerical derivatives and gradient change tracking. *Engineering Computations, 37*(4). https://doi.org/10.1108/EC-04-2019-0155

Pietrenko-Dabrowska, A., & Koziel, S. (2020a). Reliable surrogate modeling of antenna input characteristics by means of domain confinement and principal components. *Electronics, 9*(5), 1–16.

Pietrenko-Dabrowska, A., & Koziel, S. (2020b). Accelerated design optimization of miniaturized microwave passives by design reusing and kriging interpolation surrogates. *AEU International Journal of Electronics and Communications, 118*, 153165.

Pietrenko-Dabrowska, A., Koziel, S., & Al-Hasan, M. (2021). Accelerated parameter tuning of antenna structures using inverse and feature-based forward kriging surrogates. *The International Journal of Numerical Modelling, 34*(5), e2880.

Queipo, N. V., Haftka, R. T., Shyy, W., Goel, T., Vaidynathan, R., & Tucker, P. K. (2005). Surrogate-based analysis and optimization. *Progress in Aerospace Sciences, 41*(1), 1–28.

Tseng, C., & Chang, C. (2012). A rigorous design methodology for compact planar branch-line and rat-race couplers with asymmetrical T-structures. *IEEE Transactions on Microwave Theory and Techniques, 60*(7), 2085–2092.

Chapter 12
Response Features for Low-Cost Behavioral Modeling

The importance of surrogate modeling techniques in the design of modern high-frequency systems has been continuously growing over the recent years. This phenomenon is a matter of practical necessity rather than simply a fashion. On the one hand, antenna and microwave design procedures rely on full-wave electromagnetic (EM) simulation tools. On the other hand, the computational costs incurred by repetitive EM analyses involved in solving common tasks (parameter tuning, uncertainty quantification, multi-criterial design, etc.) are often prohibitive. The replacement of full-wave simulations by fast surrogates may mitigate these issues. As a matter of fact, it is the only viable option for carrying out EM-driven design in many cases. Among available modeling approaches, data-driven surrogates (Simpson et al., 2001; Queipo et al., 2005; Forrester & Keane, 2009; Yondo et al., 2019; Koziel et al., 2021) are by far the most popular due to their accessibility and versatility. At the same time, a construction of reliable models is hindered by the curse of dimensionality, high nonlinearity of system characteristics, as well as broad ranges of parameters and operating conditions that the model has to cover to ensure its design utility. Alleviating these difficulties has been the subject of extensive research, which led to the development of numerous techniques, primarily oriented toward better utilization of available training data and/or handling response nonlinearity. Another approach is a manipulation of the model domain where the focus is on restricting the modeling process to the most important parts of the parameter space, e.g., those containing high-quality designs (Koziel & Pietrenko-Dabrowska, 2020).

In this book, the exposition of response features is mainly associated with design optimization methods, both local and global, as well as statistical (Chap. 14). In this chapter, we demonstrate that feature-based approaches may also be used in the context of general-purpose surrogate modeling. By exploring the common properties of characteristic points, in particular, weakly nonlinear dependence between feature points coordinates and geometry/material parameters of the system of interest, it is possible to improve the computational efficiency of the surrogate modeling process.

A. Pietrenko-Dabrowska, S. Koziel, *Response Feature Technology for High-Frequency Electronics. Optimization, Modeling, and Design Automation*,
https://doi.org/10.1007/978-3-031-43845-5_12

This can manifest itself in a reduction of the number of the training points necessary to render a reliable model, or by improving the predictive power of the surrogate without increasing the training dataset size. Subsequent sections elaborate on specific modeling techniques following this paradigm. Chapter 13 will address utilization of response features for constrained (performance-driven) modeling.

12.1 Feature-Based Modeling Using Characteristic Point Allocation

This section discusses feature-based modeling of high-frequency structures where the surrogate model is directly constructed at the level of characteristics points. These are allocated to account for essential properties of the system responses, e.g., the resonances, and distributed either uniformly in frequency (for "flat" regions of the response) or with respect to the level (for "steep" regions). The procedure is illustrated using several examples of antenna structures.

12.1.1 Modeling Methodology

Here, we are focused on modeling reflection responses of antenna structures. The primary (expensive) model obtained from full-wave electromagnetic (EM) simulation is denoted as $\boldsymbol{R}(\boldsymbol{x})$. It represents the reflection coefficient $|S_{11}|$ of the antenna evaluated at m frequencies, ω_1 to ω_m, i.e., we have $\boldsymbol{R}(\boldsymbol{x}) = [R(\boldsymbol{x}, \omega_1) \ldots R(\boldsymbol{x}, \omega_m)]^T$. The objective is to build a replacement model (surrogate) $\boldsymbol{R}_s$. The surrogate should represent the EM model in a given region X of the design space.

The set of training samples is denoted as $X_T = \{\boldsymbol{x}^1, \boldsymbol{x}^2, \ldots, \boldsymbol{x}^N\} \subset X$. The corresponding EM model responses $\boldsymbol{R}(\boldsymbol{x}^k)$ are acquired beforehand. According to the conventional approach to data-driven surrogate construction, the responses $R(\boldsymbol{x}, \omega_j), j = 1, \ldots, m$, are approximated directly, either separately for each frequency or by treating the frequency as an additional input parameter of the model. The fundamental problem is nonlinearity of the responses, particularly for narrow-band antennas (Pan et al., 2014; Chen, 2005; Koziel et al., 2016). Typical responses of such structures are shown in Fig. 12.1 (here, reflection responses of a dielectric resonator antenna of Sect. 12.1.2). For clarity, we show the responses evaluated in the region corresponding to $7.0 \le a_x \le 9.0$ and $13.0 \le a_y \le 15.0$ (see Fig. 12.4 for parameter explanation) at three various frequencies: 5.3, 5.5, and 5.7 GHz. Accurate modeling of such nonlinear landscapes is only possible within limited parameter ranges and a large number of training samples.

Our primary objective is to reduce the number of training data samples necessary to construct an accurate surrogate model, thereby reducing the cost of training data acquisition. In the discussed approach, it is achieved by reformulating the modeling

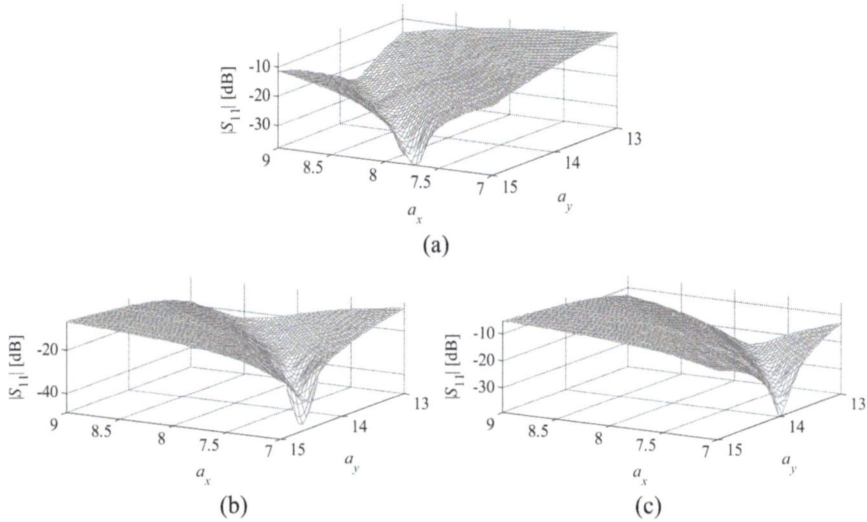

Fig. 12.1 Exemplary responses of the dielectric resonator antenna considered in Sect. 12.1.2 (here, reflection coefficient $|S_{11}|$). The responses are evaluated in the region $7.0 \le a_x \le 9.0$ and $13.0 \le a_y \le 15.0$ at the frequencies of 5.3 GHz (**a**), 5.5 GHz (**b**), and 5.7 GHz (**c**). Other variables are fixed to the following values: $a_z = 9$, $a_c = 0$, $u_s = 2$, $w_s = 10$, $y_s = 8$ (all in mm)

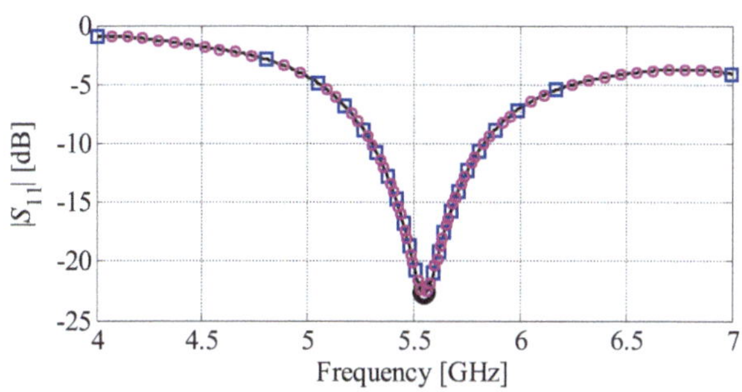

Fig. 12.2 Definition of response features in the case of a narrow-band antenna reflection characteristic: the primary point (here, corresponding to the antenna resonance) is represented as **O**; (□) represents supplemental points distributed equally with respect to response level, (○) denotes infill points distributed equally in frequency between the main and supplemental points (note that the number of points may be different for various intervals)

process and conducting it at the level of the response features (Koziel & Bandler, 2015). Figure 12.2 clarifies the definition of the feature points for narrow-band antennas in the context of surrogate modeling. The characteristic point set is constructed sequentially as follows: (1) identification of the primary point that corresponds to the center frequency (antenna resonance) and the response level at that frequency; (2) allocation of the supplemental points (here, uniformly with respect to the level and separately on the left- and right-hand-side of the primary point); (3) allocation of the infill points uniformly in frequency in between the supplemental points. Clearly, one needs to ensure that the number of characteristic points is sufficient so as to allow reliable synthesis of the antenna response (through interpolation).

On the other hand, although it is important that the major features of the response (e.g., antenna resonance) are accounted for, particular point allocation is not critical. The response features, once defined, can be easily extracted using simple post-processing of EM simulation results. For subsequent considerations, the *j*th feature point of $\boldsymbol{R}(\boldsymbol{x}^k)$ ($j = 1, \ldots, K$, $k = 1, \ldots, N$) will be denoted as $\boldsymbol{f}_k^j = [\omega_k^j \; l_k^j]$. Here, ω_k^j and l_k^j represent the frequency and the magnitude (level) components of $\boldsymbol{f}_k^j$, respectively.

For the sake of illustration, the frequency and level components of the selected feature points have been shown in Fig. 12.3. The considered design space region is the same as the one shown in Fig. 12.2. It is important that the functional landscapes of the feature points are not as nonlinear as those shown in Fig. 12.2. This is particularly the case for the frequency component. Clearly, it is to be expected that construction of a reliable surrogate model at the feature point level will require a smaller number of training samples than modeling the reflection response in a traditional manner.

Having the response features defined, the surrogate modeling process works as follows. First, the data-driven models $s_{\omega.j}(\boldsymbol{x})$ and $s_{l.j}(\boldsymbol{x})$, $j = 1, \ldots, K$, of the sets of corresponding feature points are constructed using available training designs, $\{\boldsymbol{f}_1^j, \boldsymbol{f}_2^j, \ldots, \boldsymbol{f}_N^j\}$, $j = 1, \ldots, K$ (Forrester & Keane, 2009). At this stage, we utilize kriging interpolation (Queipo et al., 2005).

The surrogate model itself is defined as

$$\boldsymbol{R}_s(\boldsymbol{x}) = [R_s(\boldsymbol{x}, \omega_1) \quad \ldots \quad R_s(\boldsymbol{x}, \omega_m)]^T \tag{12.1}$$

where its *j*th component is given as

$$R_s(\boldsymbol{x}, \omega_j) = I(\Omega(\boldsymbol{x}), L(\boldsymbol{x}), \omega_j) \tag{12.2}$$

In (12.2), $L(\boldsymbol{x}) = [s_{l.1}(\boldsymbol{x}) \; s_{l.2}(\boldsymbol{x}) \ldots s_{l.K}(\boldsymbol{x})]$ and $\Omega(\boldsymbol{x}) = [s_{\omega.1}(\boldsymbol{x}) \; s_{\omega.2}(\boldsymbol{x}) \ldots s_{\omega.K}(\boldsymbol{x})]$ are the predicted feature point locations corresponding to the evaluation design $\boldsymbol{x}$. Because we are interested in evaluating the antenna response at a discrete set of frequencies ω_1 through ω_m, it is necessary to interpolate both the level vector L and frequency vector Ω into the response at the above set of frequencies. This interpolation is represented as $I(\Omega, L, \omega)$.

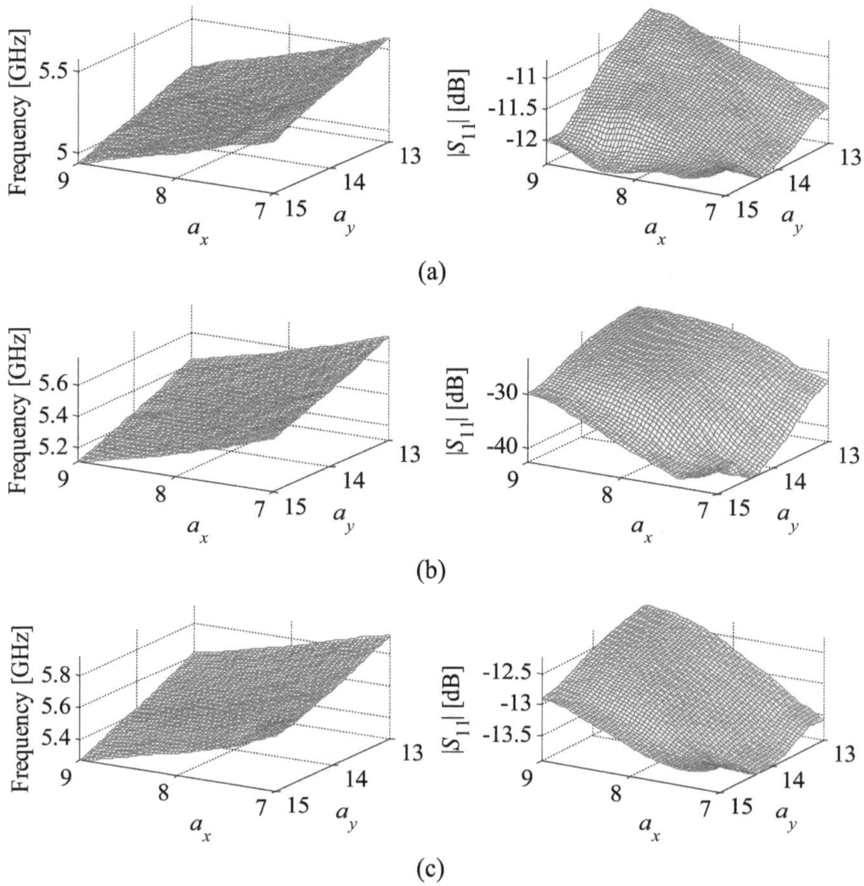

Fig. 12.3 Frequency (left panel) and level (right panel) components as functions of geometry parameters a_x and a_y (with other variables fixed) for the three selected feature points. The responses are evaluated for the same design space region as that considered in Fig. 12.2. Note that the functional landscapes of the feature point coordinates are considerably less nonlinear than those for original responses (cf. Fig. 12.2)

12.1.2 Case Study I: Dielectric Resonator Antenna

As the first verification case we consider a rectangular dielectric resonator antenna (DRA) (Sim et al., 2014). The antenna geometry is illustrated in Fig. 12.4. The dielectric resonator is implemented using a material with permittivity $\varepsilon_r = 10$. The substrate is Rogers RO4003 ($h = 0.5$ mm, $\varepsilon_r = 3.3$). The computational model of the antenna also includes polycarbonate housing with dielectric permittivity of 2.8. There are seven geometry parameters describing the DRA structure represented as $\boldsymbol{x} = [a_x\ a_y\ a_z\ a_c\ u_s\ w_s\ y_s]^T$ mm. Furthermore, there are four fixed parameters: $d_x = d_y = d_z = 1$ mm and $w_0 = 1.15$ mm.

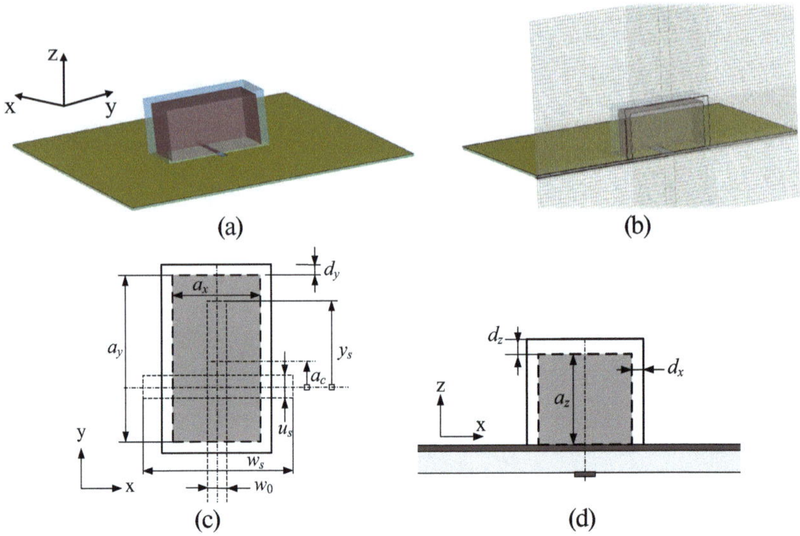

Fig. 12.4 Dielectric resonator antenna (Sim et al., 2014): (**a**) 3D view of the structure with housing, (**b**) visualization of the structure cross-section with highlight on mesh density, (**c**) top and (**d**) front views with marked dimensions

The computational model of the DRA is implemented and simulated in a full-wave electromagnetic solver of CST Microwave Studio. The physics-based model of the structure (here, denoted as $\boldsymbol{R}(\boldsymbol{x})$) has been implemented in the package and then discretized using tetrahedral mesh (~420,000 mesh cells). The discretization slice of the antenna is shown in Fig. 12.4b. The antenna response has been obtained using a time domain EM solver. The average simulation time of the model on a dual Intel Xeon E5540 machine with 6 GB RAM is 19 min. The frequency range is 4–7 GHz.

Our objective is to construct the surrogate model of the reflection coefficient $|S_{11}|$. The model domain is defined as $X = [\boldsymbol{x}^{(0)} - \boldsymbol{d}, \boldsymbol{x}^{(0)} + \boldsymbol{d}]$, with the center $\boldsymbol{x}^{(0)} = [8.0\ 14.0\ 9.0\ 0.0\ 2.0\ 10.0\ 8.0]^T$ mm and size $\boldsymbol{d} = [2.0\ 2.0\ 2.0\ 2.0\ 1.0\ 2.0\ 2.0]^T$ mm.

An important characteristic of any surrogate modeling technique is its scalability with respect to the number of training samples. Here, six training sets of various sizes are utilized for the sake of model construction. The sets contain from 20 to 800 samples. The design of experiment technique selected for sample allocation is modified Latin Hypercube Sampling (LHS) (Beachkofski & Grandhi, 2002). The error measure utilized for model verification is defined as $||\boldsymbol{R}(\boldsymbol{x}) - \boldsymbol{R}_s(\boldsymbol{x})||/||\boldsymbol{R}(\boldsymbol{x})||$ and expressed in percent (%). There is a separate set of 100 LHS-allocated testing samples for error estimation.

The model error averaged over the testing set can be found in Table 12.1. For additional verification, the feature-based surrogate has been compared to conventional modeling of S_{11}. Kriging interpolation has been selected as a representative benchmark method. The first observation is that our technique allows for considerable reduction of the training data set while ensuring comparable predictive power of

Table 12.1 Modeling results of the DRA

Model	Average error (%)					
	$N^{a} = 20$	$N = 50$	$N = 100$	$N = 200$	$N = 400$	$N = 800$
Feature-based surrogate	23.4	10.8	7.8	6.7	3.9	3.0
Kriging interpolation[b]	43.2	36.9	29.1	24.0	11.1	8.0

[a]N stands for the number of training points
[b]Direct kriging interpolation of high-fidelity model $|S_{11}|$ responses

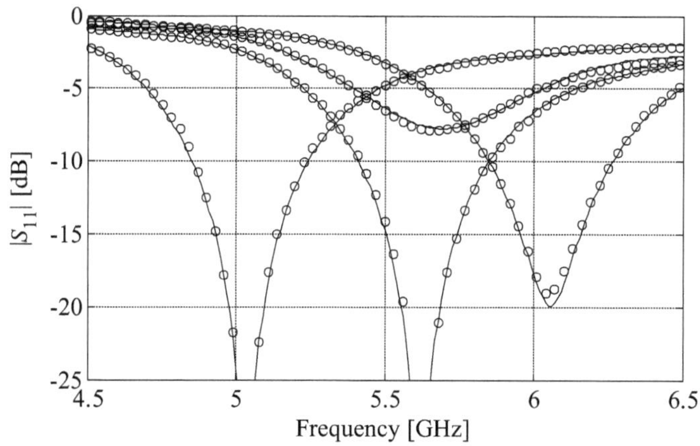

Fig. 12.5 High-fidelity (—) and feature-based model (setup with 400 high-fidelity training points) (o) at the selected test designs

the surrogate. Furthermore, even for the largest considered training set (800 samples), accuracy of the conventional model is still insufficient for conducting practical design tasks such as parametric optimization. At the same time, accuracy of the feature-based model is already satisfactory when constructed using 200 samples. Figure 12.5 shows the responses of the EM-simulated and surrogate model responses for several selected test designs. Visual inspection indicates that the alignment between the surrogate and the electromagnetic model is excellent.

An additional verification has been conducted by using the feature-based model for design optimization of the DRA. The design objective is to minimize the antenna reflection (i.e., $|S_{11}|$) in the frequency range of ± 0.2 GHz around a given operating frequency. The maximum acceptable level is -10 dB. We consider three cases with three different operating frequencies: 5.0, 5.5, and 6.0 GHz. The initial design is the same for all instances: $\boldsymbol{x}^{\text{init}} = [8.0\ 12.0\ 8.0\ 0.0\ 1.0\ 11.0\ 7.0]^T$. The responses of the optimized antenna for all three cases are shown in Fig. 12.6. It can be observed that the responses of the EM simulation model satisfy the prescribed specifications and are well aligned with the responses of the surrogate model. Therefore, no further correction is necessary.

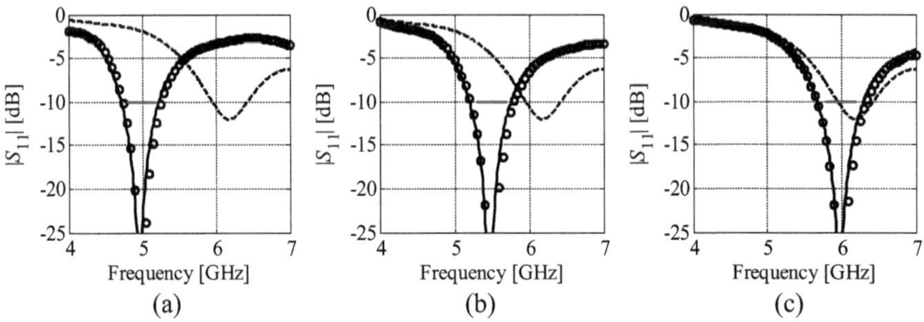

Fig. 12.6 Application of the feature-based surrogate for antenna optimization: responses of the EM model at the initial design (- - -) and at the design obtained by optimizing the surrogate (—). The surrogate model response is marked as (○). There are three cases considered corresponding to the operating frequency of 5.0 GHz (**a**), 5.5 GHz (**b**), and 6.0 GHz (**c**)

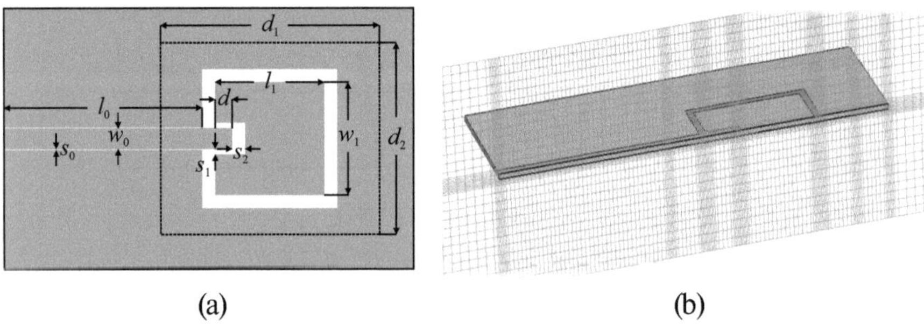

Fig. 12.7 Rectangular-slot-coupled patch antenna: (**a**) geometry with marked design parameters; (**b**) structure cross-section view with highlight on the mesh density. Gray and white colors in (**a**) represent the ground plane and slots, respectively, whereas the patch radiator (located on the opposite side of the substrate) is marked using dashed line

12.1.3 Case Study II: Patch Antenna Coupled to Rectangular Slot

Our next design example is a rectangular-slot-coupled patch antenna shown in Fig. 12.7 (Chen, 2007). The structure consists of a driven element in the form of a rectangular patch coupled to the rectangle-shaped ground plane slot localized on the other side of the substrate. The slot is coupled to a coplanar waveguide (CPW) line, which is utilized as the antenna feed. The allocation of the patch can be adjusted with respect to the center of the ground plane slot. The structure is implemented on a 0.762-mm-thick Taconic RF-35 substrate ($\varepsilon_r = 3.5$, tan $\delta = 0.0018$). Antenna parameters are represented by vector: $\boldsymbol{x} = [d\ d_1\ l_1\ s_1\ s_2\ w_1]^T$. Variable $d_2 = d_1$, whereas dimensions $w_0 = 3$, $l_0 = 30$, and $s_0 = 0.15$ remain fixed (all parameters are in mm).

Similarly as in the previous example, the antenna of Fig. 12.7 is implemented in CST Microwave Studio. It is also simulated using finite integral technique. The

Table 12.2 Modeling results of the patch antenna coupled to rectangular slot

Model	Average error (%)					
	$N^{a} = 20$	$N = 50$	$N = 100$	$N = 200$	$N = 400$	$N = 800$
Feature-based surrogate	32.3	25.9	14.4	11.7	9.3	7.5
Kriging interpolation[b]	99.5	99.2	74.2	71.9	53.9	46.4

[a]*N* stands for the number of training points
[b]Direct kriging interpolation of high-fidelity model $|S_{11}|$ responses

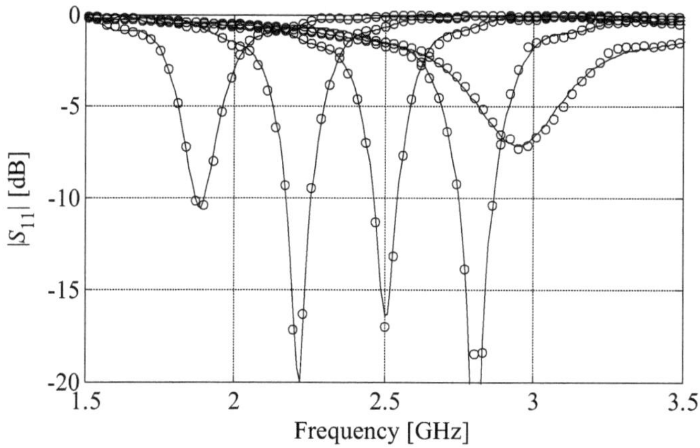

Fig. 12.8 Patch antenna: high-fidelity (—) and feature-based model (○) at the selected test designs

model discretization level is set to ~600,000 hexahedral cells. The visualization of antenna model with highlighted mesh is provided in Fig. 12.7a. Typical simulation time of the model is 12 min. The frequency range considered for modeling is from 1 to 4 GHz. The surrogate model of the reflection coefficient $|S_{11}|$ is constructed in the region $X = [\boldsymbol{x}^{(0)} - \boldsymbol{d}, \boldsymbol{x}^{(0)} + \boldsymbol{d}]$, where $\boldsymbol{x}^{(0)} = [4.5\ 4.0\ 15.0\ 1.2\ 1.2\ 15.0]^T$ mm and $\boldsymbol{d} = [2.5\ 4.0\ 5.0\ 1.0\ 1.0\ 5.0]^T$ mm.

Similarly as for the previous example, the feature-based surrogate has been constructed using six different training sets. Table 12.2 shows the average error values for particular training sets. It should be noted that because of high nonlinearity of the antenna responses and wide range of geometry parameters, conventional modeling is of poor quality (cf. Table 12.2), whereas feature-based modeling ensures practically usable prediction power. The high-fidelity and feature-based surrogate responses at the selected test designs are shown in Fig. 12.8.

For additional verification, we use the feature-based model for optimization of the antenna structure at four operating frequencies: 1.85, 2.1, 2.8, and 3.2 GHz. In all cases, the objective is to minimize reflection level $|S_{11}|$ in the frequency range of ±50 MHz around a given operating frequency. Figure 12.9 shows the antenna responses at the designs obtained by optimizing the feature-based surrogate. No further design tuning is necessary because of good accuracy of the surrogate.

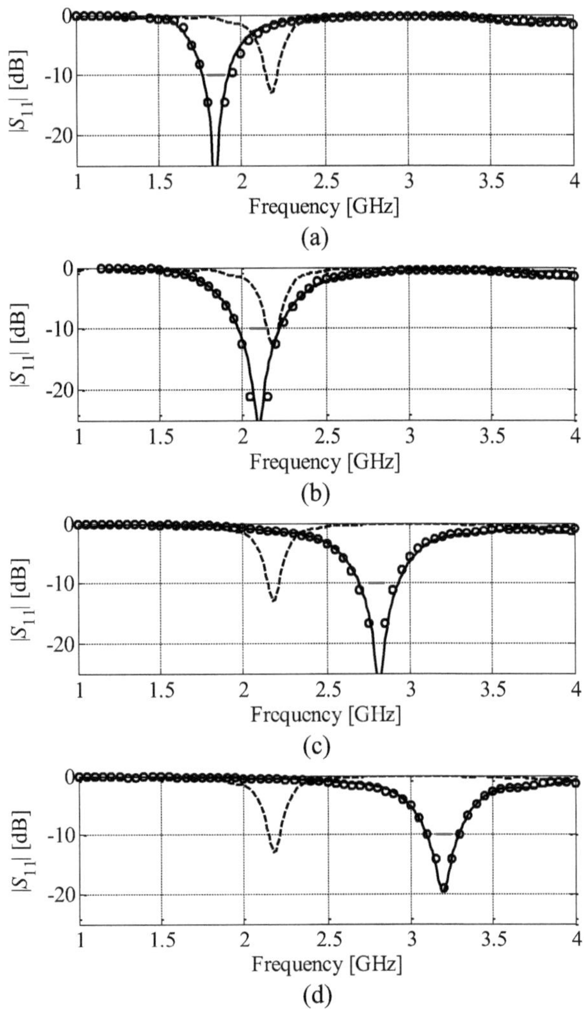

Fig. 12.9 Rectangular-slot-coupled patch antenna optimization: the high-fidelity model response at the initial design (- - -) and at the design obtained by optimizing the feature-based surrogate (—). The surrogate model response is marked as (○). Results for the operating frequency of 1.83 GHz (**a**), 2.1 GHz (**b**), 2.8 GHz (**c**), and 3.2 GHz (**d**)

12.1.4 Case Study III: Dual-Band Uniplanar Dipole Antenna

The modeling methodology outlined in Sect. 12.1.1 can be generalized for multiband antennas. Here, we illustrate it using a single-layer dual-band dipole antenna shown in Fig. 12.10 (Chen et al., 2006). The structure is implemented on a 0.762-mm-thick Taconic RF-35 dielectric substrate with relative permittivity of 3.5. Dual-band operation of the structure is ensured by two separated slots. The antenna is fed through a 50 Ω coplanar waveguide (CPW). The vector of adjustable parameters is $\boldsymbol{x} = [l_1\ l_2\ l_3\ w_1\ w_2\ w_3]^T$, whereas parameters $l_0 = 30$, $w_0 = 3$, $s_0 = 0.15$, and $o = 5$ remain fixed. The unit for all dimensions is mm. The EM antenna model $\boldsymbol{R}$

Fig. 12.10 Dual-band single-layer dipole antenna: parameterized geometry

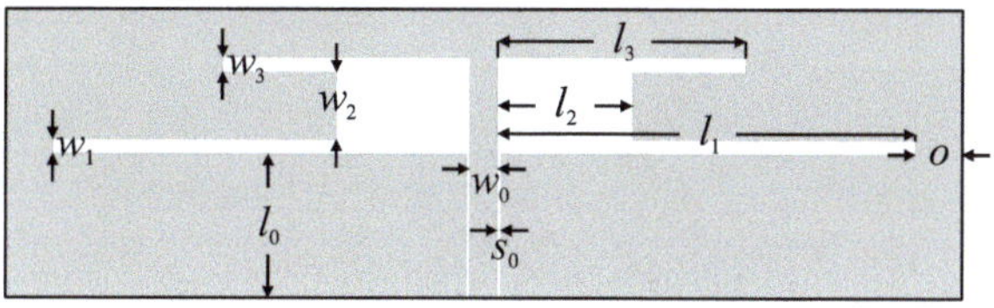

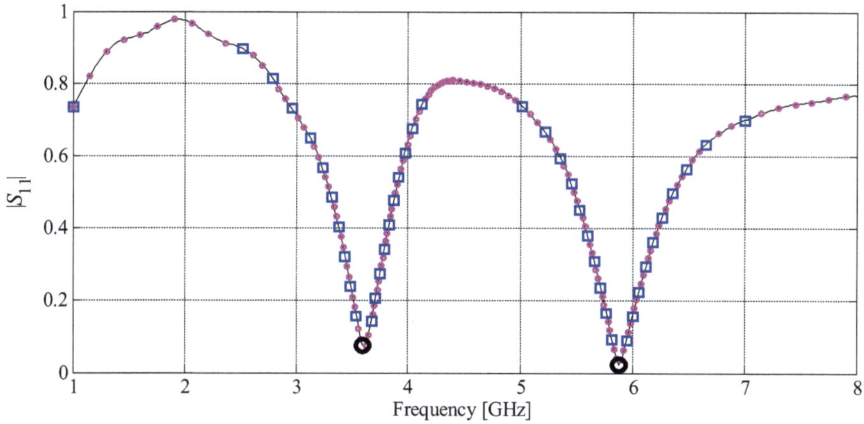

Fig. 12.11 Feature points selected for the dual-band dipole antenna: (**O**) main points (frequency and reflection level of resonances), (□) supplemental points allocated equally with respect to $|S_{11}|$ (intervals between points at the left- and the right-hand-side of the resonances are independent), (○) infill points allocated equally with respect to frequency between the remaining points (their number may vary for various intervals)

is implemented in CST Microwave Studio. It consists of about 100,000 hexahedral cells and its average simulation time is 60 s.

The feature points have been established in a similar way as described in Sect. 12.1.1 but the assignment was separate for each operating band (see also Fig. 12.11). Figure 12.12 shows the landscapes of the selected feature points derived from the antenna reflection characteristics. It should be noted that the responses are only slightly nonlinear, particularly for the frequency component, which is a close-to-linear function of the geometry parameters. Consequently, it is expected that modeling of antenna responses expressed in terms of the feature points would involve a smaller number of data samples when compared to direct modeling of the reflection characteristic.

Validation has been performed by constructing feature-based models for the training sets of 20, 50, 100, 200, 400, and 800 samples allocated using Latin Hypercube Sampling. For the sake of comparison, the same training sets have been utilized for direct modeling of antenna reflection using kriging interpolation. The region of interest is defined by the following lower and upper bounds for design variables: $\boldsymbol{l} = [24.0\ 12.0\ 16.5\ 0.2\ 0.6\ 0.5]^T$ and $\boldsymbol{u} = [32.0\ 13.0\ 19.5\ 0.6\ 3.2\ 1.0]^T$. These ranges are sufficiently wide to allow the lower operating frequency to change

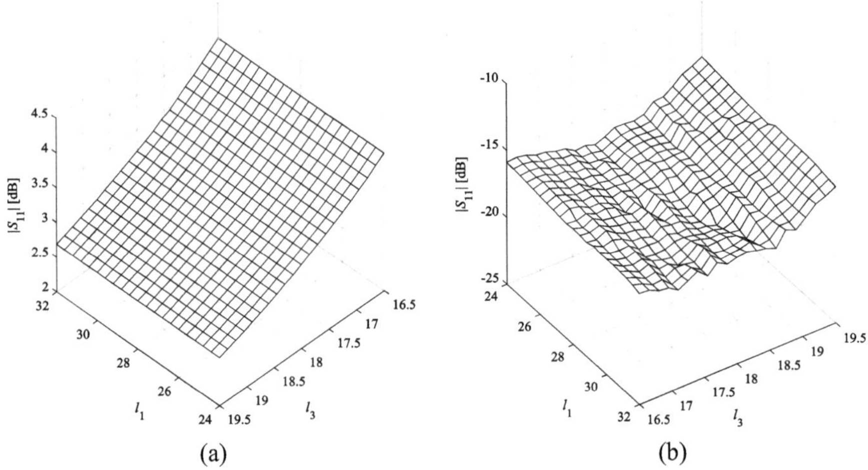

Fig. 12.12 Functional landscapes of feature points obtained for $24.0 \leq l_1 \leq 32.0$ and $16.5 \leq l_3 \leq 19.5$: (**a**) frequency and (**b**) level components of the antenna response. It should be noted that the shapes of the responses are significantly less nonlinear compared to ones obtained in the frequency domain

Table 12.3 Modeling results of the single-layer dual-band dipole

Model/cost	Average error (%)					
	$N^a = 20$	$N = 50$	$N = 100$	$N = 200$	$N = 400$	$N = 800$
Feature-based surrogate	23.3	15.2	14.8	12.4	9.2	7.4
Kriging interpolation[b]	43.2	28.4	22.1	18.2	16.9	13.3

[a]Size of the training set
[b]Direct kriging interpolation of antenna reflection

from around 2.6 to 4.2 GHz and the upper operating frequency to change from 5.2 to 6.8 GHz. The accuracy of the models is validated using the test set composed of 100 randomly generated samples. The error measure $||\boldsymbol{R}(\boldsymbol{x}) - \boldsymbol{R}_s(\boldsymbol{x})||/||\boldsymbol{R}(\boldsymbol{x})||$ is expressed in percent. When calculating the errors, dB-valued responses are considered to emphasize the importance of appropriate representation of the antenna resonances.

The comparison of conventional and feature-based models in terms of the average error is shown in Table 12.3. The results indicate that, for the same data sets, the accuracy of the feature-based modeling is 50–70% better than of the kriging-based one. Consequently, feature-based model with accuracy comparable to frequency-based model can be constructed using four- to eight-fold smaller training data set. This corresponds to reduction of the computational cost from 75% to over 85% with respect to conventional modeling.

A comparison of antenna responses obtained from EM simulation and the feature-based surrogate generated using 400 training samples is shown in Fig. 12.13. The responses of both models are well aligned, which indicates practically sufficient

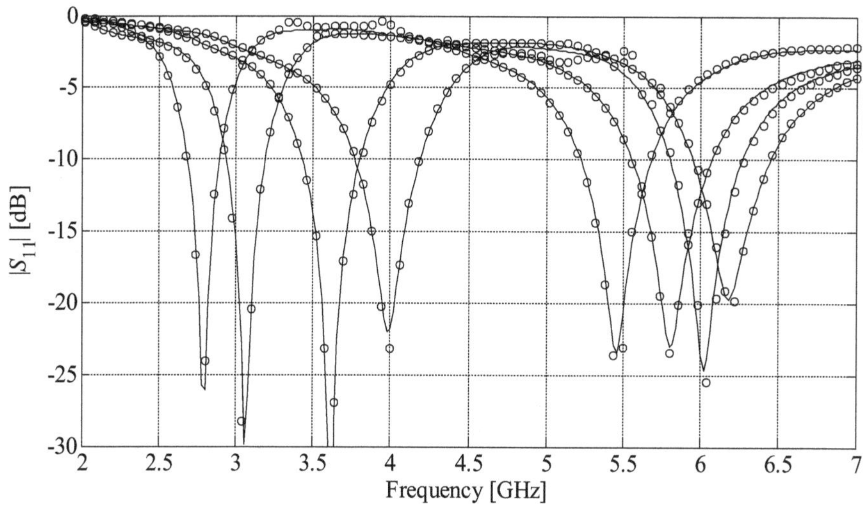

Fig. 12.13 Reflection characteristics of the high-fidelity (—) and feature-based models (400 training points) (○) at the selected test designs. Good alignment between the two datasets can be observed

accuracy of the surrogate. Resonant frequencies for the selected designs are 2.8 and 5.45 GHz, 3.05 and 5.8 GHz, 3.6 and 6.2 GHz, and 4 and 6.2 GHz.

Additional verification is provided by applying the feature-based surrogate for antenna optimization. Three different sets of operating frequencies are considered: (1) $f_1 = 3.0$ and $f_2 = 6.2$ GHz, (2) $f_1 = 3.5$ and $f_2 = 6.5$ GHz, and (3) $f_1 = 4.0$ and $f_2 = 5.5$ GHz. In all cases, the objective is to minimize $|S_{11}|$ in the frequency range of ± 0.02 GHz around the operating frequencies. The EM model responses evaluated at the designs obtained through optimization of the feature-based surrogate are shown in Fig. 12.14. Due to good predictive power of the latter, acceptable reflection responses of the antenna are obtained without further correction.

12.2 Feature-Based Modeling of Microwave Circuits

This section addresses feature-based modeling of microwave filters. The overall principles are the same as discussed in Sect. 12.1. The main difference consists in the selection of the feature points, which are associated with the circuit operating bandwidths. The modeling methodology is outlined in Sect. 12.2.1, whereas verification and application examples are provided in Sects. 12.2.2 and 12.2.3, respectively.

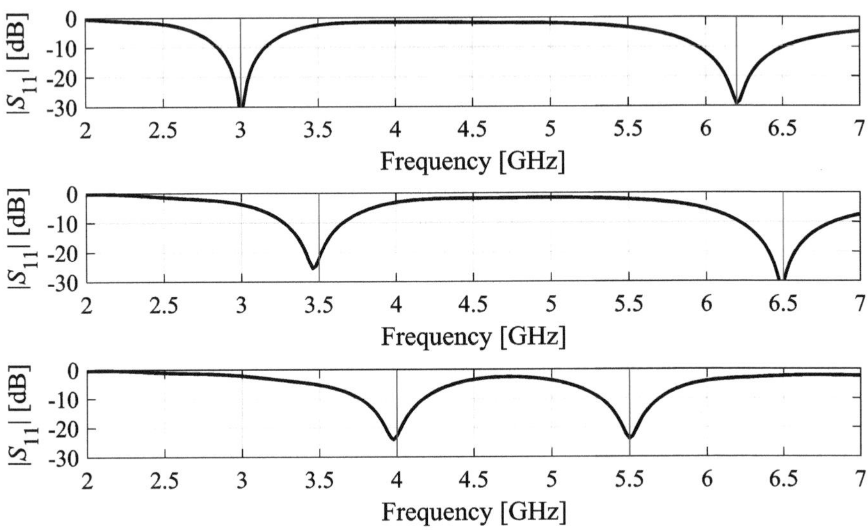

Fig. 12.14 Dual-band dipole optimization: responses of the EM model at the designs obtained through optimization of the feature-based surrogate. The selected operating frequencies: (top) 3.0/6.2 GHz, (middle) 3.5/6.5 GHz, (bottom) 4.0/5.5 GHz

12.2.1 Modeling Methodology

Let $\boldsymbol{R}_f$: $X \rightarrow R^m$, $X \subseteq R^n$ denote the response vector of the microwave device of interest. For example, $\boldsymbol{R}_f(\boldsymbol{x})$ may represent $|S_{21}|$ at m chosen frequencies, ω_1 to ω_m, i.e., $\boldsymbol{R}_f(\boldsymbol{x}) = [R_f(\boldsymbol{x}, \omega_1) \ldots R_f(\boldsymbol{x}, \omega_m)]^T$. $\boldsymbol{R}_f$ is assumed to be evaluated using high-fidelity EM simulation and, therefore, computationally expensive. The task is to build a fast surrogate model $\boldsymbol{R}_s$ that represents $\boldsymbol{R}_f$ in X. Let $X_T = \{\boldsymbol{x}^1, \boldsymbol{x}^2, \ldots, \boldsymbol{x}^N\} \subset X$ be the training set so that the responses of the high-fidelity model $\boldsymbol{R}_f(\boldsymbol{x}^k)$ at $\boldsymbol{x}^k$ are known. Conventional response surface modeling attempts to directly model $\boldsymbol{R}_f(\boldsymbol{x}, \omega_j)$, $j = 1, \ldots, m$. In many cases, the surrogate is created as an ensemble of models constructed for individual frequencies, i.e., obtained by approximating the data sets $\{(\boldsymbol{x}^k, \boldsymbol{R}_f(\boldsymbol{x}^k, \omega_j)\}_{k = 1,\ldots, N}$, for $j = 1, \ldots, m$. Sometimes (Couckuyt et al., 2010), frequency is treated as an additional designable parameter, so that the surrogate is constructed by approximating the data pairs $\{([(\boldsymbol{x}^k)^T \omega_l]^T, \boldsymbol{R}_f(\boldsymbol{x}^k, \omega_l)\}_{k = 1,\ldots, N;\ l \in \{1,\ldots,m\}}$.

Construction of data-driven models in a conventional way is challenging, particularly in the case of filter structures. One of the reasons is the nonlinearity of the responses of interest. Figure 12.15a shows the family of responses of a microstrip bandpass filter evaluated along a selected line segment in the design space $(1 - t)\boldsymbol{x}^a + t\boldsymbol{x}^b$, where $0 \leq t \leq 1$ ($\boldsymbol{x}^a$ and $\boldsymbol{x}^b$ are the two arbitrarily selected points in the design space). The circles mark the responses of design $\boldsymbol{x}^a$ at selected frequencies 1.9, 2.0, and 2.1 GHz. Now, we plot responses w.r.t. t (i.e., design value changes from $\boldsymbol{x}^a$ to $\boldsymbol{x}^b$). Figure 12.15b shows the nonlinear characteristics of $R_f((1 - t)$

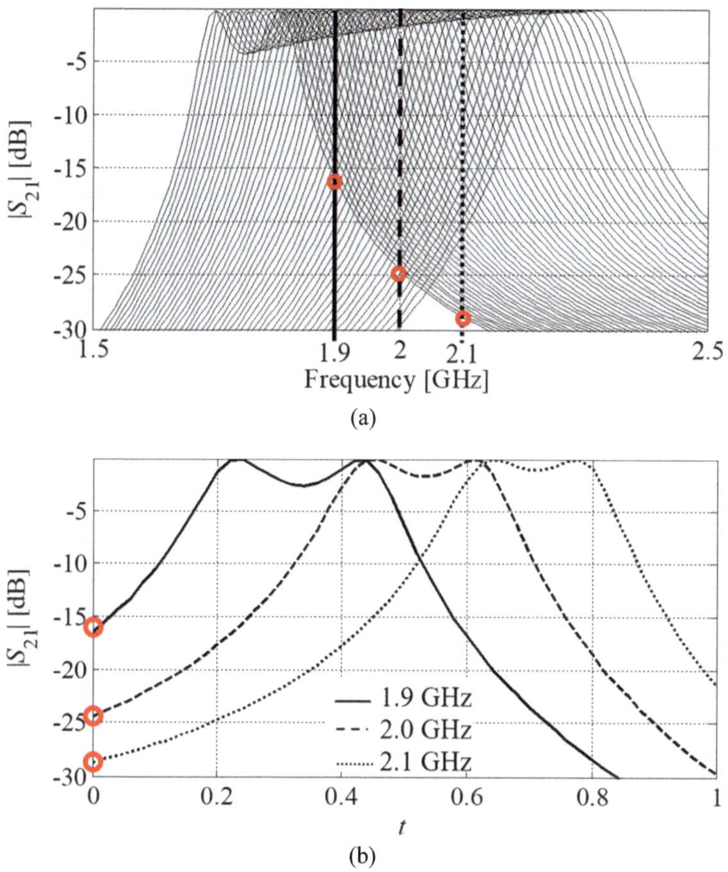

Fig. 12.15 Microstrip bandpass filter responses (**a**) family of $|S_{21}|$ responses for a microstrip bandpass filter evaluated along a selected line segment $(1 - t)\boldsymbol{x}^a + t\boldsymbol{x}^b$, $0 \le t \le 1$, circles mark the responses when $t = 0$ at selected frequencies (indicated by various vertical lines); (**b**) nonlinearity of responses w.r.t. t at selected frequencies 1.9 GHz (—), 2.0 GHz (- - -), and 2.1 GHz (····). Circles mark the responses when $t = 0$ at the selected frequencies

$\boldsymbol{x}^a + t\boldsymbol{x}^b$, ω) at the selected frequencies, 1.9, 2.0, and 2.1 GHz. Adequate modeling of such nonlinear responses across a multidimensional design space requires large quantities of training data.

Feature-based modeling considered here approximates the sets of suitably selected response features. In the case of filters, the features may include points corresponding to specific response levels (e.g., −20 dB, −10 dB, −3 dB), as well as those allocated in between fixed-level points (e.g., uniformly on the frequency scale). Figure 12.16 shows a few responses of the bandpass filter used to create the plots in Fig. 12.15, with the selected feature points marked as circles. We will use the notation $\boldsymbol{f}_k^j = [\omega_k^j \ l_k^j]^T$, $j = 1, \ldots, K$, and $k = 1, \ldots, N$ to denote the jth feature point

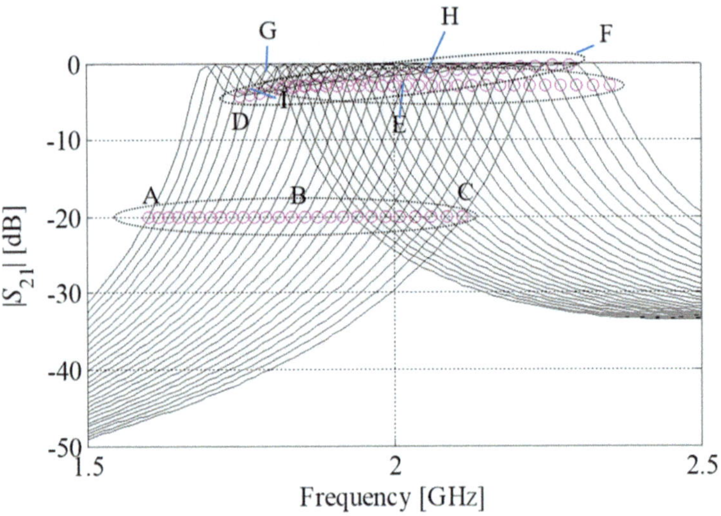

Fig. 12.16 Selected feature point progress plots between two designs $\boldsymbol{x}^a$ ($t = 0$) and $\boldsymbol{x}^b$ ($t = 1$). The lines correspond to selected responses for various values of parameter t between 0 and 1. Three selected groups of feature points (corresponding to −20 dB level on the left-hand side of the passband, right-hand side −3 dB frequency, and center frequency of the filter) are marked with circles. {A, B, C} are −20 dB level feature points corresponding to $t = 0$, 0.5, and 1; {D, E, F} are center frequency feature points corresponding to $t = 0$, 0.5, and 1; {G, H, I} are the respective −3 dB level feature points corresponding to $t = 0$, 0.5, and 1. Selected groups of corresponding features are marked with dotted-line ellipses

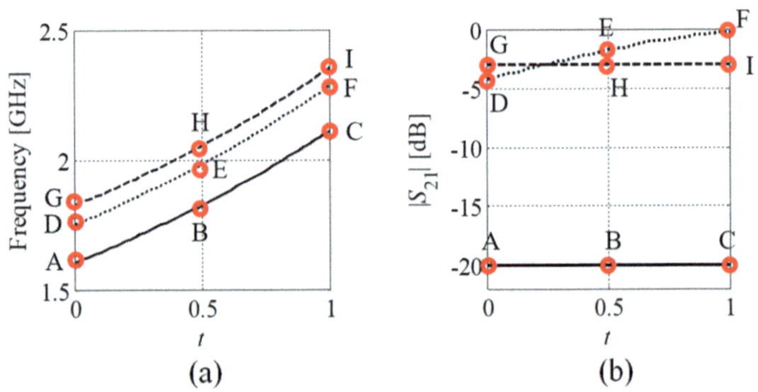

Fig. 12.17 Selected feature point progress plots between designs $\boldsymbol{x}^a$ ($t = 0$) and $\boldsymbol{x}^b$ ($t = 1$): (**a**) frequency, (**b**) levels. They correspond to three feature points: −20 dB level on the left-hand side of the passband (—), right-hand side −3 dB frequency (- - -), and center frequency of the filter (····). Letters A, B, . . ., G mark the feature points corresponding to those in Fig. 12.16

of $\boldsymbol{R}_f(\boldsymbol{x}^k)$, respectively. Here, ω_k^j and l_k^j denote the frequency and magnitude (level) components of $\boldsymbol{f}_k^j$.

Figure 12.17 shows the plots of frequency/level locations of selected feature points corresponding to the family of filter responses of Fig. 12.15. It can be observed that the functions $\omega_k^j(\boldsymbol{x})$ and $l_k^j(\boldsymbol{x})$ (i.e., dependencies of the frequencies and levels of the feature points on the designable parameters $\boldsymbol{x}$) are less nonlinear than the original responses $\boldsymbol{R}_f(\boldsymbol{x}, \omega_j)$. Similarly as discussed in Sect. 12.1, this becomes a fundamental advantage of using feature points rather than frequency-based responses $\boldsymbol{R}_f(\boldsymbol{x}, \omega)$ when it comes to modeling: a substantially smaller number of training designs is necessary to construct accurate surrogates.

The feature-based surrogate is formulated using the operator notation (Koziel et al., 2014a). We use the following two operators: (1) $\boldsymbol{C}(\cdot)$ that extracts K feature points from frequency responses of the structure of interest, and (2) $\boldsymbol{F}(\cdot)$ that carries out the inverse process, i.e., it converts the feature points back to frequency responses, by means of appropriate interpolation schemes. In practice, the operator $\boldsymbol{C}$ is implemented through "scanning" of the frequency response of the filter of interest using an auxiliary algorithm (here, implemented in Matlab). The operator $\boldsymbol{F}$ uses spline-based interpolation (also implemented in the Matlab built-in function *interp1*). Using the operator notation, the feature-based modeling procedure can be described as follows:

Step 1: Create feature points using the operator $\boldsymbol{C}$ for all training designs $\boldsymbol{x}^i$, $i = 1, \ldots, N$.

$$\{\boldsymbol{f}_i^1, \ldots, \boldsymbol{f}_i^K\} = \boldsymbol{C}\left[\boldsymbol{R}_f\left(\boldsymbol{x}^i\right), \boldsymbol{\omega}\right] \tag{12.3}$$

where $\boldsymbol{\omega} = [\omega_1 \ldots \omega_m]$ are the frequency samples corresponding to $\boldsymbol{R}_f$.

Step 2: Construct kriging interpolation models $\boldsymbol{R}_{kr.j}$, $j = 1, \ldots, K$, of the feature points using the features $\{\boldsymbol{f}_i^j\}$, $i = 1, \ldots, N$, corresponding to all training designs $\boldsymbol{x}^i$. The model $\boldsymbol{R}_{kr.j}$ is used to predict the feature point $\boldsymbol{f}^j(\boldsymbol{x})$ at the evaluation design $\boldsymbol{x}$ as follows:

$$\boldsymbol{f}^j(\boldsymbol{x}) = \boldsymbol{R}_{kr.j}(\boldsymbol{x}) \tag{12.4}$$

Step 3: The feature-based surrogate model (in the form of the corresponding frequency response) is then defined as

$$\boldsymbol{R}_s(\boldsymbol{x}) = \boldsymbol{F}\left(\{\boldsymbol{f}^1(\boldsymbol{x}), \ldots, \boldsymbol{f}^K(\boldsymbol{x})\}\right) = \boldsymbol{F}(\{\boldsymbol{R}_{kr.1}(\boldsymbol{x}), \boldsymbol{R}_{kr.2}(\boldsymbol{x}), \ldots, \boldsymbol{R}_{kr.K}(\boldsymbol{x})\}) \tag{12.5}$$

Figure 12.18 shows conceptual illustrations of the feature-based surrogate modeling process. Steps 1 and 2 correspond to surrogate model construction; they are performed only once for a given modeling problem. Step 3 corresponds to the use of the surrogate model and is performed for each model evaluation.

The feature point allocation is a two-step process: (1) identify the core feature points (e.g., those corresponding to critical levels such as −3 dB and/or −20 dB, as well as response local minima/maxima, if relevant), and (2) assign additional points, usually allocated uniformly (either in frequency or level-wise) between the core

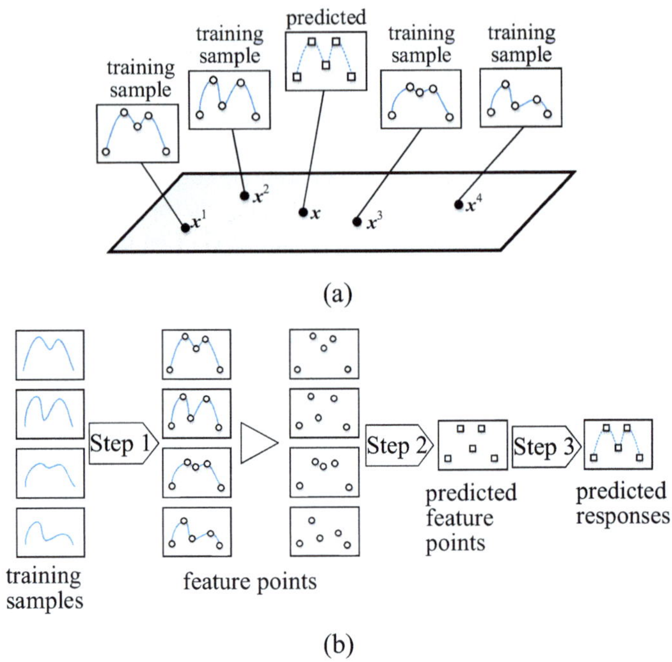

Fig. 12.18 Feature-based modeling: (**a**) conceptual illustration of the feature-based surrogate modeling: points $\boldsymbol{x}^1$ through $\boldsymbol{x}^4$ represent training designs. Their corresponding training (frequency) responses (—) are shown in the rectangular boxes. The feature points are marked by the circles in the training responses. The predicted feature points for the evaluation design $\boldsymbol{x}$ are marked using the squares. The predicted (frequency) response (---) is obtained by means of interpolating/extrapolating these predicted feature points; (**b**) illustration of the feature-based surrogate modeling steps. Step 1: extraction of the feature points from the frequency responses at the training designs; Step 2: construction of the response surface approximation model of the training feature points (here, using kriging interpolation), as well as prediction of the feature points at the evaluation design; Step 3: obtaining the frequency response at the evaluation design from the predicted feature points. Note that Steps 1 and 2 correspond to surrogate model construction (these steps are performed only once). Step 3 represents model evaluation

features. We define all possible core features in a set $\{f_1, \ldots, f_{NA}\}$ through a visual inspection of the responses. The following procedure is used to scan and find the features for all the responses $\boldsymbol{R}_f(\boldsymbol{x}^k)$, $k = 1, \ldots, N$:

```
i = 1;
while (i ≤ N_A) // Step 1
j = 1;
while ( j ≤ N)
Scan jth response R_f(x^j) for feature f_i
if feature f_i exists
j = j+1;
mark location (magnitude and frequency) of feature f_i w.r.t. R_f(x^j);
else
break;
```

```
end
if j == N // feature f_i is valid
save the locations of feature f_i for each response R_f(x^j);
end
end
i = i+1;
end
Add uniformly spaced points between the core features; //Step 2
```

It is important to note that the number of feature points should be sufficient so that the interpolation operation $\boldsymbol{F}$ has enough data for accurate response prediction. It is also important that the feature points corresponding to any two training designs are in one-to-one correspondence, and that the number of feature points is the same for all training designs. It should be mentioned, however, that if a one-to-one correspondence between the "visible" core feature points throughout the training set is not achievable, the entire concept would still formally work (in which case, the corresponding core feature points are omitted from all training designs while constructing the surrogate model). This is not the case for other feature-based methods such as SPRP (Koziel & Szczepanski, 2011) or GSPRP (Koziel & Leifsson, 2012), where the characteristic point correspondence is of fundamental importance. In Koziel and Leifsson (2012), some ways have been described of defining the feature points (or "characteristic" points in the SPRP context) to work around the possible lack of their one-to-one correspondence throughout the design space.

12.2.2 Demonstration Case Studies

The first example (Filter 1) is the wideband microstrip bandstop filter (Hsieh & Wang, 2005) shown in Fig. 12.19a. The filter is simulated in FEKO using 642 triangular meshes (simulation time: 5 s per frequency on 2.53 GHz Intel Xeon CPU with 6 GB RAM). The substrate parameters are: thickness $h = 0.508$ mm and permittivity $\varepsilon_r = 3.38$. The designable parameters are: $\boldsymbol{x} = [L_r\ W_r\ L_c\ W_c\ G_c]^T$. The region of interest is defined as an interval $[\boldsymbol{x}^0 - \boldsymbol{d}, \boldsymbol{x}^0 + \boldsymbol{d}]$ with $\boldsymbol{x}^0 = [7.0\ 1.2\ 9.0\ 0.075\ 0.10]^T$ and $\boldsymbol{d} = [1.0\ 0.4\ 1.0\ 0.025\ 0.05]^T$. The frequency range is 1–11 GHz with 100 MHz step size.

The second structure (Filter 2) is the microstrip bandpass filter with open stub inverter (Lee et al., 2000) shown in Fig. 12.19b. The filter is also simulated in FEKO using 432 triangular meshes (simulation time: 2 s per frequency). The substrate parameters are: thickness $h = 0.508$ mm and permittivity $\varepsilon_r = 2.6$. The designable parameters are: $\boldsymbol{x} = [L_1\ L_2\ L_3\ S_1\ S_2\ W_1]^T$. The region of interest is defined as an interval $[\boldsymbol{x}^0 - \boldsymbol{d}, \boldsymbol{x}^0 + \boldsymbol{d}]$ with $\boldsymbol{x}^0 = [24.0\ 10.0\ 2.0\ 0.6\ 0.2\ 0.5]^T$ and $\boldsymbol{d} = [2.0\ 2.0\ 1.0\ 0.4\ 0.1\ 0.4]^T$. The frequency range is 1.5–2.5 GHz with 12.5 MHz step size.

Filter 2 is a bandpass circuit. Therefore, for this case, the feature points are selected to capture the (frequency) location of the passband. In particular, we have: (1) two points at the -10 dB levels (i.e., determining the -10 dB passband;

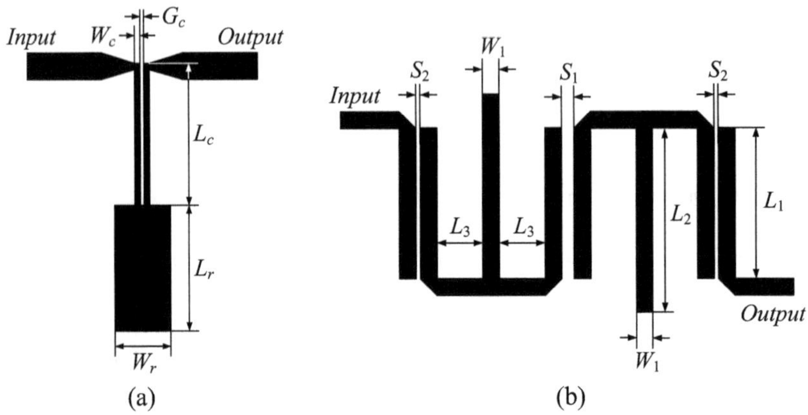

Fig. 12.19 Filter structures used for feature-based modeling verification: (**a**) wideband bandstop filter, (**b**) bandpass filter with open stub inverter

we do not use the −3 dB feature because, for certain training designs, the −3 dB level might not be achieved at all), (2) additional points uniformly distributed in frequency and placed between the −10 dB features (20–40 points depending on the example), and (3) additional points at −12 dB, −14 dB, −16 dB, and so on allocated on both sides of the −10 dB passband.

The particular choice of the −10 dB level is, of course, not critical (e.g., −11 dB will work equally well, as long as it exists in all the responses). The same principle applies to the choice of the number and levels of the additional points. For Filter 1 (bandstop filter), the feature points are selected at −20 dB level, 50 points in between these two (allocated uniformly in frequency), the two local maxima (close to the edges of the stopband), and additional points between the maxima and the −20 dB level points.

The model accuracy is verified using the relative error measure $||\boldsymbol{R}_f(\boldsymbol{x}) - \boldsymbol{R}_s(\boldsymbol{x})||/||\boldsymbol{R}_f(\boldsymbol{x})||$ expressed in percent and averaged over 100 random test designs. The feature-based model is compared to direct kriging interpolation of high-fidelity data, direct radial basis function (RBF) interpolation of $\boldsymbol{R}_f$ data, as well as GSPRP. The kriging model utilizes a Gaussian correlation function (Lophaven et al., 2002), whereas the RBF model uses Gaussian basis functions (Queipo et al., 2005). The length-scale parameter of the latter is optimized using cross-validation (Queipo et al., 2005). For all four modeling methods, five different cases are considered with the number of training points varying between $N = 20$ and $N = 400$. The results are shown in Table 12.4.

It can be observed that for all considered problems, the accuracy of the feature-based surrogate is comparable or better than GSPRP and better than the accuracy of both the kriging and RBF surrogates for the corresponding number of training points. Figure 12.20 shows the high-fidelity and feature-based model responses at selected test points for Filters 1 and 2. It should also be noted that feature-based model (similarly as GSPRP) ensures good accuracy even for a small number of

Table 12.4 Modeling results for Filters 1 and 2

Filter	Model setup cost (min)[e]	Modeling method	Average error (%)				
			$N^a = 20$	$N = 50$	$N = 100$	$N = 200$	$N = 400$
1	8·N	Feature-based[b]	9.1	5.9	5.4	4.5	4.0
		GSPRP	8.5	6.1	5.9	4.7	4.4
		Direct kriging[c]	16.0	11.8	8.8	7.1	5.9
		Direct RBFs[d]	18.9	15.3	14.9	12.2	10.5
2	3·N	Feature-based[b]	7.5	2.4	1.2	0.7	0.6
		GSPRP	7.7	2.7	1.8	1.4	1.2
		Direct kriging[c]	11.6	8.8	6.1	4.8	3.1
		Direct RBFs[d]	16.7	15.2	12.7	10.4	9.3

[a]N stands for the number of training points
[b]Feature-based surrogate constructed using the procedure outlined in Sect. 12.2.1
[c]Direct kriging interpolation of high-fidelity model $|S_{21}|$ responses
[d]Direct interpolation of high-fidelity model $|S_{21}|$ responses using Radial Basis Functions (RBFs)
[e]The cost refers to training data acquisition. The cost of surrogate model identification can be neglected

training samples ($N = 50$ or 100), which is not the case for the "direct" data-driven approaches. Our examples also indicate that the feature-based approach is capable of modeling components of various types (bandstop, bandpass, narrowband, wideband, etc.), as long as the corresponding feature sets are consistent across the design space.

12.2.3 Application Case Studies

For the sake of illustration, the feature-based model of Sect. 12.2.1 was used to carry out parametric optimization of the filter structures considered in Sect. 12.2.2. The design specifications are as follows:

- Filter 1: $|S_{21}| \leq -20$ dB for 3.0 GHz $\leq \omega \leq$ 9.00 GHz, $|S_{21}| \geq -3$ dB for $\omega \leq 2$ GHz and $\omega \geq 10$ GHz;
- Filter 2: $|S_{21}| \geq -1$ dB for 1.95 GHz $\leq \omega \leq$ 2.05 GHz, $|S_{21}| \leq -20$ dB for $\omega \leq 1.8$ GHz and $\omega \geq 2.2$ GHz.

For both cases, the feature-based surrogate created with $N = 200$ training samples was considered. Figure 12.21 shows the initial and final filter responses obtained by optimizing the feature-based surrogate model (the final design is verified by the high-fidelity model). Design specifications are marked using thick horizontal lines.

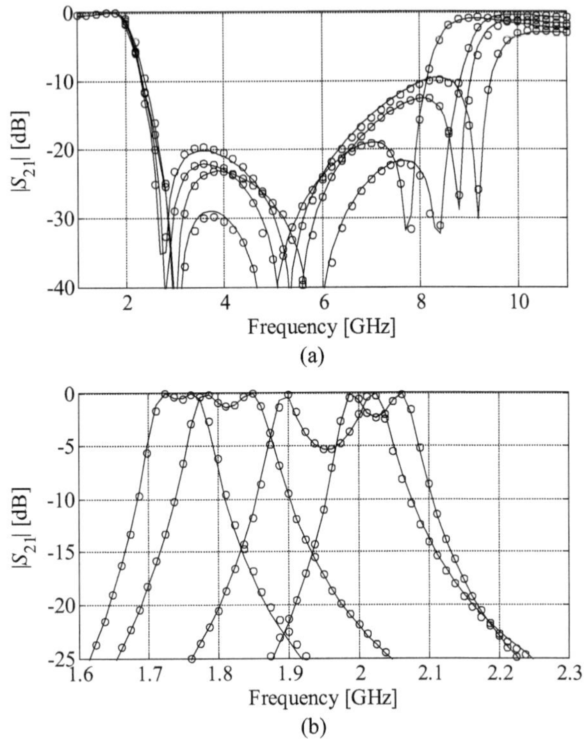

Fig. 12.20 EM-simulated (—) and feature-based surrogate responses (○) obtained for 100 base points at the selected test designs: (**a**) Filter 1, (**b**) Filter 2

Due to the very good accuracy of the surrogates, no further design tuning is necessary. In all cases, the surrogate models were optimized using a sequential quadratic programming algorithm (implemented in Matlab's *fmincon* routine).

12.3 Variable-Fidelity Feature-Based Modeling

This section discusses feature-based data-driven modeling of microwave components using variable-fidelity EM simulations. In this approach, the characteristic points of the system response are extracted from the computational models of two levels of resolution: the less accurate but faster low-fidelity representation and the accurate but more expensive high-fidelity one. Both models share the same underlying physics, which makes them well correlated. Exploring these correlations enables a rendition of reliable surrogate at a fraction of cost entailed by the procedure that exclusively relies on the high-fidelity model.

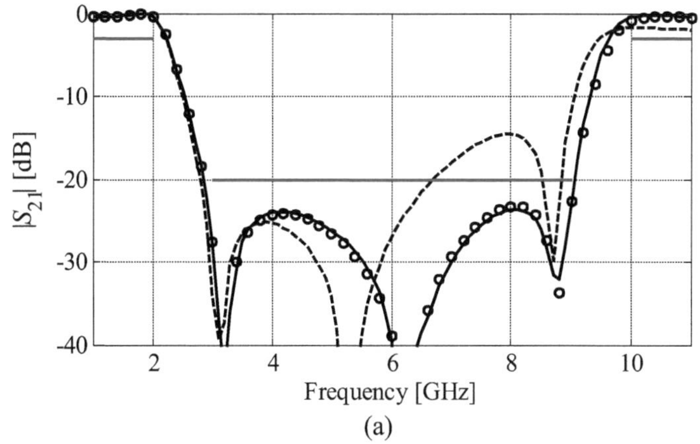

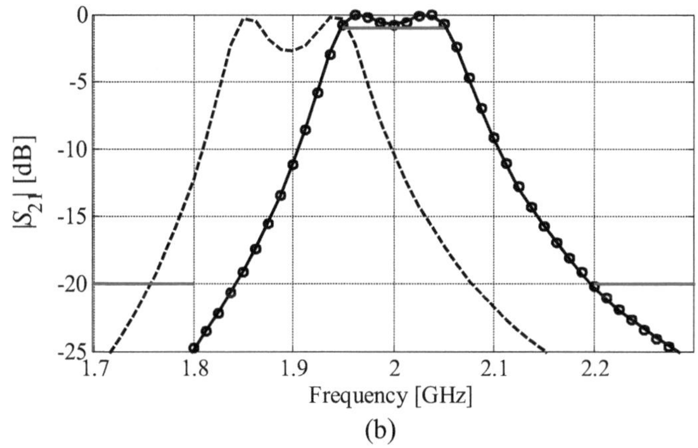

Fig. 12.21 Feature-based surrogate responses at the initial design (- - -), at the optimized design (—), and the corresponding high-fidelity model responses (○): (**a**) Filter 1, (**b**) Filter 2

12.3.1 Modeling Procedure

The modeling problem is formulated as outlined in Sect. 12.2. The construction of the surrogate model is carried out at the level of response features; however, two underlying computational (EM) models are employed to reduce the cost of the process. More specifically, we utilize training data acquired from sparsely-sampled high-fidelity model $\boldsymbol{R}_f$ (points $X_{Tf} = \{\boldsymbol{x}_f^1, \boldsymbol{x}_f^2, \ldots, \boldsymbol{x}_f^{Nf}\}$) and from densely-sampled data obtained from coarse-discretization EM simulations (low-fidelity model $\boldsymbol{R}_{cd}$), $X_{Tc} = \{\boldsymbol{x}_c^1, \boldsymbol{x}_c^2, \ldots, \boldsymbol{x}_c^{Nc}\}$. Although $\boldsymbol{R}_{cd}$ and $\boldsymbol{R}_f$ are misaligned, they are also well correlated (cf. Figs. 12.22 and 12.23) so that the initial surrogate model obtained from the $\boldsymbol{R}_{cd}$ data can be enhanced by using a few $\boldsymbol{R}_f$ points to construct the accurate, final surrogate model. We use the notation $\boldsymbol{f}_{k.f}^{\ j} = [\omega_{k.f}^{\ j}\ l_{k.f}^{\ j}]^T$, $j = 1, \ldots, K$ and $k = 1, \ldots, N_f$, to denote the jth feature point of $\boldsymbol{R}_f(\boldsymbol{x}_f^k)$, and $\boldsymbol{f}_{k.c}^{\ j} = [\omega_{k.c}^{\ j}\ l_{k.c}^{\ j}]^T$ to denote the

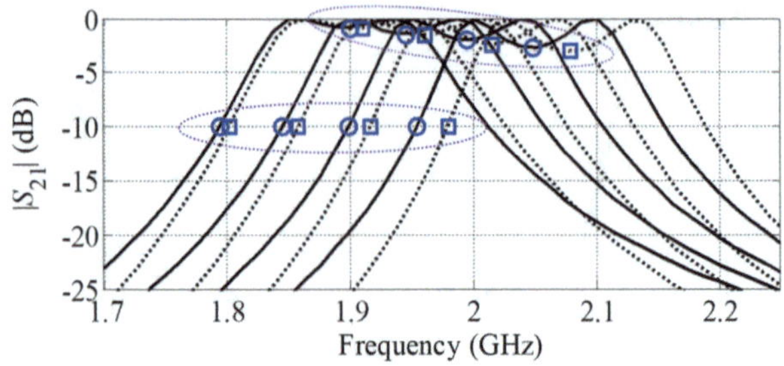

Fig. 12.22 Family of $|S_{21}|$ responses for a microstrip bandpass filter evaluated along a selected line segment $(1 - t)\boldsymbol{x}^a + t\boldsymbol{x}^b$, $0 \le t \le 1$: high-fidelity model $\boldsymbol{R}_f$ (—) and low-fidelity model $\boldsymbol{R}_{cd}$ (····). Selected feature points and groups of corresponding points marked (○) for $\boldsymbol{R}_f$ and (□) for $\boldsymbol{R}_{cd}$

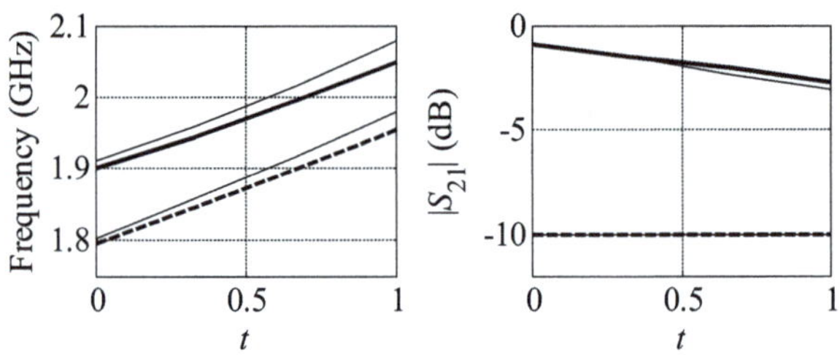

Fig. 12.23 Selected feature point plots between designs $\boldsymbol{x}^a$ ($t = 0$) and $\boldsymbol{x}^b$ ($t = 1$): (**a**) frequency, (**b**) levels. They correspond to two feature points: center frequency of the filter (- - -) and −10 dB level on the left-hand side of the passband (—); thick and thin lines are used for high- and low-fidelity model feature points, respectively

jth feature point of $\boldsymbol{R}_{cd}(\boldsymbol{x}_c^{\ k})$; $\omega_{k.f}^{\ \ j}$ and $l_{k.f}^{\ \ j}$ denote the frequency and magnitude (level) components of $\boldsymbol{f}_{k.f}^{\ \ j}$ (similarly for $\boldsymbol{f}_{k.c}^{\ \ j}$).

12.3.1.1 Variable-Fidelity Feature-Based Modeling: The Internal Approach

Multi-fidelity feature-based modeling is a two-step process. First, we construct approximation surrogates $s_{\omega.j}(\boldsymbol{x})$ and $s_{l.j}(\boldsymbol{x})$, $j = 1, \ldots, K$, corresponding to the feature points. In the internal approach presented in this section, the construction of $s_{\omega.j}(\boldsymbol{x})$ and $s_{l.j}(\boldsymbol{x})$ is based on both N_f high-fidelity and N_c low-fidelity training points and their corresponding feature points $\{\boldsymbol{f}_{k.f}^{\ \ j}, \boldsymbol{f}_{k.f}^{\ \ j}, \ldots, \boldsymbol{f}_{Nf.f}^{\ \ j}\}$, and $\{\boldsymbol{f}_{k.c}^{\ \ j}, \boldsymbol{f}_{k.c}^{\ \ j}, \ldots, \boldsymbol{f}_{Nc.c}^{\ \ j}\}$, $j = 1, \ldots, K$. Blending these two types of data sets is realized through co-kriging (Kennedy & O'Hagan, 2000) as described in the next paragraph.

Let $\boldsymbol{f}(X_{Tc})$ be the set of responses associated with the training set X_{Tc} (i.e., low-fidelity feature points $\boldsymbol{f}_{1.c}$ through $\boldsymbol{f}_{Nf.c}$). The kriging interpolant is given as (Couckuyt et al., 2010)

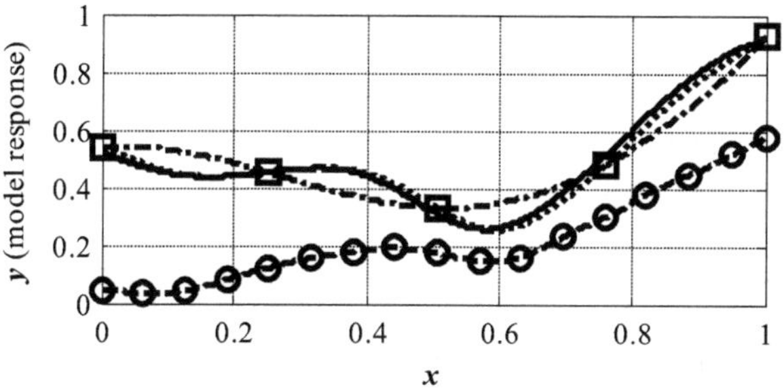

Fig. 12.24 Co-kriging concept: $\boldsymbol{R}_f$ model (—), $\boldsymbol{R}_c$ model (- - -), $\boldsymbol{R}_f$ model samples (□), $\boldsymbol{R}_c$ model samples (○). Kriging interpolation of $\boldsymbol{R}_f$ model samples (- · · -) is not an adequate representation of the $\boldsymbol{R}_f$ model (limited data set). Co-kriging interpolation (····) of blended $\boldsymbol{R}_c$ and $\boldsymbol{R}_f$ data provides better accuracy at a lower computational cost

$$s_{KR}(\boldsymbol{x}) = M\alpha + r(\boldsymbol{x}) \cdot \Psi^{-1} \cdot (f(X_{Tc}) - F\alpha) \tag{12.6}$$

where M and F are Vandermonde matrices of the test point $\boldsymbol{x}$ and the base set X_{Tc}, respectively, α is determined by generalized least squares (GLS), $r(\boldsymbol{x})$ is an $1 \times N_f$ vector of correlations between the point $\boldsymbol{x}$ and the base set X_{Tf}, where the entries are $r_i(\boldsymbol{x}) = \psi(\boldsymbol{x}, \boldsymbol{x}_f^{\,i})$, and Ψ is a $N_f \times N_f$ correlation matrix whose entries are given by $\Psi_{i,j} = \psi(\boldsymbol{x}_{KR}^{\,i}, \boldsymbol{x}_{KR}^{\,j})$. We use the exponential correlation function $\psi(\boldsymbol{x}, \boldsymbol{x}') = \exp(\sum_{k=1,\ldots,n} - \theta_k|x_k\text{-}x'_k|)$. The regression function is constant, $F = [1 \ \ldots \ 1]^T$ and $M = (1)$. Co-kriging is a type of kriging where the $\boldsymbol{R}_f$ and $\boldsymbol{R}_{cd}$ model data are combined to enhance the prediction accuracy (cf. Fig. 12.24). Co-kriging is a two-step process: first a kriging model s_{KRc} of the coarse data $(X_{Tc}, \boldsymbol{c}(X_{Tc}))$ is constructed and, on the residuals of the fine data $(X_{Tf}, \boldsymbol{R}_d)$, a second kriging model s_{KRd} is applied, where $\boldsymbol{R}_d = f(X_{Tf}) - \rho{\cdot}\boldsymbol{c}(X_{Tf})$; $\boldsymbol{c}(X_{Tf})$ can be approximated as $\boldsymbol{c}(X_{Tf}) \approx s_{KRc}(X_{Tf})$. The co-kriging interpolant is defined as (Koziel et al., 2013)

$$s_{CO}(\boldsymbol{x}) = M\alpha + r(\boldsymbol{x}) \cdot \Psi^{-1} \cdot (\boldsymbol{R}_d - F\alpha) \tag{12.7}$$

Definitions of M, F, $r(\boldsymbol{x})$, and Ψ can be found in Koziel et al. (2013).

The multi-fidelity feature-based surrogate (the internal approach) is defined as

$$\boldsymbol{R}_s(\boldsymbol{x}) = [R_s(\boldsymbol{x}, \omega_1) \quad \ldots \quad R_s(\boldsymbol{x}, \omega_K)]^T \tag{12.8}$$

with

$$R_s(\boldsymbol{x}, \omega_j) = I(\Omega(\boldsymbol{x}), L(\boldsymbol{x}), \omega_j) \tag{12.9}$$

where $L(\boldsymbol{x}) = [s_{l.1}(\boldsymbol{x}) \ldots s_{l.K}(\boldsymbol{x})]$ and $\Omega(\boldsymbol{x}) = [s_{\omega.1}(\boldsymbol{x}) \ldots s_{\omega.K}(\boldsymbol{x})]$ are predicted feature point locations corresponding to $\boldsymbol{x}$, the design being considered; $\omega_j, j = 1, \ldots, m$, is a discrete set of frequencies at which the response of the structure is being evaluated.

According to the internal approach, both $s_{\omega.j}(\boldsymbol{x})$ and $s_{l.j}(\boldsymbol{x})$ are implemented through co-kriging as in (12.7). $I(\Omega, L, \omega)$ denotes a function that interpolates the level vector L and frequency vector Ω into the response at a given frequency ω_j.

Because $\omega_k^j(\boldsymbol{x})$ and $l_k^j(\boldsymbol{x})$ (i.e., frequencies and levels of the feature points) are less nonlinear than the original responses $\boldsymbol{R}_f(\boldsymbol{x}, \omega_j)$, a substantially smaller number of training points is necessary to ensure accurate modeling. Also, good correlation between $\boldsymbol{R}_{cd}$ and $\boldsymbol{R}_f$ (cf. Fig. 12.23) allows for further reduction of the surrogate model setup cost because a very limited number of $\boldsymbol{R}_f$ samples is sufficient to elevate the $\boldsymbol{R}_{cd}$-based kriging model to high-fidelity accuracy through co-kriging.

12.3.1.2 Variable-Fidelity Feature-Based Modeling: The External Approach

The internal approach presented in Sect. 12.3.1.1 combines the low- and high-fidelity training data at the level of the feature points. In the external approach outlined below, the high-fidelity data is included through a space mapping correction of the initial feature-based surrogate model $\boldsymbol{R}_{s.c}(\boldsymbol{x})$ obtained as in (12.6) and (12.7), however, using the low-fidelity training data only. The space mapping correction is realized at the level of the original responses as follows (Bandler et al., 2006):

$$\boldsymbol{R}_s(\boldsymbol{x}) = \boldsymbol{A} \cdot \boldsymbol{R}_{s.c.F}(\boldsymbol{B} \cdot \boldsymbol{x} + \boldsymbol{c}; \boldsymbol{F}) \quad (12.10)$$

where $\boldsymbol{R}_{s.c.F}$ is a frequency scaled model $\boldsymbol{R}_{s.c}(\boldsymbol{x})$ such that $\boldsymbol{R}_{s.c.F}(\boldsymbol{x};\boldsymbol{F}) = [R_{s.c}(\boldsymbol{x}, f_0 + f_1\omega_1) \ldots R_{s.c}(\boldsymbol{x}, f_0 + f_1\omega_m)]^T$, with $R_{s.c}(\boldsymbol{x}, \omega_j)$, $j = 1, \ldots, m$, being components of $\boldsymbol{R}_{s.c}(\boldsymbol{x})$. Here, the model parameters $\boldsymbol{A}$ (diagonal matrix), $\boldsymbol{B}$ (square matrix), $\boldsymbol{c}$ (column vector), and $\boldsymbol{F}$ are obtained from the standard parameter extraction procedure needed by space mapping, namely,

$$[\boldsymbol{A}, \boldsymbol{B}, \boldsymbol{c}, \boldsymbol{F}] = \arg \min_{[\overline{\boldsymbol{A}}, \overline{\boldsymbol{B}}, \overline{\boldsymbol{c}}, \overline{\boldsymbol{F}}]} \sum\nolimits_{k=1}^{N_f} \left\| \boldsymbol{R}_f\left(\boldsymbol{x}_f^k\right) - \overline{\boldsymbol{A}} \cdot \boldsymbol{R}_{s.c.F}\left(\overline{\boldsymbol{B}} \cdot \boldsymbol{x}_f^k + \overline{\boldsymbol{c}}; \overline{\boldsymbol{F}}\right) \right\| \quad (12.11)$$

If the correlation between the low- and high-fidelity models is sufficient (which is normally the case when both models are based on EM analysis), the correction given by (12.10) and (12.11) should significantly improve the model accuracy.

It should also be noted that the computational cost of surrogate model identification (i.e., solving the parameter extraction process (12.11)) can be neglected compared to the high-fidelity data acquisition because it is realized at the level of a fast feature-based model $\boldsymbol{R}_{s.c}$.

12.3.1.3 Internal Versus External Approach

Apart from considerable conceptual differences between the internal and external approaches, i.e., blending in the high-fidelity model data at the level of the response features rather than at the level of the original responses, the external approach is simpler to implement. On the other hand, the internal approach can be more flexible because the inclusion of a larger amount of high-fidelity training data should lead to an improvement of model accuracy. This may not be the case for the external approach (beyond a certain number of points) due to the fixed number of degrees of freedom of the surrogate (12.10) (the number of model parameters depends on the problem's dimensionality but not on the cardinality of the data set).

12.3.2 Verification Examples

To illustrate the operation and performance of the considered modeling techniques, we consider two microstrip filters. Filter 1 is the stacked slotted resonators bandpass filter (Huang et al., 2008) shown in Fig. 12.25a. The high-fidelity filter model is simulated in Sonnet ***em*** using a grid of 0.05 mm × 0.05 mm. The low-fidelity model

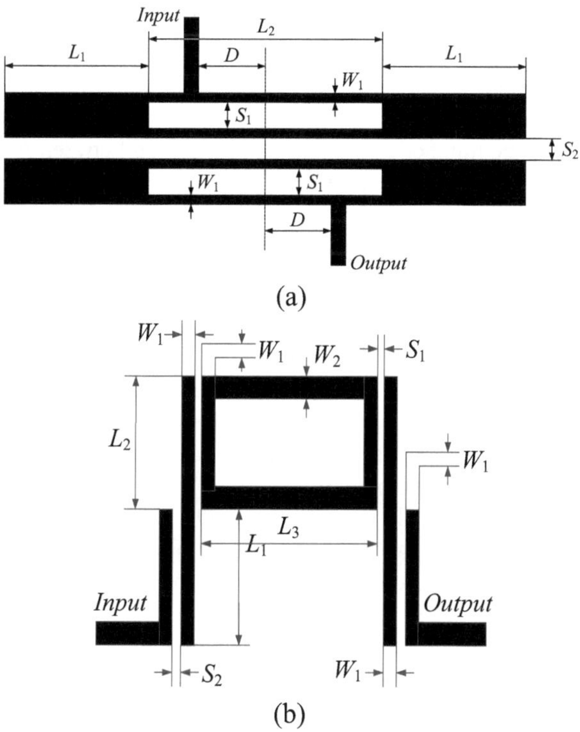

Fig. 12.25 Filter structures used for feature-based modeling verification: (**a**) stacked slotted resonators filter (Filter 1) and (**b**) fourth-order ring resonator bandpass filter (Filter 2)

is also simulated in Sonnet on a 2 mm × 2 mm grid. The substrate parameters are: thickness $h = 0.635$ mm and permittivity $\varepsilon_r = 9.7$. The designable parameters are: $\boldsymbol{x} = [L_1\ L_2\ W_1\ S_1\ S_2\ D]^T$. The region of interest is defined as the interval $[\boldsymbol{x}^0 - \boldsymbol{d}, \boldsymbol{x}^0 + \boldsymbol{d}]$ with $\boldsymbol{x}^0 = [6.0\ 9.6\ 1.0\ 1.0\ 2.0\ 2.0]^T$ and $\boldsymbol{d} = [0.8\ 0.8\ 0.2\ 0.2\ 0.4\ 0.4]^T$.

The second structure (Filter 2) is the fourth-order ring resonator bandpass filter (Salleh et al., 2008) shown in Fig. 12.25b. The high-fidelity filter model is simulated in FEKO using 952 triangular meshes. The low-fidelity FEKO model utilizes 174 meshes. The substrate parameters are: thickness $h = 1.52$ mm and permittivity $\varepsilon_r = 4.32$. The designable parameters are: $\boldsymbol{x} = [L_1\ L_2\ L_3\ S_1\ S_2\ W_1\ W_2]^T$. The region of interest is defined as the interval $[\boldsymbol{x}^0 - \boldsymbol{d}, \boldsymbol{x}^0 + \boldsymbol{d}]$ with $\boldsymbol{x}^0 = [24.0\ 20.0\ 25.0\ 0.2\ 0.2\ 1.0\ 0.5]^T$ and $\boldsymbol{d} = [1.0\ 1.0\ 1.0\ 0.1\ 0.1\ 0.2\ 0.2]^T$.

The modeling accuracy is verified using the relative error measure $||\boldsymbol{R}_f(\boldsymbol{x}) - \boldsymbol{R}_s(\boldsymbol{x})||/||\boldsymbol{R}_f(\boldsymbol{x})||$ expressed in percent and averaged over 100 random test designs. The multi-fidelity feature-based models (both the internal and external versions) are compared with the following modeling methods:

- Regular (single-fidelity) feature-based modeling (Koziel et al., 2014b);
- Generalized shape-preserving response prediction (GSPRP) (Koziel & Leifsson, 2012);
- Direct kriging interpolation of the high-fidelity data (Queipo et al., 2005);
- Response surface modeling using radial-basis functions (Buhmann & Ablowitz, 2003).

The kriging model utilizes a Gaussian correlation function, whereas the radial-basis function model uses Gaussian basis functions (Queipo et al., 2005). The length-scale parameter of the latter is optimized using cross-validation (Queipo et al., 2005). For all the above modeling methods, five different cases are considered with the number of training points varying between $N = 20$ and $N = 400$.

The results are shown in Tables 12.5 and 12.6. Figure 12.26 shows the high-fidelity and feature-based model responses at selected test points for Filters 1 and 2. The following observations can be made:

- Both the internal and external multi-fidelity feature-based models ensure excellent accuracy even with a very small number of high-fidelity training samples (specifically, 20 and 50).
- The internal feature-based modeling approach is generally superior over the external approach; however, the latter is still considerably better than a single-level feature-based surrogate for a small number of training samples and comparable or better overall.
- Asymptotically (i.e., for the number of high-fidelity training points increasing to 400), both multi-fidelity feature-based modeling methods are comparable or better than the single-fidelity feature-based models and ones based on GSPRP.
- All modeling approaches exploiting the concept of response features are considerably more accurate than conventional modeling techniques working directly with the original system responses, here, S-parameters versus frequency.

Table 12.5 Modeling results for Filter 1

Modeling method	Average error (%)				
	$N^{a} = 20$	$N = 50$	$N = 100$	$N = 200$	$N = 400$
Multi-fidelity feature-based (internal approach)[b]	0.5 [35.0][c]	0.4 [65.0][c]	0.35 [115.0][c]	0.3 [215.0][c]	0.25 [415.0][c]
Multi-fidelity feature-based (external approach)[d]	2.2	1.1	1.0	0.8	0.8
Feature-based[f]	3.9	1.8	0.9	0.6	0.4
GSPRP[e]	4.0	1.7	0.7	0.5	0.3
Direct kriging[g]	15.6	11.9	11.0	9.3	6.7
Direct RBF[h]	17.2	13.5	11.8	9.9	8.7

[a] N stands for the number of training points
[b] (Internal) multi-fidelity feature-based modeling of Sect. 12.3.1.1 (200 $\boldsymbol{R}_{cd}$ samples)
[c] Numbers in brackets refer to the total model setup cost (including $\boldsymbol{R}_{cd}$ points)
[d] (External) multi-fidelity feature-based modeling of Sect. 12.3.1.2
[e] Generalized SPRP modeling (Koziel & Leifsson, 2012)
[f] Feature-based modeling using the procedure of Koziel et al. (2014b)
[g] Direct kriging interpolation of the high-fidelity model $|S_{21}|$ responses
[h] Direct radial-basis function interpolation of the high-fidelity model data

Table 12.6 Modeling results for Filter 2

Modeling method	Average error (%)				
	$N^{a} = 20$	$N = 50$	$N = 100$	$N = 200$	$N = 400$
Multi-fidelity feature-based (internal approach)[b]	3.7 [28.3][c]	1.5 [58.3][c]	1.3 [108.3][c]	1.2 [208.3][c]	1.2 [408.3][c]
Multi-fidelity feature-based (external approach)[d]	4.7	3.6	2.7	2.6	2.5
Feature-based[f]	11.6	5.1	3.8	3.6	2.5
GSPRP[e]	8.8	5.4	4.2	4.0	3.0
Direct kriging[g]	13.5	7.8	6.3	4.6	3.6
Direct RBF[h]	21.2	19.3	15.5	13.3	11.5

[a] N stands for the number of training points
[b] (Internal) multi-fidelity feature-based modeling of Sect. 12.3.1.1 (200 $\boldsymbol{R}_{cd}$ samples)
[c] Numbers in brackets refer to the total model setup cost (including $\boldsymbol{R}_{cd}$ points)
[d] (External) multi-fidelity feature-based modeling of Sect. 12.3.1.2
[e] Generalized SPRP modeling (Koziel & Leifsson, 2012)
[f] Feature-based modeling using the procedure of Koziel et al. (2014b)
[g] Direct kriging interpolation of the high-fidelity model $|S_{21}|$ responses
[h] Direct radial-basis function interpolation of the high-fidelity model data

One should also bear in mind the limitations of the method, namely, the necessity of maintaining consistency of the feature points across the entire training set, which is a common issue for feature-based techniques. For certain structures, such as the ones considered here, as well as other structures with well-defined response "shapes" (e.g., coupling structures, narrow-band antennas, certain integrated photonic components such as microring resonators), it is straightforward to achieve. For other structures, such as high-order filters, feature-based modeling may be the method of

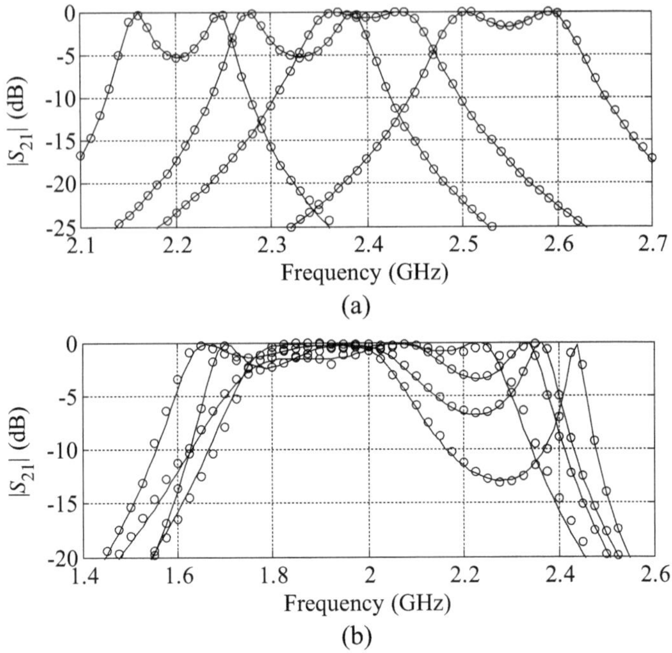

Fig. 12.26 Model responses at the selected (random) test designs: high-fidelity model (—) and multi-fidelity feature-based model (setup with 20 $\boldsymbol{R}_f$ and 200 $\boldsymbol{R}_{cd}$ points) (○): (**a**) Filter 1, (**b**) Filter 2

choice for local modeling for, e.g., statistical design purposes (Koziel & Bandler, 2015).

As an additional verification, the multi-fidelity feature-based surrogate models have been utilized for parametric optimization of the Filters 1 and 2. The following design specifications are considered:

- Filter 1: $|S_{21}| \geq -1$ dB for 2.35 GHz $\leq \omega \leq$ 2.45 GHz, $|S_{21}| \leq -20$ dB for $\omega \leq$ 2.2 GHz and $\omega \geq$ 2.6 GHz;
- Filter 2: $|S_{21}| \geq -1$ dB for 1.75 GHz $\leq \omega \leq$ 2.25 GHz, $|S_{21}| \leq -20$ dB for $\omega \leq$ 1.5 GHz and $\omega \geq$ 2.5 GHz;

In both cases, the multi-fidelity feature-based surrogate constructed using the internal approach and 50 high-fidelity training samples (200 low-fidelity samples) has been used in the process. The initial and final responses obtained by optimizing the feature-based surrogate model (the final design is verified by the high-fidelity model) are shown in Fig. 12.27. The design specifications are marked using thick horizontal lines. Because of the very good accuracy of the surrogates, no further design tuning was found necessary.

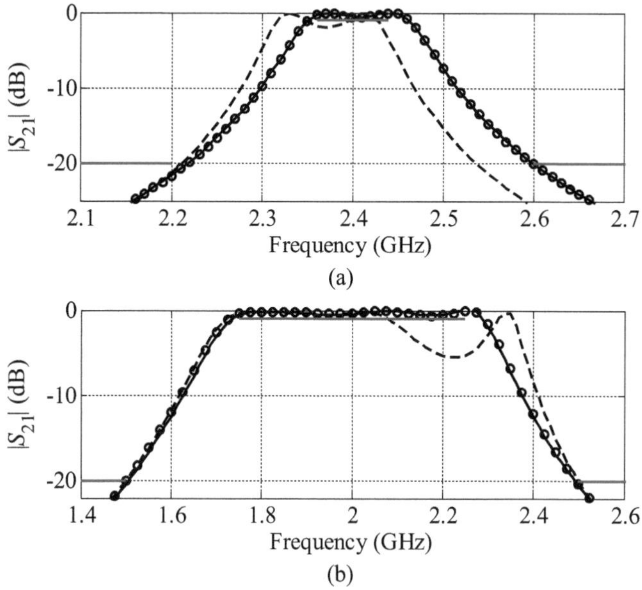

Fig. 12.27 Optimization results: (**a**) Filter 1, (**b**) Filter 2. The responses shown as (- - -) and (—) for the multi-fidelity feature-based surrogate model responses (the internal approach) at the initial design and at the optimized design; corresponding verification by the high-fidelity model responses shown as (○)

References

Bandler, J. W., Cheng, Q. S., & Koziel, S. (2006). Simplified space mapping approach to enhancement of microwave device models. *The International Journal of RF and Microwave Computer-Aided Engineering, 16*(5), 518–535.

Beachkofski, B., & Grandhi, R. (2002). *Improved distributed hypercube sampling*. American Institute of Aeronautics and Astronautics, paper AIAA 2002–1274.

Buhmann, M. D., & Ablowitz, M. J. (2003). *Radial basis functions: Theory and implementations*. Cambridge University.

Chen, J.-S. (2005). Studies of CPW-fed equilateral triangular-ring slot antennas and triangular-ring slot coupled patch antennas. *IEEE Transactions on Antennas and Propagation, 53*, 2208–2211.

Chen, J. -S. (2007). *CPW-fed compact slot-ring coupled square patch antennas*. IEEE antennas and propagation society international symposium. pp. 2297–2300.

Chen, Y. -C., Chen, S. -Y., Hsu, P. (2006). *Dual-band slot dipole antenna fed by a coplanar waveguide*. IEEE international symposium on antennas and propagation. pp. 3589–3592.

Couckuyt, I., Declercq, F., Dhaene, T., Rogier, H., & Knockaert, L. (2010). Surrogate-based infill optimization applied to electromagnetic problems. *The International Journal of RF and Microwave Computer-Aided Engineering, 20*(5), 492–501.

Forrester, A. I. J., & Keane, A. J. (2009). Recent advances in surrogate-based optimization. *Progress in Aerospace Sciences, 45*(1), 50–79.

Hsieh, M. Y., & Wang, S. M. (2005). Compact and wideband microstrip bandstop filter. *IEEE Microwave and Wireless Component Letters, 15*(7), 472–474.

Huang, C. L., Chen, Y. B., & Tasi, C. F. (2008). New compact microstrip stacked slotted resonators bandpass filter with transmission zeros using high-permittivity ceramics substrate. *Microwave and Optical Technology Letters, 50*(5), 1377–1379.

Kennedy, M. C., & O'Hagan, A. (2000). Predicting the output from complex computer code when fast approximations are available. *Biometrika, 87*, 1–13.

Koziel, S., & Bandler, J. W. (2015). Rapid yield estimation and optimization of microwave structures exploiting feature-based statistical analysis. *IEEE Transactions on Microwave Theory and Techniques, 63*(1), 107–114.

Koziel, S., & Leifsson, L. (2012). Generalized shape-preserving response prediction for accurate modeling of microwave structures. *IET Microwaves, Antennas & Propagation, 6*(12), 1332–1339.

Koziel, S., & Pietrenko-Dabrowska, A. (2020). *Performance-driven surrogate modeling of high-frequency structures*. Springer.

Koziel, S., & Szczepanski, S. (2011). Accurate modeling of microwave structures using shape-preserving response prediction. *IET Microwaves, Antennas & Propagation, 5*(9), 1116–1122.

Koziel, S., Ogurtsov, S., Couckuyt, I., & Dhaene, T. (2013). Variable-fidelity electromagnetic simulations and co-kriging for accurate modeling of antennas. *IEEE Transactions on Antennas and Propagation, 61*(3), 1301–1308.

Koziel, S., Bandler, J. W., & Cheng, Q. S. (2014a). *Low-cost feature-based modeling of microwave structures*. IEEE MTT-S international microwave symposium, Tampa, FL, pp. 1–3.

Koziel, S., Ogurtsov, S., Cheng, Q. S., & Bandler, J. W. (2014b). Rapid EM-based microwave design optimization exploiting shape-preserving response prediction and adjoint sensitivities. *IET Microwaves, Antennas & Propagation, 8*(10), 775–781.

Koziel, S., Bekasiewicz, A., & Leifsson, L. (2016). *Expedited design of dual-band antennas using feature-based optimization*. European conference on antennas and propagation, Davos. pp. 1–4.

Koziel, S., Calik, N., Mahouti, P., & Belen, M. A. (2021). Accurate modeling of antenna structures by means of domain confinement and pyramidal deep neural networks. *IEEE Transactions on Antennas and Propagation, 70*(3), 2174–2188.

Lee, J. R., Cho, J. H., & Yun, S. W. (2000). New compact bandpass filter using microstrip $\lambda/4$ resonators with open stub inverter. *IEEE Microwave and Guided Wave Letters, 10*(12), 526–527.

Lophaven, S. N., Nielsen, H. B., & Søndergaard, J. (2002). *DACE: a Matlab kriging toolbox*. Technical University of Denmark.

Pan, Y. M., Leung, K. W., & Lu, K. (2014). Compact quasi-isotropic dielectric resonator antenna with small ground plane. *IEEE Transactions on Antennas and Propagation, 62*, 577–585.

Queipo, N. V., Haftka, R. T., Shyy, W., Goel, T., Vaidynathan, R., & Tucker, P. K. (2005). Surrogate-based analysis and optimization. *Progress in Aerospace Sciences, 41*(1), 1–28.

Salleh, M. K. M., Pringent, G., Pigaglio, O., & Crampagne, R. (2008). Quarter-wavelength side-coupled ring resonator for bandpass filters. *The IEEE Transactions on Microwave Theory and Techniques, 56*(1), 156–162.

Sim, C. Y. D., Chang, M. H., & Chen, B. Y. (2014). Microstrip-fed ring slot antenna design with wideband harmonic suppression. *IEEE Transactions on Antennas and Propagation, 62*(9), 4828–4832.

Simpson, T. W., Peplinski, J., Koch, P. N., & Allen, J. K. (2001). Metamodels for computer-based engineering design: Survey and recommendations. *Engineering Computers, 17*, 129–150.

Yondo, R., Bobrowski, K., Andrés, E., & Valero, E. (2019). A review of surrogate modeling techniques for aerodynamic analysis and optimization: Current limitations and future challenges in industry. In E. Minisci, M. Vasile, J. Periaux, N. Gauger, K. Giannakoglou, & D. Quagliarella (Eds.), *Advances in evolutionary and deterministic methods for design, optimization and control in engineering and sciences, computational methods in applied sciences* (Vol. 48). Springer.

Chapter 13
Constrained Modeling with Response Features

The development of surrogate modeling techniques for high-frequency design have been instigated by excessive computational costs of direct simulation-driven optimization. Yet, it soon turned out that conventional surrogate modeling techniques face considerable difficulties in setting up reliable models caused primarily by the curse of dimensionality and high nonlinearity of system characteristics. This has encouraged the researchers to seek alternative modeling paradigms. One of them is a performance-driven modeling concept (Koziel & Pietrenko-Dabrowska, 2020a), in which the surrogate domain is contentiously delimited, so that it coincides with the most important regions of the design space, where high-quality designs reside. The two main techniques following the performance-driven (constrained) modeling paradigm include the nested kriging modeling technique (Koziel & Pietrenko-Dabrowska, 2019) and two-stage modeling approach involving random observables (Koziel & Pietrenko-Dabrowska, 2022). Both approaches permit a significant reduction of the training data acquisition cost with regard to standard modeling procedures. Subsequent sections of this chapter outline specific modeling techniques following performance-driven concept. The chapter starts from the delineation of the basic version of the nested kriging approach, which is followed by a description of its later developments: nested kriging with response features and feature-based nested kriging with explicit dimensionality reduction. These procedures require preoptimized reference designs for domain definition purposes, which in some cases could be available from the previous work with a given high-frequency structure. If, however, such a database is not readily accessible, an alternative approach offers two-stage modeling technique. Therein, the surrogate model domain is delimited using randomly acquired observables, whose quality is assessed based on the response features extracted from the respective system responses. The employment of the features for design quality estimation permitted considerable improvement of computational efficiency of performance-driven modeling frameworks. Further cost reduction has been achieved by carrying out the modeling process directly at the level of the response features rather than complete

A. Pietrenko-Dabrowska, S. Koziel, *Response Feature Technology for High-Frequency Electronics. Optimization, Modeling, and Design Automation*,
https://doi.org/10.1007/978-3-031-43845-5_13

characteristics. The last two sections of this chapter describe observable-based constrained modeling technique and its feature-based version.

13.1 Feature-Based Nested Kriging

This section presents a technique for cost-efficient design-oriented modeling of high-frequency structures. The outlined approach integrates the nested kriging modeling technique, belonging to the class of performance-driven (constrained) methods (Koziel & Pietrenko-Dabrowska, 2020a) and the response feature technology. This combination enables rendering of reliable surrogates valid within broad ranges of geometry parameters while using small training data sets. The section is organized as follows. Section 13.1 provides a brief recollection of the nested kriging modeling technique (Koziel & Pietrenko-Dabrowska, 2019). Whereas the incorporation of the response features into the constrained modeling framework is outlined in Sect. 13.1.2. Section 13.1.3 presents the results obtained using the considered approach for modeling of a dual-band antenna.

13.1.1 Nested Kriging

In performance-driven modeling (Koziel & Pietrenko-Dabrowska, 2020a), the surrogate is constructed within a parameter space region that contains high-quality designs (from the point of view of the relevant figures of interest). Approximation of such a subset is obtained from a set of preoptimized reference designs, with detailed formulation of the domain varying between implementations (Koziel, 2017; Koziel & Sigurðsson, 2018; Koziel & Pietrenko-Dabrowska, 2019). Notwithstanding, domain confinement permits construction of reliable surrogates within wide ranges of parameters and operating conditions even in spaces of relatively high dimensionality.

In the nested kriging (Koziel & Pietrenko-Dabrowska, 2019), a confined surrogate model domain is determined based on a set of reference designs $\boldsymbol{x}^{(j)} = [x_1^{(j)} \ldots x_n^{(j)}]^T$, $j = 1, \ldots, p$, preoptimized with respect to the selected performance vectors $\boldsymbol{f}^{(j)} = [f_1^{(j)} \ldots f_N^{(j)}]$. Here, f_k, $k = 1, \ldots, N$ are the performance figures of interest (e.g., target antenna operating frequencies in the case of multiband antennas). In this section, the antenna examples are utilized to present the nested kriging technique; yet, this approach has also been employed for modeling of microwave devices (Koziel & Pietrenko-Dabrowska, 2020b). The objective space F is defined by the user-defined ranges for f_k, $f_{k.\min} \le f_k^{(j)} \le f_{k.\max}$, $k = 1, \ldots, N$, which should be covered by the surrogate. It is advisable for $\{\boldsymbol{f}^{(j)}\}$ to be distributed within F in a uniform manner (Koziel & Pietrenko-Dabrowska, 2019).

Given the objective vector $\boldsymbol{f} \in F$, the optimum design corresponding to $\boldsymbol{f}$ is understood as the solution to the problem

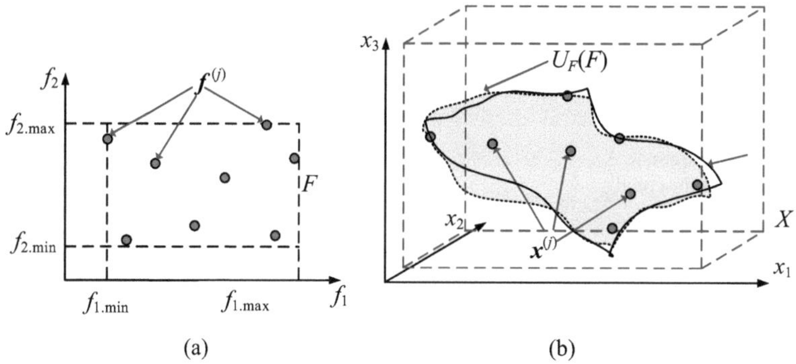

Fig. 13.1 Graphical illustration of the performance-driven modeling concepts: (**a**) objective space F (here, shown for two performance figures), (**b**) parameter space X (here, three-dimensional), the reference designs, the optimum design manifold $U_F(F)$, and the first-level surrogate image $\boldsymbol{s}_I(F)$. Because the reference designs $\boldsymbol{x}^{(j)}$ are optimal in the sense of (13.1), the manifolds $U_F(F)$ and $\boldsymbol{s}_I(F)$ coincide at all $\boldsymbol{x}^{(j)}$

$$\boldsymbol{x}^* = U_F(\boldsymbol{f}) = \arg\min_{\boldsymbol{x}} U(\boldsymbol{x},\boldsymbol{f}) \tag{13.1}$$

where U is the scalar merit function quantifying the design utility. $U_F(F)$ represents all designs that are optimum in the sense of (13.1) for all $\boldsymbol{f} \in F$. The surrogate model domain is to be established in the vicinity of this set.

According to Koziel and Pietrenko-Dabrowska (2019), $U_F(F)$ can be approximated using the reference designs $\boldsymbol{x}^{(j)} = [x_1^{(j)} \ldots x_n^{(j)}]^T j = 1, \ldots, p$, optimized with respect to the performance vectors $\boldsymbol{f}^{(j)} = [f_1^{(j)} \ldots f_N^{(j)}]$ uniformly distributed within F; $\boldsymbol{x}^{(j)}$ may be available from the previous design work on the same structure or obtained specifically for surrogate model construction, if it is justified by the intended future reuse of the framework.

Clearly, a specific definition of the merit function $U(\boldsymbol{x},\boldsymbol{f})$ is problem dependent. As an example, let us consider a multiband antenna optimized for matching improvement at the required operating frequencies $f_{0.k}$, $k = 1, \ldots, N$. In this case, the performance figures are the antenna-resonant frequencies $f_k = f_{0.k}$, and a possible formulation of the merit function is

$$U(\boldsymbol{x},\boldsymbol{f}) = \max\{|S_{11}(\boldsymbol{x},f_{0.1})|, \ldots, |S_{11}(\boldsymbol{x},f_{0.N})|\} \tag{13.2}$$

where $S_{11}(\boldsymbol{x},f)$ stands for the antenna reflection at the design $\boldsymbol{x}$ and frequency f.

The nested kriging technique utilizes two levels of models: the first one to establish the confined domain and the second-level model being the actual surrogate. Both models are implemented using kriging interpolation (Simpson et al., 2001). The first-level model $\boldsymbol{s}_I(\boldsymbol{f})$ maps the objective space F into the design space X, with $\{\boldsymbol{f}^{(j)},\boldsymbol{x}^{(j)}\}$ being the training points (cf. Fig. 13.1). The image $\boldsymbol{s}_I(F) \subset X$ of F delivers the first approximation of the region in which the surrogate is to be established. A

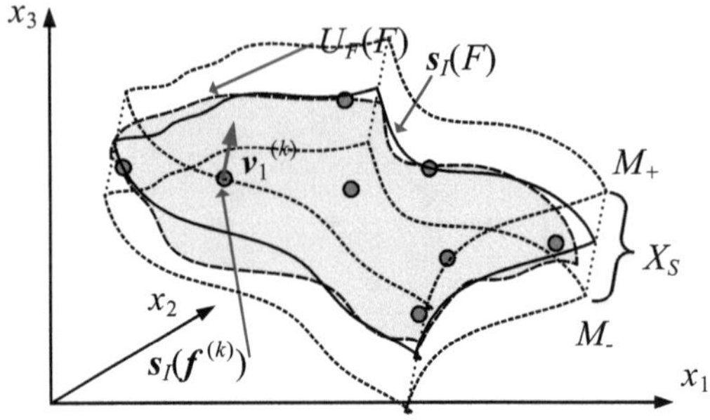

Fig. 13.2 The image $s_I(F)$ of the first-level surrogate model and the normal vector $v_1^{(k)}$ at $f^{(k)}$; the manifolds M_- and M_+ and the surrogate model domain X_S defined as the orthogonal extension of $s_I(F)$

graphical illustration of the concepts introduced in this section can be found in Fig. 13.1.

Because the first-level model is identified using limited information (in practice, only a small number of reference designs is available), it does not give a complete account for the location of the optimum designs corresponding to all $f \in F$. In order to ensure that the domain contains the entire $U_F(F)$, $s_I(F)$ has to be expanded. This is achieved through its orthogonal extension.

Let us define the following quantities: an orthonormal basis $\{v_n^{(k)}(f)\}$, $k = 1, \ldots, n - N$, of vectors normal to $s_I(F)$ at f and the parameter variations $x_d = x_{\max} - x_{\min}$ within $s_I(F)$, where $x_{\max} = \max\{x^{(k)}, k = 1, \ldots, p\}$ and $x_{\min} = \min\{x^{(k)}, k = 1, \ldots, p\}$. In addition, the vector of the coefficients of the extension is defined as

$$\alpha(f) = [\alpha_1(f) \ \ldots \ \alpha_{n-N}(f)]^T = \frac{D}{2}\left[|x_d v_n^{(1)}(f)| \ \ldots \ |x_d v_n^{(n-N)}(f)|\right]^T \quad (13.3)$$

where D is a user-defined thickness parameter which sets the "span" of the extension (i.e., the ratio of the lateral to the tangential size of the surrogate model domain). The confined domain X_S is located between the manifolds M_+ and M_- determined using the coefficients α_k (see Fig. 13.2b).

$$M_{\pm} = \left\{x \in X : x = s_I(f) \pm \sum\nolimits_{k=1}^{n-N} \alpha_k(f) v_n^{(k)}(f)\right\} \quad (13.4)$$

The formal definition of X_S takes the form of

$$X_S = \left\{\begin{array}{c} x = s_I(f) + \sum\limits_{k=1}^{n-N} \lambda_k \alpha_k(f) v_n^{(k)}(f) : f \in F, \\ -1 \le \lambda_k \le 1, \ \ k = 1, \ldots, n-N \end{array}\right\} \quad (13.5)$$

Figure 13.2 provides a graphical illustration of the manifolds M_+ and M_-, as well as the domain X_S. The second-level (i.e., the actual) surrogate is also a kriging interpolation model rendered in X_S, using $\{x_B^{(k)}, R(x_B^{(k)})\}_{k = 1, \ldots, NB}$, as the training set, where $x_B^{(k)}$ are the training samples uniformly allocated in the domain, whereas R is the EM-simulation model of the high-frequency structure of interest.

The important advantage of the above formulation is that it facilitates uniform sampling, as well as surrogate model optimization through an auxiliary mapping from the n-dimensional unit interval $[0,1]^n$ onto X_S. The mapping is a composition of the following transformations: h_1 from $[0,1]^n$ onto a Cartesian product $F \times [-1,1]^{n-N}$ defined as

$$\begin{aligned} \mathbf{y} &= h_1(\mathbf{z}) = h_1\left([z_1 \ \ldots \ z_n]^T\right) \\ &= [f_{1.\min} + z_1(f_{1.\max} - f_{1.\min}) \ \ \ldots \ \ldots \ f_{N.\min} + z_N(f_{N.\max} - f_{N.\min})] \\ &\quad \times [-1 + 2z_{N+1} \ \ \ldots \ \ -1 + 2z_n] \end{aligned} \tag{13.6}$$

where $\mathbf{z} = [z_1 \ldots z_n]^T \in [0,1]^n$, and h_2 from $F \times [-1,1]^{n-N}$ onto X_S is defined as

$$\begin{aligned} \mathbf{x} &= h_2(\mathbf{y}) = h_2\left([y_1 \ \ldots \ y_n]^T\right) = \mathbf{s}_I\left([y_1 \ \ldots \ y_N]^T\right) \\ &+ \sum_{k=1}^{n-N} y_{N+k}\alpha_k\left([y_1 \ \ldots \ y_N]^T\right)\mathbf{v}_n^{(k)}\left([y_1 \ \ldots \ y_N]^T\right) \end{aligned} \tag{13.7}$$

The overall transformation H from $[0,1]^n$ onto X_S is then a composition of h_1 and h_2

$$\mathbf{x} = H(\mathbf{z}) = h_2(h_1(\mathbf{z})) \tag{13.8}$$

As mentioned before, the mapping H facilitates uniform sampling in X_S as a set of uniformly distributed samples $\{\mathbf{x}_B{}^{(k)}\}$ in X_S can be obtained as $\{H(\mathbf{z}_B{}^{(k)})\}$, where $\{\mathbf{z}_B{}^{(k)}\}$ is the uniform data set in $[0,1]^n$. Furthermore, it is possible to optimize the surrogate model in X_S by operating within the unit hypercube and then employing the mapping H. Finally, a good initial point for parametric optimization within X_S can be conveniently generated using the first-level model as $\mathbf{x}_0 = \mathbf{s}_I(\mathbf{f}_t)$.

13.1.2 Nested Kriging with Response Features

In the technique of this section, as in all feature-based frameworks, the modeling problem is reformulated in terms of the characteristic points (features) of the system response. In the case of multiband antennas, the characteristic points of interest are frequency locations of the antenna resonances and the points corresponding to –10 dB level of reflection are of interest. These points can be extracted from the EM simulation results and then employed to allocate the resonances at the required operating frequencies and/or optimize the antenna for maximum bandwidth (see Fig. 13.3).

In the feature-based nested kriging modeling technique (Koziel & Pietrenko-Dabrowska, 2020c), the feature points referring to p antenna resonances are denoted

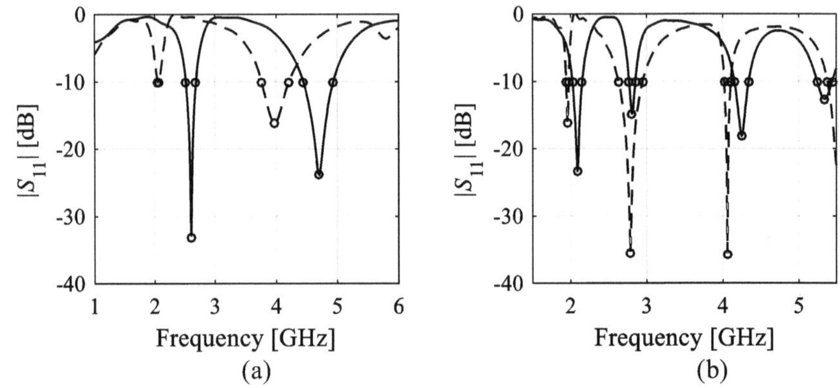

Fig. 13.3 Example input characteristics and the feature points (here, corresponding to antenna resonances and −10 dB reflection levels): (**a**) dual-band antenna, (**b**) triple-band antenna

as $\boldsymbol{R}_F(\boldsymbol{x}) = [f_1(\boldsymbol{x})\, f_2(\boldsymbol{x}) \ldots f_p(\boldsymbol{x})\; l_1(\boldsymbol{x})\; l_2(\boldsymbol{x}) \ldots l_p(\boldsymbol{x})]^T$, where f_k and l_k stand for their frequency and level coordinates, $k = 1, \ldots, p$, respectively. Here, both the first-level surrogate and the domain definition are identical to that presented in Sect. 13.1.1. Whereas the training set for the second-level surrogate consists of the pairs of the response feature vectors $\boldsymbol{R}_F(\boldsymbol{x}_B{}^{(k)})$ (rather than the original responses) and the corresponding training designs $\{\boldsymbol{x}_B{}^{(k)}, \boldsymbol{R}_F(\boldsymbol{x}_B{}^{(k)})\}_{k\,=\,1,\,\ldots,\,NB}$. Assuming that the design objectives can be uniquely quantified using the selected feature points (the surrogate merely provides a prediction about the feature coordinates at the design of interest $\boldsymbol{x} \in X_S$), modeling of $\boldsymbol{R}_F$ rather than $\boldsymbol{R}$ does not lead to any loss of information from the point of view of design purposes.

The steps undertaken to construct the surrogate model within the feature-based nested kriging framework are the following (see also Fig. 13.4). First, the objective space F is defined, for example, by the ranges of the antenna operating frequencies to be covered by the model (cf. Sect. 13.1.1). Subsequently, the reference designs $\boldsymbol{x}^{(j)}$, $j = 1, \ldots, p$, are obtained and utilized for setting up the first-level surrogate $\boldsymbol{s}_I$.

Next, the surrogate model domain is established (cf. (13.5)) with the use of the normal vectors $\boldsymbol{v}_n{}^{(k)}$, $k = 1, \ldots, n - N$, and the expansion coefficients α_k. Within that domain, the training data $\{\boldsymbol{x}_B{}^{(k)}\}_{k\,=\,1,\,\ldots,\,NB}$ are allocated by transforming the normalized training data samples $\{z_B{}^{(k)}\}_{k\,=\,1,\,\ldots,\,NB}$ from the unit hypercube $[0,1]^n$ through the mapping H (13.8). Here, the normalized samples are obtained using Latin Hypercube Sampling (Beachkofski & Grandhi, 2002). Subsequently, the EM model data $\boldsymbol{R}(\boldsymbol{x}_B{}^{(k)})$, $k = 1, \ldots, N_B$, is acquired and the response features $\boldsymbol{R}_F(\boldsymbol{x}_B{}^{(k)})$, $k = 1, \ldots, N_B$, are extracted. Finally, the surrogate is identified based on the training set $\{\boldsymbol{x}_B{}^{(k)}, \boldsymbol{R}_F(\boldsymbol{x}_B{}^{(k)})\}_{k\,=\,1,\,\ldots,\,NB}$. The surrogate model yields the feature point coordinates of the antenna response corresponding to any design $\boldsymbol{x}$ from the constrained domain X_S.

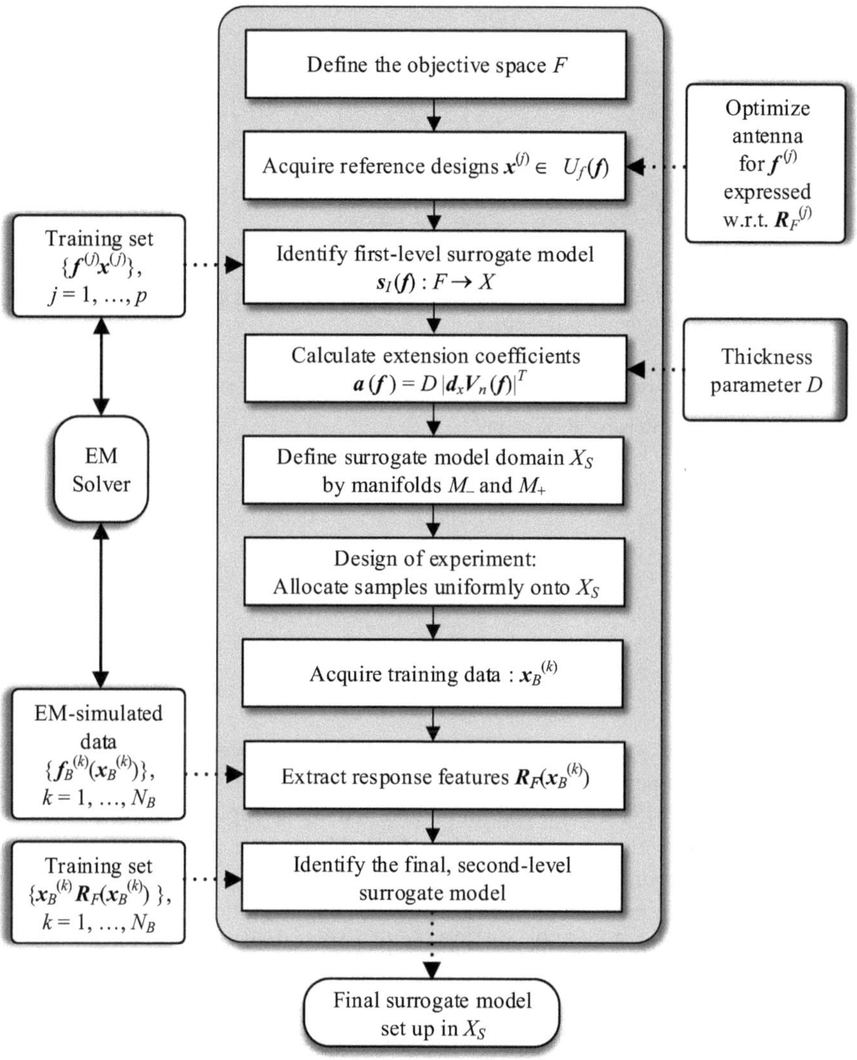

Fig. 13.4 Flow diagram of the feature-based nested kriging modeling methodology

13.1.3 Verification Case: Dual-Band Antenna

This section provides a numerical verification of the modeling framework of Sect. 13.1.2, as well as comparisons with benchmark procedures: the conventional approximation models (kriging and RBF) and the nested kriging model (Koziel & Pietrenko-Dabrowska, 2019).

The discussed methodology has been demonstrated using a dual-band antenna (Chen et al., 2006), shown in Fig. 13.5. The antenna is described by the following

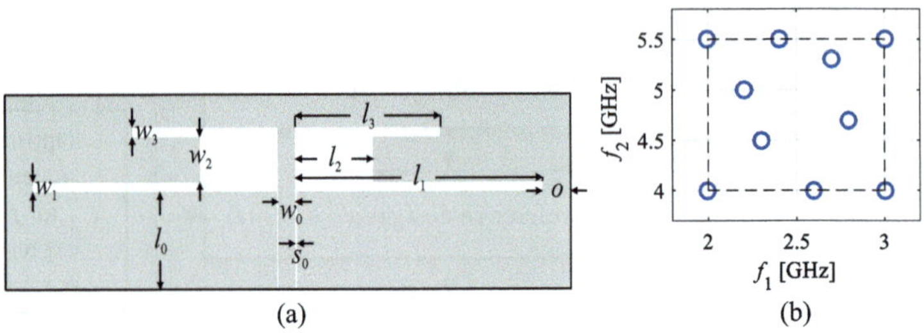

Fig. 13.5 Uniplanar dual-band dipole antenna (Chen et al., 2006): (**a**) geometry, (**b**) allocation of the reference designs in the objective space

Table 13.1 Feature-based modeling of dual-band antenna: Modeling accuracy

Number of training samples	Relative RMS error			
	Conventional kriging model	Conventional RBF	Nested kriging model	Feature-based nested kriging
20	24.5%	26.3%	19.0%	1.43%
50	21.7%	24.9%	9.9%	0.51%
100	17.3%	19.8%	6.4%	0.39%
200	12.6%	14.3%	4.4%	0.56%
400	9.3%	10.5%	3.8%	0.43%
800	7.2%	8.7%	3.4%	0.46%

design variables: $\boldsymbol{x} = [l_1\ l_2\ l_3\ w_1\ w_2\ w_3]^T$, with $l_0 = 30$, $w_0 = 3$, $s_0 = 0.15$, and $o = 5$ being fixed (all dimensions in mm). The structure is implemented on the RO4350 substrate ($\varepsilon_r = 3.48$, $h = 0.762$ mm) and fed by a coplanar waveguide. The computational model is implemented in CST Microwave Studio and simulated using its time-domain solver.

The surrogate models have been constructed using the following sizes of training data sets: 20, 50, 100, 200, 400, and 800 samples (the adopted value of the thickness parameter was $D = 0.05$). Table 13.1 reports the modeling errors (average relative RMS), estimated based on 100 independent random test points. It should be noted that in the case of the feature-based nested kriging, the errors of the frequency coordinates of the characteristic points are presented.

The goal is to construct the surrogate model valid for the following ranges of operating frequencies: 2.0 GHz $\leq f_1 \leq$ 3.0 GHz (lower band) and 4.0 GHz $\leq f_2 \leq$ 5.5 GHz (upper band). The reference designs $\boldsymbol{x}^{(j)}$, $j = 1, \ldots, 10$, are allocated as shown in Fig. 13.5b. The lower and upper bounds for design variables, derived from $\boldsymbol{x}^{(j)}$, are $\boldsymbol{l} = [27\ 5.0\ 16.5\ 0.22\ 1.8\ 0.5]^T$ and $\boldsymbol{u} = [40\ 12\ 23.5\ 0.6\ 4.3\ 2.7]^T$.

The results presented in Table 13.1 reveal that the feature-based surrogate is far more accurate than the benchmark models. A remark should be made that the model predictive power fluctuates for training data set sizes comprising more than 100 samples. This is due to the fact that the variance of the model error estimation and the error itself, here, around a fraction of percent, are at a similar level.

Table 13.2 Optimization results for dual-band antenna

Target operating conditions		Geometry parameter values [mm]					
f_1 [GHz]	f_2 [GHz]	l_1	l_2	l_3	w_1	w_2	w_3
2.45	5.30	33.78	8.63	18.16	0.30	2.11	1.50
2.20	4.50	34.99	6.62	18.60	0.44	3.89	1.83
3.00	5.00	28.69	9.75	19.80	0.37	2.75	1.09
2.10	4.20	36.42	6.32	19.79	0.45	3.97	1.81

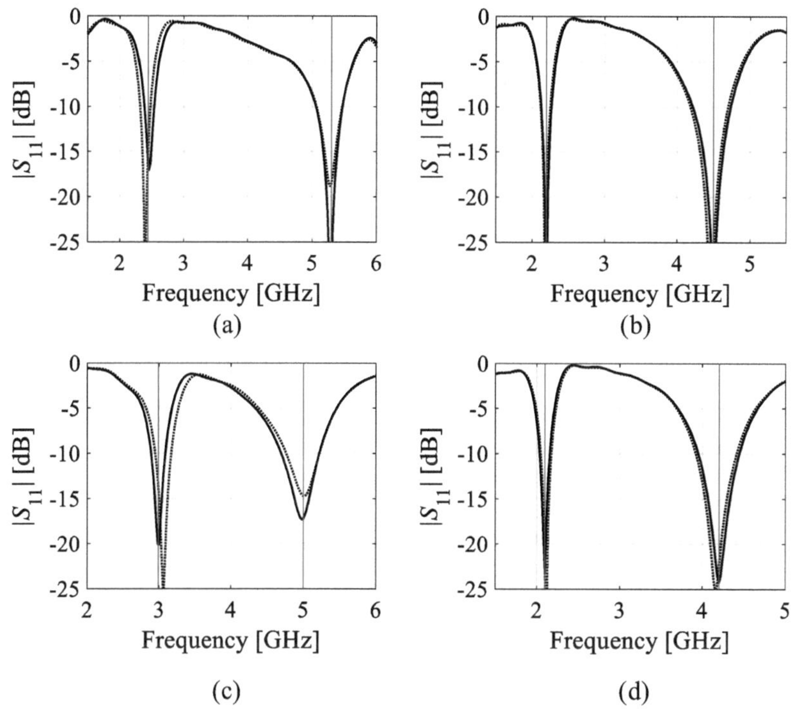

Fig. 13.6 Optimization results for dual-band antenna of Fig. 13.5a using feature-based nested kriging: EM-simulated antenna responses at the initial designs (.....) obtained with the surrogate and the optimized responses (—). Required operating frequencies are marked using vertical lines: (**a**) $f_1 = 2.45$ GHz, $f_2 = 5.3$ GHz, (**b**) $f_1 = 2.2$ GHz, $f_2 = 4.5$ GHz, (**c**) $f_1 = 3.0$ GHz, $f_2 = 5.0$ GHz, and (**d**) $f_1 = 2.1$ GHz, $f_2 = 4.2$ GHz

In the following, the examples of design applications of the feature-based nested kriging model for antenna optimization are presented. Table 13.2 and Fig. 13.6 show the results obtained for the surrogate set up with 50 training samples and four selected pairs of the target operating frequencies. It should be emphasized that the predictions yielded by the surrogate are indeed very accurate. As a matter of fact, even the initial designs obtained for the target vector $\boldsymbol{f}_t$ with the first-level model as $\boldsymbol{x}^{(0)} = \boldsymbol{s}_I(\boldsymbol{f}_t)$ are of high quality (see Fig. 13.6). Figure 13.7 shows the realized gain as well as radiation patterns of the antenna of Fig. 13.5a at the design of Fig. 13.6a.

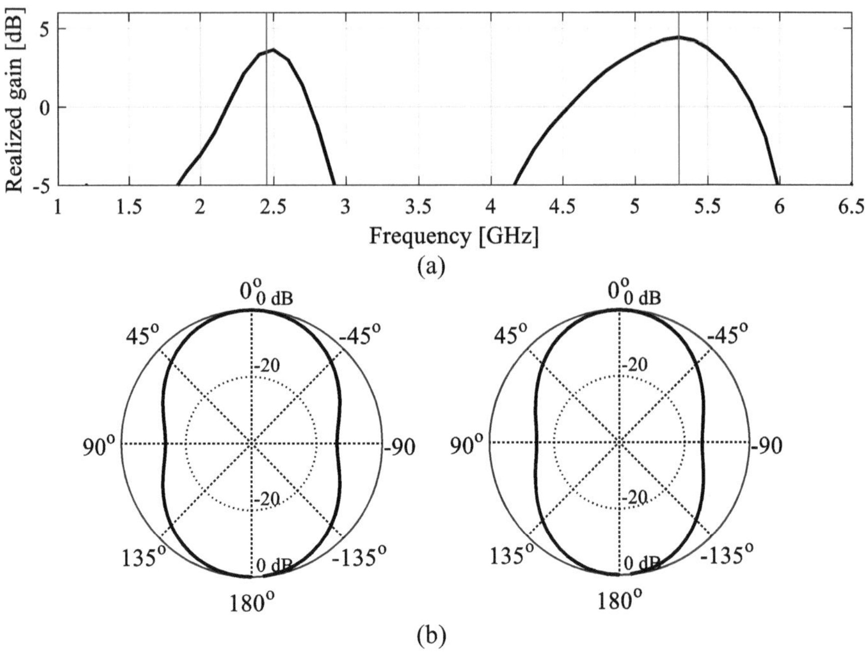

Fig. 13.7 Simulated characteristics of antenna of Fig. 13.4a at the design shown in Fig. 13.5a: (**a**) realized gain, (**b**) H plane radiation patterns at 2.45 GHz and 5.3 GHz, respectively. The target operating frequencies are shown using vertical lines

The obtained results allow us to draw the following conclusions: Modeling the feature-point coordinates instead of the complete responses allows for achieving a considerably better predictive power of the surrogate. At the same time, a very small number of training samples is required to set up an accurate model. Clearly, the surrogates constructed at the level of selected feature points do not carry full information about the relevant antenna characteristics. On the other hand, focusing the scope of the model on the coordinates of important portions of the responses (here, resonance allocation) leads to a dramatic reduction of the EM simulation data that need to be acquired to secure the appropriate predictive power of the surrogate. The adequacy of such a model depends on its intended use. Having in mind that parametric optimization is one of the most commonly executed tasks, the provided application examples do demonstrate that the presented framework might be a useful antenna design tool.

13.2 Feature-Based Nested Kriging with Dimensionality Reduction

This section outlines the nested kriging modeling technique with response features and dimensionality reduction (Pietrenko-Dabrowska & Koziel, 2020), which constitutes an advancement over the procedure described in Sect. 13.1. Similarly as in the previous section, here, also, only characteristic (feature) points are handled instead of complete approaches. This permits smoothening out the functional landscape to be processed, which facilitates the rendition of the surrogate model and leads to considerable reduction of the computational expenses associated with the acquisition of the training data. Further cost savings are obtained as a result of surrogate domain dimensionality reduction.

The section is structured as follows. Section 13.2.1 directly delineates the nested kriging modeling technique with explicit dimensionality reduction, as the delineation of the core modeling technique, that is, the nested kriging, has been already provided in Sect. 13.1.1. The incorporation of the response features technology into the dimensionality-reduced nested kriging modeling framework is outlined in Sect. 13.2.2. Section 13.2.3 provides verification example corroborating suitability of the considered methodology for antenna modeling purposes using the example of a triple-band antenna.

13.2.1 Nested Kriging with Explicit Dimensionality Reduction

The dimensionality-reduced performance-driven modeling framework (Koziel et al., 2020) follows a general paradigm of the constrained modeling (Koziel & Pietrenko-Dabrowska, 2020a). In the latter (see also Sect. 13.1.1), the surrogate is constructed within a confined domain—a subset of the original parameter space—that encompasses high-quality designs with respect to the assumed figures of interest. The domain definition requires setting up an auxiliary inverse model (the first-level surrogate).

In the original nested kriging formulation modeling (Koziel & Pietrenko-Dabrowska, 2019), it is subsequently extended in the directions orthogonal to the objective space image through the first-level model. Whereas the dimensionality-reduced domain of Koziel et al. (2020) is defined by orthogonally extending the objective space image merely in a limited number of directions selected based on the principal component analysis of the reference design set. The volume of the confined domain (both of the original and the dimensionality-reduced one) is significantly smaller than the volume of the typically used box-constrained domain, delimited by the lower and upper bands on the design variables. This allows for constructing reliable surrogates at a fraction of the cost required by conventional modeling techniques.

In this section, we use the notation utilized in the performance-modeling techniques, which has been defined in Sect. 13.1. Whereas the basic definitions pertaining to the principal component analysis of the reference design sets are provided below. Let $\boldsymbol{x}_m$ refer to a center of gravity of the set $\{\boldsymbol{x}^{(j)}\}_{j = 1,..., p}$:

$$\boldsymbol{x}_m = \frac{1}{p}\sum_{j=1}^{p}\boldsymbol{x}^{(p)} \tag{13.9}$$

Further, let $\boldsymbol{S}_p$ denote the covariance matrix of $\{\boldsymbol{x}^{(j)}\}$ defined as

$$\boldsymbol{S}_p = \frac{1}{p-1}\sum_{j=1}^{p}\left(\boldsymbol{x}^{(j)} - \boldsymbol{x}_m\right)\left(\boldsymbol{x}^{(j)} - \boldsymbol{x}_m\right)^T \tag{13.10}$$

The principal components of the reference design set (Jolliffe, 2002), that is, the eigenvectors $\boldsymbol{a}_i$, $i = 1, \ldots, n$ of $\boldsymbol{S}_p$, indicate the most significant directions of correlations between the optimized design parameters and the target vectors within the objective space F. Whereas the variance of the reference set in the eigenspace is described by the eigenvalues λ_i, which are assumed to be arranged in the descending order, that is, $\lambda_1 \geq \lambda_2 \geq \ldots \geq \lambda_n \geq 0$. We will also use the following matrix:

$$\boldsymbol{A}_i = [\boldsymbol{a}_1 \quad \ldots \quad \boldsymbol{a}_i] \tag{13.11}$$

whose columns are the first i eigenvectors $\boldsymbol{a}_i$. In addition, the matrix $\boldsymbol{A} = \boldsymbol{A}_n$ will stand for the matrix containing all eigenvectors.

13.2.1.1 Domain Definition

This subsection provides a description of the surrogate model domain definition according to the methodology proposed in (Koziel et al., 2020), which is employed in the discussed technique. Toward this end, the inverse surrogate $\boldsymbol{s}_I(\boldsymbol{f})$: $F \rightarrow X$, (i.e., the first-level model) is rendered using the training data set $\{\boldsymbol{f}^{(j)}, \boldsymbol{x}^{(j)}\}$, $j = 1, \ldots, p$. Here, similarly as in Sect. 13.1, the model $\mathbf{s}_I$ is utilized for approximating the optimum-design manifold $U_F(F)$. As the reference design set is of a limited size, $s_I(F)$ is an imperfect rendition of $U_F(F)$. In order to accommodate the discrepancies between the two sets, $s_I(F)$ has to be somewhat extended. In the original nested kriging formulation (Koziel & Pietrenko-Dabrowska, 2019) (see also Sect. 13.1), the extension is performed in all orthogonal directions $\{\boldsymbol{v}_n^{(k)}(\boldsymbol{f})\}$, $k = 1, \ldots, n - N$, whereas in PCA-based constrained modeling (Koziel et al., 2020), the extension is applied in a selective manner. Accordingly, the manifold $\boldsymbol{s}_I(F)$ is outstretched along the vectors constructed is spanned by K most dominant eigenvectors $\boldsymbol{a}_i$ of the reference set.

On the one hand, the number K of the principal components, in which the extension is to be carried out and has to be greater than the number N of the objectives (otherwise, the extension would be trivial). On the other hand, only the directions that carry meaningful information about the antenna response variability should be used, which is decided upon using the eigenvalue analysis. Once the exact value of K is chosen, the extension vectors are constructed as follows. First, the representation of $\{t_j(f)\}_{j=1,\ldots,N}$, in the basis $\{a_i\}_{i=1,\ldots,K}$, has to be found as

$$[\bar{t}_1(f) \;\; \ldots \;\; \bar{t}_N(f)] = A_K^T [t_1(f) \;\; \ldots \;\; t_N(f)] \tag{13.12}$$

where $t_j(f)$, $j = 1, \ldots, N$ denote the vectors tangent to $s_I(F)$ at f. The size of the vectors $\bar{t}_j(f), j = 1, \ldots, N$, is $K \times 1$. In finding the vectors normal to $s_I(F)$ within the subspace spanned by the columns of the matrix A_K, the following matrix $T(f)$, being a complement of $[\bar{t}_1(f) \;\; \ldots \;\; \bar{t}_N(f)]$ to a square matrix of size $K \times K$, is utilized:

$$T(f) = [\bar{t}_1(f) \;\; \ldots \;\; \bar{t}_N(f) \;\; e_{N+1} \;\; e_{N+2} \;\; \ldots \;\; e_K] \tag{13.13}$$

In (13.13), $e_j = [0 \ldots 0\ 1\ 0 \ldots 0]^T$, that is, the entries of the vector e_j are equal to zero, except for the jth position containing 1. Employing the Gram-Schmidt orthogonalization (Bau III & Trefethen, 1997) to $T(f)$ allows for rendering the orthonormal basis of K vectors:

$$T_{GS}(f) = [\tilde{t}_1(f) \;\; \ldots \;\; \tilde{t}_N(f) \;\; w_1(f) \;\; \ldots \;\; w_{K-N}(f)] \tag{13.14}$$

The second part of the matrix T_{GS} consists of the vectors $w_j(f), j = 1, \ldots, K - N$. These will be used to extend $s_I(F)$. The dependence of the vectors $w_j(f)$ on the objective vector f implicates that for each $f \in F$, a separate calculation is required. A remark should be made that the vectors $\tilde{t}_j(f)$of (13.12) are close to the vectors $\bar{t}_j(f)$of (13.13), due to a typically good alignment between the tangent vectors $t_j(f)$ and the eigenvectors a_j, $j = 1, \ldots, N$.

Let us now address the issue of an appropriate choice of the number K of the principal components. As mentioned before, one needs $K > N$. At the same time, the eigenvalues are normally quickly decreasing, which indicates that the majority of information about antenna response variability is contained within the subspace spanned by the first few eigenvectors. In practice, it suffices to appoint $K = N + 1$ or $N + 2$. The actual choice of K is problem specific.

In the last step, the first-level surrogate $s_I(f)$ and the vectors w_j are used to define the surrogate model domain X_S. We have (cf. (13.5))

$$X_S = \left\{ \begin{array}{c} x = s_I(f) + T \sum_{k=1}^{K-N} \lambda_k w_n^{(k)}(f) : f \in F, \\ -1 \leq \lambda_k \leq 1, \;\; k = 1, \ldots, n - N \end{array} \right\} \tag{13.15}$$

In other words, X_S contains all points of the form

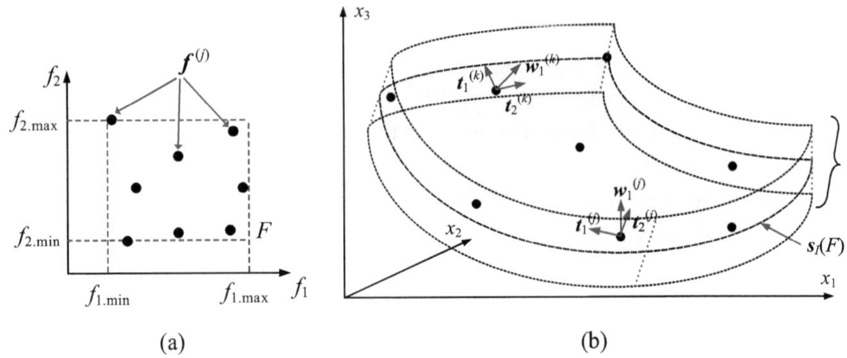

Fig. 13.8 Conceptual illustration of the surrogate domain confinement with explicit dimensionality reduction: (**a**) two-dimensional objective space F, (**b**) three-dimensional parameter space X, the first-level model image $s_I(F)$, along with two exemplary points $s_I(f)$ and their respective tangent vectors t_1 and t_2 as well as the normal vector w_1; the reference designs are marked with black circles. It should be noted that $K < n$, that is, the intended dimensionality of the confined domain X_s should be smaller than the dimensionality of the original parameter space X; yet shown is the sole case that could be presented graphically, that is, $K = n$

$$\boldsymbol{x} = \boldsymbol{s}_I(\boldsymbol{f}) + T\sum_{k=1}^{K-N} \lambda_k \boldsymbol{w}_n^{(k)}(\boldsymbol{f}) \tag{13.16}$$

for all $\boldsymbol{f} \in F$ and all combinations of $-1 \leq \lambda_k \leq 1$ for $k = 1, \ldots, K-N$. The parameter T of (13.15) refers to the lateral dimension of the domain X_S, and its value is chosen to be a fraction (between 5% and 10%) of the reference set spread along the most dominant eigenvector. Also, it accommodates the relationships between the subsequent eigenvalues λ_k. A modification to this scheme is possible, in which individual coefficient values T_k are adopted for each vector $\boldsymbol{w}_k$, instead of a joint value T. This would enable to account for the relative contribution of particular directions. The surrogate model domain dimensionality is appointed by selecting the parameter K. A graphical illustration of the considered concepts can be found in Fig. 13.8, here presented for a two-dimensional objective space F and a three-dimensional parameter space X. Figure 13.8 shows the case for the dimensionality K of the domain X_S equal to the dimensionality n of X (to enable proper illustration). Whereas in practice, one should assume $K < n$.

13.2.2 PCA-Based Constrained Modeling with Response Features

As argued on numerous occasions, the motivation behind reformulating the design task in terms of the features has come from the scrutiny of the dependence of the feature coordinates on the design variables, which appears to be much less nonlinear (in fact, often close to linear) than a similar dependence of the original responses.

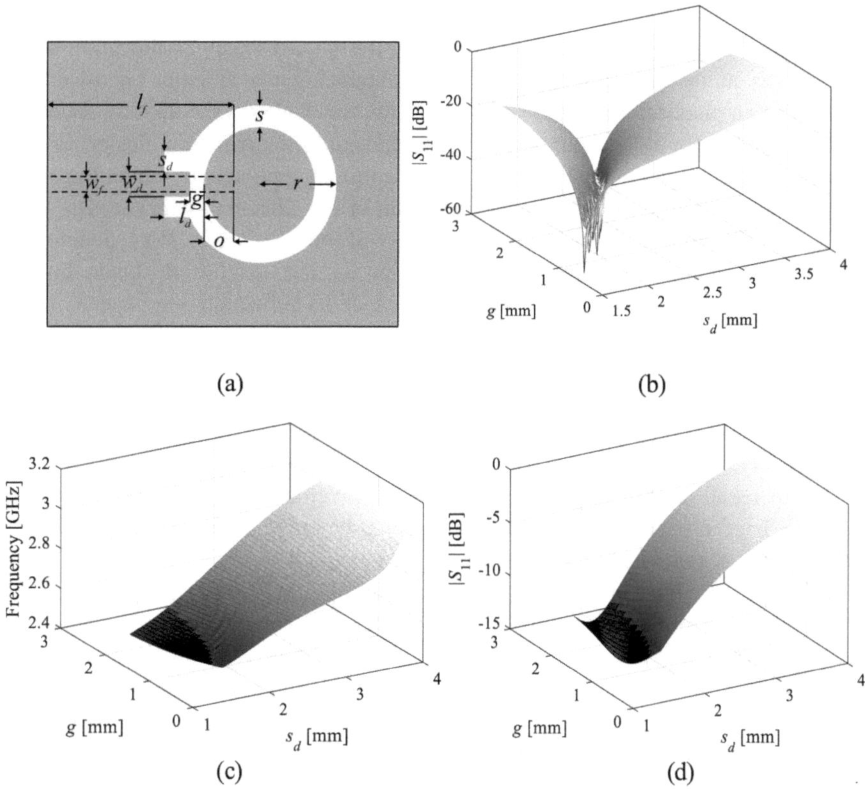

Fig. 13.9 Graphical illustration of the benefits of response feature approach: (**a**) geometry of a ring slot antenna, (**b**) and its reflection responses evaluated for $1.5 \le s_d \le 4.0$ and for $0.3 \le g \le 2.3$ at $f = 2.4$ GHz. Other geometry parameters are set to: $l_f = 27$, $l_d = 6.5$, $w_d = 2.2$, $r = 14.5$, $s = 5.3$, $o = 5.1$ (all dimensions in mm). The feature point components evaluated within the same design space region: frequency (**c**) and level (**d**). It can be observed that the dependence of the feature point coordinates on antenna geometry parameters is significantly less nonlinear than for the entire characteristics. At the same time, the information carried by the feature points is sufficient for handling the design problems the points were defined for (here, the matching improvement task of (13.2))

The example of such a relationship is shown in Fig. 13.9, where graphical illustration of the benefits coming from employing response feature approach: The dependence of the feature point coordinates on the antenna parameters is shown in Fig. 13.9c, d, whereas the respective dependence for the entire characteristics is shown in Fig. 13.9b.

Technically, the features have to be selected within a given design context and have to reflect the design goals. The actual selection of the response features requires the analysis of the system outputs and identification of easily distinguishable characteristic points. An illustrative example of the concepts outlined above constitute narrow- and multiband antennas. Here, a natural choice of response features are the points corresponding to the antenna resonances and -10 dB level of reflection. This selection of response features allows handling design tasks such as allocation of the resonances at the intended frequencies or enhancing the antenna bandwidth. It should be emphasized that the feature-based approach allows for directly encoding the knowledge about the figures of interest relevant to the particular design optimization problem at hand. Whereas in the conventional approach, in which the entire responses are processed, subsequent extraction of this information is necessary.

In this section, the following notation will be used. Let $\boldsymbol{R}(\boldsymbol{x})$ denote the EM-simulated antenna response at the design $\boldsymbol{x}$. The symbol $\boldsymbol{R}$ stands for the aggregated antenna outputs and may consists of its reflection response $S_{11}(\boldsymbol{x},f)$, gain $G(\boldsymbol{x},f)$, etc., where f is the frequency within a certain simulation range as provided by the model. Let also $\boldsymbol{\varphi} = [\varphi_1{}^T \ldots \varphi_P{}^T]^T$ be the vector whose entries are the characteristic points of the response (features): $\varphi_k = [\omega_k\ \lambda_k]^T$, $k = 1, \ldots, P$, with ω_k and λ_k being the frequency and the level (magnitude) coordinates, respectively.

Let us recall the design task of matching improvement of multiband antennas of Sect. 13.1 (cf. (13.2)). Here, the feature vector $\boldsymbol{\varphi}$ comprises the frequency and level coordinates of the antenna resonances: $\omega_k = f_{0.k}$ and $\lambda_k = S_{11}(f_{0.k})$, $k = 1, \ldots, P$, respectively. We have $\boldsymbol{\varphi} = [\varphi_1{}^T \ldots \varphi_P{}^T]^T = [f_{0.1}\ S_{11}(f_{0.k}) \ldots f_{0.P}\ S_{11}(f_{0.P})]^T$, where P refers to the number of antenna bands. In the case of bandwidth enhancement, the feature vector is $\boldsymbol{\varphi} = [\varphi_1{}^T \ldots \varphi_P{}^T]^T$, where each consecutive vector $\varphi_k{}^T = [f_{0.k}\ S_{11}(f_{0.k})\ f_{L.k}\ S_{11}(f_{L.k})\ f_{H.k}\ S_{11}(f_{H.k})]^T$, that is, it contains the frequency and level coordinates of the k-th antenna resonance, along with the coordinates of the points of -10 dB level of the antenna response around it, that is, $\lambda_{L.k} = |S_{11}(f_{L.k})| = -10$ dB and $\lambda_{H.k} = |S_{11}(f_{H.k})| = -10$ dB, $k = 1, \ldots, k$.

In the feature-based framework, the objective function (13.1) has to be reformulated in terms of the response features and will be denoted as $U_f\ (\boldsymbol{\varphi}(\boldsymbol{x}))$. Accordingly, the design task is transformed into

$$\boldsymbol{x}^* = \arg\min_{\boldsymbol{x}} U_\varphi(\boldsymbol{\varphi}(\boldsymbol{x})) \tag{13.17}$$

In the case of the multiband antennas, the merit function (13.2) of Sect. 13.1 may therefore be defined as

$$\begin{aligned} U_\varphi(\boldsymbol{x},\boldsymbol{\varphi}) &= \max\{|\lambda(\boldsymbol{x},f_{0.1}(\boldsymbol{x}))|, \ldots, |\lambda(\boldsymbol{x},f_{0.P}(\boldsymbol{x}))|\} \\ &\quad + \beta \cdot \left\| [f_{0.1}(\boldsymbol{x}) \ldots f_{0.P}(\boldsymbol{x})]^T - [f_{0.1}{}^t \ldots f_{0.P}{}^t]^T \right\|^2 \end{aligned} \tag{13.18}$$

where $f_{0.k}{}^t$ denotes the k-th target operating frequency and β is a penalty coefficient. In (13.18), the primary objective is minimization of antenna reflection, and the second term of (13.18) permits allocation of the antenna operating frequencies. In the case of bandwidth enhancement, the merit function may be defined as

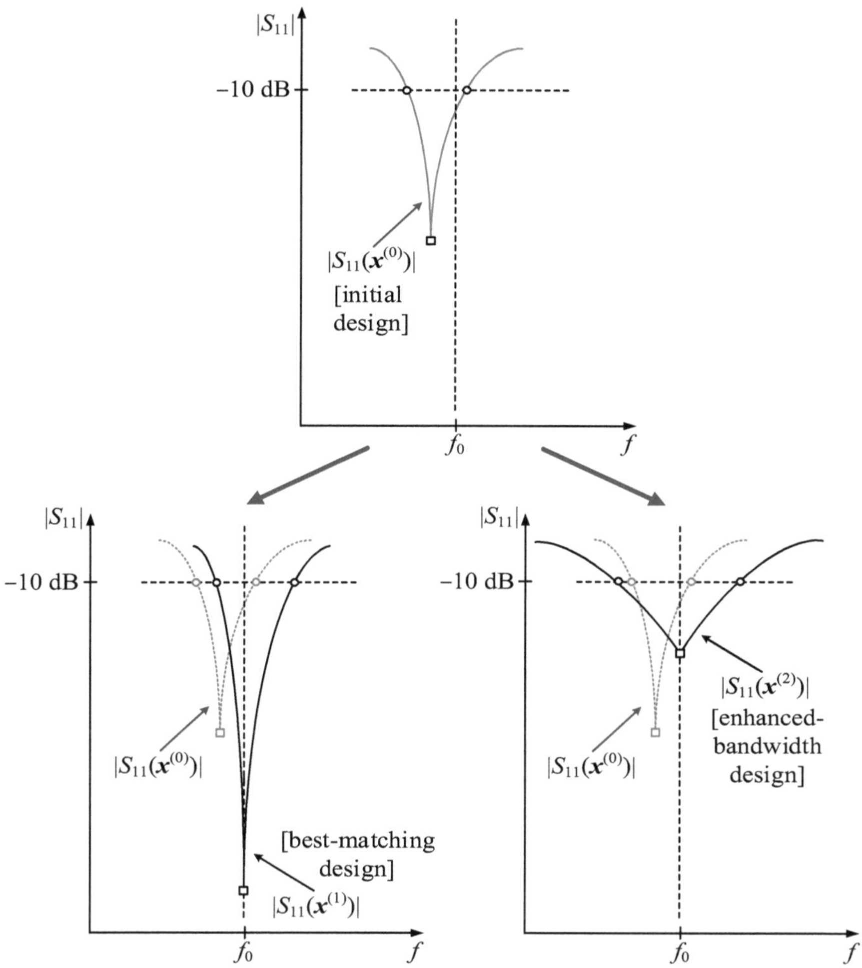

Fig. 13.10 Graphical explanation of the response features selection for a particular design task: the features corresponding to −10 dB level (°) (for bandwidth enhancement at a target frequency f_0 – bottom-left panel) and the point corresponding to f_0 (□) (for antenna matching improvement at f_0 – bottom-right panel)

$$U_\varphi(\boldsymbol{x}, \boldsymbol{\varphi}) = -\min\{B_1(\boldsymbol{x}), \ldots, B_P(\boldsymbol{x})\} \tag{13.19}$$

where the kth relative (fractional) bandwidth B_k is defined as

$$B_k(\boldsymbol{x}) = \frac{2 \cdot \min\{f_{0.k}(\boldsymbol{x}) - f_{L.k}(\boldsymbol{x}), f_{H.k}(\boldsymbol{x}) - f_{L.k}(\boldsymbol{x})\}}{f_{0.k}} \tag{13.20}$$

The graphical illustration of the response features selection depending on the optimization goals is shown in Fig. 13.10. The top panel of Fig. 13.10 presents the exemplary antenna reflection characteristic at the initial design $\boldsymbol{x}^{(0)}$, along with the response features corresponding to the antenna resonance and -10 dB reflection levels. The bottom-left panel of Fig. 13.10 shows the antenna response optimized for minimum in-band reflection with the use of the "resonance" feature point (marked with square). Whereas the bottom-right panel of Fig. 13.10 illustrates the response of the same structure optimized for maximum symmetrical bandwidth, obtained using the -10 dB characteristic points (marked with circles).

The discussed feature-based modeling procedure over dimensionality-reduced constrained domain consists of the following steps:

1. Objective space F definition by selecting relevant performance figures and their ranges
2. Reference designs acquisition $\boldsymbol{x}^{(j)}$, $j = 1, \ldots, p$, (Sect. 13.1)
3. Application of the principal component analysis of the reference design set to yield the eigenvectors $\boldsymbol{a}_k$ (Sect. 13.2)
4. Construction of the first-level surrogate $\boldsymbol{s}_I$
5. Definition of the dimensionality-reduced constrained domain X_S
6. Allocation of the training data samples $\boldsymbol{x}_B^{(k)}$, $k = 1, \ldots, N_B$, within X_S (Sect. 13.2.2.1)
7. Training data acquisition $\boldsymbol{R}(\boldsymbol{x}_B^{(k)})$, $k = 1, \ldots, N_B$
8. Response features extraction $\boldsymbol{\varphi}(\boldsymbol{R}(\boldsymbol{x}_B^{(k)}))$, $k = 1, \ldots, N_B$;
9. Surrogate model identification using the data pairs $\{\boldsymbol{x}_B^{(k)}, \boldsymbol{\varphi}(\boldsymbol{x}_B^{(k)})\}_{k = 1, \ldots, NB}$

The surrogate model is set up within X_S using kriging interpolation (Simpson et al., 2001) with $\{\boldsymbol{x}_B^{(k)}, \boldsymbol{\varphi}(\boldsymbol{x}_B^{(k)})\}_{k = 1, \ldots, NB}$ being the training data set. The samples $\boldsymbol{x}_B^{(k)}$ are uniformly allocated within X_S; $\boldsymbol{\varphi}(\boldsymbol{x}_B^{(k)})$ refer to the extracted characteristic points of the corresponding responses. The surrogate provides a prediction of the response feature coordinates for any given design $\boldsymbol{x}$ within X_S.

The details concerning the design experiments are delineated in the next subsection. Figure 13.11 shows the flow diagram of the overall modeling framework: exploiting the response feature technology and operating within the constrained domain of reduced dimensionality.

13.2.2.1 Design of Experiments. Surrogate Model Optimization

The geometry of the surrogate model domain is generally complex (in particular, its lateral dimensions significantly exceed the tangential ones); therefore, its handling may be inconvenient from the point of view of design of experiments (DoE). In particular, uniform allocation of the training data samples in X_S is a nontrivial task. In order to facilitate DoE within the considered framework, a surjective mapping H between the unit hypercube $[0,1]^K$ (normalized domain) and X_S is defined (see also Sect. 13.1). Using this transformation, it suffices to generate uniformly distributed data samples $\{z^{(k)}\}$, $k = 1, \ldots, N_B$, within $[0,1]^K$ using a standard procedure (e.g., Latin Hypercube Sampling (Beachkofski & Grandhi, 2002)) and subsequently transform them into X_S using H as

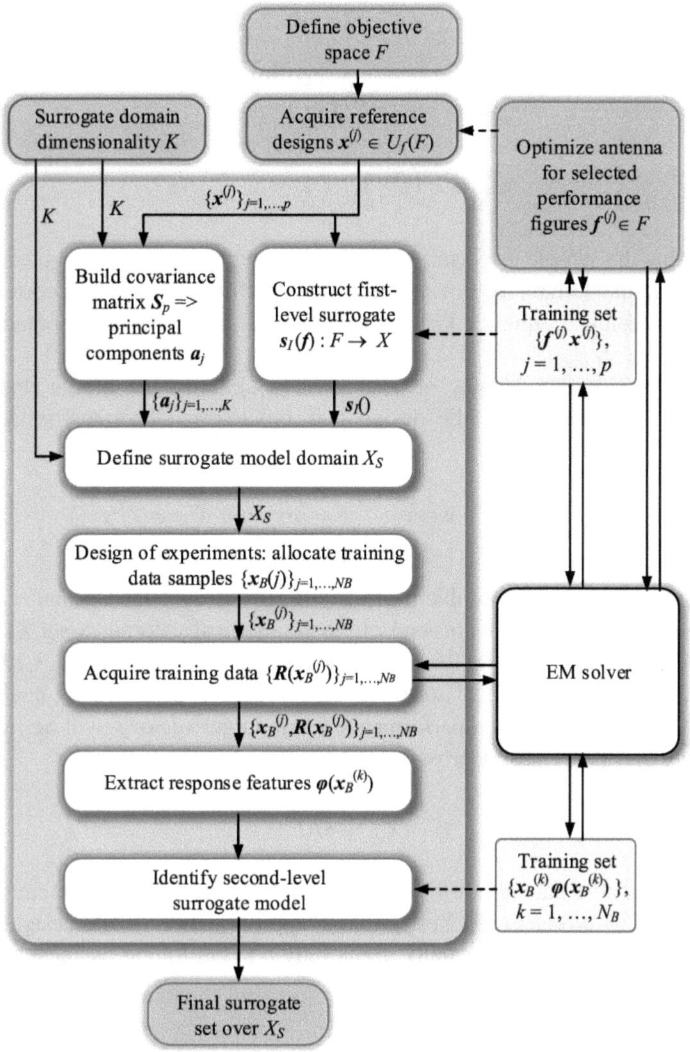

Fig. 13.11 Flow diagram of the discussed surrogate modeling procedure exploiting the response feature technology and operating on the constrained domain of reduced dimensionality

$$\begin{aligned} \boldsymbol{x} &= H(z) = H\left([z_1 \ \ldots \ z_n]^T\right) \\ &= \boldsymbol{s}_I\left(\boldsymbol{f}_z\right) + T\sum_{k=1}^{K-N}(-1 + 2z_{N+k})\boldsymbol{w}_n^{(k)}\left(\boldsymbol{f}_z\right) \end{aligned} \tag{13.21}$$

where

$$\boldsymbol{f}_z = [f_{1.\min} + z_1(f_{1.\max} - f_{1.\min}) \dots f_{N.\min} + z_N(f_{N.\max} - f_{N.\min})]^T \tag{13.22}$$

The procedure (13.21), (13.22) is repeated for all vectors $z^{(k)}$ to yield the uniformly distributed training data set $\{\boldsymbol{x}_B{}^{(k)}\}$ in X_S as

$$\boldsymbol{x}_B^{(k)} = H\left(\boldsymbol{z}^{(k)}\right), \qquad k = 1, \dots, N_B \tag{13.23}$$

Note that the sample uniformity is understood with respect to the objective space F rather than the domain X_S. For example, if the performance figure f_1 corresponds to the antenna center frequency $f_{0.1}$, the antenna designs are equally spread over the interval $[f_{1.\min}, f_{1.\max}]$.

The transformation H may be also used to simplify the design optimization process. More specifically, the design task (13.17) can be transformed into an equivalent problem

$$\boldsymbol{x}^* = H(\boldsymbol{z}^*) \qquad \text{where} \qquad \boldsymbol{z}^* = \arg \min_{\boldsymbol{z} \in [0,\,1]^n} U_\varphi(\boldsymbol{\varphi}(H(\boldsymbol{z}))) \tag{13.24}$$

The benefit is that the problem (13.24) is solved over the normalized domain delimited by box constraints instead of the geometrically complex set X_S. Possible formulations of the objective function U_φ are given by (13.18) and (13.19).

Furthermore, the first-level surrogate can be employed to yield a quality initial design, which, for any given performance figure target vector $\boldsymbol{f}^t$ will be (for details, see Koziel and Pietrenko-Dabrowska (2019))

$$\boldsymbol{x}^{(0)} = \boldsymbol{s}_I(\boldsymbol{f}^t) \tag{13.25}$$

Given available data about the antenna at hand encoded in the reference design set, Eq. (13.25) produces the best possible approximation of the design optimum with respect to $\boldsymbol{f}^t$, which normally needs only a slight tuning by means of a local optimization algorithm.

13.2.3 *Verification Case: Triple-Band Antenna*

This section discusses validation of the modeling procedure outlined in Sect. 13.2, as well as provides comparisons with five benchmark procedures: the conventional data-driven surrogates: RBF and kriging interpolation (both set up within the unconstrained domain delimited by the lower and upper bounds on the parameters $\boldsymbol{l}$, $\boldsymbol{u}$, respectively), as well as three variants of the nested kriging technique: the basic procedure (Koziel & Pietrenko-Dabrowska, 2019), the feature-based version of Sect. 13.1 (Koziel & Pietrenko-Dabrowska, 2020c), and the dimensionality-reduced nested kriging (Koziel et al., 2020). The numerical experiments are carried out on

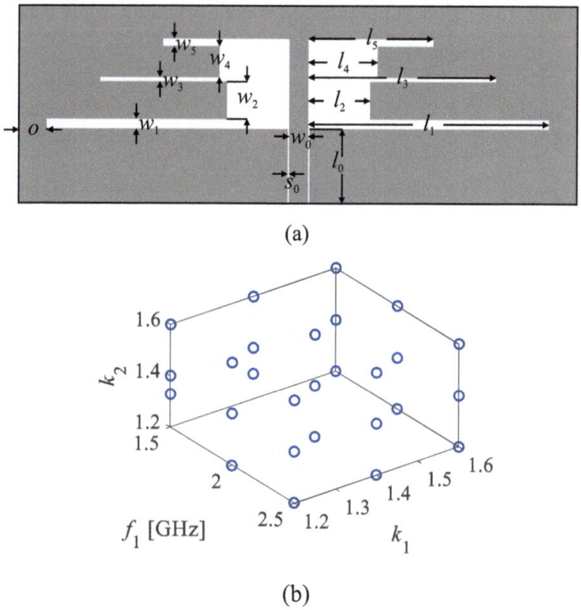

Fig. 13.12 Uniplanar triple-band dipole (Chen et al., 2006): (**a**) geometry, (**b**) allocation of the reference de-signs in the objective space

the example of a triple-band uniplanar dipole. The application of the surrogates rendered within the considered technique for antenna optimization is also investigated.

As a verification case study, we use a triple-band uniplanar antenna built as a stack of three ground plane slits separated by two slots that is fed by a coplanar waveguide (Chen et al., 2006). Figure 13.12 shows the antenna geometry described by the following vector of the design variables: $\boldsymbol{x} = [l_1\ l_2\ l_3\ l_4\ l_5\ w_1\ w_2\ w_3\ w_4\ w_5]^T$, with $l_0 = 30$, $w_0 = 3$, $s_0 = 0.15$, and $o = 5$ being fixed (all dimensions in mm). The structure is implemented on the RO4350 substrate. The antenna computational model is simulated using CST Microwave Studio transient solver (~180,000 cells; simulation time ~80 seconds).

The following setup for the validation process has been adopted. The surrogate models have been constructed using the training data sets of various sizes: 20, 50, 100, 200, 400, and 800 samples. The modeling error (average relative RMS) has been estimated through the split sample method (Jiang et al., 2020), with the test set consisting of 100 random samples.

The number of the principal directions $K = 4$ has been adopted. The lateral dimension of the constrained domain X_S has been set as $T = 0.25$ mm (this value corresponds to around 5% of the lateral span of the unconstrained domain X; (see Koziel and Pietrenko-Dabrowska (2019) for details).

The goal is to build the surrogate valid for the antenna operating frequencies f_k, $k = 1, 2, 3$, with $f_2 = f_1k_1$ and $f_3 = f_2k_2$ (the actual operating frequencies are calculated based on the values of the ratios k_1 and k_2). The intended frequency ranges are: $f_1 \in [1.5, 2.5]$ GHz and $k_1, k_2 \in [1.2, 1.6]$. The objective space comprises

Table 13.3 Modeling results and benchmarking for triple-band antenna

Number of training samples	Relative RMS error					
	Conventional kriging model	Conventional RBF	Nested kriging model	Feature-based nested kriging	PCA-based nested kriging modeling	PCA-based nested kriging modeling with response features
20	28.5%	30.1%	38.9%	2.65%	44.8%	1.71%
50	22.7%	23.5%	16.0%	0.25%	16.5%	0.66%
100	19.9%	19.8%	11.2%	0.22%	10.9%	0.24%
200	18.6%	19.2%	9.9%	0.19%	8.7%	0.89%
400	17.2%	18.8%	9.7%	0.14%	6.4%	0.43%
800	16.8%	17.4%	7.8%	0.20%	4.3%	0.19%

the vectors $[f_1\ k_1\ k_2]^T$. The geometry parameter lower and upper bounds are: $\boldsymbol{l} = [30\ 5.0\ 20\ 5.0\ 15\ 0.2\ 0.2\ 0.2\ 0.2\ 0.2]^T$ and $\boldsymbol{u} = [50\ 15\ 30\ 15\ 21\ 2.2\ 4.2\ 2.2\ 4.2\ 2.2]^T$, respectively. Table 13.3 gathers the modeling errors for the discussed and benchmark frameworks. Similarly as in Sect. 13.1.3, for feature-based modeling frameworks, the errors of the frequency coordinates of the characteristic points are reported.

The design utility of the considered surrogate has been demonstrated by applying the model for antenna optimization. The results obtained by the surrogate constructed using 50 training samples for four different sets of the target operating frequencies are provided in Fig. 13.13 and Table 13.4. Figure 13.13 presents the EM-simulated initial and optimized responses of a triple-band antenna. The results corroborate suitability of the discussed feature-based modeling framework with dimensionality reduction for antenna design purposes. Despite using such a low number of training data samples, the surrogate yields the optimized designs of acceptable quality with the target operating frequencies allocated according to the design specifications, which is corroborated by the full-wave simulations. Despite the fact that the operating frequencies of the initial designs are not accurately allocated, the optimized designs meet the requirements.

The section discusses a modeling framework that combines the nested kriging technique with explicit dimensionality reduction through principal component analysis with the response feature technology. Conducting the modeling process at the level of the feature points of the antenna responses while operating within the constrained domain of reduced dimensionality requires remarkably smaller training data sets—as compared to benchmark methods—to ensure usable predictive power of the surrogate. The formulation of the considered technique incorporates the mechanisms for efficient design of experiments, and for expedited surrogate model optimization. The discussed methodology has been validated using a triple-band antenna described by ten parameters. Thorough benchmarking against two conventional data-driven modeling techniques and three variations of the performance-driven surrogates constructed within constrained domain corroborate reliability of

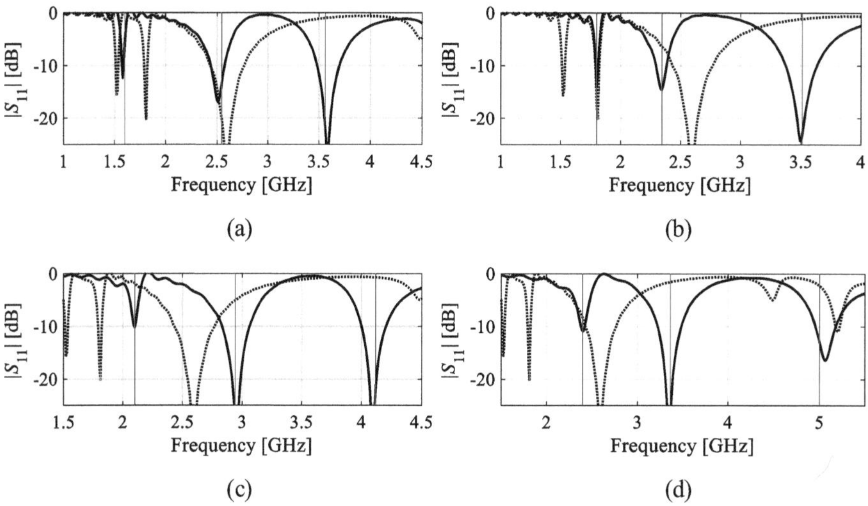

Fig. 13.13 EM-simulated antenna responses of Antenna III: initial designs (. . . ..) yielded by the PCA-and feature-based surrogate, and the optimized designs (—). Vertical lines mark target operating frequencies: (**a**) $f_1 = 1.6$ GHz, $k_1 = 1.6$, $k_2 = 1.4$ ($f_2 = 2.56$ GHz, $f_3 = 3.58$ GHz), (**b**) $f_1 = 1.8$ GHz, $k_1 = 1.3$, $k_2 = 1.5$ ($f_2 = 2.34$ GHz, $f_3 = 3.51$ GHz), (**c**) $f_1 = 2.1$ GHz, $k_1 = 1.4$, $k_2 = 1.4$ ($f_2 = 2.94$ GHz, $f_3 = 4.12$ GHz), (**d**) $f_1 = 2.4$ GHz, $k_1 = 1.4$, $k_2 = 1.5$ ($f_2 = 3.36$ GHz, $f_3 = 5.04$ GHz)

Table 13.4 Optimization results for triple-band antenna

Target operating frequencies [GHz]			Geometry parameters [mm]									
f_1	f_2	f_3	l_1	l_2	l_3	l_4	l_5	w_1	w_2	w_3	w_4	w_5
1.6	2.56	3.58	40.20	4.85	33.89	9.63	24.66	0.25	0.95	0.78	2.43	0.61
1.8	2.34	3.51	39.05	7.98	34.08	7.80	24.67	0.43	1.23	0.49	2.37	0.82
2.1	2.94	4.12	37.24	9.98	31.24	10.32	22.58	0.50	1.29	0.91	1.09	0.55
2.4	3.36	5.04	36.42	12.08	29.03	11.21	20.48	0.57	0.94	1.04	0.59	0.67

the considered approach. The design utility of the surrogate has been demonstrated through design optimization of the considered antenna for various combinations of performance figures.

13.3 Two-Stage Observable-Based Constrained Modeling

The nested kriging modeling framework offers a work-around the dimensionality issues affecting conventional modeling procedures utilized in high-frequency design by restricting the model domain to the parameter space regions that contain high-quality designs with respect to the assumed performance metrics. However, the domain determination requires acquisition of a set of preoptimized reference designs, which adds to the overall computational cost of the surrogate model setup in a significant manner. This section discusses an alternative approach, where the confined domain definition does not require any reference designs whatsoever (Koziel et al., 2021; Koziel & Pietrenko-Dabrowska, 2022). Instead, a preselected set of random observables is employed. The selection of the observables is carried out based on the performance figure values (response features) extracted from the EM-simulated component responses across the observable set. Numerical validation carried out with the use of antenna and microwave structures indicates that two-stage observable-based constrained modeling offers computational benefits similar to those of nested kriging technique while considerably reducing the initial setup cost by a factor of around 70% on the average. This cost reduction has been achieved owing to the exploitation of the response features in the observable selection process.

The necessity of acquiring the reference designs is the most serious practical issue related to performance-driven modeling methods for two primary reasons: (i) Depending on the objective space dimensionality, the number of reference designs varies between 10 and 20 (Koziel & Pietrenko-Dabrowska, 2020a); assuming that each design is obtained through local (e.g., gradient-based) search and the average parameter space dimensionality is 10, the cost of reference point acquisition may be as high as a thousand EM analyses of the component at hand; this cost adds to the surrogate model setup cost; (ii) acquisition of the reference vectors is a numerically challenging task because the structure of interest has to be redesigned over broad ranges of operating conditions; in practice, it may entail several optimization attempts to obtain a single design, including manual adjustment of the starting point.

It is possible to reduce the number of reference points using advanced techniques, such as the establishment of the first-level surrogate using gradient-enhanced kriging (Pietrenko-Dabrowska et al., 2020), or attempt to automate the acquisition process (Koziel & Pietrenko-Dabrowska, 2020d). Nevertheless, the main problem, that is, the necessity of gathering at least several designs allocated within the optimum design manifold, remains.

13.3.1 Inverse Regression Surrogate

The reference designs, as elaborated on in Sects. 13.1 and 13.2, provide a great deal of information about the spatial allocation of the optimum design manifold, which subsequently becomes the core of the surrogate model definition (Koziel & Pietrenko-Dabrowska, 2020a). In the discussed reference-design-free modeling framework, the reference designs are replaced by a preselected set of random observables generated in the parameter space X. Using the response features extracted from the observables, an inverse regression model is constructed that replaces the first-level interpolative surrogate of the nested kriging approach. The inverse model will be then used to define the domain of the final surrogate.

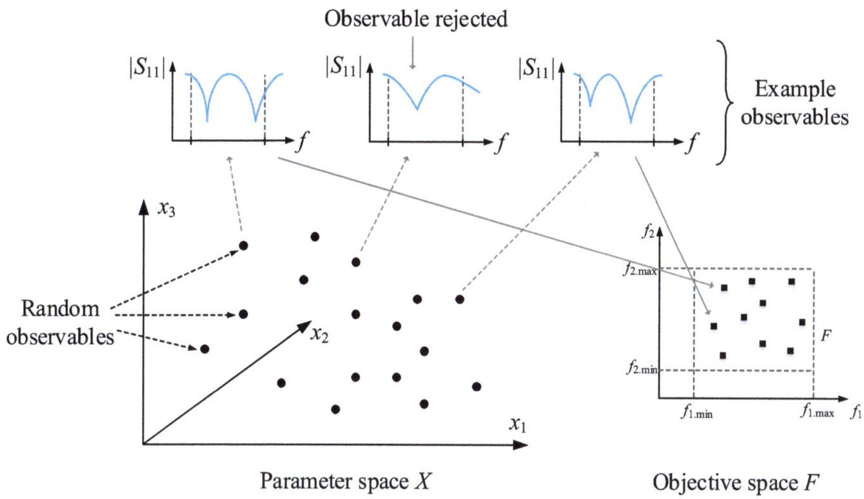

Fig. 13.14 Graphical illustration of the random observable generation procedure using an example of a dual-band antenna. The antenna is to be modelled over a two-dimensional objective space and within a three-dimensional parameter space. The random vectors that correspond to designs featuring operating frequencies within the assumed objective space are accepted, others are rejected. The final observable set $\{\boldsymbol{x}_r^{(j)}\}_{j=1,\ldots,N_r}$ will be used to generate the inverse regression model $s_r(\cdot)$

Let $\boldsymbol{x}_r^{(j)}$, $j = 1, 2, \ldots$, be a sequence of random parameter vectors sequentially generated in X using a uniform probability distribution. Let $\boldsymbol{f}_r^{(j)}$ be the response feature vector extracted from the EM-simulated antenna response at $\boldsymbol{x}_r^{(j)}$. For example, in the case of a dual-band antenna with the figures of interest being the target operating frequencies, the vector $\boldsymbol{f}_r^{(j)}$ contains the actual operating frequencies of the structure (regardless of the resonance depths). If $\boldsymbol{f}_r^{(j)} \in F$, the observable is accepted; otherwise (also, if the components of $\boldsymbol{f}_r^{(j)}$ fall beyond the assumed lower or upper limits for the figures of interest or cannot be identified at all), it is rejected. Figure 13.14 shows a graphical illustration for the discussed case of a dual-band antenna. The sampling process is concluded when the prescribed number of observables has been found, here denoted as N_r (typically 50 to 100).

The data set $\{\boldsymbol{x}_r^{(j)}, \boldsymbol{f}_r^{(j)}\}_{j=1,\ldots,\mathrm{Nr}}$, is used to establish the inverse regression model s_r, which provides a rough allocation of the optimum design manifold. The inverse model is defined over the objective space F with the values in the parameter space X, or s_r: $F \rightarrow X$. In the discussed framework, the following analytical form of s_r is assumed:

$$s_r(f) = s_r([f_1 \ \dots \ f_N]^T) = \begin{bmatrix} s_{r.1}(f) \\ \dots \\ s_{r.n}(f) \end{bmatrix}$$
$$= \begin{bmatrix} a_{1.0} + a_{1.1}\exp\left(\sum_{k=1}^{N} a_{1.k+1} f_k\right) \\ \dots \\ a_{n.0} + a_{n.1}\exp\left(\sum_{k=1}^{N} a_{n.k+1} f_k\right) \end{bmatrix} \tag{13.26}$$

which is motivated by a normally weakly nonlinear dependence between the geometry parameters and operating figures such as resonant frequencies, bandwidths, etc.

The model is identified by solving the nonlinear regression problems of the form

$$[a_{j.0} \ a_{j.1} \ \dots \ a_{j.K+1}] = \arg \min_{[b_0 \ b_1 \ \dots \ b_{K+1}]} \sum_{k=1}^{N_r} w_k \left[s_{r.j}\left(f_r^{(k)}\right) - x_{r.j}^{(k)} \right]^2, j$$
$$= 1, \dots, n \tag{13.27}$$

where $x_{r.j}^{(k)}$ is the jth component of the observable vector $x_r^{(k)}$. The weighting factors w_k are introduced to discriminate between "good" and "poor" observables. This is decided upon using the supplementary vectors $p_r^{(j)} = [p_{r.1}^{(j)} \dots p_{r.N}^{(j)}]^T$, also extracted from EM-simulated outputs of the component under design. The entries of the vector $p_r^{(j)}$ contain relevant data concerning the design quality. For example, in the discussed case of a dual band antenna, these could be the reflection levels at the operating frequencies, the lower the better. The weighing factors are assigned as

$$w_k = \left[M - \max\left\{ p_1\left(x^{(j)}\right), \dots, p_N\left(x^{(j)}\right) \right\} \right]^2, \quad k = 1, \dots, N_r \tag{13.28}$$

where $M = \max\{k = 1, \dots, N_r, j = 1, \dots, N : p_j^{(k)}\}$. Here, it is assumed that $p_j^{(k)}$ is nonnegative and lower values correspond to a better design. Again, the previously considered example of reflection levels at the operating frequencies conform to this assumption, as their moduli may change between zero and one.

The idea behind the weighting factors is to put more emphasis on high-quality observables, while still incorporating information contained in the remaining vectors. In particular, the designs characterized by higher weights are allocated closer to the optimum design manifold; therefore, they should determine the inverse surrogate in a more significant manner than the lower-quality ones. The aforementioned feature-based assessment of the quality of the observables allows for a more efficient and reliable rendition of the inverse surrogate. A graphical illustration of the inverse regression model, corresponding to the example of Fig. 13.14, has been shown in Fig. 13.15.

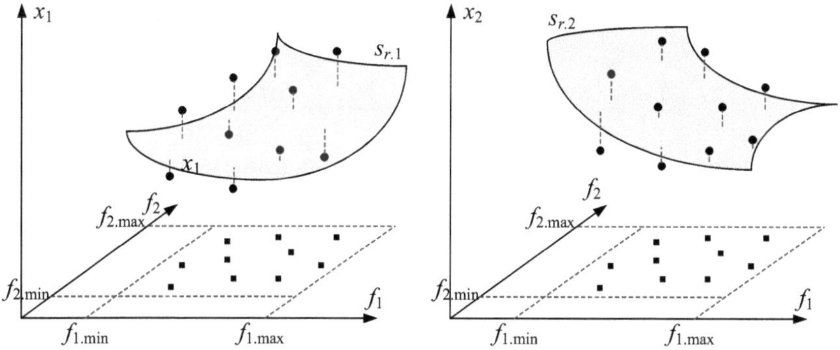

Fig. 13.15 Graphical illustration of the inverse regression model s_r established using the preselected observables $\boldsymbol{x}_r^{(j)}$ and the corresponding objective vectors $\boldsymbol{f}_r^{(j)}$. The components $s_{r.j}$ of the inverse model are obtained as in (13.26)–(13.28), and their images are visualized as the gray-shaded surfaces for antenna parameters x_1 (left), and x_2 (right), respectively

13.3.2 Domain Definition and Final Surrogate

The image $s_r(F)$ of the inverse model provides a rough approximation of the spatial allocation of the optimum design manifold $U_F(F)$. In order to encapsulate most of $U_F(F)$ into the domain of the surrogate model, $s_r(F)$ needs to be extended, which is realized using the procedure similar to the one developed for the nested kriging framework (Koziel & Pietrenko-Dabrowska, 2019). Let $\{\boldsymbol{v}_n^{(k)}(\boldsymbol{f})\}$, $k = 1, \ldots, n - N$, denote the orthonormal basis of vectors normal to $s_r(F)$ at the objective vector $\boldsymbol{f}$. Further, let $\boldsymbol{T} = [T_1 \ldots T_n]^T$ be a vector of positive real numbers determining the amount of extension for the structure parameters. Using these, we calculate the extension coefficients

$$\boldsymbol{\alpha}(\boldsymbol{f}) = [\alpha_1(\boldsymbol{f}) \;\; \ldots \;\; \alpha_{n-N}(\boldsymbol{f})]^T = \left[|\boldsymbol{T}\boldsymbol{v}_n^{(1)}(\boldsymbol{f})| \;\; \ldots \;\; |\boldsymbol{T}\boldsymbol{v}_n^{(n-N)}(\boldsymbol{f})| \right]^T \tag{13.29}$$

The domain X_S of the final surrogate model is then defined similarly as in the nested kriging framework of Sect. 13.1 (Koziel & Pietrenko-Dabrowska, 2019):

$$X_S = \left\{ \begin{array}{c} \boldsymbol{x} = s_r(\boldsymbol{f}) + \sum_{k=1}^{n-N} \lambda_k \alpha_k(\boldsymbol{f}) \boldsymbol{v}_n^{(k)}(\boldsymbol{f}) : \boldsymbol{f} \in F, \\ -1 \le \lambda_k \le 1, \;\; k = 1, \ldots, n - N \end{array} \right\} \tag{13.30}$$

which is a set of vectors of the form of (13.30) obtained for all $\boldsymbol{f} \in F$, and all $\lambda_k \in [-1, 1]$, $k = 1, \ldots, n - N$. It can be noted that the extension boundaries of the domain are obtained for the extreme values of the coefficients λ_k, which are the surfaces defined by

$$S_+ = \left\{ \boldsymbol{x} \in X : \boldsymbol{x} = s_r(\boldsymbol{f}) + \sum\nolimits_{k=1}^{n-N} \alpha_k(\boldsymbol{f}) \boldsymbol{v}_n^{(k)}(\boldsymbol{f}) \right\} \tag{13.31}$$

$$S_- = \left\{ \boldsymbol{x} \in X : \boldsymbol{x} = s_r(\boldsymbol{f}) - \sum\nolimits_{k=1}^{n-N} \alpha_k(\boldsymbol{f}) \boldsymbol{v}_n^{(k)}(\boldsymbol{f}) \right\} \tag{13.32}$$

It should be emphasized that the extension factors T_j (the entries of the vector $\boldsymbol{T}$) are individualized for all geometry parameters (e.g., in the nested kriging framework of (Koziel & Pietrenko-Dabrowska, 2019), it was a scalar coefficient). Also, the amount of extension can be estimated using the available observable set (in (Koziel & Pietrenko-Dabrowska, 2019), it was arbitrarily chosen by the user). The assessment of the extension factors is arranged as follows. Let, for any observable pair $\{\boldsymbol{x}_r^{(j)}, \boldsymbol{f}_r^{(j)}\}$, the vector $P_k(\boldsymbol{x}_r^{(j)})$ be the element of the space $[l_k\ u_k] \times F$ (recall that l_k and u_k are parameter-space-defining lower and upper bounds for the kth parameter) that minimizes the distance between $[x_{r.k}^{(j)}\ (\boldsymbol{f}_r^{(j)})^T]^T$ and $[s_{r.k}(\boldsymbol{f})\ \boldsymbol{f}^T]^T, \boldsymbol{f} \in F$ (here, $x_{r.k}^{(j)}$ is the kth component of the observable vector $\boldsymbol{x}_r^{(j)}$), that is,

$$P_k\left(\boldsymbol{x}_r^{(j)}\right) = \arg\min_{\boldsymbol{f} \in F} \left\| \left[x_{r.k}^{(j)}\ \left(\boldsymbol{f}_r^{(j)}\right)^T \right]^T - \left[s_r(\boldsymbol{f})\ \boldsymbol{f}^T \right]^T \right\| \tag{13.33}$$

In other words, $P_k(\boldsymbol{x}_r^{(j)})$ determines the orthogonal projection of $[x_{r.k}^{(j)}\ (\boldsymbol{f}_r^{(j)})^T]^T$ onto the image of the kth component of the inverse regression model within the space $[l_k\ u_k] \times F$. Then,

$$d_{r.k}\left(\boldsymbol{x}_r^{(j)}\right) = \left\| \left[x_{r.k}^{(j)}\ \left(\boldsymbol{f}_r^{(j)}\right)^T \right]^T - \left[s_r\left(P\left(\boldsymbol{x}_r^{(j)}\right)\right)\ P\left(\boldsymbol{x}_r^{(j)}\right)^T \right]^T \right\| \tag{13.34}$$

is the minimum distance between $[x_{r.k}^{(j)}\ (\boldsymbol{f}_r^{(j)})^T]^T$ and the said image. Graphically, $d_{r.k}$ can be interpreted as the distances between the respective observable components and the gray-colored surfaces in Fig. 13.15.

Using the above, the extension factor T_k can be then defined as

$$T_k = \frac{1}{2N_r} \sum_{j=1}^{N_r} d_{r.k}\left(\boldsymbol{x}_r^{(j)}\right) \tag{13.35}$$

which is half of the average distance between the observable component and the respective inverse regression model surface. It should be recalled that high-quality observables are allocated close to the optimum design manifold $U_F(F)$ (they also contribute more to the determination of the inverse model through the associated weighting factors w_k, cf. (13.27)), whereas poor observables are farther away from $U_F(F)$. Consequently, the average distance between the inverse regression model surface and the observables gives an overly pessimistic estimate of the necessary domain extension (i.e., if used, the domain would also contain many designs that are

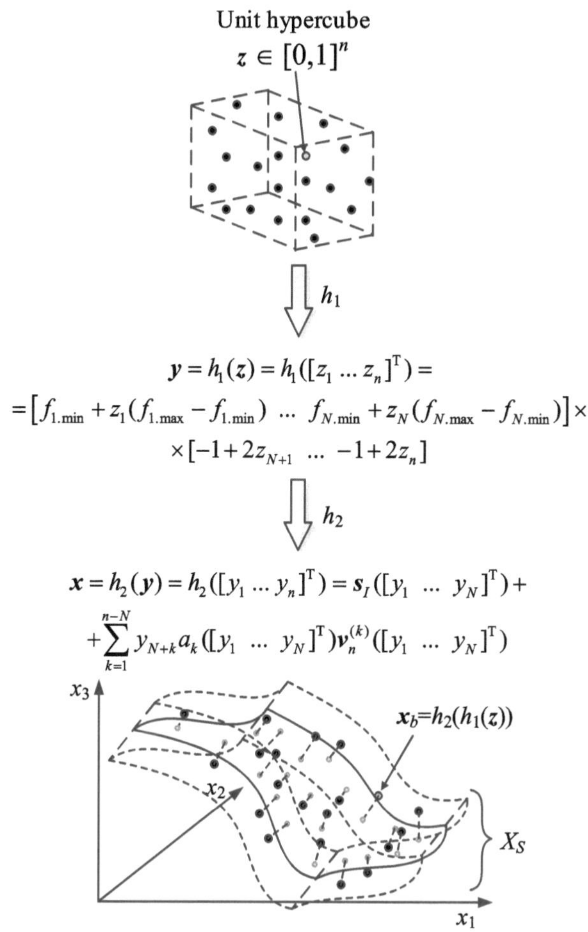

Fig. 13.16 Conceptual illustration of the employed design experiments procedure: (i) samples allocated using Latin Hypercube Sampling (LHS) (Beachkofski & Grandhi, 2002) (top), (ii) transformation with the use of function h_1 onto the Cartesian product of the objective space F and $[-1,1]^{n-N}$ (middle), (iii) samples (black circles) mapped onto the constrained domain X_S using transformation h_2 (bottom); whereas the samples allocated onto the image of F before orthogonal shift are marked with light gray circles

of not-so-good quality). To account for this, the definition (13.35) uses half of the average distance instead.

Having the domain X_S, the final surrogate $\boldsymbol{s}(\boldsymbol{x})$ is generated therein using kriging interpolation (Simpson et al., 2001). The training data set is denoted as $\{\boldsymbol{x}_B^{(k)}, \boldsymbol{R}(\boldsymbol{x}_B^{(k)})\}_{k = 1, \ldots, NB}$, where $\boldsymbol{x}_B^{(k)} \in X_S$, whereas $\boldsymbol{R}$ represents the relevant antenna characteristics obtained through EM analysis. Figure 13.16 presents graphical illustration of the design of experiments procedure similar to that proposed in (Koziel & Pietrenko-Dabrowska, 2019) (see also Sect. 13.1), which capitalizes on the one-to-one transformation of the unit interval $[0,1]^n$ onto X_S.

Observe that the data samples $\boldsymbol{x}_B^{(k)} = h_2(h_1(z^{(k)}))$ allocated in X_S are uniformly distributed with respect to the objective space F and not necessarily with respect to the parameter space X. Here, $\{z^{(k)}\}$, $k = 1, \ldots, N_B$, denotes a uniformly distributed set of samples in $[0,1]^n$ (obtained using Latin Hypercube Sampling, LHS (Beachkofski & Grandhi, 2002)). Note that the same mapping can be also employed to carry out design procedures (e.g., parametric optimization) using the two-stage surrogate model, similarly as it has been described in Sect. 13.2.2.1. Although a straightforward application of the model (i.e., operating with the domain X_S) might be challenging due to a relatively complex geometry of the domain, all operations can be executed within the unity interval $[0, 1]^n$, with the parameters mapped to X_S for the purpose of surrogate model evaluation. Furthermore, the vector $s_r(f)$ provides a good initial design for any target objective vector $\boldsymbol{f} \in F$.

The computational benefits of restricting the model domain, as described in this section, are expected to be similar to those of other performance-driven procedures (Koziel & Pietrenko-Dabrowska, 2020a), in particular, the nested kriging framework. By focusing on a volume-reduced region of the space, the surrogate can be constructed using small training data sets without formally limiting range of validity of the model, both in terms of the ranges of geometry parameters and the operating conditions. On the other hand, the approach discussed here does not require any preoptimized reference designs, which has a significant impact on the model setup cost. An additional benefit is that defining the surrogate domain using random observables provides a convenient way to determine its lateral dimensions, which is a nontrivial problem for prior performance-driven techniques.

13.3.3 Complete Two-Stage Observable-Based Modeling Framework

This section provides a concise summary of the discussed two-stage observable-based modeling procedure. The only user-defined control parameters are the number of (accepted) random observables N_r (typically 50 to 100) and the number N_B of training data samples for rendering the final surrogate model. The input data consist of the definition of the parameter space X and the objective space F. The modeling process works as follows:

1. Generate random observables $\boldsymbol{x}_r^{(j)} \in X$ until N_r samples are found such that their corresponding objective vectors $\boldsymbol{f}_r^{(j)}$ are in the assumed objective space F (cf. Sect. 13.3.2); for these samples, also evaluate supplementary performance vectors $\boldsymbol{p}_r^{(j)}$.
2. Construct the inverse regression model s_r using $\{\boldsymbol{x}_r^{(j)},\boldsymbol{f}_r^{(j)}\}_{j = 1,\ldots,Nr}$, as the training data set (cf. (13.26)) and supplementary vectors $\boldsymbol{p}_r^{(j)}$ to compute the weighting factors for the observables (cf. (13.28)).
3. Calculate the extension vector $\boldsymbol{T}$ as in (13.33)–(13.35) and the extension coefficients $\boldsymbol{\alpha}$ (cf. (13.29)).
4. Define the surrogate model domain X_S using (13.30).
5. Perform design of experiments (see Fig. 13.16) to find N_B training data samples $\{\boldsymbol{x}_B^{(k)},\boldsymbol{R}(\boldsymbol{x}_B^{(k)})\}_{k = 1, \ldots, NB}$.

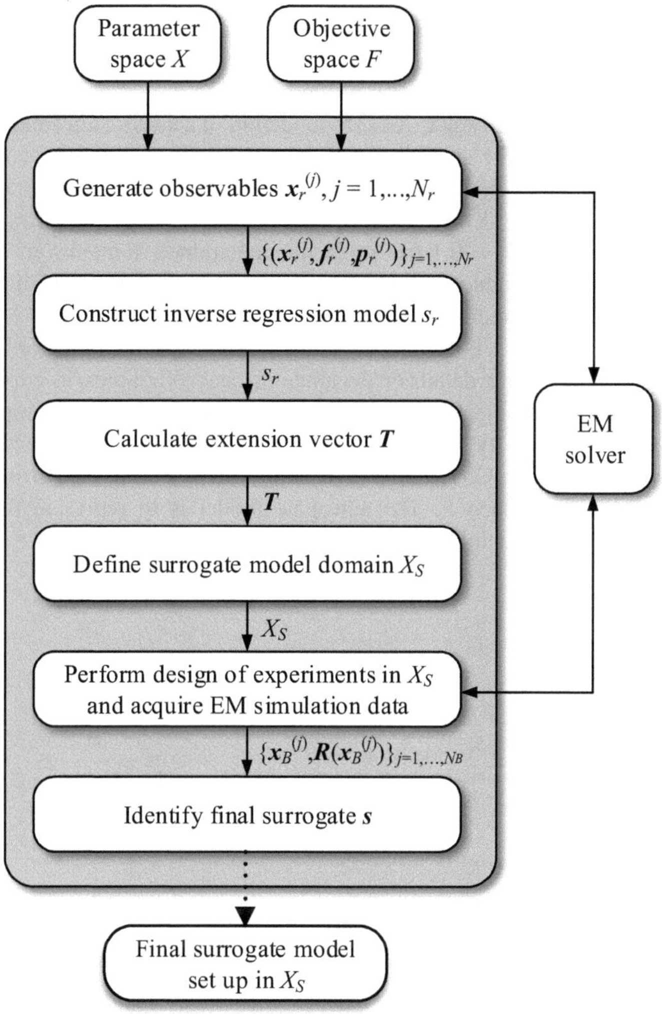

Fig. 13.17 Flow diagram of the two-stage observable-based modeling procedure involving inverse regression surrogates

6. Identify the final surrogate model $\boldsymbol{s}$ using kriging interpolation of the training samples obtained in Step 5, extended by the observable set $\{\boldsymbol{x}_r^{(k)}, \boldsymbol{R}(\boldsymbol{x}_r^{(k)})\}_{k = 1, \ldots, N}$.

Figure 13.17 shows the flow diagram of the modeling process. It should be noted that adding the observable set to the training data is expected to considerably improve the predictive power of the considered model for small values of N_B, which is an additional advantage of the presented approach.

13.3.4 Verification Case Study I: Quasi-Yagi Antenna

This section demonstrates application of the two-stage observable-based procedure (Koziel & Pietrenko-Dabrowska, 2022) to modeling of a quasi-Yagi antenna. For the sake of benchmarking, the considered technique is compared to conventional surrogates: kriging interpolation model and RBF model (both set up in the box-constrained domain X), as well as the nested kriging framework with the thickness parameter of $T = 0.1$ (set up in the constrained domain X_S). The nested kriging model utilizes eight reference designs; the cost of their acquisition is 1899 EM simulations of the antenna.

Figure 13.18 presents a quasi-Yagi antenna with a parabolic reflector (Hua et al., 2015). Table 13.5 gives the details on the antenna design variables, its computational model, and the design objective ranges. The feed line width W_1 is computed for a given substrate permittivity to ensure 50-ohm input impedance. The computational model is implemented in CST Microwave Studio. In this case, the permittivity is a part of the objective space F. The surrogate model is to represent the antenna reflection and realized gain characteristics over the objective space F. Note that

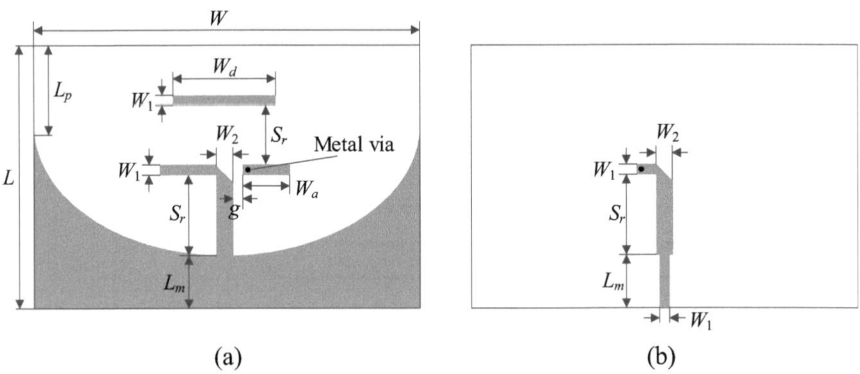

Fig. 13.18 Quasi-Yagi antenna geometry (Hua et al., 2015): (**a**) top layer, (**b**) bottom layer

Table 13.5 Quasi-Yagi antenna of Fig. 13.18: Relevant details

Parameters	
Substrate thickness	$h = 1.5$ mm
Designable parameters	$\boldsymbol{x} = [W\ L\ L_m\ L_p\ S_d\ S_r\ W_2\ W_a\ W_d\ g]^T$
Lower bounds	$\boldsymbol{l} = [22.0\ 3.5\ 0.3\ 6.5\ 3.0\ 0.5\ 3.5\ 0.2]^T$
Upper bounds	$\boldsymbol{u} = [27.0\ 8.0\ 2.3\ 16.0\ 7.0\ 5.5\ 6.0\ 2.3]^T$
Computational model	~300,000 cells
Simulation time	90 s
Design objective ranges	
Substrate permittivity	$2.0 \leq \varepsilon_r \leq 5.0$
Operating frequency	2.5 GHz $\leq f_0 \leq$ 6.5 GHz

the dimensionality of the design space X is 10, which makes this case challenging from the point of view of surrogate modeling.

The discussed modeling procedure has been validated with the number of random (accepted) observables (cf. Sect. 13.3.2) set to $N_r = 50$. These were obtained using 192 antenna simulations. The surrogates were constructed for $N_B = 50$, 100, 200, 400, and 800 training samples so that the scalability of the modeling error can be investigated as well. Furthermore, these numbers are considered typical for modeling of high-frequency components in engineering practice. The training data were allocated using modified Latin Hypercube Sampling (LHS) (Beachkofski & Grandhi, 2002).

Table 13.6 shows the numerical results for the two-stage surrogate and the benchmark techniques. Visualization of the inverse regression model for the selected antenna parameters can be found in Fig. 13.19, whereas Fig. 13.20 shows a comparison of EM-simulated and surrogate-predicted responses for selected test points. The results indicate that the predictive power of the two-stage surrogate is considerably better than for the conventional models, and it is comparable with the nested kriging for training data set of 200 samples and larger, but it is superior for $N_B = 50$ and 100. The computational savings with respect to nested kriging are substantial due to high cost of reference design acquisition for the latter (1899 EM simulations). For example, the cost reduction of the surrogate model setup is 88 percent for $N_B = 50$ and 63% for $N_B = 800$.

The design utility of the model has been demonstrated by optimizing the antenna for the target operating frequency f_0 assuming a given substrate permittivity ε_r (both within the assumed objective space). The optimization process aimed at ensuring at least 8 percent fractional bandwidth (symmetric with respect to f_0) and maximizing the average realized gain within the same bandwidth. The results are shown in Table 13.7 and Fig. 13.21. It can be noticed that the operating bandwidth of the antenna is well controlled and the inverse regression model yields good initial designs for further tuning. Also, the agreement between the EM-simulated antenna responses and the predictions of the two-stage surrogate is excellent.

13.3.5 Verification Case Study II: Compact Branch-Line Coupler

The considered modeling approach (Koziel et al., 2021) is demonstrated here using a microstrip coupler. The modeling accuracy is compared to conventional surrogates (kriging, radial basis functions, and artificial neural networks), as well as the nested kriging framework. Furthermore, application examples are considered in the form of circuit optimization for selected target values of the figures of interest. Experimental validation is also provided.

The compact branch-line coupler (BLC) (Tseng & Chang, 2012) is shown in Fig. 13.22. The circuit is implemented on a 0.76-mm-thick substrate of the

Table 13.6 Quasi-Yagi antenna: Modeling results and benchmarking

Number of training samples	Modeling technique							
	Kriging		RBF		Nested kriging		Two-stage surrogate	
	Modeling error	Model setup cost	Modeling error	Model setup cost	Modeling error	Model setup cost[a]	Modeling error	Model setup cost[b]
50	61.4%	50	65.3%	50	17.9%	1949	10.8%	242
100	50.7%	100	51.8%	100	13.3%	1999	8.4%	292
200	39.8%	200	43.2%	200	7.5%	2099	7.1%	392
400	32.8%	400	37.1%	400	5.4%	2299	5.9%	592
800	31.8%	800	33.6%	800	4.5%	2699	5.0%	992

[a]The cost includes acquisition of the reference designs, which is 1899 EM simulations of the antenna
[b]The cost includes generation of random observables; here, 192 simulations in total to yield $N_r = 50$ accepted samples

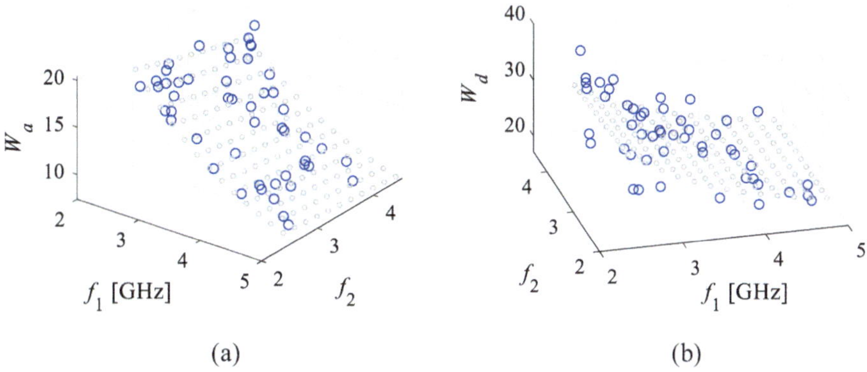

Fig. 13.19 Quasi-Yagi antenna: inverse regression model for selected antenna parameters: (**a**) W_a, and (**b**) W_d. Inverse model surfaces and random observables shown using gray and blue circles, respectively

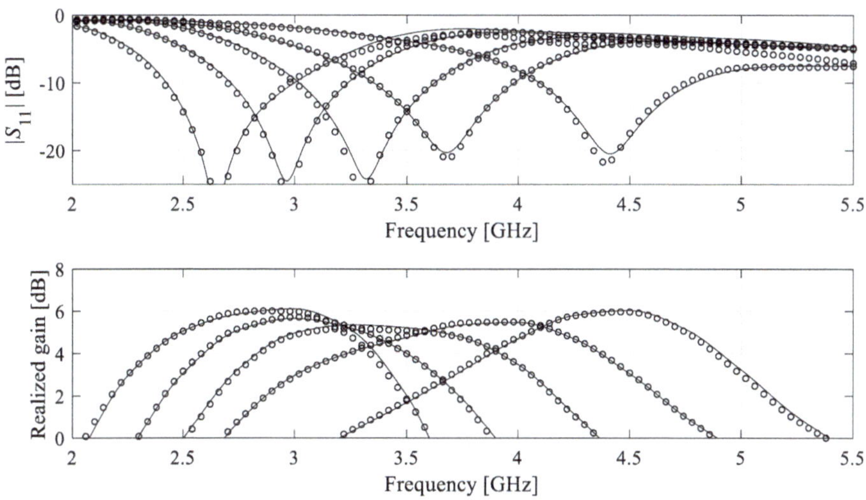

Fig. 13.20 Quasi-Yagi antenna: reflection (top) and realized gain (bottom) characteristics at the selected test designs: EM model (—) and the two-stage surrogate (o). The surrogate setup using $N_B = 400$ training samples

Table 13.7 Quasi-Yagi antenna: Optimization results

Target operating conditions		Geometry parameter values [mm]									
f_0 [GHz]	ε_r	W	L	L_m	L_p	S_d	S_r	W_2	W_a	W_d	g
3.6	3.0	118.4	71.8	22.4	22.3	13.5	13.7	3.96	13.3	23.7	0.69
4.1	3.0	118.5	71.8	22.9	22.3	14.1	13.3	3.72	11.2	21.8	0.68
2.7	4.4	114.3	66.3	19.4	21.4	13.8	16.6	3.96	16.6	28.6	0.83
4.2	4.4	117.2	65.0	24.0	22.4	15.8	12.4	3.86	8.22	18.8	0.69

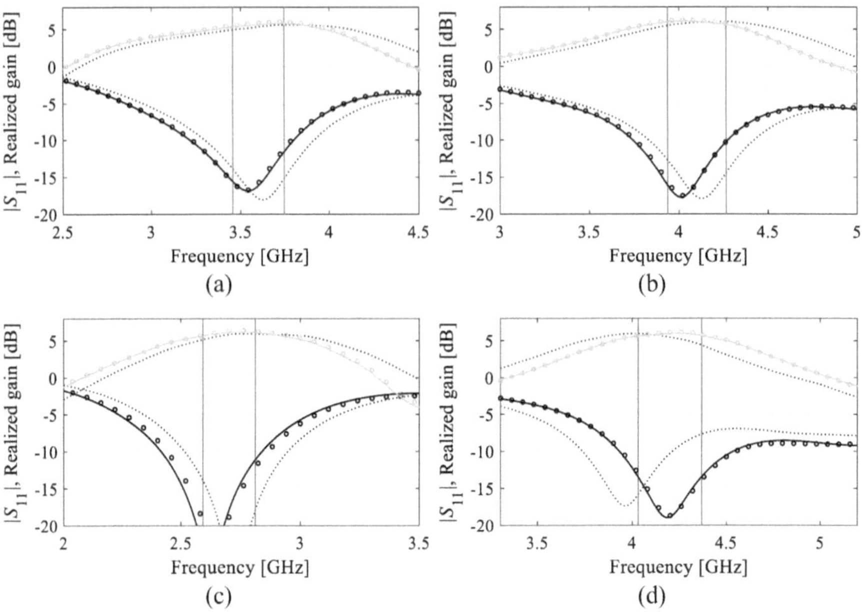

Fig. 13.21 Quasi-Yagi antenna: EM-simulated reflection responses at the initial design (····) obtained using the inverse regression model, as well as surrogate (o) and EM-simulated response (—) at the design obtained by optimizing the two-stage surrogate set up using $N_B = 800$ training samples. The designs obtained for the following objective vectors: (**a**) $f_0 = 3.6$ GHz, $\varepsilon_r = 3.0$, (**b**) $f_0 = 4.1$ GHz, $\varepsilon_r = 3.0$, (**c**) $f_0 = 2.7$ GHz, $\varepsilon_r = 4.4$, (**d**) $f_0 = 4.2$ GHz, $\varepsilon_r = 4.4$. Vertical lines denote the target operating band of the antenna. Reflection and realized gain characteristics shown using black and gray lines, respectively

Fig. 13.22 Compact branch-line coupler: Geometry (Tseng & Chang, 2012)

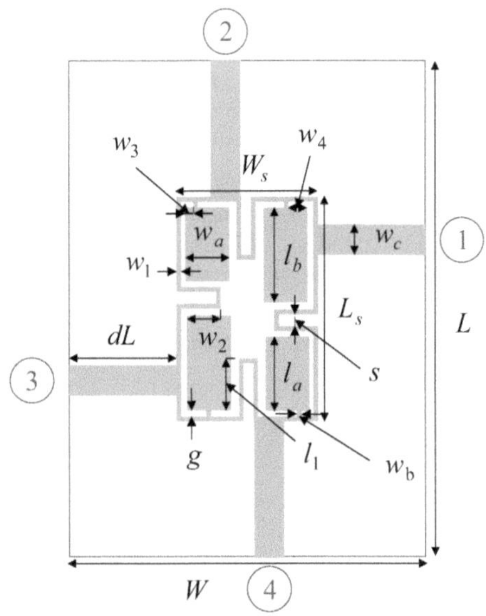

permittivity ε_r, being one of the components of the objective space. The independent parameters are $\boldsymbol{x} = [g\ l_{1r}\ l_a\ l_b\ w_1\ w_{2r}\ w_{3r}\ w_{4r}\ w_a\ w_b]^T$. Other variables are described by the following relations: $L = 2dL + L_s$, $L_s = 4w_1 + 4\,g + s + l_a + l_b$, $W = 2dL + W_s$, $W_s = 4w_1 + 4\,g + s + 2w_a$, $l_1 = l_b l_{1r}$, $w_2 = w_a w_{2r}$, $w_3 = w_{3r} w_a$, and $w_4 = w_{4r} w_a$. The EM model is implemented and simulated in CST Microwave Studio.

The objective space is defined by the operating frequency f_0 of the coupler within the range from 1.0 GHz to 2.0 GHz and the relative permittivity ε_r of the substrate the circuit is to be implemented on; the range of interest is $2.0 \le \varepsilon_r \le 5.0$. The conventional parameter space X is defined by the lower bounds $\boldsymbol{l} = [0.4\ 0.43\ 5.9\ 7.7\ 0.68\ 0.28\ 0.1\ 0.1\ 2.0\ 0.2]^T$ and the upper bounds $\boldsymbol{u} = [1.0\ 0.86\ 14.0\ 16.5\ 1.5\ 0.99\ 0.65\ 0.25\ 5.5\ 0.8]^T$. The surrogate model of the BLC S-parameters ($|S_{11}|$, $|S_{21}|$, $|S_{31}|$, and $|S_{41}|$) is to be valid over the entire objective space. The design optimality is understood as minimization of the matching and isolation characteristics at f_0, as well as maintaining the required power split ratio K_p also at f_0 (here, equal, i.e., $|S_{21}(\boldsymbol{x}, f)| = |S_{31}(\boldsymbol{x},f)|$). The objective function is defined as

$$\begin{aligned} U(\boldsymbol{x},\boldsymbol{f}) = U\left(\boldsymbol{x}, [f_0\ \ K_P]^T\right) &= \max\{|S_{11}(\boldsymbol{x},f_0)|, |S_{41}(\boldsymbol{x},f_0)|\} \\ &+\beta[K_P - [|S_{21}(\boldsymbol{x},f_0)| - |S_{31}(\boldsymbol{x},f_0)|]]^2 \end{aligned} \tag{13.36}$$

where $S_{k1}(\boldsymbol{x},f)$ is the respective S-parameter at the design $\boldsymbol{x}$ and frequency f. The second term in (13.36) is a penalty function enforcing the required power split.

The modeling framework is validated as follows: $N_r = 100$ is used as the number of random observables, acquisition of which required 226 EM simulations of the BLC. The surrogates were constructed assuming $N_B = 50$, 100, 200, 400, and 800 samples. In this case, due to the challenging character of the problem, the extension parameters T_k were set to half of the values obtained from (13.35).

The following benchmark methods were used: kriging, RBF, ANN (all in conventional space X), and nested kriging (Koziel & Pietrenko-Dabrowska, 2019), with the thickness parameter of $T = 0.025$ (established in the domain X_S). The nested kriging model utilizes nine reference designs; the cost of their acquisition is 1014 EM simulations.

Table 13.8 shows the numerical results for all considered modeling techniques. The inverse model plots for selected coupler parameter can be found in Fig. 13.23. Note that for this example, the observables are scattered and do not follow the trends determined by the inverse surrogate, which indicates a certain parameter redundancy, typical for compact components realized using the slow-wave phenomenon. Figure 13.24 shows a comparison of the coupler scattering parameters predicted by the considered surrogate and EM simulation.

The two-stage inverse/forward surrogate outperforms all considered conventional surrogates. Its performance is similar to that of nested kriging, yet better for smaller training data sets, which can be explained as follows: The incorporation of the observable data yields additional benefits. Finally, the computational efficiency of the discussed approach is significantly improved over the nested kriging framework due to avoiding the usage of the reference designs. The extra cost associated with

Table 13.8 Compact BLC: Modeling results and benchmarking

Number of training samples	Modeling technique									
	Kriging		RBF		ANN[a]		Nested kriging		Two-stage surrogate	
	Error [%]	Setup cost	Error [%]	Setup cost	Error [%]	Setup cost	Error [%]	Setup cost[c]	Error [%]	Setup cost[b]
50	52.3	50	51.8	50	29.9	50	10.0	1064	7.6	276
100	38.3	100	40.5	100	22.2	100	7.4	1114	6.2	326
200	31.0	200	37.4	200	15.2	200	6.8	1214	4.7	426
400	27.3	400	32.8	400	10.5	400	5.1	1414	4.5	626
800	23.3	800	27.2	800	9.8	800	4.9	1814	3.4	1026

[a]Feedforward network with two hidden layers, trained using Levenberg-Marquardt algorithm
[b]The cost includes generation of random observables; here, 226 simulations in total to yield $N_r = 50$ accepted samples
[c]The cost includes reference design acquisition, i.e., 1014 EM simulations of the BLC

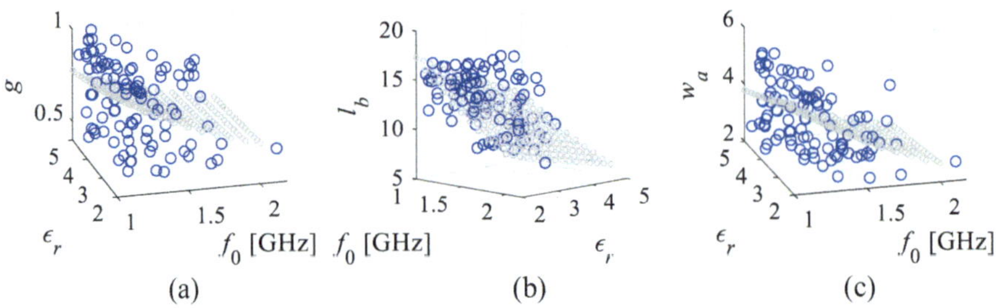

Fig. 13.23 Compact BLC of Fig. 13.22: Inverse regression model for selected parameters: (**a**) g, (**b**) l_b, (**c**) w_a. Inverse model surfaces and random observables shown using gray and blue circles, respectively

random observable generation is only 226 EM simulations, whereas similar cost for nested kriging (reference design acquisition) is as high as the already mentioned 1014 simulations. This translates into computational savings: 74% for $N_B = 50$, 70% for $N_B = 100$, and over 43% for $N_B = 800$.

The relevance of the considered modeling framework for design purposes has been demonstrated through several application case studies. The BLC has been optimized for several target operating frequencies and substrate permittivity, as shown in Table 13.9 and Fig. 13.25. In all considered cases, the initial design obtained from the inverse model is of good quality, and the optimization process further improves the circuit responses. At the same time, the agreement between the surrogate predictions and EM-simulated data is satisfactory.

The designs considered in Fig. 13.25 were fabricated and experimentally validated. Figure 13.26 shows the comparison between the simulated and measured S-parameters. The agreement between the two data sets is good; the minor discrepancies result from the effects of connectors not included in the computational model of the coupler.

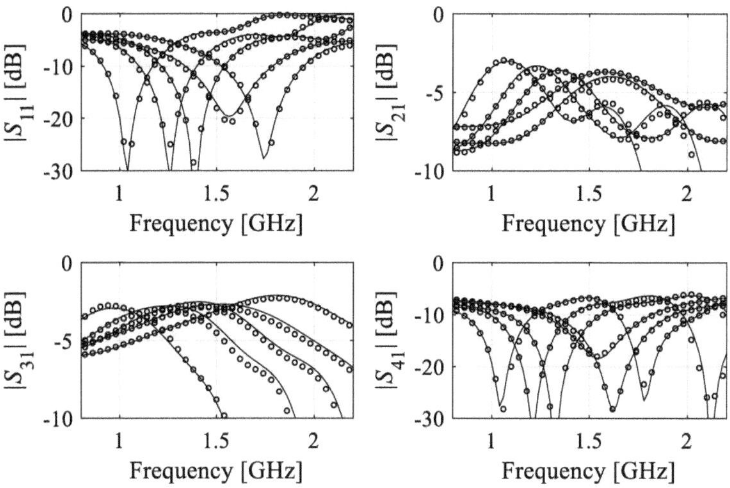

Fig. 13.24 Compact BLC of Fig. 13.22: Scattering parameters at the selected test designs: EM model (—), and the two-stage surrogate (o). The surrogate set up using $N_B = 400$ training samples

Table 13.9 Compact BLC: Optimization results

Target operating conditions		Geometry parameter values									
f_0 [GHz]	ε_r	g	l_{1r}	l_a	l_b	w_1	w_{2r}	w_{3r}	w_{4r}	w_a	w_b
1.0	3.0	0.85	0.60	9.05	15.2	1.06	0.71	0.40	0.20	4.76	0.54
1.2	3.0	0.77	0.60	8.84	13.4	1.17	0.56	0.39	0.18	3.90	0.54
1.5	2.5	0.69	0.61	9.03	11.7	1.20	0.65	0.41	0.18	3.54	0.53
1.7	2.5	0.64	0.61	8.94	10.9	1.21	0.61	0.41	0.19	3.18	0.55

This section outlines a procedure for low-cost surrogate modeling of high-frequency structures. The considered two-stage observable-based methodology capitalizes on the performance-driven modeling concept. The confined domain of the final surrogate is defined using an auxiliary inverse regression model whose construction does not require any preoptimized reference designs. Instead, the inverse model is rendered based on a set of random observables generated within the assumed parameter space. This has a profound effect on the computational cost of the surrogate model setup, which is reduced by up to 70%, as compared to the nested kriging framework. As demonstrated using antenna and microwave examples, the considered technique ensures superior predictive power of the surrogate while using limited numbers of training data samples. The accuracy-related advantages are particularly pronounced for smaller data sets as the random observables are also taken into account for model identification. The employed definition of the confined domain also provides means for straightforward uniform sampling as well as model optimization. Finally, the definition of the model domain permits a convenient determination of its lateral size. Design utility of the two-stage approach has been

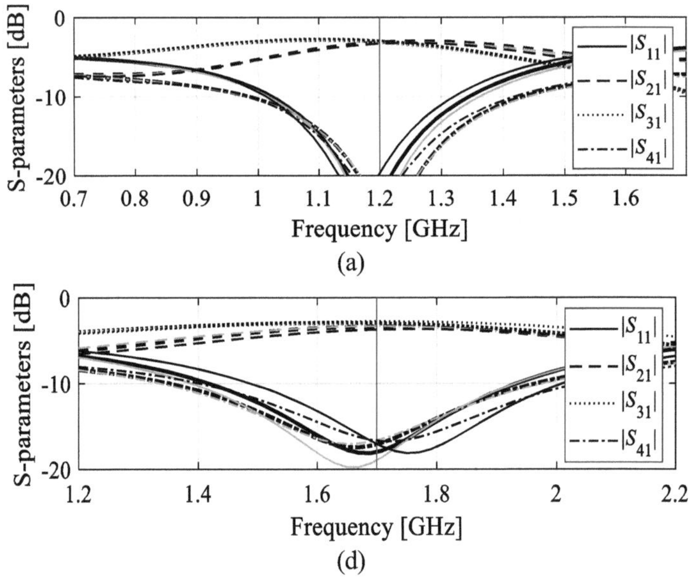

Fig. 13.25 Compact BLC: EM-simulated S-parameters at the initial design (thin black lines) obtained using the inverse regression model, as well as surrogate (gray lines) and EM-simulated response (thick black lines) at the design obtained by optimizing the two-stage surrogate setup using $N_B = 800$ training samples. The vertical lines denote the target operating frequencies: (**a**) $f_0 = 1.2$ GHz, $\varepsilon_r = 3.0$ and (**b**) $f_0 = 1.7$ GHz, $\varepsilon_r = 2.5$

corroborated through application case studies (design optimization) and experimental validation.

13.4 Two-Stage Feature-Based Modeling

The last modeling framework discussed in this chapter is a two-stage observable-based modeling technique with response features (Pietrenko-Dabrowska et al., 2022). The considered approach further improves the efficacy of the two-stage observable-based modeling method of Sect. 13.3 (Koziel & Pietrenko-Dabrowska, 2022) by carrying out the modeling process at the level of response features, similarly as in the case of the feature-based nested kriging technique of Sect. 13.2. The examples of this section pertain to modeling of antenna responses; yet, applying the discussed approach for modeling of characteristics of microwave components is also possible.

In the discussed approach, the surrogate domain is defined in a cost-efficient manner following the methodology outlined in Sect. 13.3 (Koziel & Pietrenko-Dabrowska, 2022), whereas the surrogate is set up at the level of the antenna response features, which allows for smoothing out the functional landscape to be

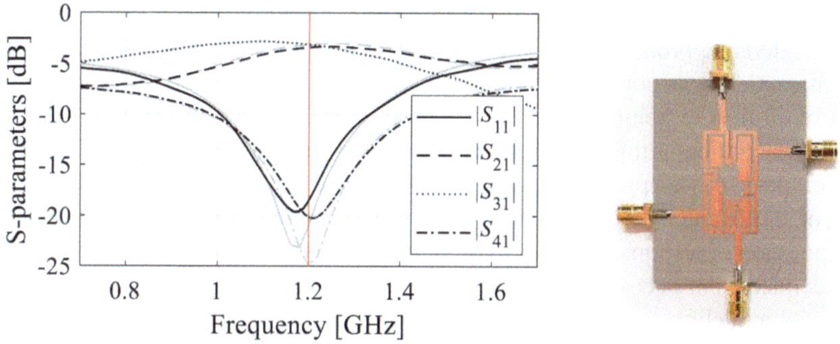

(a)

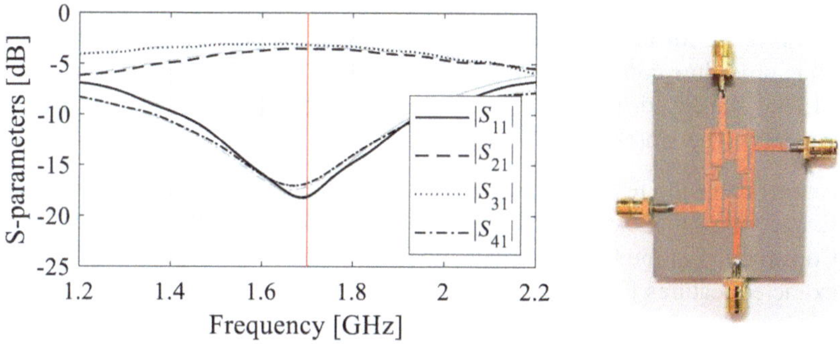

(b)

Fig. 13.26 Experimental validation of the BLC designs considered in Fig. 13.25. The EM simulation and measurement data shown using gray and black lines, respectively: (**a**) design optimized for $f_0 = 1.2$ GHz, $\varepsilon_r = 3.0$; (**b**) design optimized for $f_0 = 1.7$ GHz, $\varepsilon_r = 2.5$

approximated. Joint exploitation of both methodologies allows for constructing metamodels of antenna input characteristics using data sets of truly low cardinalities without compromising the modeling accuracy.

In the considered modeling technique, the second-level metamodel is set up using the training data pairs $\{\boldsymbol{x}_B^{(k)}, \boldsymbol{F_R}(\boldsymbol{x}_B^{(k)})\}_{k=1,\ldots,N_B}$ (along with the observables $\{\boldsymbol{x}_r^{(l)}, \boldsymbol{R}(\boldsymbol{x}_r^{(l)})\}_{l=1,\ldots,N_r}$). Here, $\boldsymbol{F_R}(\boldsymbol{x}) = [f_1(\boldsymbol{x})\, f_2(\boldsymbol{x}) \ldots f_p(\boldsymbol{x})\, \lambda_1(\boldsymbol{x})\, \lambda_2(\boldsymbol{x}) \ldots \lambda_p(\boldsymbol{x})]^T$ denotes the response feature vector corresponding to a given design $\boldsymbol{x}$. The entries of the vector $\boldsymbol{F_R}$ are the frequency f_j and level coordinates λ_j, $j = 1, \ldots, p$, of p antenna resonances. In other words, the response of the second-level surrogate yields predictions only about the feature point coordinates rather than the entire antenna characteristic at a given design $\boldsymbol{x} \in X_S$ (in the same manner as in the feature-based nested kriging of Sect. 13.1). Naturally, focusing only on the response features does lead to some unavoidable loss of information. Still, this loss is irrelevant from the point of view of the design goals: As mentioned earlier, the features are defined so as to allow for quantifying the design objectives unequivocally. In general, some of the feature points may not be distinguishable (e.g., -10 dB reflection points do not exist if the antenna resonance level is above that limit). However, this is not an issue for the considered approach, because the very definition of the domain X_S ensures that the antenna designs contained therein are of high quality, which ensures the existence of all feature points.

The description of this section is very brief to avoid repetition, as both constituent parts of the discussed procedure have been already described in detail in the previous parts of the chapter: two-stage observable-based modeling in Sect. 13.3, whereas incorporation of response features into performance-driven modeling framework has been delineated in Sect. 13.1.2. Figure 13.27 shows the conceptual illustration of the discussed modeling procedure. The user needs to define the parameter space and the objective space (by providing the respective lower and upper bounds) and also decide upon the number of the observables N_r (typically several dozen or so) that are to be used for domain definition purposes. Moreover, the number of the training data samples N_B for the construction of the final surrogate has to be selected. The modeling procedure consists of virtually the same steps as the procedure described in Sect. 13.3.4, except for the last stages, which are performed after Step 5 (design of experiments). These are:

6. Extraction of the features $\boldsymbol{F_R}(\boldsymbol{x}_B{}^{(k)})\}_{k = 1, \ldots, NB}$ from the training designs $\boldsymbol{x}_B{}^{(k)}$
7. Extraction of the features $\boldsymbol{F_R}(\boldsymbol{x}_r{}^{(k)})\}_{k = 1, \ldots, Nr}$ from the observables $\boldsymbol{x}_r{}^{(k)}$
8. Construction of the final surrogate model $\boldsymbol{s}_f$, with the training data set being the extracted features from the training samples and the observable set

13.4.1 Verification Case Study I: Ring-Slot Antenna

The discussed modeling framework has been demonstrated using a ring-slot antenna (Sim et al., 2014) presented in Fig. 13.28. Table 13.10 provides the details concerning the design variables and objectives, as well as simulation model of the antenna. In the considered example, the substrate relative permittivity ε_r is one of the performance figures; therefore, its ranges are provided under the design objective ranges of Table 13.10. The computational model is evaluated in CST Microwave Studio and simulated using its time-domain solver. The design objective is to minimize the antenna reflection at its operating frequency f_0.

The surrogate models have been constructed within the region of validity given in Table 13.10, using the following sizes of the training data sets: 20, 50, 100, 200, 400, and 800 samples. The benchmark techniques include: (i) conventional kriging in an unconstrained domain, (ii) basic nested kriging technique (Koziel & Pietrenko-Dabrowska, 2019), two-stage observable-based modeling technique (Koziel & Pietrenko-Dabrowska, 2022), and feature-based nested kriging technique (Koziel & Pietrenko-Dabrowska, 2020c). Table 13.11 provides the modeling results: the computational costs of model setup and its accuracy. Observe that in the case of the feature-based performance-driven techniques (i.e., feature-based nested kriging and

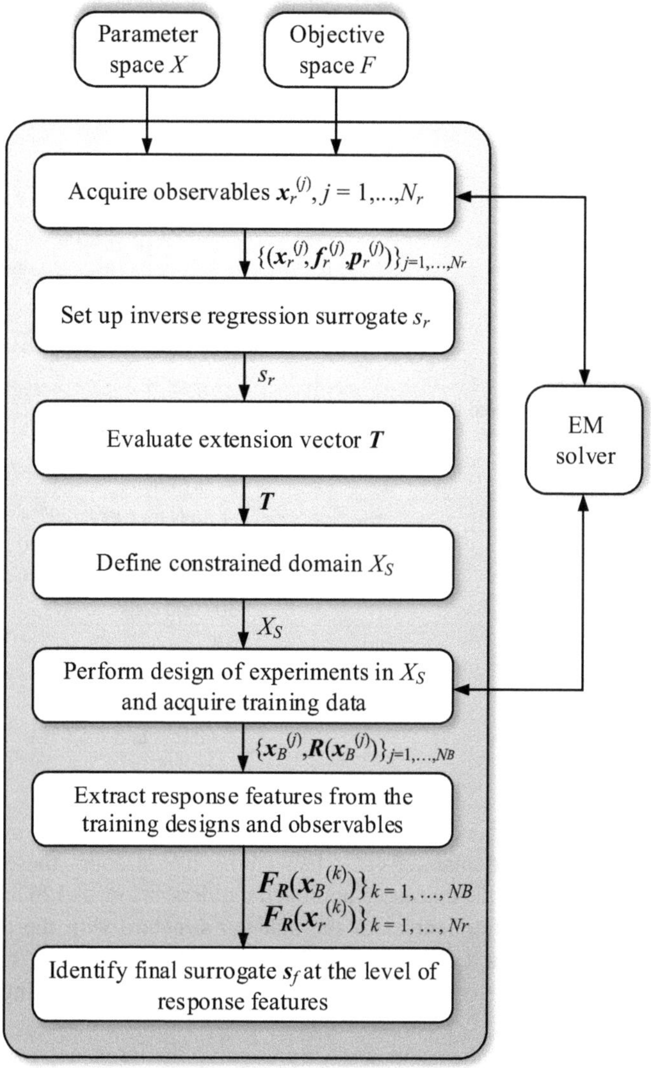

Fig. 13.27 Flow diagram of the feature-based observable-based constrained modeling technique

the discussed approach), the modeling accuracies of the frequency and level coordinates of the response features are provided.

The results of Table 13.11 demonstrate that carrying out the modeling process at the level of the response features allows for achieving superior accuracy of representing the characteristic points relevant to the assumed design objectives at a remarkably small computational cost. The considered feature-based surrogate allows for achieving accuracy of less than 1% for training data sizes containing merely 50 samples from the constrained domain. Observe that the total cost of training data

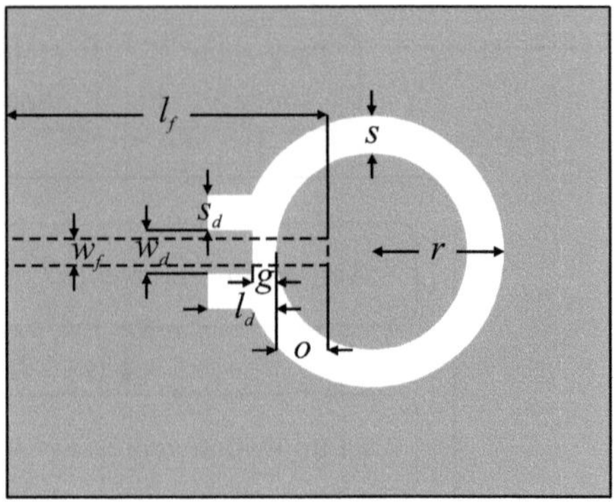

Fig. 13.28 Geometry of a ring slot antenna (Sim et al., 2014) with microstrip feed marked using a dashed line

Table 13.10 Ring-slot antenna: Details

Parameters	
Substrate thickness	$h = 0.76$ mm
Designable parameters	$\boldsymbol{x} = [l_f\ l_d\ w_d\ r\ s\ s_d\ o\ g]^T$
Lower bounds	$\boldsymbol{l} = [22.0\ 3.5\ 0.3\ 6.5\ 3.0\ 0.5\ 3.5\ 0.2]^T$
Upper bounds	$\boldsymbol{u} = [27.0\ 8.0\ 2.3\ 16.0\ 7.0\ 5.5\ 6.0\ 2.3]^T$
Computational model	~300,000 cells
Simulation time	90 s
Design objective ranges	
Substrate permittivity	$2.0 \leq \varepsilon_r \leq 5.0$
Operating frequency	2.5 GHz $\leq f_0 \leq$ 6.5 GHz

acquisition includes also the cost of generating 106 observables, which are necessary to define the surrogate domain. Even when taking into account these additional computational expenses, the discussed approach requires as low as 126 data samples to assess the operating frequencies of the antenna structure with the accuracy of around 1%, which makes it a very cost-efficient modeling technique. The conventional surrogate does not ensure satisfactory accuracy even for the training data set of 800 samples.

It should be noted that the accuracy of representing the frequency coordinates of the feature points is exceptionally good even for small data sets, whereas it is not as good for the level coordinates. On the one hand, this is of little practical significance because, for antenna design procedures, it is the frequency allocation that is of primary importance; reflection level, as long as it is below −10 dB is of secondary relevance. On the other hand, the reason for degraded level rendition are low values of reflection coefficients at the antenna resonances (typically, −20 dB or less), which implies a considerable amount of numerical noise caused by the EM simulation process itself (related to adaptive meshing techniques as well as terminating the time-domain simulation at relatively high levels of residual energy).

For better visualization, Fig. 13.29 shows the scatter plots of the relevant feature points. Observe that correlation between the surrogate-model-predicted and

Table 13.11 Ring-slot antenna: Modeling results and benchmarking

Number of training samples	Modeling technique: Kriging Error [%]	Kriging Setup cost	Nested kriging Error [%]	Nested kriging Setup cost	Observable-based modeling Error [%]	Observable-based modeling Setup cost	Feature-based nested kriging Error of f [%]	Feature-based nested kriging Error of l [%]	Feature-based nested kriging Setup cost[a]	Observable-based modeling with features Error of f [%]	Observable-based modeling with features Error of l [%]	Observable-based modeling with features Setup cost[b]
20	93.8	20	59.5	884	45.1	126	3.91	32.8	884	0.71	12.8	126
50	56.9	50	19.4	914	13.4	156	3.29	27.8	914	0.43	8.5	156
100	50.8	100	12.9	964	9.9	206	0.38	20.6	964	0.26	7.7	206
200	35.8	200	7.7	1064	6.9	306	0.32	22.8	1064	0.18	6.6	306
400	31.5	400	5.1	1264	5.4	506	0.19	13.4	1264	0.17	6.8	506
800	25.6	800	3.7	1664	4.4	906	0.23	11.9	1664	0.15	6.2	906

[a]The cost includes reference design acquisition, i.e., 864 EM simulations of the antenna

[b]The cost includes generation of random observables; here, 106 simulations in total to yield $N_r = 50$ accepted samples

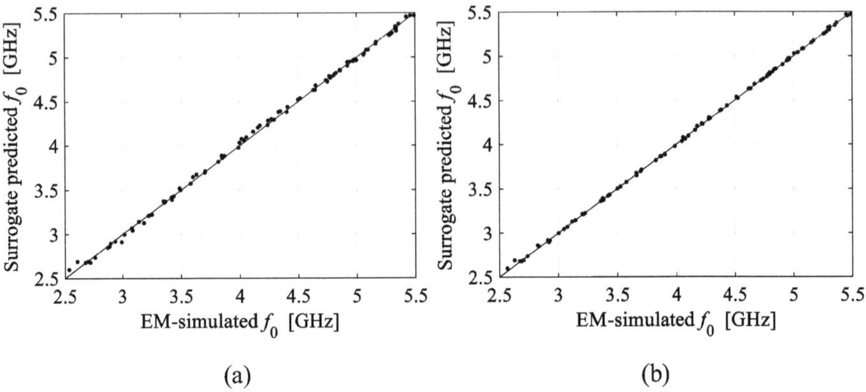

Fig. 13.29 Ring-slot antenna of Fig. 13.28: Scatter plots of the center frequency f_0 [GHz] yielded by the feature-based surrogate against their EM-simulated counterparts; surrogates constructed using (**a**) $N_B = 20$ and (**b**) $N_B = 100$ samples

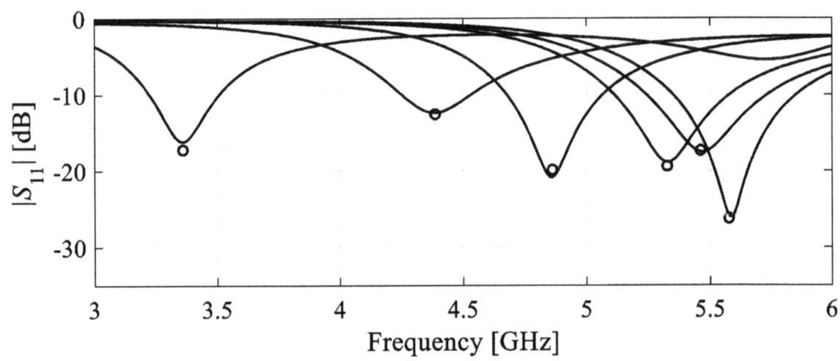

Fig. 13.30 Ring-slot antenna of Fig. 13.28: EM-simulated antenna responses at the selected test designs (—), along with the surrogate-predicted relevant antenna feature points (o); the surrogates set up using $N = 200$ training samples

EM-simulated results is good, even in the case of the model set up using only 20 training data samples. For one hundred data samples, the said correlation is excellent. In addition, Fig. 13.30 provides EM-simulated antenna reflection characteristics at the selected test locations, along with the characteristic points yielded by the surrogate. The accuracy of predicting the frequencies that are relevant from the point of view of the assumed design objectives is remarkable.

13.4.2 Verification Case Study II: Quasi-Yagi Antenna

For the second example, let us consider a quasi-Yagi antenna shown in Fig. 13.31 (Hua et al., 2015), which is implemented on RO4350 substrate. The details on the antenna designable and fixed parameters, the antenna computational model, and the

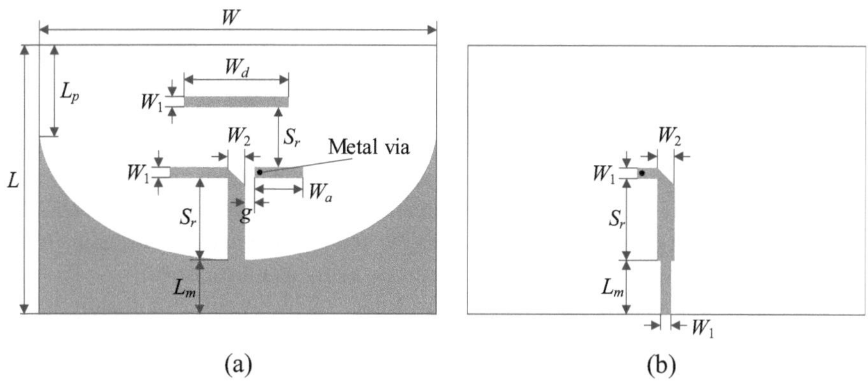

Fig. 13.31 Quasi-Yagi antenna geometry (Hua et al., 2015): (**a**) Top layer, (**b**) bottom layer

Table 13.12 Quasi-Yagi antenna of Fig. 13.31: Relevant details

Parameters	
Substrate thickness	$h = 1.5$ mm
Designable parameters	$\boldsymbol{x} = [W\ L\ L_m\ L_p\ S_d\ S_r\ W_2\ W_a\ W_d\ g]^T$
Lower bounds	$\boldsymbol{l} = [22.0\ 3.5\ 0.3\ 6.5\ 3.0\ 0.5\ 3.5\ 0.2]^T$
Upper bounds	$\boldsymbol{u} = [27.0\ 8.0\ 2.3\ 16.0\ 7.0\ 5.5\ 6.0\ 2.3]^T$
Computational model	~600,000 cells
Simulation time	240 s
Design objective ranges	
Substrate permittivity	$2.0 \le \varepsilon_r \le 5.0$
Operating frequency	2.5 GHz $\le f_0 \le$ 6.5 GHz

design objectives are given in Table 13.12. The antenna has been already utilized as a verification example in Sect. 13.3.4; yet, for convenience of the reader, here, also, the most important details on antenna structure are reiterated. As before, the permittivity is a part of the objective space F. In this case, dimensionality of the design space is ten.

The considered two-stage feature-based modeling framework has been validated in a similar way as for the first case study. The assumed number of random observables is $N_r = 50$, here, obtained using 192 EM simulations of the antenna structure. Similarly, as in the previous case, the surrogates were constructed for $N_B = 50$, 100, 200, 400, and 800 training samples. The benchmark methods are the same as in Sect. 13.4.1: kriging interpolation and RBF, nested kriging surrogate, two-stage observable based modeling, and feature-based nested kriging. The nested kriging model utilizes eight reference designs; the cost of their acquisition is 1899 EM simulations of the antenna using feature-based optimization. This cost is added to the overall expenses pertinent to the surrogate model setup, and it is much higher than for the previous example due to problem dimensionality and more involved formulation of the optimization task.

For this antenna structure, the design goal is to minimize the antenna reflection and enhance gain within 8% fractional bandwidth around f_0. To handle this task, the following set of characteristic points has been appointed: the lower and upper frequencies for which the reflection response assumes −10 dB levels, as well as the corresponding values of the realized gain characteristic, supplemented by five additional points equally distributed in frequency in between these points. The supplementary points are required to adequately reconstruct the gain characteristics within the operating band in order to assess the antenna average in-band gain. Thus, in this case, the response feature set contains as many as nine entries.

The numerical results are provided in Table 13.13. Again, for feature-based modeling techniques, the accuracies of the frequency and level coordinates of the response features are given. As in Sect. 13.4.1, the results of Table 13.13 show that performing the modeling process using response features ensures excellent accuracy of representing the characteristic points of interest at a truly low cost. The cost of generating 192 observables, necessary to define the model domain, is included in the overall cost of training data acquisition. Yet, it requires only 212 training samples to assess the antenna operating frequency with the error below 1%. This accuracy is beyond capability of the conventional surrogate even when 800 samples are used. As in the previous example, the accuracy of representing the level coordinates of the characteristic points is not as good as those of the frequency coordinates.

Figure 13.32 presents the scatter plots of the antenna feature points corresponding to lower and upper frequencies corresponding to −10 dB reflection levels. The correlation between the modelled and the EM-evaluated results is good even for 20 training data samples. Figure 13.33 shows EM-simulated antenna reflection and gain characteristics at the selected test locations with surrogate-predicted features. The accuracy of predicting of all the characteristic points is superb.

This section outlined a technique for low-cost feature-based surrogate modeling of antenna input characteristics, with the surrogate constructed in the constrained domain determined cost-efficiently using a set of random observables. The considered approach enhances the original observable-based modeling framework by incorporating the response features technology, thereby allowing for further reduction of the training data acquisition cost. At the same time, it improves the surrogate model predictive power. The discussed modeling procedure has been favorably compared to several benchmark techniques: conventional data-driven model and performance-driven methods operating on the complete antenna responses. The obtained results demonstrate that combining two algorithmic approaches: (i) the observable-based model domain definition, and (ii) reformulating the modeling task in terms of characteristic points of antenna responses, enables notable computational savings without compromising surrogate model accuracy. Owing to its generic formulation, it can also find applications in various engineering fields that rely on costly simulation models, including microwave design.

Table 13.13 Quasi-Yagi antenna: Modeling results and benchmarking

Number of training samples	Modeling technique												
	Kriging		Nested kriging		Observable-based modeling		Feature-based nested kriging			Observable-based modeling with features			
	Error [%]	Setup cost	Error [%]	Setup cost	Error [%]	Setup cost	Error of f [%]	Error of l [%]	Setup cost[a]	Error of f [%]	Error of l [%]	Setup cost[b]	
20	69.3	20	39.4	1919	17.0	212	3.3	6.33	1919	1.28	5.2	212	
50	61.4	50	17.9	1949	10.8	242	2.8	6.93	1949	0.89	3.6	242	
100	50.7	100	13.3	1999	8.4	292	2.8	6.55	1999	0.74	3.8	292	
200	39.8	200	7.5	2099	7.1	392	2.3	6.10	2099	0.40	2.4	392	
400	32.8	400	5.4	2299	5.9	592	2.4	6.53	2299	0.40	2.2	592	
800	31.8	800	4.5	2699	5.0	992	2.3	5.96	2699	0.34	1.8	992	

[a]The cost includes reference design acquisition, i.e., 1899 EM simulations of the antenna
[b]The cost includes generation of random observables; here, 192 simulations in total to yield $N_r = 50$ accepted samples

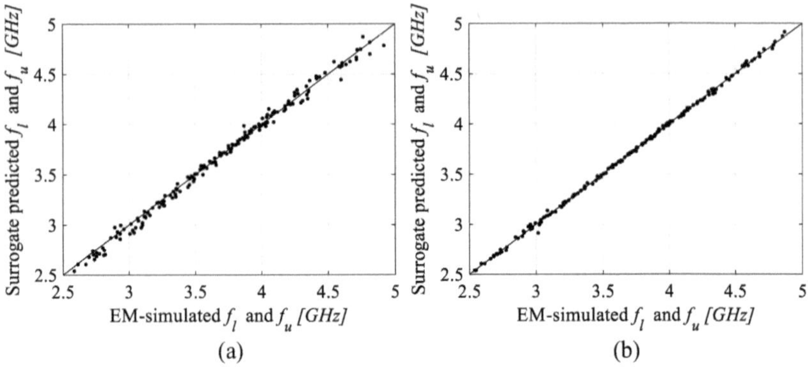

Fig. 13.32 Quasi-Yagi antenna of Fig. 13.31: Scatter plots of the lower and upper frequencies corresponding to −10 dB reflection levels [GHz] yielded by the surrogate against their EM-simulated counterparts; surrogates constructed using (**a**) $N_B = 20$ and (**b**) $N_B = 100$ samples

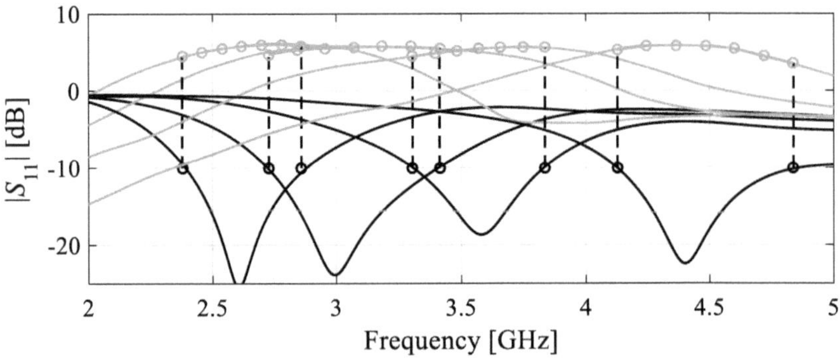

Fig. 13.33 Quasi-Yagi antenna of Fig. 13.31: EM-simulated antenna responses at the selected test designs (—), along with the surrogate-predicted relevant antenna feature points (o) (lower and upper frequencies corresponding to −10 dB reflection levels, as well as corresponding values of the realized gain characteristic with five infill points are marked with circles); the surrogates set up using $N = 200$ training samples

13.5 Summary

This chapter addressed performance-driven modeling of high-frequency structures enhanced with the use of the response features. In constrained modeling, the domain of the constrained surrogate is restricted to cover the most important parts of the entire design space. This allows for a significant reduction of training data acquisition cost in comparison to conventional modeling procedures. Two initial sections of this chapter outline two nested kriging modeling techniques, in which the modeling process is carried out at the level of response features: a basic version and its variation with dimensionality reduction. In the nested kriging approach, the surrogate domain definition involves preoptimized reference designs. Needless to say, the process of reference design acquisition is neither simple nor cheap. The remaining part of the chapter has been devoted to the more recent developments of the nested kriging technique not requiring reference designs, that is, two-stage observable-based modeling. Therein, the surrogate model domain is identified in a more cost-efficient manner using random observables, the quality of which is assessed based on their respective response features. Excellent efficacy of two-stage observable-based modeling is a result of leveraging the properties of response features. Yet, even a more efficient technique has been developed, where also the modeling process itself is conducted at the level of the response features. Exposition of that approach has been the focus of the last section of this chapter.

References

Bau, D., III, & Trefethen, L. N. (1997). *Numerical linear algebra*. Society for Industrial and Applied Mathematics.

Beachkofski, B., & Grandhi, R. (2002). Improved distributed hypercube sampling. In *American Institute of Aeronautics and Astronautics* (pp. 2002–1274). Paper AIAA.

Chen, Y. C., Chen, S. Y., Hsu, P. (2006) Dual-band slot dipole antenna fed by a coplanar waveguide. In *IEEE Antennas and Propagation Society International Symposium*, Albuquerque, NM, USA, 2006, pp. 3589–3592.

Hua, Z., Haichuan, G., Hongmei, L., Beijia, L., Guanjun, L., & Qun, W. (2015). A novel high-gain quasi-Yagi antenna with a parabolic reflector. In *International Symposium on Antennas and Propagation (ISAP)*, Hobart, Australia, 2015.

Jiang, P., Zhou, Q., & Shao, X. (2020). Verification methods for surrogate models. In *Surrogate model-based engineering design and optimization* (Springer tracts in mechanical engineering). Springer.

Jolliffe, I. T. (2002). *Principal component analysis* (2nd ed.). Springer.

Koziel, S. (2017). Low-cost data-driven surrogate modeling of antenna structures by constrained sampling. *IEEE Antennas and Wireless Propagation Letters, 16*, 461–464.

Koziel, S., & Pietrenko-Dabrowska, A. (2019). Performance-based nested surrogate modeling of antenna input characteristics. *IEEE Transactions on Antennas and Propagation, 67*(5), 2904–2912.

Koziel, S., & Pietrenko-Dabrowska, A. (2020a). *Performance-driven surrogate modeling of high-frequency structures*. Springer.

Koziel, S., & Pietrenko-Dabrowska, A. (2020b). Reduced-cost surrogate modelling of compact microwave components by two-level kriging interpolation. *Engineering Optimization, 52*(6), 960–972.

Koziel, S., & Pietrenko-Dabrowska, A. (2020c). Design-oriented computationally-efficient feature-based surrogate modelling of multi-band antennas with nested kriging. *AEU – International Journal of Electronics and Communications, 120*, 153202.

Koziel, S., & Pietrenko-Dabrowska, A. (2020d). On computationally-efficient reference design acquisition for reduced-cost constrained modeling and re-design of compact microwave passives. *IEEE Access, 8*, 203317–203330.

Koziel, S., & Pietrenko-Dabrowska, A. (2022). Knowledge-based performance-driven modeling of antenna structures. *Knowledge-Based Systems, 237*, 107698.

Koziel, S., & Sigurðsson, A. T. (2018). Triangulation-based constrained surrogate modeling of antennas. *IEEE Antennas and Wireless Propagation Letters, 66*(8), 4170–4179.

Koziel, S., Pietrenko-Dabrowska, A., & Al-Hasan, M. (2020). Design-oriented two-stage surrogate modeling of miniaturized microstrip circuits with dimensionality reduction. *IEEE Access, 8*, 121744–121754.

Koziel, S., Pietrenko-Dabrowska, A., & Ullah, U. (2021). Low-cost modeling of microwave components by means of two-stage inverse/forward surrogates and domain confinement. *IEEE Transactions on Microwave Theory and Techniques, 69*(12), 5189–5202.

Pietrenko-Dabrowska, A., & Koziel, S. (2020). Simulation-driven antenna modeling by means of response features and confined domains of reduced dimensionality. *IEEE Access, 8*, 228942–228954.

Pietrenko-Dabrowska, A., Koziel, S., & Al-Hasan, M. (2020). Cost-efficient bi-layer modeling of antenna input characteristics using gradient kriging surrogates. *IEEE Access, 8*, 140831–140839.

Pietrenko-Dabrowska, A., Koziel, S., & Ullah, U. (2022). Reduced-cost two-level surrogate antenna modeling using domain confinement and response features. *Scientific Reports, 12*, 4667.

Sim, C.-Y.-D., Chang, M.-H., & Chen, B.-Y. (2014). Microstrip-fed ring slot antenna design with wideband harmonic suppression. *IEEE Transactions on Antennas and Propagation, 62*, 4828–4832.

Simpson, T. W., Peplinski, J. D., Koch, P. N., & Allen, J. K. (2001). Metamodels for computer-based engineering design: Survey and recommendations. *Engineering Computers, 17*, 129–150.

Tseng, C., & Chang, C. (2012). A rigorous design methodology for compact planar branch-line and rat-race couplers with asymmetrical T-structures. *IEEE Transactions on Microwave Theory and Techniques, 60*(7), 2085–2092.

Chapter 14
Feature-Based Uncertainty Quantification

Fabrication tolerances and other types of uncertainties, e.g., the lack of precise knowledge of material parameters or operating conditions, have detrimental effects on electrical and field performance of high-frequency systems. Mitigating their impact requires accounting for possible parameter deviations already at the design stage. This involves optimization of appropriately defined statistical figures of merit such as yield. Although important, robust (or tolerance-aware) design is an intricate endeavor because manufacturing inaccuracies are normally described using probability distributions, and their quantification has to be based on statistical analysis. The major bottleneck here is High computational cost: for reliability reasons, high-frequency components are evaluated using full-wave electromagnetic (EM) models, whereas conventionally utilized analysis methods (e.g., Monte Carlo simulation) are associated with massive circuit evaluations. A practical approach that allows for circumventing aforementioned obstacles offer surrogate modeling techniques, which has been a dominant trend over the recent years. Notwithstanding, a construction of accurate metamodels may require considerable computational investments, especially for higher-dimensional cases. The response feature technology can be used to alleviate these difficulties by capitalizing on weakly-nonlinear dependence between the characteristic point coordinates and geometry/material parameters of the structure of interest, which has been discussed and explored in the previous chapters.

This chapter demonstrates how the response feature approach can be incorporated into uncertainty quantification procedures, especially for estimation of the fabrication yield and its optimization. We start by recalling formulation of the yield estimation and enhancement task. Subsequent sections describe algorithms for yield optimization, where response features are employed to reformulate the statistical analysis process, thereby accelerating it, as well as discuss more involved approaches where robust design procedures are further expedited by incorporating variable-resolution EM models. Furthermore, we investigate an alternative formulation of the robust design task, which is tolerance optimization along with the feature-based methods of solving at low computational expenses. Finally, the use

A. Pietrenko-Dabrowska, S. Koziel, *Response Feature Technology for High-Frequency Electronics. Optimization, Modeling, and Design Automation*,
https://doi.org/10.1007/978-3-031-43845-5_14

of inverse surrogates is discussed in the context of rapid yield optimization. The presented design techniques are illustrated using a number of high-frequency components such as antennas and microwave structures (couplers, filters, etc.), and compared to non-feature-based methods such as direct EM-driven statistical analysis and surrogate-assisted algorithms.

14.1 Motivation

The development of contemporary high-frequency systems is mostly concerned with the so-called nominal design. Therein, possible uncertainties pertinent to geometry and material parameters as well as operating conditions (e.g., bending, proximity of environmental components) are neglected. In particular, it is assumed that the dimensions of the fabricated prototype coincide with those obtained at the design stage, typically, by means of the optimization-based design closure. Notwithstanding, the presence of uncertainties may degrade the system characteristics by reducing the likelihood of satisfying performance specifications for a particular physical realization of the device (Biernacki et al., 2012). Among the two major types of uncertainties, the most important are the aleatory ones (Prasad & Roy, 2019), e.g., deviations of the design parameters from their nominal values; these include variations of geometry parameters due to fabrication tolerances or changes of permittivity/height between particular sheets of the dielectric substrate. Aleatory uncertainties can be characterized by probability distributions. Needless to say, accurate assessment of the effects of tolerances is instrumental in ensuring design robustness (Li et al., 2014), which requires carrying out statistical analysis (Rossi et al., 2014; Du & Roblin, 2017). Furthermore, enhancing the design robustness entails the improvement of appropriately defined statistical performance metrics, typically, the yield (Hassan et al., 2015). The second class of uncertainties, referred to as epistemic ones (Zeng et al., 2017), are systematic deviations induced by a limited knowledge concerning the operating conditions (e.g., temperature, geometrical distortion such as bending, etc.). Their handling requires the designer to ensure acceptable system performance within the specified ranges of the aforementioned operating parameters.

For reliability reasons, uncertainty quantification of high-frequency systems has to be performed at the level of full-wave electromagnetic (EM) analysis, which makes the process computationally inefficient. In particular, Monte Carlo simulation is a daunting task as its slow convergence entails excessive numbers (hundreds or even thousands) of EM simulations. Simplified approaches such as worst-case analysis (Sengupta et al., 2005) are computationally tractable but do not give accurate account for the effects of the tolerances. It seems that the most promising techniques of mitigating the high-cost issue are based on surrogate modeling methods. The replacement of EM analysis by a fast (usually, data-driven (Koziel et al., 2005)) metamodel allows for rapid execution of all sorts of uncertainty quantification tasks, including statistical analysis, yield estimation, and robust design. Popular modeling methods utilized in this context include response surface approximation (Matoglu et al., 2004), neural networks (Sharma et al., 2020), and polynomial chaos expansion (PCE) (Boeykens et al., 2014; Austin et al., 2013; Cheng et al., 2019; Acikgoz & Mittra, 2018). The latter is particularly attractive because it allows for estimating the statistical moments of the performance figures based on the expansion coefficients (i.e., without running Monte Carlo simulation). The bottleneck of surrogate-assisted techniques is the construction of the metamodel itself, which becomes troublesome for higher-dimensional cases. The mitigation methods include hybridization (e.g., PC kriging (Leifsson et al., 2020)), dimensionality reduction (e.g., by means of principal component analysis (Ochoa & Cangellaris, 2013)), variable-resolution methods (e.g., space mapping (Abdel-Malek et al., 2006); co-kriging (Kennedy & O'Hagan, 2000)), model order reduction (Sommer et al., 2015), and, recently, performance-driven modeling (Pietrenko-Dabrowska, 2021). These developments have been in line with rapid advancements of generic optimization techniques, including nature-inspired methods (Agushaka et al., 2022; Abualigah et al., 2021; Oyelade et al., 2022), as well as in the context of global and multi-objective design (Li et al., 2022; Deng et al., 2022).

While quantifying the effects of uncertainties is important, diminishing them is even more vital. Given the manufacturing process and the assumed ranges of operating conditions, not much can be done about the tolerances themselves; however, it might be possible to improve the design robustness by increasing the likelihood that any particular realization (prototype) of the device will satisfy the performance requirements. The associated design procedures are referred to as tolerance-aware design, yield-driven design, or design centering (Ma et al., 2018; Scotti et al., 2005; Ren et al., 2016). Typically, a merit function that undergoes maximization is the yield (Kouassi et al., 2016), but it is also possible to relocate the design to place it—as much as possible—in the middle of the feasible space (geometrical design centering; Laxminidhi & Pavan, 2007). Unfortunately, yield-driven design is a computationally expensive process because the statistical figures of merit have to be calculated at each stage thereof. In particular, direct EM-based yield improvement is computationally prohibitive in most practical cases. Again, the problem can be alleviated using surrogate modeling methods (Bandler et al., 2002; Hassan & Mohamed, 2013; Easum et al., 2018) with the popular techniques including response surface approximations, neural networks (Kim et al., 2020), PCE (Kersaudy et al., 2014), or space mapping (Zhang et al., 2020). On the other hand, optimization of yield typically requires quite a considerable adjustment of the design (as compared to the parameter tolerances); therefore, construction of the surrogate may be hindered by the curse of dimensionality. A possible solution is sequential approximate optimization (SAO) (Koziel & Bekasiewicz, 2018), with the metamodel rendered locally (in the vicinity of the current design), then moved along the optimization path. The approach reduces computational overhead of training data acquisition but requires iterative refinement of the model. The response feature technique (Koziel, 2015) offers another approach, where the surrogate is constructed at the level of characteristic points extracted from the system responses rather than the entire outputs. This method leverages a less nonlinear dependence between the feature point coordinates and geometry parameters, thereby leading to a reduction of the number of necessary training data points (Koziel & Bandler, 2015). Yet another possibility is utilization of performance-driven modeling concept, where the surrogate is only constructed along the specified directions within the parameter space, which are the most influential in terms of affecting the performance figures of the system (Pietrenko-Dabrowska et al., 2020).

Despite the advancements described in the previous paragraph, the existing robust design techniques, including the surrogate-assisted procedures, still lack computational efficiency, especially in the case of structures described by larger numbers of parameters (i.e., more than a few). The incorporation of response features allows for improving both the reliability and computational efficiency of robust design procedures. The remaining sections of this chapter describe specific feature-based uncertainty quantification approaches and demonstrate them using real-world antenna and microwave components.

14.2 Fabrication Yield and Design Centering: Problem Formulation

The objective of robust design is to maximize—under the assumed tolerances concerning geometry and material parameters—the likelihood of satisfying performance requirements imposed upon the system of interest. For high-frequency components, the specifications are often expressed in a minimax form with acceptance thresholds set up for specific characteristics, e.g., reflection response over target frequency intervals. Let $\boldsymbol{x} = [x_1 \ldots x_n]^T$ represent the vector of n adjustable parameters of the system under design. The variables are typically geometry parameters, i.e., physical dimensions of the structural details, such as transmission line lengths, widths, spacing between elements, patch sizes, gap widths, and so on, but these may also be material parameters (e.g., dielectric permittivity of the substrate).

For the sake of example, consider a multiband antenna, for which f_{0k}, $k = 1, \ldots, N$, are the target operating frequencies and B is the target fractional bandwidth. We say that the antenna at the design $\boldsymbol{x}$ fulfills the matching requirements if

$$\max\left\{f \in \cup_{k=1}^{N}[(1-B/2)f_{0k}, (1+B/2)f_{0k}] \ : \ |S_{11}(\boldsymbol{x},f)|\right\} \leq S_{\max} \tag{14.1}$$

where $S_{\max}$ is the acceptance threshold for $|S_{11}|$, normally set to $S_{\max} = -10$ dB. In (14.1), f denotes the frequency and S_{11} is marked as an explicit function of both $\boldsymbol{x}$ and f. Satisfaction of the condition (14.1) is equivalent to antenna matching being at the level of $S_{\max}$ or better, simultaneously at all target bandwidths.

We denote by $\boldsymbol{x}^{(0)}$ the nominal design, e.g., obtained by minimizing the antenna reflection at all operating bands, i.e., by solving the minimax problem of the form

$$x^{(0)} = \arg\min_{x} \left\{ \max\left\{ f \in \cup_{k=1}^{N} [(1-B/2)f_{0k}, (1+B/2)f_{0k}] \ : \ |S_{11}(x,f)| \right\} \right\} \tag{14.2}$$

The solution to (14.2) corresponds to the design featuring the best possible antenna matching within the bands of interest.

As a second example, consider an N-band coupler, for which $[f_{L.k} f_{R.k}]$ denote the kth operating band, $k = 1, \ldots, N$, with $f_{L.k}$ and $f_{R.k}$ being the lower and upper ends of the band, respectively. Within that band, the matching and isolation characteristics are not to exceed a user-defined value S_{max} (e.g., -20 dB). Finally, let D_k be the tolerance thresholds for the power split deviations at the center frequency $f_{0k} = [f_{R.k} + f_{L.k}]/2$ and S_k stands for the intended power split ratio. The following conditions are to be fulfilled for the design x to satisfy the aforementioned specifications, where $S_{k1}(x, f)$ stands for the respective S-parameter response, $k = 1, \ldots, 4$:

$$\max\left\{ f \in \cup_{k=1}^{N} [f_{L.k}, f_{R.k}] \ : \ |S_{11}(x,f)| \right\} \le S_{max} \tag{14.3}$$

$$\max\left\{ f \in \cup_{k=1}^{N} [f_{L.k}, f_{R.k}] \ : \ |S_{41}(x,f)| \right\} \le S_{max} \tag{14.4}$$

$$\left| |S_{31}(x,f_{0.k})| - |S_{21}(x,f_{0.k})| - S_k \right| \le D_k \qquad k = 1, \ldots, N \tag{14.5}$$

The fulfillment of (14.3)–(14.5) is analogous to the circuit operating at the required bandwidth and ensuring assumed power division for all frequency intervals $[f_{L.k} f_{R.k}]$. Although the conditions (14.3)–(14.5) define the minimum requirements, it is normally possible, through optimization, to improve the circuit performance further. This can be done, e.g., by improving both matching and isolation over the target bandwidths or to broaden the bandwidths at the S_{max} level. The design that is optimized in the former sense will be referred to as the nominal one and denoted as $x^{(0)}$. It is obtained by solving

$$x^{(0)} = \arg\min_{x} \left\{ \max\left\{ f \in \cup_{k=1}^{N} [f_{L.k}, f_{R.k}] \ : \ \max\{|S_{11}(x,f)|, |S_{41}(x,f)|\} \right\} \right\} \tag{14.6}$$

subject to the equality constraint

$$|\, S_{31}(x,f_{0.k}) \,| - |\, S_{21}(x,f_{0.k}) \,| = S_k \qquad k = 1, \ldots, N \tag{14.7}$$

The design $\boldsymbol{x}^{(0)}$ refers to the best achievable design that can be attained without considering manufacturing tolerances. Here, the nominal design such as (14.2) or (14.6), (14.7) will be used as a starting point for yield optimization.

The deviations from the nominal design, e.g., due to manufacturing tolerances, will be denoted as $d\boldsymbol{x}$. In the case of geometry parameters, parameter deviations are normally a result of fabrication inaccuracies (e.g., related to chemical etching or mechanical milling), whereas in the case of material parameters, these are typically due to incomplete knowledge of the material properties, e.g., permittivity differences between different substrate material sheets. The deviations follow the assumed probability distribution, typically joint Gaussian with zero mean and a fabrication-process-specific variance σ, or uniform with maximum deviation $d_{\max}$ (also specific to a particular fabrication process).

The likelihood of satisfying the design specifications can be quantified using yield Y as a performance metric (Kim et al., 2020), which is defined as

$$Y\left(\boldsymbol{x}^{(0)}\right) = \int_{X_f} p\left(\boldsymbol{x}, \boldsymbol{x}^{(0)}\right) d\boldsymbol{x} \tag{14.8}$$

where $p(\boldsymbol{x}, \boldsymbol{x}^{(0)})$ is a joint probability density function describing statistical variations of the design $\boldsymbol{x}$ with respect to the nominal design $\boldsymbol{x}^{(0)}$, whereas X_f is the feasible space, i.e., the set of parameter vectors that satisfy the prescribed design specifications. Onc of practical ways of cstimating thc yicld is Montc Carlo simulation, in which Y is computed as

$$Y\left(\boldsymbol{x}^{(0)}\right) = p^{-1} \sum\nolimits_{k=1}^{p} H\left(\boldsymbol{x}^{(k)}\right) \tag{14.9}$$

where $\boldsymbol{x}^{(k)} = \boldsymbol{x}^{(0)} + d\boldsymbol{x}^{(k)}$, $k = 1, \ldots, N_r$, are random observables with $d\boldsymbol{x}^{(k)}$ being deviations drawn according to the probability distribution determined by $p(\cdot)$. In (14.9), the function $H(\boldsymbol{x})$ is defined to assume a value 1 if the performance requirements are satisfied, and 0 otherwise.

When it comes to posing the robust design task, it might be understood as direct maximization of the yield

$$\boldsymbol{x}^* = \arg\min_{\boldsymbol{x}} \{-Y(\boldsymbol{x})\} \tag{14.10}$$

where the yield estimation has to be carried out throughout the entire optimization process.

An alternative to (14.10) is design centering, where we strive to allocate design as far away from the boundaries of the feasible region as possible, thereby reducing the chances that the specifications will be violated for any particular realization of the deviation vector $d\boldsymbol{x}$. A conceptual explanation of design centering can be found in Fig. 14.1.

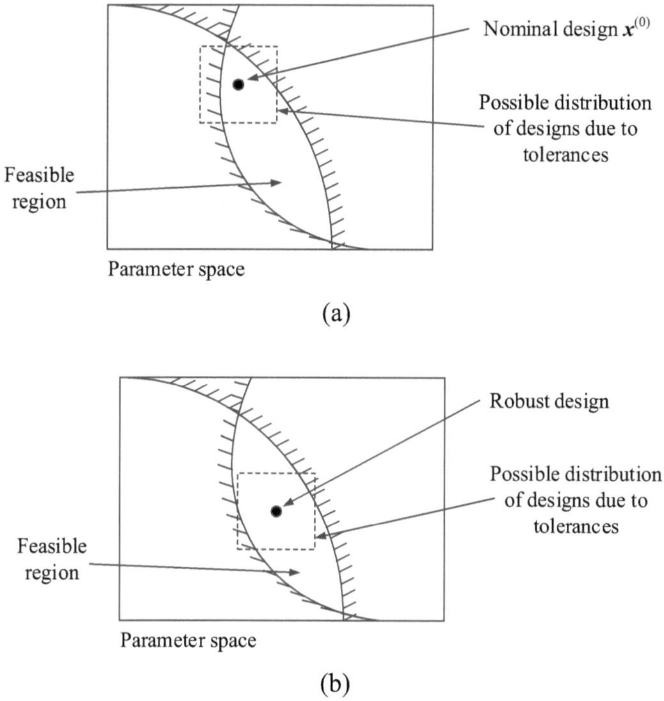

Fig. 14.1 Graphical illustration of design centering: (**a**) nominal design $\boldsymbol{x}^{(0)}$ allocated in the feasible region along with the area marked with dashed line that corresponds to possible design distribution due to tolerances (here, independent uniform probability distributions assumed for all parameters); (**b**) robust design allocated so that most of the dashed area resides in the feasible region. The fabrication yield is higher for the robust design than for the nominal design, although the (nominal) objective function value (e.g., minimax) may be better at $\boldsymbol{x}^{(0)}$

14.3 Yield Optimization Using Response Features

Here, design centering is realized as an iterative process embedded in the trust region framework (Conn et al., 2000), where estimation of the yield is realized using forward surrogate models established at the level of appropriately defined response features. Formulation of the yield optimization problem in terms of characteristic points is followed by a description of the design algorithm, as well as numerical results and benchmarking.

14.3.1 Yield Optimization Task Using Response Features

Figure 14.2 shows an exemplary frequency characteristics of a microwave coupler with the response features defined to account for the design specifications formulated

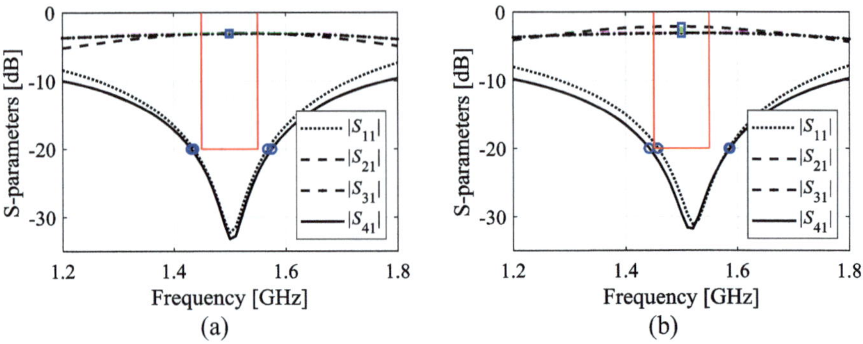

Fig. 14.2 Scattering parameters of an exemplary microwave coupler along with the response features corresponding to −20 dB level of matching ($|S_{11}|$) and isolation ($|S_{41}|$) characteristics (○) and transmission ($|S_{21}|$, $|S_{31}|$) at the target operating frequency of 1.5 GHz (□). The feature points allow for determining satisfaction of the performance requirements (here, the matching/isolation bandwidth of 1.45–1.55 GHz and maximum power split error of 0.5 dB at 1.5 GHz): (**a**) design satisfying specifications, (**b**) design violating both bandwidth and power split requirements

in (14.3)–(14.5). It can be observed that for certain points, specifically those corresponding to −20 dB levels of matching and isolation response, we are only interested in the frequency coordinates, whereas for others (here, those corresponding to transmission responses) we only need the level coordinates. The aggregated feature vector can be therefore written as

$$\boldsymbol{P}(\boldsymbol{x}) = [p_1(\boldsymbol{x})\ p_2(\boldsymbol{x})\ \ldots\ p_6(\boldsymbol{x})]^T = [f_1(\boldsymbol{x})\ f_2(\boldsymbol{x})\ f_3(\boldsymbol{x})\ f_4(\boldsymbol{x})\ l_1(\boldsymbol{x})\ l_2(\boldsymbol{x})]^T \tag{14.11}$$

where f_1 and f_2 are the frequencies corresponding to −20 dB level of $|S_{11}|$, f_3 and f_4 are the frequencies corresponding to −20 dB level of $|S_{41}|$, and l_1 and l_2 are the levels of $|S_{21}|$ and $|S_{31}|$, respectively, at the target operating frequency of the coupler. For a multiband coupler, we will have

$$\begin{aligned}\boldsymbol{P}(\boldsymbol{x}) = [p_1(\boldsymbol{x})\ p_2(\boldsymbol{x})\ \ldots\ p_{6N}(\boldsymbol{x})]^T = [&f_{1.1}(\boldsymbol{x})\ f_{2.1}(\boldsymbol{x})\ f_{3.1}(\boldsymbol{x})\ f_{4.1}(\boldsymbol{x})\ l_{1.1}(\boldsymbol{x})\ l_{2.1}(\boldsymbol{x})\ \ldots\ \ldots\\ &f_{1.N}(\boldsymbol{x})\ f_{2.N}(\boldsymbol{x})\ f_{3.N}(\boldsymbol{x})\ f_{4.N}(\boldsymbol{x})\ l_{1.N}(\boldsymbol{x})\ l_{2.N}(\boldsymbol{x})]^T\end{aligned} \tag{14.12}$$

where the second subscript indicates the operating band (from 1 to N). Clearly, a particular definition of the vector $\boldsymbol{P}$ depends on the design specifications. The feature points themselves are readily extracted from EM-simulated circuit characteristics.

Using the feature vector $\boldsymbol{P}$, the design specification conditions can be reformulated as

$$f_{1.k}(\boldsymbol{x}) \le f_{L.k}, \quad f_{3.k}(\boldsymbol{x}) \le f_{L.k}, \quad k = 1, \ldots, N \tag{14.13}$$

$$f_{2.k}(\boldsymbol{x}) \geq f_{R.k}, \quad f_{4.k}(\boldsymbol{x}) \geq f_{R.k}, \quad k = 1, \ldots, N \tag{14.14}$$

$$|\ l_{1.k}(\boldsymbol{x}) - l_{2.k}(\boldsymbol{x})\ | \leq D_k, \quad k = 1, \ldots, N \tag{14.15}$$

The primary advantage of working at the level of response features rather than the entire circuit responses is weakly-nonlinear dependence of the feature points on geometry parameters of the system, which allows for constructing accurate surrogates using smaller training data sets. In the case of design centering, this would translate into improved reliability of the optimization process as well as its computational efficiency.

14.3.2 Design Centering by Feature-Based Surrogates and Trust Regions

The design centering procedure solves the yield maximization problem (14.10) using a trust-region (TR)-based algorithm that renders approximations $\boldsymbol{x}^{(i)}$, $i = 0, 1, \ldots$, to $\boldsymbol{x}^*$ (Koziel & Pietrenko-Dabrowska, 2019a).

In each iteration, the new approximation $\boldsymbol{x}^{(i+1)}$ is obtained by solving

$$\boldsymbol{x}^{(i+1)} = \arg \min_{\left\|\boldsymbol{x} - \boldsymbol{x}^{(i)}\right\| \leq d^{(i)}} \{-Y_P(\boldsymbol{x})\} \tag{14.16}$$

where the predicted yield Y_P is obtained from the conditions (14.13)–(14.15) evaluated using the auxiliary linear regression model $\boldsymbol{L}^{(i)}$ of the feature vector $\boldsymbol{P}(\boldsymbol{x})$:

$$\boldsymbol{L}^{(i)}(\boldsymbol{x}) = \boldsymbol{P}\left(\boldsymbol{x}^{(i)}\right) + \boldsymbol{J}_P\left(\boldsymbol{x}^{(i)}\right) \cdot \left(\boldsymbol{x} - \boldsymbol{x}^{(i)}\right) \tag{14.17}$$

in which the Jacobian matrix $\boldsymbol{J}_P$ is defined as

$$\boldsymbol{J}_P\left(\boldsymbol{x}^{(i)}\right) = \begin{bmatrix} \dfrac{\partial p_1\left(\boldsymbol{x}^{(i)}\right)}{\partial x_1} & \cdots & \dfrac{\partial p_1\left(\boldsymbol{x}^{(i)}\right)}{\partial x_n} \\ \vdots & \ddots & \vdots \\ \dfrac{\partial p_{2N}\left(\boldsymbol{x}^{(i)}\right)}{\partial x_1} & \cdots & \dfrac{\partial p_{2N}\left(\boldsymbol{x}^{(i)}\right)}{\partial x_n} \end{bmatrix}^T \tag{14.18}$$

The partial derivatives in (14.18) can be computed based on the sensitivities of the circuit S-parameters (here, obtained using finite differentiation). The trust region size $d^{(i)}$ is adjusted adaptively using the standard TR rules. The vector $\boldsymbol{x}^{(i+1)}$ is accepted if the gain ratio, computed as $r = [Y_{P\#}(\boldsymbol{x}^{(i+1)}) - Y_P(\boldsymbol{x}^{(i)})]/[Y_P(\boldsymbol{x}^{(i+1)}) - Y_P(\boldsymbol{x}^{(i)})]$, is positive. Here, $Y_{P\#}$ is computed using a linear model similar to (14.17) but with $\boldsymbol{P}(\boldsymbol{x}^{(i)})$ replaced by $\boldsymbol{P}(\boldsymbol{x}^{(i+1)})$ extracted from EM-simulated circuit response at $\boldsymbol{x}^{(i+1)}$. It should be noted that the employment of $Y_{P\#}$ is a computationally efficient way of validating the candidate design (at the cost of only one EM analysis of the circuit at hand), and it assumes that the feature point sensitivities do not alter considerably while relocating the design from $\boldsymbol{x}^{(i)}$ to $\boldsymbol{x}^{(i+1)}$. This assumption is reasonable as the expected relocations are small in the norm sense (i.e., $||\boldsymbol{x}^{(i+1)} - \boldsymbol{x}^{(i)}||$), more specifically, comparable to $d_{\max}$.

Estimation of the yield using the feature-based regression model $\boldsymbol{L}^{(i)}$ is carried out through Monte Carlo integration (Robert & Casella, 2004) of (14.8). It is arranged using a large number of random samples (in our numerical experiments, 100,000) generated according to the joint probability distribution $p(\cdot)$. For each sample $\boldsymbol{x}_r$, the regression model response $\boldsymbol{L}^{(i)}(\boldsymbol{x}_r)$ is tested against the conditions (14.13)–(14.15). The process is vectorized to ensure rapid evaluation of (14.8) (a typical evaluation time is 40 ms in Matlab, which is negligible compared to EM simulation time of the circuit). At the same time, using a large random set leads to a very small variance of yield estimation (less than 0.1%).

Figure 14.3 shows the flow diagram of the design centering procedure. The input parameter is the nominal design $\boldsymbol{x}^{(0)}$. In the first two steps, the EM analysis of the circuit is performed along with estimation of its response sensitivities, followed by extraction of the feature vectors and construction of the feature-based regression model. The latter is then applied to render a candidate design $\boldsymbol{x}^{(i+1)}$, which is accepted if it leads to a positive gain ratio. At the same time, the trust region size is adjusted as well. Termination condition is based on convergence in argument $||\boldsymbol{x}^{(i+1)} - \boldsymbol{x}^{(i)}|| < \varepsilon$ or shrinking the trust region size beyond the termination threshold $d^{(i+1)} < \varepsilon$. In the numerical experiments of Sect. 14.3.3, we use $\varepsilon = 10^{-3}$.

14.3.3 Demonstration Examples

This section provides numerical verification of the design centering procedure discussed above using two examples of microstrip couplers. The computational efficiency of our technique is compared to several surrogate-based yield optimization methods. Furthermore, direct EM-based Monte Carlo analysis is employed to validate reliability of yield estimation using feature-based regression model at both the nominal and optimized designs.

Consider a compact rat-race coupler (RRC; Circuit I) (Koziel & Pietrenko-Dabrowska, 2019b) shown in Fig. 14.4a and a dual-band branch-line coupler (Circuit II) (Xia et al., 2020) shown in Fig. 14.4b. The relevant information concerning the three structures, including the substrate parameters, geometry parameters, nominal design, and performance requirements, can be found in Table 14.1. The EM-simulation models are implemented in CST Microwave Studio and evaluated using the time-domain solver.

Fig. 14.3 Operating flow of the discussed design centering algorithm

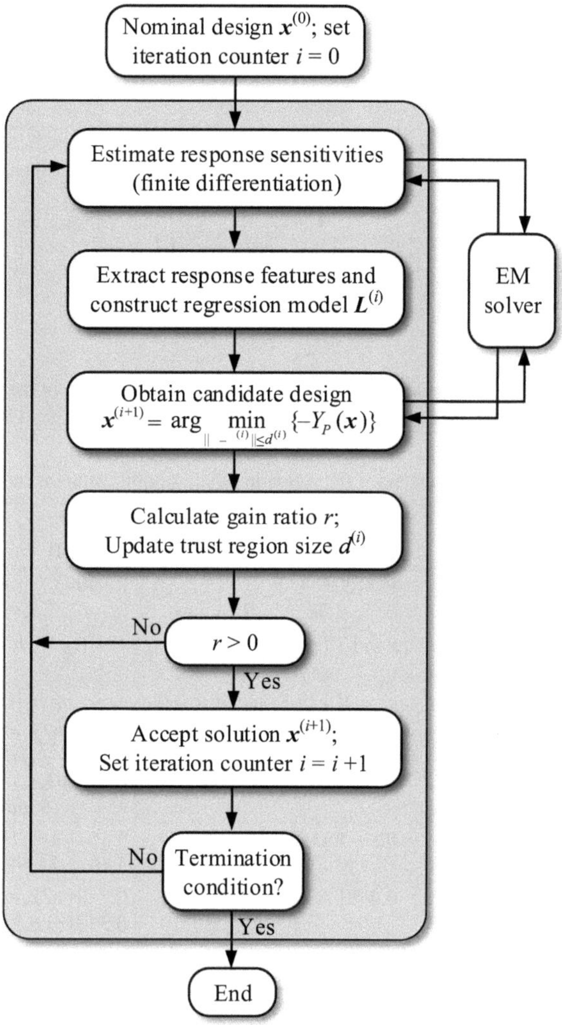

Design centering is carried out assuming that geometry parameter deviations follow independent Gaussian probability distribution with variance 0.017 mm with maximum deviations limited to 0.05 mm (about triple variance).

For the sake of benchmarking, the yield optimization results obtained using the discussed procedure are compared to those produced by several surrogate-based methods. The reference techniques are briefly characterized below.

Reference algorithm 1: This is a surrogate-assisted approach incorporating a metamodel established in a relatively large vicinity of the nominal design to enable sufficient relocation of the design during yield optimization. The problem (14.10) is solved using local optimization algorithm. Here, the metamodel is constructed using kriging interpolation (Queipo et al., 2005), whereas the domain is defined as an

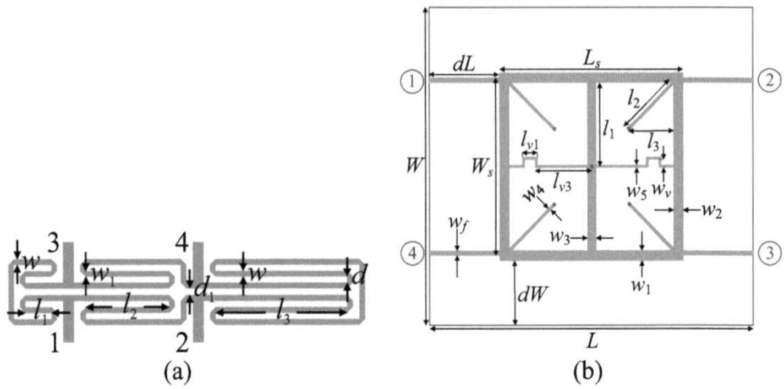

Fig. 14.4 Microwave structures employed for validation of the design centering algorithm of Sect. 14.3.2: (**a**) compact rat-race coupler (Circuit I), (**b**) dual-band branch-line coupler (Circuit II)

Table 14.1 Test cases for validating yield optimization procedure of Sect. 14.3.2

	Case study[a]	
	Circuit I	Circuit II
Substrate	RO4003 ($\varepsilon_r = 3.38$, $h = 0.76$ mm)	RO4003 ($\varepsilon_r = 3.5$, $h = 0.51$ mm)
Design parameters	$\boldsymbol{x} = [l_1\ l_2\ l_3\ d\ w\ w_1]^T$	$\boldsymbol{x} = [L_s\ W_s\ l_{3r}\ w_1\ w_2\ w_3\ w_4\ w_5\ w_v]^T$
Other parameters	$d_1 = d + \lvert w - w_1 \rvert$, $d = 1.0$, $w_0 = 1.7$, and $l_0 = 15$	$d_L = d_W = 10$ mm, $L = 2d_L + L_s$, $W = 2d_W + 2w_1 + (W_s - 2w_f)$, $l_1 = W_s/2$, $l_2 = l_3 2^{1/2}$, $l_3 = l_{3r}((L_s - w_3)/2 - w_4/2^{1/2})$, $l_{v1} = l_3/3$, $l_{v3} = L_s/2 - w_3/2 - l_3 + l_{v1}$; $w_f = 1.15$ mm
Operating bands	0.89–1.11 GHz	2.36–2.44 GHz 5.16–5.24 GHz
Maximum power split error	0.4 dB at 1 GHz	0.5 dB at 2.4 GHz 0.5 dB at 5.2 GHz
Nominal design	$\boldsymbol{x}^{(0)} = [4.50\ 11.08\ 21.81\ 0.65\ 0.94\ 0.86]^T$	$\boldsymbol{x}^{(0)} = [25.05\ 0.85\ 0.76\ 1.90\ 1.23\ 0.36\ 0.71\ 0.30\ 0.30]^T$

[a] Parameters with subscript r are relative, and their deviations are recalculated accordingly in order to have the corresponding absolute parameters following the assumed probability distribution (here, Gaussian with variance of 0.017 mm)

interval $[\boldsymbol{x}^{(0)} - \boldsymbol{d}, \boldsymbol{x}^{(0)} + \boldsymbol{d}]$. The entries of the size vector $\boldsymbol{d} = [d_1 \ldots d_n]^T$ are set to $d_k = 10d_{\max}$, $k = 1, \ldots, n$ (recall that $d_{\max}$ is the maximum parameter deviation, which, in our experiments is 0.05 mm). This size of domain normally creates a sufficient room for design relocation in the yield maximization process. The advantage of this procedure is implementation simplicity. The drawback is a potentially high cost of training data acquisition, which is especially troublesome to structures described by larger numbers of geometry parameters.

Reference algorithm 2: This procedure adopts the sequential approximate optimization (SAO) approach (Koziel & Bekasiewicz, 2018) in which the original problem (14.10) is solved iteratively as

$$\boldsymbol{x}^{(i+1)} = \arg\min_{\boldsymbol{x}} \left\{ -Y_s^{(i)}(\boldsymbol{x}) \right\} \tag{14.19}$$

where $\boldsymbol{x}^{(i)}$, $i = 0, 1, \ldots$, are approximations to the optimum design $\boldsymbol{x}^*$. The yield estimated in the *i*th iteration, $Y_s^{(i)}$, is computed based on the surrogate model established in the current domain being the interval $[\boldsymbol{x}^{(i)} - \boldsymbol{d}_l, \boldsymbol{x}^{(i)} - \boldsymbol{d}_l]$, with $\boldsymbol{x}^{(i)} = [x_1^{(i)} \ldots x_n^{(i)}]^T$, i.e., the domain is always centered at the current iteration point. Because the surrogate models are constructed along the optimization path, their domains can be considerably smaller than for the first reference algorithm (here, $\boldsymbol{d}_l = [d_{l.1} \ldots d_{l.n}]^T$ is set to $d_{l.k} = 3d_{\max}$, $k = 1, \ldots, n$). The problem (14.19) is constrained to satisfy $x_k^{(i)} - d_{l.k} + d_{\max} \le x_k \le x_k^{(i)} + d_{l.k} - d_{\max}$, $k = 1, \ldots, n$, which ensures that the solution is located in the domain interior and at the distance at least $d_{\max}$ from its boundary. The latter allows to carry out Monte Carlo simulation within the region of validity of the metamodel. The advantage of this procedure is a lower cost of setting up the surrogate as compared to the previous method. However, several iterations are normally required to approach the optimum solution.

Reference algorithm 3: This procedure incorporates the performance-driven modeling concept (Koziel & Pietrenko-Dabrowska, 2020). It relies on a single metamodel. The latter is rendered in the domain spanned by the most relevant directions within the parameter space, specifically those that affect the likelihood of satisfying the assumed design requirements in the most significant manner. The relevant directions are found through auxiliary local optimizations (Pietrenko-Dabrowska, 2021). The principal advantage of this method is low volume of the surrogate model domain, which is however of sufficient size wherever necessary. Consequently, the third algorithm effectively combines the advantages of the first and the second reference methods. The conceptual illustration of the metamodel domain definition can be found in Fig. 14.5.

The numerical results obtained using the discussed design centering approach as well as the reference algorithms outlined above have been gathered in Tables 14.2 and 14.3 for Circuit I and II, respectively. The following final (yield-optimized) designs were obtained:

- Circuit I: $\boldsymbol{x}^* = [4.65\ 11.22\ 21.73\ 0.73\ 0.94\ 0.86]^T$
- Circuit II: $\boldsymbol{x}^* = [25.16\ 0.80\ 0.77\ 1.89\ 1.25\ 0.42\ 0.75\ 0.30\ 0.30]^T$

The computational cost of the design centering process was very low and corresponded to only 25 and 32 EM simulations of Circuit I and II, respectively. It can be observed that this cost is considerably lower than that required by the reference algorithms. In particular, the average computational savings over the reference algorithm 1 are as high as 95%, almost 90% over the reference algorithm 2, and almost 70% over the reference algorithm 3.

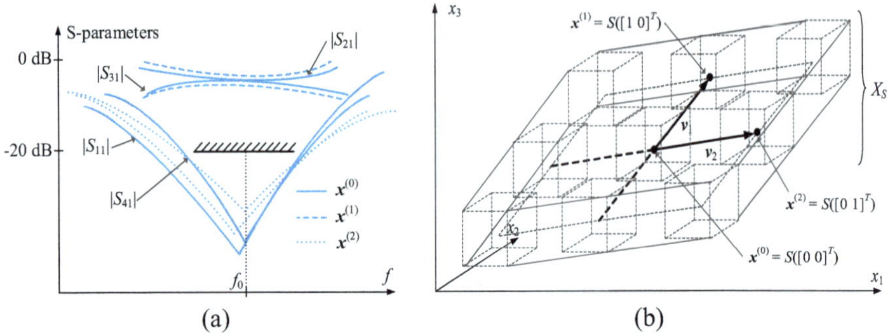

Fig. 14.5 Yield optimization using performance-driven surrogate modeling concept (Koziel & Pietrenko-Dabrowska, 2020): (**a**) S-parameters of a microwave coupler at the nominal design $\boldsymbol{x}^{(0)}$, design $\boldsymbol{x}^{(1)}$ (spoiled power split), and design $\boldsymbol{x}^{(2)}$ (improved -20 dB bandwidth); only the selected S-parameters are shown for $\boldsymbol{x}^{(1)}$ ($|S_{21}|$, $|S_{31}|$) and $\boldsymbol{x}^{(2)}$ ($|S_{11}|$, $|S_{41}|$) for clarity. The designs $\boldsymbol{x}^{(1)}$ and $\boldsymbol{x}^{(2)}$ determine the important directions from the point of view of yield manipulation; (**b**) Designs $\boldsymbol{x}^{(0)}$, $\boldsymbol{x}^{(1)}$, and $\boldsymbol{x}^{(2)}$ form surface $S(\boldsymbol{t})$ parameterized by vector $\boldsymbol{t} = [t_1\ t_2]^T$. The union of intervals $S_I(\boldsymbol{t})$ for $-1 \le t_1, t_2 \le 1$ becomes the domain X_S of the surrogate model

Table 14.2 Design centering results for Circuit I

Optimization algorithm	Initial yield (%) Estimated by surrogate model	Initial yield (%) EM-based	Optimized yield (%) Estimated by surrogate model	Optimized yield (%) EM-based	CPU cost[a]
Reference algorithm 1	50	42	100	97	400
Reference algorithm 2	45	42	97	97	200[b]
Reference algorithm 3	44	42	98	98	82
Feature-based (Sect. 14.3.2)	45	42	99	98	25

[a]Optimization cost in number of EM simulations of the considered circuit
[b]The algorithm convergence after three iterations (surrogate setup cost 50 training samples per iteration)

The improvements in terms of computational efficiency are primarily due to the usage of response features and incorporation of a simple (here, linear) surrogate. The accuracy of the latter is a result of constructing the metamodel at the level of feature point coordinates, which—as mentioned before—are in weakly-nonlinear relationship with the geometry parameters of the circuit.

The quality of the designs obtained using the considered approach regarding the final yield values are similar to those produced by other methods. On the other hand, the predictions of the feature-based surrogate are comparable or better than those of the reference algorithms. This is corroborated through direct EM-driven Monte Carlo (MC) simulations at the nominal and the yield-optimized designs as indicated in Tables 14.2 and 14.3. Limited predictive power is particularly pertinent to the

Table 14.3 Design centering results for Circuit II

Optimization algorithm	Initial yield (%) Estimated by surrogate model	Initial yield (%) EM-based	Optimized yield (%) Estimated by surrogate model	Optimized yield (%) EM-based	CPU cost[a]
Reference algorithm 1	80	71	99	93	800
Reference algorithm 2	88	71	96	91	500[b]
Reference algorithm 3	74	71	94	92	123
Feature-based (Sect. 14.3.2)	71	71	93	89	32

[a]Optimization cost in number of EM simulations of the considered circuit
[b]The algorithm convergence after four iterations (surrogate setup cost 100 training samples per iteration)

reference Algorithm 1: although the surrogate models therein were constructed using a relatively large number of samples (400 for Circuit I and 800 for Circuit II), the relative RMS error of the metamodels is about 3.5% (Circuit I) and 6.1% (Circuit II), which is insufficient to render accurate yield estimation. At the same time, it can be noticed that the accuracy of yield estimation using feature-based surrogate is not as good as for other techniques for Circuit II. This indicates that the relationships between the geometry parameters of this circuit and response feature coordinates is more nonlinear than for other considered structures. This does lead to slightly inferior results in terms of the final yield (as confirmed by EM-driven Monte Carlo simulation), and further improvements are needed. Graphical illustrations of EM-driven MC for the nominal and yield-optimized designs can be found in Figs. 14.6 and 14.7 for Circuit I and II, respectively. The analysis has been performed using 500 random samples. A comparison between the left- and the right-hand-side panels indicates noticeable increase in the number of samples satisfying the prescribed specifications at the optimized design as compared to the nominal one.

14.4 Yield Optimization of Microwave Filters Using Feature-Based Linear Regression Models

This section illustrates utilization of response features for estimation and improvement of fabrication yield for microwave filters (Koziel & Bandler, 2015). The presented technique uses a feature-based linear regression model to enable fast Monte Carlo analysis in the vicinity of the current design and relocates it along the optimization path in the course of the search process.

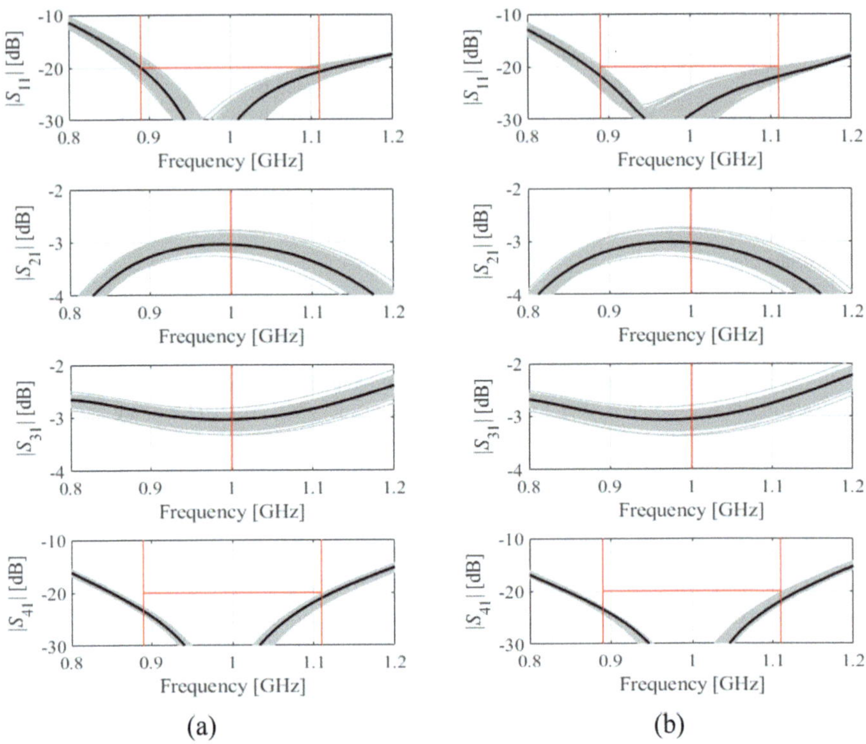

Fig. 14.6 EM-driven Monte Carlo simulation of Circuit I at (**a**) the nominal design, and (**b**) the final design obtained using the algorithm considered in this section. MC carried out using 500 random samples. EM simulation data shown as gray curves. Black plots show the circuit responses at the nominal (**a**) and the optimized design (**b**)

14.4.1 Feature-Based Approximation Model

Consider $|S_{11}|$ responses of the bandpass filter considered in Sect.14.3.2, shown in Fig. 14.8. The plot shows the response at the nominal design (i.e., a typically desired minimax optimum w.r.t. the design specifications marked with horizontal lines, here, $|S_{11}| \leq -20$ dB for 10.55–11.45 GHz and $|S_{11}| \geq -1$ dB for frequencies lower than 10.3 GHz and higher than 11.7 GHz), as well as a set of so-called feature points, here, represented by -1 and -20 dB levels as well as the peaks of the response in the passband. The location of these points is sufficient to determine whether the response violates or satisfies the given design specifications. In particular, assuming small design perturbations, the feature points corresponding to -1 and -20 dB may move toward lower or higher frequencies violating (in some cases) the specifications regarding passband and/or stopband frequencies; the feature points corresponding to $|S_{11}|$ maxima in the passband may move up leading to violation of the $|S_{11}| \leq -20$ requirement. Figure 14.8 also shows the response and the corresponding feature points at a perturbed design (which, in this case, violates the specifications).

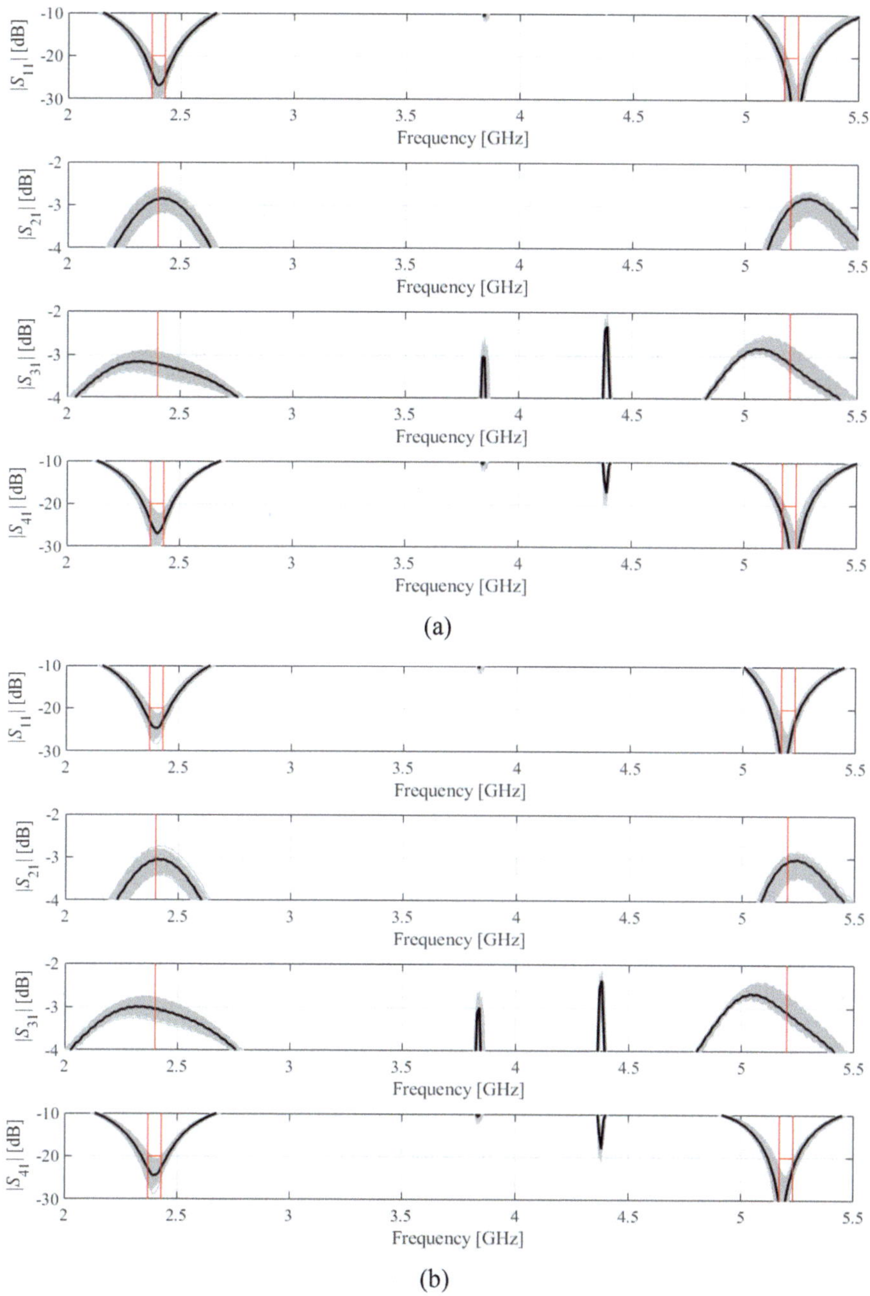

Fig. 14.7 EM-driven Monte Carlo simulation of Circuit II at (**a**) the nominal design, and (**b**) the final design obtained using the algorithm considered here. MC carried out using 500 random samples. EM simulation data shown as gray curves. Black plots show the circuit responses at the nominal (**a**) and the optimized design (**b**)

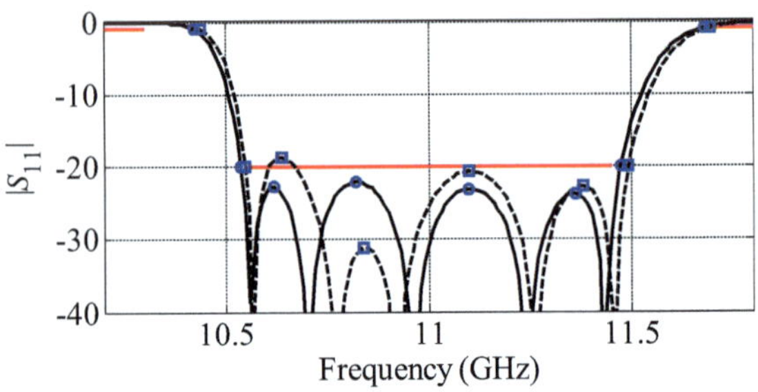

Fig. 14.8 Reflection response of the bandpass filter (—) at the optimum design with respect to given minimax specifications (marked with horizontal lines), as well as the response at a perturbed design (- - -). Circles and squares denote feature points for both responses, here, corresponding to the -1 and -20 dB levels as well as the response maxima in the passband. Design specifications are: $|S_{11}| \leq -20$ dB for 10.55–11.45 GHz and $|S_{11}| \geq -1$ dB for frequencies lower than 10.3 GHz and higher than 11.7 GHz

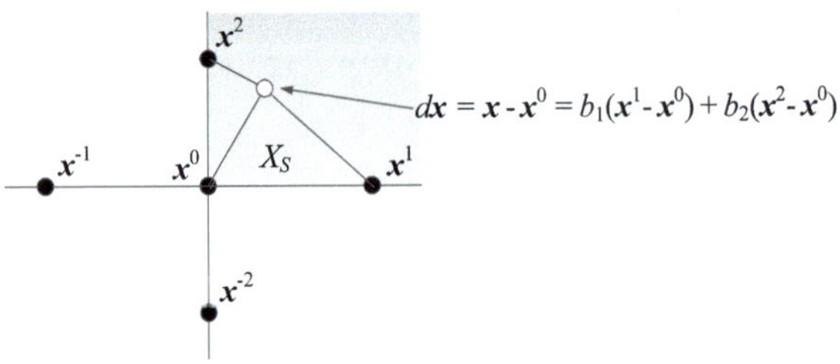

Fig. 14.9 Deviation vector dx and its expansion using star-distributed training vectors x^0 and x^k, $k = -2, -1, 1, 2$ (denoted as •). The shaded area denotes an area defined by a subset X_S of points being the closest to $x^0 + dx$, which is represented as a linear combination of vectors $x^k - x^0$. The feature points at $x^0 + dx$ are calculated using the coefficients of this linear combination and the feature points of $R(x^k)$ for $x^k \in X_S$. Here, we have $p^j = p_0^j + \beta_1(p_1^j - p_0^j) + \beta_2(p_2^j - p_0^j)$

In order to construct the feature-based model, we consider $2n + 1$ evaluations of the original model R at the nominal design, $R(x^0)$, and at the perturbed designs $x^k = [x_{0.1} \ldots x_{0.k} + \text{sign}(k)\cdot\delta_k \ldots x_{0.n}]^T$, $k = -n, \ldots, -1, 1, \ldots, n$, where δ_k may be, e.g., a maximum assumed deviation of the kth parameter from its nominal value. The feature points of the response vector $R(x^k)$ are denoted as $p_k^j = [f_k^j \; r_k^j]^T, j = 1, \ldots, K$, where f and r are the frequency and magnitude components of the respective point, and where K is the total number of feature points.

The aim is to predict the position of the feature points corresponding to a perturbed vector $x^0 + dx$, using the available training set $\{x^k; p_{-n}^j, \ldots, p_{-1}^j, p_0^j, p_1^j, \ldots, p_n^j\}$. For any given dx, we find a subset X_S of the base set $\{x^k\}$ that defines an area containing $x^0 + dx$. The surrogate model is set up using all the points from X_S, as shown in Fig. 14.9 for $n = 2$. Without loss of generality, we can assume that $X_S = \{x^0, x^1, \ldots, x^n\}$. We define

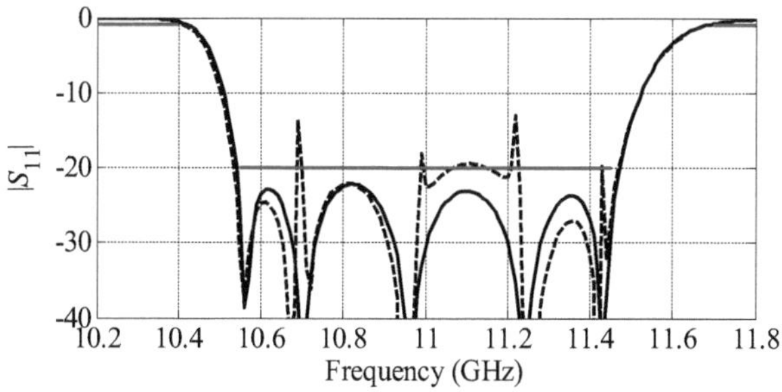

Fig. 14.10 Responses of the fifth-order bandpass filter at the nominal design (—) and the response obtained from a linear model (14.23) constructed using a perturbed design (- - -) (at the selected reference design). The spikes that appear due to the linear modeling of sharp responses lead to considerable yield underestimation

$$\boldsymbol{p}^j = \boldsymbol{p}_0^j + \beta_1(\boldsymbol{p}_1^j - \boldsymbol{p}_0^j) + \beta_2(\boldsymbol{p}_2^j - \boldsymbol{p}_0^j) + \cdots + \beta_n(\boldsymbol{p}_n^j - \boldsymbol{p}_0^j) \tag{14.20}$$

where $\beta_1, \beta_2, \ldots, \beta_n$, determines a unique representation of $d\boldsymbol{x}$ using vectors $\boldsymbol{v}_i = \boldsymbol{x}^k - \boldsymbol{x}^0$, $i = 1, \ldots, n$. Coefficients β_i can be explicitly found as

$$[\beta_1 \ \ \beta_2 \ \ \ldots \ \ \beta_n]^T = [\boldsymbol{v}_1 \ \ \boldsymbol{v}_2 \ \ \ldots \ \ \boldsymbol{v}_n]^{-1} \cdot (\boldsymbol{x} - \boldsymbol{x}^0) \tag{14.21}$$

Using (14.20) and (14.21), we can define an approximation model of the feature points as

$$s(\boldsymbol{x}^0 + d\boldsymbol{x}) = [\boldsymbol{p}^1(\boldsymbol{x}^0 + d\boldsymbol{x}) \ \ \cdots \ \ \boldsymbol{p}^K(\boldsymbol{x}^0 + d\boldsymbol{x})] \tag{14.22}$$

where $\boldsymbol{p}^j(\boldsymbol{x}^0 + d\boldsymbol{x}) = [f^j(\boldsymbol{x}^0 + d\boldsymbol{x})\ r^j(\boldsymbol{x}^0 + d\boldsymbol{x})]^T$, $j = 1, \ldots, K$. Having $s(\boldsymbol{x}^0 + d\boldsymbol{x})$, one can estimate yield through numerical integration of the probability density function determining the parameter deviations, cf. (14.9). The fundamental difference is that the satisfaction/violation of the design specification frequencies/levels is verified for the feature points only rather than for the entire responses.

It should be emphasized that using response features for estimating yield rather than constructing, e.g., a linear model of the entire *S*-parameter response, is critical to accuracy. Let us consider a simple, first-order Taylor expansion of the filter model around the nominal design $\boldsymbol{x}^0$

$$\boldsymbol{R}_L(\boldsymbol{x}) = \boldsymbol{R}(\boldsymbol{x}^0) + \boldsymbol{J}_{\boldsymbol{R}}(\boldsymbol{x}^0) \cdot (\boldsymbol{x} - \boldsymbol{x}^0) \tag{14.23}$$

where $\boldsymbol{J}_{\boldsymbol{R}}(\boldsymbol{x}^0)$ is an estimated Jacobian of $\boldsymbol{R}$ at the nominal design. The estimate can be obtained using evaluations of $\boldsymbol{R}$ at the perturbed designs $\boldsymbol{x}^k$.

Figure 14.10 shows an *S*-parameter prediction obtained by evaluating a linear surrogate $\boldsymbol{R}_L$ constructed from the filter responses evaluated for the same training set used for the feature-based model. The lack of accuracy coming from the very sharp responses (as functions of frequency) is reflected in underestimated yield predictions. This indicates that the feature-based yield estimation, although based on the same data set, is fundamentally different from simple linear modeling.

14.4.2 Yield Estimation Examples

Consider the X-band waveguide filter with nonsymmetrical irises (Hauth et al., 1993) shown in Fig. 14.11. The design variables are $\boldsymbol{x} = [z_1\ z_2\ z_3\ d_1\ d_2\ d_3\ t_1\ t_2\ t_3]^T$. The filter is simulated in CST Microwave Studio. The nominal design, $\boldsymbol{x}^0 = [12.0836\ 14.2069\ 14.6875\ 13.9841\ 11.6864\ 10.8076\ 1.5455\ 3.0671\ 2.4557]^T$ mm, is a minimax optimum with respect to the following design specifications: $|S_{11}| \leq -20$ dB for 10.55 GHz $\leq \omega \leq$ 11.45 GHz, and $|S_{11}| \geq -1$ dB for $\omega \leq$ 10.4 GHz and $\omega \geq$ 11.7 GHz. The minimax optimization is understood as minimization of a maximum violation of the aforementioned design specifications within the respective frequency sub-bands.

Yield estimation has been carried out using four scenarios for geometry parameter deviations, including a uniform probability distribution with a maximum deviation equal to 0.01 and 0.02 mm (Cases 1 and 2) and a normal distribution with zero mean and standard deviation 0.01 and 0.02 mm (Cases 3 and 4). The deviations are taken as uncorrelated. The yield has been estimated with the methodology described in Sect. 14.4.1 using the eight feature points shown in Fig. 14.8. For comparison, the yield was also estimated using conventional Monte Carlo analysis with 500 random samples (the number of samples is limited due to the computational cost of the EM simulation). The results are shown in Table 14.4. Figure 14.12 presents a visualization of the yield estimation for Case 2. The agreement between the yield estimation obtained using the considered method and conventional Monte Carlo analysis is excellent. As a matter of fact, the results obtained using the considered approach are

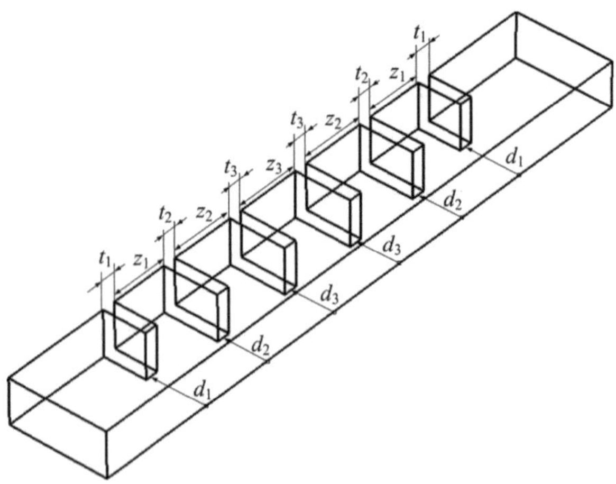

Fig. 14.11 Fifth-order waveguide bandpass filter (Hauth et al., 1993)

Table 14.4 Yield estimation for fifth-order waveguide filter

Case	Distribution	Yield estimation method	Estimated yield	CPU cost[a]
1	Uniform (max. dev. 0.01 mm)	Feature points	0.97	19
		EM-based Monte Carlo	0.97	500
		Linear modeling[b]	0.63	19
2	Uniform (max. dev. 0.02 mm)	Feature points	0.56	19
		EM-based Monte Carlo	0.55	500
		Linear modeling[b]	0.12	19
3	Gaussian (std. dev. 0.01 mm)	Feature points	0.69	19
		EM-based Monte Carlo	0.69	500
		Linear modeling[b]	0.19	19
4	Gaussian (std. dev. 0.02 mm)	Feature points	0.25	19
		EM-based Monte Carlo	0.24	500
		Linear modeling[b]	0.02	19

[a]Estimation cost in number of EM analyses. Feature-based yield estimation utilizes $N = 5000$ random samples

[b]Estimation based on a linear model of the S-parameter response around the nominal design

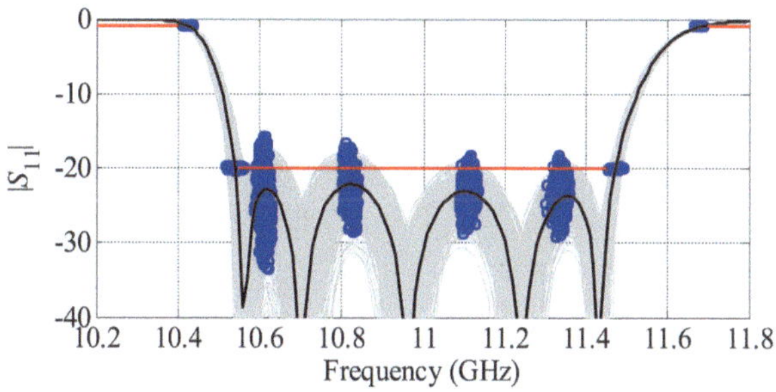

Fig. 14.12 Fifth-order waveguide bandpass filter: yield estimation for Case 2. Gray lines correspond to 500 EM-simulated random samples for Monte Carlo analysis, circles represent corresponding feature points calculated using the feature-based approximation model

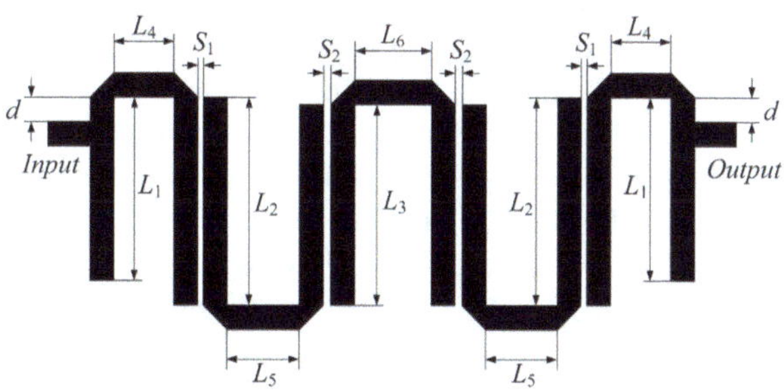

Fig. 14.13 Microstrip hairpin filter: geometry (Brady, 2002)

more reliable than MC: the uncertainty in the latter is relatively large due to the small number of samples used in the process to keep the cost low. Feature-based yield estimation was executed for $N = 5000$.

As a second example, consider the microstrip hairpin filter (Brady, 2002) shown in Fig. 14.13. The design variables are $\boldsymbol{x} = [L_1\ L_2\ L_3\ L_4\ L_5\ L_6\ s_1\ s_2\ d]^T$. The microstrip width is fixed to $w = 1.5$ mm. The substrate parameters are

Table 14.5 Yield estimation for microstrip hairpin filter

Case	Distribution	Yield estimation method	Estimated yield	CPU cost[a]
1	Uniform (max. dev. 0.01 mm)	Feature points	0.18	19
		EM-based Monte Carlo	0.19	500
		Linear modeling[b]	0.07	19
2	Uniform (max. dev. 0.02 mm)	Feature points	0.032	19
		EM-based Monte Carlo	0.040	500
		Linear modeling[b]	0.014	19
3	Gaussian (std. dev. 0.01 mm)	Feature points	0.102	19
		EM-based Monte Carlo	0.110	500
		Linear modeling[b]	0.002	19
4	Gaussian (std. dev. 0.02 mm)	Feature points	0.021	19
		EM-based Monte Carlo	0.026	500
		Linear modeling[b]	0.000	19

[a]Estimation cost in number of EM analyses. Feature-based yield estimation utilizes $N = 5000$ random samples

[b]Estimation based on a linear model of the S-parameter response around the nominal design

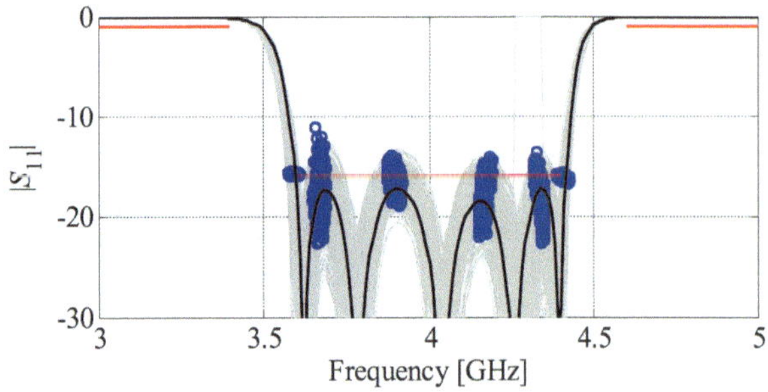

Fig. 14.14 Microstrip hairpin filter: yield estimation for Case 1. Gray lines correspond to 500 EM-simulated random samples for Monte Carlo analysis, circles represent corresponding feature points calculated using the feature-based approximation model

$h = 0.787$ mm (substrate height) and $\varepsilon_r = 3.38$ (dielectric permittivity). The filter is simulated in FEKO. The nominal design, $\boldsymbol{x}^0 = [9.357\ 11.684\ 9.814\ 0.402\ 0.507\ 1.514\ 0.053\ 0.130\ 0.182]^T$ mm, is a design optimized with respect to the following specifications: $|S_{11}| \leq -16$ dB for 3.6 GHz $\leq \omega \leq$ 4.4 GHz and $|S_{11}| \geq -1$ dB for $\omega \leq$ 3.4 GHz and $\omega \geq$ 4.6 GHz.

For this example, the yield has been estimated using four scenarios of uncertainties in the geometry parameters, as shown in Table 14.5. Figure 14.14 shows a visualization of yield-estimation for the selected case. The results are consistent with the previous examples in terms of agreement with the Monte Carlo analysis; however, the prediction of yield given by the discussed approach is lower than for Monte Carlo. It should be emphasized that in this case, the yield estimation with Monte Carlo based on 500 samples is not particularly reliable due to low overall values of the yield. For example, in Case 2 (a Gaussian distribution with standard deviation of 0.02 mm), the standard deviation of the estimated yield exceeds 50% of the mean value (based on multiple yield estimates obtained using the method considered in this section with 500 samples). To the contrary, because of its speed, the feature-based approach allows any number of samples, which greatly improves the yield estimation accuracy.

14.4.3 Yield Optimization Examples

Here, the feature-based approximation model of Sect. 14.4.1 is used for tolerance-aware design. Let $Y_s(\boldsymbol{x}^0)$ denotes the yield estimated at the nominal design $\boldsymbol{x}^0$ using multiple evaluations of the feature model s. The yield estimation is based on multiple evaluations of this model, i.e., $s(\boldsymbol{x}^0 + d\boldsymbol{x}^k)$, where $d\boldsymbol{x}^k$, $k = 1, \ldots, N$, are random vectors sampled according to the assumed probability distribution. Let us define a function $Y_s(\boldsymbol{x}; \boldsymbol{x}^0)$ realizing a yield estimation for any design $\boldsymbol{x}$ (in practice, $\boldsymbol{x}$ should be in the vicinity of $\boldsymbol{x}^0$), where the yield is calculated using evaluations of $s(\boldsymbol{x} + d\boldsymbol{x}^k)$. However, because s is defined with respect to $\boldsymbol{x}^0$ rather than an arbitrary $\boldsymbol{x}$, the yield estimation has to be performed using the following set of evaluations: $s(\boldsymbol{x}^0 + d\boldsymbol{x}'^k)$, where $d\boldsymbol{x}'^k = d\boldsymbol{x}^k + [\boldsymbol{x} - \boldsymbol{x}^0]$, $k = 1, \ldots, N$.

The design maximizing the yield is then found as

$$\boldsymbol{x}^* = \arg\min_{\boldsymbol{x}} \left\{ -Y_s\left(\boldsymbol{x};\boldsymbol{x}^0\right) \right\} \tag{14.24}$$

To ensure reliability (i.e., to minimize the yield estimation variance), a large number of samples is used, here, $N = 5000$. Also, the same sample set of perturbations $\{d\boldsymbol{x}^k\}$ should be utilized for all evaluations of $Y_s(\boldsymbol{x}; \boldsymbol{x}^0)$ in order to avoid numerical noise related to yield estimation variance.

In practice, the accuracy of a yield estimate using $Y_s(\boldsymbol{x}; \boldsymbol{x}^0)$ will degrade as $\boldsymbol{x}$ moves away from $\boldsymbol{x}^0$, so yield optimization should preferably be implemented as an iterative process, namely,

$$\boldsymbol{x}^{i+1} = \arg\min_{\boldsymbol{x}} \left\{ -Y_s\left(\boldsymbol{x};\boldsymbol{x}^i\right) \right\} \tag{14.25}$$

where $\boldsymbol{x}^i$, $i = 0, 1, \ldots$, are approximations of $\boldsymbol{x}^*$ obtained by optimizing $Y_s(\boldsymbol{x}; \boldsymbol{x}^i)$, i.e., a yield estimation function setup similarly to $Y_s(\boldsymbol{x}; \boldsymbol{x}^0)$ but centered around $\boldsymbol{x}^i$ rather than $\boldsymbol{x}^0$ and using corresponding perturbations. The cost of each iteration is $2n + 1$ EM simulations, where n is the number of designable parameters. The procedure is terminated when the current iteration does not lead to yield improvement, which is understood as $Y_s(\boldsymbol{x}^{i+1}; \boldsymbol{x}^{i+1}) > Y_s(\boldsymbol{x}^i; \boldsymbol{x}^i)$. The feature-based model Y_s is optimized using a pattern search algorithm (Koziel, 2010).

Consider again the waveguide filter of Fig. 14.11, assuming a Gaussian probability distribution with a standard deviation equal to 0.02 mm (Case 4). The optimization is started from the nominal design $\boldsymbol{x}^0$ with an estimated yield of 0.25. The optimized design is $\boldsymbol{x}^* = [12.05\ 14.18\ 14.67\ 14.06\ 11.74\ 10.85\ 1.49\ 3.03\ 2.46]^T$ with an estimated yield of 0.46. This design was obtained in four iterations of the yield optimization procedure at a total cost of $4 \times 19 = 76$ filter evaluations.

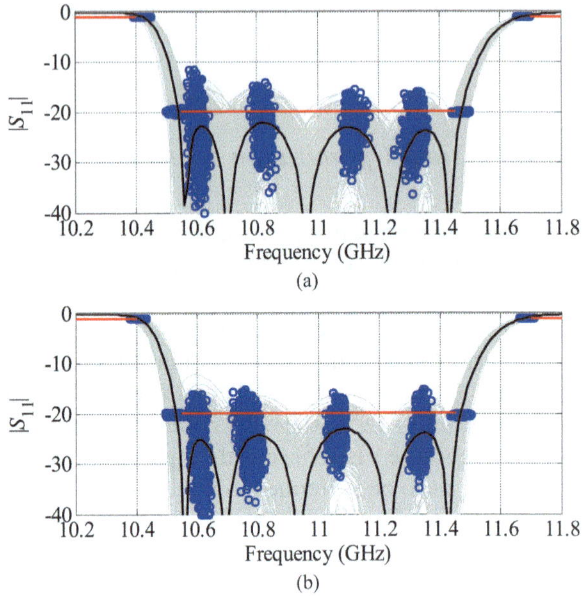

Fig. 14.15 Fifth-order waveguide filter: yield estimation assuming a Gaussian probability distribution with standard deviations of 0.02 mm (Case 4) at (**a**) the nominal design ($Y = 0.25$) and (**b**) the optimized design ($Y = 0.46$). Gray lines correspond to 500 EM-simulated random samples for Monte Carlo analysis, circles represent corresponding feature points calculated using the feature-based approximation model

Table 14.6 Yield optimization of the fifth-order waveguide filter

Case[a]	Yield estimation method	Estimated yield	CPU cost[b]
Nominal design (minimax optimum)	Feature points	0.25	19
	EM-based Monte Carlo	0.24	500
Yield-optimized design	Feature points	0.46	19
	EM-based Monte Carlo	0.47	500

[a]Estimation for a uniform distribution with maximum deviations of 0.01 mm
[b]Estimation cost in number of EM analyses

Figure 14.15 shows a visualization of the yield estimate at the optimized design. Table 14.6 indicates the yield estimated using the discussed method and conventional Monte Carlo analysis. Good agreement between the two estimations is observed.

Yield optimization for the hairpin filter of Fig. 14.13 was executed for Case 1 (a uniform probability distribution with a maximum deviation equal to 0.01 mm). The optimized design, $\boldsymbol{x}^* = [9.141\ 11.725\ 9.84\ 0.436\ 0.607\ 1.576\ 0.030\ 0.107\ 0.266]^T$ with an estimated yield of 0.25, is obtained in two iterations (total cost $2 \times 19 = 38$ filter simulations). Table 14.7 indicates the yield estimated using both the considered method and conventional Monte Carlo analysis.

Table 14.7 Yield optimization of the microstrip hairpin filter

Case[a]	Yield estimation method	Estimated yield	CPU cost[b]
Nominal design (minimax optimum)	Feature points	0.18	19
	EM-based Monte Carlo	0.19	500
Yield-optimized design	Feature points	0.25	19
	EM-based Monte Carlo	0.28	500

[a]Estimation for a uniform distribution with maximum deviations of 0.01 mm
[b]Estimation cost in number of EM analyses

14.5 Expedited Feature-Based Uncertainty Quantification Using Variable-Fidelity EM Simulations

This section illustrated expedited feature-based robust design employing variable-fidelity EM simulations. In Sections 14.5.1 and 14.5.2, we briefly outline two methods capitalizing on this idea and exploring correlations between computational models of different resolutions.

14.5.1 Variable-Fidelity Yield Optimization Using Second-Order Feature-Based Surrogates

For the purpose of this section, we denote the aggregated vector of feature points, i.e., containing both the frequency and the level coordinates, by $\boldsymbol{F}_f(\boldsymbol{x})$ (a vector of the size $2p \times 1$, where p is the number of feature points). A second-order surrogate model of the feature points is constructed as

$$s(z) = \boldsymbol{a}_0 + \boldsymbol{A}_1\left(\boldsymbol{x}^{(0)} + z\right) + \boldsymbol{A}_2\left[\boldsymbol{x}^{(0)} + z\right]^2 \quad (14.26)$$

in which $[\cdot]^2$ is a component-wise operation. In (14.26), $\boldsymbol{a}_0$ is a $2p \times 1$ vector, whereas $\boldsymbol{A}_1$ and $\boldsymbol{A}_2$ are $2p \times n$ matrices, where n is the number of coupler parameters. The coefficient matrix $\boldsymbol{A} = [\boldsymbol{a}_0 \boldsymbol{A}_1{}^T \boldsymbol{A}_2{}^T]^T$ is found by solving the regression problems

$$s\left(\boldsymbol{x}^{(0)} + \boldsymbol{x}_B{}^{(j)}\right) = \boldsymbol{F}\left(\boldsymbol{x}^{(0)} + \boldsymbol{x}_B{}^{(j)}\right), j = 1, \ldots, N_B \quad (14.27)$$

where $\boldsymbol{x}_B{}^{(j)}$ are the base designs. In this section, the design of experiments technique of choice is star distribution, with $\boldsymbol{x}_B{}^{(0)}$ being a zero vector and $\boldsymbol{x}_B{}^{(j)} = (-1)^j d\boldsymbol{e}^{(\lceil j/2 \rceil)}$ (here, $\boldsymbol{e}^{(j)}$ are the standard R^n basis vectors and $\lceil\cdot\rceil$ is the ceiling function); d is the maximum assumed parameter deviation (e.g., 0.05 mm). Note that the training set contains $2n + 1$ points, which means that the model coefficients are determined uniquely and the surrogate itself is interpolative.

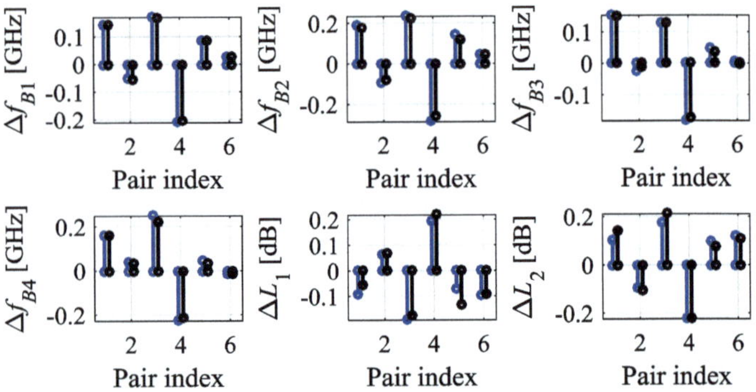

Fig. 14.16 Correlation between response features of the high- (left bars) and low-fidelity model (right bars). Shown are differences between selected feature point coordinates (both frequency and level) for several pairs of randomly generated designs

We denote by $\boldsymbol{F}_c(\boldsymbol{x})$ the response feature vector of the auxiliary low-fidelity EM model of the structure under design. The low-fidelity model is simulated using the same solver as the one used for the high-fidelity model but with coarser discretization of the circuit geometry. Typically, it is possible to reduce the evaluation time of the low-fidelity model by a factor at least 3–10, which is obtained at the expense of compromising the accuracy. Typical discrepancies between the high- and low-fidelity models include the frequency and level shifts. At the same time, the model correlation is usually good as shown in Fig. 14.16 for the microwave coupler considered later in this section.

To use the low-fidelity-based surrogate for yield estimation, it needs to be corrected. Here, the correction is based on a single high-fidelity evaluation (at the center design). It is implemented as

$$\boldsymbol{s}(\boldsymbol{z}) = \boldsymbol{s}_c(\boldsymbol{z}) + \left[\boldsymbol{F}_f\left(\boldsymbol{x}^{(0)}\right) - \boldsymbol{F}_c\left(\boldsymbol{x}^{(0)}\right)\right] \tag{14.28}$$

The correction is of a zero-order type; the example correction vectors have been shown in Fig. 14.17. Note that the computational cost of constructing the surrogate (14.28) is one high-fidelity and $2n + 1$ low-fidelity model evaluations.

Let $\boldsymbol{x}^{(0)}$ be the initial design (typically, the nominal design obtained assuming no parameter deviations) and d be the maximum parameter deviation. The latter is user-defined and would typically be equal to the maximum parameter deviation (for uniform probability distributions) or equal to 3σ (for Gaussian distributions). It should be noted that the probability distributions for all parameters are assumed independent; in general, a covariance matrix can be used to determine dependence along with individual maximum deviations for individual variables.

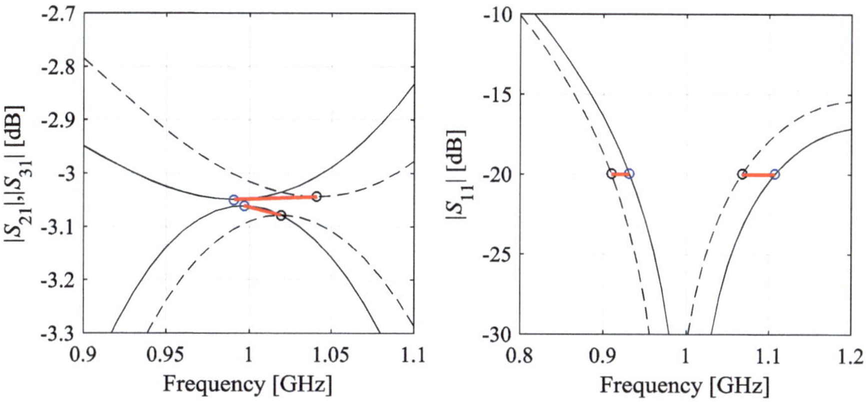

Fig. 14.17 Example responses of the low- (- - -) and high-fidelity model (—), corresponding feature points (○), as well as feature point correction vectors $\boldsymbol{F}_f(\boldsymbol{x}) - \boldsymbol{F}_c(\boldsymbol{x})$ marked using thick lines

The operation of the yield optimization procedure can be described as follows:

1. Set $i = 0$;
2. Acquire training data: $\boldsymbol{F}_f(\boldsymbol{x}^{(i)})$, $\boldsymbol{F}_c(\boldsymbol{x}^{(i)})$, $\boldsymbol{F}_c(\boldsymbol{x}^{(i)} + \boldsymbol{x}_B{}^{(j)})$, $j = 1, \ldots, 2n$, with $\boldsymbol{x}_B{}^{(j)} = (-1)^j 2d\boldsymbol{e}^{(\lceil j/2 \rceil)}$;
3. Construct the surrogate $\boldsymbol{s}^{(i)}(\cdot)$ as in (14.28) (i indicates the center design for surrogate construction);
4. Solve the subproblem:

$$z^* = \arg\min_{z} \left\{ -Y\left(\boldsymbol{s}^{(i)}(z)\right) \right\} \tag{14.29}$$

subject to

$$-d \le x_k \le d, \quad k = 1, \ldots, n \tag{14.30}$$

5. Set $\boldsymbol{x}^{(i+1)} = \boldsymbol{x}^{(i)} + z^*$ and $i = i + 1$;
6. Acquire training data: $\boldsymbol{F}_f(\boldsymbol{x}^{(i)})$, $\boldsymbol{F}_c(\boldsymbol{x}^{(i)})$, $\boldsymbol{F}_c(\boldsymbol{x}^{(i)} + \boldsymbol{x}_B{}^{(j)})$, $j = 1, \ldots, 2n$, with $\boldsymbol{x}_B{}^{(j)} = (-1)^j 2d\boldsymbol{e}^{(\lceil j/2 \rceil)}$;
7. Construct the surrogate $\boldsymbol{s}^{(i)}(\cdot)$ as in (14.28) (i indicates the center design for surrogate construction);
8. If $Y(\boldsymbol{s}^{(i)}(0)) > Y(\boldsymbol{s}^{(i-1)}(0))$ go to 4;
9. END.

The variable-fidelity feature-based surrogate is constructed at the beginning of each iteration at the cost of $2n + 1$ low-fidelity and one high-fidelity model evaluation. The surrogate domain is set to be twice the maximum parameter deviation, which gives a room for relocating the design. More specifically, the yield optimization problem (14.29) and (14.30) is solved within the domain of the size equal to d. Note that each evaluation of the objective function in (14.29) requires running Monte Carlo simulation within the interval of the size determined by d around the

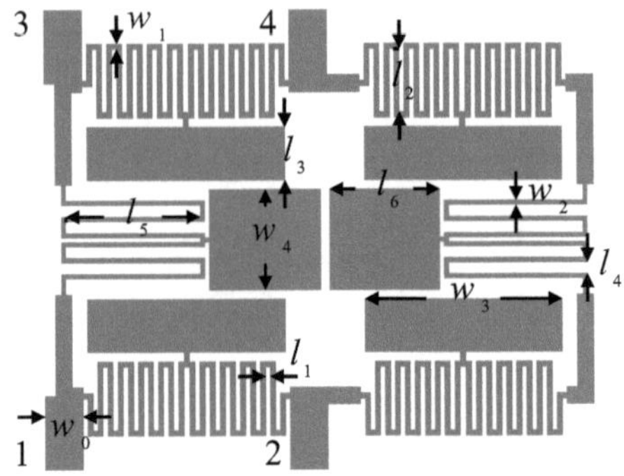

Fig. 14.18 Geometry of the compact microstrip rat-race coupler

evaluation point z. Yield estimation is conducted using the same set of 2000 pre-allocated samples that are remapped into the current model domain. This allows us to reduce the numerical noise related to a nonzero variance of the Monte-Carlo-evaluated yield. Procedure termination condition (Step 8) involves comparison of the yield estimated at the new design (using the new feature-based surrogate, more accurate at that design than the previous one) and the previous design. Lack of improvement terminates the process. In practice, a few iterations are sufficient. The problem (14.29) and (14.30) is solved using a pattern search algorithm (Koziel, 2010) as the objective function is non-smooth.

The yield optimization framework is demonstrated using an equal-split rat-race coupler (RRC). The circuit geometry has been shown in Fig. 14.18. The RRC is implemented on RO4003 substrate ($h = 0.813$ mm, $\varepsilon_r = 3.38$). The design parameters are $x = [w_1\ l_1\ l_2\ l_3\ w_2\ l_4\ l_5\ l_6]^T$; $w_0 = 1.7$ is fixed to ensure 50-Ω input impedance. The dependent variables are $y = [w_3\ w_4]^T$, where $w_3 = 20w_1 + 19\ l_1$ and $w_4 = 6w_2 + 7\ l_5$. All dimensions are in mm. The EM models are implemented in Keysight Momentum: high-fidelity (5600 mesh cells, simulation time ~5 min), low-fidelity (450 cells, ~25 s). The nominal design is $x^{(0)} = [0.41\ 0.20\ 2.19\ 1.88\ 0.22\ 0.43\ 5.01\ 7.18]^T$ mm. Design specifications are set as $dS_{\max} = 0.2$ dB and $B_{\min} = 100$ MHz.

The verification set consists of four cases, which are different with respect to the probability distributions describing geometry parameter deviations:

- Case I: uniform distribution with maximum deviation of 0.03 mm;
- Case II: uniform distribution, maximum deviation 0.05 mm;
- Case III: Gaussian distribution with zero mean and variance $\sigma = 0.01$ mm;
- Case IV: Gaussian distribution with zero mean and variance $\sigma = 0.017$ mm.

For validation purposes, the initial and final yield estimations obtained with feature-based surrogates have been compared to those found using direct Monte Carlo simulations at the high-fidelity EM model level. The results are gathered in Tables 14.8 and 14.9. Figure 14.19 provides visualization of yield estimation (using both direct Monte Carlo and feature-based surrogates) for the selected case. It can be observed that variable-fidelity feature-based framework allows for reliable yield

Table 14.8 Estimated yield at nominal and optimized designs for coupler of Fig. 14.18

Case	Design specifications/probability distribution[a]	Estimated yield		CPU cost[b]
		Nominal design	Yield-optimized design	
I	$dS_{max} = 0.2$ dB, $B_{min} = 100$ MHz U(0.03 mm)	0.56	0.77	13
II	$dS_{max} = 0.2$ dB, $B_{min} = 100$ MHz U(0.05 mm)	0.18	0.44	13
III	$dS_{max} = 0.2$ dB, $B_{min} = 100$ MHz G(0.01 mm)	0.83	0.96	8
IV	$dS_{max} = 0.2$ dB, $B_{min} = 100$ MHz G(0.017 mm)	0.59	0.80	10

[a]Probability distribution of geometry parameter deviations: $U(d)$—uniform with maximum deviation d, $G(\sigma)$—Gaussian with variance σ; dS_{max} stands for the acceptance threshold for the power split ration, B_{min} is the minimum −20 dB bandwidth for coupler matching and isolation responses
[b] Yield optimization cost in the number of equivalent high-fidelity EM analyses

Table 14.9 Feature-based surrogate model verification for coupler of Fig. 14.18

Case	Design	Yield estimation method	Estimated yield	CPU cost[a]
I	Nominal	Variable-fidelity feature-based	0.56	2.5
		EM-based Monte Carlo	0.59	500
	Optimized	Variable-fidelity feature-based	0.77	2.5
		EM-based Monte Carlo	0.77	500
II	Nominal	Variable-fidelity feature-based	0.18	2.5
		EM-based Monte Carlo	0.28	500
	Optimized	Variable-fidelity feature-based	0.44	2.5
		EM-based Monte Carlo	0.46	500
III	Nominal	Variable-fidelity feature-based	0.83	2.5
		EM-based Monte Carlo	0.81	500
	Optimized	Variable-fidelity feature-based	0.96	2.5
		EM-based Monte Carlo	0.94	500
IV	Nominal	Variable-fidelity feature-based	0.59	2.5
		EM-based Monte Carlo	0.63	500
	Optimized	Variable-fidelity feature-based	0.80	2.5
		EM-based Monte Carlo	0.78	500

[a]Estimation cost in the number of equivalent high-fidelity EM analyses. Feature-based yield estimation utilizes $N = 5000$ random samples

optimization at extremely low computational cost corresponding to about 11 high-fidelity coupler simulations on average. It is worth mentioning that a single yield estimation (equal to the cost of constructing the surrogate) is less than three high-fidelity EM analyses.

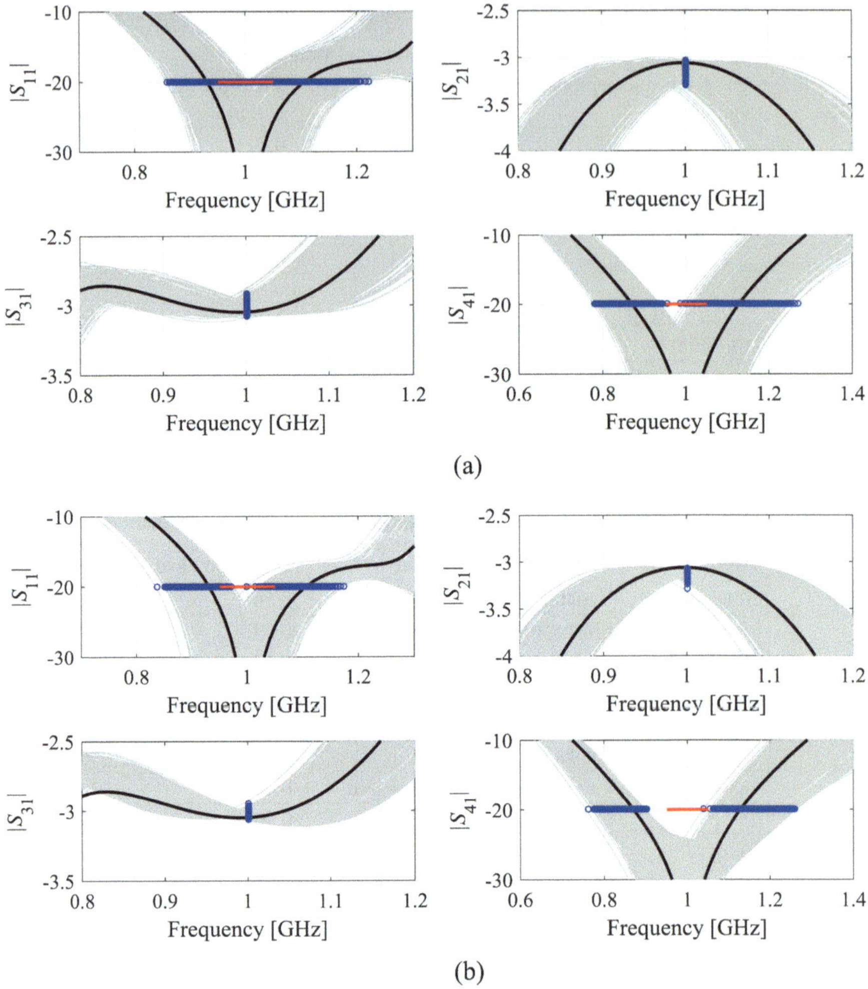

Fig. 14.19 Monte Carlo simulation of the coupler of Fig. 14.18 using EM simulations (gray plots) and variable-fidelity response feature surrogate model (circles) for Case I: (**a**) at the nominal design, (**b**) at yield-optimized design. Required operational bandwidth shown using a horizontal line; the nominal and yield-optimized design response shown in black in figures (**a**) and (**b**), respectively

14.5.2 Variable-Fidelity Yield Optimization Using Linear Feature-Based Surrogates and Trust Regions

In this section, we again use variable-fidelity EM simulations to accelerate optimization of the fabrication yield (Pietrenko-Dabrowska & Koziel, 2022). The properties of computational models of different resolutions are briefly recalled, followed by an exposition of the optimization procedure and an illustration example.

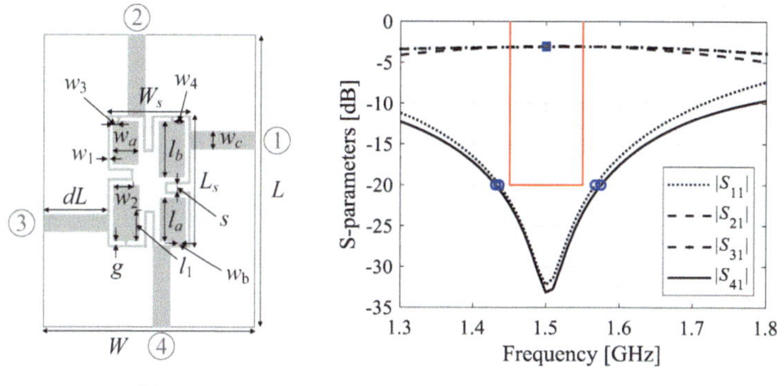

Fig. 14.20 Conceptual illustration of response features for a microwave coupler: S-parameters and feature points representing −20 dB level of $|S_{11}|$ (matching) and $|S_{41}|$ (isolation) responses (○), along with $|S_{21}|$ and $|S_{31}|$ (transmission) at the intended operating frequency $f_0 = 1.5$ GHz (□). The characteristic point coordinates permit assessment of performance requirements satisfaction (100 MHz matching/isolation bandwidth centered around f_0), and maximal power split deviation of 0.5 dB at f_0): (**a**) coupler geometry, (**b**) design fulfilling specifications and the response features selected to account for the circuit bandwidth and its power split ratio

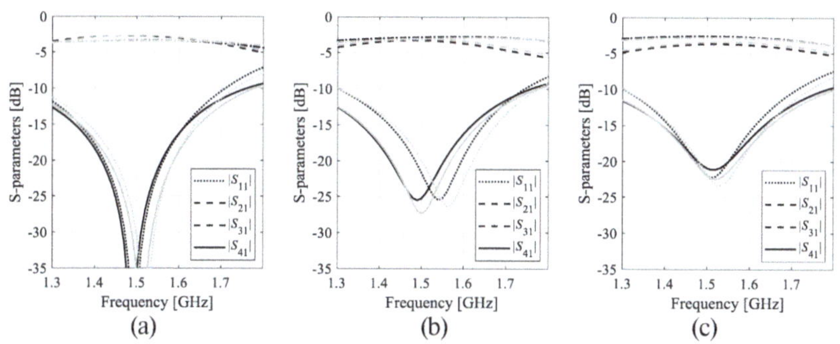

Fig. 14.21 S-parameters for the coupler considered in Fig. 14.20 evaluated using the low- (gray) and high-fidelity (black) EM simulation. Observe considerable misalignment, especially in terms of the frequency shifts, yet the overall shape and the amount of misalignment is consistent for different designs shown in the pictures (plots (**a**)–(**c**))

Consider a miniaturized branch-line coupler shown in Fig. 14.20, and its EM-simulated scattering parameters evaluated for the low- and high-fidelity model at three randomly generated designs, shown in Fig. 14.21. In this case, the high-fidelity model features about 80,000 mesh cells, whereas the low-fidelity model is set up with approximately 20,000 cells. As the robust design task is solved at the level of response features (cf. Fig. 14.20b), we are mainly interested in correlation between the coordinates of the feature points of low- and high-fidelity models, or, more specifically, the feature point sensitivities. Figure 14.22 shows the differentials $\Delta p_j = p_j(\boldsymbol{x}_k) - p_j(\boldsymbol{x}_l)$ of the high- and low-fidelity model computed for all pairs of

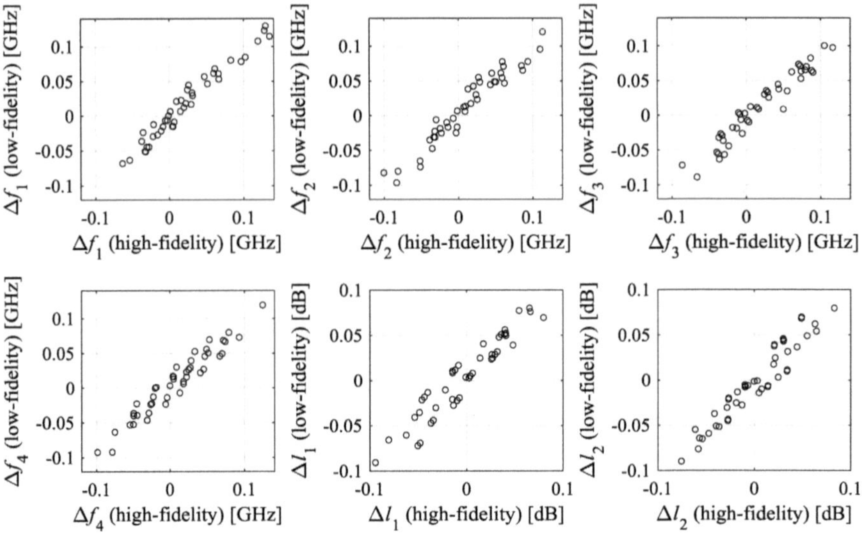

Fig. 14.22 Scatter plots of feature coordinate differentials $\Delta p_j = p_j(\boldsymbol{x}_k) - p_j(\boldsymbol{x}_l)$ computed for all pairs of ten designs $\{\boldsymbol{x}_k\}_{k=1,...,10}$, generated in the vicinity of a nominal vector $\boldsymbol{x}^{(0)}$ for a microwave coupler considered in Fig. 14.20. Horizontal and vertical axis correspond to the differentials of the high- and low-fidelity model, respectively

designs selected from a ten-element random set $\{\boldsymbol{x}_k\}_{k=1,\ldots,10}$, generated in the vicinity of a nominal vector $x^{(0)}$ for a coupler of Fig. 14.20a. It can be observed that the correlation between the models is excellent with the linear correlation coefficients equal to 0.99, 0.97, 0.97, 0.97, 0.96, and 0.96 for f_1, f_2, f_3, f_4, l_1, and l_2, respectively. This means that response feature sensitivity predicted by the low-fidelity model will be in a good agreement with sensitivity evaluated with the high-fidelity one. Still, the computational expenditures associated with sensitivity estimation is considerably lower, typically, by a factor 3–4.

The core of the yield optimization procedure follows the technique introduced in Pietrenko-Dabrowska and Koziel (2021), which is enhanced by utilization of the variable-resolution EM models. The yield is maximized in an iterative fashion using the trust-region (TR) procedure (Conn et al., 2000).

In every iteration, a new design $\boldsymbol{x}^{(i+1)}$ being an approximation of the optimum vector $\boldsymbol{x}^*$ is produced, $i = 0, 1, \ldots$ (here, the nominal design $\boldsymbol{x}^{(0)}$ is the starting point), by solving

$$\boldsymbol{x}^{(i+1)} = \arg \min_{\left\| \boldsymbol{x} - \boldsymbol{x}^{(i)} \right\| \le d^{(i)}} \{ -Y_P(\boldsymbol{x}) \} \tag{14.31}$$

The yield Y_P is computed at the level or response features as in (14.13)–(14.15), with the feature point coordinates predicted using a linear (regression) model $\boldsymbol{L}^{(i)}$ of $\boldsymbol{P}(\boldsymbol{x})$ comprising the response features

$$L^{(i)}(x) = P\left(x^{(i)}\right) + J_{P.L}\left(x^{(i)}\right) \cdot \left(x - x^{(i)}\right) \tag{14.32}$$

in which $J_{P.L}$ is the approximation of the Jacobian matrix J_P

$$J_P\left(x^{(i)}\right) = \left[\frac{\partial p_j\left(x^{(i)}\right)}{\partial x_k}\right]^T_{\substack{j=1,\ldots,2N \\ k=1,\ldots,n}} \tag{14.33}$$

estimated by finite differentiation with the use of the low-fidelity EM model. Let us denote the time evaluation ratio M between the high- and low-fidelity model. The cost of constructing the model $L^{(i)}$ equals to $1 + n/M$, where n is the parameter space dimensionality. Assuming conservatively that $M = 3$, the computational savings reach 60% for $n = 10$.

Adjusting the size parameter $d^{(i)}$ is an important consideration in TR frameworks. It is normally done based on the gain ratio r, which is defined as a ratio between the observed improvement of the merit function (here, the yield) and the improvement predicted by the linear model.

It should be noted that evaluation of the yield at the candidate design $x^{(i+1)}$ requires rebuilding of the linear model at $x^{(i+1)}$, which would turn to be a waste of computational resources if $r < 0$, i.e., if the candidate design is rejected; according to the TR principles (Conn et al., 2000), the design $x^{(i+1)}$ is only retained if $r > 0$). For the sake of cost savings, in this section, the gain factor is evaluated as $r = [Y_{P\#}(x^{(j+1)}) - Y_P(x^{(j)})]/[Y_P(x^{(j+1)}) - Y_P(x^{(j)})]$, where $Y_{P\#}$ is obtained from a linear model defined as in (14.32); however, instead of $P(x^{(i)})$ the vector $P(x^{(i+1)})$ derived from EM-evaluated system output for $x^{(i+1)}$. This only requires one EM analysis. The approximation is due to using the same Jacobian matrix for both $x^{(i)}$ and $x^{(i+1)}$, which is tenable as the distance between these two vectors is normally small (comparable to or smaller than the maximum assumed parameter deviation).

The actual yield estimation with the linear predictor $L^{(i)}$ is executed as Monte-Carlo-based integrating of (14.8), with the use of a large number of randomly allocated observables to reduce the estimation variance. For this purpose, the feature-based performance specifications (14.13)–(14.15) are verified for the output of the model $L^{(i)}$ obtained for each observable generated using the probability distribution assumed for input tolerances. The cost of this process is small in comparison to EM simulation of the circuit.

The flow diagram of the yield enhancement procedure considered in this section has been presented in Fig. 14.23. As emphasized before, we use the nominal design $x^{(0)}$ as the starting point for robust design process. The high-resolution EM model is only utilized to evaluate the circuit response at the current design, whereas response feature sensitivities are estimated using the low-fidelity model. The linear regression model is constructed upon extracting feature point coordinates, and utilized to obtain the candidate design $x^{(i+1)}$. The gain ratio is evaluated as explained above. It is used to decide about the acceptance of the candidate design and to adjust the search region

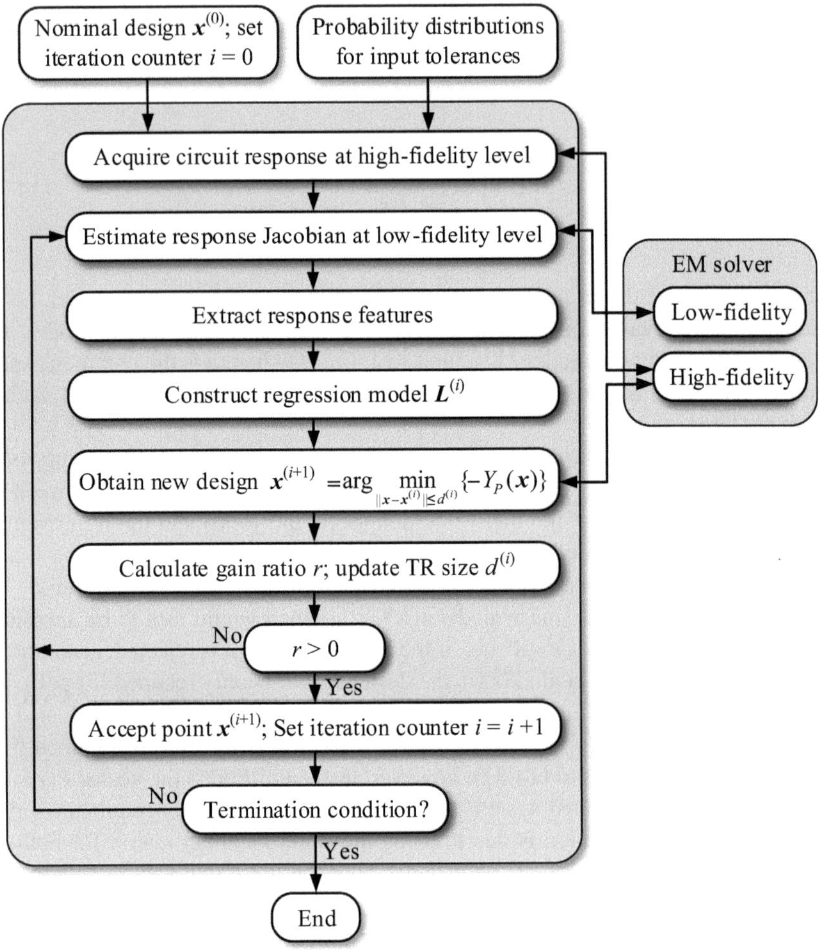

Fig. 14.23 Flow diagram of the considered yield optimization procedure

size parameter $d^{(i)}$. The algorithm is terminated if either $\|x^{(i+1)} - x^{(i)}\| < \varepsilon$ (convergence in argument) or $d^{(i+1)} < \varepsilon$ (reduction of the TR size). In the numerical experiments discussed below, we employ the termination threshold $\varepsilon = 10^{-3}$.

For the sake of illustration, consider the coupler of Fig. 14.20. Its essential parameters including design specifications have been gathered in Table 14.10. The input parameter tolerances are assumed to follow independent zero-mean Gaussian distributions with 0.017 mm variance, and maximum deviations limited to $d_{max} = 0.05$ mm. The circuit yield has been optimized using the variable-fidelity feature-based procedure described in this section, as well as four benchmark algorithms. The first three have been described in Sect. 14.3.3 (Algorithms 1–3), whereas the fourth one is the feature-based procedure as described above but using the high-

Table 14.10 Essential parameters of the coupler circuit of Fig. 14.20

Parameter	Value
Substrate	AD300 ($\varepsilon_r = 2.97$, $h = 0.76$ mm)
Design parameters	$\boldsymbol{x} = [g\ l_{1r}\ l_a\ l_b\ w_1\ w_{2r}\ w_{3r}\ w_{4r}\ w_a\ w_b]^T$
High-fidelity model[a]	~85,000 mesh cells; simulation time 90 s
Low-fidelity model[a]	~27,000 mesh cells; simulation time 33 s
Other parameters	$L = 2dL + L_s$, $L_s = 4w_1 + 4\ g + s + l_a + l_b$, $W = 2dL + W_s$, $W_s = 4w_1 + 4\ g + s + 2w_a$, $l_1 = l_b l_{1r}$, $w_2 = w_a w_{2r}$, $w_3 = w_{3r} w_a$, $w_4 = w_{4r} w_a$
Operating bands	1.45–1.55 GHz
Maximum power split error	0.5 dB at 1.5 GHz
Nominal design	$\boldsymbol{x}^{(0)} = [0.63\ 5.90\ 9.34\ 12.45\ 1.29\ 2.02\ 0.99\ 0.32\ 2.81\ 0.22]^T$

[a]EM models are implemented in CST Microwave Studio and evaluated using the time domain solver

[b]Parameters with subscript r are relative, and their deviations are recalculated accordingly in order to have the corresponding absolute parameters following the assumed probability distribution (here, Gaussian with variance of 0.017 mm)

Table 14.11 Design centering results for the coupler of Fig. 14.20

	Initial yield (%)		Optimized yield (%)		
Optimization algorithm	Estimated by surrogate model	EM-based	Estimated by surrogate model	EM-based	CPU cost[a]
Reference algorithm 1	82	77	93	88	800
Reference algorithm 2	76	77	94	93	320
Reference algorithm 3	79	77	92	93	112
Reference algorithm 4	79	77	90	92	37
Variable-fidelity feature-based algorithm	79	77	94	92	23.1[b]

[a]Optimization cost in number of EM simulations of the considered circuit

[b]Equivalent number of high-fidelity EM simulations (actual number of analyses was 7 high-fidelity and 44 low-fidelity)

fidelity EM model. The yield-optimized design obtained using the variable-fidelity approach is $\boldsymbol{x}^* = [0.63\ 5.98\ 9.21\ 12.53\ 1.29\ 2.08\ 0.99\ 0.31\ 2.82\ 0.28]^T$.

Table 14.11 gathers the comparison data. As expected, the incorporation of variable-resolution models leads to significant improvement of the cost efficacy of the yield optimization routine. It is already remarkably low for Algorithms 3 and 4, with the average number of high-fidelity EM simulations being 112 and 37, respectively. Yet, the variable-fidelity approach brings the cost even lower, to about 23 the average of 20, which corresponds to 36% saving over Algorithm 4 and 89% savings over Algorithm 3.

In terms of design quality, the solution obtained using the variable-fidelity approach are comparable to that identified using the benchmark methods. The same can be said about reliability, as confirmed through EM-based Monte Carlo simulations (see Fig. 14.24). It should be reiterated that the variance of the MC-estimated yield is relatively high (up to 2%), as emphasized before, due to a

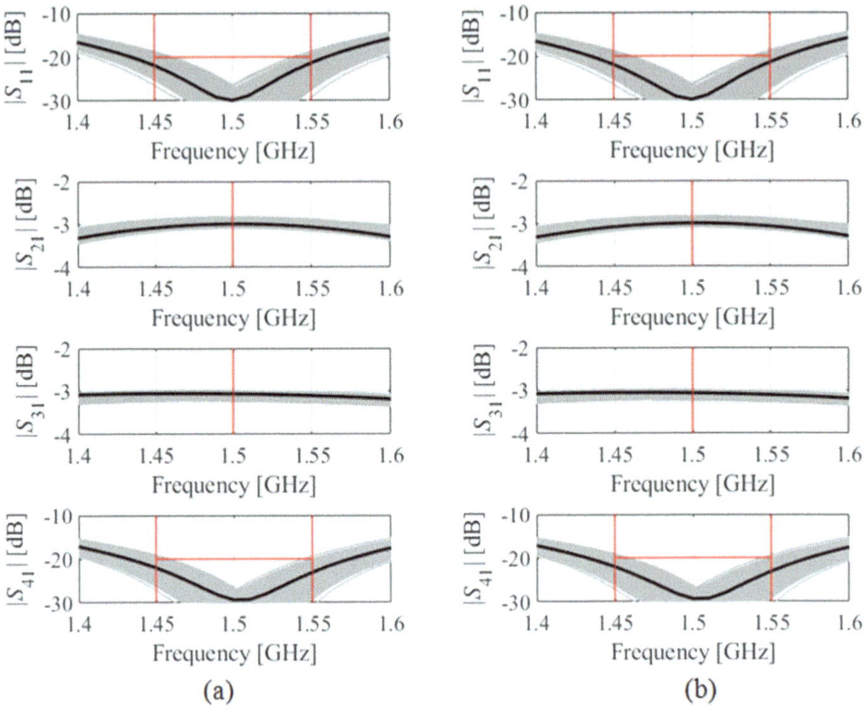

Fig. 14.24 Visualization of EM-based Monte Carlo analysis for the circuit of Fig. 14.20 at the (**a**) nominal design and (**b**) optimal design rendered with the use of the procedure considered in this section. MC is executed using 500 random data points. Gray curves represent EM simulations; circuit characteristics at the nominal (**a**) and optimal design (**b**) are shown black

relatively low number of observables used in the process. This means that the yield differences of up to 2% or 3% are statistically insignificant.

14.6 Feature-Based Tolerance Optimization

The formulation of the robust design problem considered in this chapter so far was based on maximizing the fabrication yield given the probability density function that describes the deviations of geometry parameters from their nominal values. In this section, we investigate a different approach, which is somehow related to the maximum input tolerance hypervolume (MITH) improvement (Wu et al., 2021). More specifically, we aim at maximizing the input parameter deviations as much as possible while maintaining 100% yield. In the numerical experiments shown below, we assume that the deviations are allocated according to independent Gaussian distributions with common variance σ; therefore, the task is formulated as

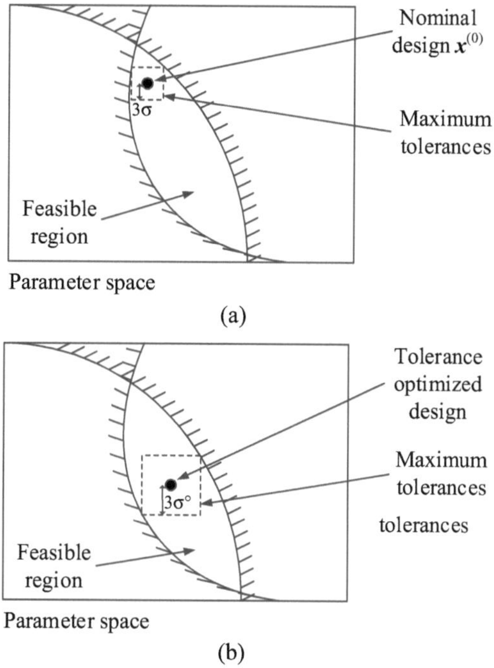

Fig. 14.25 Graphical illustration of (input) tolerance optimization: (**a**) (feasible) nominal design $x^{(0)}$ and the area marked using the dashed line, corresponding to maximum values of parameter deviations that still ensure 100% yield (here, independent Gaussian distributions assumed with common variance σ); (**b**) optimized design, featuring larger value of variance, for which 100% yield is possible. Note that the nominal objective function value may be lower at $x^{(0)}$ than at the optimized design

$$x^* = \arg\min_{x} U_Y(x) \tag{14.34}$$

where the objective function U_Y is defined as

$$U_Y(x) = -\arg\max_{\sigma} \{Y(x, \sigma) = 1\} \tag{14.35}$$

In other words, the value of the objective function at the design x is the maximum value of variance σ for which the yield (here, marked as depending on both x and σ) is still equal to 1. This can be readily generalized to other types of probability distributions controlled by one or more coefficients (variances, maximum deviations, etc.). More details about U_Y and its numerical evaluation will be provided in Sect. 14.6.2. Figure 14.25 provides a graphical illustration of the tolerance optimization process according to (14.34) and (14.35).

14.6.1 Feature-Based Regression Model

To ensure computational efficiency and reliability, in this section, yield estimation will be realized using a response feature approach, specifically, feature-based regression models. The approach employed here will be explained using a case of multiband antennas optimized for impedance bandwidth. In such a case, the feature

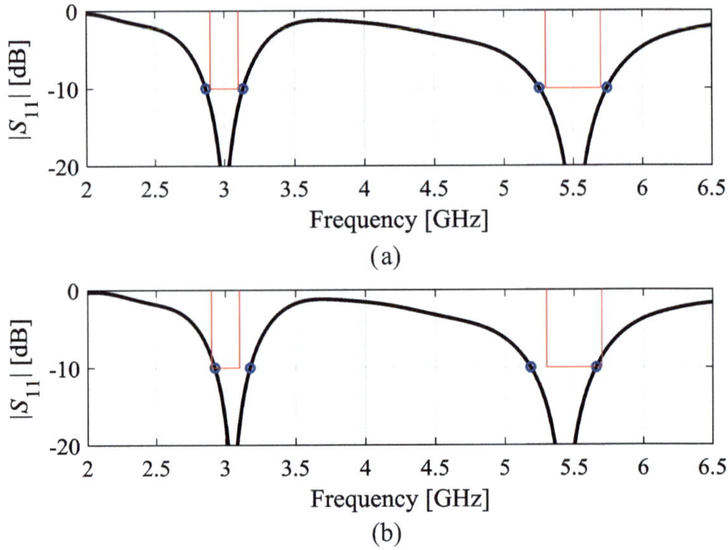

Fig. 14.26 Reflection response of a dual-band antenna (—) and the corresponding response features (○), along with exemplary design specifications (thin lines). The frequency coordinates of the feature points are sufficient to determine satisfaction of design specifications on impedance matching: (**a**) design satisfying specifications, (**b**) design violating specifications

points are selected to be the frequencies corresponding to -10 dB level of $|S_{11}|$. Examples of reflection response of a dual-band antenna along with its characteristic points have been shown in Fig. 14.26. As mentioned before, for other design scenarios, a different feature point setup might be necessary. Let f_{0k}, $k = 1, \ldots, N$, be the target operating frequencies, and let B_k denote the target fractional bandwidth over which the reflection coefficient should not exceed $S_{\max}$ (typically $S_{\max} = -10$ dB).

The matching requirements are satisfied as

$$\max\left\{f \in \cup_{k=1}^{N}[(1-B_k/2)f_{0k}, (1+B_k/2)f_{0k}] \quad : |S_{11}(\boldsymbol{x},f)| \le S_{\max}\right\} \tag{14.36}$$

where f denotes the frequency and S_{11} is marked as explicitly depending on $\boldsymbol{x}$ and f. The feature vector $\boldsymbol{P}$ at the design $\boldsymbol{x}$ is defined as

$$\boldsymbol{P}(\boldsymbol{x}) = [p_1(\boldsymbol{x}) \ p_2(\boldsymbol{x}) \ \ldots \ p_{2N}(\boldsymbol{x})]^T = [f_1(\boldsymbol{x}) \ f_2(\boldsymbol{x}) \ \ldots \ f_{2N}(\boldsymbol{x})]^T \tag{14.37}$$

where f_{2k-1} and f_{2k} are the frequencies corresponding to -10 dB level of $|S_{11}|$ for the kth operating band of the antenna, $k = 1, \ldots, N$. The feature points can be extracted from the results of EM analysis of the structure.

The performance specifications (14.36) can be expressed in terms of the feature vector $\boldsymbol{P}$ as follows:

$$p_{2k-1}(\boldsymbol{x}) \leq f_{0.k} - B_k/2, \qquad p_{2k}(\boldsymbol{x}) \geq f_{0.k} + B_k/2, k = 1, \ldots, N \tag{14.38}$$

Let $\boldsymbol{x}^{(i)}$ be the current design produced by the optimization algorithm. The regression model of the feature vector $\boldsymbol{P}$ is established to be a linear model $\boldsymbol{L}_P{}^{(i)}(\boldsymbol{x})$, defined as

$$L_P^{(i)}(\boldsymbol{x}) = [p_{L.1}(\boldsymbol{x}) \ \ldots \ p_{L.2N}(\boldsymbol{x})]^T = \begin{bmatrix} l_{0.1} + \boldsymbol{L}_1^T(\boldsymbol{x} - \boldsymbol{x}^{(i)}) \\ \vdots \\ l_{0.2N} + \boldsymbol{L}_{2N}^T(\boldsymbol{x} - \boldsymbol{x}^{(i)}) \end{bmatrix} \tag{14.39}$$

The model is identified using $n + 1$ pairs $\{\boldsymbol{x}_B{}^{(j)}, \boldsymbol{P}(\boldsymbol{x}_B{}^{(j)})\}_{j = 1, \ldots, n+1}$, with $\boldsymbol{x}_B{}^{(1)} = \boldsymbol{x}^{(i)}$ and $\boldsymbol{x}_B{}^{(j)} = \boldsymbol{x}^{(i)} + [0 \ldots 0 \; d \; 0 \ldots 0]^T$ with d on the $(j - 1)$th position. In the numerical experiments, $d = 3\sigma$, i.e., it (approximately) corresponds to the maximum parameter deviation. Using these, we get

$$\begin{bmatrix} l_{0.j} \\ \boldsymbol{L}_j \end{bmatrix} = \begin{bmatrix} 1 & \left(\boldsymbol{x}_B^{(1)} - \boldsymbol{x}^{(i)}\right)^T \\ & \vdots \\ 1 & \left(\boldsymbol{x}_B^{(n+1)} - \boldsymbol{x}^{(i)}\right)^T \end{bmatrix}^{-1} \begin{bmatrix} \left[\boldsymbol{P}\left(\boldsymbol{x}_B^{(1)}\right)\right]_j \\ \vdots \\ \left[\boldsymbol{P}\left(\boldsymbol{x}_B^{(n+1)}\right)\right]_j \end{bmatrix}, \quad j = 1, \ldots, 2N \tag{14.40}$$

Here, $[\boldsymbol{P}(\boldsymbol{x}_B{}^{(k)})]_j$ is the jth component of the feature vector $\boldsymbol{P}(\boldsymbol{x}_B{}^{(k)})$.

14.6.2 Evaluating Objective Function: Tolerance Optimization Algorithm

The objective function $U_Y(\boldsymbol{x})$ for tolerance analysis has been defined in (14.35). For the purpose of the optimization process, it is evaluated using numerical integration of (14.8) with the aid of the feature-based regression model (14.39). In the ith iteration of the optimization algorithm, we use the model $\boldsymbol{L}_P{}^{(i)}(\boldsymbol{x})$, which provides predictions of the response features and allows for verifying the conditions (14.38). The yield $Y(\boldsymbol{x}, \sigma)$ is estimated using a large number of random samples $\boldsymbol{x}_r{}^{(j)}$ (here, 10,000) generated according to the assumed probability distribution characterized by the variance σ. The yield evaluation procedure works as follows:

1. Given the variance σ, generate random set $\{\boldsymbol{x}_r{}^{(j)}\}_{j = 1, \ldots, Nr}$;
2. Evaluate regression model $\boldsymbol{L}_P{}^{(i)}(\boldsymbol{x}_r{}^{(j)})$ for $j = 1, \ldots, N_r$;
3. Evaluate conditions (14.38) for all random observables using predicted feature points $p_{L.k}(\boldsymbol{x}_r{}^{(j)})$, $j = 1, \ldots, N_r$;
4. Estimate the yield $Y(\boldsymbol{x}, \sigma)$ according to (14.10) (the function H therein assumes the value of 1 if conditions (14.38) are satisfied, and 0 otherwise).

Due to using a large number of random samples, the variance of yield estimation is low. At the same time, as all the steps in the above procedure are vectorized, the yield computation is extremely fast. In particular, evaluation of regression model for all random observables can be arranged as a single matrix multiplication due to the regression model being a linear function of its coefficients.

To evaluate the objective function, the problem (14.35) is solved using a golden ratio search procedure (Press et al., 2007) (other methods, e.g., gradient-based algorithms, can be used if the probability distribution is characterized by multiple parameters, e.g., a covariance matrix). When evaluating the yield for different values of variance σ, the same set of random observables is used, generated using a unity variance and scaled accordingly. For the purpose of subsequent considerations, the objective function $U_Y(\boldsymbol{x})$ evaluated at the ith iteration of the tolerance optimization procedure with the use of the regression model $\boldsymbol{L}_P^{(i)}$ will be identified as $U_Y^{(i)}(\boldsymbol{x})$.

In this section, tolerance optimization is arranged as an iterative process following the trust-region (TR) principles (Conn et al., 2000). At each iteration, a new approximation $\boldsymbol{x}^{(i)}$ of the optimum design $\boldsymbol{x}^*$ is generated as

$$\boldsymbol{x}^{(i+1)} = \arg \min_{\left\|\boldsymbol{x}-\boldsymbol{x}^{(i)}\right\| \leq d^{(i)}} U_Y^{(i)}(\boldsymbol{x}) \tag{14.41}$$

The search process is subject to constraint $||\boldsymbol{x} - \boldsymbol{x}^{(i)}|| \leq d^{(i)}$, where the trust region size is adaptively altered using the standard TR procedures (Conn et al., 2000). Having $\boldsymbol{x}^{(i+1)}$, the gain ratio is calculated as

$$r = \frac{U_Y^{\#(i)}\left(\boldsymbol{x}^{(i+1)}\right) - U_Y^{(i)}\left(\boldsymbol{x}^{(i)}\right)}{U_Y^{(i)}\left(\boldsymbol{x}^{(i+1)}\right) - U_Y^{(i)}\left(\boldsymbol{x}^{(i)}\right)} \tag{14.42}$$

where $U_Y^{\#(i)}$ is calculated the same way as U_Y; however, with the regression model $\boldsymbol{L}_P^{(i)}$ replaced by the model $\boldsymbol{L}_P^{\#(i)}$, which is constructed as in (14.39) and (14.40) but with the coefficient vector $[l_{0.1} \ldots l_{0.2N}]^T$ replaced by $\boldsymbol{P}(\boldsymbol{x}^{(i+1)})$. The latter is extracted from EM-simulated responses at $\boldsymbol{x}^{(i+1)}$. Note that using $U_Y^{\#(i)}$ is a CPU-efficient way of verifying the candidate design $\boldsymbol{x}^{(i+1)}$ as it only requires a single EM analysis. The assumption made here is that the feature point gradients are relatively stable around the current design (i.e., they do not change considerably between $\boldsymbol{x}^{(i)}$ and $\boldsymbol{x}^{(i+1)}$). This is justifiable because $||\boldsymbol{x}^{(i+1)} - \boldsymbol{x}^{(i)}||$ is normally comparable to the variance σ.

The candidate design $\boldsymbol{x}^{(i+1)}$ is accepted if $r > 0$, otherwise it is rejected and the iteration is repeated with reduced TR size $d^{(i)}$. The algorithm termination is based on the requested resolution of the optimization process. Here, we use the conditions $||\boldsymbol{x}^{(i+1)} - \boldsymbol{x}^{(i)}|| < \varepsilon$ or $d(i) < \varepsilon$, with $\varepsilon = 10^{-3}$. The flow diagram of the tolerance optimization procedure has been shown in Fig. 14.27.

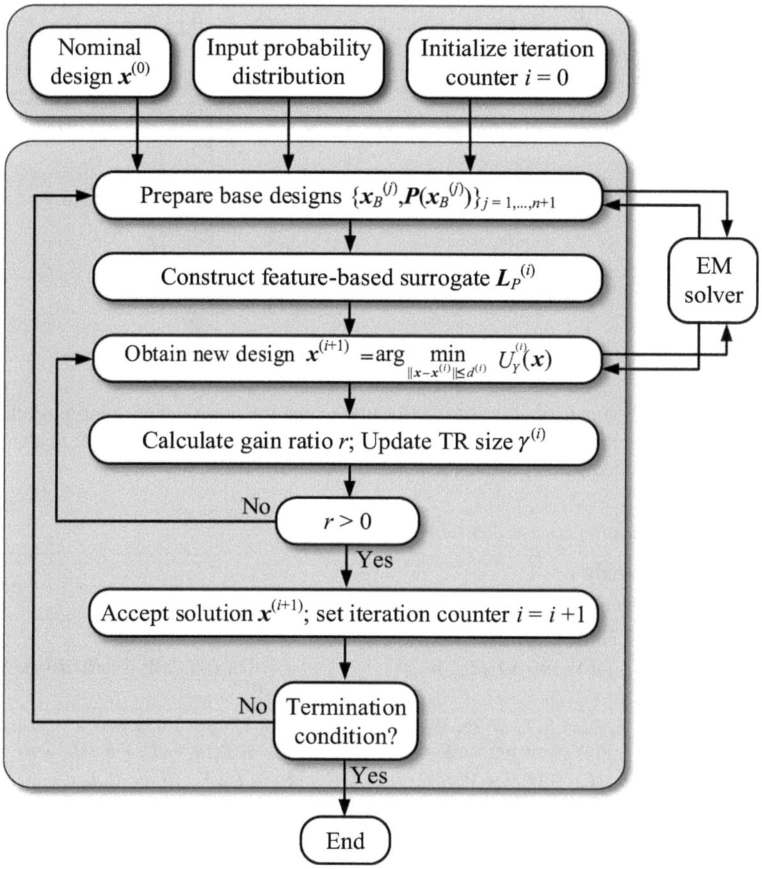

Fig. 14.27 Flow diagram of the feature-based algorithm for input tolerance optimization

14.6.3 Demonstration Examples 1: Antenna Design

Consider a dual-band dipole antenna with a truncated substrate (Qudrat-E-Maula & Shafai, 2014), and a triple-band U-slotted patch with L-slot defected ground structure (Consul, 2015) shown in Fig. 14.28. The relevant information about the structures (dielectric substrate, design variables, nominal design, target operating bands) has been gathered in Table 14.12. The EM simulation models of the considered antennas are implemented in CST Microwave Studio and simulated using the time-domain solver. The fabrication tolerances are modeled using independent Gaussian distributions with zero mean and variance σ. The maximum parameter deviations are restricted to 3σ. Furthermore, in the case of relative parameters, the tolerances are scaled to adhere to the assumed input distributions at the level of the corresponding absolute variables (according to the parameter relationships as described in Table 14.12).

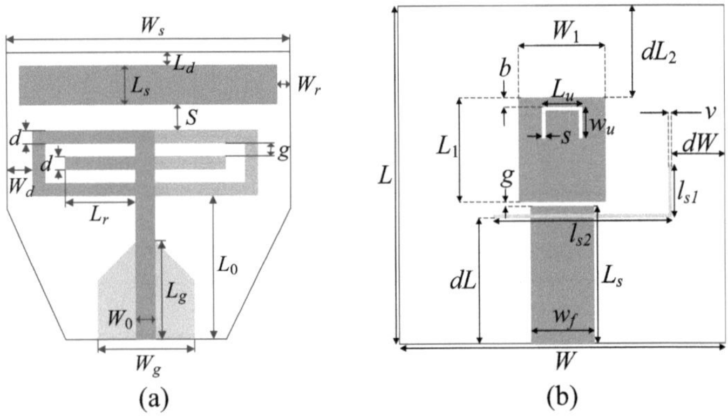

Fig. 14.28 Antenna structures used for verification of the tolerance optimization procedure: (**a**) dual-band dipole antenna with truncated substrate (Antenna I) (Quadrat-E-Maula & Shafai, 2014), (**b**) triple-band U-slotted patch antenna (Antenna II) (Consul, 2015)

Table 14.12 Verification case studies (antenna design)

	Case study	
	Antenna I	Antenna II
Substrate	$\varepsilon_r = 3.38$, $h = 0.81$ mm	$\varepsilon_r = 3.2$, $h = 3.1$ mm
Design parameters	$\boldsymbol{x} = [L_{rr}\ d\ W_s\ W_d\ S\ L_d\ L_{gr}\ W_{gr}]^T$	$\boldsymbol{x} = [L_1\ L_s\ L_{ur}\ W\ W_1\ dL_r\ dW_r\ g\ l_{s1r}\ l_{s2r}\ w_{ur}]^T$
Other parameters	$W_r = 5$, $L_s = 5$, $L_0 = 25$, $W_0 = 1.9$, $L_r = L_{rr}((W_s - W_0)/2 - W_d - d)$, $L_g = L_{gr}(L_0 - W_g/2 + W_0/2)$, $W_g = W_{gr}W_s$, $g = W_d$	$b = 1$, $w_f = 7.4$, $s = 0.5$, $w = 0.5$, $dL_2 = L_1$; $L = L_s + g + L_1 + dL_2$, $L_u = L_{ur}W_1$, $dL = dL_rL$, $dW = dW_rW$, $l_{s1} = l_{s1r}(L - dL)$, $l_{s2} = l_{s2r}(W - dW)$, $w_u = w_{ur}(L_1 - b - s)$
Operating bands	3.35–3.65–4.05–4.35 GHz	120 MHz bandwidth centered at operating frequencies 3.5, 5.8, and 7.5 GHz
Nominal design	$\boldsymbol{x}^{(0)} = [0.91\ 1.33\ 48.11\ 3.45\ 2.29\ 5.36\ 0.99\ 0.38]^T$	$\boldsymbol{x}^{(0)} = [11.52\ 19.03\ 0.60\ 52.47\ 10.81\ 0.44\ 0.20\ 0.32\ 0.21\ 0.55\ 0.20]^T$

The numerical results have been provided in Table 14.13. The optimized parameter vectors for the three considered antennas are:

- Antenna I: $\boldsymbol{x}^* = [0.91\ 1.22\ 48.06\ 3.47\ 2.46\ 5.27\ 1.00\ 0.39]^T$
- Antenna II: $\boldsymbol{x}^* = [11.53\ 19.04\ 0.59\ 52.48\ 10.79\ 0.44\ 0.19\ 0.30\ 0.21\ 0.55\ 0.20]^T$

The computational cost of yielding the final designs corresponds to only 58 and 69 full-wave simulations of Antenna I and II, respectively. This level of computational efficiency is possible due to utilization of the response feature approach, which allows for setting up simple (i.e., linear) yet reliable predictors for uncertainty quantification of the structures.

The design relocation is the largest for Antenna I and quite small for Antenna II, whose operating bands are the most sensitive to geometry parameter changes. Also,

Table 14.13 Tolerance optimization results for the considered antenna structures

Antenna structure		Antenna I	Antenna II
Nominal design	Maximum variance σ ensuring 100% yield[a]	7.5 μm	13.1 μm
	Maximum parameter deviation ensuring 100% yield	22.5 μm	39.2 μm
	EM-based yield estimation[b]	100%	100%
Tolerance-optimized design	Maximum variance σ ensuring 100% yield[a]	27.8 μm	17.4 μm
	Maximum parameter deviation ensuring 100% yield	83.5 μm	52.2 μm
	EM-based yield estimation[b]	99.4%	100%
Optimization cost[c]		58	69

[a] σ refers to the variance of the independent zero-mean Gaussian distributions assumed to describe the fabrication tolerances. The maximum assumed parameter deviations are 3σ
[b] Estimation obtained using Monte Carlo simulation based on 500 random samples
[c] Optimization cost in terms of the number of EM analyses of the antenna structure

Antenna II is a three-band structure, meaning that we have a larger number of conditions that need to be satisfied to fulfill design requirements.

Notwithstanding, the increase of the input tolerance levels ensuring 100% yield is significant, from 22.5 to 83.5 μm (Antenna I) and from 39.2 to 52.2 μm (Antenna II), which corresponds to multiplication factors of almost 4 and 1.3 (average of 2.5) for Antennas I and II, respectively.

The reliability of the procedure has been validated using direct EM-driven Monte Carlo analysis at the nominal and tolerance-optimized designs, as reported in Table 14.13 and graphically illustrated in Figs. 14.29 and 14.30. In both cases, the estimated yield is 100% or close to it, whereas the plots in Figs. 14.29 and 14.30 indicate that the optimized designs exhibit considerably improved robustness: the spread of the antenna responses corresponding to the random observables generated during the Monte Carlo analysis is much larger for optimized designs than for the nominal ones.

14.6.4 Demonstration Examples 1: Microwave Design

In this section, the discussed tolerance optimization methodology is demonstrated using two microstrip filters: (1) an upper UWB-band (5.8–10.6 GHz) microstrip filter using a stepped-impedance resonator (SIR) (Circuit I, Zukocinski, 2018), and (2) a compact wideband filter using shunt resonators coupled through admittance inverters (Circuit II, Sans et al., 2014), both shown in Fig. 14.31. Figure 14.32 shows the definition of the feature points. The necessary data pertaining to the considered structures can be found in Table 14.14. The computational models of all circuits are evaluated using the time-domain solver of CST Microwave Studio. The input (manufacturing) tolerances are represented by means of independent normal

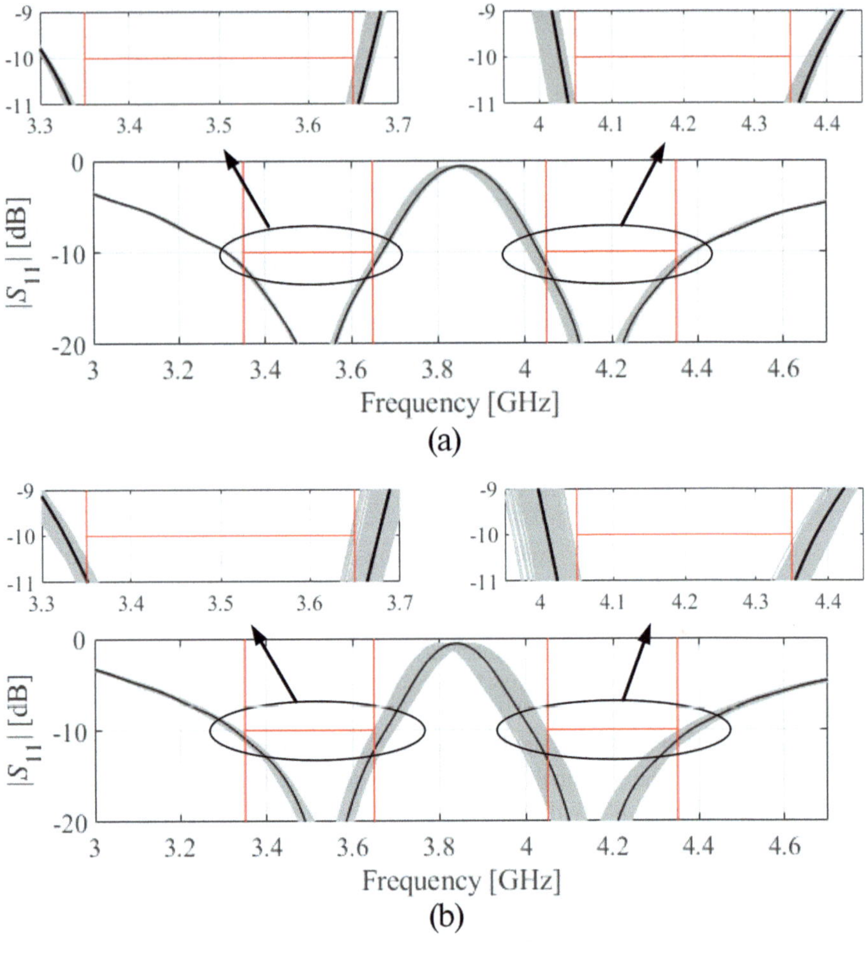

Fig. 14.29 Monte Carlo analysis of Antenna I (Fig. 14.28a) using EM simulations (gray plots) at: (**a**) the nominal design and (**b**) the design found using the discussed procedure. Black plots show the antenna response at the nominal and optimized designs, respectively

distributions with zero mean and a common variance σ. Further, the maximum parameter deviations are limited to 3σ. It should be noted that some of the circuit parameters are relative, in which case their parameter deviations are recalculated (scaled) to comply with the assumed input tolerances for the corresponding absolute parameters (based on the functional dependencies provided in Table 14.14).

Table 14.15 gathers the numerical results, in particular, the variance and maximum parameter deviation ensuring 100% yield at the nominal design and after applying the robustness enhancement procedure. The obtained final designs for the considered circuits are:

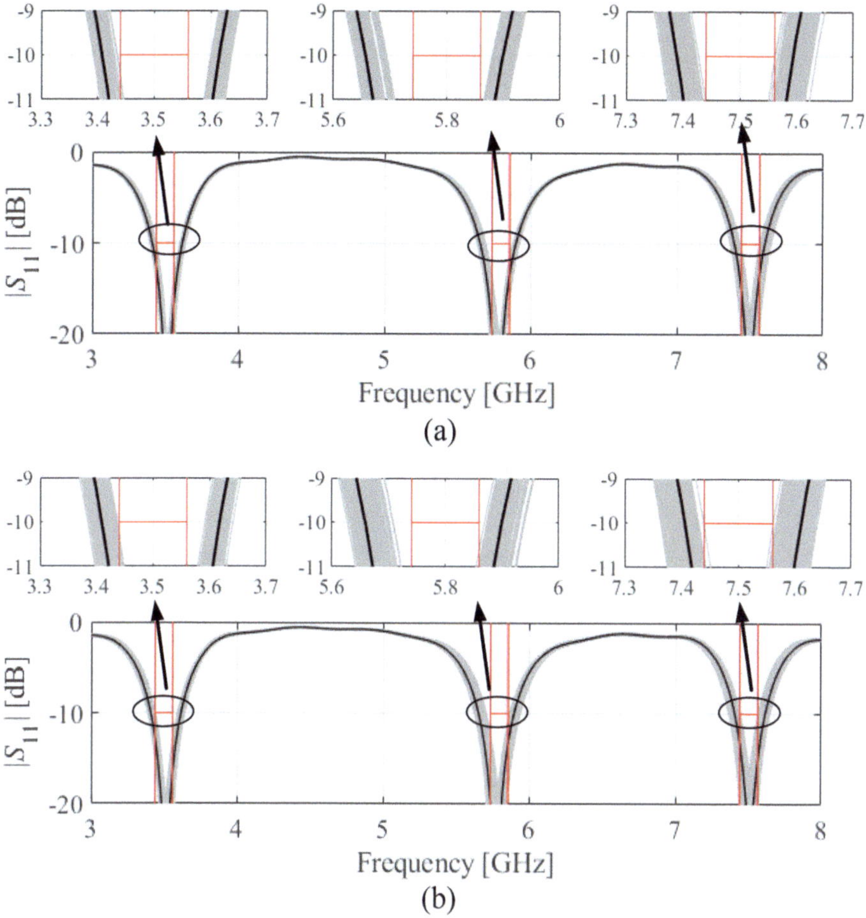

Fig. 14.30 Monte Carlo analysis of Antenna II (Fig. 14.28b) using EM simulations (gray plots) at: (**a**) the nominal design and (**b**) the design found using the discussed procedure. Black plots show the antenna response at the nominal and optimized designs, respectively

- Circuit I: $\boldsymbol{x}^* = [4.24\ 5.17\ 4.02\ 6.71\ 1.06\ 0.47]^T$
- Circuit II: $\boldsymbol{x}^* = [2.84\ 3.53\ 1.77\ 10.53\ 11.62\ 10.81\ 0.74\ 2.4\ 0.53\ 3.67\ 2.3\ 3.88]^T$

It should be noted that the CPU cost of finding the robust designs is low. It corresponds to just 40 and 118 EM simulations of the respective circuits. Although the cost for Circuit II is considerably higher than for other structures, the dimensionality of the parameters is also higher (12 parameters versus 6 for Circuit I). The primary factor enabling computational speedup is utilization of feature-based surrogates, which permits rapid uncertainty quantification.

The design relocations are generally small, which is due to the fact that the responses of all circuits are highly dependent on their geometry parameters. Nevertheless, the discussed robustness enhancement procedure allows for a considerable

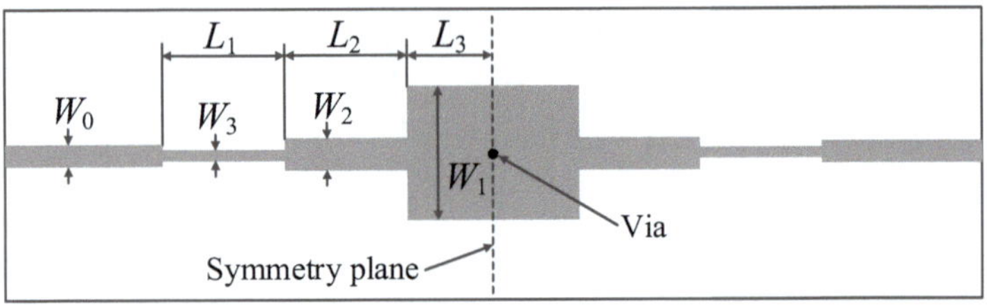

(a)

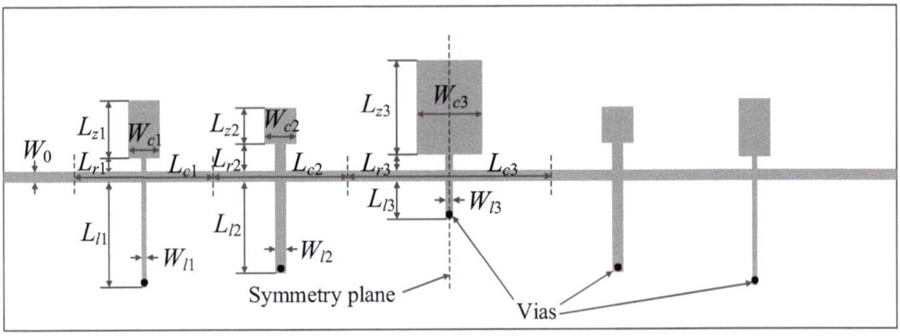

(b)

Fig. 14.31 Microwave circuits utilized for demonstrating the tolerance optimization procedure: (**a**) UWB filter using stepped-impedance resonator (Circuit I), and (**b**) wideband filter using shunt resonators (Circuit II)

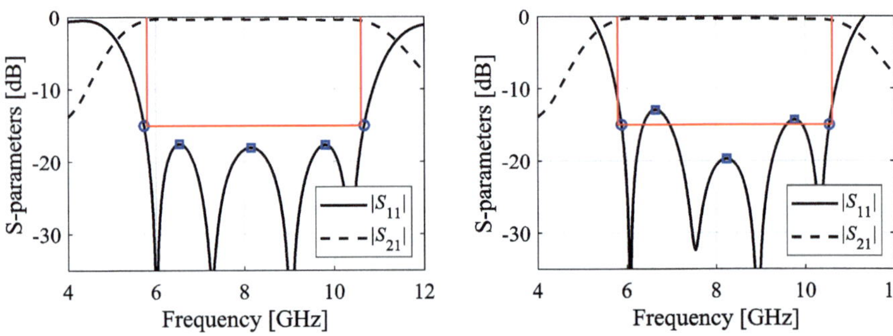

Fig. 14.32 Response features for an exemplary microwave filter. Shown are scattering parameters of a filter with the feature points corresponding to -15 dB level of $|S_{11}|$ and local maxima of $|S_{11}|$ within the filter operating band; the feature points permit determination of whether the return loss characteristic satisfies the matching conditions over the upper UWB bandwidth (5.9–10.6 GHz). The left- and right-hand-side panels illustrate designs satisfying and violating the performance specifications, respectively

Table 14.14 Verification case studies

	Case study[a]	
	Circuit I	Circuit II
Substrate	RO4003C ($\varepsilon_r = 3.55$, $h = 0.305$ mm)	RO3010 ($\varepsilon_r = 10.2$, $h = 0.63$ mm)
Design parameters	$\boldsymbol{x} = [L_1\ L_2\ L_3\ W_1\ W_2\ W_3]^T$	$\boldsymbol{x} = [L_{l1}\ L_{l2}\ L_{l3}\ L_{c1}\ L_{c2}\ L_{c3}\ L_{r1}\ L_{r2}\ L_{r3}\ L_{z1}\ L_{z2}\ L_{z3}]^T$
Other parameters	$W_0 = 0.66$ mm	$W_0 = 0.6$ mm, $W_{c1} = W_{c2} = 2$ mm, $W_{c3} = 3$ mm, $W_{l1} = W_{l3} = 0.3$ mm, $W_{l2} = 0.5$ mm
Performance specifications	$\lvert S_{11}\rvert \le -15$ dB within the operating band of 5.8–10.6 GHz	$\lvert S_{11}\rvert \le -15$ dB within the operating band of 2.1–2.9 GHz
Nominal design	$\boldsymbol{x}^{(0)} = [4.25\ 5.20\ 4.04\ 6.69\ 1.07\ 0.47]^T$	$\boldsymbol{x}^{(0)} = [2.86\ 3.53\ 1.77\ 10.55\ 11.62\ 10.82\ 0.74\ 2.42\ 0.54\ 3.67\ 2.30\ 3.89]^T$

[a]Parameters with subscript r are relative, and their deviations are recalculated accordingly in order to have the corresponding absolute parameters following the assumed probability distribution

Table 14.15 Robustness enhancement results for Circuits I and II

Verification structure		Circuit I	Circuit II
Nominal design	Maximum variance σ ensuring 100% yield[a]	5.6 μm	1.2 μm
	Maximum parameter deviation ensuring 100% yield	16.7 μm	3.7 μm
	EM-based yield estimation[b]	100%	98%
Tolerance-optimized design	Maximum variance σ ensuring 100% yield[a]	8.7 μm	1.8 μm
	Maximum parameter deviation ensuring 100% yield	26.1 μm	5.4 μm
	EM-based yield estimation[b]	100%	99%
Optimization cost[c]		40	118

[a]σ refers to the variance of the independent zero-mean Gaussian distributions assumed to describe the fabrication tolerances. The maximum parameter deviations are assumed to be 3σ
[b]Estimation obtained using Monte Carlo simulation based on 500 random samples
[c]Optimization cost in terms of the number of EM analyses of the circuit under design

increase in the input tolerance levels ensuring 100% fabrication yield, by a factor of 1.6 and 1.5 for Circuit I and II, respectively. In absolute terms, the maximum parameter deviations increase from 16.7 to 26.1 μm (Circuit I) and from 3.7 to 5.4 μm (Circuit II).

For the sake of verifying the reliability, EM-based Monte Carlo analysis using 500 random samples has been executed for both the nominal and the robust design. The purpose was to verify whether 100% yield is indeed preserved, which would be a confirmation of a sufficient predictive power of the feature-based surrogates utilized for uncertainty quantification. It can be observed (Table 14.15) that the MC-validated yield is 100% for Circuit I and Circuit II, and it is slightly lower for Circuit II. It should be emphasized that EM-driven MC exhibits relatively large standard deviation of yield estimation, which is inversely proportional to $N_r^{1/2}$; at the

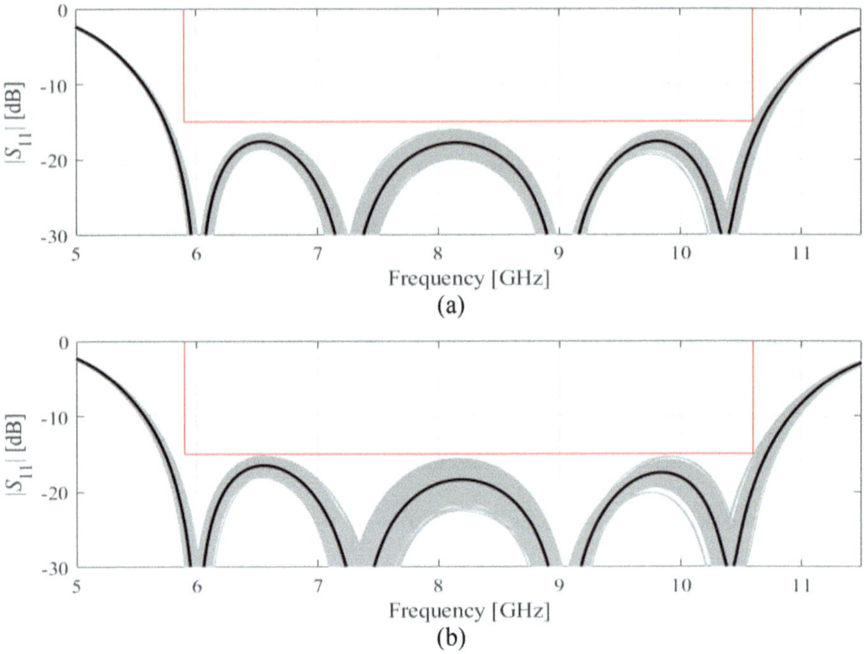

Fig. 14.33 EM-driven Monte Carlo analysis of Circuit I (Fig. 14.31a) at (**a**) the nominal design and (**b**) the design found using the robustness enhancement algorithm. Black plots indicate the circuit responses at the nominal and the optimized designs, respectively; gray plots represent EM simulation data at the random observables generated during the MC analysis

same time, the number of samples was limited to 500 due to the high CPU cost of the procedure.

Graphical illustrations of MC can be found in Figs. 14.33 and 14.34. It can be observed that the allowable tolerance levels are much higher at the optimized designs than at the nominal ones, which is indicated by the broader spread of the circuit characteristics (gray plots). The improvement obtained by relocating the design manifests itself by slight broadening of the bandwidth at the expense of other characteristics (e.g., the increase of the in-band maxima). Also, in all cases, the families of circuit responses are essentially touching the specification lines, which is a visual confirmation of both the fact that the yield at both the nominal and the robust design is close to 100%, and that the tolerance levels cannot be improved further.

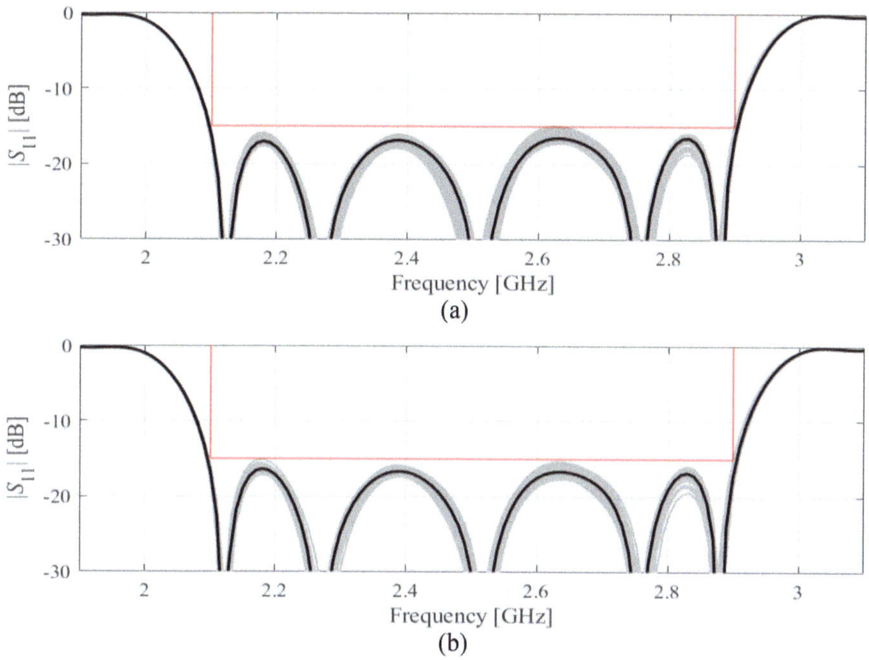

Fig. 14.34 EM-driven Monte Carlo analysis of Circuit II (Fig. 14.31b) at (**a**) the nominal design and (**b**) the design found using the robustness enhancement algorithm. Black plots indicate the circuit responses at the nominal and the optimized designs, respectively; gray plots represent EM simulation data at the random observables generated during the MC analysis

14.7 Design Centering Using Response Features and Inverse Regression

This section discussed rapid design centering procedure using response features and inverse regression surrogates (Koziel & Pietrenko-Dabrowska, 2022). A brief outline of the problem statement (Sect. 14.7.1) is followed by a description of the optimization approach (Sect. 14.7.2) and the demonstration case studies (Sect. 14.7.3).

14.7.1 Task Formulation and Models

The method is explained using the case of multiband antenna optimized for the best in-band matching. Let f_{0k}, $k = 1, \ldots, N$, be the target operating frequencies, and B be the target fractional bandwidth. We say that the antenna at the design $\boldsymbol{x}$ fulfills the matching requirements if

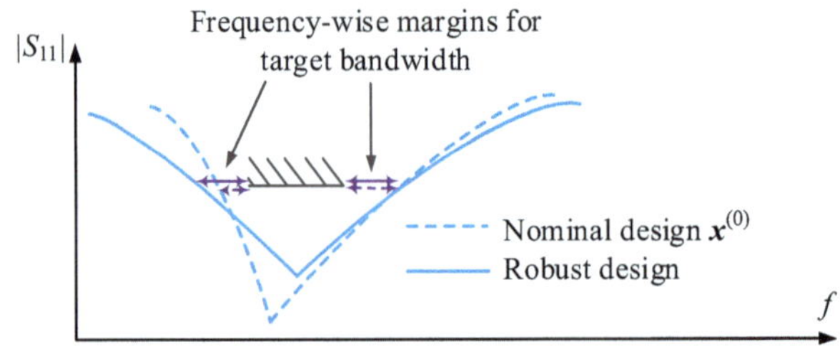

Fig. 14.35 Nominal design versus robust design: although $x^{(0)}$ may feature lower value of the (here, minimax) objective function, robust design ensures a wider margin in terms of satisfying specifications due to the frequency shifts

$$\max\{f \in \cup_{k=1}^{N}[(1-B/2)f_{0k}, (1+B/2)f_{0k}] \ : \ |S_{11}(\boldsymbol{x},f)|\} \leq S_{\max} \tag{14.43}$$

where $S_{\max}$ is the acceptance threshold for $|S_{11}|$, normally set to $S_{\max} = -10$ dB. In (14.43), f denotes the frequency and S_{11} is marked as an explicit function of both $\boldsymbol{x}$ and f. Satisfaction of the condition (14.43) is equivalent to antenna matching being at the level of $S_{\max}$ or better, simultaneously at all target bandwidths.

The objective of design centering is to relocate the design so that it is as away from the boundaries of the feasible region as possible. This, implicitly, results in increasing the fabrication yield. Here, we approach the problem from the perspective of antenna responses. While in the parameter space, the robust design is allocated possibly away from the boundaries of the feasible region, from the point of view of antenna characteristics, we want to have a possibly large margin, safeguarding from a violation of the design specifications. In the case of multiband antennas, this margin would be considered in terms of frequencies corresponding to -10 dB level of reflection response and how far are those from the frequencies determining the target operating bandwidth, as indicated in Fig. 14.35. At the same time, extending the bandwidth in this sense should take into account the problem-specific knowledge in the form of the sensitivity of -10 dB frequencies to geometry parameters. In particular, higher sensitivity requires more margin, whereas lower sensitivity allows for a reduced safeguard. A rigorous formulation of the design centering procedure that realizes robust design in this sense in a computationally efficient manner has been described below.

We aim at increasing the margin for satisfying the condition (14.43) under the assumed geometry parameter tolerances. For the kth operating band of the antenna at hand, the lower- and the upper-end margins are defined as

$$M_{L.k} = s_{L.k}[(1-B/2)f_{0.k} - f_{L.k}] \tag{14.44}$$

$$M_{R.k} = s_{R.k}[f_{R.k} - (1+B/2)f_{0.k}] \tag{14.45}$$

for $k = 1, \ldots, N$, where $f_{L.k}$ and $f_{R.k}$ are the frequency coordinates of the feature points corresponding to -10 dB reflection levels determining the kth operating band, whereas $s_{L.k}$ and $s_{R.k}$ are the scaling factors that account for the sensitivity of $f_{L.k}$ and $f_{R.k}$ to antenna geometry parameters at the vicinity of the nominal design. In other words, $M_{L.k}$ and $M_{R.k}$ are sensitivity-normalized margins for the current antenna design to satisfy the condition (14.43). The scaling factors are calculated as

$$s_{L.k} = \frac{\left\| \nabla_{f_{L.k}}\left(\boldsymbol{x}^{(0)}\right) \right\|}{M_J} \tag{14.46}$$

$$s_{R.k} = \frac{\left\| \nabla_{f_{R.k}}\left(\boldsymbol{x}^{(0)}\right) \right\|}{M_J} \tag{14.47}$$

for $k = 1, \ldots, N$, where

$$M_J = \begin{matrix} \max\left\{ \max\left\{k = 1, \ldots N : \left\| \nabla_{f_{L.k}}\left(\boldsymbol{x}^{(0)}\right) \right\| \right\}, \right. \\ \left. \max\left\{k = 1, \ldots N : \left\| \nabla_{f_{R.k}}\left(\boldsymbol{x}^{(0)}\right) \right\| \right\} \right\} \end{matrix} \tag{14.48}$$

with $\nabla_{fL.k}$ and $\nabla_{fR.k}$ being gradients of $f_{L.k}$ and $f_{R.k}$, respectively, i.e.,

$$\nabla_{f_{L.k}}\left(\boldsymbol{x}^{(0)}\right) = \left[\frac{\partial f_{L.k}}{\partial x_1}\left(\boldsymbol{x}^{(0)}\right) \quad \ldots \quad \frac{\partial f_{L.k}}{\partial x_n}\left(\boldsymbol{x}^{(0)}\right) \right]^T \tag{14.49}$$

$$\nabla_{f_{R.k}}\left(\boldsymbol{x}^{(0)}\right) = \left[\frac{\partial f_{R.k}}{\partial x_1}\left(\boldsymbol{x}^{(0)}\right) \quad \ldots \quad \frac{\partial f_{R.k}}{\partial x_n}\left(\boldsymbol{x}^{(0)}\right) \right]^T \tag{14.50}$$

These gradients can be readily obtained from the Jacobian matrix of the antenna reflection at the nominal design, which is normally available as a by-product of obtaining $\boldsymbol{x}^{(0)}$ (using, e.g., gradient-based procedures) or can be estimated by means of finite differentiation.

The objective of design centering in our formulation would be to solve the problem

$$\boldsymbol{x}^* = \arg\min_{\boldsymbol{x}} \{-C(\boldsymbol{x})\} \tag{14.51}$$

where the objective function $C(\boldsymbol{x})$ is defined as

$$C(\boldsymbol{x}) = \min\{\min\{k = 1, \ldots, N : M_{L.k}\}, \min\{k = 1, \ldots, N : M_{R.k}\}\} \tag{14.52}$$

In other words, we attempt to increase the worst-case normalized margin for satisfying condition (14.43). Clearly, the formulation should be modified if design specifications are defined differently, e.g., with individual target fractional bandwidths set for different operating bands.

The problem (14.51) and (14.52) can be solved directly through EM-based minimization of the cost function $C(\boldsymbol{x})$, using, e.g., gradient-based algorithms. Here, seeking computational savings, we use the inverse modeling approach as described below. For that purpose, we first define the feature vector

$$\boldsymbol{F}(\boldsymbol{x}) = [F_1(\boldsymbol{x}) \ \ldots \ F_{2N}(\boldsymbol{x})]^T = [f_{L.1}(\boldsymbol{x}) \ f_{R.1}(\boldsymbol{x}) \ \ldots \ f_{L.N}(\boldsymbol{x}) \ f_{R.N}(\boldsymbol{x})]^T \quad (14.53)$$

Let $\boldsymbol{x}^{(i)}$ be the ith approximation of $\boldsymbol{x}^*$ in (14.51), obtained using an iterative optimization procedure. We aim at constructing an inverse model $F_I(\boldsymbol{F})$ with the values in the geometry parameter space of the antenna under design. Due to weakly-nonlinear dependence between the feature vector entries and geometry parameters, as well as the local nature of the search, we assume that the model is linear and established using $2N + 1$ pairs $\{\boldsymbol{x}_B^{(j)}, \boldsymbol{F}(\boldsymbol{x}_B^{(j)})\}_{j = 1, \ldots, 2N + 1}$. We have $\boldsymbol{x}_B^{(1)} = \boldsymbol{x}^{(i)}$ and $\boldsymbol{x}_B^{(j)} = \boldsymbol{x}^{(i)} + \delta_r^{(i)}\boldsymbol{x}_r$, where $\boldsymbol{x}_r$ is a random vector generated in the interval $[-1, 1]^n$, whereas $\delta_r^{(i)}$ is a size parameter of the procedure, adaptively adjusted as described below.

The inverse model takes the form of

$$F_I(\boldsymbol{F}) = \begin{bmatrix} l_{0.1} + \boldsymbol{L}_1^T \boldsymbol{F} \\ \\ l_{0.n} + \boldsymbol{L}_n^T \boldsymbol{F} \end{bmatrix} \quad (14.54)$$

The model coefficients have to ensure that $\boldsymbol{x}_B^{(j)} = F_I(\boldsymbol{F}(\boldsymbol{x}_B^{(j)}))$ for $j = 1, \ldots, 2N + 1$, i.e., we get

$$\begin{bmatrix} l_{0.j} \\ \boldsymbol{L}_j \end{bmatrix} = \begin{bmatrix} 1 & \ldots & \boldsymbol{F}\left(\boldsymbol{x}_B^{(1)}\right)^T \\ \\ 1 & \ldots & \boldsymbol{F}\left(\boldsymbol{x}_B^{(2N+1)}\right)^T \end{bmatrix}^{-1} \begin{bmatrix} x_{B.j}^{(1)} \\ \vdots \\ x_{B.j}^{(2N+1)} \end{bmatrix} \quad (14.55)$$

where $x_{B.j}^{(k)}$ is the jth entry of the vector $\boldsymbol{x}_B^{(k)}$. The inverse model will be used to provide predictions of antenna geometry parameters given for a given feature vector $\boldsymbol{F}$.

1. Set the iteration index $i = 0$;
2. Generate the training set $\{\boldsymbol{x}_B^{(j)}\}_{j=1,\ldots,2N+1}$ using the current size vector $\delta_r^{(i)}$;
3. Acquire EM simulation data and extract $\{\boldsymbol{F}(\boldsymbol{x}_B^{(j)})\}_{j=1,\ldots,2N+1}$;
4. Construct the inverse model F_I;
5. Find $\boldsymbol{F}_{tmp} = \mathrm{argmin}\{F : -C_F(\boldsymbol{F})\}$ subject to $|F_j - F_j(\boldsymbol{x}^{(i)})| \le r_j^{(i)}$, $j = 1, \ldots, 2N$;
6. Evaluate EM model at $\boldsymbol{x}_{tmp} = F_I(\boldsymbol{F}_{tmp})$, and extract the feature vector $\boldsymbol{F}(\boldsymbol{x}_{tmp})$;
7. Calculate gain ratio $\rho = [C(\boldsymbol{x}_{tmp}) - C(\boldsymbol{x}^{(i)})] \,/\, [C_F(\boldsymbol{F}_{tmp}) - C(\boldsymbol{x}^{(i)})]$ where C_F is the objective function calculated directly using the feature vector $\boldsymbol{F}_{tmp}$;
8. **if** $\rho > \rho_{incr}$
   $\boldsymbol{r}^{(i+1)} = 2\boldsymbol{r}^{(i+1)}$;
   elseif $\rho < \rho_{decr}$
   $\boldsymbol{r}^{(i+1)} = \boldsymbol{r}^{(i+1)}/3$; $\delta_r^{(i+1)} = \delta_r^{(i)}/2$;
   end
9. **if** $\rho > 0$
   $\boldsymbol{x}^{(i+1)} = \boldsymbol{x}_{tmp}$; Set $i = i + 1$;
   end
10. **if** $\delta_r^{(i)} < \varepsilon$ OR $||\boldsymbol{r}^{(i)}|| < \varepsilon$
    Terminate the algorithm and return $\boldsymbol{x}^* = \boldsymbol{x}^{(i)}$;
    else
    Go to 2;
    end

Fig. 14.36 Pseudocode of the discussed design centering algorithm

14.7.2 Design Centering Algorithm

The problem (14.51) is solved using the inverse regression model and the iterative trust-region procedure described below. The initial solution is the nominal design $\boldsymbol{x}^{(0)}$, whereas the control parameters are the size parameter $\delta_r^{(i)}$ for generating the inverse model training data, as well as the size vector $\boldsymbol{r}^{(i)} = [r_1^{(i)} \ldots r_{2N}^{(i)}]^T$, determining the search region for candidate design identification. Both are adjusted adaptively in the course of the optimization process. The initial value for $\delta_r^{(i)}$ is the maximum assumed geometry tolerance, whereas the initial values for $r_j^{(i)}$, $j = 1, \ldots, 2N$, are set to be standard deviations of the vector $[F_j(\boldsymbol{x}_B^{(1)}) \ldots F_j(\boldsymbol{x}_B^{(2N+1)})]^T$.

We also define an auxiliary objective function C_F as

$$C_F(\boldsymbol{F}) = \min\{\min\{k=1,\ldots,N : M_{L.k}(\boldsymbol{F})\}, \min\{k=1,\ldots,N : M_{R.k}(\boldsymbol{F})\}\} \quad (14.56)$$

which is the same as (14.52), except that it is defined directly on the feature vector $\boldsymbol{F}$. Note that $C(\boldsymbol{x}) = C_F(\boldsymbol{F})$ for $\boldsymbol{F} = \boldsymbol{F}(\boldsymbol{x})$, i.e., if $\boldsymbol{F}$ is the specific feature vector extracted from EM simulation results at the design $\boldsymbol{x}$.

The operation of the design centering algorithm has been shown in Fig. 14.36. In the algorithm, Steps 2–4 are for preparation of inverse model. The critical stage is Step 5, which makes a prediction about the design featuring improved value of the objective function (14.52). This design is verified through the EM analysis and the extraction of the problem-specific knowledge in the form of the corresponding feature points in Steps 6 and 7. The size parameters δ_r and $\boldsymbol{r}$ are then adaptively adjusted in Steps 8 and 9, where we use the standard trust-region thresholds $\rho_{incr} = 0.75$ and $\rho_{decr} = 0.25$ (Conn et al., 2000). The termination condition is based on diminishing the size parameters beyond a user-defined threshold ε, here, set to 10^{-3}.

It should be noted that the inverse model is a prediction tool giving an indication of the promising directions in the parameter space, which are then tested and validated through EM analysis. The most important advantage of this approach is its computational efficiency: the cost of constructing the new inverse model is only $2N$ EM simulations (e.g., two for a single-band antenna and six for a triple-band antenna), regardless of the number of geometry parameters, which is normally considerably higher than $2N$. At the same time, it should be emphasized that the discussed approach does not directly optimize the yield; however, as argued before, improved values of the objective function C are expected to translate into higher yield values.

It should also be noted that the procedure does not have any control parameters that would have to be tuned for a particular problem at hand. As a matter of fact, the only control parameters are those related to the updating of the trust region size, which are selected according to the general recommendations for the trust region procedures, as well as those related to the termination condition. The former are not critical because by the very nature of the TR process. The latter are only used to determine the required resolution of the search process. Using $\varepsilon = 10^{-3}$ is a typical value for antenna structures with typical variable ranges being from fractions of mm to tens of mm as in the case of the problems considered in Sect. 14.7.3.

14.7.3 Demonstration Case Studies

The design centering procedure of Sect. 14.7.2 is demonstrated using two examples of microstrip antennas, a single- and a triple-band one. The efficacy as well as the computational efficiency of the algorithm is compared to three benchmark methods, all of which are state-of-the-art surrogate-assisted frameworks. The reliability is also validated by means of EM-driven Monte Carlo simulation performed at the nominal and final designs.

Consider a ring-slot antenna shown in Fig. 14.37 (Antenna I) and a triple-band dipole antenna shown in Fig. 14.38 (Antenna II). Computational models of the antennas are implemented in CST Microwave Studio and simulated using its time-domain solver. In all cases, for the purpose of design centering, the geometry parameter deviations are assumed to be described by independent uniform probability distributions with the maximum deviation $d_{\max} = 0.05$ mm.

The presented algorithm is compared to three state-of-the-art surrogate-assisted procedures whose taxonomy is provided in Table 14.16. In the first technique

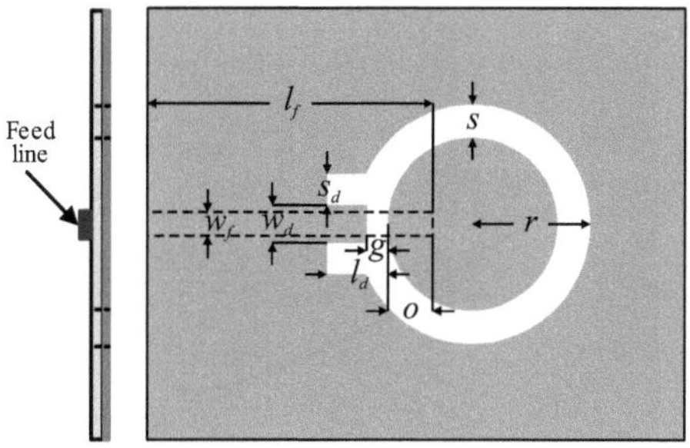

Substrate	$\varepsilon_r = 2.0$, $h = 0.76$ mm
Design parameters	$\boldsymbol{x} = [l_f\ l_d\ w_d\ r\ s\ s_d\ o\ g]^T$
Other parameters	-
Operating bands	4.15 GHz – 4.85 GHz
Nominal design	$\boldsymbol{x}^{(0)} = [20.28\ 6.54\ 0.24\ 11.83\ 2.95\ 6.77\ 7.85\ 2.23]^T$

(a) (b)

Fig. 14.37 Antenna I: (**a**) ring-slot antenna with a microstrip feed marked using a dashed line (Koziel & Pietrenko-Dabrowska, 2020), (**b**) the details pertaining to the antenna structure

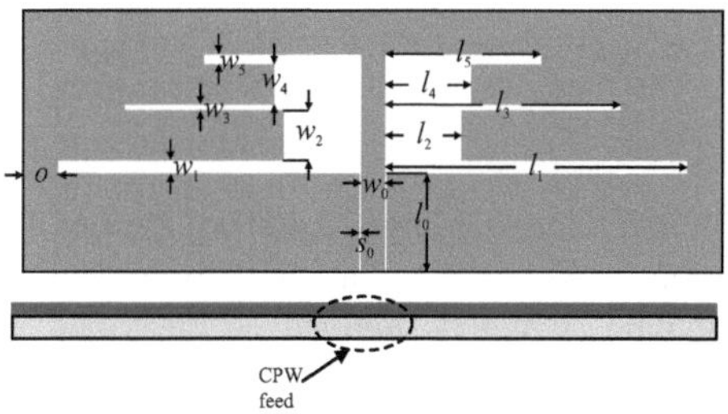

Substrate	$\varepsilon_r = 3.5$, $h = 0.76$ mm
Design parameters	$\boldsymbol{x} = [l_1\ l_2\ l_3\ l_4\ l_5\ w_1\ w_2\ w_3\ w_4\ w_5]^T$
Other parameters	$l_0 = 30$, $w_0 = 3$, $s_0 = 0.15$, and $o = 5$
Operating bands	4-percent fractional bandwidth w.r.t. center frequencies 2.45 GHz, 3.6 GHz, and 5.3 GHz
Nominal design	$\boldsymbol{x}^{(0)} = [35.42\ 11.54\ 26.07\ 8.09\ 17.14\ 0.60\ 0.99\ 1.44\ 0.78\ 1.17]^T$

(a) (b)

Fig. 14.38 Antenna II: (**a**) triple-band uniplanar dipole antenna (Chen et al., 2006), (**b**) the details pertaining to the antenna structure

Table 14.16 Taxonomy of benchmark surrogate-assisted yield estimation procedures

Algorithm		Number of surrogates	Surrogate domain	Pros	Cons
1		One	$[\boldsymbol{x}^{(0)} - \boldsymbol{d}, \boldsymbol{x}^{(0)} + \boldsymbol{d}]$ $\boldsymbol{d} = [d_1 \ldots d_n]^T$ $d_k = 10d_{\max}$ [a]	Simple implementation	High training data acquisition cost
2	Sequential approximate optimization	Several	$[\boldsymbol{x}^{(i)} - \boldsymbol{d}_l, \boldsymbol{x}^{(i)} - \boldsymbol{d}_l]$ $\boldsymbol{x}^{(i)} = [x_1^{(i)} \ldots x_n^{(i)}]^T$ $\boldsymbol{d}_l = [d_{l.1} \ldots d_{l.n}]^T$ $d_k = 3d_{\max}$ [a]	Low training data acquisition cost	Necessity of reiterating surrogate construction
3	Performance-driven surrogate modeling	One	Constrained domain spanned by the directions of maximum response changes w.r.t. performance figures	Low training data acquisition cost of constructing single surrogate model	Considerable cost of acquiring reference designs for domain definition

[a] $d_{\max}$ denotes the maximum parameter deviation

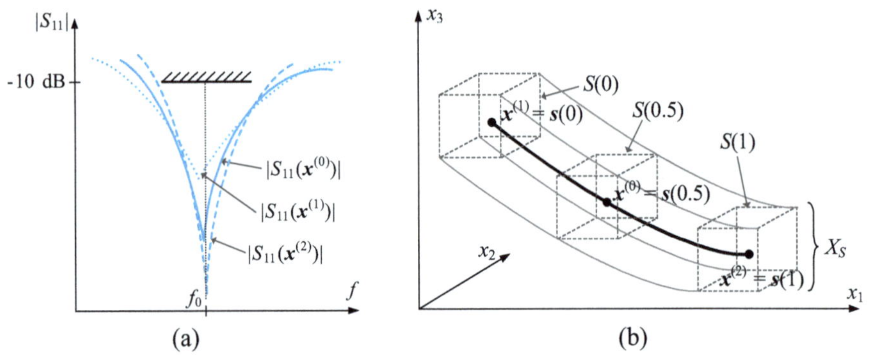

Fig. 14.39 Surrogate-assisted yield optimization by means of performance-driven surrogates (Pietrenko-Dabrowska et al., 2020): (**a**) $|S_{11}|$ responses an antenna at the nominal design $x^{(0)}$, design optimized for maximum bandwidth $x^{(1)}$, and design optimized for best matching (at the center frequency f_0) $x^{(2)}$. These designs determine the directions of the most significant response changes and span the surrogate model domain; (**b**) the reference designs $x^{(0)}$ through $x^{(2)}$ form a path (parameterized curve $s(t)$). The union of intervals $S(t)$ forms the domain of the surrogate model

(Algorithm 1), a single metamodel is built in a relatively large region adjacent to the nominal design as a kriging interpolation model. Next, the yield optimization problem is solved within the model domain, which is defined as described in Table 14.16. The second benchmark technique (Algorithm 2) is a variation of sequential approximate optimization (SAO) (Koziel & Ogurtsov, 2013). Therein, the yield optimization task is transformed into an iterative process of solving $x^{(i+1)} = \arg\min\{x : -Y_s^{(i)}(x)\}$, where $x^{(i)}$, $i = 0, 1, \ldots$, is a sequence of designs approximating the optimum x^*, whereas $Y_s^{(i)}$ is the estimated yield in the ith iteration. Each consecutively built surrogate is constructed in the domain that is centered at $x^{(i)}$; this domain is shifted between the algorithm iterations (see Table 14.16). Here, the surrogate domain is considerably smaller than that of Algorithm 1. The yield optimization problem is solved subject to $x_k^{(i)} - d_{l.k} + d_{\max} \leq x_k \leq x_k^{(i)} + d_{l.k} - d_{\max}$, $k = 1, \ldots, n$, which is to ensure that the point x is always located in the interior of the current domain, and at least at the distance $d_{\max}$ from its boundary (in all directions). This is necessary to conduct Monte Carlo simulation within the domain. The last benchmark algorithm is a surrogate-assisted technique involving performance-driven modeling (Pietrenko-Dabrowska et al., 2020), where a single surrogate model is constructed (no further iterations). In this technique, the surrogate domain is spanned by the directions corresponding to the most significant changes of antenna responses from the point of view of satisfying the design specifications. These are determined using auxiliary local optimization runs, leading to the so-called reference designs (Pietrenko-Dabrowska et al., 2020). This allows for ensuring sufficient size of the domain while maintaining its low overall volume. In other words, Algorithm 3 combines the benefits of Algorithms 1 and 2, while mitigating their drawbacks (see also Table 14.16). Figure 14.39 shows the conceptual illustration of the surrogate model domain definition.

Table 14.17 Yield optimization results for Antenna I (Fig. 14.37)

	Initial yield (%)		Optimized yield (%)		
Optimization algorithm	Estimated by surrogate model	EM-based	Estimated by surrogate model	EM-based	CPU cost[a]
Algorithm 1	81	81	92	93	400
Algorithm 2	81	81	91	91	150[b]
Algorithm 3	81	81	91	91	62[c]
Feature-based inverse surrogates	–	81	–	91	9

[a]Optimization cost in number of EM analyses of the antenna structure
[b]The algorithm convergence after three iterations (surrogate setup cost 50 training samples per iteration)
[c]The cost includes training data acquisition (35 EM analyses) and generation of the reference designs $\boldsymbol{x}^{(1)}$ and $\boldsymbol{x}^{(2)}$ (27 EM simulations in total)

Table 14.18 Yield optimization results for Antenna II (Fig. 14.38)

	Initial yield (%)		Optimized yield (%)		
Optimization algorithm	Estimated by surrogate model	EM-based	Estimated by surrogate model	EM-based	CPU cost[a]
Algorithm 1	63	58	75	66	1600
Algorithm 2	62	58	72	69	400[b]
Algorithm 3	60	58	72	69	132[c]
Feature-based inverse surrogates	–	58	–	67	36

[a]Optimization cost in number of EM analyses of the antenna structure
[b]The algorithm convergence after four iterations (surrogate setup cost 100 training samples per iteration)
[c]The cost includes training data acquisition (100 EM analyses) and generation of the reference designs $\boldsymbol{x}^{(1)}$ and $\boldsymbol{x}^{(2)}$ (32 EM simulations in total)

Tables 14.17 and 14.18 show the results obtained using the discussed and the benchmark algorithms for Antennas I and II, respectively. The two evaluation metrics that are of interest here are: (1) the quality of the optimization process, measured in terms of the optimized value of the fabrication yield, and (2) the computational cost of the optimization process, measured in terms of the number of full-wave EM simulations of the antenna structure at hand. The final designs obtained using the presented approach are as follows: $\boldsymbol{x}^* = [20.23\ 6.61\ 0.20\ 11.80\ 3.02\ 6.80\ 7.92\ 2.29]^T$ (Antenna I) and $\boldsymbol{x}^* = [35.40\ 11.55\ 26.08\ 8.09\ 17.13\ 0.62\ 0.97\ 1.40\ 0.79\ 1.19]^T$ (Antenna II). These designs were obtained at a remarkably low computational cost of only 9 and 36 EM analyses of the respective antennas, which is significantly lower than the cost of all benchmark techniques, including Algorithm 3. At the same time, the design quality is similar to the benchmark in terms of the optimized yield values. The computational efficiency can be mainly attributed to exploiting the problem-specific knowledge, specifically, the usage of response features and the inverse surrogate. The former allows for flattening the functional landscape considered during the optimization process, whereas the latter mitigates

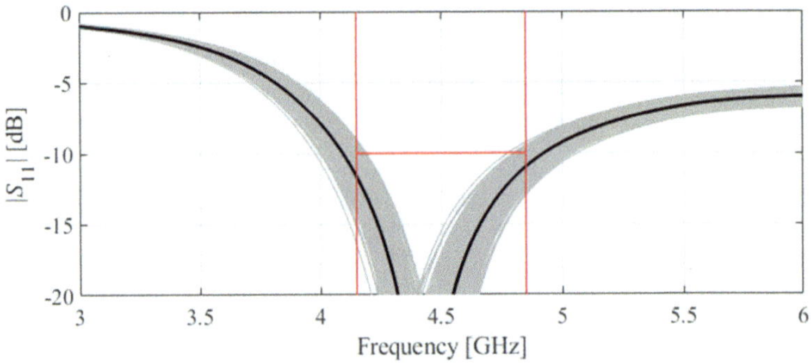

Fig. 14.40 Monte Carlo analysis of Antenna I using EM simulations (gray plots) at the design found using the methodology considered in this section. Black plots show the antenna response at the nominal and optimized designs, respectively

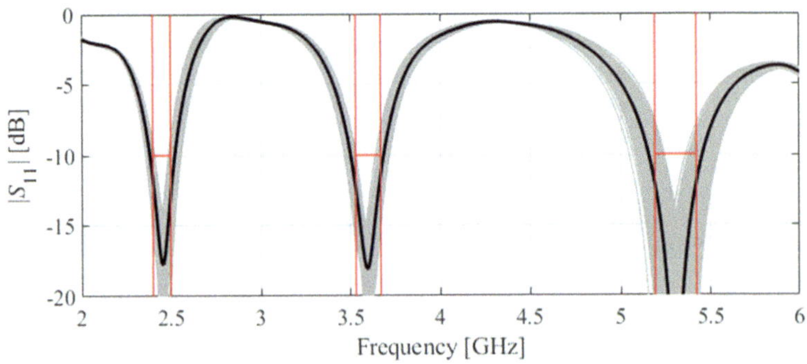

Fig. 14.41 Monte Carlo analysis of Antenna II using EM simulations (gray plots) at the design found using the methodology considered in this section. Black plots show the antenna response at the nominal and optimized designs, respectively

the dimensionality issue due to mapping low-dimensional feature space into the geometry parameter space of the antenna at hand. As a matter of fact, the computational expenses scale according to the number of antenna bands rather than the number of antenna geometry parameters, which is an attractive feature of the presented methodology.

The yield estimation in Tables 14.17 and 14.18 has been obtained using EM-based Monte Carlo simulation involving 500 random samples. Figures 14.40 and 14.41 provide visualization of the Monte Carlo analysis at the optimized designs for the three considered antenna structures. Note that there is no surrogate-based yield estimation provided for the considered method because it does not directly handle the yield as the objective function.

14.8 Summary

This chapter investigated utilization of response features for computationally efficient uncertainty quantification of high-frequency structures. We considered a number of feature-based techniques for statistical analysis and robust design of antenna and microwave components, where the underlying optimization task is reformulated in terms of the characteristic points of the system response at hand. This leads to inherent regularization of the problems and facilitates a construction of reliable replacement models, here, used for rapid numerical integration of the probability density functions describing the deviations of geometry parameters. The specific design tasks addressed here included optimization of the fabrication yield, as well as optimization of the input parameter tolerances (so that the yield can be kept at 100%). We also demonstrated application of feature-based inverse surrogates for low-cost design centering.

Extensive numerical experiments and benchmarking that involved a large number of antennas and microwave components demonstrated the computational advantages of the considered techniques, with cost savings being up to 90% with respect to state-of-the-art surrogate-assisted methods. As usual, the limitation of feature-based algorithms is that the characteristic points of the system are extractable in the entire domain of interest. In the case of statistical analysis, satisfaction of this condition is somehow facilitated by relative small design relocation in the course of the optimization process, which make the presented methods viable alternatives to the approaches reported in the literature so far.

References

Abdel-Malek, H. L., Hassan, A. S. O., Soliman, E. A., & Dakroury, S. A. (2006). The ellipsoidal technique for design centering of microwave circuits exploiting space-mapping interpolating surrogates. *IEEE Transactions on Microwave Theory and Techniques, 54*(10), 3731–3738.

Acikgoz, H., & Mittra, R. (2018). Stochastic polynomial chaos expansion analysis of a split-ring resonator at terahertz frequencies. *IEEE Transactions on Antennas and Propagation, 66*(4), 2131–2134.

Agushaka, J. O., Ezugwu, A. E., & Abualigah, L. (2022). Dwarf mongoose optimization algorithm. *Computer Methods in Applied Mechanics and Engineering, 391*, 114570.

Austin, A. C. M., Sood, N., Siu, J., & Sarris, C. D. (2013). Application of polynomial chaos to quantify uncertainty in deterministic channel models. *IEEE Transactions on Antennas and Propagation, 61*(11), 5754–5761.

Bandler, J. W., Rayas-Sánchez, J. E., & Zhang, Q. J. (2002). Yield-driven electromagnetic optimization via space mapping-based neuromodels. *International Journal of RF and Microwave Computer-Aided Engineering, 12*, 79–89.

Biernacki, R., Chen, S., Estep, G., Rousset, J., & Sifri, J. (2012). Statistical analysis and yield optimization in practical RF and microwave systems. In *IEEE MTT-S international microwave symposium digest*, Montreal (pp. 1–3).

Boeykens, F., Rogier, H., & Vallozzi, L. (2014). An efficient technique based on polynomial chaos to model the uncertainty in the resonance frequency of textile antennas due to bending. *IEEE Transactions on Antennas and Propagation, 62*(3), 1253–1260.

Brady, D. (2002). The design, fabrication and measurement of microstrip filter and coupler circuits. *High Frequency Electronics, 1*(1), 22–30.

Chen, Y.-C., Chen, S.-Y., & Hsu, P. (2006). Dual-band slot dipole antenna fed by a coplanar waveguide. In *IEEE Antennas and Propagation Society international symposium* (pp. 3589–3592).

Cheng, X., Shao, W., Wang, K., & Wang, B.-Z. (2019). Uncertainty analysis in dispersive and lossy media for ground-penetrating radar modeling. *IEEE Antennas and Wireless Propagation Letters, 18*(9), 1931–1935.

Conn, A.R., Gould, N.I.M., & Toint, P.L. (2000). *Trust region methods.* MPS-SIAM Series on Optimization.

Consul, P. (2015). Triple band gap coupled microstrip U-slotted patch antenna using L-slot DGS for wireless applications, In *Communication, Control and Intelligent Systems (CCIS)*, Mathura, India (pp. 31–34).

Deng, W., Zhang, X., Zhou, Y., Liu, Y., Zhou, X., Chen, H., & Zhao, H. (2022). An enhanced fast non-dominated solution sorting genetic algorithm for multi-objective problems. *Information Sciences, 585*, 441–453.

Du, J., & Roblin, C. (2017). Statistical modeling of disturbed antennas based on the polynomial chaos expansion. *IEEE Antennas and Wireless Propagation Letters, 16*, 1843–1847.

Easum, J. A., Nagar, J., Werner, P. L., & Werner, D. H. (2018). Efficient multiobjective antenna optimization with tolerance analysis through the use of surrogate models. *IEEE Transactions on Antennas and Propagation, 66*(12), 6706–6715.

Hassan, A. S. O., & Mohamed, A. S. A. (2013). Surrogate-based circuit design centering. In S. Koziel & L. Leifsson (Eds.), *Surrogate-based modeling and optimization* (pp. 27–49). Springer.

Hassan, A.S.O., Abdel-Malek, H.L., Mohamed, A.S.A., Abuelfadl, T.M., & Elqenawy, A.E. (2015). Statistical design centering of RF cavity linear accelerator via non-derivative trust region optimization. In *IEEE MTT-S international conference on Numerical Electromagnetic and Multiphysics Modeling and Optimization (NEMO)* (pp. 1–3).

Hauth, W., Keller, R., Papziner, U., Ihmels, R., Sieverding, T., & Arndt, F. (1993) Rigorous CAD of multipost coupled rectangular waveguide components. In *Proceedings of the 23rd European Microwave conference*, Madrid, Spain (pp. 611–614).

Kennedy, M. C., & O'Hagan, A. (2000). Predicting the output from complex computer code when fast approximations are available. *Biometrika, 87*, 1–13.

Kersaudy, P., Mostarshedi, S., Sudret, B., Picon, O., & Wiart, J. (2014). Stochastic analysis of scattered field by building facades using polynomial chaos. *IEEE Transactions on Antennas and Propagation, 62*(12), 6382–6393.

Kim, D., Kim, M., & Kim, W. (2020). Wafer edge yield prediction using a combined long short-term memory and feed- forward neural network model for semiconductor manufacturing. *IEEE Access, 8*, 215125–215132.

Kouassi, A., Nguyen-Trong, N., Kaufmann, T., Lallechere, S., Bonnet, P., & Fumeaux, C. (2016). Reliability-aware optimization of a wideband antenna. *IEEE Transactions on Antennas and Propagation, 64*(2), 450–460.

Koziel, S. (2010). Computationally efficient multi-fidelity multi-grid design optimization of microwave structures. *Applied Computational Electromagnetics Society Journal, 25*(7), 578–586.

Koziel, S. (2015). Fast simulation-driven antenna design using response-feature surrogates. *International Journal of RF and Microwave Computer-Aided Engineering, 25*(5), 394–402.

Koziel, S., & Bekasiewicz, A. (2018). Sequential approximate optimization for statistical analysis and yield optimization of circularly polarized antennas. *IET Microwaves, Antennas & Propagation, 12*(13), 2060–2064.

Koziel, S., & Bandler, J. W. (2015). Rapid yield estimation and optimization of microwave structures exploiting feature-based statistical analysis. *IEEE Transactions on Microwave Theory and Techniques, 63*(1), 107–114.

Koziel, S., & Ogurtsov, S. (2013). Multi-level microwave design optimization with automated model fidelity adjustment. *International Journal of RF and Microwave Computer-Aided Engineering, 24*(3), 281–288.

Koziel, S., & Pietrenko-Dabrowska, A. (2019a). Reduced-cost electromagnetic-driven optimization of antenna structures by means of trust-region gradient-search with sparse Jacobian updates. *IET Microwaves, Antennas & Propagation, 13*(10), 1646–1652.

Koziel, S., & Pietrenko-Dabrowska, A. (2019b). Reduced-cost surrogate modeling of compact microwave components by two-level kriging interpolation. *Engineering Optimization, 52*(6), 960–972.

Koziel, S., & Pietrenko-Dabrowska, A. (2020). *Performance-driven surrogate modeling of high-frequency structures*. Springer.

Koziel, S., & Pietrenko-Dabrowska, A. (2022). Rapid design centering of multi-band antennas using knowledge-based inverse models and response features. *Knowledge Based Systems, 252*, paper no. 109360.

Koziel, S., Bandler, J., Mohamed, A., & Madsen, K. (2005). Enhanced surrogate models for statistical design exploiting space mapping technology. In *IEEE MTT-S international microwave symposium digest*, Long Beach, CA (pp. 1–4).

Laxminidhi, T., & Pavan, S. (2007). Efficient design centering of high-frequency integrated continuous-time filters. *IEEE Transactions on Circuits and Systems I, 54*(7), 1481–1488.

Leifsson, L., Du, X., & Koziel, S. (2020). Efficient yield estimation of multi-band patch antennas by polynomial chaos-based kriging. *International Journal of Numerical Modeling, 33*(6), paper No. e2722.

Li, Y., Ding, Y., & Zio, E. (2014). Random fuzzy extension of the universal generating function approach for the reliability assessment of multi-state systems under aleatory and epistemic uncertainties. *IEEE Transactions on Reliability, 63*(1), 13–25.

Li, G., Li, Y., Chen, H., & Deng, W. (2022). Fractional-order controller for course-keeping of underactuated surface vessels based on frequency domain specification and improved particle swarm optimization algorithm. *Applied Sciences, 12*(6), 3139.

Ma, B., Lei, G., Liu, C., Zhu, J., & Guo, Y. (2018). Robust tolerance design optimization of a PM claw pole motor with soft magnetic composite cores. *IEEE Transactions on Magnetics, 54*(3), paper No. 8102404.

Matoglu, E., Pham, N., De Araujo, D., Cases, M., & Swaminathan, M. (2004). Statistical signal integrity analysis and diagnosis methodology for high-speed systems. *IEEE Transactions on Advanced Packaging, 27*(4), 611–629.

Ochoa, J. S., & Cangellaris, A. C. (2013). Random-space dimensionality reduction for expedient yield estimation of passive microwave structures. *IEEE Transactions on Microwave Theory and Techniques, 61*(12), 4313–4321.

Oyelade, O. N., Ezugwu, A. E.-S., Mohamed, T. I. A., & Abualigah, L. (2022). Ebola optimization search algorithm: A new nature-inspired metaheuristic optimization algorithm. *IEEE Access, 10*, 16150–16177.

Pietrenko-Dabrowska, A. (2021). Rapid tolerance-aware design of miniaturized microwave passives by means of confined-domain surrogates. *International Journal of Numerical Modeling, 33*(6), e2779.

Pietrenko-Dabrowska, A., & Koziel, S. (2021). Design centering of compact microwave components using response features and trust regions. *Enerties, 14*(24), paper No. 8550.

Pietrenko-Dabrowska, A., & Koziel, S. (2022). Rapid yield optimization of miniaturized microwave passives by response features and variable-fidelity EM simulations. *Scientific Reports, 12*, 22440.

Pietrenko-Dabrowska, A., Koziel, S., & Al-Hasan, M. (2020). Expedited yield optimization of narrow- and multi-band antennas using performance-driven surrogates. *IEEE Access*, 143104–143113.

Prasad, A. K., & Roy, S. (2019). Reduced dimensional Chebyshev-polynomial chaos approach for fast mixed epistemic-aleatory uncertainty quantification of transmission line networks. *IEEE Transactions on Components, Packaging and Manufacturing Technology, 9*(6), 1119–1132.

Press, W. H., Teukolsky, S. A., Vetterling, W. T., & Flannery, B. P. (2007). *Golden section search in one dimension, in numerical recipes: The art of scientific computing* (3rd ed.). Cambridge University Press.

Qudrat-E-Maula, M., & Shafai, L. (2014). A dual band microstrip dipole antenna. In *International symposium on antenna technology and applied electromagnetics (ANTEM)*, Victoria, BC, Canada (pp. 1–2).

Queipo, N. V., Haftka, R. T., Shyy, W., Goel, T., Vaidynathan, R., & Tucker, P. K. (2005). Surrogate-based analysis and optimization. *Progress in Aerospace Sciences, 41*(1), 1–28.

Ren, Z., He, S., Zhang, D., Zhang, Y., & Koh, C. S. (2016). A possibility-based robust optimal design algorithm in preliminary design state of electromagnetic devices. *IEEE Transactions on Magnetics, 52*(3), paper No. 7001504.

Robert, C. P., & Casella, G. (2004). *Monte Carlo statistical methods* (2nd ed.). Springer.

Rossi, M., Dierck, A., Rogier, H., & Vande Ginste, D. (2014). A stochastic framework for the variability analysis of textile antennas. *IEEE Transactions on Antennas and Propagation, 62*(16), 6510–6514.

Scotti, G., Tommasino, P., & Trifiletti, A. (2005). MMIC yield optimization by design centering and off-chip controllers. *IET Proceedings – Circuits Devices and Systems, 152*(1), 54–60.

Sengupta, M., Saxena, S., Daldoss, L., Kramer, G., Minehane, S., & Cheng, J. (2005). Application-specific worst case corners using response surfaces and statistical models. *IEEE Transactions on Computer-Aided Design of Integrated Circuits and Systems, 24*(9), 1372–1380.

Sharma, Y., Zhang, H. H., & Xin, H. (2020). Machine learning techniques for optimizing design of double T-shaped monopole antenna. *IEEE Transactions on Antennas and Propagation, 68*(7), 5658–5663.

Sommer, A., Farle, O., & Dyczij-Edlinger, R. (2015). A new method for accurate and efficient residual computation in adaptive model-order reduction. *IEEE Transactions on Magnetics, 51*(3), 1–4. Art no. 1100104.

Wu, Q., Chen, W., Yu, C., Wang, H., & Hong, W. (2021). Multilayer machine learning-assisted optimization-based robust design and its applications to antennas and arrays. *IEEE Transactions on Antennas and Propagation, 69*(9), 6052–6057.

Xia, L., Li, J., Twumasi, B. A., Liu, P., & Gao, S. (2020). Planar dual-band branch-line coupler with large frequency ratio. *IEEE Access, 8*, 33188–33195.

Zeng, Z., Kang, R., Wen, M., & Zio, E. (2017). A model-based reliability metric considering aleatory and epistemic uncertainty. *IEEE Access, 5*, 15505–15515.

Zhang, J., Feng, F., Jin, J., & Zhang, Q.-J. (2020). Efficient yield estimation of microwave structures using mesh deformation-incorporated space mapping surrogates. *IEEE Microwave and Wireless Components Letters, 30*(10), 937–940.

Chapter 15
Generalized Response Features

Solving simulation-driven design problems using response features has numerous advantages that include improved computational efficiency and reliability. These and other benefits have been extensively discussed in Chaps. 5 through 14 in the context of local and global parametric optimization, general-purpose surrogate modeling, multicriterial design, and uncertainty quantification. At the same time, it has been mentioned that the response feature technology comes with certain limitations. The major issues are related to the very definition, existence, and extractability of the characteristic points of the system outputs. On the one hand, the feature points are strongly related to the response shape. Consequently, their definition is problem-dependent and normally requires case-to-case extraction procedures. On the other hand, in some cases of severely distorted characteristics, the feature points may not exist, which entails applicability limitations. For this reason, the response feature technique is more suitable for problems such as local parameter tuning, statistical analysis, robust design, or performance-driven modeling, where the likelihood of degenerated responses is reasonably small.

In any case, application of response feature requires a customized approach, with the characteristic points individually defined for each type of the system outputs and tailored to a particular type of design specifications. This requires user experience and hinders a widespread application of FBO. This chapter discusses a generalized and unified feature point definition in the context of antenna design. It is suitable for majority of typical antenna input characteristics (narrow-, multiband, enhanced bandwidth, wideband) and performance specifications (matching improvement, bandwidth enhancement, mixture thereof). The presented framework allows for an automated definition of the feature points given the performance specifications, along with their extraction from EM-simulated responses. Its operation is illustrated using a range of planar antennas and favorably compared to conventional (nonfeature-based) design closure task formulation.

A. Pietrenko-Dabrowska, S. Koziel, *Response Feature Technology for High-Frequency Electronics. Optimization, Modeling, and Design Automation*,
https://doi.org/10.1007/978-3-031-43845-5_15

15.1 Motivation

As mentioned on multiple occasions throughout this book, EM-driven optimization has nowadays become a commonplace in high-frequency electronics, including antenna design. Beyond elementary cases (e.g., a simple patch antenna), it is virtually infeasible to arrive at the optimum set of dimensions using theoretical models or interactive parameter sweeping, especially when the antenna topology is complex and has to conform to multiple performance requirements pertinent to its electrical and field properties (Rokunuzzaman et al., 2017). Accounting for variable interactions and their effects on antenna characteristics requires simultaneous adjustment of all parameters, which can only be achieved using numerical procedures (Yang et al., 2017). For reliability, antenna evaluation has to be carried out through EM analysis (Wang et al., 2019). EM-driven optimization may be computationally expensive even if only local (e.g., gradient-based (Zhao et al., 2015)) search is needed. Global optimization, uncertainty quantification, or multiobjective design entail considerably higher expenses (Al-Azza et al., 2016; Tang et al., 2018; Du & Roblin, 2017; De Villiers et al., 2017). The research efforts undertaken to address the aforementioned difficulties led to the development of a variety of techniques that aim at making EM-driven design procedures faster and more reliable. Available methods include the incorporation of adjoint sensitivities (Maghrabi et al., 2020), sparse sensitivity updates (Koziel & Pietrenko-Dabrowska, 2019; Pietrenko-Dabrowska & Koziel, 2020), variable-fidelity models (Wu et al., 2020), and a wide range of surrogate-assisted methods. The latter involve both physics-based techniques (space mapping, Tu et al., 2013, response correction, Koziel & Unnsteinsson, 2018) and data-driven ones (Hassan et al., 2018; Tak et al., 2018; Dong et al., 2019; Jamshidi et al., 2019). Recently, integration of global optimization with machine learning has been fostered as a way of efficient handling of multidimensional parameter spaces (Liu et al., 2019).

The methods mentioned in the previous paragraph are conventional in the sense of the design problem formulation, in particular, they operate on the original system outputs (e.g., the frequency characteristics) and use appropriately defined merit functions defined over these outputs. Response features methodologies being the subject of this book (Koziel, 2015; Koziel & Pietrenko-Dabrowska, 2020a) attempt to exploit a particular structure of the system outputs and the fact that the dependence of the frequency and level coordinates of suitably defined characteristic points of these responses exhibit less nonlinear dependence on the design parameters than the original ones, when considered in their entirety (Koziel, 2015). Reformulation of the design task (e.g., parametric optimization) in terms of response features leads to flattening the objective function landscape and allows for improving the reliability and computational efficiency of EM-driven procedures (Koziel, 2015; Koziel & Pietrenko-Dabrowska, 2020b). In some cases, for example, microwave devices such as filters, the response features may be selected as zeros and poles of the transfer function (Feng et al., 2019) or the entries of the coupling matrix (Glubokov & Koziel, 2014). However, the most generic approach (e.g., Koziel, 2015), more suitable for antennas, assumes that the characteristics points are defined on and directly extracted from the simulated frequency characteristics of the system under design. These may be the location of the antenna resonances (Koziel, 2015), the minima of the coupler matching/isolation responses, the local (in-band) maxima of the filter return loss characteristic (Koziel & Bandler, 2015), or the points corresponding to particular response levels (e.g., −10 dB when determining the antenna impedance bandwidth, Koziel, 2015). In other words, the feature definition depends on the design problem (i.e., considered performance figures and requirements) and the type of response (narrow-band, multiband, broadband, etc.). These issues have been extensively illustrated in the earlier chapters of this book, and make the technique less accessible for inexperienced users.

This chapter discusses a generalized formulation of the response feature technology for optimization of antenna input characteristics. In particular, we discuss a unified definition of the feature points, which is suitable for the majority of antenna responses, including the narrow-band, multiband, enhanced-bandwidth, and broadband responses. Furthermore, this approach enables handling of various types of performance requirements (matching improvement, bandwidth enhancement) and allows for an automated definition of the feature points given the specifications and their extraction from EM-simulated responses. The numerical results obtained for several antenna structures of different input characteristics corroborate versatility of the approach and the computational benefits it provides when compared to the standard (nonfeature-based) formulation of the design closure problem. It should be emphasized that the scope of the chapter is limited to input characteristics. Generalization for other types of responses (gain, axial ratio, etc.) will be addressed elsewhere.

15.2 Response Features: Recollection of the Concept and Advantages

This section recalls the concept of feature-based optimization, here, considered for antenna input characteristics. We begin by formulating the antenna design closure task, which is followed by discussing conventional feature-based optimization and its benefits but also limitations. Although it is a repetition of the material already included earlier in the book, it allows for making this chapter self-contained for the convenience of the readers.

The primary computational representation of the antenna under design is a full-wave EM model $\boldsymbol{R}(\boldsymbol{x})$ that returns all relevant characteristics (reflection $S_{11}(\boldsymbol{x})$, gain $G(\boldsymbol{x})$, etc.). It is a function of a designable variable vector $\boldsymbol{x}$ that contains all independent (mostly geometry) parameters undergoing the adjustment process. In the numerical optimization nomenclature, the design closure task can be formulated as

$$x^* = \arg\min_{x} U(\boldsymbol{R}(\boldsymbol{x})) \tag{15.1}$$

where $\boldsymbol{x}^*$ is the optimum design to be identified, whereas U is the objective function that encodes the design requirements imposed upon the structure. Often, the objective function is of a minimax type, that is, a set of lower and/or upper acceptance levels is defined over the specified frequencies or frequency bands. For example, when optimizing a wideband antenna to improve its matching over the frequency range $f \in [f_L, f_U]$, the function U would be defined as

$$U(\boldsymbol{R}(x)) = \max\{f_L \le f \le f_U : |S_{11}(\boldsymbol{x},f)|\} \tag{15.2}$$

If the antenna is optimized to reduce the gain variation over the same frequency range while satisfying the condition $|S_{11}(\boldsymbol{x},f)| \le -10$ dB for $f_L \le f \le f_R$, the objective function could be defined as

$$U(\boldsymbol{R}(\boldsymbol{x})) = \max\{f_L \le f \le f_R : G(\boldsymbol{x})\} - \min\{f_L \le f \le f_R : G(\boldsymbol{x})\} + \beta c(\boldsymbol{x})^2 \tag{15.3}$$

where $c(\boldsymbol{x}) = \max\{\max\{f_L \le f \le f_R: |S_{11}(\boldsymbol{x},f)| + 10\},0\}$ is a penalty function enforcing the matching condition. The latter may be also handled explicitly, as a design constraint. Similarly, optimizing a multiband antenna to be matched at the target operating frequencies $f_{0.j}$, $j = 1, ..., N$ would imply the following function U:

$$U(\boldsymbol{R}(\boldsymbol{x})) = \max\{|S_{11}(\boldsymbol{x},f_{0.1})|, \ldots, |S_{11}(\boldsymbol{x},f_{0.N})|\} \tag{15.4}$$

The particular cases mentioned above constitute merely selected examples that are supposed to give a flavor of a large variety of possible options. Other design scenarios will be discussed later in the chapter.

Feature-based optimization (FBO) has been originally introduced in Koziel (2015) as a way of improving computational efficiency and reliability of EM-driven design procedures. FBO departs from the original formulation of the design problem and expresses the performance requirements in terms of appropriately defined characteristic points of the antenna responses. For the sake of further discussion, we denote the feature points as $\boldsymbol{p}_k = [f_k\ l_k]$, $k = 1, \ldots, N_p$, where f_k and l_k stand for the frequency and level coordinate of the kth point and N_p is the number of features defined for a given antenna structure. The aggregated feature vector will be denoted as $\boldsymbol{P}(\boldsymbol{x}) = [\boldsymbol{p}_1(\boldsymbol{x}) \ldots \boldsymbol{p}_{Np}(\boldsymbol{x})]^T$. FBO capitalizes on the fact that the dependence between the design parameters $\boldsymbol{x}$ and the coordinates f_k and l_k is significantly less nonlinear than a similar dependence for the original responses, that is, $\boldsymbol{R}(\boldsymbol{x})$ (Koziel, 2015), as illustrated in Fig. 15.1 for a dual-band antenna, with the feature points being the locations of the antenna resonances. The following consequences of the aforementioned property can be observed:

- Reformulating the original design problem (15.1) in terms of the feature points results in "flattening" the functional landscape to be handled by the optimization

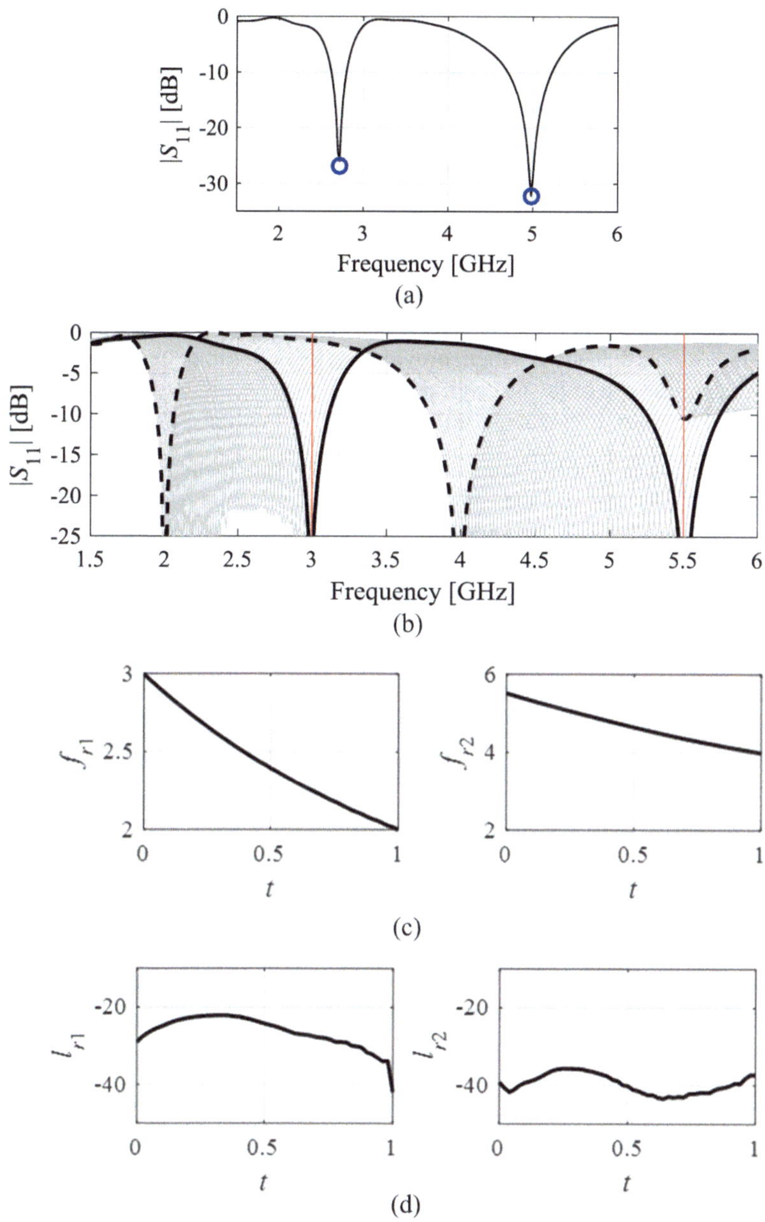

Fig. 15.1 Response feature concept: (**a**) example reflection characteristic of a dual-band antenna and the feature points corresponding to the antenna resonances (o); (**b**) exemplary initial design (- - -), the design optimized for target frequencies 3.0 GHz and 5.5 GHz (—), and the family of reflection responses along the line segment connecting these two designs, parameterized by $0 \leq t \leq 1$ (gray lines); (**c**) frequency, and (**d**) level coordinates of the feature points as functions of the parameter t.

algorithm, which leads to improved reliability and a faster convergence (Koziel, 2015).

- When used in the modelling context, feature-based approach may enable considerable computational savings in terms of training data acquisition, that is, less training samples are required (as compared to operating at the level of frequency characteristics) to render the surrogates of similar predictive power (Koziel & Pietrenko-Dabrowska, 2020a).
- Operating at the level of response features facilitates frequency manipulation of the antenna characteristics (e.g., shifting the resonances to different operating frequencies), which is usually challenging when handling the entire responses through, for example, minimax objective functions.
- Using response features often provides a leeway for poor initial designs, in particular, the use of local search procedures may be sufficient even in situations normally requiring global algorithms (Koziel & Pietrenko-Dabrowska, 2020b).

The advantages of FBO (and response feature approach in general) have been demonstrated in the context of design optimization of multiband (Koziel & Pietrenko-Dabrowska, 2020a) and broadband antennas, microwave couplers, surrogate modeling (Koziel & Pietrenko-Dabrowska, 2020a), uncertainty quantification (Koziel & Bandler, 2015), and globalized search (Koziel & Pietrenko-Dabrowska, 2020b); see also Chaps. 5 through 14.

Selection of the feature points is generally problem dependent. It might be related to the circuit transfer function (e.g., pole and zero location) (Feng et al., 2019) or may be directly extracted from the circuit responses, for example, locations of the resonances (Koziel, 2015) or local maxima of the filter return loss characteristic in the passband (Zhang et al., 2015). In either case, the prerequisite is that the information carried by the characteristic points is sufficient to account for the design specifications. For illustration, let us consider the minimax objective function (15.4), defined for the problem of allocating the antenna resonances at the target operating frequencies $f_{0.j}$, $j = 1, \ldots, N$ and improving the matching therein. In this case, it is sufficient to select the response features to be the locations of the antenna resonances so that we have $\boldsymbol{p}_k(\boldsymbol{x}) = [f_{r.k}(\boldsymbol{x})\ l_{r.k}(\boldsymbol{x})]$, $k = 1, \ldots, N$, where $f_{r.k}(\boldsymbol{x})$ and $l_{r.k}(\boldsymbol{x})$ are the resonant frequencies and the corresponding reflection levels. Using these, the feature-based objective function U_F can be defined as

$$U_F(\boldsymbol{P}(\boldsymbol{x})) = \max\{l_{r.1}(\boldsymbol{x}), \ldots, l_{r.N}(\boldsymbol{x})\} + \beta \left\| \begin{bmatrix} f_{0.1} \\ \vdots \\ f_{0.N} \end{bmatrix} - \begin{bmatrix} f_{r.1}(\boldsymbol{x}) \\ \vdots \\ f_{r.2}(\boldsymbol{x}) \end{bmatrix} \right\|^2 \quad (15.5)$$

where β is a multiplication factor determining the contribution of the frequency misalignment term into the objective function. It should be noted that the optimum designs according to (15.4) and (15.5) coincide (assuming that it is possible to perfectly match the target frequencies), yet the profile of (15.5) along the line section

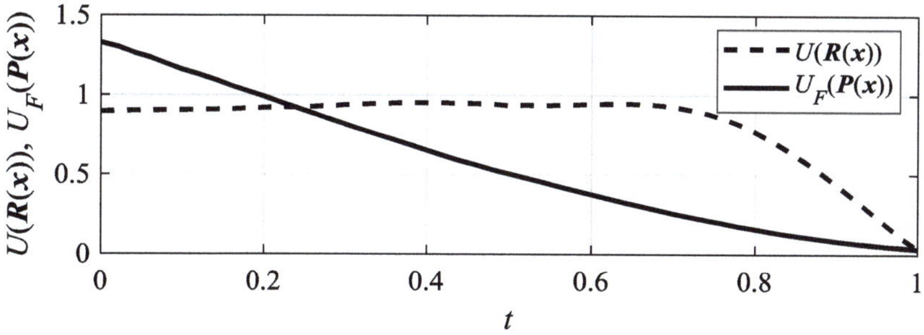

Fig. 15.2 Minimax objective function (15.4) (- - -) and feature-based objective function (15.5) (—) versus parameter t (cf. Fig. 15.1). The feature-based objective function is smoother and monotonic, and the target design is reachable from the given initial point through local search

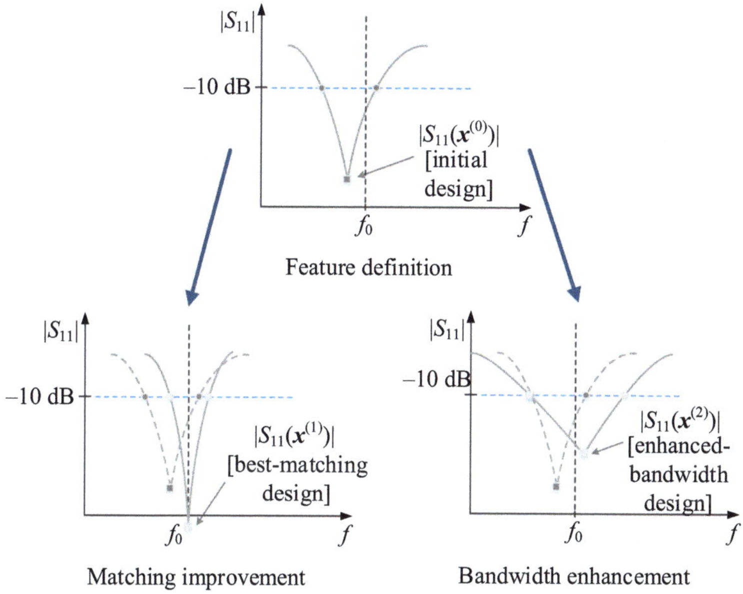

Fig. 15.3 Selecting response features for particular design tasks. The feature point corresponding to the antenna resonance can be used for matching improvement at the target frequency f_0 (left plot). The points corresponding to –10 dB levels can be employed for bandwidth enhancement (right plot)

considered in Fig. 15.1 is monotonic as opposed to that of (15.4), cf. Figure 15.2, which makes the problem easier to handle.

As mentioned earlier, the response features need to be defined so that the information therein allows us to account for performance specifications imposed on the system at hand. This is illustrated in Fig. 15.3, where—for the purpose of matching improvement at the target frequency—the right choice of the feature point would be the location of the resonance, whereas for the sake of bandwidth enhancement, it is more appropriate to select the points corresponding to −10 dB levels of the reflection characteristic.

15.3 Generalized Response Features

This section introduces the concept of generalized response features. The material is organized as follows. Section 15.3.1 provides a unified definition of characteristics points suitable for various types of antenna input characteristics (multi-band, wide-band, etc.). Feature-based objective function definition is dealt with in Sect. 15.3.2, whereas Sect. 15.3.3 discusses the overall algorithm operation and illustrates it using a flow diagram.

15.3.1 Concept and Feature Point Definition

In general, the feature definition, its extraction from EM-simulated antenna responses, and a definition of the feature-based objective function has to be tailored to a particular type of antenna response and a specific design task. This is perhaps the most serious disadvantage of the FBO technique in terms of its automation, because the setup of the optimization framework requires a certain amount of user experience and interaction. Here, a unified definition of response features is discussed so that the characteristic points can be defined and extracted automatically regardless of a specific set of performance specifications, and subsequently used to formulate a feature-based objective function (which will be dealt with in Sect. 15.3.2). As mentioned earlier, in this chapter, our considerations are restricted to antenna input characteristics (reflection response). Generalization to other types of responses such as gain or axial ratio will be given elsewhere.

Design specifications concerning the antenna reflection response are either related to matching improvement (or satisfying given acceptance thresholds) at a specific frequency band (or bands in the case of multi-band structures) or to bandwidth enhancement at, and in the vicinity of, the specific operating frequencies. Thus, when considering the characteristic points capable of accounting for the aforementioned types of requirements, one needs to include the following:

- A point corresponding to the reflection minimum within the bandwidth; the left-most local minimum is to be considered
- Bandwidth-defining points, which for the reflection characteristic would be those corresponding to −10 dB level of $|S_{11}|$
- Points corresponding to the target operating frequencies
- Points corresponding to the levels in the middle (in dB scale) between the reflection minimum and 0 dB
- Points corresponding to the local maxima within the bandwidth

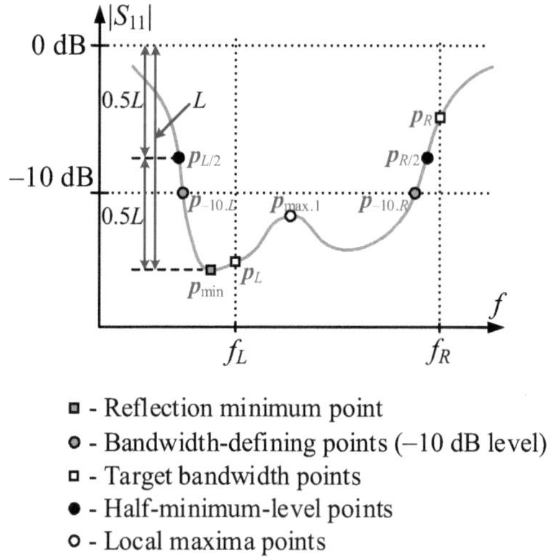

Fig. 15.4 Example reflection response, target bandwidth (frequencies f_L and f_R), and the feature points: $\boldsymbol{p}_{\min}$—the left-most local reflection minimum; $\boldsymbol{p}_{-10.L}, \boldsymbol{p}_{-10.H}$—the points corresponding to –10 dB levels; $\boldsymbol{p}_L, \boldsymbol{p}_H$—the points corresponding to the target frequencies f_L and f_H; and $\boldsymbol{p}_{L/2}, \boldsymbol{p}_{H/2}$—the points corresponding to the levels determined by half of the distance between 0 dB and the level of the minimum point (in dB), and the local maxima ($\boldsymbol{p}_{\max.k}$). In this case, the characteristic only contains one in-band local maximum (i.e., $k = 1$). The picture shows a generic situation where all types of feature points actually exist. Degenerate situations will be discussed later in the section, see also Fig. 15.5

The rationale behind selecting these particular points will be elaborated on further in this section. The above set of points will be considered the maximum feature set, which may be reduced for some cases by merging the points that belong to different categories, depending on the type of antenna response, the design specifications, or simply the fact that some of the points may be nonexistent at particular designs. Furthermore, in the case of multiband antennas, there will be a separate set of feature points corresponding to each band. However, for the sake of simplifying the discussion, we assume the case of a single-band antenna at the moment. Figure 15.4 illustrates the definition and meaning of the particular categories of feature points. In the exemplary situation shown in the picture, all types of points are present. Notwithstanding, in many practical situations, one will often encounter degenerate cases. These will be discussed later in the section.

The following notation will be used throughout (for the sake of brevity, henceforth, the dependence of the entries of the vector $\boldsymbol{P}(\boldsymbol{x})$ on the design variables will be omitted):

$$\boldsymbol{P}(\boldsymbol{x}) = \left[\boldsymbol{p}_{\min}\ \boldsymbol{p}_{-10.L}\ \boldsymbol{p}_{-10.R}\ \boldsymbol{p}_L\ \boldsymbol{p}_R\ \boldsymbol{p}_{L/2}\ \boldsymbol{p}_{R/2}\ \boldsymbol{p}_{\max.1}\ \cdots\right] \tag{15.6}$$

where (f and l are the frequency and the level coordinates of the feature points):

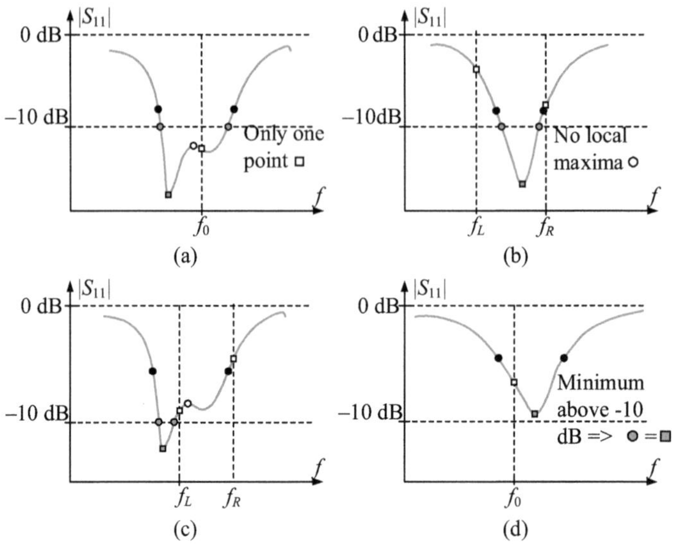

Fig. 15.5 Response features for several specific scenarios (explanations of the feature points can be found in Fig. 15.4): (**a**) the presence of a single operating frequency f_0 (instead of the target bandwidth defined by the frequencies f_L and f_R); (**b**) the lack of local maxima; (**c**) part of the operating bandwidth below −10 dB, in which case half-level points are more appropriate to estimate the antenna bandwidth than −10 dB ones; (**d**) reflection minimum above −10 dB, in which case the–10 dB points are assigned to be identical to the minimum point

- $\boldsymbol{p}_{\min} = [f_{\min}\ l_{\min}]^T$—feature point corresponding to the left-most in-band minimum.
- $\boldsymbol{p}_{-10.L} = [f_{-10.L}\ l_{-10.L}]^T, \boldsymbol{p}_{-10.R} = [f_{-10.R}\ l_{-10.R}]^T$—feature points corresponding to – 10 dB reflection levels (left- and right-hand side, respectively).
- $\boldsymbol{p}_L = [f_L\ l_L]^T, \boldsymbol{p}_R = [f_R\ l_R]^T$—feature points corresponding to the target bandwidth frequencies f_L and f_H, respectively.
- $\boldsymbol{p}_{L/2} = [f_{L/2}\ l_{L/2}]^T, \boldsymbol{p}_{R/2} = [f_{R/2}\ l_{R/2}]^T$—feature points corresponding to the level in the middle (in dB scale) between the reflection minimum and 0 dB.
- $\boldsymbol{p}_{\max.k} = [f_{\max.k}\ l_{\max.k}]^T$—feature points corresponding to the kth local in-band reflection maximum. For the purpose of extracting $\boldsymbol{p}_{\max.k}$, the bandwidth is understood as the frequency range between $\min\{f_{-10.L}, f_{L/2}\}$ and $\max\{f_{-10.R}, f_{R/2}\}$.

The reason for defining these particular types of points is that they are necessary but also sufficient to account for the various types of specifications that one may encounter while handling the reflection characteristics of antennas. In particular:

- $\boldsymbol{p}_{\min}$ can be used to control the antenna matching at the operating frequency (in the cases of narrow- and multiband antennas) and to relocate the antenna resonances toward the target frequencies.
- $\boldsymbol{p}_{-10.L}$ and $\boldsymbol{p}_{-10.R}$ can be used to control the antenna bandwidth (e.g., when optimizing the antenna for maximum bandwidth).

- $\boldsymbol{p}_L$ and $\boldsymbol{p}_R$ will be useful to implement minimax type of specifications (e.g., matching improvement over a specific frequency range).
- $\boldsymbol{p}_{L/2}$ and $\boldsymbol{p}_{R/2}$ are auxiliary points that will be useful to approximately determine antenna bandwidth in some degenerate cases (e.g., when the overall in-band reflection level is above −10 dB);
- $\boldsymbol{p}_{\max.k}$ will be employed to control antenna matching for minimax type of specifications (e.g., in the cases of wideband structures).

As mentioned before, some of the features may or may not exist for a given instance of the reflection characteristic:

- If the minimum reflection level in the vicinity of the antenna operating bandwidth is higher than −10 dB, neither $\boldsymbol{p}_{-10.L}$ nor $\boldsymbol{p}_{-10.R}$ exist and we assign $\boldsymbol{p}_{-10.L} = \boldsymbol{p}_{-10.R} = \boldsymbol{p}_{\min}$.
- If the reflection characteristic does not have any local maxima in the operating band, the points $\boldsymbol{p}_{\max.k}$ are not assigned.
- In some cases, one may have $f_L = f_R = f_0$, i.e., the design specifications are defined with respect to an operating frequency f_0 rather than the operating bandwidth $[f_L\ f_R]$. In this situation, we assign $\boldsymbol{p}_L = \boldsymbol{p}_R = \boldsymbol{p}_0 = [f_0\ l_0]^T$, where l_0 is the response level at f_0.

The reason for the above assignments is that for the sake of practical handling of the response features (e.g., through numerical procedures), it is more convenient to maintain a constant number of points (apart from the local maxima, which are at the end of the response feature vector), so that their identification within $\boldsymbol{P}(\boldsymbol{x})$ is straightforward. Figure 15.5 shows several degenerate situations that may be encountered in practice. Depending on the case, some of the points may not be defined as described above.

It should also be noted that some of the feature points may not be well-defined simply because of a limited frequency range of antenna simulation. This would be typically the case for ultra-wideband antennas, which are well matched even for the highest simulation frequency. Therein, the feature points $\boldsymbol{p}_R$ and $\boldsymbol{p}_{R/2}$ may not exists.

In the case of multiband antenna, each band will have an independent set of feature points extracted according to the same rules as discussed in this section. The overall feature point structure will consist of vectors

$$\boldsymbol{P}_j(\boldsymbol{x}) = \left[\boldsymbol{p}_{j.\min}\ \boldsymbol{p}_{j.-10.L}\ \boldsymbol{p}_{j.-10.R}\ \boldsymbol{p}_{j.L}\ \boldsymbol{p}_{j.R}\ \boldsymbol{p}_{j.L/2}\ \boldsymbol{p}_{j.R/2}\ \boldsymbol{p}_{j.\max.1}\ \cdots\right] \quad (15.7)$$

with $j = 1, \ldots, N_B$, where N_B is the number of antenna operating bands.

15.3.2 Objective Function Definition

The primary purpose of the feature-based technology is to improve the reliability and to expedite the design optimization procedures. Having a unified definition of response features, the next goal is to develop a versatile formulation of the feature-based objective function that would express the original (often, minimax) objectives imposed on the antenna reflection characteristic in terms of the feature point coordinates. To this end, we will consider several types of specifications, which essentially cover most of the practical possibilities in terms of antenna reflection response manipulation, the designer might want to pursue. For the sake of notational simplicity, the list below concerns single-band antennas, but generalization for multiband cases is straightforward. As before, f_0 will be referred to as a (single) target frequency, whereas f_L and f_R will denote the target operating bandwidth:

- Matching improvement within the frequency range $[f_L\ f_R]$ (which implies relocation of the antenna operating bandwidth near $[f_L\ f_R]$)
- Matching improvement at f_0 (which implies relocation of the antenna resonance to f_0)
- Maximization of bandwidth that includes the frequency range $[f_L\ f_R]$ (relocation of the antenna operating bandwidth near $[f_L\ f_R]$ is implied)
- Maximization of (symmetric) bandwidth around the center frequency f_0 (relocation of antenna operating bandwidth towards f_0 is implied)

However, as mentioned before, the second and the fourth case are special cases of the first and the third one, respectively (with the assignment of $f_L = f_R = f_0$, and $\boldsymbol{p}_L = \boldsymbol{p}_R = \boldsymbol{p}_0 = [f_0\ l_0]^T$). Thus, in the following, only these two cases (matching improvement and bandwidth maximization) will be considered.

15.3.2.1 Objective Function for In-Band Matching Improvement

The objective function for matching improvement will be defined as (single-band case)

$$U_F(\boldsymbol{P}(\boldsymbol{x})) = \max\{l_L(\boldsymbol{x}), l_R(\boldsymbol{x}), l_{\max.1}(\boldsymbol{x}), \ldots\} + \beta_m\left[c_{m.L}(\boldsymbol{x})^2 + c_{m.R}(\boldsymbol{x})^2\right] \quad (15.8)$$

where

$$c_{m.L}(\boldsymbol{x}) = \max\left\{\min\{f_{-10.L}(\boldsymbol{x}), f_{L/2}(\boldsymbol{x})\} - f_L, 0\right\} \quad (15.9)$$

$$c_{m.R}(\boldsymbol{x}) = \max\left\{f_R - \min\{f_{-10.R}(\boldsymbol{x}), f_{R/2}(\boldsymbol{x})\}, 0\right\} \tag{15.10}$$

The first term in (15.8), which is the primary objective, represents the maximum in-band reflection to be minimized. Note that for the case $f_L = f_R = f_0$ (single operating frequency instead of a range $[f_L\, f_R]$), this term reduces to $l_0(\boldsymbol{x})$. The second term is a penalty factor that allows us to relocate the antenna bandwidth to $[f_L\, f_R]$. The penalty term has a nonzero contribution if either the left-hand-side edge of the actual operating band is higher than the target f_L or the right-hand-side edge of the operating band is lower than the target f_R.

Generalization of (15.8)–(15.10) for the case of multiband antennas will take the form of

$$U_F(\boldsymbol{P}(\boldsymbol{x})) = \max\{l_{1.L}(\boldsymbol{x}), \ldots, l_{N_B.L}(\boldsymbol{x}), l_{1.R}(\boldsymbol{x}), \ldots l_{N_B.R}(\boldsymbol{x}), l_{1.\max.1}(\boldsymbol{x}), \ldots, l_{N_B.\max.1}(\boldsymbol{x}), \ldots\} + \beta_m\left[\sum_{j=1}^{N_B} c_{j.m.L}(\boldsymbol{x})^2 + \sum_{j=1}^{N_B} c_{j.m.R}(\boldsymbol{x})^2\right] \tag{15.11}$$

with

$$c_{j.m.L}(\boldsymbol{x}) = \max\left\{\min\{f_{j.-10.L}(\boldsymbol{x}), f_{j.L/2}(\boldsymbol{x})\} - f_{j.L}, 0\right\} \tag{15.12}$$

$$c_{j.m.R}(\boldsymbol{x}) = \max\left\{f_{j.R} - \min\{f_{j.-10.R}(\boldsymbol{x}), f_{j.R/2}(\boldsymbol{x})\}, 0\right\} \tag{15.13}$$

The above definitions are consistent with the "traditional" feature-based objective functions. For example, it can be observed that the objective function (15.5) considered in Sect. 15.2 is a special case of (15.11)–(15.13) because we have: $l_{j.L}(\boldsymbol{x}) = l_{j.R}(\boldsymbol{x}) = l_{j.\min}(\boldsymbol{x}) = l_{r.j}$ for $j = 1, \ldots, N_B = N$, that is, the first term in (15.11) becomes $\max\{l_{r.1}(\boldsymbol{x}), \ldots, l_{r.N}(\boldsymbol{x})\}$. Also, $f_{j.-10.L}(\boldsymbol{x}) = f_{j.-10.R}(\boldsymbol{x}) = f_{j.\min}(\boldsymbol{x}) = f_{r.j}$; hence, $f_{j.L/2}(\boldsymbol{x}) < f_{j.-10.L}(\boldsymbol{x})$ and $f_{j.R/2}(\boldsymbol{x}) > f_{j.-10.R}(\boldsymbol{x})$. Due to these relations, we have $\min\{f_{j.-10.L}(\boldsymbol{x}), f_{j.L/2}(\boldsymbol{x})\} - f_{j.L}\} = f_{j.L/2}(\boldsymbol{x}) - f_{0.j}$ and $f_{j.R} - \min\{f_{j.-10.R}(\boldsymbol{x}), f_{j.R/2}(\boldsymbol{x})\} = f_{0.j} - f_{j.R/2}(\boldsymbol{x})$. Now, if $f_{j.L/2}(\boldsymbol{x}) < f_{0.j}$, then $c_{j.m.L}(\boldsymbol{x}) = 0$; otherwise, $c_{j.m.L}(\boldsymbol{x}) = f_{j.L/2}(\boldsymbol{x}) - f_{0.j}$. Similarly, if $f_{0.j} < f_{j.R/2}(\boldsymbol{x})$, then $c_{j.m.R}(\boldsymbol{x}) = 0$; otherwise, $c_{j.m.R}(\boldsymbol{x}) = f_{0.j} - f_{j.R/2}(\boldsymbol{x})$. Thus, the second term in (15.11) becomes

$$\sum_{j=1}^{N_B} c_{j.m.L}(\boldsymbol{x})^2 + \sum_{j=1}^{N_B} c_{j.m.R}(\boldsymbol{x})^2 = \sum_{j=1}^{N_B}\left(\max\left\{f_{j.L/2}(\boldsymbol{x}) - f_{0.j}, 0\right\}\right)^2 + \sum_{j=1}^{N_B}\left(\min\left\{f_{j.R/2}(\boldsymbol{x}) - f_{0.j}, 0\right\}\right)^2 \tag{15.14}$$

The term (15.14) becomes zero if the antenna operating frequencies are sufficiently close to the target frequencies, that is, $f_{j.L/2} < f_{0.j} < f_{j.R/2}$ for all j, and it is proportional to the second power of the respective frequency differences otherwise. On the other hand, the penalty term in (15.5) approaches zero under the same conditions, that is, $f_{r.j} \approx f_{0.j}$ for all j, and it has quadratic dependence between the operating frequency deviations with respect to the target otherwise.

15.3.2.2 Objective Function for Bandwidth Maximization

The objective function for bandwidth maximization is more complex. For a single-band case, it takes the form of

$$U_F(\boldsymbol{P}(\boldsymbol{x})) = -B(\boldsymbol{x}) + \beta_b c_b(\boldsymbol{x})^2 \tag{15.15}$$

where the bandwidth (here, symmetric with respect to the center frequency $f_c = (f_L + f_R)/2$) is defined as

$$B(\boldsymbol{x}) = 2\min\{f_c - f_{B.L}, f_{B.R} - f_c\} \tag{15.16}$$

The bandwidth-defining frequencies $f_{B.L}$ and $f_{B.R}$ are

$$f_{B.L} = \begin{cases} f_{L/2} & \text{if} \quad l_{\min} > -10 \text{ dB} \quad \text{or} \quad \dfrac{f_{-10.L} - f_{L/2}}{f_{R/2} - f_{L/2}} > \dfrac{1}{3} \\ f_{-10.L} & \text{otherwise} \end{cases} \tag{15.17}$$

$$f_{B.R} = \begin{cases} f_{R/2} & \text{if} \quad l_{\min} > -10 \text{ dB} \quad \text{or} \quad \dfrac{f_{R/2} - f_{-10.R}}{f_{R/2} - f_{L/2}} > \dfrac{1}{3} \\ f_{-10.R} & \text{otherwise} \end{cases} \tag{15.18}$$

The reflection levels at $f_{B.L}$ and $f_{B.R}$ are $l_{B.L}$ and $l_{B.R}$, respectively (and equal to either -10 dB or to $L_{L/2}/L_{R/2}$ depending on the case in (15.17) and (15.18)). The penalty factor $c_b(\boldsymbol{x})$ in (15.15) is to ensure that the antenna reflection does not exceed -10 dB within its operating bandwidth. We have

$$c_b(\boldsymbol{x}) = \max\{\max\{l_{B.L}(\boldsymbol{x}), l_{B.R}(\boldsymbol{x}), l_{\max.1}(\boldsymbol{x}), \ldots\} + 10, 0\} \tag{15.19}$$

The reason for the particular definition of the bandwidth-defining frequencies in (15.17) and (15.18) is that—in some cases—the -10-dB points $\boldsymbol{p}_{-10.L}$ and $\boldsymbol{p}_{-10.R}$ do not give a proper account for the bandwidth, either because these points may not exist (if $l_{\min} > -10$) or because of a specific shape of the reflection characteristic (cf. Fig. 15.6). Eventually, that is, when close to the optimum, the bandwidth will be determined by $\boldsymbol{p}_{-10.L}$ and $\boldsymbol{p}_{-10.R}$.

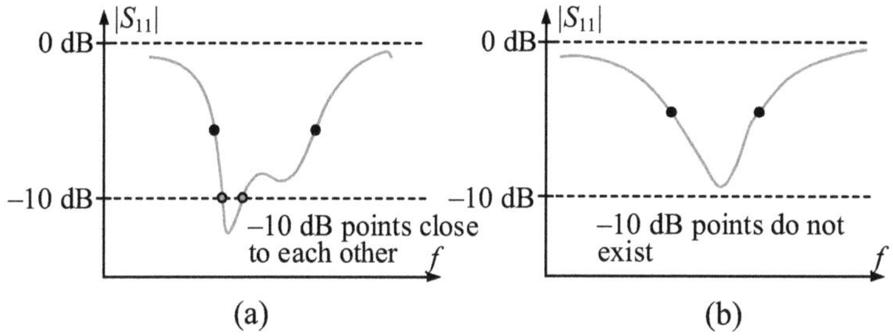

Fig. 15.6 Exemplary situations when a better account for the impedance bandwidth can be obtained using the frequencies $f_{L/2}$ and $f_{R/2}$ rather than −10-dB points: (**a**) the −10-dB points are very close to each other; (**b**) the −10-dB points do not exist. Both cases motivate the definitions (15.17) and (15.18) of the frequencies $f_{B.L}$ and $f_{B.R}$

A practical problem related to (15.17) and (15.18) is that $f_{B.L}$ and $f_{B.R}$ are discontinuous with respect to $\boldsymbol{x}$ whenever there is a jump from −10 dB frequencies to $f_{L/2}$ or $f_{R/2}$. This would be troublesome from the point of view of numerical optimization. The following formulas provide alternative versions, which are smooth with respect to $\boldsymbol{x}$, therefore more suitable for being incorporated into numerical procedures, especially gradient-based algorithms:

$$f_{B.L} = \begin{cases} f_{-10.L} & \text{if} \quad g_L(\boldsymbol{x}) \le 1/3 \\ f_{-10.L} + 0.5\left(f_{L/2} - f_{-10.L}\right)[6(g_L(\boldsymbol{x}) - 1/3)]^2 \\ \quad \text{if} \quad 1/3 < g_L(\boldsymbol{x}) \le 1/2 \\ f_{L/2} - 0.5\left(f_{L/2} - f_{-10.L}\right)[6(2/3 - g_L(\boldsymbol{x}))]^2 \\ \text{if} \quad 1/2 < g_L(\boldsymbol{x}) \le 2/3 \\ f_{L/2} \quad \text{if} \quad 2/3 < g_L(\boldsymbol{x}) \end{cases} \tag{15.20}$$

where $g_L(\boldsymbol{x}) = [f_{-10.\,L} - f_{L/2}]/[f_{R/2} - f_{L/2}]$ (here, $f_{-10.L} = f_{min}$ if $l_{min} > -10$ dB) and

$$f_{B.R} = \begin{cases} f_{-10.R} & \text{if} \quad g_R(\boldsymbol{x}) \le 1/3 \\ f_{-10.R} + 0.5\left(f_{R/2} - f_{-10.R}\right)[6(g_R(\boldsymbol{x}) - 1/3)]^2 \\ \text{if} \quad 1/3 < g_R(\boldsymbol{x}) \le 1/2 \\ f_{R/2} - 0.5\left(f_{R/2} - f_{-10.R}\right)[6(2/3 - g_R(\boldsymbol{x}))]^2 \\ \text{if} \quad 1/2 < g_R(\boldsymbol{x}) \le 2/3 \\ f_{R/2} \quad \text{if} \quad 2/3 < g_R(\boldsymbol{x}) \end{cases} \tag{15.21}$$

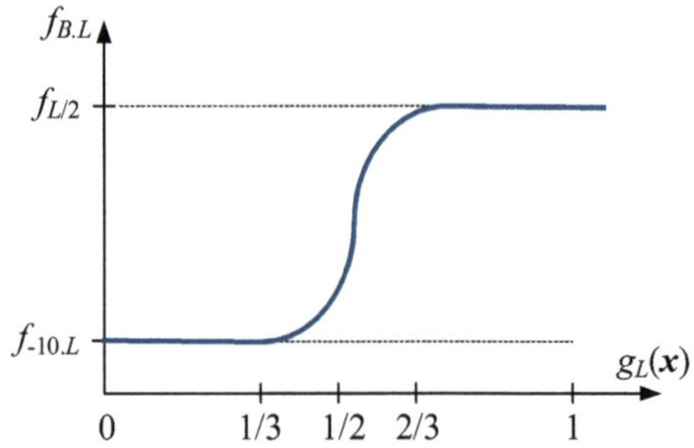

Fig. 15.7 Functional dependence of $f_{B.L}$ on $g_L(\boldsymbol{x})$ according to (15.20)

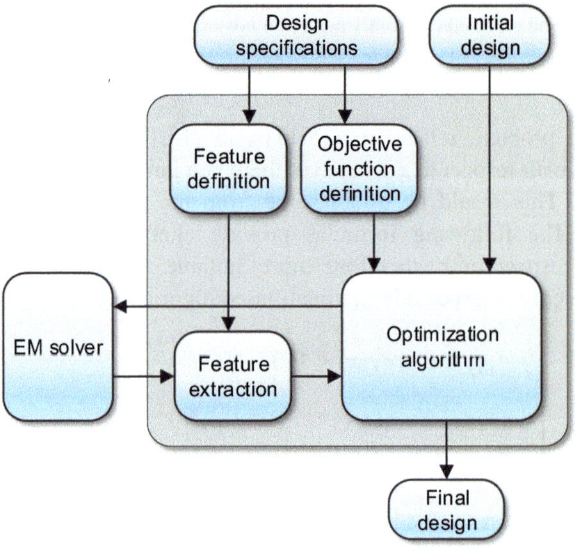

Fig. 15.8 Optimization using generalized response features: algorithm flow. The response feature part is pertinent to the design task formulation and EM simulation results postprocessing, as discussed in Sects. 15.3.1 and 15.3.2. The optimization algorithm is an independent part of the process. In this chapter, a trust-region (TR) gradient-based procedure is employed for all verification cases of Sect. 15.4

where $g_R(\boldsymbol{x}) = [f_{R/2} - f_{-10.\ R}]/[f_{R/2} - f_{L/2}]$ (here, $f_{-10.R} = f_{\min}$ if $l_{\min} > -10$ dB). Figure 15.7 shows the functional dependence of $f_{B.L}$ on $g_L(\boldsymbol{x})$. Generalization of (15.15)–(15.21) for the case of multiband antennas is straightforward. The details are omitted for the sake of brevity.

15.3.3 Optimization Algorithm Flow

This section briefly summarizes the operation of the optimization procedure involving generalized response features. The algorithm flow has been shown in Fig. 15.8. Design specifications are used to define the feature points (cf. Sect. 15.3.1) and the feature-based objective function (cf. Sect. 15.3.2).

The feature points are extracted from EM simulation responses each time a specific design is to be evaluated by the optimization algorithm. Using this data, the objective function value is computed. The optimization algorithm itself is an independent part of the procedure. In the numerical experiments discussed in Sect. 15.4, the trust-region (TR) gradient-search (Conn et al., 2000) is employed.

It should be noted that although the unified definition of the feature points (Sect. 15.3.1) and the defining details of the feature-based objective functions (Sect. 15.3.2) may seem complex, both the feature extraction and objective function technicalities, once implemented, are hidden from the user.

As mentioned earlier, such a generalization is the main purpose of this chapter so that the same routines can be used regardless of the particular type of antenna characteristics (narrow-band, broadband, multiband) to both extract the feature points and to translate the performance specifications into mathematically rigorous cost functions that adequately assess the design quality and can be directly employed by the optimization algorithms.

15.4 Illustration Examples

This section provides a numerical verification of the unified feature based optimization (FBO) technique in the context of design optimization of antenna input characteristics of various types. The numerical studies are carried out using four planar structures of distinct reflection responses: a ring slot antenna, a patch antenna with enhanced bandwidth, a multiband dipole antenna, and an ultra-wideband monopole.

All verification structures have been optimized for best in-band matching, whereas the ring-slot antenna has been also optimized for bandwidth enhancement. The same routines implementing the feature point extraction and objective functions are employed in each case showing the versatility of the presented approach. The feature-based formulation is compared to the conventional ways of representing the performance specifications (mostly minimax), so that the benefits of generalized FBO in terms of design reliability but also computational efficiency can be identified.

The presented methodology has been benchmarked against the following methods: (i) gradient-based TR algorithm (Conn et al., 2000), (ii) derivative-free pattern search algorithm (Conn et al., 2009) (the specific implementation used here, described in (Koziel, 2010), is a stencil-based procedure with variable grid size

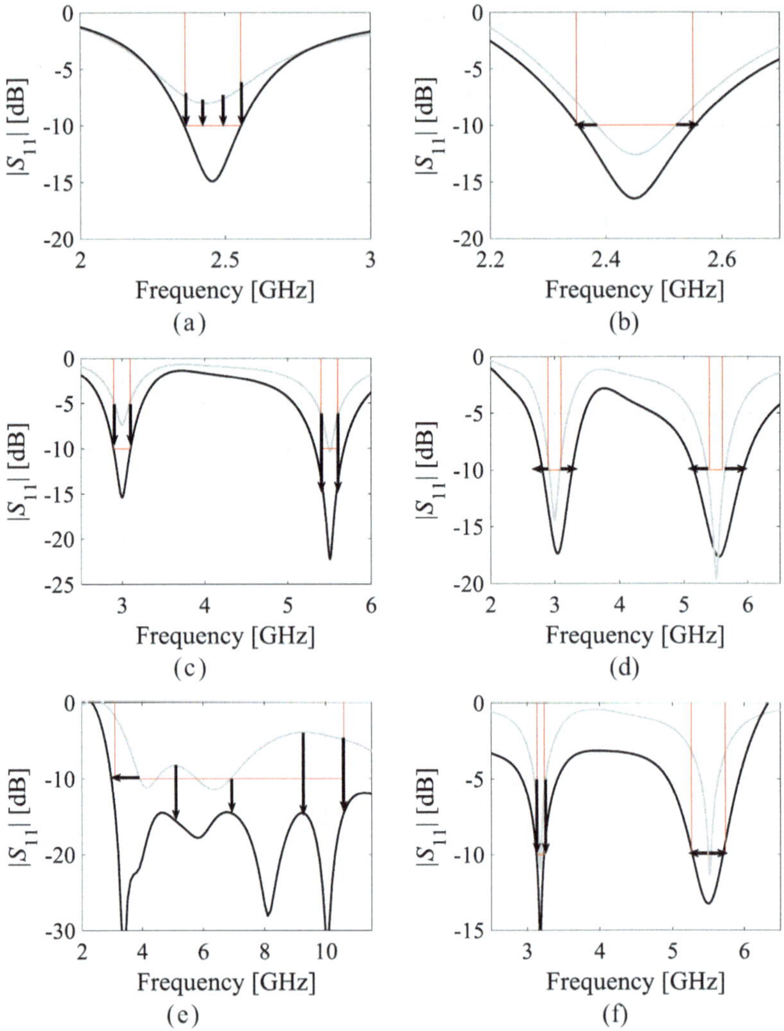

Fig. 15.9 A variety of the optimization tasks that may be handled in a unified manner regardless of the particular type of antenna characteristics within the considered optimization approach using the generalized formulation of response features: (**a**) a single-band antenna optimized for improved matching, (**b**) a single-band antenna optimized for bandwidth enhancement, (**c**) a dual-band antenna optimized for improved matching in both operating bands, (**d**) a dual-band antenna optimized for bandwidth enhancement in both bands, (**e**) a wide-band antenna optimized for improved matching within the target operating bandwidth, (**f**) a dual-band antenna optimized for improved matching at and around the first operating frequency and, at the same time, optimized for bandwidth enhancement at and around the second operating frequency; the initial and optimal designs are marked gray and black, respectively, horizontal and vertical lines indicate the design specifications

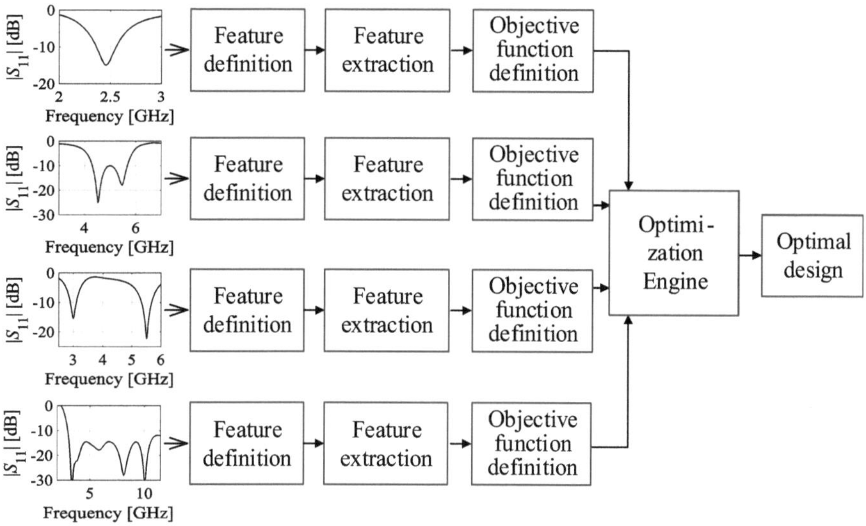

Fig. 15.10 Conceptual illustration of the optimization process flow according to the conventional response features approach (from left to right): (i) optimized antenna input characteristics of various types; (ii) definition of individual response features; (iii) feature extraction procedure ((ii) and (iii) are mandatory in the conventional response feature approach or for bandwidth enhancement within the conventional approach); (iv) definition of the objective function that encodes design specifications (has to be carried out separately for each design task and each antenna type); (v) design optimization routine of choice; and (vi) the optimal solution

enhanced by grid-constrained line search), and (iii) particle swarm optimizer (PSO) (Jin & Rahmat-Samii, 2007). The first two algorithms are local search procedures, whereas the third is a global nature-inspired population-based routine.

Figure 15.9 presents a selection of a variety of optimization tasks of the antenna input characteristics that the considered approach is capable of handling without introducing any adjustments pertaining to the objective function formulation otherwise unavoidable in the conventional approach. These include:

- Optimization for the best in-band matching in the assumed target operating band of a single-band antenna (Fig. 15.9a)
- Optimization for bandwidth enhancement (Fig. 15.9b)

- Minimization of a dual-band antenna reflection coefficient in both operating bands (Fig. 15.9c)
- Maximization of the bandwidths of the same antenna (Fig. 15.9d)
- Optimization for improved matching of a wide-band antenna within the target operation bandwidth (Fig. 15.9e)
- Optimization of a dual-band antenna for improved matching at and around the first operating frequency and, at the same time, for bandwidth enhancement at and around the second operating frequency (Fig. 15.9f)

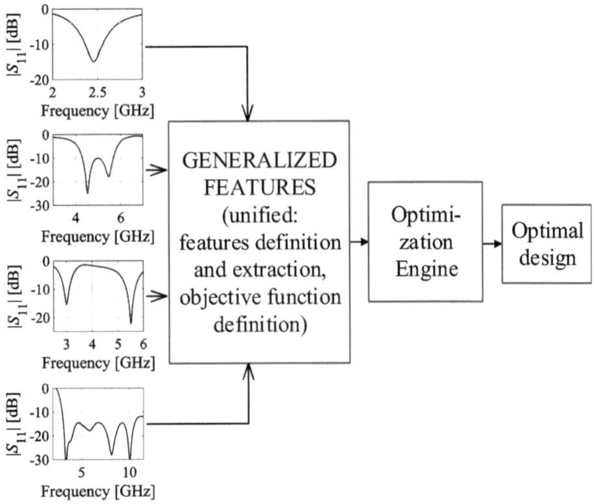

Fig. 15.11 Conceptual illustration of the optimization process flow in the generalized response features approach (from left to right): (i) optimized characteristics of various types, (ii) generalized features procedure that handles: response features definition and extraction and unified objective function definition, (iii) design optimization routine of choice, and (iv) the optimal solution

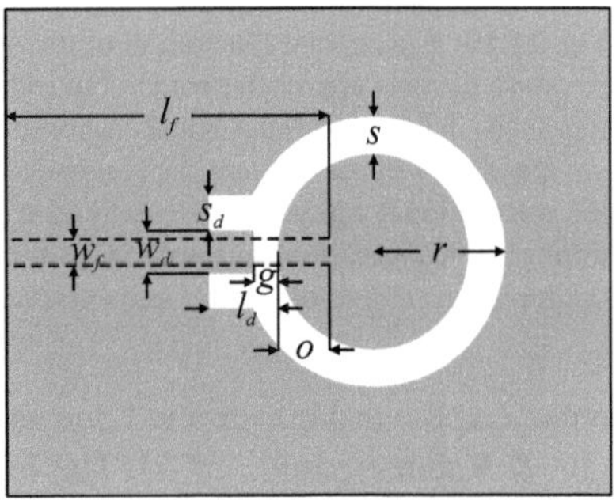

Fig. 15.12 Geometry of a ring-slot antenna; dashed lines mark feeding line (Sim et al., 2014)

In order to demonstrate the benefits offered by a unified definition of response features, a comparison of the optimization flow arrangement is discussed for the conventional FBO approach involving individualized response features definition of (Koziel, 2015) (Fig. 15.10), and the approach discussed in this chapter (Fig. 15.11). The former requires that the designer separately handles any specific type of antenna structure and its corresponding response, which entails individual definition and extraction of the feature points (including implementation of the relevant Matlab codes, etc.), cf. Figure 15.10. Subsequently, the designer needs to define the objective function encoding the design specifications. Whereas in the considered approach, as shown in Fig. 15.11, the generalized features procedure handles all the response types and the design tasks in a unified and automated manner. Thus, the response features definition and extraction and the definition of the merit function is carried out without the designer's interaction, which streamlines the entire design process.

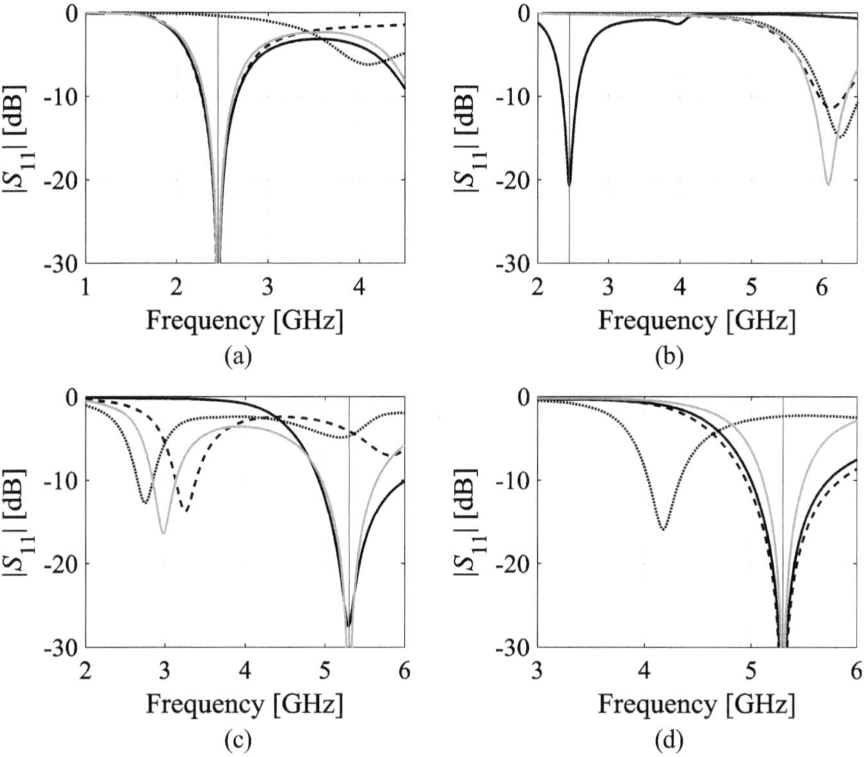

Fig. 15.13 Ring-slot antenna of Fig. 15.12, optimization for best in-band matching. The following responses are shown: at the initial design (····), the design optimized using the algorithm exploiting generalized formulation of response features (—), the design optimized using classic TR algorithm (- - -), and the design optimized using the conventional pattern search algorithm (gray). The presented designs correspond to the following target operating frequencies: (**a**), (**b**) f_0 = 2.45 GHz, (**c**), (**d**) f_0 = 5.3 GHz and have been optimized starting from different initial designs. Vertical lines indicate the design specifications

15.4.1 Case I: Ring Slot Antenna

The first example is a slot antenna shown in Fig. 15.12 (Sim et al., 2014). The structure is implemented on a 0.76-mm-thick substrate of relative permittivity $\varepsilon_r = 3.5$. The antenna is excited through a microstrip line, which feeds a circular ground plane slot with defected ground structure (Sim et al., 2014). The designable variables are $\boldsymbol{x} = [l_f\ l_d\ w_d\ r\ s\ s_d\ o\ g]^T$. The unit for all the parameters is millimeter. The feed line width w_f is adjusted to achieve 50 ohm input impedance. The lower and upper bounds for the design variables are: $\boldsymbol{l} = [20\ 1\ 0.2\ 6.5\ 0.2\ 0.2\ 1\ 0.2]^T$ and $\boldsymbol{u} = [35\ 10\ 2.2\ 15.5\ 7.0\ 7.0\ 12\ 4]^T$, respectively. The EM simulation model is evaluated in CST Microwave Studio (~300,000 cells, simulation 90 s). The antenna

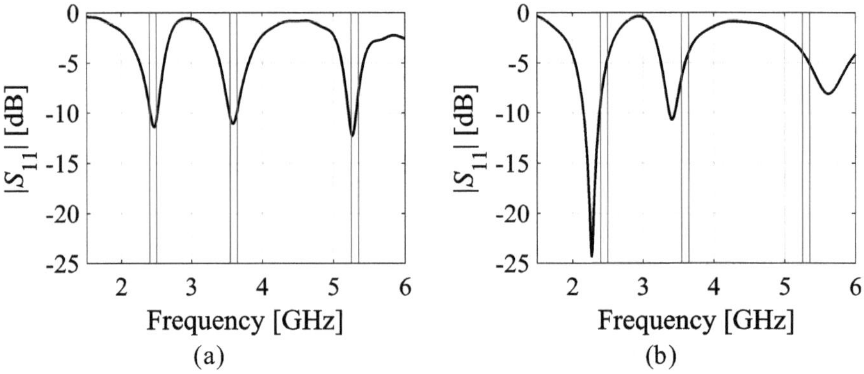

Fig. 15.14 Ring-slot antenna of Fig. 15.12, optimization for bandwidth enhancement (observe that this kind of optimization task cannot be directly carried out within the classical approach). Responses of ring-slot antenna at the designs optimized using the algorithm exploiting generalized formulation of response features: the initial design (····), the design optimized for enhanced bandwidth (—), and the design optimized for the best in-band matching (- - -). Shown are the designs corresponding the target operating frequency $f_0 = 2.45$ GHz and $f_0 = 5.3$ GHz. Horizontal lines mark the design specifications

has been optimized for: (i) the best in-band matching around the center frequency f_0 and (ii) enhanced impedance bandwidth around f_0.

The first design task is to improve antenna matching at the target operating frequency f_0. The numerical experiments have been performed for the following two values of the operating frequency: $f_0 = 2.45$ GHz and $f_0 = 5.3$ GHz. Figure 15.13 shows the designs optimized within the considered approach, as well those rendered using the conventional local search procedures: trust region and pattern search algorithm, starting from four different initial designs. The optimal designs shown in Fig. 15.13a, c satisfy the assumed specifications for both the approaches. Whereas the conventional TR local search starting from the initial designs distant from the required target operating frequencies fails to find satisfactory designs (see Fig. 15.13b, d). As for the pattern search algorithm, it has been capable of appropriately allocating the operating frequency in three out of four cases, and failed only in the case of the design of Fig. 15.13b.

Notwithstanding, the generalized feature approach is capable of accurately allocating target operating frequencies in all considered cases. As the objective function is defined differently in both approaches, for comparison purposes, we quantify the design quality as the maximum reflection level at f_0. For the designs yielded in our approach (shown in Fig. 15.13), the average value is -17.1 dB, whereas for the TR algorithm, it is -14.0 dB, and -12.5 dB in the case of pattern search algorithm.

The respective computational costs equal 152, 72, and 338 EM analyses (i.e., the cost of design optimization using the pattern search algorithm is the highest among the compared local procedures, over twice as much as the considered technique). Yet, perhaps the most important factor that allows for comparing both the approaches is the accuracy of the operating frequency allocation. In our approach,

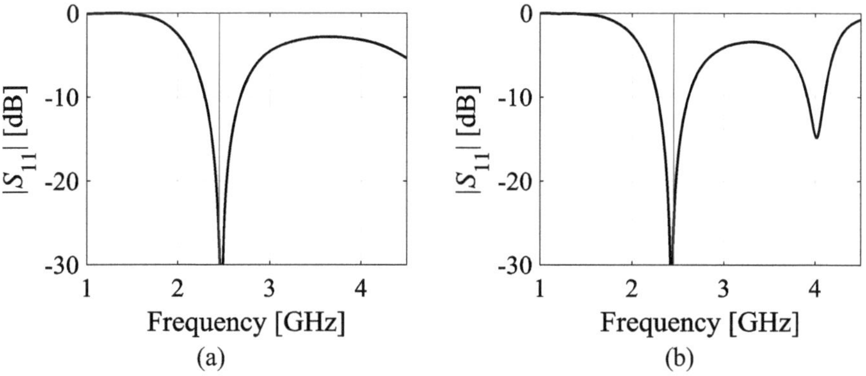

Fig. 15.15 Ring-slot antenna of Fig. 15.12, optimization for the best in-band matching at $f_0 = 2.45$ GHz using PSO benchmark algorithm (—). Representative optimal designs are shown. Vertical lines indicate design specifications

Fig. 15.16 Geometry of the triple-band dipole antenna (Koziel & Bekasiewicz, 2018)

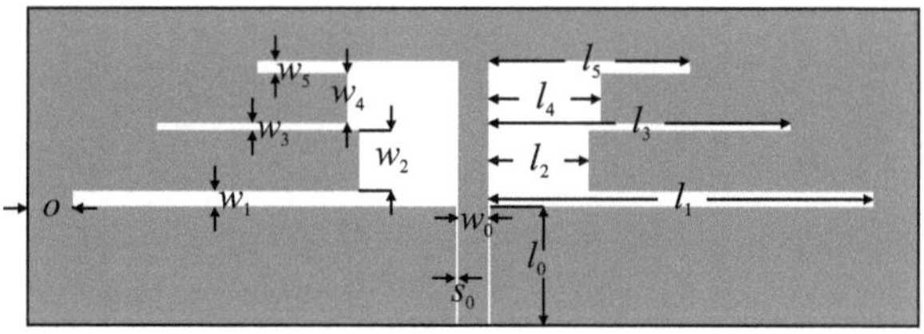

the average error of the operating frequency misalignment (for the designs of Fig. 15.13) equals merely 2.6 percent, whereas in the conventional approaches, it is as high as 91.9 and 91.0 percent, for TR and pattern search algorithm, respectively. This corroborates suitability of the generalized features optimization framework for reliable antenna design.

Figure 15.14 shows the results of maximizing the bandwidth for the ring-slot antenna of Fig. 15.12 using our methodology. As before, the optimal designs meet the assumed design specifications. It should be observed that—within the conventional local approach (both TR and pattern search algorithms)—the bandwidth enhancement cannot be performed without some kind of feature definition (as well as extraction) for assessing the antenna bandwidth at each iteration of the optimization algorithm run.

For comparison purposes, Fig. 15.15 provides the selected optimal designs for best in-band matching at $f_0 = 2.45$ GHz rendered by multiple runs (10) of the benchmark particle swarm optimization (PSO) procedure (population size of 20, maximum number of iterations equal to 50). The quality of the PSO-optimized designs is very good (the maximum reflection at f_0 is -27.7 dB on average, with the standard deviation of 2.2 dB across 10 algorithm runs). The accuracy of the operating frequency is excellent and equals 1.0 percent. Yet, this is achieved at the computational cost that equals 1000 EM analyses, which is significantly higher (by a

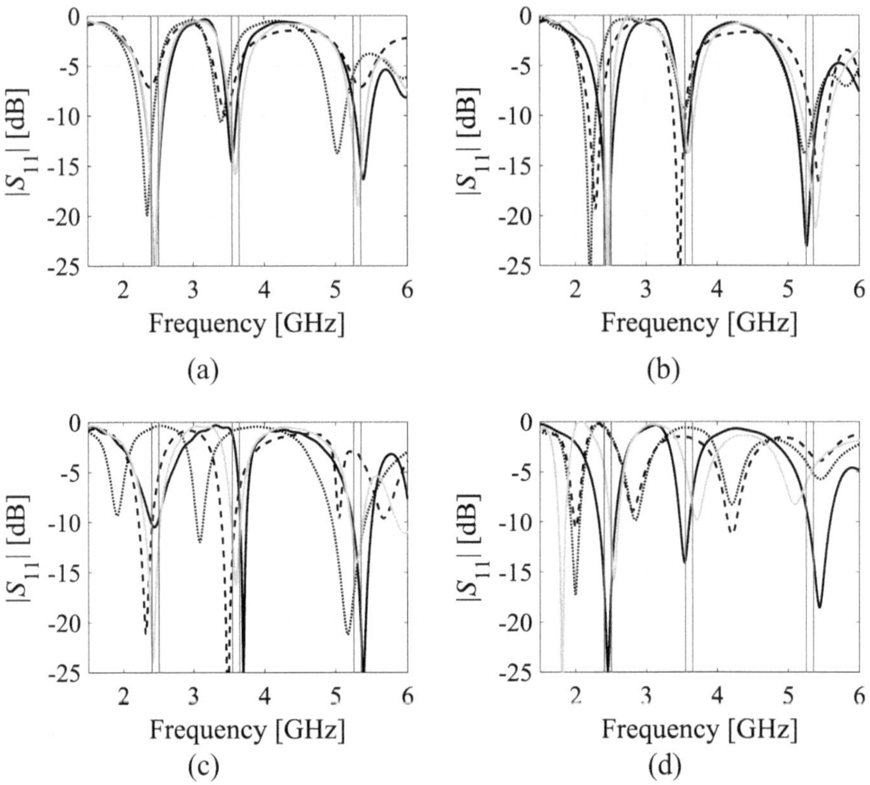

Fig. 15.17 Triple-band antenna of Fig. 15.16, optimization for the best in-band matching at the ±50 MHz bands centered at 2.45 GHz, 3.6 GHz, and 5.3 GHz. The following responses are shown: At four different initial designs (**a**) through (**d**) (····), the designs optimized using the algorithm exploiting generalized formulation of response features (—), the designs optimized using classic TR algorithm (- - -), and the designs optimized using the conventional pattern search algorithm (gray). Vertical lines indicate design specifications

factor of about ten) than that of the local search (both classical and the generalized feature one).

15.4.2 *Case II: Triple-Band Dipole Antenna*

The second verification case is a triple-band dipole antenna shown in Fig. 15.16 (Koziel & Bekasiewicz, 2018). The structure is fed by a coplanar waveguide and implemented on RO4350 substrate (ε_r = 3.48, h = 0.762 mm). The geometry parameters are: $\boldsymbol{x} = [l_1\ l_2\ l_3\ l_4\ l_5\ w_1\ w_2\ w_3\ w_4\ w_5]^T$. The following dimensions $l_0 = 30$, $w_0 = 3$, $s_0 = 0.15$, and $o = 5$ remain fixed. All dimensions are expressed in mm. The lower and upper bounds on the design variables are: $\boldsymbol{l}$ = [30 5 20 5 15 0.2

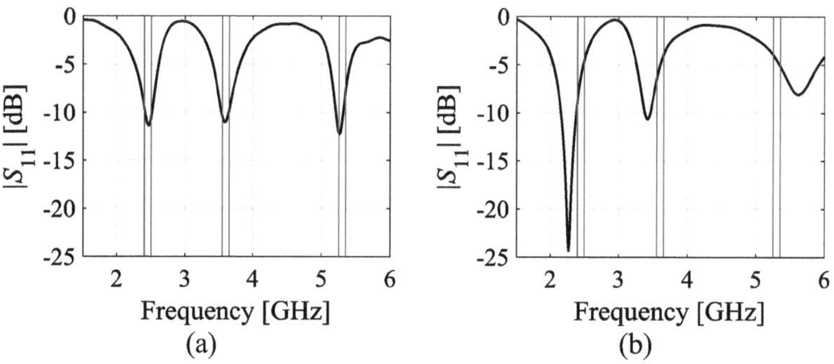

Fig. 15.18 Triple-band antenna of Fig. 15.16, optimization for the best in-band matching at the ±50 MHz bands centered at 2.45 GHz, 3.6 GHz, and 5.3 GHz using PSO benchmark algorithm (—). Shown are two selected PSO-optimized designs featuring: (**a**) correctly allocated operating frequencies; (**b**) operating frequencies misaligned with respect to the assumed design specifications (marked with vertical lines)

0.2 0.2 0.2 0.2]^T and $\boldsymbol{u}$ = [50 15 30 14.9 21 2.2 4.2 2.2 4.2 2.2] ^{T}, respectively. The computational model is simulated in CST Microwave Studio using its time-domain solver. The antenna is intended to operate within 50 MHz bandwidths around the following center frequencies: 2.45 GHz, 3.6 GHZ, and 5.3 GHz.

Figure 15.17 shows the designs optimized within the classical and the considered approach for the triple-band antenna of Fig. 15.16. The results of Fig. 15.17 show superiority of the considered algorithm over the benchmark: The target operating frequencies of all the optimized designs are allocated accurately. Whereas the optimal designs rendered by the classical local search TR procedure are considerably misaligned with the target in all presented cases. As far as pattern search algorithm is concerned, it has been capable of allocating the operating frequencies in three out of four cases. The respective errors of the target frequency allocation equal 1.6 percent (our approach), 26.2 percent (TR approach), and 27.4 percent (pattern search algorithm).

The design quality (quantified as the maximum reflection level at the center frequencies) is −11.8 dB, −3.9 dB, and −12.0 dB (generalized features, TR and pattern search approaches, respectively) on the average, across the design set of Fig. 15.17. The corresponding CPU cost is 76.3, 85.7, and 730 EM analyses, respectively. As before, the cost of pattern search algorithm is the highest (around 10 times higher than that of the generalized feature procedure) among local routines.

For additional verification, Fig. 15.18 presents the designs optimized with the use of PSO (selected from the ten algorithm runs). The operating frequencies of the antenna of PSO-rendered designs are allocated accurately for the design of Fig. 15.18a; yet, they are considerably misaligned for the design shown in Fig. 15.18b. The average maximum reflection around at the operating frequencies is −7.0 dB, with the standard deviation of 2.5 dB. The average accuracy of the operating frequencies allocation is slightly worse than in the generalized feature approach and equals to 1.8 percent. Still, one has to bear in mind the tremendous cost

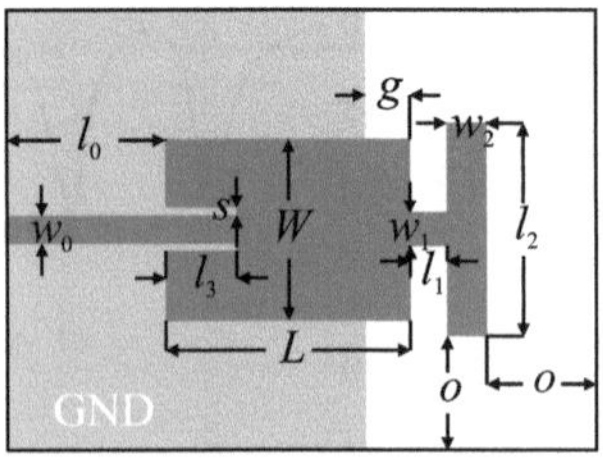

Fig. 15.19 Geometry of the planar antenna with enhanced bandwidth (Bekasiewicz & Koziel, 2016). Ground plane marked using light-gray shade

of optimizing these designs (1000 EM simulations), considerably (over ten times) higher than that of the generalized feature technique.

15.4.3 Case III: Bandwidth-Enhanced Patch Antenna

The third example is a bandwidth-enhanced planar antenna, presented in Fig. 15.19 (Bekasiewicz & Koziel, 2016). The structure comprises two radiating elements in the form of a quasi-microstrip patch with inset feed excited through a microstrip line. The antenna is implemented on 0.76-mm-thick substrate of relative permittivity $\varepsilon_r = 3.0$, $\tan\delta = 0.0018$. The independent geometry parameters are $\boldsymbol{x} = [L\ l_1\ l_2\ l_3\ W\ w_1\ w_2\ g]^T$. Whereas the following parameters are fixed: $o = 7$, $l_0 = 10$, and $s = 0.5$. The lower and upper bounds on design variables are: $\boldsymbol{l} = [10\ 0.1\ 5\ 0.1\ 2\ 0.2\ 0.1\ {-6}]^T$ and $\boldsymbol{u} = [40\ 10.0\ 30\ 15\ 30\ 8\ 15\ 15]^T$, respectively. The unit for all dimensions is mm. The feed line width w_0 is adjusted to ensure 50-ohm impedance. The EM antenna model is implemented in CST Microwave Studio.

The antenna has been optimized for the best in-band matching within the frequency range 4.5 GHz $\leq f \leq$ 5.5 GHz. Figure 15.20 shows the designs optimized using the presented methodology, along with those optimized using classical TR gradient-based search and pattern search procedure. In three out of four cases, the generalized feature approach outperforms both benchmark local procedures in terms of allocating the target bandwidth. It should be emphasized that the initial designs are considerably misaligned with respect to the target bandwidth; yet, the generalized feature technique has been capable of finding the optimal solutions satisfying the design specifications. The average value of the maximum of the input reflection characteristic within the target bandwidth is −15.8 dB for the presented approach, as well as −9.9 dB and −1.1 dB for the local benchmark procedures, respectively (across the set of designs of Fig. 15.20). The computational cost of rendering the optimal designs is comparable for generalized features and the TR approach, and it equals to 113.8 and 110.5 EM simulations per design, respectively. Whereas in the case of pattern search algorithm. it is significantly higher and equals to 585 EM analyses (i.e., it is over five times higher).

As for the previous examples, Fig. 15.21 provides the exemplary optimal designs rendered with PSO optimization algorithm. One of them accurately allocates the antenna operating bandwidth (from 4.5 GHz to 5.5 GHz), whereas the other is

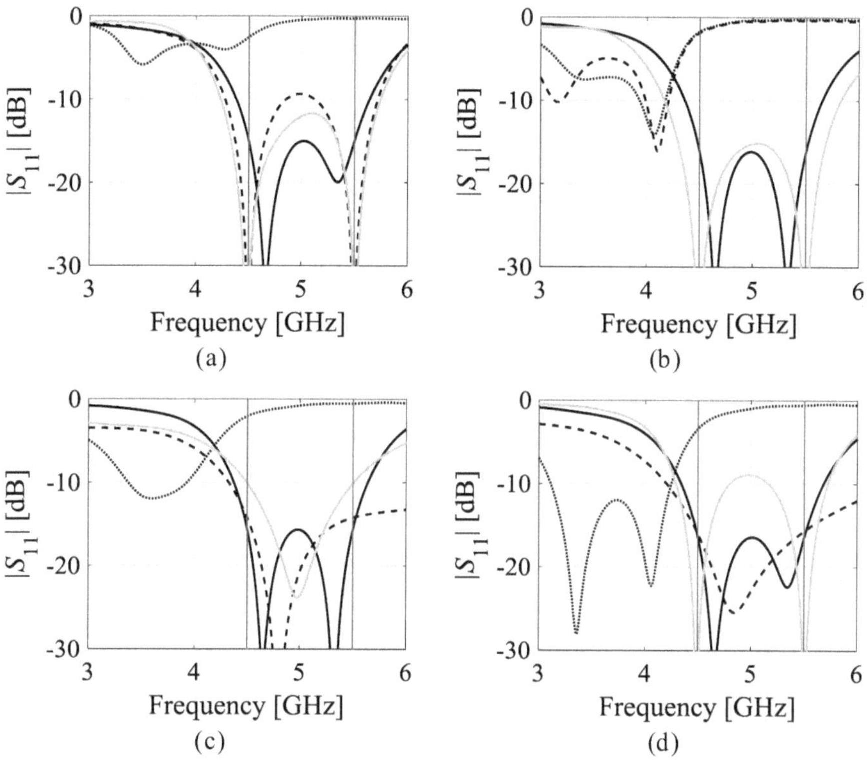

Fig. 15.20 Bandwidth-enhanced patch antenna of Fig. 15.19, optimization for the best in-band matching within the frequency range 4.5 GHz $\leq f \leq$ 5.5 GHz. Shown are the responses: At four different initial designs considerably misaligned with the target bandwidth (····), the designs optimized using the algorithm exploiting generalized formulation of response features (—), the design optimized using classic TR algorithm (- - -), and the design optimized using the conventional pattern search algorithm (gray). Vertical lines indicate design specifications

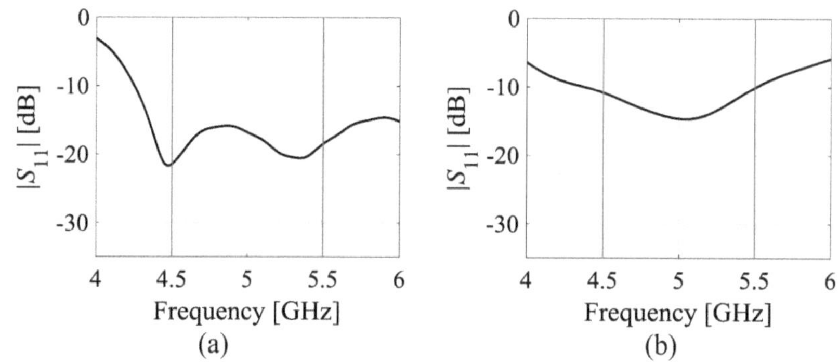

Fig. 15.21 Bandwidth-enhanced patch antenna of Fig. 15.19, optimization for the best in-band matching within the frequency range 4.5 GHz $\leq f \leq$ 5.5 GHz using PSO benchmark algorithm (—). Representative optimized designs are shown. Vertical lines indicate design specifications

considerably misaligned. The average quality of PSO-rendered designs equals –12.9 dB, with the standard deviation of 4.1 dB (across 10 algorithm runs). The operational bandwidth has been properly allocated at 6 out of 10 optimal designs. Nevertheless, the cost of rendering these designs is equal to 1000 EM analyses, which is—again—around 10 times higher than that of the local search.

15.4.4 Case IV: Ultra-Wideband Antenna

The last verification case is an ultra-wideband antenna implemented on Taconic RF-35 substrate ($h = 0.762$ mm, $\varepsilon_r = 3.5$, $\tan\delta = 0.0018$), shown in Fig. 15.22 (Alsath & Kanagasabai, 2015). The structure employs a quasi-circular radiator, along with a modified ground plane for bandwidth enhancement. The antenna geometry parameter vector is $\boldsymbol{x} = [L_0\ dR\ R\ r_{\text{rel}}\ dL\ dw\ L_g\ L_1\ R_1\ dr\ c_{\text{rel}}]^T$. The lower and upper bounds on the parameters are: $\boldsymbol{l} = [4.0\ 0\ 3.0\ 0.1\ 0\ 0\ 4.0\ 0\ 2.0\ 0.2\ 0.2]^T$ and $\boldsymbol{u} = [15.0\ 6.0\ 8.0\ 0.9\ 5.0\ 8.0\ 15.0\ 6.0\ 5.0\ 1.0\ 0.9]^T$, all dimensions in mm. The antenna is to be optimized for minimum reflection within the UWB frequency range (3.1–10.6 GHz).

Figure 15.23 shows the optimal designs for antenna of Fig. 15.22 rendered in the presented approach, as well as those obtained using the benchmark local TR and pattern search algorithms. The results indicate that the generalized feature approach allows for meeting the design specifications in all the considered cases, whereas the TR algorithm is not capable of assigning the antenna bandwidth accurately, especially when the initial design is allocated above the intended frequency range (see Fig. 15.23c, d); the same pertains to the pattern search procedure. The average computational cost of rendering the designs of Fig. 15.23 is 125.5, 83.5, and 598 EM simulations for our, TR, and pattern search algorithms, respectively. The corresponding design quality (maximum value of the reflection characteristic within the UWB range) is –14.7 dB, –7.51 dB, and –6.7 dB, respectively. Again, the cost of the pattern search procedure is the highest one (around five times higher than that of the feature-based technique).

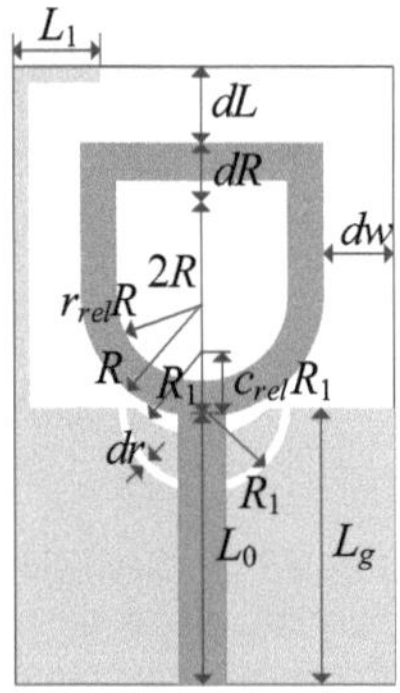

Fig. 15.22 Geometry of the ultra-wideband antenna (Alsath & Kanagasabai, 2015). Ground plane marked using light gray shade

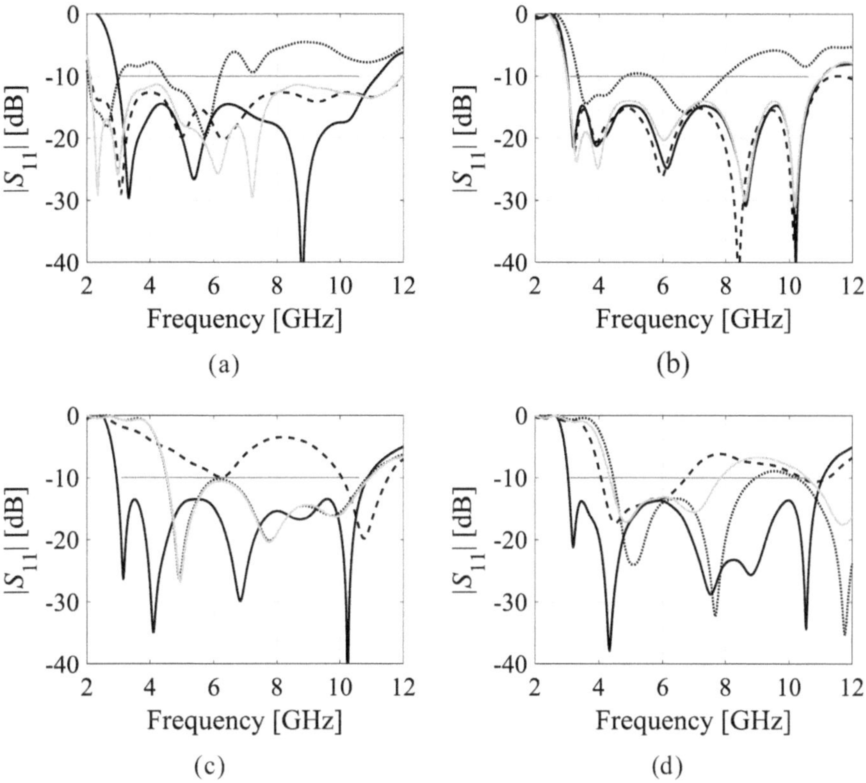

Fig. 15.23 Broadband antenna of Fig. 15.22, optimization for best in-band matching within UWB band from 3.1 GHz to 10.6 GHz. The following responses are shown: at the initial designs (····), the designs optimized using the algorithm exploiting generalized formulation of response features (—), the designs optimized using classic TR algorithm (- - -), and the designs optimized using the conventional pattern search algorithm (gray). Horizontal line indicates the design specifications

As a supplementary verification, the broadband antenna of Fig. 15.22 has also been optimized with the PSO algorithm (see Fig. 15.24). One of the presented PSO-optimal designs is of high quality: The intended UWB bandwidth is allocated accurately, and the reflection does not exceed −10 dB level (Fig. 15.24a). However, this is not the case for the design shown in Fig. 15.24b, which exceeds the assumed reflection level. The cost of rendering both the designs shown in Fig. 15.24 is as high as 1000 EM analyses. The average design quality (maximum in-band reflection) is −10.9 dB (i.e., it is poorer than within our approach), with standard deviation of 1.8 dB across 10 algorithm runs. The maximum reflection level of the PSO-rendered designs does not exceed −10 dB level for 7 out of 10 designs.

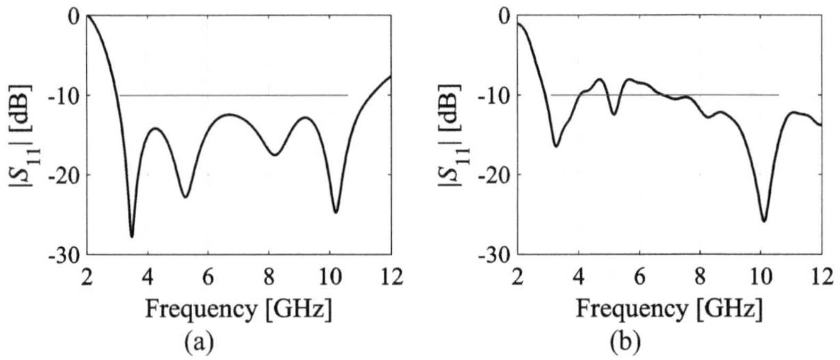

Fig. 15.24 Broadband antenna of Fig. 15.22, optimization for best in-band matching within UWB band from 3.1 GHz to 10.6 GHz using PSO benchmark algorithm (—).Two designs are shown: (**a**) meeting the assumed design specifications (marked with horizontal line) and (**b**) with $|S_{11}|$ exceeding the level of –10 dB in a part of the UWB frequency range

15.4.5 Discussion

The analysis of the results presented in Sects. 15.4.1 through 15.4.4 allows us to draw several conclusions concerning the advantages offered by the optimization algorithm exploiting the unified and automated definition of response features, as well as the performance of the procedure when compared to that of the conventional local search and particle swarm optimization algorithm.

The observations can be summarized as follows:

- The generalized feature approach has been demonstrated to successfully handle a variety of antenna structures featuring different types of input characteristics while using identical setup with no need of user interaction.
- The allocation and extraction of the feature points is automated with the only information provided by the user being design specifications.
- Based on the same information, the procedure selects and sets up the appropriate feature-based objective function, again, without engaging the expert knowledge of the user.
- The results confirm the reliability of the presented approach: For all of the verification structures, the optimal designs satisfy the assumed specifications, even for severely misaligned initial designs.
- The conventional local trust-region gradient search algorithm frequently fails to find the optimal designs meeting the target operating frequencies, and it is very much dependent on the quality of the start point.
- The pattern search algorithm (the second local benchmark procedure) also is incapable of yielding the designs that satisfy the specifications in some of the presented cases, and its outcome heavily depends on the antenna operating frequencies at the initial design. In addition, it incurs the highest computational cost among the three local optimizers (from 5 to even 10 times higher, depending on the structure).

- The considered problems are generally challenging even for state-of-the-art nature-inspired procedures (here, represented by PSO). The PSO-rendered optimal designs for some structures do not meet the assumed design specifications, which puts its reliability in question.
- The quality of the optimal designs rendered by all the compared procedures (the generalized features, conventional local algorithms: TR and pattern search, as well as PSO optimizer) is similar across the set of four benchmark antenna structures of different types of input characteristics (provided the local and PSO optimizer have been able to identify these).
- The computational cost of the presented procedure is—on the average—comparable (or slightly higher) to that of the conventional TR approach and considerably lower than that of the pattern search algorithm. Still, even though the TR local search is in some cases faster, its reliability is significantly inferior to that of our approach.
- The computational cost of the presented procedure is around 10 times smaller than for the PSO algorithm while providing optimal designs of similar quality.

The overall conclusion is that the performance of the optimization process is greatly improved by employing the generalized definition of the response features. The execution of the entire procedure is facilitated by the unified feature definition and extraction, but also unified objective function formulation. Comprehensive validation studies demonstrate that generalized response feature technology allows for obtaining high-quality designs, even when starting from poor initial parameter vectors, which normally call the employment of global procedures. Moreover, the procedure is computationally efficient with the cost comparable to the conventional local TR procedure, a fraction (10–20 percent) of the cost of pattern search algorithm, and about 10 percent of that of the particle swarm optimizer. The most important observation is that the generalization performs up to the expectations and the purpose it was developed for, especially in terms of handling a variety of design tasks and types of antenna input characteristics within a single and automated algorithmic framework.

15.5 Summary

This chapter discussed a generalized formulation of the response feature technology for efficient and reliable optimization of antenna input characteristics of various types. The presented approach unifies different definitions used so far in the literature to determine the characteristic points of antenna responses and provides a single framework, suitable for handling narrow-, multiband, and enhanced bandwidth, as well as broadband (also ultra-wideband) structures. The generalized definition of response features is accompanied by a unified formulation of the feature-based objective functions that cover the design tasks related to in-band matching improvement and bandwidth enhancement. The major advantage of this methodology is that it eliminates the need for defining both the response features and the cost functions individually for a particular antenna structure. Thus, it makes the feature-based optimization (FBO) more accessible for an inexperienced user.

A comprehensive numerical verification demonstrates the performance of the method over a range of antenna structures featuring different types of input characteristics (ring-slot, triple-band dipole, enhanced-bandwidth patch antenna, and ultra-wideband monopole antenna) but also corroborates other advantages of FBO, which include the improved reliability (in particular, good performance when facing poor initial design) and reduced computational cost as compared to the standard design problem definitions. The considered formulation has been specifically developed for handling antenna input characteristics. Generalizations for antenna responses such as gain, axial ratio, etc. (i.e., other than input characteristics), as well as for compound problems (e.g., gain maximization under reflection constraints, or simultaneous impedance matching and axial ratio bandwidth improvement) is yet to be developed and implemented.

All the aforementioned benefits are achieved at the expense of limiting the scope of applicability of the modelling method to structures whose responses contain well-distinguished characteristic points, for example, the aforementioned multiband antennas. As a result, the methodology is not as versatile as other frameworks that do not impose any restraints on the response structure of the component under design. Yet, the characteristics of many real-world antennas are inherently structured (e.g., narrow-band or multiband antennas). Consequently, the employment of the feature-based techniques is not hindered by the aforementioned factors. Having said that, by unifying the definition of response features, the generalized approach overcomes some of the limitations of the prior versions (cf. Chaps. 5 through 14), all of which exploited individual (i.e., problem-specific) definitions of the characteristic points. The application of the latter has been significantly limited to antenna structures whose responses featured readily discernible characteristic points (e.g., multiband antennas). Therefore, such frameworks have been not as versatile as other procedures capable of handling antenna responses of arbitrary shapes.

The technique discussed in this chapter can be considered a step toward extending the applicability of the response feature technology for antenna design. It can be of interest, among others, for the readers dealing with simulation-driven optimization tasks that are difficult to handle by means of conventional local optimization algorithms, yet, for which the use of global search routines may be impractical from the point of view of their computational complexity. It is also useful whenever the antennas need to be redesigned with respect to their operating frequencies because FBO is a practical tool for manipulating the antenna resonances/bandwidth in a convenient manner. Finally, it should be mentioned that the feature-based optimization concept can also be generalized for handling other types of high-frequency structures, including microwave components. Notwithstanding, such a generalization is a more challenging endeavor due to a larger variety of possible system outputs (e.g., pertinent to filters, couplers, and power dividers) that need to be taken into account by a respective framework.

References

Al-Azza, A. A., Al-Jodah, A. A., & Harackiewicz, F. J. (2016). Spider monkey optimization: A novel technique for antenna optimization. *IEEE Antennas and Wireless Propagation Letters, 15*, 1016–1019.

Alsath, M. G. N., & Kanagasabai, M. (2015). Compact UWB monopole antenna for automotive communications. *IEEE Transactions on Antennas and Propagation, 63*(9), 4204–4208.

Bekasiewicz, A., & Koziel, S. (2016). Precise control of reflection response in bandwidth-enhanced planar antennas. *The International Journal of RF and Microwave Computer-Aided Engineering, 26*(8), 653–659.

Conn, A. R., Gould, N. I. M., & Toint, P. L. (2000). Trust region methods, MPS-SIAM series on optimization, SIAM, Philadelphia, PA.

Conn, A. R., Scheinberg K., & Vicente, L. N. (2009). Introduction to derivative-free optimization, MPS-SIAM series on optimization, SIAM, Philadelphia, PA.

De Villiers, D. I. L., Couckuyt, I., & Dhaene, T. (2017). *Multi-objective optimization of reflector antennas using kriging and probability of improvement.* IEEE international symposium on antennas and propagation, San Diego, USA. pp. 985–986.

Dong, J., Qin, W., & Wang, M. (2019). Fast multi-objective optimization of multi-parameter antenna structures based on improved BPNN surrogate model. *IEEE Access, 7*, 77692–77701.

Du, J., & Roblin, C. (2017). Statistical modeling of disturbed antennas based on the polynomial chaos expansion. *IEEE Antennas and Wireless Propagation Letters, 16*, 1843–1847.

Feng, F., Zhang, C., Na, W., Zhang, J., Zhang, W., & Zhang, Q. J. (2019). Adaptive feature zero assisted surrogate-based EM optimization for microwave filter design. *IEEE Microwave and Wireless Components Letters, 29*(1), 2–4.

Glubokov, O., & Koziel, S. (2014) *EM-driven tuning of substrate integrated waveguide filters exploiting feature-space surrogates.* IEEE international microwave symposium.

Hassan, A. K., Etman, A. S., & Soliman, E. A. (2018). Optimization of a novel nano antenna with two radiation modes using kriging surrogate models. *IEEE Photonics Journal, 10*(4), paper no. 4800807.

Jamshidi, M. B., Lalbakhsh, A., Lotfi, S., Siahkamari, H., Mohamadzade, B., & Jalilian, J. (2019). A neuro-based approach to designing a Wilkinson power divider. *The International Journal of RF and Microwave Computer-Aided Engineering, 30*, article no. e22091.

Jin, N., & Rahmat-Samii, Y. (2007). Advances in particle swarm optimization for antenna designs: Real-number, binary, single-objective and multiobjective implementations. *IEEE Transactions on Antennas and Propagation, 55*(3), 556–567.

Koziel, S. (2010). Computationally efficient multi-fidelity multi-grid design optimization of microwave structures. *The Applied Computational Electromagnetics Society Journal, 25*, 578–586.

Koziel, S. (2015). Fast simulation-driven antenna design using response-feature surrogates. *The International Journal of RF and Microwave Computer-Aided Engineering, 25*(5), 394–402.

Koziel, S., & Bandler, J. W. (2015). Rapid yield estimation and optimization of microwave structures exploiting feature-based statistical analysis. *IEEE Transactions on Microwave Theory and Techniques, 63*(1), 107–114.

Koziel, S., & Bekasiewicz, A. (2018). Fast re-design and geometry scaling of multi-band antennas using inverse surrogate modeling techniques. *International Journal of Numerical Modelling, 31*(3), paper no. 2287.

Koziel, S., & Pietrenko-Dabrowska, A. (2019). Variable-fidelity simulation models and sparse gradient updates for cost-efficient optimization of compact antenna input characteristics. *Sensors, 19*(8), paper no. 1806.

Koziel, S., & Pietrenko-Dabrowska, A. (2020a). Design-oriented computationally-efficient feature-based surrogate modelling of multi-band antennas with nested kriging. *AEU -International Journal of Electronics and Communications, 120*, paper no. 153202.

Koziel, S., & Pietrenko-Dabrowska, A. (2020b). Expedited feature-based quasi-global optimization of multi-band antennas with Jacobian variability tracking. *IEEE Access, 8*, 83907–83915.

Koziel, S., & Unnsteinsson, S. D. (2018). Expedited design closure of antennas by means of trust-region-based adaptive response scaling. *IEEE Antennas and Wireless Propagation Letters, 17*(6), 1099–1103.

Liu, B., Akinsolu, M. O., Ali, N., & Abd-Alhameed, R. (2019). Efficient global optimisation of microwave antennas based on a parallel surrogate model-assisted evolutionary algorithm. *IET Microwaves, Antennas and Propagation, 13*(2), 149–155.

Maghrabi, M. M. T., Bakr, M. H., Kumar, S., Elsherbeni, A. Z., & Demir, V. (2020). FDTD-based adjoint sensitivity analysis of high-frequency nonlinear structures. *IEEE Transactions on Antennas and Propagation, 68*(6), 4727–4737.

Pietrenko-Dabrowska, A., & Koziel, S. (2020). Computationally-efficient design optimization of antennas by accelerated gradient search with sensitivity and design change monitoring. *IET Microwaves, Antennas & Propagation, 14*(2), 165–170.

Rokunuzzaman, M., Samsuzzaman, M., & Islam, M. T. (2017). Unidirectional wideband 3-D antenna for human head-imaging application. *IEEE Antennas and Wireless Propagation Letters, 16*, 169–172.

Sim, C., Chang, M., & Chen, B. (2014). Microstrip-fed ring slot antenna design with wideband harmonic suppression. *IEEE Transactions on Antennas and Propagation, 62*(9), 4828–4832.

Tak, J., Kantemur, A., Sharma, Y., & Xin, H. (2018). A 3-D-printed W-band slotted waveguide array antenna optimized using machine learning. *IEEE Antennas and Wireless Propagation Letters, 17*(11), 2008–2012.

Tang, M., Chen, X., Li, M., & Ziolkowski, R. W. (2018). Particle swarm optimized, 3-D-printed, wideband, compact hemispherical antenna. *IEEE Antennas and Wireless Propagation Letters, 17*(11), 2031–2035.

Tu, S., Cheng, Q. S., Zhang, Y., Bandler, J. W., & Nikolova, N. K. (2013). Space mapping optimization of handset antennas exploiting thin-wire models. *IEEE Transactions on Antennas and Propagation, 61*(7), 3797–3807.

Wang, Y., Zhang, J., Peng, F., & Wu, S. (2019). A glasses frame antenna for the applications in internet of things. *IEEE Internet of Things Journal, 6*(5), 8911–8918.

Wu, Q., Wang, H., & Hong, W. (2020). Multistage collaborative machine learning and its application to antenna modeling and optimization. *IEEE Transactions on Antennas and Propagation, 68*(5), 3397–3409.

Yang, X. S., Wang, J., & Wang, B. Z. (2017). Topology optimization of patch antennas with specified polarization and beam characteristics. In *6th Asia-Pacific Conf. Ant. Propag., Xian, China.*

Zhang, C., Feng, F., Gongal-Reddy, V. M. R., Zhang, Q. J., & Bandler, J. W. (2015). Cognition-driven formulation of space mapping for equal-ripple optimization of microwave filters. *IEEE Transactions on Microwave Theory and Techniques, 63*(7), 2154–2165.

Zhao, Z., Pichot, C., & Dedeban, C. (2015). *Topological shape gradient and Level-Set method for optimizing planar antennas.* IEEE Conference on Antenna Measurements & Applications (CAMA), Chiang Mai, Thailand, pp. 1–4.

Chapter 16
Summary

Contemporary high-frequency engineering relies heavily on computational models. Although analytical and circuit-theory-based tools are ubiquitous, they are most often used to develop initial designs that require further tuning. This is because equivalent networks and similar means are unable to account for mutual coupling, dielectric and radiation losses, substrate anisotropy, or the presence of environmental components (connectors, human tissue, housing, etc.). In many cases, especially antenna structures, parameterized network models are simply unavailable. These and other phenomena affecting the operation of high-frequency devices can be quantified using full-wave electromagnetic (EM) analysis, which has become the ultimate evaluation tool. Modern EM simulation packages (CST Microwave Studio, Ansys HFSS, FEKO, Sonnet ***em***, Keysight Momentum, etc.) permit implementation of realistic models that adequately represent the operation of antennas or microwave components with the accuracy comparable to physical measurements. Consequently, any design obtained using either analytical methods or equivalent networks has to be validated through EM analysis. In general, due to limited accuracy of traditional models, the system parameters have to be tuned so that performance requirements imposed on the operating parameters (bandwidth, gain, isolation, axial ratio, phase response, etc.) are satisfied at the EM level of description.

EM-based models are indispensable, yet they tend to be computationally expensive. A long simulation time is not an issue for a one-time system evaluation or even parametric studies often undertaken at various stages of the design process. However, it becomes a serious problem whenever repetitive analyses are required. The flagship example thereof is optimization-based parameter adjustment, which may require many dozens or hundreds of EM simulations even in the case of local tuning. Other EM-driven procedures such as global or multi-criterial design or uncertainty quantification (statistical analysis, yield optimization) incur considerably higher costs, often unmanageable. At the same time, as mentioned earlier, the employment of full-wave computational models is a practical necessity fostered by reliability requirements.

A. Pietrenko-Dabrowska, S. Koziel, *Response Feature Technology for High-Frequency Electronics. Optimization, Modeling, and Design Automation*,
https://doi.org/10.1007/978-3-031-43845-5_16

Expediting simulation-based design procedures have been addressed by researchers for more than two decades. In the realm of gradient-based algorithms, a notable example is the incorporation of adjoint sensitivities, which is, however, an intrusive technique of limited accessibility through commercial simulation packages. Non-intrusive methods include the algorithms with sparse sensitivity updates, typically based on detecting the stability patterns for the system response Jacobians or employing selectively applied updating formulas. Another group of techniques is those exploring variable-fidelity models, which, in the case of antennas, are mostly based on variable-resolution EM simulations. The main idea is to shift the computational burden onto the lower-fidelity representation, often in the form of a physics-based surrogate, with only occasional reference to the high-fidelity EM analysis (for design verification and surrogate model enhancement). Popular implementations include space mapping (with its various implementations), as well as response correction techniques (e.g., manifold mapping, adaptive response scaling, shape-preserving response prediction). Surrogate-assisted approaches involving data-driven models have been drawing an increasing attention over the recent years due to their versatility. These methods often interleave the surrogate model construction and the prediction stage (in local or global optimization) through sequential design of experiments involving various infill criteria, e.g., maximization of the expected improvement or minimization of the statistical lower bound. Among the techniques of this group that are popular in high-frequency design, kriging, radial basis functions, neural networks, support vector regression, or polynomial chaos expansion may be listed. Recently, integration of global optimization with machine learning has been fostered as a way of efficient handling of multidimensional parameter spaces. Notwithstanding, surrogate-assisted methods are affected by a number of issues, related to parameter space dimensionality, reliability, as well as appropriate selection of a low-fidelity model. As mentioned, for many practical high-frequency structures, a reasonably reliable lower-fidelity model can only be obtained through a coarse-mesh EM analysis, which is detrimental for computational efficiency of the optimization process. At the same time, data-driven modeling of responses featuring highly nonlinear outputs is limited to structures described by a few parameters within relatively narrow ranges thereof.

The response feature methodologies, being the main topic of this book, offer a qualitatively different approach to addressing the common issues of EM-driven design. They attempt to exploit a particular structure of the system outputs as well as the fact that the frequency and level coordinates of suitably defined characteristic points of these responses are less nonlinearly dependent on the design parameters than the original (complete) characteristics. Reformulation of the design task (e.g., parametric optimization) in terms of response features leads to flattening the objective function landscape and allows for improving the reliability as well as the computational efficiency of simulation-based procedures. In the case of surrogate modeling problems (including surrogate-assisted optimization or uncertainty quantification), feature-based methods allow for reducing the computational cost of model construction and improving its predictive power as compared to traditional formulations. A critical aspect of the response feature technology is a definition of the characteristics points. In some cases, e.g., microwave devices such as filters, the response features may be selected as zeros and poles of the transfer function, or the entries of the coupling matrix. However, the most generic approach assumes that the characteristic points are defined on and directly extracted from the simulated frequency characteristics of the system under design. These may be the locations of the device resonances, the minima of the coupler matching/isolation responses, the local (in-band) maxima of the filter return loss characteristics, or the points corresponding to particular response levels (e.g., -10 dB when determining the antenna impedance bandwidth). In other words, the feature definition depends on the design problem (i.e., considered performance figures and requirements) and the type of response (narrow-band, multi-band, broadband, etc.).

The main purpose of this book was the exposition of the response feature technology. In particular, we discussed the concept of characteristic points and their properties in the context of high-frequency devices, outlined feature-based procedures, as well as their applications to parameter tuning, global and multi-criterial optimization, and uncertainty quantification of antenna and microwave structures. The presented algorithms have been illustrated using a large number of real-world design scenarios and benchmarked against conventional (non-feature-based) approaches to emphasize the potential benefits of the incorporation of response features from the perspective of computational efficiency and reliability of EM-driven design.

The secondary purpose of this book was to provide an overall introduction to numerical optimization and surrogate modeling as well as a review of the most popular techniques. This was mainly realized in the first four chapters covering simulation-based design (Chap. 2), local and global optimization (Chap. 3), as well as data-driven and physics-based modeling, along with surrogate-assisted optimization (Chap. 4). Although the majority of this material was placed in a generic context, particular aspects pertinent to high-frequency structures have been emphasized as well. Exposition of the theoretical background was supplemented by examples of antenna and microwave engineering case studies.

As mentioned earlier, response feature methodologies bring unique opportunities in terms of improving the computational efficiency and reliability of EM-driven design procedures. Chapter 5 summarized the fundamentals of this technology in the context of local parameter tuning of antennas and microwave components. In particular, we discussed the definition of the characteristic points and the associated notation, as well as reformulation of the design problem in terms of the feature points along with inherent regularization of the objective function and computational benefits that are a consequence thereof. Chapter 6 elaborated on utilization of response features for local parameter tuning. The emphasis was on the appropriate selection of characteristic points for different types of high-frequency structures such as single- and multi-band antennas, broadband antennas, antenna arrays, microwave couplers, or bandpass filters. The case studies considered in this chapter were selected to illustrate a variety of possible scenarios and the ways of defining feature points so as to capitalize on their properties, thereby enabling computational speedup. Chapters 7 and 8 discussed accelerated local optimization, where the response feature methodologies were coupled with variable-fidelity EM models (Chap. 7), as well as adjoint sensitivities or sparse sensitivity updating schemes (Chap. 8). In all cases, the incorporation of the mentioned algorithmic tools brings additional speedup without being detrimental to design quality, as demonstrated using several antennas and microwave devices.

One of the most appealing properties of the response feature methodology is that it allows for improving the reliability of the optimization procedures, in particular, for immunizing local search algorithms against poor initial designs. These issues are of fundamental importance in the design of multi-band structures or system re-design over broad ranges of operating frequencies. Chapter 9 discussed two classes of techniques where the response features were employed to detect the operating parameters of the structure at hand and to facilitate optimum identification. The first type of algorithms iteratively adjusts design specifications to ensure that they are reachable using local tuning at each stage of the optimization process. In the second type of procedures, the objective function is regularized to alter its profile and make it monotonic with respect to the distance between the actual and target operating frequencies. A generalization of the mentioned method to variable-fidelity settings has been discussed as well.

Applications of response features for global and multi-objective optimization have been covered in Chap. 10. Therein, we discussed globalized search procedures, where partial or complete reformulation of the design task in terms of characteristic points allows for a considerable reduction of the computational costs of the search process, either conducted directly, or within surrogate-assisted frameworks. In a related vein, Chap. 11 investigated feature-based warm-start optimization of high-frequency structures, where the goal is to enable rapid parameter tuning over broad ranges of geometry/material parameters as well as operating frequencies by re-using information encoded in pre-optimized reference designs acquired beforehand. The pivotal role in the discussed frameworks is played by feature-based inverse surrogate models that allow fast identification of close-to-optimum sets of design variables corresponding to the target operating parameters of choice.

Surrogate modeling using response features has been the subject of Chaps. 12 and 13. In Chap. 12, we covered a construction of feature-based models representing complete frequency responses of high-frequency components, where characteristics points are allocated to account for the most important parts of the responses such as resonances or passbands (in the case of filtering circuits). The same chapter also discussed variable-resolution modeling, where the data acquired using low- and high-fidelity EM analysis is combined using co-kriging or space mapping. On the other hand, Chap. 13 elaborated on performance-driven modeling, where feature-based surrogates are established in confined domains defined using auxiliary reference designs or randomized parameter space pre-screening. Extensions oriented toward further improvement of computational efficiency, e.g., dimensionality reduction, were investigated as well. The modeling procedures described in both chapters capitalize on inherent regularization of the functional landscapes being handled in the course of the modeling process, enabled by the weakly nonlinear relationship between the operating parameters of the structure of interest and its geometry/material parameters.

Chapter 14 focused on uncertainty quantification using response features. In particular, we discussed reformulation of the yield estimation and optimization task in terms of characteristic points of the system outputs, as well as cost-efficient solution algorithms. Furthermore, procedures for rapid input tolerance optimization have been described, where the objective is to maximize allowed parameter deviations for which the perfect fabrication yield of 100% is still ensured. Expedited versions of the aforementioned algorithms involving variable-fidelity EM simulations and inverse feature-based surrogates were also discussed and illustrated using microwave and antenna components. Chapter 15 introduced a generalized formulation of response features in the context of antenna design, specifically, optimization of the input characteristics. Therein, the goal was to provide a versatile definition of characteristic points, which is applicable to a broad range of antenna responses, including single- and multi-band, as well as broadband ones. An associated definition of the objective function expressed in terms of generalized features has been discussed as well. The fundamental advantage of this sort of formalism is that it mitigates one of the main issues related to response feature methodologies, which is the necessity to individually define and extract features on a case-to-case basis.

The analysis of the feature-based techniques presented in this book, including the verification results, benchmarking, and applications for solving a variety of design tasks, allows us to draw several conclusions that can be summarized as follows:

- Reformulating the design task in terms of suitably chosen characteristic points of the system responses (as opposed to handling complete outputs, e.g., frequency characteristics) allows for a considerable improvement of both reliability and computational efficiency of EM-driven design procedures. The underlying property is a weakly nonlinear dependence between design variables and frequency/level coordinates of the feature points.
- In the case of local optimization, feature-based approaches typically enable acceleration, yet the most important property is regularization, which often leads to quasi-global search capabilities while using (formally) local optimization algorithms, especially gradient-based procedures.
- Response features are also suitable for enhancing global and multi-objective optimization algorithms, as well as warm-start optimization routines. A limiting factor here is the existence of the assigned feature points across the entire design space. In this context, feature-based inverse surrogates are particularly suitable due to being able to yield reliable predictions on the optimum design location at extremely low computational expenses.
- Utilization of response features for general-purpose modeling is conditional upon the same factors as mentioned above, i.e., the existence of the characteristic points across the design space. From this perspective, response features are particularly suitable for performance-driven (constrained) modeling, where confining the surrogate domain to regions containing high-quality designs facilitates satisfaction of the said condition.

- Feature-based approaches seem to be especially attractive for uncertainty quantification (statistical analysis, robust design) tasks. This is because the expected design relocation is limited when solving problems such as yield optimization, which helps preserving the consistency of the characteristic point structure.
- Typical gains offered by feature-based approaches, either in terms of computational savings or reliability improvements, are considerable, as illustrated by numerous examples discussed throughout this book. Optimization process speedup of 30–50% is not uncommon, just as the two- or threefold reduction of the surrogate modeling error.
- The response feature techniques can be conveniently combined with other algorithmic tools and mechanisms such as local/global search routines, variable-fidelity modeling approaches, and domain confinement. This is because feature-based methods are generally non-intrusive, i.e., often operate at the level of formulating the problem (e.g., the objective function) instead of affecting the underlying algorithms.

As mentioned on multiple occasions, the most important limitation of the response feature technology is the need to identify, define, and extract the set of characteristic points for a given structure and design task. This means, in particular, that generic and versatile implementation of feature-based optimization or modeling procedures is difficult. The generalized response feature approach described in Chap. 15 illustrates this issue: even when restricted to the input characteristics of antennas, the mathematical formalism becomes relatively complex. Also, it cannot be translated into other cases (e.g., microwave couplers or filters) in a straightforward manner. On the other hand, if the class of problems is well defined and relatively narrow (e.g., sidelobe level reduction of antenna arrays, or resonant frequency allocation of multi-band antennas), feature-based methods may become indispensable. In terms of implementation, the most important aspect is the extraction of response features from EM-simulated characteristics of the device under design, which requires auxiliary algorithms that identify the frequency and level coordinates thereof. The process is typically carried out by scanning the frequency response in search for, e.g., resonances, local minima/maxima, or frequencies corresponding to specific levels of the response, depending on the assumed definitions of the feature points. In this book, all auxiliary codes were implemented in Matlab.

It should be reiterated that simulation-driven design, including modeling and optimization, is a challenging endeavor for the reasons mentioned on several occasions throughout this book. Efficient procedures tend to be involved in terms of the algorithmic tools. More often than not, the design frameworks combine several mechanisms, which are selected or custom-developed to address particular issues pertinent to the task at hand. The response feature technology described in this book is one of such methodologies. The authors believe that the presented material might be helpful for the readers interested in EM-driven design in all its aspects, but also researchers and practitioners who might adapt the concepts outlined here to other engineering disciplines, as well as incorporate them into design frameworks.

We conclude with a few remarks concerning possible extensions of the discussed methods and their design applications. One of such directions is the further generalization of the response feature concept, e.g., oriented toward defining and handling system response properties that are not necessarily directly related to "visible" characteristic points such as resonances, but can still be used to account for design specifications. Perhaps a good conceptual illustration is representing frequency responses by means of zeros and poles of rational function approximations, which is commonly used especially in the design of microwave components. Another direction would be automation of the feature extraction process, similar to but more generic than what was discussed in Chap. 15. The objective would be to implement versatile extraction procedures, which use the definitions of characteristic points as their input arguments and are able to properly operate for broad ranges/classes of such definitions. Yet another option is the development of feature-based procedures for handling specific classes of high-frequency devices (e.g., couplers, filters, or multi-band antennas), which share the main properties of their outputs in a structural sense, thereby ensuring feature point existence and extractability for all instances of a given class. A realization of the aforementioned goals would make the response feature technology more accessible to engineers who may not have specialized knowledge in numerical modeling/optimization yet are interested in applying these methods in their design work.

Index

A
Accelerated design, 184
Accelerated optimization, 294
Adaptively adjusted
design specifications, 257, 282
Adjoint sensitivities, 5, 205, 232–241, 245, 254, 550, 584, 586
Algorithm convergence, 50, 73, 500, 501, 543
Antenna design, 2, 28, 115, 158, 178, 182, 197, 205, 243–245, 254, 274, 310, 326, 351, 352, 444, 454, 456, 466, 476, 478, 527–529, 537, 549–551, 571, 580, 587
Antenna modeling, 445
Approximation model management optimization (AMMO), 108, 111
Artificial neural networks (ANNs), 94, 95, 467, 471, 472

B
Bandwidth enhancement, 163, 206, 211–213, 307, 317, 358, 450, 451, 549, 551, 555, 556, 565–567, 570, 571, 576, 579
Bandwidth maximization, 150, 159, 160, 199, 560, 562–564

C
Characteristic point allocation, 404–415
Characteristics points, 309, 315, 404, 551, 556, 585, 586
Circuit decomposition, 174, 184–190, 200
Computational efficiency, 4, 10, 12, 37, 78, 86, 98, 118, 156, 164, 253, 274–276, 292, 294, 321, 324, 344, 346, 358, 403, 435, 471, 490, 495, 496, 500, 523, 528, 540, 543, 549, 550, 552, 565, 584–587
Computer-aided design, v
Constrained optimization, 45, 55–59
Correlation analysis, 208, 209, 215, 225

D
Database acquisition, 355–401
Data-driven surrogates, 6–8, 35, 36, 88–98, 103–108, 403, 404, 454
Derivative-free optimization, 59–62
Descent methods, 48–52, 54, 55
Design automation, 9, 11, 85
Design centering, 147, 489–493, 495–501, 521, 535–545
Design of experiments (DOEs), 8, 35, 52, 59, 64, 69, 71, 75, 89–91, 131, 160, 167, 177, 225, 309, 321, 322, 340, 347, 360, 388, 390, 397, 399, 400, 408, 438, 440, 452–454, 456, 458, 463, 464, 473, 475, 476, 478, 511, 559, 562, 577, 584
Design optimization, 6, 8, 10, 12, 86, 108, 128, 136, 137, 139, 149–151, 159, 160, 165, 174, 175, 177, 183, 184, 193, 199, 208, 220, 238, 242, 258, 273, 302, 305, 325, 326, 345, 352, 362, 363, 370, 373, 379, 390, 393, 403, 409, 450, 454, 457, 474, 554, 560, 565, 567, 568, 570
Domain confinement, 436, 448

A. Pietrenko-Dabrowska, S. Koziel, *Response Feature Technology for High-Frequency Electronics. Optimization, Modeling, and Design Automation*,
https://doi.org/10.1007/978-3-031-43845-5

E
Electromagnetic (EM) analysis, 2, 3, 25, 37, 355, 488, 583
EM-driven design, 2, 11, 27, 78, 127, 236, 403, 550, 552, 584, 585, 587, 588
EM-driven optimization, 270, 358, 550
External approach, 428–431

F
Fast optimization, 129, 186, 365
Feature-based design, 78, 169, 172, 173, 177, 190
Feature-based design automation, 78
Feature-based modeling, 403–432, 442, 452, 456, 474–482
Feature-based objective function, 143, 358, 359, 376, 554–556, 560, 561, 565, 578, 579
Feature-based optimization (FBO), 5, 10, 11, 128, 135–146, 149–190, 193, 199, 200, 206–215, 217–254, 305, 347, 348, 350, 376, 379, 551, 552, 565, 580
Feature-based problem formulation, 191, 206–208, 234–236
Feature-based statistical analysis, 488, 545, 588
Feature point definition, 128, 549, 556–559
Filter design, 100, 163, 165, 394, 398
Forward models, 356, 360, 364, 384
Frequency-based regularization, 11, 257, 284–292, 294, 298, 300–302
Frequency scaling, 101–103, 114, 134, 137

G
Generalization error, 89, 96, 97
Generalized response features, 556–565, 568, 579, 588
Globalized search, 147, 307, 308, 554, 586
Global optimization, 6, 9, 11, 25–27, 45, 59, 62–78, 95, 97, 103, 107, 127, 155, 243, 257, 259, 305–345, 352, 550, 584, 585
Gradient-based optimization, 4, 47–59, 153, 234, 236, 286, 319, 392
Gradient-based search, 52, 59, 153–154, 259, 277, 281, 283, 298, 356, 574
Gradient variability, 233, 294

H
High-frequency design, 4, 78, 98, 108, 127, 355, 435, 457, 584

I
Internal approach, 426–429, 431–433
Inverse modeling, 8, 11, 324, 538
Inverse models, 306, 309, 312–315, 317, 342, 359–363, 371, 379, 380, 384, 387, 398–399, 445, 458, 459, 461, 462, 469, 471–473, 538–540
Inverse sensitivity, 377–383, 385
Inverse surrogates, 305–324, 341, 344, 352, 355, 356, 360, 362, 364, 367, 371, 372, 380–382, 384, 386, 446, 460, 471, 488, 543, 545, 586, 587

K
Kriging interpolation, 97, 309, 360, 386, 388, 406, 408, 409, 411, 413, 414, 419, 420, 422, 423, 427, 430, 431, 437, 438, 452, 454, 463, 465, 466, 481, 497, 542

L
Local optimization, 5, 7, 87, 88, 104, 146, 149–200, 205, 233, 243, 244, 257–259, 265, 269, 271, 284, 292, 302, 305, 306, 308, 318, 319, 324, 338, 340, 341, 344, 362, 371, 454, 497, 499, 542, 580, 585, 587
Low-fidelity models, 7, 9, 35, 87, 98–103, 108–110, 112–118, 132–138, 160, 165, 166, 176, 189, 205, 206, 208–212, 214–216, 221–225, 228–231, 274, 346, 348, 397, 425, 426, 429, 512, 517–519, 584

M
Matching improvement, 206, 213–215, 272, 276, 294, 307, 317, 326, 437, 449–451, 549, 551, 555, 556, 559–562, 579
Microwave design, 29, 33, 38, 205, 326, 386–393, 403, 482, 529–534
Microwave modeling, 415–424

Model validation, 35, 89, 97
Multi-objective design, 11, 33, 34, 127, 128, 305, 327, 345, 346, 352, 489

N
Nature-inspired algorithms, 27, 33, 74, 78, 318, 341, 344
Nested kriging, 435–458, 461, 462, 464, 466–468, 471–476, 479, 481, 483–485
Numerical optimization, 27, 45, 55, 196, 284, 286, 355, 377, 551, 563, 585

O
Optimization process reliability, 200, 338

P
Parameter tuning, 4, 26–28, 127–139, 146, 258, 259, 272, 282, 292, 324, 325, 355–371, 383, 403, 549, 585, 586
Physics-based surrogates, 7, 10, 35, 98–103, 108–119, 584

Q
Quasi-global optimization, 243, 254, 301
Quasi Newton method, 52–54

R
Radial basis functions (RBFs), 6, 92, 93, 422, 423, 441, 442, 454, 456, 466–468, 471, 472, 481, 584
Reference designs, 100, 103, 113, 114, 131, 133, 208–210, 222, 309, 364, 366, 371–389, 393, 435–438, 440, 442, 445, 446, 448, 452, 454, 457, 458, 464, 466–468, 471–473, 479, 481, 483–485, 505, 541–543, 586
Regression models, 10, 93, 94, 209, 310, 312, 315, 458–462, 464, 467, 469, 470, 472–474, 495, 496, 501–510, 519, 523–526, 539
Response correction, 7, 98–100, 111, 115, 134, 137, 184, 185, 274, 550, 584
Response features, 10, 11, 25, 38, 45, 78, 127–147, 149–200, 207, 208, 218–219, 235, 242–245, 247, 249, 251–254, 257, 284, 305–352, 355–401, 403–432, 435, 436, 439–440, 445, 448–454, 456, 458, 459, 474–477, 482, 484, 485, 487, 489, 490, 493–501, 505, 512, 516–519, 523–525, 528, 532, 535–545, 549–556, 558–560, 564, 566–570, 572, 575, 577–580, 584–589

S
Sequential sampling, 6, 8, 35, 36, 86, 91, 92, 97, 107
Shape-preserving response prediction (SPRP), 8, 108, 112–113, 421, 430, 431, 584
Side-lobe level (SLL) reduction, 167, 172
Simplex, 61–63, 92, 325, 328, 332–336, 338, 344
Simplex algorithm, 61–63, 342, 344
Simulation-driven design, 2, 3, 5, 10, 25–39, 45, 86, 127, 306, 549, 588
Size reduction, 4, 32, 150, 161–163, 174, 178, 182, 190–200, 290, 294, 370
Space filling designs, 35, 91
Space mapping (SM), 5, 7, 37, 88, 90, 100–102, 108–111, 113, 115, 128, 134, 136, 137, 139, 274, 346, 428, 489, 550, 584, 586
Sparse sensitivity updates, 205, 232, 233, 241–254, 284, 292–302, 316, 550, 584
Statistical analysis, 5, 8, 11, 37, 38, 128, 147, 208, 487, 488, 545, 549, 583, 588
Support vector regression (SVR), 6, 92, 95, 96, 584
Surrogate-assisted optimization, 11, 78, 85–88, 104, 106, 107, 113, 127, 137, 208, 584, 585
Surrogate modeling, 11, 25, 27, 34–36, 45, 85–119, 127, 147, 403, 405, 406, 408, 419, 420, 435, 453, 467, 473, 482, 487–489, 500, 541, 549, 554, 584–586, 588

T
Tolerance analysis, 525
Tolerance hypervolume optimization, 4
Trust-region algorithm, 116, 143, 144, 153, 262, 264–266, 340
Two-stage modeling, 435

U
Uncertainty quantification, 3–5, 8, 11, 25, 34, 37–39, 127, 403, 487–545, 549, 550, 554, 583–585, 587, 588

V
Variable-fidelity EM simulations, 183, 185, 190, 205, 206, 292, 302, 424, 511–522, 587
Variable-fidelity feature-based modeling, 424–432
Variable-fidelity robust design, 511
Variable-resolution modeling, 586

W
Warm-start optimization, 11, 355–401, 586, 587

Y
Yield-driven design, 37, 489

MIX
Papier aus verantwortungsvollen Quellen
Paper from responsible sources
FSC® C105338

If you have any concerns about our products,
you can contact us on
ProductSafety@springernature.com

In case Publisher is established outside the EU,
the EU authorized representative is:
Springer Nature Customer Service Center GmbH
Europaplatz 3, 69115 Heidelberg, Germany

Printed by Libri Plureos GmbH
in Hamburg, Germany